Impact of Engineered Nanomaterials in Genomics and Epigenomics

Impact of Engineered Nanomaterials in Genomics and Epigenomics

Edited by

Saura C. Sahu
Formerly of Center for Food Safety and Applied Nutrition, US Food and Drug Administration, MD, USA

WILEY

This edition first published 2023

Registered Offices
John Wiley & Sons, Inc., 111 River Street, Hoboken, NJ 07030, USA
John Wiley & Sons Ltd, The Atrium, Southern Gate, Chichester, West Sussex, PO19 8SQ, UK

For details of our global editorial offices, customer services, and more information about Wiley products visit us at http://www.wiley.com.

Wiley also publishes its books in a variety of electronic formats and by print-on-demand. Some content that appears in standard print versions of this book may not be available in other formats.

A catalogue record for this book is available from the Library of Congress

Hardback ISBN: 9781119896227; ePub ISBN: 9781119896241; ePDF ISBN: 9781119896234;
oBook ISBN: 9781119896258

Cover Image: © Gio_tto/Getty Images
Cover Design: Wiley

Set in 9.5/12.5pt STIXTwoText by Integra Software Services Pvt. Ltd, Pondicherry, India

Dedicated to my family:

My late parents, Gopinath and Ichhamoni, for their gifts of life and love, and for being living examples.

My wife, Jharana, for her life-long friendship, love and support as well as for her patience and understanding of the long hours spent at home on planning, writing, and editing this book.

My children, Meghamala, Sudhir, Subir, and their spouses for their love and care.

My eight-year-old granddaughter, Naomi, and six-year-old grandson, Jonah, for their unconditional love, faith, and trust in me.

Saura C. Sahu, Ph.D.
Columbia, Maryland, USA
June 5, 2022

Contents

List of Contributors

Salma M.S. Ahmad
Department of Biological and Environmental Sciences
College of Arts and Sciences
Qatar University
Doha, Qatar

Saleh Alfuraih
Department of Pharmaceutical Sciences
College of Pharmacy
Nova Southeastern University
Fort Lauderdale, FL, USA

Eid Alshammari
Department of Biochemistry, Microbiology, and Immunology
Wayne State University School of Medicine
Detroit, MI, USA

Maryam Al-Mansoob
Department of Biological and Environmental Sciences
College of Arts and Sciences
Qatar University
Doha, Qatar

Najlaa Al-Thani
Research and Development Department
Barzan Holdings
Doha, Qatar

Rais A. Ansari
Department of Pharmaceutical Sciences
College of Pharmacy
Nova Southeastern University
Fort Lauderdale, FL, USA

Piyoosh Kumar Babele
College of Agriculture
Rani Lakshmi Bai Central Agricultural University
Jhansi, Uttar Pradesh, India

Ravi Kant Bhatia
Department of Biotechnology
Himachal Pradesh University
Shimla, Himachal Pradesh, India

Delia Cavallo
Department of Occupational and Environmental Medicine, Epidemiology and Hygiene
Italian Workers' Compensation Authority-INAIL
Monte Porzio Catone, Rome, Italy

Ramesh Chaudhari
Biological and Life Sciences, School of Arts and Sciences
Ahmedabad University (Central Campus)
Ahmedabad, Gujarat, India

Pieranna Chiarella
Department of Occupational and Environmental Medicine, Epidemiology and Hygiene
Italian Workers' Compensation Authority-INAIL
Monte Porzio Catone, Rome, Italy

Aureliano Ciervo
Department of Occupational and Environmental Medicine, Epidemiology and Hygiene
Italian Workers' Compensation Authority-INAIL
Monte Porzio Catone, Rome, Italy

Nicholas Day
Biological Sciences Division
Pacific Northwest National Laboratory
Richland, WA, USA

Valentina Del Frate
Department of Occupational and Environmental Medicine, Epidemiology and Hygiene
Italian Workers' Compensation Authority-INAIL
Monte Porzio Catone, Rome, Italy

Anna Maria Fresegna
Department of Occupational and Environmental Medicine, Epidemiology and Hygiene
Italian Workers' Compensation Authority-INAIL
Monte Porzio Catone, Rome, Italy

Matthew J. Gaffrey
Biological Sciences Division
Pacific Northwest National Laboratory
Richland, WA, USA

Mario Ganau
Neuroscience Division
Oxford University Hospitals
Oxford, UK

Mohammad Z. Haider
Department of Basic Medical Sciences
College of Medicine, QU Health
Qatar University
Doha, Qatar

George Hammons
Division of Biochemical Toxicology
National Center for Toxicological Research
U.S. Food & Drug Administration
Jefferson, AR, USA

Vratko Himič
Neuroscience Division
Oxford University Hospitals
Oxford, UK

and

University of Oxford Medical School
Oxford, UK

Saleem Javed
Department of Biochemistry
Aligarh Muslim University
Aligarh, India

Hafiz Muhammad Jhanzab
Department of Agronomy
The University of Agriculture Dera Ismail Khan
Khyber Pakhtunkhwa, Pakistan

Pius Joseph
Molecular Carcinogenesis Laboratory
Toxicology and Molecular Biology Branch
Health Effects Laboratory Division
National Institute for Occupational Safety and Health (NIOSH)
Morgantown, WV, USA

Firas Kobeissy
Program for Neurotrauma, Neuroproteomics & Biomarkers Research
Departments of Emergency Medicine
University of Florida
Gainesville, FL, USA

Setsuko Komatsu
Faculty of Environment and Information Sciences
Fukui University of Technology
Gakuen, Fukui, Japan

Ashutosh Kumar
Biological & Life Sciences, School of Arts & Sciences
Ahmedabad University (Central Campus)
Ahmedabad, Gujarat, India

Gianfranco K.I. Ligarotti
Institute of AeroSpace Medicine
Milan, Italy

Beverly Lyn-Cook
Division of Biochemical Toxicology
National Center for Toxicological Research
U.S. Food & Drug Administration
Jefferson, AR, USA

Ghazala Mustafa
Department of Plant Sciences
Quaid-i-Azam University
Islamabad, Pakistan

Yadollah Omidi
Department of Pharmaceutical Sciences
College of Pharmacy
Nova Southeastern University
Fort Lauderdale, FL, USA

Stuti Patel
Department of Biology
University of Florida
Gainesville, FL, USA

Vishva Patel
Biological and Life Sciences, School of Arts and Sciences
Ahmedabad University (Central Campus)
Ahmedabad, Gujarat, India

Marta Pogribna
Division of Biochemical Toxicology
National Center for Toxicological Research
U.S. Food & Drug Administration
Jefferson, AR, USA

Wei-Jun Qian
Biological Sciences Division
Pacific Northwest National Laboratory
Richland, WA, USA

Saura C. Sahu
FORMER (retired) emplyee of the Food and Dug Administaration
Columbia, MD, USA

Abdullah Shaito
Biomedical Research Center
College of Medicine and Department of Biomedical Sciences at College of Health Sciences
Qatar University
Doha, Qatar

Saghir A. Shakil
ToxInternational Inc.
Hilliard, OH, USA

and

Department of Biomedical and Biological Sciences
Aga Khan University
Karachi, Pakistan

Kamran Shekh
Toxicology Consultant
Yordas Group (Canada Office)
Hamilton, ON, Canada

Nikolaos Syrmos
School of Medicine
Aristotle University of Thessaloniki
Greece

Brian D. Thrall
Biological Sciences Division
Pacific Northwest National Laboratory
Richland, WA, USA

Cinzia Lucia Ursini
Department of Occupational and Environmental Medicine, Epidemiology and Hygiene
Italian Workers' Compensation Authority-INAIL
Monte Porzio Catone, Rome, Italy

Carlos Wells
Division of Biochemical Toxicology
National Center for Toxicological Research
U.S. Food & Drug Administration
Jefferson, AR, USA

Alexander Yang
Department of Biochemistry, Microbiology, and Immunology
Wayne State University School of Medicine
Detroit, MI, USA

Nouran Yonis
Department of Biochemistry, Microbiology, and Immunology
Wayne State University School of Medicine
Detroit, MI, USA

Zhe Yang
Department of Biochemistry, Microbiology, and Immunology
Wayne State University School of Medicine
Detroit, MI, USA

Farhat Yasmeen
Department of Biosciences
University of Wah
Wah Cantt, Pakistan

Tong Zhang
Biological Sciences Division
Pacific Northwest National Laboratory
Richland, WA, USA

Yingxue Zhang
Department of Biochemistry, Microbiology, and Immunology
Wayne State University School of Medicine
Detroit, MI, USA

Preface

Engineered nanomaterials are products of the twenty-first century. Because of their superior physiochemical properties, they are applied in different areas of development, such as agriculture, energy, environment, medicine, biotechnology, and material science. They have great potential for improving human life. They are used in a wide range of consumer products, such as cosmetics, drugs, medical devices, paints, nanofabrics, and electronics. Humans are exposed to engineered nanomaterials every day. The effect of their long-term exposure is of public concern. In spite of their beneficial impacts on human life, much remains to be known about their safety. They appear to impact human life both ways, good and bad. Humans benefit from them in many ways, but at the same time, public concern about their safety continues to exist. At the moment, it is up to us to trust nanoscience or avoid it as much as possible.

It is my sincere hope that the up-to-date information presented in this monograph will serve as a stimulus to investigators interested in the impact of engineered nanomaterials on genomics and epigenomics.

The importance of research in this area of scientific discipline is evidenced by the increasing number of contributions published each year. It becomes increasingly clear that developments in this field are moving so rapidly that new means are needed to report the status of ongoing research activities. The contributions presented in this monograph represent a collaborative effort by international experts working in this emerging field of science.

The main purpose of this book is to assemble up-to-date, state-of-the-art information on the impact of engineered nanomaterials in genomics and epigenomics presented by internationally recognized experts in a single edition. Therefore, I sincerely hope that this book will provide an authoritative source of current information on this area of nanoscience and prove useful to the scientists interested in this scientific discipline throughout the world. However, it should be of interest to a variety of other scientific disciplines, including toxicology, medicine, and pharmacology, as well as drug and food sciences.

Saura C. Sahu, Ph.D.
Columbia, Maryland, USA

Acknowledgments

I am indebted to the internationally recognized experts who shared my enthusiasm for the field of nanoscience and contributed generously to this book. They were selected from academia, industry, and government for their expertise in their areas of research. Their work speaks for itself, and I am grateful to them for their strong commitment, cooperation, and excellent contributions in their areas of expertise.

I thank John Wiley & Sons Ltd, particularly Jenny Cossham and Elke Morice-Atkinson, for their excellent help, cooperation, support, and assistance for the timely publication of this book. Wiley published my first book, *Hepatotoxicity: From Genomics to in vitro and in vivo Models*, 682 pages, in 2007. Since then, Wiley has published more than a dozen of my books in different, new, emerging, and developing areas of modern toxicology and medicine (i.e. nanotoxicology, systems toxicology, toxicogenomics, epigenomics, microRNAs, stem cells). Wiley published my last book, *Genomic and Epigenomic Biomarkers of Toxicology and Disease: Clinical and Therapeutic Actions*, in 2022, and is currently publishing my book titled *Impact of Engineered Nanomaterials in Genomics and Epigenomics*. I express my gratitude to John Wiley & Sons Ltd.

Saura C. Sahu, Ph.D.
Columbia, Maryland, USA

1

Impact of Engineered Nanomaterials in Genomics and Epigenomics

Saura C. Sahu

FORMER (retired) emplyee of the Food and Dug Administaration Columbia, MD, USA

Nanotechnology: A Technological Advancement of the Twenty-First Century

Nanotechnology is a new technological development of the twenty-first century. The US National Nanotechnology Initiative (NNI) defines nanotechnology as "the understanding and control of matter at dimensions between approximately 1 and 100 nm, where unique phenomena enable novel applications" (NNI 2014; NSTC 2011). Nanotechnology is a new and developing technology. The ratio of the surface area to volume of a nanoparticle is high compared with its larger counterpart (Roduner 2006). This property makes them more reactive compared with the larger particles. Engineered nanomaterials are used in a wide range of consumer products (Shen et al. 2013), such as cosmetics, drugs, medical devices, paints, nanofabric clothes, and electronics because of their superior physiochemical properties. They demonstrate better magnetic, electrical, optical, and thermal properties compared with their larger counterparts. Therefore, they have found useful applications in different areas of development, such as agriculture, energy, environment, medicine, biotechnology, and material science. They have shown great potential for impacting human life because of their beneficial properties.

Genomics and Epigenomics

The National Institute of Health (NIH) defines genomics as "the study of all of a person's genes (the genome), including interactions of those genes with each other and with the person's environment". The National Cancer Institute (NCI) defines genomics as an interdisciplinary field of biology focusing on the structure, function, evolution, mapping, and editing of genomes. A genome is defined as a complete set of DNA, including all of its genes in an organism. The genome contains all the information needed for an organism to develop and grow. The global analysis of gene expression profiles provides a comprehensive view of toxicity and disease.

Impact of Engineered Nanomaterials in Genomics and Epigenomics, First Edition. Edited by Saura C. Sahu.

In genomic mechanisms of toxicity and disease, the genomic DNA sequence is altered by the chemical exposure. Such modified genomic DNA sequences are not cell and tissue specific. However, in some cases, toxicity and diseases are caused by DNA modifications due to chemical exposure, but in the absence of any direct alteration in genomic DNA sequence. Such DNA modifications without direct alterations in genomic DNA sequences are known as epigenomics, where DNA methylation regulates gene expression without direct alteration in the DNA sequence. In DNA methylation, gene expression occurs at the cytosine dinucleotide when a methyl group is added at position-5 producing methylcytosine (de Gannes et al. 2020). Unlike genomic changes, the epigenetic changes are cell and tissue specific. The epigenetic changes may be heritable and nonheritable. DNA methylation is associated with several human diseases including cancer.

The epigenome is defined as heritable biological information contained outside the DNA sequence (Dolinoy and Jirtle 2008). It consists of DNA methylation, histone modifications, and microRNAs. Noncoding RNAs (ncRNAs) regulate gene expression at the transcriptional or post-translational levels without changing the genomic DNA sequence.

Beneficial Impacts of Engineered Nanomaterials on Human Life

Engineered nanomaterials demonstrate a huge potential to transform human life for the better. Their use in consumer products is increasing rapidly. They are used in our food, cosmetics, medicine, and agriculture (Sahu and Hayes 2017). They are used in our water filters to remove microorganisms, such as bacteria from drinking water. They are used in water treatment systems. They are used to make our fabrics fire resistant and to prepare plastic bottles for daily use. They are used in cosmetics, sunscreens, pharmaceuticals, medicine, and medical devices. They are used for drug delivery in chemotherapy and as nanosensors for patients. They are used in computer circuits and for fuel efficiency in vehicles. Engineered nanomaterials are used in vehicles and sports equipment to make them lighter, stronger, and chemical resistant. They are used in solar plastics to collect solar energy. They are used to clean up chemical spills and airborne pollutants.

Potential Adverse Health Effects of Engineered Nanomaterials

Humans are exposed to engineered nanomaterials every day. Therefore, the health effects of these nanomaterials are of public concern. In spite of the various beneficial impacts of engineered nanomaterials on human life, our knowledge of engineered nanomaterials is not complete. We must keep in mind that nanoscience is a developing new science. Many things remain unknown. We do not know much about long-term effects engineered nanomaterials. We do not know much about their safety. Many questions about their potential effects on our health, planet, and ecosystems come to our mind. At the moment, they are unregulated. There are no recognized standards for producing and handling them. Are they safe? Are they double-edged swords? Such concerns will continue to exist in our minds until more is known about them. At the moment, it is up to us to trust nanoscience or avoid it as much as possible.

The molecular mechanisms of gene–environment interactions have attracted widespread interest in recent years. These effects may be of genomic and/or epigenomic in nature, highlighting potential molecular targets following the exposure of engineered nanomaterials.

Thai et al. (2016) published the first report on genomic effects of titanium dioxide nanomaterials in an in vitro study using human liver HepG2 cells. This study linked some of the in vitro canonical pathways to in vivo adverse outcomes: NRF2-mediated response pathways to oxidative stress, acute phase response to inflammation, cholesterol biosynthesis to steroid hormones alteration, fatty acid metabolism changes to lipid homeostasis alteration, G2/M cell checkpoint regulation to apoptosis, and hepatic fibrosis/stellate cell activation to liver fibrosis.

Bicho et al. (2020) in a multigenerational study demonstrated epigenetic effects of copper oxide nanomaterials in environmental species *Enchytraeus crypticus*. Using gene expression analyses, they showed changes in the epigenetic gene targets, depending on the generation and form of copper. Also, they showed its transgenerational effects in postexposure generations. They observed nanoparticle-specific effects indicating differences between organisms exposed to different forms of copper.

Lu et al. (2016) and Sierra et al. (2016) reported the effect of nanomaterial exposure on the mammalian epigenome.

Conclusions

Currently, engineered nanomaterials appear to be double-edged swords. They impact our lives both ways, good and bad. We benefit from them in many ways, but at the same time we are concerned about their adverse health effects. Public concern about their safety will continue until we understand them completely. At the moment, it is up to us to trust nanoscience or avoid it as much as possible.

With regard to the need for a book on the impact of engineered nanomaterials in genomics and epigenomics, the rate of publications during the past few years has demonstrated that the impact of engineered nanomaterials in genomics and epigenomics has attracted widespread interest and, therefore, there is a need for new means to report the updated current status of this developing area of research. As the editor of this monograph *Impact of Engineered Nanomaterials in Genomics and Epigenomics*, it gives me great pride, pleasure, and honor to introduce this unique book that encompasses many aspects of genomic and epigenomic research never published together before.

References

Bicho, R.C., Roelofs, D., Mariën, J., Scott-Fordsmand, J.J., and Amorim, M.J.B. (2020 March). Epigenetic effects of (nano)materials in environmental species – Cu case study in *Enchytraeus crypticus*. *Environ. Int.* 136: 105447.

de Gannes, M., Ko, C., Zhang, X., Biesiada, J., Niu, L., Koch, S.E., Medvedovic, M., Rubinstein, J., and Pugak, A. (2020). Dioxin disrupts dynamic DNA methylation patterns in genes that govern cardiomyocyte maturation. *Toxicol. Sci.* 178 (2): 325–337.

Dolinoy, D.C. and Jirtle, R.L. (2008). Environmental epigenomics in human health and disease. *Enviiron. Mol. Mutagen.* 49 (1): 4–8.

Lu, X., Miousse, I.R., Pirela, S.V., Melnyk, S., Koturbash, I., and Demokritou, P. (2016). Short-term exposure to engineered nanomaterials affects cellular epigenome. *Nanotoxicology* 10 (2): 140–150. doi: 10.3109/17435390.2015.1025115.

NNI, National Nanotechnology Initiative Strategic Plan (2014 February). http://nano.gov/sites/default/files/pub_resource/2014_nni_strategic_plan.pdf.

NSTC, National Science and Technology Council, Committee on Technology, Subcommittee on Nanoscale Science (2011). National Technology Initiative Strategic Plan. https://www.nano.gov.

Roduner, E. (2006). Size matters: why nanomaterials are different. *Chem. Soc. Rev.* 35: 583–592.

Sahu, S.C. and Hayes, A.W. (2017). Toxicity of nanomaterials found in human environment: a literature review. *Toxicol. Res. Appl.* 1: 1–13.

Shen, C., James, S.A., de Jonge, M.D., Turney, T.W., Wright, P.F. A., and Feltis, B.N. (2013). Relating cytotoxicity, zinc ions, and reactive oxygen in ZnO nanoparticle–exposed human immune cells. *Toxicol. Sci.* 136 (1): 120–130.

Sierra, M.I., Valdés, A., Fernández, A.F., Torrecillas, R., and Fraga, M.F. (2016). The effect of exposure to nanoparticles and nanomaterials on the mammalian epigenome. *Int. J. Nanomed.* 11: 6297–6306. doi: 10.2147/IJN.S120104.

Thai, S., Wallace, K.A., Jones, C.P., Ren, G., Grulke, E., Castellon, B.T., Crooks, J. and Kitchin, K.T. (2016). Differential genomic effects of six different TiO2 nanomaterials on. *J. Biochem. Mol. Toxicol.* 30 (7): 331–341. doi: 10.1002/jbt.21798. Epub 2016 Feb 26.

2

Molecular Impacts of Advanced Nanomaterials at Genomic and Epigenomic Levels

Kamran Shekh[1], Rais A. Ansari[2,], Yadollah Omidi[3], and Saghir A. Shakil[4,5,6]*

[1] *Yordas Group (Canada Office), Hamilton, ON, Canada*
[2] *Department of Pharmaceutical Sciences, College of Pharmacy, Nova Southeastern University, Fort Lauderdale, FL, USA*
[3] *Department of Pharmaceutical Sciences, College of Pharmacy, Nova Southeastern University, Fort Lauderdale, FL, USA*
[4] *ToxInternational Inc, Hilliard, OH, USA*
[5] *Department of Biomedical and Biological Sciences, Aga Khan University, Karachi, Pakistan*
[6] *Institute of Environmental Science and Meteorology, University of the Philippines-Diliman, Quezon City, Philippines*
[*] *Corresponding author*

Introduction

Nanomaterials (NMs) are generally defined as nanoscale materials that have at least one of their dimensions in the range of 1–100 nm. The terms nanomaterials and nanoparticles (NPs) are often used interchangeably even though these terms are distinct. Nanoparticles, as the name indicates, are spherical particles in the nano-size range, typically 1–100 nm, whereas, NMs include NPs along with other nano-sized objects, such as nanofibers and nanorods (Auffan et al. 2009; Jeevanandam et al. 2018). NMs could be naturally occurring, such as different types of biomolecular particles in the human body that make the foundation of many biomacromolecular structures. Some examples of natural NMs are glucose particles (size approximately 1 nm), DNA (2.2–2.6 nm), ribosomes (25 nm), and antibodies (2–200 nm) (Ménétret et al. 2007; Milo and Phillips 2015; Papazoglou and Parthasarathy 2007; Schaefer 2010; Sinden 2012). Another subtype of naturally occurring NMs, which are often categorized separately, are incidental NMs, which are produced either unintentionally from anthropogenic activities or produced from purely natural processes. Examples of incidental NMs are cosmic dust, volcanic eruptions, forest fires, engine exhaust, building demolition, cigarette smoke, and NMs formed in the environment from various types of plastic waste (Jeevanandam et al. 2018; Kumar et al. 2020). Nanoscale materials that are intentionally produced due to their desirable properties or their application in a specific area are designated as engineered nanomaterials (ENMs) (Yokel and Macphail 2011). ENMs have huge applications in a variety of sectors, including medicine, specialty chemicals, solar batteries, aviation and construction industries, paint, food industry, and electronics, just to name a few (Ali 2020). Due to such huge applications, the use of NMs has risen dramatically in recent years and with high usage, the exposure of humans and the environment to NMs has also significantly increased.

Impact of Engineered Nanomaterials in Genomics and Epigenomics, First Edition. Edited by Saura C. Sahu.

The use of NMs increased tremendously after findings that several key physicochemical properties of matter change significantly when they are produced at the nano-sized range. Some of the key physicochemical changes that happen in the properties of matter at the nanoscale level include increased strength and hardness, decreased melting point, increased heat capacity, and solubility. The key change in the properties of matter at the nanoscale is the significant increase in the surface area. Remarkably, the majority of unique physicochemical properties of NMs can be explained by an increased surface area (Trotta and Mele 2019). Interestingly, the increased surface area of NMs along with their other unique physicochemical properties are also responsible for their unique biological, pharmacological, and toxicological properties.

Classification of NMs

Intentionally produced NMs can be classified in several ways. The most common way to classify NMs is based on their nature, size/dimension, and morphology. Upon the constituent materials, NMs are commonly classified into four types: (i) carbon-based NMs; (ii) inorganic-based NMs; (iii) organic-based NMs; and (iv) composite-based NMs (Majhi and Yadav 2021). Carbon-based NMs are available in different types of morphologies, such as fullerenes, carbon nanotubes, nanofibers, carbon black, and graphene. Inorganic-based NMs include metal and metal oxide NMs, such as AgNPs, AuNPs, FeONPs, SiO_2NPs, and ZnONPs. Organic-based NMs are made from organic materials, they do not include carbon-based NMs (Jeevanandam et al. 2018). Composite-based NMs, which are multiphase solid materials with one of the phases, have one, two, or three dimensions of less than 100 nm or structures with nanoscale repeat distances between the phases making up the materials. NMs are also classified based on their size/dimension and they could be zero-dimensional, where all three dimensions of the NMs are in the nanoscale range. Most NPs fall into this category. For one-dimensional NMs, one of the dimensions is in the nanoscale range and the other two dimensions are not. Nanorods and nanotubes fall within this category. The two-dimensional NMs, as the name indicates have two of the three dimensions in the nanoscale range and the third out of the nanoscale range; major examples of this type of NMs are nanofilms and nanolayers. Finally, three-dimensional (3D) NMs have none of its three dimensions in the nanoscale range. Materials such as nanocomposites and bundles of nanotubes are examples of 3D NMs (Jeevanandam et al. 2018).

Absorption and Distribution of NMs

Inhalation Exposure

Inhalation is considered the most relevant route of possible exposure for NMs. Depending on the size and physicochemical properties, NMs can be deposited at different locations throughout the respiratory tract. Nanoscale particles can reach the deepest part of the lungs, i.e., gas exchange surfaces, and larger particles are deposited further up. Similarly, large-diameter fibers are deposited in the respiratory branches whereas smaller-diameter fibers move further down and may reach up to the alveoli. Very long fibers (high aspect ratio) with smaller diameters may, however, remain stuck in upper airways (Hoet et al. 2004; Lippmann 1990; Oberdörster 2001). NMs deposited in the lungs are cleared through physical translocation as well as chemical clearance. NMs that are soluble in physiological fluids are chemically dissolved and the process is much faster compared to the physical translocation. NMs, which are relatively insoluble in the lung environment, are cleared through physical transition via mucociliary function.

The first step after the inhalation of NMs is the NMs' interaction with the phospholipids and proteins in the lining fluid of the respiratory epithelium. Upon the mucociliary action, the ciliated epithelial cells in the bronchi continuously move the NMs-associated mucosal entity into the pharyngeal region resulting in the possible clearance of NMs from the lungs (Semmler et al. 2004; Takenaka et al. 2001). Another route of lung clearance for NPs of particle size <10 microns is phagocytosis in the alveolar region, where macrophage cells can phagocyte the accumulated NPs. During the process of phagocytosis, the NPs are digested in the lysosomes of phagocytes, and the digested material is removed by exocytosis (Buzea et al. 2007).

Notably, the inhaled NMs can also be transported to the nervous system through the olfactory nerve system. The nasal, tracheal, and bronchial regions are rich in nerve endings; thus, such routes can be a potential transportation path for NMs (Oberdörster et al. 2005a). In monkeys, silver-coated gold nanoparticles were demonstrated to migrate to the olfactory nerves when instilled nasally (Oberdörster et al. 2005a). Similarly, magnesium oxide (MgO) NPs and carbon NPs have also been shown to be translocated to the olfactory nerves when given to rats through the inhalation route (Elder et al. 2006; Oberdörster et al. 2002). The olfactory region is the only part of the body where the central nervous system and peripheral environment are in direct contact with each other, and chemicals absorbed through the olfactory system circumvent the blood-brain barrier (BBB) and directly reach the brain (Selvaraj et al. 2018). Alternatively, NMs can access the central nervous system by crossing the BBB which selectively controls the inward and outward traverse of substances in the brain. In this line, BBB has been shown to possess selectivity toward cationic NPs (Lockman et al. 2004). For example, AgNPs injected subcutaneously in rats have been shown to reach the brain through BBB (Hirst et al. 2013). Similarly, titanium dioxide (TiO_2)NPs when administered intragastrically, accumulate in the hippocampus region of the brain (Hu et al. 2011).

Oral Exposure

Two NMs that are highly used in products for dietary consumption are TiO_2NPs and silicates. NMs used in pharmaceuticals and oral cosmetics are also subjected to exposure through the oral route. Additionally, NMs cleared through mucociliary clearance also eventually reach the gastrointestinal (GI) tract. The absorption of particles through the GI tract is a size-dependent phenomenon that decreases with the increasing size of the NMs (Jani et al. 1990). The time to cross the mucosal layer of the intestinal epithelium seems to be also dependent on the size (Hoet et al. 2004). Only NPs, which cross the mucus layer and reach enterocytes, can be absorbed. The mucus layer is in general negatively charged; therefore, diffusion of NMs through the mucus layer is charge dependent as well, where positively charged materials are trapped and negatively charged materials are diffused with relative ease and reach epithelial enterocytes (Hoet et al. 2004).

Dermal Exposure

Studies have shown that NMs are capable of penetrating through the stratum corneum, the outer portion of the epidermis, which is composed of a keratinized layer of dead cells (Borm et al. 2006; Oberdörster et al. 2005b). *In vitro* experiments have also proved intracellular penetration of NPs (Monteiro-Riviere et al. 2005). Although it is still debatable whether NMs can penetrate through unbroken skin, the penetration of NMs through hair follicles and broken skin are well-known pathways of their uptake. For example, TiO_2NPs used in commercially available sunscreen products showed higher penetration in hairy skin, indicating that hair follicles possibly provide an additional route for the absorption of TiO_2NPs (Tsuji et al. 2006).

Circulatory Distribution

Studies have shown that the circulatory distribution of NMs is highly size- and material-dependent, in which metallic NPs of size less than <30 nm can most readily enter the circulatory system (Donaldson et al. 2005; Liu et al. 2006; Oberdörster et al. 2005b; Takenaka et al. 2001). When the exposure happens through inhalation, NMs of <20 nm diameter are cleared rapidly from the lungs compared to larger particles and hence these smaller particles enter rapidly into the circulatory system (Takenaka et al. 2001). AuNPs, TiO_2NPs, and AgNPs all with a diameter ≤30 nm have been shown to translocate rapidly through the blood to organs, such as the brain, liver, heart, kidney, and spleen (Geiser et al. 2005; Oberdörster et al. 2005a; Takenaka et al. 2001). On the other hand, very little information regarding the rapid circulatory translocation of nonmetallic NMs are available and the results are contradictory. For example, some studies show no translocation of carbon NPs of various sizes ranging between 5 and 100 nm, but some studies show rapid and extensive translocation of nonmetallic NPs of sizes ranging from 5 to 30 nm (Brown et al. 2002; Mills et al. 2006; Nemmar et al. 2002a,b; Oberdörster et al. 2002; Wiebert et al. 2006a,b). In terms of the mechanism of NMs' circulatory distribution, it is known that NMs are taken up by all major blood cells. The uptake of NPs by red blood cells (RBC) is primarily determined by the size of NPs, whereas the charge seems to be the primary determinant of NPs uptake in platelets (Nemmar et al. 2008; Peters et al. 2006).

Accumulation of NMs in Organs

Although endothelial cells of the vascular system are supposed to be a physical barrier to nanoscale particles, they have been shown to cross different types of barriers. For example, ferritin molecules of ~10 nm were detected in the deep brain when injected into the cerebrospinal fluids of rats (Buzea et al. 2007). AgNPs (4–10 nm in diameter) were accumulated in critical organs (e.g., the liver, kidney, and heart) of rats through blood circulation following inhalation exposure (Takenaka et al. 2001). Similarly, NMs generated from wearing dental bridges were accumulated in the liver and kidney through circulatory distribution after absorption from the GI tract; particles accumulated in the liver and kidney were 20 and <6 μm, respectively. The accumulation of NMs has also been shown to occur after oral exposure. For example, when rats were fed with polystyrene, spheres ranging from a few nanometers to a few micrometers in size were distributed to organs, including the liver and spleen when the particle size was <300 nm (Jani et al. 1990).

Major Adverse Effects of NMs

As discussed above, inhalation is the most relevant route of exposure to NMs and the clearance of NMs from the lungs is the key factor in determining the adverse effects of NMs in the lungs as a high lung NM burden may lead to toxicological effects. Upon continuous exposure, the insoluble NMs accumulate rapidly following the saturation of the mucociliary elimination mechanisms. When the NM accumulation reaches a level that is beyond the saturation point of the phagocytosis in the lungs, it leads to adverse effects. The most common effect of NMs exposure to the lungs is the induction of inflammation, epithelial cell proliferation, fibrosis, emphysema, and tumor (Borm et al. 2004; Dasenbrock 1996; Driscoll et al. 1996; Ferin 1994; Nikula et al. 1995; Oberdörster et al. 1994). The mechanisms behind the effects of NMs on the lung are not well known but the generation of reactive oxygen species (ROS) has been indicated to play a key role. For example, carbon nanotubes (CNTs) produce ROS both *in vitro* and *in vivo* (Oberdörster et al. 2005a). In general, it is suggested that NMs generate ROS either by directly interfering with the cellular respiratory pathways or by the activation of macrophages. The enhanced oxidative stress leads to alteration in intracellular calcium signaling that eventually leads to the depletion of calcium levels impairing macrophages. Reduced phagocytosis in the lungs leads to interaction between NMs and epithelial cells resulting in inflammation, further enhancing oxidative stress, cell injury, and other morphological effects (Donaldson and Stone 2003; Risom et al. 2005).

Similar to the lungs, many of the adverse effects on the brain (e.g., various neurodegenerative diseases) have been shown to originate from excessive oxidative stress or due to alteration in the homeostasis of trace elements. Several NMs have been shown to cause damage to the nervous system. Although mechanisms of these toxic effects are not well known, oxidative stress and inflammation could be potential mechanisms. AgNPs can damage BBB and cause neuronal harm when injected subcutaneously into rats (Tang et al. 2009). TiO_2NPs can also cause brain damage, specifically imposing detrimental impacts on glial cells (Ma et al. 2010).

Under normal conditions, blood is in a balanced state between clotting and bleeding and the balance is critical for human health. The tendency of blood to maintain a balance between clotting and lysing of such clots is known as "hemostatic balance". The hemostatic balance is dictated by a balance between procoagulant and anticoagulant factors. Any disturbance in the hemostatic balance leads to either bleeding or clotting. Platelets, RBCs, various protein factors, and enzymes are the components that play key roles in hemostatic balance. When NMs reach the bloodstream, they may interact with coagulation, anticoagulation, and fibrinolytic systems, which may lead to dysregulation of the hemostatic balance. CNTs have shown to induce platelet activation and aggregation in some *in vitro* studies and it has also shown to cause vascular thrombosis in an *in vivo* study in rats (Radomski et al. 2005; Semberova et al. 2009). CNT has also shown procoagulant activities in some *in vitro* studies (Luyts et al. 2014). AgNPs have been shown to cause endothelial dysfunction, reduced platelet adhesion and aggregation, and enhanced venous thrombosis formation in a couple of *in vitro* studies (Ragaseema et al. 2012; Sun et al. 2016). Ag nanowires have been shown to reduce RBC aggregation in some *in vitro* studies (Kim and Shin 2014). Au nanoparticles have demonstrated a size-dependent effect where 68 nm NPs showed no effect on platelet, whereas 20 nm NPs showed platelet activation (Deb et al. 2011). FeONPs induced RBC aggregation and platelet aggregation in some *in vitro* studies (Achilli et al. 2016; Ran et al. 2015); however, contradictory results are also found in the literature, where FeONPs showed suppression of platelet aggregation in an *in vitro* study (Cabrera et al. 2020). Several *in vitro* studies with silica have shown a vast variety of effects on the circulatory system including NO imbalance, increased coagulation factors, decreased anticoagulant factors, shortened coagulation time, increased platelet activation, and aggregation (Corbalan et al. 2012; Feng et al. 2019; Gryshchuk and Galagan 2016; Guo et al. 2015). TiO_2NPs have also shown size-dependent effects on the circulatory system. TiO_2NPs (4–6 nm in diameter) have shown increased platelet aggregation both *in vitro* and *in vivo*, whereas particles ~40 nm showed no effects (Bihari et al. 2010; Nemmar et al. 2008). ZnONPs have shown procoagulant activity and increased factor VIII levels (Luyts et al. 2014).

Inhalation of NMs also causes cardiorespiratory diseases. TiO_2NPs have been shown to accumulate in the heart of zebrafish (Chen et al. 2011) causing inflammation and necrosis of heart cells. Molecular effects induced by TiO_2 exposure in cardiac muscles are increased ROS levels and increased DNA peroxidation. Additionally, TiO_2 reduced the expression of antioxidant enzymes (Sheng et al. 2013). Cardiac injury biomarkers were also shown to increase after oral administration of TiO_2. These biomarkers included creatine kinase and lactate dehydrogenase (Bu et al. 2010). Several other NMs, such as ZnONP, AgNP, carbon NPs, and silica NPs, have also been shown to cause several myocardial damage (Bostan et al. 2016). ZnONPs have been suggested to cause myocardial toxicity through their effect on the calcium metabolism in cells. AgNP accumulates in

Table 2.1 Some examples of the toxicity of nanomaterials.

Nanomaterial	Effects	Experimental model	References
MWCNT	1) DNA damage 2) Cytotoxicity 3) Spleen toxicity 4) Inflammatory response	1) Lung cells 2) Rat glioma cells 3) Mice 4) Rats	(Deng et al. 2009; Guo et al. 2012; Han et al. 2012; Muller et al. 2009)
SWCNT	1) Neuronal toxicity 2) Inflammatory response 3) Reduced cell proliferation and adhesive ability	1) Chicken embryo 2) Human macrophage cells 3) HEK293 cells	(Belyanskaya et al. 2009; Cui et al. 2005; Fiorito et al. 2006)
C60 fullerene particles	1) Pulmonary toxicity 2) Genotoxicity	1) Rats 2) Bacterial cells	(Fujita et al. 2009; Zhang et al. 2009)
ZNONPs	1) Cytotoxicity 2) Delayed apoptosis 3) Changes in the expression of cell adhesion molecules	1) Human pulmonary cell line 2) Human neutrophils 3) Human lymphocytes	(Goncalves and Girard 2014; Lozano-Fernández et al. 2014; Wang et al. 2015)
TiO_2NPs	1) Oxidative stress 2) DNA damage 3) Changes in the expression of cell adhesion molecules 4) Cytotoxicity	1) Human gastric cells 2) Human gastric cells 3) Human lymphocytes 4) Mice skin cells	(Botelho et al. 2014; Jebali and Kazemi 2013; Lozano-Fernández et al. 2014)
AgNPs	1) Oxidative stress and cytotoxicity 2) Cell injury and dysfunction	1) Human colon carcinoma cells 2) Human umbilical endothelial cells	(Miethling-Graff et al. 2014; Shi et al. 2014)

myocardial cells after exposure, causing induction in oxidative stress. Carbon NMs cause histological damage in the left ventricle by increasing the level of injury biomarkers and a decrease in protective thiols in rodents (Ge et al. 2012; Shvedova et al. 2014). Several detailed reviews are available, covering the toxicity of various NMs (Sharifi et al. 2012; Srivastava et al. 2015). Table 2.1 presents some of the main toxicity of various NMs.

Known Cellular and Nuclear Uptake Mechanisms for Nanoparticles

A mechanistic understanding of cellular and nuclear uptake is essential for interpreting several aspects of NMs such as therapeutic efficacy and toxicity as well as the mechanism of action of toxicity. The cellular uptake as well as uptake within different cellular compartments/organelles depends on several physicochemical characteristics of the NMs. Some of these characteristics are size, shape, charge, partition coefficient, and the nature of surface coatings. The predominant mechanism of NM cellular uptake is vesicular trafficking through endocytosis and/or transcytosis. Several types of endocytosis mechanisms operate in cells. These types of endocytic pathways are clathrin-mediated endocytosis, caveolae-mediated endocytosis, phagocytosis, macropinocytosis, and pinocytosis.

Both clathrin- and caveolae-mediated endocytosis are receptor-based mechanisms and are operated in many types of cells to internalize different substrates, including viruses and NPs. Different types of cell receptors are involved in clathrin-mediated endocytosis, such as transferrin receptors, lipoprotein receptors, epidermal growth factor receptors, and beta-adrenergic receptors (Park and Oh 2014). The process of clathrin-mediated endocytosis is complex and involves the formation of the coated pit by nucleation of relevant cytoplasmic proteins followed by the invagination of the plasma membrane. Thereafter, an intracellular vesicle is formed by cutting and separating the invagination from the plasma membrane. The clathrin-mediated endocytosis proceeds through entrapment and internalization of the NPs in the vesicles (Muñoz and Costa 2012). Caveolae-mediated endocytosis proceeds through the formation of flask-shaped vesicles that are stabilized by a caveolin protein-based coat (Kaksonen and Roux 2018). Phagocytosis processes are undertaken mainly in immune cells, such as macrophages, dendritic cells, and neutrophils. Phagocytosis is primarily meant to clear pathogens and diseased cells. Nanoparticles have also been shown to be cleared through phagocytosis. The process of phagocytosis begins with the interaction between the foreign body and the phagocyte receptors such as Fc receptor, which leads to opsonization and adsorption of immunoglobulin (Sahay et al. 2010; Stuart and Ezekowitz 2005). Macropinocytosis is an uptake mechanism in many types of cells for extracellular fluid and solutes. This process is mediated through actin in which NPs become entrapped in vesicular structures and hence are transported intracellularly (Palm 2019).

The size of an NM and the type of the cell are crucial determinants of the endocytic mechanism in cells exposed to it. Nanoparticles (e.g., SiO_2NPs and CNTs) ranging between 120 and 150 nm are predominantly taken up via clathrin- or caveolin-mediated endocytosis. However, NPs of 250 nm to 3 μm have been shown *in vitro* to be taken up through phagocytosis, for example, polystyrene, PEGelyated gold nanorods [caviston from fbioe] (Foroozandeh and Aziz 2018). NMs with a size of a few to several hundred nanometers are internalized via macropinocytosis, which, as discussed above, is a generalized process to take up fluids along with particles into the cells from the extracellular environment (Lu et al. 2009; Park and Oh 2014). Some NMs internalized through macropinocytosis are micron-sized polystyrene and lipid NMs (Behzadi et al. 2017).

In comparison with the mechanistic understanding of the cellular uptake of NMs, the uptake into the nucleus is not that well understood. ENMs usually utilize nuclear localization sequences (NLS) to achieve the nuclear localization of NMs. An NLS is an amino acid sequence that naturally acts as a tag for proteins found in cells for their transport to the nucleus (Lu et al. 2021). Molecules that are larger than 10 nm are selectively transported through nuclear pore complexes after they are tagged with NLS (De Robertis et al. 1978; Dingwall et al. 1982). It has been shown that AuNPs coated with NLS are transported into the nucleus via NPC, as long as the diameter is up to 26 nm (Dworetzky et al. 1988).

Epigenetic Mechanisms and the Effect of NMs

Epigenetics alter the activity of genes without making any changes in the DNA sequence, and the changes induced by epigenetics can be transmitted to daughter cells (Simmons 2008). It has become clear in recent years that the role played by epigenetics in controlling the phenotype is very critical and should be considered for the adverse effects of xenobiotics. There are three major types of epigenetic modifications: (i) DNA methylation; (ii) modification of histone; and (iii) regulation of noncoding RNAs. NMs have shown to induce all major types of epigenetic effects. Therefore, it is essential to understand each of these mechanisms individually and how NMs affect them.

DNA Methylation

One of the major epigenetic DNA modification mechanisms is the covalent modification of the cytosine residues within the DNA. Fundamentally, DNA methylation is the addition of a methyl group on the fifth carbon position of the cytosine residues of the DNA (Figure 2.1). The methyl donor in this reaction is *S*-adenosyl-l-methionine and as a result of this reaction, cytosine is converted into 5-methylcytosine (5-meC). The enzymes responsible for DNA methylation belong to DNA methyltransferase 3 (DNMT3). One of the most interesting features of the DNA methylation process is its dynamics within the DNA. It is well known that DNA methylation is a dynamic process, which means that methylation of DNA can be seen as a balance between methylation and demethylation. The three main DNMTs are DNMT1, 3A, and 3B. Another interesting feature of DNA methylation is its location within the DNA. In mammalian cells, DNA methylation is known to occur only in the CpG dinucleotide regions of the DNA (Moore et al. 2013). As discussed above, DNA methylation is a dynamic process between methylation and demethylation. Two processes through which DNA demethylation is achieved are active and passive mechanisms. The active mechanism of DNA demethylation is more complicated than the passive one. In the active mechanism, 5-meC is sequentially converted into cytosine. First, 5-meC is oxidized to 5-hydroxymethlcytosine, which in turn is further oxidized to 5-formylcytosine, which is finally oxidized to 5-carboxylcytosine. Thymine DNA glycosylase can remove 5-formylcytosine and 5-carboxylcytosine to finally lead back to cytosine (Maiti et al. 2013). Interestingly, based on recent experimental evidence, it is suggested that 5-hydroxymethlcytosine, 5-formylcytosine, and 5-carboxylcytosine could potentially not just be intermediates but may have stable epigenetic roles in cells (Kohli and Zhang 2013; Moore et al. 2013). "Passive" demethylation mechanism, on the other hand, is replication dependent and operates during cell division. This pathway does not involve any enzyme-mediated reactions. It is known that during replication, the DNA methylation maintenance system is absent, which leads to a loss of methylation from cytosine during rounds of replication (Moore et al. 2013).

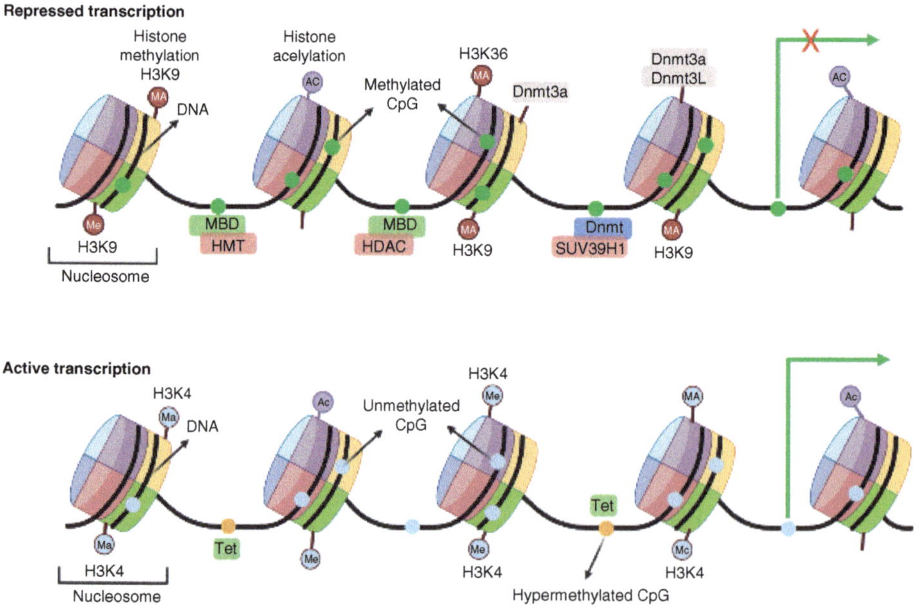

Figure 2.1 DNA methylation and demethylation. Created in BioRender.com.

Nanomaterial Effects on DNA Methylation

Extensive research has been done regarding nanomaterial effects on cellular targets. Although the effects of NMs on DNA methylation have not been studied in as great detail as other endpoints such as genotoxicity, several NMs of biomedical importance have been shown to alter the DNA methylation process (Table 2.2). For example, AgNP causes changes in DNA methylation patterns in mouse hippocampal neuronal cells and the mechanism of these changes is an alteration in the expression of DNA methyltransferase enzyme (Mytych et al. 2017). *In vivo* experiments with NMs have also demonstrated the epigenetic effects of NMs in organisms. For example, citrate-coated AuNPs of various sizes caused hypomethylation in some genes and hypermethylation in many others (Tabish et al. 2017). Many genes affected by AuNPs, such as ATM, CDK, GPX, and GSR, play key roles in regulating the cell cycle process and maintaining cellular homeostasis.

TiO_2NPs have been well studied for their genotoxic effects, while little is known about their epigenetic effects. Notably, global DNA hypomethylation was noted in a few studies. For example, global DNA hypomethylation and total DNA methyltransferase activity were noted in MRC5 lung

Table 2.2 Summary of selected nanomaterial-induced epigenetic changes via DNA modifications.

Nanomaterial	Experimental model	Epigenetic effect	References
ZnONPs	Various human and other mammalian cells	Reduction in 5-mC and enhanced expression of TET gene; DNA hypomethylation and decreased DNMT activity; increased mutation in HGPRT	(Choudhury et al. 2017; Deobagkar et al. 2016; Jain et al. 2019)
AgNPs	*In vitro* and *in vivo* mammalian data	Alteration in nucleosome assembly and DNA methylation; decrease in DNA methylation in placenta tissue	(Gurunathan et al. 2018; Zhang et al. 2015b)
AuNPs	*In vivo* mammalian data	Hypo and hypermethylation of several genes	(Tabish et al. 2017)
CuONPs	*In vitro* data	L1 and Alu hypermethylation; changes in the expression of several genes	(Lu et al. 2016b)
TiO_2NPs	Several *in vitro* studies	Enhanced methylation of SINE1; enhanced expression of TET2; PARP1 methylation increase; hypomethylation of DNA; decreased expression of DNMT activity; methionine deficiency; decreased global DNA methylation	(Lu et al. 2016a; Ma et al. 2017; Patil et al. 2016; Tucci et al. 2013)
SiO_2NPs	*In vitro*	Decreased gene and protein expression of DNMT; global DNA hypomethylation; hypermethylation of PARP1; differential methylation in several promoters	(Gong et al. 2010, Gong et al. 2012; Zou et al. 2016)
MWCNTs	*In vitro* and *in vivo*	Increased global DNA methylation; hypomethylation and increased IFN and TNF gene expression; hypermethylation and decreased expression of Thy-1 gene	(Brown et al. 2016; Li et al. 2016)
SWCNT	*In vivo*	Hypomethylation of the promoter of ATM gene	(Tabish et al. 2017)
Cadmium telluride quantum dots	*In vitro*	Condensation of chromatin	(Lovrić et al. 2005)

fibroblast cells when exposed to TiO_2NP for 24 and 48 h (Stoccoro et al. 2017). In a separate study with A549 cells, exposure to TiO_2NP caused global hypomethylation; similar demethylation was observed in human small airway epithelial cells (SAEC) on the SINE B1 repetitive elements after 24 h exposure to TiO_2NP (Lu et al. 2016a; Stoccoro et al. 2017). Experiments in several other toxicologically relevant cell lines, such as HepG2, A-431, and Caco-2, have also demonstrated a TiO_2 NP-induced global DNA hypomethylation (Pogribna et al. 2020). It is also common to observe gene-specific hypermethylation in conjunction with global DNA hypomethylation. For example, hypermethylation in several genes (e.g., SCARA3 and CDKN1A) was observed in all three cell lines studied, i.e., HepG2, Caco-2, and A-431, when treated with TiO_2NP (Pogribna et al. 2020). Changes in the expression of the genes, which are involved in maintaining DNA methylation patterns, were affected by TiNP treatment (Pogribna and Hammons 2021).

Similar to other NMs, silica (SiO_2)NM induces hypomethylation in human cells accompanied by a reduced expression of DNMT proteins (Gong et al. 2010). However, prolonged treatment of human bronchial epithelial BEAS-2B cells with SiO_2 nanoparticles promoted DNA hypermethylation (Zou et al. 2016). Zinc nanoparticles have also demonstrated a loss in DNA methylation. For example, MRC5 lung fibroblast cells, after treating with ZnONPs for 24 and 48 h demonstrated a significant reduction in DNA global methylation, and similar to many other nanoparticles, the DNA methylation changes were accompanied by a marked reduction in the expression of DNMT (Patil et al. 2016). Similar results were also recorded in other human cell lines such as human embryonic kidney HEK293 cells (Choudhury et al. 2017).

Carbon nanotubes exposure to human THP-1 cells has shown to cause promoter hypomethylation in more than a thousand genes (Öner et al. 2017). CNT-induced DNA hypomethylation also occurs in other cell lines such as 16HBE14 bronchial epithelial cells: both hypomethylation and hypermethylation at different CpG sites (Ghosh et al. 2018). However, another carbon nanomaterial known as carbon dot does not affect the methylation status of the DNA when tested in human embryonic lung fibroblast HEL 12469 cells (Sima et al. 2020). DNA methylation changes in blood cells have also been demonstrated in the workers exposed to multiwall carbon nanotubes (MWCNT); these methylation changes were present in the promoters of several critical genes (Ghosh et al. 2016).

The evidence of most of the DNA methylation from NMs discussed above comes from *in vitro* studies; DNA methylation has also been reported in animal models and human monitoring studies. Carbon-based NMs as well as other NMs, such as AuNPs and CuNPs, have been shown to affect DNA methylation status in animal studies. DNA hypermethylation has also been observed in some animal organs as a result of carbon-based nanomaterial exposure. For example, a seven-day-long exposure to MWCNTs caused significant global hypomethylation of DNA in the lung tissue and white blood cells in mice (Brown et al. 2016). Black carbon nanomaterial caused a significant increase in global cytosine DNA methylation status in the heart tissue of zebrafish (Zhou et al. 2019). Other nanoparticles such as a single dose of AuNP caused hypermethylation of several genes and hypomethylation of several other genes in the mouse lung tissues (Tabish et al. 2017). On the other hand, CuONP has shown contradictory findings. CuONP induced DNA hypermethylation in the lung tissue of male BALB/c mice after a single intratracheal dose but no changes in DNA methylation in the female ICR mice after inhalation exposure of varying lengths, ranging from 3 days to 3 months (Lu et al. 2016b; Rossner et al. 2020).

Titanium NP also caused DNA methylation changes in rats, as evidenced by a study in which rats were exposed orally to TiNP for six weeks resulted in significant DNA hypomethylation in exposed animals (Badr El Dine et al. 2018). In another study, mice were exposed to TiNPs through nose-only inhalation for 30 days. In this study, both young and adult mice were used and a decrease in global methylation was only observed in the young group. Moreover, as expected, RNA sequencing (RNA-seq) revealed that the expression of 197 genes was upregulated. This study also demonstrated that there could be age-specific effects of NMs with the younger population may be more sensitive to exposure to NMs (Ma et al. 2019).

Although limited, available evidence from human monitoring indicates that various NMs have the potential to induce DNA methylation changes in humans. For example, monitoring of DNA methylation in the white blood cells of the workers from the SiO_2NP manufacturing factory revealed that there was a significant hypomethylation in the workers compared with the controls (Liou et al. 2017). In contrast, there were no effects on the DNA methylation status in the blood cells of workers that worked in the MWCNT manufacturing factory (Ghosh et al. 2017), which was an interesting observation because MWCNT showed significant changes in the global DNA methylation status in some *in vitro* studies. Nonetheless, methylation changes were indeed observed in the individual CpG sites in the promoter region of several genes.

Histone Modification

Five forms of histone proteins that make nucleosomes are H1, H2A, H2B, H3, and H4. The N-terminal tails of the nucleosome histones are subject to extensive post-translational modifications (Figure 2.2). These post-translational modifications include methylation, acetylation, phosphorylation, and ubiquitination (Jenuwein 2001; Shilatifard 2006; Wade et al. 1997). Post-translational modification of

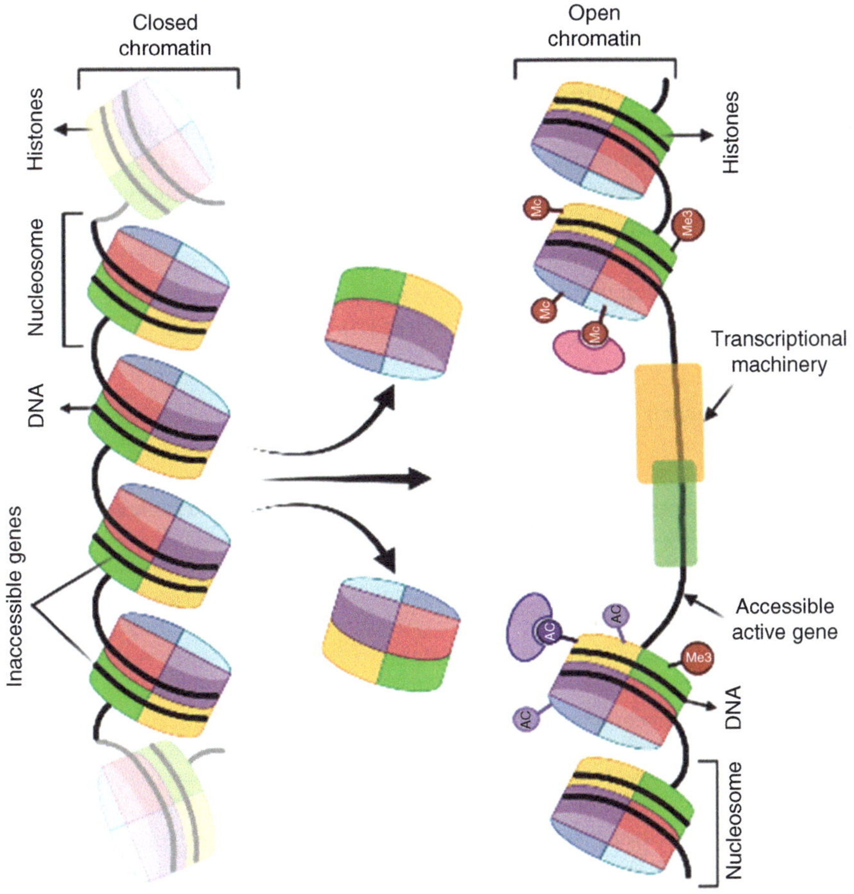

Figure 2.2 Process of the modification of histone. Created in BioRender.com.

histones regulates several important aspects of nucleic acid functionality, including regulation of chromatin structure, accessibility of regulatory proteins to the DNA, and regulation of gene expression. Post-translational modification of histones could lead to either relaxation or compression of the chromatin, thus leading to profound changes in the expression of genes. Therefore, any kind of disruption that impedes the proper functioning of histone may have significant consequences. It is well known that various types of chemical exposure could impact histone post-translational modification thus causing an effect on the phenotype (Galligan 2019). Two major enzyme families responsible for histone modification are histone acetyltransferases (HATs) and histone deacetylases (HDACs). As their name suggests, HAT acetylates histone (mainly H3 and H4), whereas HDACs deacetylate histones. It is generally observed that acetylation is responsible for keeping the genes in "active" mode, whereas deacetylated histones are normally associated with inactive genes. Histone methylation of lysine residues is also an important post-translational modification that can affect the expression of genes. As opposed to acetylation, methylation of histones can have both silencing or activating effects on genes depending on the gene in question. Some histone lysines are methylated as a result of post-translational modification and therefore lysine of histone exists in four forms, i.e., unmethylated, mono-, di-, and tri-methylated forms. The lysine methylation is mediated by s-adenosyl methionine-dependent histone methyltransferases (HMT) (Mentch et al. 2015).

All histone tails have acceptor sites that can be phosphorylated and dephosphorylated by kinases and phosphatases, respectively. Amino acid residues that can be subjected to phosphorylation are serine, threonine, and tyrosine. Several proteins are capable of recognizing phosphorylated histones and performing their downstream functions depending on the status of the phosphorylation of histones.

Nanomaterial Effects on Histone Modification

ENM-induced histone effects are not as well studied as DNA methylation and most studies are from *in vitro* models; nonetheless, available evidence still suggests that histones are targets for several NMs (Table 2.3). One of the first known effects of NMs on histones was that of cadmium telluride quantum dots in human MCF-7 breast cancer cells. Exposure to cadmium telluride caused a significant reduction in the acetylation of the histone H3. Moreover, inhibition of HDAC with a selective HDAC inhibitor reversed the reduction in acetylation of histone, suggesting that the cadmium telluride quantum dots-induced effects on histones were due to the involvement of HDACs (Choi et al. 2008).

Table 2.3 Summary of selected nanomaterials-induced epigenetic changes via histone modifications.

Nanomaterial	Experimental model	Epigenetic effect	References
AgNPs	*In vitro*	Decreased histone methylation on b-globin locus; induced phosphorylation of histone H3 via activation of aurora kinase	(Qian et al. 2015; Zhao et al. 2017)
ZnO	*In vitro*	Increased methylation of H3K9 and a decreased acetylation of H4K5; chromatin condensation	(Gao et al. 2016)
AuNPs	*In vitro*	Heterochromatin modulation; dephosphorylation and deacetylation of H3 (negatively charged NPs); decreased phosphorylation and acetylation of H3 (positively charged NPs)	(Mazumder and Shivashankar 2007; Surapaneni et al. 2018)
SiO_2NPs	*In vitro*	Decreased deacetylation of H3K9 and H3K56	(Zhang et al. 2018)
As_2O_3NPs	*In vitro*	Reduced global histone acetylation	(Liu et al. 2015)

Exposure to AgNPs has shown to decrease methylation at the lysine residues on the b-globin, decreased HMT, and decreased hemoglobin production in erythroleukemia cells. AgNPs have also shown to affect histone phosphorylation in human lung and breast adenocarcinoma cells, HaCaT, A549, and MCF-7 (Gedda et al. 2019). Increased histone deacetylation has also been shown to be one of the effects of AgNP exposure to cells such as in A549 cells (Zhao et al. 2016a).

Zinc oxide (ZnO) NPs have also shown histone alteration in HaCaT cells, specifically, enhanced histone methylation and decreased histone acetylation (Gao et al. 2016). In contrast, ZnONP exposure caused a decrease in global methylation in the human bladder cancer T24 cells. At the enzymatic levels, ZnONPs exposure in HaCaT cells resulted in the upregulation of methyltransferase genes and a downregulation of histone acetyltransferase genes (Gao et al. 2016).

Silica NPs exposure has been reported to decrease the deacetylation of histone in A549 cell lines, leading to a decreased expression of SIRT6 genes (Zhang et al. 2018). SiO_2NPs also increased the acetylation of histones H3 and H4 in Bhas 42 cells, which also enhanced the level of HDAC2 protein but reduced the levels of HDAC1 and 6 proteins (Seidel et al. 2017).

The exposure of human adipose-derived stem cells (hASCs) with TiO_2 nanotubes increased the methylation of histones at the promoter region of two different osteogenic genes with the result of osteogenic differentiation of hASCs. Moreover, histone demethylase was also inhibited (Lv et al. 2015).

Gold NPs are also known to cause several types of histone modification such as decreased histone methylation in SAEC and increased deacetylation and dephosphorylation in MDA-MB-231 cells. Interestingly, the effects of AuNPs in MDA-MB-231 cells were charge specific, where negatively charged AuNPs showed increased deacetylation and dephosphorylation as described above. However, exposure to positively charged AuNPs resulted in opposite effects, i.e., increased histone acetylation and phosphorylation (Surapaneni et al. 2018). The As_2O_3-NPs have also shown similar effects in cellular systems including HEK293 or HeLa cells, where exposure to As_2O_3-NPs reduced global histone acetylation (Liu et al. 2015). Similar to the effects of many NPs at the enzymatic level, CuONPs exposure to A549 cells caused a decrease in HDAC activity that could have significant epigenetic effects (Kalaiarasi et al. 2018).

Increased phosphorylation of histone, H2A histone family member X (abbreviated as H2AX) at serine-139, forming γH2AX, is also a very commonly reported effect of various NMs. Several studies have demonstrated that AgNP, TiO_2NPs, SiO_2NPs, AuNPs, CuONPs, As_2O_3NPs, and ZnONPs induce the formation of γH2AX in cells such as in MCF-7, A549, Caco-2, and human skin fibroblast. One of the most common observations in the studies demonstrating NPs-induced γH2AX is the induction of oxidative stress demonstrating the possible role of NP-induced γH2AX in adverse effects and disease (Gedda et al. 2019; Stoccoro et al. 2017). However, it is noteworthy that oxidative stress is not always observed in the presence of NPs-induced γH2AX.

Noncoding RNAs

Noncoding RNAs are important epigenetic modifiers. As the name indicates, noncoding RNAs do not translate to proteins. Therefore, noncoding RNAs were initially considered junk DNAs. Structurally, noncoding RNAs have been divided into two groups based on length: small and long noncoding RNAs. Small noncoding RNAs are shorter than 200 nucleotides; whereas, noncoding RNAs longer than 200 nucleotides are designated as long noncoding RNAs (lncRNAs). There are three main functionally distinct subtypes of small noncoding RNAs, namely short-interfering RNAs (siRNAs), microRNAs (miRNAs), and piwi-interacting RNAs (piRNAs). Among the noncoding RNAs, microRNAs and lncRNAs are the most toxicologically relevant entities because they are known for their involvement in the initiation and progression of various diseases (Miguel et al. 2020).

The miRNA family consists of a large number of small noncoding RNAs of approximately 22 nucleotide lengths. miRNAs play a significant role in the regulation of protein-coding genes by reducing the translation or promoting the degradation of mRNAs. The miRNAs can bind to the complementary sequences in the 3′ untranslated regions (UTR) of mRNAs and block the translation. miRNAs have a huge influence on the biological function of the cells because a single miRNA can target a large number of mRNAs affecting some critical functions in a variety of biological processes (Miguel et al. 2020). As described earlier, lncRNA are longer than 200 nucleotides untranslated transcripts, which is estimated to be more than 100,000. Similar to miRNAs, lncRNAs are also involved in the activation or repression of transcripts and regulation of the stability of mRNA. The *cis*-acting lncRNAs regulate genes located nearby. Some examples include X-inactive specific transcript (Xist), antisense Igfr2 noncoding RNA (AIRN), and lincRNA-p21. On the other hand, the *trans*-acting lncRNAs do not act at the site of transcription; rather, they influence the chromatin organization (Miguel et al. 2020).

Role of Noncoding RNAs in Disease and Environmental Health

It is well recognized by now that miRNAs and lncRNAs are involved in the pathophysiology of several diseases. For example, miRNAs play a key role in many of the biological processes that are directly linked with cancer like resistance to apoptosis, angiogenesis, and metastasis. Establishing a general rule in terms of the role of noncoding RNAs in cancer is difficult because noncoding RNAs may contribute to the promotion or inhibition of tumorigenesis depending on the tissue, cell type, and genes in question. Many cardiovascular diseases, including cardiomyopathy, myocardial infarction, and heart failure, are a result of chronic activation of myocardium remodeling. The miRNA expressions might be altered in various heart diseases. Moreover, different heart diseases have shown unique signature miRNA expression patterns. The miRNAs have been shown to regulate the expression of various genes that have well-known roles in the pathogenesis of cardiac diseases. Neurodegenerative diseases affect motor or cognitive functions depending on the type of cells that are affected; the hallmark feature of neurodegenerative diseases is the progressive death of neuronal cells. The most common neurodegenerative disorders are Alzheimer's, Parkinson's, Huntington's, and amyotrophic lateral sclerosis. The miRNAs have been shown to play key roles in several imperative processes within neurodevelopment and neurotransmission. Therefore, it is not surprising that the dysregulation of miRNAs is linked to neurodegenerative disorders. Since members of the miRNAs family have been demonstrated to play a role in alpha-synuclein, which is a key protein involved in the origin of Parkinson's disease, the reduced expression of such miRNA family members could have a significant role in Parkinson's disease. Similarly, reduced expressions of the members of the miRNA family are detected in the brains of Alzheimer's patients. Moreover, the miR-29a/b-1 cluster affects an enzyme, BACE-1 secretase, which results in the generation of Amyloid ("plague") beta species (Lekka and Hall 2018). Noncoding RNAs have also been implicated directly in other neuronal disorders, including Huntington's disease and spinal muscular atrophy. The overexpression of a particular miRNA, miR-196a, has shown to cause a reduction in mutant huntingtin and thus lessen the formation of pathological aggregates (Lekka and Hall 2018).

One of the epigenetic triggers of the environmental agents is noncoding RNAs. Abnormality in the expression of noncoding RNAs has been linked to various diseases. Notably, exposure to various types of environmental agents has been shown to affect the expression of various noncoding RNAs. For example, several lncRNAs have been shown to affect the Nrf2 pathway. Further, environmental agents, which affect redox homeostasis, have also been shown to alter the expression of lncRNAs such as Metastasis Associated Lung Adenocarcinoma Transcript 1 (MALAT1), demonstrating a link between environmental agents and perturbation of biochemical pathways through noncoding RNA. Similarly, another lncRNA, H19, is also sensitive to redox-affecting agents, such as hydrogen peroxide. The expression of miRNAs is reported to affect human keratinocytes in response to UV radiation suggesting the role of noncoding RNAs in UV-mediated effects on the skin. Several lncRNAs have also been shown to involve in UV-mediated damages (Miguel et al. 2020).

NMs and Their Effects on Noncoding RNAs

The effect of NMs on noncoding RNAs has not been researched in as much detail as other epigenetic mechanisms, such as DNA methylation and histone modification. However, limited evidence indicates that NMs could impact noncoding RNAs (Table 2.4). In a large-scale study, four different NMs were studied in mice: CuONP, MWCNT, spherical TiO_2, and rod-shaped TiO_2. Moreover, four different surface chemistries (pristine, COOH, NH_2, and PEG) for each of the NPs were also included in the study to assess the impact of the surface chemistry of the NPs on their biological effects and a multi-omic approach was utilized to assess which of the adverse effects caused by the NPs were mediated through noncoding RNAs. Mice were exposed to NPs for four days and the study revealed that both the NPs and coating types impact the expression of several lncRNA transcripts with CuONPs affecting the largest number of lncRNA transcripts. All NPs affected various key lncRNA-dependent biological processes such as mitotic cell division. The study also suggested that the immunomodulatory effects of spherical TiO_2 could be due to the alterations in genes that are affected by altered lncRNA expression (Ndika et al. 2021).

In a study, NiNPs-induced apoptosis in Sertoli cells was directly linked to lncRNA type LOC102551356. LOC102551356 was shown to be upregulated when primary cultured Sertoli-germ cells were treated with NiNPs. The upregulation in LOC102551356 was correlated to Igfbp3 gene in the P53 pathway, suggesting that lncRNA could have a role in the apoptosis of germ cells (Kong et al. 2019). In another

Table 2.4 Summary of selected nanomaterials-induced epigenetic changes via miRNA alterations.

Nanomaterial	Experimental model	Epigenetic effect	References
AgNP	*In vitro*	Alteration of up to 63 miRNAs with corresponding changes in the expression of several genes including genes of bone formation	(Eom et al. 2014; Mahmood et al. 2011)
AuNP	*In vitro* and *in vivo*	Up to 28 miRNA dysregulated including miR-298, miR-155, and miR-183; chromatin condensation	(Balansky et al. 2013; Chew et al. 2012; Ng et al. 2011)
Fe_2O_3NPs	*In vitro*	miRNA expression profile changed at the entire genome-wide level	(Li et al. 2011a)
CO_3O_4NPs	*In vitro*	Changes in several miRNA levels such as miR-21, miR-30a, miR-155	(Alinovi et al. 2017)
SiO_2NPs	*In vivo*	miR-122 and miR-192 upregulated	(Nagano et al. 2013)
MWCNTs	*In vitro*	Dysregulation of miRNA expression profile	(Li et al. 2011a)
TiO_2NPs	*In vitro* and *in vivo*	Upregulation of several miRNA such as miR-1, miR-449a, miR-135B and downregulation of several miRNA such as miR-21 and miR-30a	(Alinovi et al. 2017; Halappanavar et al. 2011)
Cd–Te quantum dots	*In vitro*	Alteration in the expression of miRNA at the global level	(Li et al. 2011b)

study, Al_2O_3NPs, consisting of a well-known pulmonary inflammatory agent, were also shown to affect miR-297 (Yun et al. 2020). The effects were observed in human bronchial epithelial cells, and it was demonstrated that miR-297 regulated the inflammatory response by reducing NF-κB-activating protein. A very similar miRNA-mediated inflammatory response was observed in neuronal cells after exposure to MnNPs and the type of miRNA responsible for affecting miR-155. It was demonstrated that the exposure to MnNPs caused a decrease in the levels of miR-155, which in turn caused an increase in mRNA levels of TNFa and IL-6 (Grogg et al. 2016).

Nonhuman species also provide important clues on the potential impact of NMs on noncoding RNAs. For example, a very commonly used cosmetic ingredient nano polystyrene has been shown to cause alteration in the expression of miRNAs, such as miR-35, miR-38, miR-76, miR-354, and miR-794 in *Caenorhabditis elegans* (Qu et al. 2019). ZnNps causes alteration in miRNA expression of granulosa cells of hens (Zhao et al. 2016b).

The SiO_2NPs are one of the well-studied NPs in terms of their effects on noncoding RNAs. SiO_2-induced lung damage in mice has been suggested to be mediated, at least in part, through miR-212 and miR-18a (Yang et al. 2017). SiO_2NP-induced apoptosis in rodent spermatocyte cells was suggested to be due to the repression of miR-2861, which in turn upregulates key apoptosis-contributing genes such as *fas* and *fadd* (Ren et al. 2019). In zebrafish, the effects of SiO_2NPs and a connection between miR-7147, miR-26a, and miR-375 and cardiovascular toxicity were shown by Hu et al. (2017). Although not conclusive, there is also some evidence that AgNPs could cause oxidative stress and genotoxicity, at least partially, through alteration in miRNA-linked pathways. For example, in an *in vitro* study with human neural stem/progenitor cells, exposure to AgNPs altered miRNAs that are known to involve in oxidative stress and inflammation (Oh et al. 2016). Similar to *in vitro* experiments, exposure to AgNP in ciliates *Euplotes vannus* caused altered expression of several miRNAs in addition to adverse effects such as oxidative stress and cell cycle alteration (Pan et al. 2018). AuNPs also affected several miRNAs expressed in human cells (Huang et al. 2015). FeO nanoparticles (superparamagnetic) have been shown to affect miRNAs expression in PC12 and HepG2, the miRNAs that are involved in the process of cell proliferation in cancer (Brzóska et al. 2019; Sun et al. 2015).

Genetic and Genomic Effects of NMs

Direct damage to DNA and/or altered transcription govern, to a great extent, the response of a cell to environmental exposure. The ability of an environmental agent to cause genotoxicity is considered a serious adverse event because of the potentially inheritable nature of the effect that genotoxicity may cause. Therefore, genotoxicity evaluation is mandatory for the majority of chemicals in all major regulatory regimes. Another important aspect of the toxicity of environmental agents is their genomic effects at the messenger RNA level, which provide great detail on the overall effect of an agent on the whole system level. Responses at the genetic and genomic levels provide a more sensitive approach to understanding the effects of chemicals compared with traditional toxicity endpoints such as carcinogenicity and reproductive toxicity. Moreover, genomic effects also provide a better mechanistic understanding of the adverse effects caused by a chemical (Nymark et al. 2018).

Genetic Damage (Genotoxicity)

Many chemical substances have the potential to interact with genetic material. The ability of a chemical substance to harm genetic information is called "genotoxicity," which encompasses two types of mechanisms of DNA damage, namely mutagenicity and clastogenicity. Mutagenicity refers to the ability of a substance to cause mutations or changes in genetic code that are permanent and inheritable.

On the other hand, clastogenicity refers to the ability of a chemical to cause breaks in DNA strands that could be single- or double-stranded (Ren et al. 2017). The molecular mechanisms through which genotoxicity is caused by chemicals are not fully understood but it is well-recognized that DNA repair pathways that operate after damage to nucleotides in the DNA strands often lead to mutations. These repair pathways include base and nucleotide excision repair pathways and mismatch repair pathways (Ren et al. 2017). Chemicals cause damage to the nucleotide through mechanisms such as the alkylation of DNA bases, interacting with nucleotides and forming DNA adducts, and oxidation of bases through generating ROS. Overall, from the regulatory perspective, two types of genotoxicity endpoints are relevant: mutagenicity and structural chromosomal aberration, and a battery of tests are employed to detect the two categories of genotoxicity endpoints. The term battery is used here in the type of tests to emphasize the fact that one test is often not sufficient to provide information on all types of genotoxicity endpoints. In recent decades, significant progress has been made in the development of *in vitro* methods for genotoxicity assessments and a battery of several *in vitro* genotoxicity methods is usually sufficient to address genotoxicity endpoints and, in some cases, where the results of *in vitro* methods are contradictory or if conclusion on genotoxicity cannot be reached with *in vitro* tests, *in vivo* genotoxicity assays in rats and/or mouse are used to complement the data. Among *in vitro* tests, the most commonly used studies to investigate gene (point) mutation are bacterial reverse mutation assay (OECD Test Guideline (TG) 471) and *in vitro* gene mutation assay in mammalian cells (OECD TG 476), and the most commonly used *in vitro* studies for investigating chromosome aberrations are *in vitro* chromosomal aberration assay (OECD TG 473) and micronucleus assay (OECD TG 487). Among *in vivo* tests, gene mutation assays in transgenic models (draft OECD TG) is most commonly used to investigate gene mutations; whereas, mammalian erythrocyte micronucleus test (OECD TG 474) and mammalian bone marrow chromosomal aberration test (OECD TG 475) are used to investigate chromosome damage. In addition, two *in vivo* methods are often used to investigate primary DNA damage: comet assay and mammalian unscheduled DNA synthesis assay (OECD TG 486). Comet assay has also been utilized in the format.

Many NMs have been shown to induce genotoxicity in a variety of models, in which the mechanisms of NM-induced genotoxicity have been investigated extensively (Doak and Dusinska 2017; Evans et al. 2017; Magdolenova et al. 2015; Nelson et al. 2017; Pfuhler et al. 2017, 2020). Genotoxicity, which is induced by the direct interaction of NMs with the genome, depends on whether or not NMs have the ability to enter the cell and reach the nucleus. Studies have indicated that NMs may enter the nucleus via nuclear pores or through endocytosis (Kazimirova et al. 2020; Zhang et al. 2019). Once inside the nucleus, NMs may have different targets to cause direct genotoxicity, depending on the cell cycle stage. For example, NMs may interact with DNA molecules directly and interfere with replication if the cell is in the interphase; whereas, during mitosis, NMs may interfere with the spindle apparatus and/or centrioles and thus potentially causing effects, such as clastogenicity or aneugenicity (Kisin et al. 2011). Indirect genotoxicity, as the name indicates, is not due to the direct interaction of NMs with DNA. Instead, it is mediated by ROS or the release of ions from NMs (Doak and Dusinska 2017; Oecd 2014). Indirect mechanisms are considered to be the predominant mode of genotoxicity for NMs. Similar to other toxicity endpoints, genotoxicity by NMs also depends on size, surface properties, and shape in addition to exposure duration and dose.

In the literature, it is apparent that the Ames test is not as widely conducted for NMs as other *in vitro* genotoxicity tests. Some of the NPs that have been studied in the Ames assay are AgNPs, TiO_2NPs, Al_2O_3NPs, ZnONPs, SWCNTs, and MWCNTs (Guo and Chen 2015). It is interesting to note that a literature review of the NPs tested in the Ames test revealed that the majority of the results were negative (Guo and Chen 2015). Among 86 publications that included Ames test data available for 126 NMs, it was shown that only 27% of the assays demonstrated positive outcomes in at least one or more strains used in the test. In addition, the positive responses were weak in the majority of the cases. One of the reasons for the lack of positive results for NPs in the Ames test could be the insensitivity of the bacterial strains to NPs due to the lack of their capability to internalize or process the NPs. A literature review has revealed that the comet assay is one of the most commonly used assays for the evaluation of the genotoxicity of NMs. The *in vitro* comet format utilizes mammalian cells, providing more biologically relevant information compared with the Ames test which utilizes bacterial cells. Carbon-based NMs, TiO_2NPs, and Fe- and Ag-based NMs are among the most commonly tested NMs for genotoxicity using comet assay (Guo and Chen 2015).

Interestingly, 78% of the total reported comet assays with NPs in 168 reports showed positive results. One of the possible reasons for the high positivity rate for NMs in the comet assay is the ability of the assay to detect various types of DNA damage, such as single and double strand breaks, DNA cross-links, and oxidative damage to DNA bases (Mei et al. 2012; Olive and Banáth 2006; Tice et al. 2000). The standard alkaline comet assay detects the breaks in DNA but DNA modifications such as oxidative damage to bases and cross-links are not always detected by standard tests, and hence, a variation of the comet assay designated as the enzyme-modified comet assay is used, which utilizes lesion-specific endonucleases to detect base modifications, such as oxidized nucleotides. The most common modifying enzymes used in the modified comet assay are based on amidopyrimidine DNA glycosylase (FPG) and endonuclease III (EndoIII). NMs-induced oxidative lesions in DNA have been observed with Fe_3O_4 in SWCNTs in mammalian cells (Jacobsen et al. 2008; Karlsson et al. 2008).

Micronucleus assay, which is considered a gold standard for detecting clastogenic and aneugenic activities of genotoxicants, is also widely utilized in the *in vitro* format (OECD 487). With NMs, the literature review revealed that *in vitro* micronucleus assay is the second most widely used genotoxicity assay and the rate of positive response was high but significantly lower than the comet assay (60% compared with 78%). NMs have shown positive results in TiO_2NP, SiO_2NP, CeO_2NP, MWCNT, AgNP, etc. (Guo and Chen 2015). Chromosomal aberration (CA) assay, another extremely useful, although a bit less definitive than the micronucleus assay, for the detection of structural damage in the DNA, is also widely used for the detection of environmental genotoxicants in the *in vitro* format (OECD 473). One of the disadvantages of the CA assay is that it is very labor-intensive and much more prone to manual error. Nonetheless, the *in vitro* CA assay was revealed to be the third most commonly used assay for NMs after the *in vitro* comet and micronucleus assays. The positivity rate for the CA assay was found to be lower than the micronucleus and comet assay (Guo and Chen 2015).

Other *in vitro* assays that have been used for testing of NMs are sister chromatid exchange, mouse lymphoma assay (MLA), and hypoxanthine phosphoribosyl transferase (HPRT) assay, although to a significantly lesser extent. A recent literature review showed that there were only two MLA assays and six HPRT assays with NMs. In the MLA assay, both AgNP and WC-CoNPs demonstrated positive responses (Mei et al. 2012; Moche et al. 2014); whereas, TiO_2NP, SiO_2NPs, and ultrafine quartz showed positive responses in the HPRT assay (Wang et al. 2007a, 2007b; Wang et al. 2007). Moreover, SWCNT of 1–3 μm dimension also indicated a positive response as opposed to SWCNT of 400–800 nm or 5–30 μm size (Manshian et al. 2013). Chronic exposure to TiO_2NPs demonstrated negative results in the HPRT assay in CHO-K1 cells (Wang et al. 2011).

The use of *in vivo* assays for the genotoxicity assessment has been reduced significantly in recent years because the combination of several *in vitro* assays provides an accurate prediction of the genotoxicity hazard of chemicals. Nonetheless, *in vivo* genotoxicity assays are still considered the gold standard and are still used if *in vitro* results are contradictory or if a conclusion on genotoxicity cannot be reached using *in vitro* assays. A recent literature review revealed that similar to the *in vitro* assay, the *in vivo* comet assay is the most used assay for the determination of genotoxicity of NMs, which is followed by micronucleus and CA assays (Guo and Chen 2015). Surprisingly, the rate of positive outcomes in the *in vivo* genotoxicity assays was found to be significantly lower compared with the *in vitro* genotoxicity assays. This could have been due to the higher repair capacity of in the intact animals. Some of the NMs that showed a positive response in terms of chromosomal damage and strand breaks in the *in vivo* assays are MWCNTs, TiO_2NPs, CdSeNPs, AgNPs, and carbon black. As discussed above, sometimes the standard comet assay could show a negative response despite the presence of DNA base modifications and such modifications can be revealed by the modified comet assay. For example, mice exposed to PVP or silicon-coated AgNPs showed a negative result in the standard comet assay; however, the presence of oxidative damage was revealed by the EndoIII and hOOG1-modified comet assay (Guo and Chen 2015).

Genomic Changes on the Messenger RNA Level

Measurement of the effect of environmental pollution on the entire genome level falls within the domain of "omics". When these effects are measured at the gene level (i.e., up and downregulation of mRNA), it is known as transcriptomics. The measurement at the protein expression level is regarded as "proteomics". It is noteworthy that the evaluation of mRNA expression after exposure to a toxicant does not provide the full picture of the actual biological effects because information at the level of mRNA does not take into account the fact that many proteins are modified post-transcriptionally and their activities are significantly affected.

The pattern of gene expressions is extremely sensitive to external stimuli, including environmental toxicants. These effects could be direct, for example, the binding of an environmental toxicant to a transcription factor, and this complex could in turn modulate the transcription of a gene or several genes. The effects of toxicants on the expression of genes could also be indirect or compensatory, which results in a response to cellular damage by the toxicant. The power of mRNA analysis comes from the fact that a pattern of expression for multiple genes may provide information about the mode of action (MOA) of a toxicant (Ankley et al. 2006). As more knowledge is gained about the link between the pattern of gene expressions and adverse outcomes, mRNA analysis at the transcriptome level could also be a great tool for predicting the adverse effects of the less well-studied toxicants. The mRNA data can be integrated and used to build adverse outcome pathways (AOP) that can help in the prediction of effects at higher levels of biological organizations. When enough mRNA and adverse effect data is available for multiple species with multiple toxicants, the information can be integrated to build models that can predict the impact of toxicants on a broader range of species in a variety of ecosystems since species within classes and families often demonstrate conservation in AOPs (Kunin et al. 2005; Luo et al. 2015; Siepel et al. 2005). Therefore, it is clear that the information on the effect of NMs on the genomic changes at the mRNA level is of critical importance.

The fundamental principle of the toxicogenomics technology is that the cells exposed to an environmental stressor will express protein-coding mRNA differently compared with an unexposed cell thus providing a type of genetic fingerprint for the stressor in question. Two major techniques utilized to evaluate genome-wide mRNA expression levels are RNA-seq and microarray. Microarrays are also referred to as DNA chips, on which DNA molecules are attached to act as probes. In microarray analysis, mRNA molecules are collected from treatment and control groups and converted into cDNA, and samples are then labeled with a fluorescent probe of a different color. Samples are then mixed and added to the microarray chip to allow binding, in a process called hybridization. Following hybridization, the chip is scanned and the expression of genes is evaluated. Hence, the microarray chip provides a method to evaluate the effect of a stressor on the expression of a large number of genes simultaneously (Matson 2009). RNA-seq is a newer technique than microarray. In RNA-seq technology, the first step is the separation of a subset of RNA (most common mRNA from the total RNA). Once the mRNA is separated, cDNA is synthesized using reverse transcription. The next step is the addition of adapters to the end of fragments. These adapters facilitate amplification and once the amplification is complete, the cDNA library is sequenced. In general, RNA-seq is considered superior to microarray. This is due to several reasons; the main reason is that RNA-seq can be used to even detect mRNA from organisms for which genome sequence is not yet available. Moreover, RNA-seq data is more sensitive at detecting very low and very high transcription and it also has a low background signal (Corchete et al. 2020).

Evaluation of gene expression as a result of NM exposure has been very valuable in revealing the mechanisms behind the toxicity of various NMs (Foldbjerg et al. 2012; Kedziorek et al. 2010; Li et al. 2014; Liu and Wang 2013). Therefore, the application of toxicogenomics to the toxicology evaluation of NMs could lead to great progress in revealing the mechanisms of their toxicity. For example, etoposide, which is a widely used cancer medication that induces apoptotic cell death, when loaded as NPs and dosed to retinoblastoma cell line (Y-79), showed differential expression of genes in the microarray analysis, with an upregulation of 171 genes, and downregulation of 280 genes. Interestingly, upregulated genes belonged to apoptosis pathways and cell cycle; thus, supporting the mechanistic understanding of the cellular effects of etoposide (Mitra et al. 2011).

Microarray analysis with AgNPs-exposed zebrafish embryos for 24 h showed 35% of genes were downregulated and 65% of genes were upregulated (van Aerle et al. 2013). Microarray analysis has also revealed important mechanistic details about adverse effects caused by SWCNT. When human macrophages were exposed to SWCNT, 23 vital cellular pathways were shown to be affected and it was demonstrated that SWCNT induced oxidative stress, cytokines activation, apoptosis, and induction of T cells due to the activation of nuclear factor κB (NF-κB) and activator protein 1 (AP-1) (Chou et al. 2008). Similarly, macrophages were shown to be affected drastically by AgNP exposure in microarray analysis, as revealed by effects on up to 45 genes, including critical stress genes such as heat shock protein and metallothionein (Lim et al. 2012). Silica NPs, for which an increasing body of evidence has suggested that they cause inflammation in the lungs (McCarthy et al. 2012), were reported to affect genes that play a key role in lung inflammation, such as Cxcl, Ccl4, and Cxcr4 in macrophages (Waters et al. 2009). Interestingly, the particle size of silica nanoparticles did not have any meaningful difference in their effects on genes. FeNPs showed induction of a few common genes in two different cell lines using the microarray experiments in the THP-1 and HepG2 cell lines. A set of common genes between two cells induced by FeNPs was found to be dose-dependent, while a set of 11 genes was induced at a low dose of FeNPs and a separate set of 10 genes was induced at the high dose. The overall conclusion is that FeNP exposure enriched the hepatitis C pathway in both cell lines (Zhang et al. 2015a). Microarray analysis has also helped greatly in establishing the mechanism of toxicity of MWCNT. Data for MWCNT exposure-induced gene expression is available for two different cell lines, i.e., human skin fibroblasts and human embryonic lung fibroblasts with various genes of critical cellular pathways are shown to alter at both low and high doses. These pathways include but are not limited to protein metabolism, Golgi transport pathways, cell cycle, and fatty acid biosynthesis (Ding et al. 2005). TiO_2NPs are one of the most intensely studied materials in terms of gene expression analysis. Exposure to TiO_2NPs in BV2 microglia has shown to have widespread effects, ranging from apoptosis to inflammation (Long et al. 2007). As reviewed elsewhere, other types of NPs that have demonstrated wide-ranging effects on gene expression using microarray analysis are ZnONPs, WCNPs, $CoCl_2NPs$, CeO_2NPs, and Al_2O_3NPs in cellular and animal models (Dwivedi et al. 2018).

The use of RNA-seq technology with NMs has only started recently; therefore, the available data on the effects of NMs on gene expression is not exhaustive as it is with microarray technology. Nonetheless, this situation is changing rapidly and almost all newer information on NM-induced global gene expression is now generated using RNA-seq. As discussed earlier, one of the major strengths of RNA-seq technology is that it can be used to even detect mRNA from organisms for which the genome sequence is not yet available. Hence, RNA-seq has proven to be a valuable tool for scientists investigating NM effects on a variety of species, and available data in nonhuman species could provide valuable information related to molecular mechanisms of toxicity of NMs. For example, available data with TiO_2 and ZnONPs exposure to nonhuman species reveal vital clues related to the MOA of these NPs in human diseases. Exposure to TiO_2 and ZnONPs in *Chlamydomonas reinhardtii* revealed inhibition of proteasome pathways hence a possibility of serious diseases, including Alzheimer's disease (Simon et al. 2013). Fullerene exposure in MCF-7 cells revealed alteration in a large number of pathways in the cell cycle, cell proliferation, protein synthesis, and maturation (Lucafò et al. 2013). Exposure to SiO_2NP in *Hydra vulgaris* caused differential expression in 45 genes with 29 genes being upregulated and 16 downregulated. However, more pathway-related information is needed to understand the mechanism better (Ambrosone et al. 2014). TiO_2NP exposure in fifth-instar silkworm (*Bombyx mori*) larvae caused alterations in a whopping 11,268 genes, many of which were related to metabolic, and cellular pathways that mediate growth, development, and nutrient metabolism (Tian et al. 2016). Earthworm, *Eisenia fetida*, when exposed to AgNPs showed a large number of upregulated and downregulated genes compared with the control, and the pathways that were affected were the metabolism of sugar and proteins and molecular stress, among many others (Novo et al. 2015).

Conclusion

Engineered NMs have huge applications in a variety of sectors such as medicine, specialty chemicals, solar batteries, aviation and construction industries, paint, food industry, and electronics. Due to such a huge number of applications, the use of NMs has risen dramatically in recent years, and with high usage, the exposure of humans and the environment to the NMs has also increased significantly. Humans are exposed to ENMs through all major routes of exposure, i.e., inhalation, oral, and dermal. ENMs have widespread toxicity in a variety of organs; therefore, understanding the hazard profile of ENMs is of paramount importance. Mechanisms of cellular uptake have been well studied showing that several types of endocytosis mechanisms that normally operate in cells are involved in the cellular uptake of ENMs. These types of endocytic pathways are clathrin-mediated endocytosis, caveolae-mediated endocytosis, phagocytosis, macropinocytosis, and pinocytosis. ENMs have also shown to penetrate nuclei; however, the mechanisms of nuclear uptake have not been studied well as those of cellular uptake. Nonetheless, available evidence indicates that ENMs usually utilize specific amino acid sequences known as NLS for their uptake into the nucleus. There could be many mechanisms of toxicity of ENMs. Collectively, in this chapter, we have reviewed the evidence available for the ENM-induced alteration in epigenetic and genomic effects and the data indicate that all major ENMs have far-reaching effects on both epigenetic and genetic levels.

References

Achilli, C., Grandi, S., Guidetti, G.F., Ciana, A., Tomasi, C., Capsoni, D., and Minetti, G. (2016). Fe_3O_4@SiO2 core–shell nanoparticles for biomedical purposes: adverse effects on blood cells. *Biomater. Sci.* 4 (10): 1417–1421. https://doi.org/10.1039/c6bm00374e.

Ali, A.S. (2020). Application of nanomaterials in environmental improvement. In: *Nanotechnology and the Environment, M. Sen (ed.).* IntechOpen.

Alinovi, R., Goldoni, M., Pinelli, S., Ravanetti, F., Galetti, M., Pelosi, G., De Palma, G., Apostoli, P., Cacchioli, A., Mutti, A., and Mozzoni, P. (2017). Titanium dioxide aggregating nanoparticles induce autophagy and under-expression of microRNA 21 and 30a in A549 cell line: a comparative study with cobalt (II, III) oxide nanoparticles. *Toxicol. Vitro: Int. J. Publ. Assoc. BIBRA* 42: 76–85.

Ambrosone, A., Scotto Di Vettimo, M.R., Malvindi, M.A., Roopin, M., Levy, O., Marchesano, V., Pompa, P.P., Tortiglione, C., and Tino, A. (2014). Impact of amorphous SiO2 nanoparticles on a living organism: morphological, behavioral, and molecular biology implications. *Front. Bioeng. Biotechnol.* 2: 37.

Ankley, G.T., Daston, G.P., Degitz, S.J., Denslow, N.D., Hoke, R.A., Kennedy, S.W., Miracle, A.L., Perkins, E.J., Snape, J., Tillitt, D.E., Tyler, C.R., and Versteeg, D. (2006). Toxicogenomics in regulatory ecotoxicology. *Environ. Sci. Technol.* 40 (13): 4055–4065.

Auffan, M., Rose, J., Bottero, J.-Y., Lowry, G.V., Jolivet, J.-P., and Wiesner, M.R. (2009). Towards a definition of inorganic nanoparticles from an environmental, health and safety perspective. *Nat. Nanotechnol.* 4 (10): 634–641.

Badr El Dine, F.M.M., Mohamed Hussein, H.A.S., and Ghazala, R.A. (2018). Evaluation of epigenetic changes of liver tissue induced by oral administration of titanium dioxide nanoparticles and possible protective role of *Nigella sativa* oil, in adult male albino rats. *Nanomed. J.* 5 (4): 192–198.

Balansky, R., Longobardi, M., Ganchev, G., Iltcheva, M., Nedyalkov, N., Atanasov, P., Toshkova, R., De Flora, S., and Izzotti, A. (2013). Transplacental clastogenic and epigenetic effects of gold nanoparticles in mice. *Mutat. Res.* 751–752: 42–48.

Behzadi, S., Serpooshan, V., Tao, W., Hamaly, M.A., Alkawareek, M.Y., Dreaden, E.C., Brown, D., Alkilany, A.M., Farokhzad, O.C., and Mahmoudi, M. (2017). Cellular uptake of nanoparticles: journey inside the cell. *Chem. Soc. Rev.* 46 (14): 4218–4244.

Belyanskaya, L., Weigel, S., Hirsch, C., Tobler, U., Krug, H.F., and Wick, P. (2009). Effects of carbon nanotubes on primary neurons and glial cells. *NeuroToxicology* 30 (4): 702–711. https://doi.org/10.1016/j.neuro.2009.05.005.

Bihari, P., Holzer, M., Praetner, M., Fent, J., Lerchenberger, M., Reichel, C.A., Rehberg, M., Lakatos, S., and Krombach, F. (2010). Single-walled carbon nanotubes activate platelets and accelerate thrombus formation in the microcirculation. *Toxicology* 269 (2–3): 148–154.

Borm, P.J.A., Robbins, D., Haubold, S., Kuhlbusch, T., Fissan, H., Donaldson, K., Schins, R., Stone, V., Kreyling, W., Lademann, J., Krutmann, J., Warheit, D., and Oberdorster, E. (2006). The potential risks of nanomaterials: a review carried out for ECETOC. *Part. Fibre Toxicol.* 3: 11.

Borm, P.J.A., Schins, R.P.F., and Albrecht, C. (2004). Inhaled particles and lung cancer, part B: paradigms and risk assessment. *Int. J. Cancer* 110 (1): 3–14. https://doi.org/10.1002/ijc.20064.

Bostan, H.B., Rezaee, R., Valokala, M.G., Tsarouhas, K., Golokhvast, K., Tsatsakis, A.M., and Karimi, G. (2016). Cardiotoxicity of nano-particles. *Life Sci.* 165: 91–99.

Botelho, M.C., Costa, C., Silva, S., Costa, S., Dhawan, A., Oliveira, P.A., and Teixeira, J.P. (2014). Effects of titanium dioxide nanoparticles in human gastric epithelial cells in vitro. *Biomed. Pharmacother.:* 68 (1): 59–64.

Brown, J.S., Zeman, K.L., and Bennett, W.D. (2002). Ultrafine particle deposition and clearance in the healthy and obstructed lung. *Am. J. Respir. Crit. Care Med.* 166 (9): 1240–1247.

Brown, T.A., Lee, J.W., Holian, A., Porter, V., Fredriksen, H., Kim, M., and Cho, Y.H. (2016). Alterations in DNA methylation corresponding with lung inflammation and as a biomarker for disease development after MWCNT exposure. *Nanotoxicology* 10 (4): 453–461.

Brzóska, K., Grądzka, I., and Kruszewski, M. (2019). Silver, gold, and iron oxide nanoparticles alter miRNA expression but do not affect DNA methylation in HepG2 cells. *Materials* 12 (7). https://doi.org/10.3390/ma12071038.

Bu, Q., Yan, G., Deng, P., Peng, F., Lin, H., Xu, Y., Cao, Z., Zhou, T., Xue, A., Wang, Y., Cen, X., and Zhao, Y.-L. (2010). NMR-based metabonomic study of the sub-acute toxicity of titanium dioxide nanoparticles in rats after oral administration. *Nanotechnology* 21 (12): 125105.

Buzea, C., Pacheco, I.I., and Robbie, K. (2007). Nanomaterials and nanoparticles: sources and toxicity. *Biointerphases* 2 (4): MR17–MR71.

Cabrera, D., Walker, K., Moise, S., Telling, N.D., and Harper, A.G.S. (2020). Controlling human platelet activation with calcium-binding nanoparticles. *Nano Res.* 13 (10): 2697–2705.

Chen, J., Dong, X., Xin, Y., and Zhao, M. (2011). Effects of titanium dioxide nano-particles on growth and some histological parameters of zebrafish (*Danio rerio*) after a long-term exposure. *Aquat. Toxicol.* 101 (3–4): 493–499.

Chew, W.-S., Poh, K.-W., Siddiqi, N.J., Alhomida, A.S., Yu, L.E., and Ong, W.-Y. (2012). Short- and long-term changes in blood miRNA levels after nanogold injection in rats—potential biomarkers of nanoparticle exposure. *Biomarkers: Biochem. Indicators Exposure Response Susceptibility Chem.* 17 (8): 750–757.

Choi, A.O., Brown, S.E., Szyf, M., and Maysinger, D. (2008). Quantum dot-induced epigenetic and genotoxic changes in human breast cancer cells. *J. Mol. Med.* 86 (3): 291–302.

Chou, -C.-C., Hsiao, H.-Y., Hong, Q.-S., Chen, C.-H., Peng, Y.-W., Chen, H.-W., and Yang, P.-C. (2008). Single-walled carbon nanotubes can induce pulmonary injury in mouse model. *Nano Lett.* 8 (2): 437–445. https://doi.org/10.1021/nl0723634.

Choudhury, S.R., Ordaz, J., Lo, C.-L., Damayanti, N.P., Zhou, F., and Irudayaraj, J. (2017). From the cover: zinc oxide nanoparticles-induced reactive oxygen species promotes multimodal cyto- and epigenetic toxicity. *Toxicol. Sci.: Offic. J. Soc. Toxicol.* 156 (1): 261–274.

Corbalan, J.J., Medina, C., Jacoby, A., Malinski, T., and Radomski, M.W. (2012). Amorphous silica nanoparticles aggregate human platelets: potential implications for vascular homeostasis. *Int. J. Nanomed.* 7: 631–639.

Corchete, L.A., Rojas, E.A., Alonso-López, D., De Las Rivas, J., Gutiérrez, N.C., and Burguillo, F.J. (2020). Systematic comparison and assessment of RNA-seq procedures for gene expression quantitative analysis. *Sci. Rep.* 10 (1): 19737.

Cui, D., Tian, F., Ozkan, C.S., Wang, M., and Gao, H. (2005). Effect of single wall carbon nanotubes on human HEK293 cells. *Toxicol. Lett.* 155 (1): 73–85.

Dasenbrock, C. (1996). The carcinogenic potency of carbon particles with and without PAH after repeated intratracheal administration in the rat. *Toxicol. Lett.* 88 (1–3): 15–21. https://doi.org/10.1016/0378-4274(96)03712-5.

De Robertis, E.M., Longthorne, R.F., and Gurdon, J.B. (1978). Intracellular migration of nuclear proteins in Xenopus oocytes. *Nature* 272 (5650): 254–256.

Deb, S., Patra, H.K., Lahiri, P., Dasgupta, A.K., Chakrabarti, K., and Chaudhuri, U. (2011). Multistability in platelets and their response to gold nanoparticles. *Nanomedicine: Nanotechnol. Biol. Med.* 7 (4): 376–384.

Deng, X., Wu, F., Liu, Z., Luo, M., Li, L., Ni, Q., Jiao, Z., Wu, M., and Liu, Y. (2009). The splenic toxicity of water soluble multi-walled carbon nanotubes in mice. *Carbon* 47 (6): 1421–1428. https://doi.org/10.1016/j.carbon.2008.12.032.

Deobagkar, D.D., Patil, N., and Gade, W.N. (2016). Epigenetic modulation upon exposure of lung fibroblasts to TiO2 and ZnO nanoparticles: alterations in DNA methylation. *Int. J. Nanomed.* 11: 4509–4519. https://doi.org/10.2147/ijn.s110390.

Ding, L., Stilwell, J., Zhang, T., Elboudwarej, O., Jiang, H., Selegue, J.P., Cooke, P.A., Gray, J.W., and Chen, F.F. (2005). Molecular characterization of the cytotoxic mechanism of multiwall carbon nanotubes and nano-onions on human skin fibroblast. *Nano Lett.* 5 (12): 2448–2464.

Dingwall, C., Sharnick, S.V., and Laskey, R.A. (1982). A polypeptide domain that specifies migration of nucleoplasmin into the nucleus. *Cell* 30 (2): 449–458.

Doak, S.H. and Dusinska, M. (2017). NanoGenotoxicology: present and the future. *Mutagenesis* 32 (1): 1–4.

Donaldson, K. and Stone, V. (2003). Current hypotheses on the mechanisms of toxicity of ultrafine particles. *Annali dell'Istituto Superiore Di Sanita* 39 (3): 405–410.

Donaldson, K., Tran, L., Jimenez, L.A., Duffin, R., Newby, D.E., Mills, N., MacNee, W., and Stone, V. (2005). Combustion-derived nanoparticles: a review of their toxicology following inhalation exposure. *Part. Fibre Toxicol.* 2: 10.

Driscoll, K.E., Carter, J.M., Howard, B.W., Hassenbein, D.G., Pepelko, W., Baggs, R.B., and Oberdörster, G. (1996). Pulmonary inflammatory, chemokine, and mutagenic responses in rats after subchronic inhalation of carbon black. *Toxicol. Appl. Pharmacol.* 136 (2): 372–380.

Dwivedi, S., Saquib, Q., Ahmad, B., Ansari, S.M., Azam, A., and Musarrat, J. (2018). Toxicogenomics: a new paradigm for nanotoxicity evaluation. *Adv. Exp. Med. Biol.* 1048: 143–161.

Dworetzky, S.I., Lanford, R.E., and Feldherr, C.M. (1988). The effects of variations in the number and sequence of targeting signals on nuclear uptake. *J. Cell Biol.* 107 (4): 1279–1287.

Elder, A., Gelein, R., Silva, V., Feikert, T., Opanashuk, L., Carter, J., Potter, R., Maynard, A., Ito, Y., Finkelstein, J., and Oberdörster, G. (2006). Translocation of inhaled ultrafine manganese oxide particles to the centralnervous system. *Environ. Health Perspect.* 114 (8): 1172–1178. https://doi.org/10.1289/ehp.9030.

Eom, H.-J., Chatterjee, N., Lee, J., and Choi, J. (2014). Integrated mRNA and micro RNA profiling reveals epigenetic mechanism of differential sensitivity of Jurkat T cells to AgNPs and Ag ions. *Toxicol. Lett.* 229 (1): 311–318.

Evans, S.J., Clift, M.J.D., Singh, N., de Oliveira Mallia, J., Burgum, M., Wills, J.W., Wilkinson, T.S., Jenkins, G.J.S., and Doak, S.H. (2017). Critical review of the current and future challenges associated with advanced in vitro systems towards the study of nanoparticle (secondary) genotoxicity. *Mutagenesis* 32 (1): 233–241.

Feng, L., Yang, X., Liang, S., Xu, Q., Miller, M.R., Duan, J., and Sun, Z. (2019). Silica nanoparticles trigger the vascular endothelial dysfunction and prethrombotic state via miR-451 directly regulating the IL6R signaling pathway. *Part. Fibre Toxicol.* 16 (1): 16.

Ferin, J. (1994). Pulmonary retention and clearance of particles. *Toxicol. Lett.* 72 (1–3): 121–125. https://doi.org/10.1016/0378-4274(94)90018-3.

Fiorito, S., Serafino, A., Andreola, F., and Bernier, P. (2006). Effects of fullerenes and single-wall carbon nanotubes on murine and human macrophages. *Carbon* 44 (6): 1100–1105. https://doi.org/10.1016/j.carbon.2005.11.009.

Foldbjerg, R., Irving, E.S., Hayashi, Y., Sutherland, D.S., Thorsen, K., Autrup, H., and Beer, C. (2012). Global gene expression profiling of human lung epithelial cells after exposure to nanosilver. *Toxicol. Sci.: Offic. J. Soc. Toxicol.* 130 (1): 145–157.

Foroozandeh, P. and Aziz, A.A. (2018). Insight into cellular uptake and intracellular trafficking of nanoparticles. *Nanoscale Res. Lett.* 13 (1): 339.

Fujita, K., Morimoto, Y., Ogami, A., Myojyo, T., Tanaka, I., Shimada, M., Wang, W.-N., Endoh, S., Uchida, K., Nakazato, T., Yamamoto, K., Fukui, H., Horie, M., Yoshida, Y., Iwahashi, H., and

Nakanishi, J. (2009). Gene expression profiles in rat lung after inhalation exposure to C60 fullerene particles. *Toxicology* 258 (1): 47–55.

Galligan, J.J. (2019). Chromatin modifications in toxicology. *Chem. Res. Toxicol.* 32 (5): 794–795.

Gao, F., Ma, N., Zhou, H., Wang, Q., Zhang, H., Wang, P., Hou, H., Wen, H., and Li, L. (2016). Zinc oxide nanoparticles-induced epigenetic change and G2/M arrest are associated with apoptosis in human epidermal keratinocytes. *Int. J. Nanomed.* 11: 3859–3874.

Ge, C., Meng, L., Xu, L., Bai, R., Du, J., Zhang, L., Li, Y., Chang, Y., Zhao, Y., and Chen, C. (2012). Acute pulmonary and moderate cardiovascular responses of spontaneously hypertensive rats after exposure to single-wall carbon nanotubes. *Nanotoxicology* 6 (5): 526–542. https://doi.org/10.3109/17435390.2011.587905.

Gedda, M.R., Babele, P.K., Zahra, K., and Madhukar, P. (2019). Epigenetic aspects of engineered nanomaterials: is the collateral damage inevitable? *Front. Bioeng. Biotechnol.* 7: 228.

Geiser, M., Rothen-Rutishauser, B., Kapp, N., Schürch, S., Kreyling, W., Schulz, H., Semmler, M., Im Hof, V., Heyder, J., and Gehr, P. (2005). Ultrafine particles cross cellular membranes by nonphagocytic mechanisms in lungs and in cultured cells. *Environ. Health Perspect.* 113 (11): 1555–1560.

Ghosh, M., Öner, D., Duca, R.C., Bekaert, B., Vanoirbeek, J.A.J., Godderis, L., and Hoet, P.H.M. (2018). Single-walled and multi-walled carbon nanotubes induce sequence-specific epigenetic alterations in 16 HBE cells. *Oncotarget* 9 (29): 20351–20365.

Ghosh, M., Öner, D., Poels, K., Tabish, A.M., Vlaanderen, J., Pronk, A., Kuijpers, E., Lan, Q., Vermeulen, R., Bekaert, B., Hoet, P.H., and Godderis, L. (2017). Changes in DNA methylation induced by multi-walled carbon nanotube exposure in the workplace. *Nanotoxicology* 11 (9–10): 1195–1210.

Ghosh, M., Öner, D., Tabis, A., Poels, K., Hoet, P., Pronk, A., Vermeulen, R., Bekaert, B., and Godderis, L. (2016). DNA methylation changes in workers occupational exposed to carbon nanotubes. *Eur. Respir. J.: Offic. J. Eur. Soc. Clin. Respir. Physiol.* 48 (suppl 60). https://doi.org/10.1183/13993003.congress-2016.PA4275.

Goncalves, D.M. and Girard, D. (2014). Zinc oxide nanoparticles delay human neutrophil apoptosis by a de novo protein synthesis-dependent and reactive oxygen species-independent mechanism. *Toxicol. Vitro: Int. J. Publ. Assoc. BIBRA* 28 (5): 926–931.

Gong, C., Tao, G., Yang, L., Liu, J., Liu, Q., Li, W., and Zhuang, Z. (2012). Methylation of PARP-1 promoter involved in the regulation of nano-SiO2-induced decrease of PARP-1 mRNA expression. *Toxicol. Lett.* 209 (3): 264–269.

Gong, C., Tao, G., Yang, L., Liu, J., Liu, Q., and Zhuang, Z. (2010). SiO2 nanoparticles induce global genomic hypomethylation in HaCaT cells. *Biochem. Biophys. Res. Commun.* 397 (3): 397–400.

Grogg, M.W., Braydich-Stolle, L.K., Maurer-Gardner, E.I., Hill, N.T., Sakaram, S., Kadakia, M.P., and Hussain, S.M. (2016). Modulation of miRNA-155 alters manganese nanoparticle-induced inflammatory response. *Toxicol. Res.* 5 (6): 1733–1743.

Gryshchuk, V. and Galagan, N. (2016). Silica nanoparticles effects on blood coagulation proteins and platelets. *Biochem. Res. Int.* 2016: 2959414.

Guo, C., Xia, Y., Niu, P., Jiang, L., Duan, J., Yu, Y., Zhou, X., Li, Y., and Sun, Z. (2015). Silica nanoparticles induce oxidative stress, inflammation, and endothelial dysfunction in vitro via activation of the MAPK/Nrf2 pathway and nuclear factor-κB signaling. *Int. J. Nanomed.* 10: 1463–1477.

Guo, F., Ma, N., Horibe, Y., Kawanishi, S., Murata, M., and Hiraku, Y. (2012). Nitrative DNA damage induced by multi-walled carbon nanotube via endocytosis in human lung epithelial cells. *Toxicol. Appl. Pharmacol.* 260 (2): 183–192.

Guo, X. and Chen, T. (2015). Progress in genotoxicity evaluation of engineered nanomaterials. *Nanomaterials – Toxicity Risk Assess.* https://doi.org/10.5772/61013.

Gurunathan, S., Qasim, M., Park, C., Yoo, H., Choi, D.Y., Song, H., Park, C., Kim, J.-H., and Hong, K. (2018). Cytotoxicity and transcriptomic analysis of silver nanoparticles in mouse embryonic fibroblast cells. *Int. J. Mol. Sci.* 19 (11). https://doi.org/10.3390/ijms19113618.

Halappanavar, S., Jackson, P., Williams, A., Jensen, K.A., Hougaard, K.S., Vogel, U., Yauk, C.L., and Wallin, H. (2011). Pulmonary response to surface-coated nanotitanium dioxide particles includes induction of acute phase response genes, inflammatory cascades, and changes in microRNAs: a toxicogenomic study. *Environ. Mol. Mutagen.* 52 (6): 425–439.

Han, Y.-G., Xu, J., Li, Z.-G., Ren, -G.-G., and Yang, Z. (2012). In vitro toxicity of multi-walled carbon nanotubes in C6 rat glioma cells. *Neurotoxicology* 33 (5): 1128–1134.

Hirst, S.M., Karakoti, A., Singh, S., Self, W., Tyler, R., Seal, S., and Reilly, C.M. (2013). Bio-distribution and in vivo antioxidant effects of cerium oxide nanoparticles in mice. *Environ. Toxicol.* 28 (2): 107–118.

Hoet, P.H.M., Brüske-Hohlfeld, I., and Salata, O.V. (2004). Nanoparticles–known and unknown health risks. *J. Nanobiotechnol.* 2 (1): 1–15.

Hu, H., Shi, Y., Zhang, Y., Wu, J., Asweto, C.O., Feng, L., Yang, X., Duan, J., and Sun, Z. (2017). Comprehensive gene and microRNA expression profiling on cardiovascular system in zebrafish co-exposured of SiNPs and MeHg. *Sci. Total Environ.* 607-608: 795–805.

Hu, R., Zheng, L., Zhang, T., Gao, G., Cui, Y., Cheng, Z., Cheng, J., Hong, M., Tang, M., and Hong, F. (2011). Molecular mechanism of hippocampal apoptosis of mice following exposure to titanium dioxide nanoparticles. *J. Hazard. Mater.* 191 (1–3): 32–40.

Huang, Y., Lü, X., Qu, Y., Yang, Y., and Wu, S. (2015). MicroRNA sequencing and molecular mechanisms analysis of the effects of gold nanoparticles on human dermal fibroblasts. *Biomaterials* 37: 13–24.

Jacobsen, N.R., Pojana, G., White, P., Møller, P., Cohn, C.A., Korsholm, K.S., Vogel, U., Marcomini, A., Loft, S., and Wallin, H. (2008). Genotoxicity, cytotoxicity, and reactive oxygen species induced by single-walled carbon nanotubes and C(60) fullerenes in the FE1-Mutatrade markMouse lung epithelial cells. *Environ. Mol. Mutagen.* 49 (6): 476–487.

Jain, A.K., Singh, D., Dubey, K., Maurya, R., and Pandey, A.K. (2019). Zinc oxide nanoparticles induced gene mutation at the HGPRT locus and cell cycle arrest associated with apoptosis in V-79 cells. *J. Appl. Toxicol.* 39 (5): 735–750.

Jani, P., Halbert, G.W., Langridge, J., and Florence, A.T. (1990). Nanoparticle uptake by the rat gastrointestinal mucosa: quantitation and particle size dependency. *J. Pharm. Pharmacol.* 42 (12): 821–826.

Jebali, A. and Kazemi, B. (2013). Triglyceride-coated nanoparticles: skin toxicity and effect of UV/IR irradiation on them. *Toxicol. Vitro: Int. J. Publ. Assoc. BIBRA* 27 (6): 1847–1854.

Jeevanandam, J., Barhoum, A., Chan, Y.S., Dufresne, A., and Danquah, M.K. (2018). Review on nanoparticles and nanostructured materials: history, sources, toxicity and regulations. *Beilstein J. Nanotechnol.* 9: 1050–1074.

Jenuwein, T. (2001). Re-SET-ting heterochromatin by histone methyltransferases. *Trends Cell Biol.* 11 (6): 266–273.

Kaksonen, M. and Roux, A. (2018). Mechanisms of clathrin-mediated endocytosis. *Nat. Rev. Mol. Cell Biol.* 19 (5): 313–326.

Kalaiarasi, A., Sankar, R., Anusha, C., Saravanan, K., Aarthy, K., Karthic, S., Mathuram, T.L., and Ravikumar, V. (2018). Copper oxide nanoparticles induce anticancer activity in A549 lung cancer cells by inhibition of histone deacetylase. *Biotechnol. Lett.* 40 (2): 249–256.

Karlsson, H.L., Cronholm, P., Gustafsson, J., and Möller, L. (2008). Copper oxide nanoparticles are highly toxic: a comparison between metal oxide nanoparticles and carbon nanotubes. *Chem. Res. Toxicol.* 21 (9): 1726–1732.

Kazimirova, A., El Yamani, N., Rubio, L., García-Rodríguez, A., Barancokova, M., Marcos, R., and Dusinska, M. (2020). Effects of titanium dioxide nanoparticles on the Hprt gene mutations in V79 hamster cells. *Nanomaterials (Basel, Switzerland)* 10 (3). https://doi.org/10.3390/nano10030465.

Kedziorek, D.A., Muja, N., Walczak, P., Ruiz-Cabello, J., Gilad, A.A., Jie, C.C., and Bulte, J.W.M. (2010). Gene expression profiling reveals early cellular responses to intracellular magnetic labeling with superparamagnetic iron oxide nanoparticles. *Magn. Reson. Med.: Offic. J. S. Magn. Reson. Med./ Soc. Magn. Reson. Med.* 63 (4): 1031–1043.

Kim, M.J. and Shin, S. (2014). Toxic effects of silver nanoparticles and nanowires on erythrocyte rheology. *Food Chem. Toxicol.: Int. J. Publ. Br. Ind. Biol. Res. Assoc.* 67: 80–86.

Kisin, E.R., Murray, A.R., Sargent, L., Lowry, D., Chirila, M., Siegrist, K.J., Schwegler-Berry, D., Leonard, S., Castranova, V., Fadeel, B., Kagan, V.E., and Shvedova, A.A. (2011). Genotoxicity of carbon nanofibers: are they potentially more or less dangerous than carbon nanotubes or asbestos? *Toxicol. Appl. Pharmacol.* 252 (1): 1–10.

Kohli, R.M. and Zhang, Y. (2013). TET enzymes, TDG and the dynamics of DNA demethylation. *Nature* 502 (7472): 472–479.

Kong, L., Hu, W., Gao, X., Wu, Y., Xue, Y., Cheng, K., and Tang, M. (2019). Molecular mechanisms underlying nickel nanoparticle induced rat Sertoli-germ cells apoptosis. *Sci. Total Environ.* 692: 240–248.

Kumar, A.G., Anjana, K., Hinduja, M., Sujitha, K., and Dharani, G. (2020). Review on plastic wastes in marine environment – Biodegradation and biotechnological solutions. *Mar. Pollut. Bull.* 150: 110733.

Kunin, V., Ahren, D., Goldovsky, L., Janssen, P., and Ouzounis, C.A. (2005). Measuring genome conservation across taxa: divided strains and united kingdoms. *Nucleic Acids Res.* 33 (2): 616–621.

Lekka, E. and Hall, J. (2018). Noncoding RNAs in disease. *FEBS Lett.* 592 (17): 2884–2900.

Li, J., Tian, M., Cui, L., Dwyer, J., Fullwood, N.J., Shen, H., and Martin, F.L. (2016). Low-dose carbon-based nanoparticle-induced effects in A549 lung cells determined by biospectroscopy are associated with increases in genomic methylation. *Sci. Rep.* 6 (1). https://doi.org/10.1038/srep20207.

Li, S., Wang, H., Qi, Y., Tu, J., Bai, Y., Tian, T., Huang, N., Wang, Y., Xiong, F., Lu, Z., and Xiao, Z. (2011a). Assessment of nanomaterial cytotoxicity with SOLiD sequencing-based microRNA expression profiling. *Biomaterials* 32 (34): 9021–9030.

Li, S., Wang, Y., Wang, H., Bai, Y., Liang, G., Wang, Y., Huang, N., and Xiao, Z. (2011b). MicroRNAs as participants in cytotoxicity of CdTe quantum dots in NIH/3T3 cells. *Biomaterials* 32 (15): 3807–3814.

Li, X., He, Q., and Shi, J. (2014). Global gene expression analysis of cellular death mechanisms induced by mesoporous silica nanoparticle-based drug delivery system. *ACS Nano* 8 (2): 1309–1320.

Lim, D.-H., Jang, J., Kim, S., Kang, T., Lee, K., and Choi, I.-H. (2012). The effects of sub-lethal concentrations of silver nanoparticles on inflammatory and stress genes in human macrophages using cDNA microarray analysis. *Biomaterials* 33 (18): 4690–4699.

Liou, S.-H., Wu, W.-T., Liao, H.-Y., Chen, C.-Y., Tsai, C.-Y., Jung, W.-T., and Lee, H.-L. (2017). Global DNA methylation and oxidative stress biomarkers in workers exposed to metal oxide nanoparticles. *J. Hazard. Mater.* 331: 329–335.

Lippmann, M. (1990). Effects of fiber characteristics on lung deposition, retention, and disease. *Environ. Health Perspect.* 88: 311–317.

Liu, D., Wu, D., Zhao, L., Yang, Y., Ding, J., Dong, L., Hu, L., Wang, F., Zhao, X., Cai, Y., and Jin, J. (2015). Arsenic trioxide reduces global histone H4 acetylation at lysine 16 through direct binding to histone acetyltransferase hMOF in human cells. *PLoS One* 10 (10): e0141014.

Liu, J., Wong, H.-L., Moselhy, J., Bowen, B., Wu, X.Y., and Johnston, M.R. (2006). Targeting colloidal particulates to thoracic lymph nodes. *Lung Cancer* 51 (3): 377–386.

Liu, Y. and Wang, J. (2013). Effects of DMSA-coated Fe3O4 nanoparticles on the transcription of genes related to iron and osmosis homeostasis. *Toxicol. Sci.: Offic. J. Soc. Toxicol.* 131 (2): 521–536.

Lockman, P.R., Koziara, J.M., Mumper, R.J., and Allen, D.D. (2004). Nanoparticle surface charges alter blood–brain barrier integrity and permeability. *J. Drug Target.* 12 (9–10): 635–641.

Long, T.C., Tajuba, J., Sama, P., Saleh, N., Swartz, C., Parker, J., Hester, S., Lowry, G.V., and Veronesi, B. (2007). Nanosize titanium dioxide stimulates reactive oxygen species in brain microglia and damages neurons in vitro. *Environ. Health Perspect.* 115 (11): 1631–1637.

Lovrić, J., Bazzi, H.S., Cuie, Y., Fortin, G.R.A., Winnik, F.M., and Maysinger, D. (2005). Differences in subcellular distribution and toxicity of green and red emitting CdTe quantum dots. *J. Mol. Med.* 83 (5): 377–385.

Lozano-Fernández, T., Ballester-Antxordoki, L., Pérez-Temprano, N., Rojas, E., Sanz, D., Iglesias-Gaspar, M., Moya, S., González-Fernández, Á., and Rey, M. (2014). Potential impact of metal oxide nanoparticles on the immune system: the role of integrins, L-selectin and the chemokine receptor CXCR4. *Nanomedicine: Nanotechnol. Biol. Med.* 10 (6): 1301–1310.

Lu, F., Wu, S.-H., Hung, Y., and Mou, C.-Y. (2009). Size effect on cell uptake in well-suspended, uniform mesoporous silica nanoparticles. *Small* 5 (12): 1408–1413.

Lu, J., Wu, T., Zhang, B., Liu, S., Song, W., Qiao, J., and Ruan, H. (2021). Types of nuclear localization signals and mechanisms of protein import into the nucleus. *Cell Commun. Signal.* 19 (1): 60.

Lu, X., Miousse, I.R., Pirela, S.V., Melnyk, S., Koturbash, I., and Demokritou, P. (2016a). Short-term exposure to engineered nanomaterials affects cellular epigenome. *Nanotoxicology* 10 (2): 140–150.

Lu, X., Miousse, I.R., Pirela, S.V., Moore, J.K., Melnyk, S., Koturbash, I., and Demokritou, P. (2016b). In vivo epigenetic effects induced by engineered nanomaterials: a case study of copper oxide and laser printer-emitted engineered nanoparticles. *Nanotoxicology* 10 (5): 629–639.

Lucafò, M., Gerdol, M., Pallavicini, A., Pacor, S., Zorzet, S., Da Ros, T., Prato, M., and Sava, G. (2013). Profiling the molecular mechanism of fullerene cytotoxicity on tumor cells by RNA-seq. *Toxicology* 314 (1): 183–192.

Luo, H., Gao, F., and Lin, Y. (2015). Evolutionary conservation analysis between the essential and nonessential genes in bacterial genomes. *Sci. Rep.* 5: 13210.

Luyts, K., Smulders, S., Napierska, D., Van Kerckhoven, S., Poels, K., Scheers, H., Hemmeryckx, B., Nemery, B., Hoylaerts, M.F., and Hoet, P.H.M. (2014). Pulmonary and hemostatic toxicity of multi-walled carbon nanotubes and zinc oxide nanoparticles after pulmonary exposure in Bmal1 knockout mice. *Part. Fibre Toxicol.* 11 (1). https://doi.org/10.1186/s12989-014-0061-5.

Lv, L., Liu, Y., Zhang, P., Zhang, X., Liu, J., Chen, T., Su, P., Li, H., and Zhou, Y. (2015). The nanoscale geometry of TiO2 nanotubes influences the osteogenic differentiation of human adipose-derived stem cells by modulating H3K4 trimethylation. *Biomaterials* 39: 193–205.

Ma, L., Liu, J., Li, N., Wang, J., Duan, Y., Yan, J., Liu, H., Wang, H., and Hong, F. (2010). Oxidative stress in the brain of mice caused by translocated nanoparticulate TiO2 delivered to the abdominal cavity. *Biomaterials* 31 (1): 99–105. https://doi.org/10.1016/j.biomaterials.2009.09.028.

Ma, Y., Guo, Y., Wu, S., Lv, Z., Zhang, Q., and Ke, Y. (2017). Titanium dioxide nanoparticles induce size-dependent cytotoxicity and genomic DNA hypomethylation in human respiratory cells. *RSC Adv.* 7 (38): 23560–23572.

Ma, Y., Guo, Y., Ye, H., Huang, K., Lv, Z., and Ke, Y. (2019). Different effects of titanium dioxide nanoparticles instillation in young and adult mice on DNA methylation related with lung inflammation and fibrosis. *Ecotoxicol. Environ. Saf.* 176: 1–10.

Magdolenova, Z., Drlickova, M., Henjum, K., Rundén-Pran, E., Tulinska, J., Bilanicova, D., Pojana, G., Kazimirova, A., Barancokova, M., Kuricova, M., Liskova, A., Staruchova, M., Ciampor, F., Vavra, I., Lorenzo, Y., Collins, A., Rinna, A., Fjellsbø, L., Volkovova, K., and Dusinska, M. (2015). Coating-dependent induction of cytotoxicity and genotoxicity of iron oxide nanoparticles. *Nanotoxicology* 9 (Suppl 1): 44–56.

Mahmood, M., Li, Z., Casciano, D., Khodakovskaya, M.V., Chen, T., Karmakar, A., Dervishi, E., Xu, Y., Mustafa, T., Watanabe, F., Fejleh, A., Whitlow, M., Al-Adami, M., Ghosh, A., and Biris, A.S. (2011). Nanostructural materials increase mineralization in bone cells and affect gene expression through miRNA regulation. *J. Cell. Mol. Med.* 15 (11): 2297–2306.

Maiti, A., Michelson, A.Z., Armwood, C.J., Lee, J.K., and Drohat, A.C. (2013). Divergent mechanisms for enzymatic excision of 5-formylcytosine and 5-carboxylcytosine from DNA. *J. Am. Chem. Soc.* 135 (42): 15813–15822.

Majhi, K.C. and Yadav, M. (2021). Synthesis of inorganic nanomaterials using carbohydrates. In: *Green Sustainable Process for Chemical and Environmental Engineering and Science* (ed. B.R. Inamuddin, M.I. Ahamed, and A.M. Asiri), 109–135. Elsevier.

Manshian, B.B., Jenkins, G.J., Williams, P.M., Wright, C., Barron, A.R., Brown, A.P., Hondow, N., Dunstan, P.R., Rickman, R., Brady, K., and Doak, S.H. (2013). Single-walled carbon nanotubes: differential genotoxic potential associated with physico-chemical properties. *Nanotoxicology* 7 (2): 144–156.

Matson, R. (2009). Introduction to microarray technologies. In: *Microarray Methods and Protocols*, 1–11. CRC Press. https://doi.org/10.1201/9781420046663.

Mazumder, A. and Shivashankar, G.V. (2007). Gold-nanoparticle-assisted laser perturbation of chromatin assembly reveals unusual aspects of nuclear architecture within living cells. *Biophys. J.* 93 (6): 2209–2216.

McCarthy, J., Inkielewicz-Stępniak, I., Corbalan, J.J., and Radomski, M.W. (2012). Mechanisms of toxicity of amorphous silica nanoparticles on human lung submucosal cells in vitro: protective effects of fisetin. *Chem. Res. Toxicol.* 25 (10): 2227–2235.

Mei, N., Zhang, Y., Chen, Y., Guo, X., Ding, W., Ali, S.F., Biris, A.S., Rice, P., Moore, M.M., and Chen, T. (2012). Silver nanoparticle-induced mutations and oxidative stress in mouse lymphoma cells. *Environ. Mol. Mutagen.* 53 (6): 409–419.

Ménétret, J.-F., Schaletzky, J., Clemons, W.M., Osborne, A.R., Skånland, S.S., Denison, C., Gygi, S.P., Kirkpatrick, D.S., Park, E., Ludtke, S.J., Rapoport, T.A., and Akey, C.W. (2007). Ribosome binding of a single copy of the SecY complex: implications for protein translocation. *Mol. Cell* 28 (6): 1083–1092. https://doi.org/10.1016/j.molcel.2007.10.034.

Mentch, S.J., Mehrmohamadi, M., Huang, L., Liu, X., Gupta, D., Mattocks, D., Gómez Padilla, P., Ables, G., Bamman, M.M., Thalacker-Mercer, A.E., Nichenametla, S.N., and Locasale, J.W. (2015). Histone methylation dynamics and gene regulation occur through the sensing of one-carbon metabolism. *Cell Metab.* 22 (5): 861–873.

Miethling-Graff, R., Rumpker, R., Richter, M., Verano-Braga, T., Kjeldsen, F., Brewer, J., Hoyland, J., Rubahn, H.-G., and Erdmann, H. (2014). Exposure to silver nanoparticles induces size- and dose-dependent oxidative stress and cytotoxicity in human colon carcinoma cells. *Toxicol. Vitro* 28 (7): 1280–1289. https://doi.org/10.1016/j.tiv.2014.06.005.

Miguel, V., Lamas, S., and Espinosa-Diez, C. (2020). Role of non-coding-RNAs in response to environmental stressors and consequences on human health. *Redox Biol.* 37: 101580.

Mills, N.L., Amin, N., Robinson, S.D., Anand, A., Davies, J., Patel, D., de la Fuente, J.M., Cassee, F.R., Boon, N.A., Macnee, W., Millar, A.M., Donaldson, K., and Newby, D.E. (2006). Do inhaled carbon nanoparticles translocate directly into the circulation in humans? *Am. J. Respir. Crit. Care Med.* 173 (4): 426–431.

Milo, R. and Phillips, R. (2015). *Cell Biology by the Numbers*. Garland Science.

Mitra, M., Dilnawaz, F., Misra, R., Harilal, A., Verma, R.S., Sahoo, S.K., and Krishnakumar, S. (2011). Toxicogenomics of nanoparticulate delivery of etoposide: potential impact on nanotechnology in retinoblastoma therapy. *Cancer Nanotechnol.* 2 (1–6): 21–36.

Moche, H., Chevalier, D., Barois, N., Lorge, E., Claude, N., and Nesslany, F. (2014). Tungsten carbide-cobalt as a nanoparticulate reference positive control in in vitro genotoxicity assays. *Toxicol. Sci.: Offic. J. Soc. Toxicol.* 137 (1): 125–134.

Monteiro-Riviere, N.A., Nemanich, R.J., Inman, A.O., Wang, Y.Y., and Riviere, J.E. (2005). Multi-walled carbon nanotube interactions with human epidermal keratinocytes. *Toxicol. Lett.* 155 (3): 377–384.

Moore, L.D., Le, T., and Fan, G. (2013). DNA methylation and its basic function. *Neuropsychopharmacology: Offic. Publ. Am. College Neuropsychopharmacol.* 38 (1): 23–38.

Muller, J., Delos, M., Panin, N., Rabolli, V., Huaux, F., and Lison, D. (2009). Absence of carcinogenic response to multiwall carbon nanotubes in a 2-year bioassay in the peritoneal cavity of the rat. *Toxicol. Sci.* 110 (2): 442–448. https://doi.org/10.1093/toxsci/kfp100.

Muñoz, A. and Costa, M. (2012). Elucidating the mechanisms of nickel compound uptake: a review of particulate and nano-nickel endocytosis and toxicity. *Toxicol. Appl. Pharmacol.* 260 (1): 1–16.

Mytych, J., Zebrowski, J., Lewinska, A., and Wnuk, M. (2017). Prolonged effects of silver nanoparticles on p53/p21 pathway-mediated proliferation, DNA damage response, and methylation parameters in HT22 hippocampal neuronal cells. *Mol. Neurobiol.* 54 (2): 1285–1300.

Nagano, T., Higashisaka, K., Kunieda, A., Iwahara, Y., Tanaka, K., Nagano, K., Abe, Y., Kamada, H., Tsunoda, S.-I., Nabeshi, H., Yoshikawa, T., Yoshioka, Y., and Tsutsumi, Y. (2013). Liver-specific microRNAs as biomarkers of nanomaterial-induced liver damage. *Nanotechnology* 24 (40): 405102.

Ndika, J., Karisola, P., Kinaret, P., Ilves, M., and Alenius, H. (2021). Profiling non-coding RNA changes associated with 16 different engineered nanomaterials in a mouse airway exposure model. *Cells* 10 (5). https://doi.org/10.3390/cells10051085.

Nelson, B.C., Wright, C.W., Ibuki, Y., Moreno-Villanueva, M., Karlsson, H.L., Hendriks, G., Sims, C.M., Singh, N., and Doak, S.H. (2017). Emerging metrology for high-throughput nanomaterial genotoxicology. *Mutagenesis* 32 (1): 215–232.

Nemmar, A., Hoet, P.H.M., Vanquickenborne, B., Dinsdale, D., Thomeer, M., Hoylaerts, M.F., Vanbilloen, H., Mortelmans, L., and Nemery, B. (2002a). Passage of inhaled particles into the blood circulation in humans. *Circulation* 105 (4): 411–414.

Nemmar, A., Hoylaerts, M.F., Hoet, P.H.M., Dinsdale, D., Smith, T., Xu, H., Vermylen, J., and Nemery, B. (2002b). Ultrafine particles affect experimental thrombosis in an in vivo hamster model. *Am. J. Respir. Crit. Care Med.* 166 (7): 998–1004. https://doi.org/10.1164/rccm.200110-026oc.

Nemmar, A., Melghit, K., and Ali, B.H. (2008). The acute proinflammatory and prothrombotic effects of pulmonary exposure to rutile TiO2 nanorods in rats. *Exp. Biol. Med.* 233 (5): 610–619.

Ng, C.-T., Dheen, S.T., Yip, W.-C.G., Ong, C.-N., Bay, B.-H., and Lanry Yung, L.-Y. (2011). The induction of epigenetic regulation of PROS1 gene in lung fibroblasts by gold nanoparticles and implications for potential lung injury. *Biomaterials* 32 (30): 7609–7615.

Nikula, K.J., Snipes, M.B., Barr, E.B., Griffith, W.C., Henderson, R.F., and Mauderly, J.L. (1995). Comparative pulmonary toxicities and carcinogenicities of chronically inhaled diesel exhaust and carbon black in F344 rats. *Fundam. Appl. Toxicol.: Offic. J. Soc. Toxicol.* 25 (1): 80–94.

Novo, M., Lahive, E., Díez-Ortiz, M., Matzke, M., Morgan, A.J., Spurgeon, D.J., Svendsen, C., and Kille, P. (2015). Different routes, same pathways: molecular mechanisms under silver ion and nanoparticle exposures in the soil sentinel *Eisenia fetida*. *Environ. Pollut.* 205: 385–393.

Nymark, P., Kohonen, P., Hongisto, V., and Grafström, R.C. (2018). Toxic and genomic influences of inhaled nanomaterials as a basis for predicting adverse outcome. *Ann. Am. Thorac. Soc.* 15 (Suppl 2): S91–S97.

Oberdörster, G. (2001). Pulmonary effects of inhaled ultrafine particles. *Int. Arch. Occup. Environ. Health* 74 (1): 1–8.

Oberdörster, G., Ferin, J., and Lehnert, B.E. (1994). Correlation between particle size, in vivo particle persistence, and lung injury. *Environ. Health Perspect.* 102 (Suppl 5): 173–179.

Oberdörster, G., Oberdörster, E., and Oberdörster, J. (2005a). Nanotoxicology: an emerging discipline evolving from studies of ultrafine particles. *Environ. Health Perspect.* 113: 823–839. http://citeseerx.ist.psu.edu/viewdoc/summary?doi=10.1.1.276.8630.

Oberdörster, G., Oberdörster, E., and Oberdörster, J. (2005b). Nanotoxicology: an emerging discipline evolving from studies of ultrafine particles. *Environ. Health Perspect.* 113 (7): 823–839. https://doi.org/10.1289/ehp.7339.

Oberdörster, G., Sharp, Z., Atudorei, V., Elder, A., Gelein, R., Lunts, A., Kreyling, W., and Cox, C. (2002). Extrapulmonary translocation of ultrafine carbon particles following whole-body inhalation exposure of rats. *J. Toxicol. Environ. Health A* 65 (20): 1531–1543.

Oecd, T.N. (2014). 43: Genotoxicity of manufactured nanomaterials: report of the OECD expert meeting. OECD Publishing.

Oh, J.-H., Son, M.-Y., Choi, M.-S., Kim, S., Choi, A.-Y., Lee, H.-A., Kim, K.-S., Kim, J., Song, C.W., and Yoon, S. (2016). Integrative analysis of genes and miRNA alterations in human embryonic stem cells-derived neural cells after exposure to silver nanoparticles. *Toxicol. Appl. Pharmacol.* 299: 8–23.

Olive, P.L. and Banáth, J.P. (2006). The comet assay: a method to measure DNA damage in individual cells. *Nat. Protoc.* 1 (1): 23–29.

Öner, D., Moisse, M., Ghosh, M., Duca, R.C., Poels, K., Luyts, K., Putzeys, E., Cokic, S.M., Van Landuyt, K., Vanoirbeek, J., Lambrechts, D., Godderis, L., and Hoet, P.H.M. (2017). Epigenetic effects of carbon nanotubes in human monocytic cells. *Mutagenesis* 32 (1): 181–191.

Palm, W. (2019). Metabolic functions of macropinocytosis. *Philos. Trans. R. Soc. B* 374 (1765): 20180285.

Pan, Y., Zhang, W., and Lin, S. (2018). Transcriptomic and microRNAomic profiling reveals molecular mechanisms to cope with silver nanoparticle exposure in the ciliate Euplotes vannus. *Environ. Sci.: Nano* 5 (12): 2921–2935.

Papazoglou, E.S. and Parthasarathy, A. (2007). BioNanotechnology. *Synth. Lect. Biomed. Eng.* 2 (1): 1–139. https://doi.org/10.2200/s00051ed1v01y200610bme007.

Park, J.H. and Oh, N. (2014). Endocytosis and exocytosis of nanoparticles in mammalian cells. *Int. J. Nanomed.* 51. https://doi.org/10.2147/ijn.s26592.

Patil, N.A., Gade, W.N., and Deobagkar, D.D. (2016). Epigenetic modulation upon exposure of lung fibroblasts to TiO2 and ZnO nanoparticles: alterations in DNA methylation. *Int. J. Nanomed.* 11: 4509–4519.

Peters, A., Veronesi, B., Calderón-Garcidueñas, L., Gehr, P., Chen, L.C., Geiser, M., Reed, W., Rothen-Rutishauser, B., Schürch, S., and Schulz, H. (2006). Translocation and potential neurological effects of fine and ultrafine particles a critical update. *Part. Fibre Toxicol.* 3: 13.

Pfuhler, S., Downs, T.R., Allemang, A.J., Shan, Y., and Crosby, M.E. (2017). Weak silica nanomaterial-induced genotoxicity can be explained by indirect DNA damage as shown by the OGG1-modified comet assay and genomic analysis. *Mutagenesis* 32 (1): 5–12.

Pfuhler, S., van Benthem, J., Curren, R., Doak, S.H., Dusinska, M., Hayashi, M., Heflich, R.H., Kidd, D., Kirkland, D., Luan, Y., Ouedraogo, G., Reisinger, K., Sofuni, T., van Acker, F., Yang, Y., and Corvi, R. (2020). Use of in vitro 3D tissue models in genotoxicity testing: strategic fit, validation status and way forward. Report of the working group from the 7th International Workshop on Genotoxicity Testing (IWGT). *Mutat. Res. Genet. Toxicol. Environ. Mutagen.* 850–851: 503135. https://doi.org/10.1016/j.mrgentox.2020.503135.

Pogribna, M. and Hammons, G. (2021). Epigenetic effects of nanomaterials and nanoparticles. *J. Nanobiotechnol.* 19 (1): 2.

Pogribna, M., Koonce, N.A., Mathew, A., Word, B., Patri, A.K., Lyn-Cook, B., and Hammons, G. (2020). Effect of titanium dioxide nanoparticles on DNA methylation in multiple human cell lines. *Nanotoxicology* 14 (4): 534–553.

Qian, Y., Zhang, J., Hu, Q., Xu, M., Chen, Y., Hu, G., Zhao, M., and Liu, S. (2015). Silver nanoparticle-induced hemoglobin decrease involves alteration of histone 3 methylation status. *Biomaterials* 70: 12–22.

Qu, M., Luo, L., Yang, Y., Kong, Y., and Wang, D. (2019). Nanopolystyrene-induced microRNAs response in *Caenorhabditis elegans* after long-term and lose-dose exposure. *Sci. Total Environ.* 697: 134131.

Radomski, A., Jurasz, P., Alonso-Escolano, D., Drews, M., Morandi, M., Malinski, T., and Radomski, M.W. (2005). Nanoparticle-induced platelet aggregation and vascular thrombosis. *Br. J. Pharmacol.* 146 (6): 882–893.

Ragaseema, V.M., Unnikrishnan, S., Kalliyana Krishnan, V., and Krishnan, L.K. (2012). The antithrombotic and antimicrobial properties of PEG-protected silver nanoparticle coated surfaces. *Biomaterials* 33 (11): 3083–3092.

Ran, Q., Xiang, Y., Liu, Y., Xiang, L., Li, F., Deng, X., Xiao, Y., Chen, L., Chen, L., and Li, Z. (2015). Eryptosis indices as a novel predictive parameter for biocompatibility of Fe3O4 magnetic nanoparticles on erythrocytes. *Sci. Rep.* 5 (1). https://doi.org/10.1038/srep16209.

Ren, L., Zhang, J., Wang, J., Wei, J., Liu, J., Li, X., Zhu, Y., Li, Y., Guo, C., Duan, J., Sun, Z., and Zhou, X. (2019). Silica nanoparticles induce spermatocyte cell apoptosis through microRNA-2861 targeting death receptor pathway. *Chemosphere* 228: 709–720.

Ren, N., Atyah, M., Chen, W.-Y., and Zhou, C.-H. (2017). The various aspects of genetic and epigenetic toxicology: testing methods and clinical applications. *J. Transl. Med.* 15 (1): 110.

Risom, L., Møller, P., and Loft, S. (2005). Oxidative stress-induced DNA damage by particulate air pollution. *Mutat. Res.* 592 (1–2): 119–137.

Rossner, P., Jr, Vrbova, K., Rossnerova, A., Zavodna, T., Milcova, A., Klema, J., Vecera, Z., Mikuska, P., Coufalik, P., Capka, L., Krumal, K., Docekal, B., Holan, V., Machala, M., and Topinka, J. (2020). Gene expression and epigenetic changes in mice following inhalation of copper(II) oxide nanoparticles. *Nanomaterials (Basel, Switzerland)* 10 (3). https://doi.org/10.3390/nano10030550.

Sahay, G., Alakhova, D.Y., and Kabanov, A.V. (2010). Endocytosis of nanomedicines. *J. Control. Release* 145 (3): 182–195.

Schaefer, H.-E. (2010). *Nanoscience: The Science of the Small in Physics, Engineering, Chemistry, Biology and Medicine*. Springer Science & Business Media.

Seidel, C., Kirsch, A., Fontana, C., Visvikis, A., Remy, A., Gaté, L., Darne, C., and Guichard, Y. (2017). Epigenetic changes in the early stage of silica-induced cell transformation. *Nanotoxicology* 11 (7): 923–935.

Selvaraj, K., Gowthamarajan, K., and Karri, V.V.S.R. (2018). Nose to brain transport pathways an overview: potential of nanostructured lipid carriers in nose to brain targeting. *Artif. Cells Nanomed. Biotechnol.* 46 (8): 2088–2095.

Semberova, J., De Paoli Lacerda, S.H., Simakova, O., Holada, K., Gelderman, M.P., and Simak, J. (2009). Carbon nanotubes activate blood platelets by inducing extracellular Ca2+ influx sensitive to calcium entry inhibitors. *Nano Lett.* 9 (9): 3312–3317.

Semmler, M., Seitz, J., Erbe, F., Mayer, P., Heyder, J., Oberdörster, G., and Kreyling, W.G. (2004). Long-term clearance kinetics of inhaled ultrafine insoluble iridium particles from the rat lung, including transient translocation into secondary organs. *Inhal. Toxicol.* 16 (6–7): 453–459.

Sharifi, S., Behzadi, S., Laurent, S., Laird Forrest, M., Stroeve, P., and Mahmoudi, M. (2012). Toxicity of nanomaterials. *Chem. Soc. Rev.* 41 (6): 2323–2343. https://doi.org/10.1039/c1cs15188f.

Sheng, L., Wang, X., Sang, X., Ze, Y., Zhao, X., Liu, D., Gui, S., Sun, Q., Cheng, J., Cheng, Z., Hu, R., Wang, L., and Hong, F. (2013). Cardiac oxidative damage in mice following exposure to nanoparticulate titanium dioxide. *J. Biomed. Mater. Res. A* 101 (11): 3238–3246.

Shi, J., Sun, X., Lin, Y., Zou, X., Li, Z., Liao, Y., Du, M., and Zhang, H. (2014). Endothelial cell injury and dysfunction induced by silver nanoparticles through oxidative stress via IKK/NF-κB pathways. *Biomaterials* 35 (24): 6657–6666.

Shilatifard, A. (2006). Chromatin modifications by methylation and ubiquitination: implications in the regulation of gene expression. *Annu. Rev. Biochem.* 75: 243–269.

Shvedova, A.A., Kisin, E.R., Murray, A.R., Mouithys-Mickalad, A., Stadler, K., Mason, R.P., and Kadiiska, M. (2014). ESR evidence for in vivo formation of free radicals in tissue of mice exposed to single-walled carbon nanotubes. *Free Radic. Biol. Med.* 73: 154–165.

Siepel, A., Bejerano, G., Pedersen, J.S., Hinrichs, A.S., Hou, M., Rosenbloom, K., Clawson, H., Spieth, J., Hillier, L.W., Richards, S., Weinstock, G.M., Wilson, R.K., Gibbs, R.A., Kent, W.J., Miller, W., and Haussler, D. (2005). Evolutionarily conserved elements in vertebrate, insect, worm, and yeast genomes. *Genome Res.* 15 (8): 1034–1050.

Sima, M., Vrbova, K., Zavodna, T., Honkova, K., Chvojkova, I., Ambroz, A., Klema, J., Rossnerova, A., Polakova, K., Malina, T., Belza, J., Topinka, J., and Rossner, P., Jr. (2020). The differential effect of carbon dots on gene expression and DNA methylation of human embryonic lung fibroblasts as a function of surface charge and dose. *Int. J. Mol. Sci.* 21 (13). https://doi.org/10.3390/ijms21134763.

Simmons, D. (2008). Epigenetic influence and disease. *Nat. Educ.* 1 (6). https://tahomaclinic.com/Private/Articles2/Lithium/Simmons%202008%20-%20Epigenetic%20influences%20and%20disease.pdf.

Simon, D.F., Domingos, R.F., Hauser, C., Hutchins, C.M., Zerges, W., and Wilkinson, K.J. (2013). Transcriptome sequencing (RNA-seq) analysis of the effects of metal nanoparticle exposure on the transcriptome of *Chlamydomonas reinhardtii*. *Appl. Environ. Microbiol.* 79 (16): 4774–4785. https://doi.org/10.1128/aem.00998-13.

Sinden, R.R. (2012). *DNA Structure and Function*. Elsevier.

Srivastava, V., Gusain, D., and Sharma, Y.C. (2015). Critical review on the toxicity of some widely used engineered nanoparticles. *Ind. Eng. Chem. Res.* 54 (24): 6209–6233. https://doi.org/10.1021/acs.iecr.5b01610.

Stoccoro, A., Di Bucchianico, S., Coppedè, F., Ponti, J., Uboldi, C., Blosi, M., Delpivo, C., Ortelli, S., Costa, A.L., and Migliore, L. (2017). Multiple endpoints to evaluate pristine and remediated titanium dioxide nanoparticles genotoxicity in lung epithelial A549 cells. *Toxicol. Lett.* 276: 48–61.

Stuart, L.M. and Ezekowitz, R.A.B. (2005). Phagocytosis: elegant complexity. *Immunity* 22 (5): 539–550.

Sun, B., Liu, R., Ye, N., and Xiao, Z.-D. (2015). Comprehensive evaluation of microRNA expression profiling reveals the neural signaling specific cytotoxicity of superparamagnetic iron oxide nanoparticles (SPIONs) through N-methyl-D-aspartate receptor. *PLoS One* 10 (3): e0121671. https://doi.org/10.1371/journal.pone.0121671.

Sun, X., Shi, J., Zou, X., Wang, C., Yang, Y., and Zhang, H. (2016). Silver nanoparticles interact with the cell membrane and increase endothelial permeability by promoting VE-cadherin internalization. *J. Hazard. Mater.* 317: 570–578.

Surapaneni, S.K., Bashir, S., and Tikoo, K. (2018). Gold nanoparticles-induced cytotoxicity in triple negative breast cancer involves different epigenetic alterations depending upon the surface charge. *Sci. Rep.* 8 (1): 12295.

Tabish, A.M., Poels, K., Byun, H.-M., Luyts, K., Baccarelli, A.A., Martens, J., Kerkhofs, S., Seys, S., Hoet, P., and Godderis, L. (2017). Changes in DNA methylation in mouse lungs after a single intra-tracheal administration of nanomaterials. *PLoS One* 12 (1): e0169886.

Takenaka, S., Karg, E., Roth, C., Schulz, H., Ziesenis, A., Heinzmann, U., Schramel, P., and Heyder, J. (2001). Pulmonary and systemic distribution of inhaled ultrafine silver particles in rats. *Environ. Health Perspect.* 109 (Suppl 4): 547–551.

Tang, J., Xiong, L., Wang, S., Wang, J., Liu, L., Li, J., Yuan, F., and Xi, T. (2009). Distribution, translocation and accumulation of silver nanoparticles in rats. *J. Nanosci. Nanotechnol.* 9 (8): 4924–4932.

Tian, J.H., Hu, J.S., Li, F.C., Ni, M., Li, Y.Y., Wang, B.B., Xu, K.Z., Shen, W.D., and Li, B. (2016). Effects of TiO2 nanoparticles on nutrition metabolism in silkworm fat body. *Biol. Open* 5 (6): 764–769.

Tice, R.R., Agurell, E., Anderson, D., Burlinson, B., Hartmann, A., Kobayashi, H., Miyamae, Y., Rojas, E., Ryu, J.C., and Sasaki, Y.F. (2000). Single cell gel/comet assay: guidelines for in vitro and in vivo genetic toxicology testing. *Environ. Mol. Mutagen.* 35 (3): 206–221.

Trotta, F. and Mele, A. (2019). Nanomaterials: classification and properties. *Nanosponges* 1–26. https://doi.org/10.1002/9783527341009.ch1.

Tsuji, J.S., Maynard, A.D., Howard, P.C., James, J.T., Lam, C.-W., Warheit, D.B., and Santamaria, A.B. (2006). Research strategies for safety evaluation of nanomaterials, part IV: risk assessment of nanoparticles. *Toxicol. Sci.: Offic. J. Soc. Toxicol.* 89 (1): 42–50.

Tucci, P., Porta, G., Agostini, M., Dinsdale, D., Iavicoli, I., Cain, K., Finazzi-Agró, A., Melino, G., and Willis, A. (2013). Metabolic effects of TiO2 nanoparticles, a common component of sunscreens and cosmetics, on human keratinocytes. *Cell Death Dis.* 4: e549.

van Aerle, R., Lange, A., Moorhouse, A., Paszkiewicz, K., Ball, K., Johnston, B.D., De-bastos, E., Booth, T., Tyler, C.R., and Santos, E.M. (2013). Molecular mechanisms of toxicity of silver nanoparticles in zebrafish embryos. *Environ. Sci. Technol.* 47 (14): 8005–8014.

Wade, P.A., Pruss, D., and Wolffe, A.P. (1997). Histone acetylation: chromatin in action. *Trends Biochem. Sci.* 22 (4): 128–132.

Wang, C., Wang, H., Lin, M., and Hu, X. (2015). ZnO nanoparticles induced cytotoxicity on human pulmonary adenocarcinoma cell line LTEP-a-2. *Process Saf. Environ. Prot.* 93: 265–273. https://doi.org/10.1016/j.psep.2014.05.007.

Wang, J.J., Sanderson, B.J.S., and Wang, H. (2007a). Cytotoxicity and genotoxicity of ultrafine crystalline SiO2 particulate in cultured human lymphoblastoid cells. *Environ. Mol. Mutagen.* 48 (2): 151–157.

Wang, J.J., Sanderson, B.J.S., and Wang, H. (2007b). Cyto- and genotoxicity of ultrafine TiO2 particles in cultured human lymphoblastoid cells. *Mutat. Res.* 628 (2): 99–106.

Wang, J.J., Wang, H., and Sanderson, B.J.S. (2007). Ultrafine quartz-induced damage in human lymphoblastoid cells in vitro using three genetic damage end-points. *Toxicol. Mech. Methods* 17 (4): 223–232. https://doi.org/10.1080/15376510600943775.

Wang, S., Hunter, L.A., Arslan, Z., Wilkerson, M.G., and Wickliffe, J.K. (2011). Chronic exposure to nanosized, anatase titanium dioxide is not cyto- or genotoxic to Chinese hamster ovary cells. *Environ. Mol. Mutagen.* 52 (8): 614–622. https://doi.org/10.1002/em.20660.

Waters, K.M., Masiello, L.M., Zangar, R.C., Tarasevich, B.J., Karin, N.J., Quesenberry, R.D., Bandyopadhyay, S., Teeguarden, J.G., Pounds, J.G., and Thrall, B.D. (2009). Macrophage responses to silica nanoparticles are highly conserved across particle sizes. *Toxicol. Sci.: Offic. J. Soc. Toxicol.* 107 (2): 553–569.

Wiebert, P., Sanchez-Crespo, A., Falk, R., Philipson, K., Lundin, A., Larsson, S., Möller, W., Kreyling, W.G., and Svartengren, M. (2006a). No significant translocation of inhaled 35-nm carbon particles to the circulation in humans. *Inhal. Toxicol.* 18 (10): 741–747. https://doi.org/10.1080/08958370600748455.

Wiebert, P., Sanchez-Crespo, A., Seitz, J., Falk, R., Philipson, K., Kreyling, W.G., Möller, W., Sommerer, K., Larsson, S., and Svartengren, M. (2006b). Negligible clearance of ultrafine particles retained in healthy and affected human lungs. *Eur. Respir. J.: Offic. J. Eur. Soc. Clin. Respir. Physiol.* 28 (2): 286–290.

Yang, H., Zhang, Y., Li, W., Lao, C., Li, M., and Zheng, Y. (2017). Altered microRNA expression profiles in lung damage induced by nanosized SiO2. *Bioengineered* 8 (1): 45–54.

Yokel, R.A. and Macphail, R.C. (2011). Engineered nanomaterials: exposures, hazards, and risk prevention. *J. Occup. Med. Toxicol.* 6: 7.

Yun, J., Yang, H., Li, X., Sun, H., Xu, J., Meng, Q., Wu, S., Zhang, X., Yang, X., Li, B., and Chen, R. (2020). Up-regulation of miR-297 mediates aluminum oxide nanoparticle-induced lung inflammation through activation of Notch pathway. *Environ. Pollut.* 259: 113839.

Zhang, L., Becton, M.D., Liu, N., Averett, R.D., Pidaparti, R.M., and Wang, X. (2019). Physiochemical effects of nanoparticles on cell nuclear complex pore transport: a coarse-grained computational model. *J. Chem. Theory Comput.* 15 (11): 6382–6392.

Zhang, L., Han, B., Xiang, J., Liu, K., Dong, H., and Gao, X. (2018). Silica nanoparticle releases SIRT6-induced epigenetic silencing of follistatin. *Int. J. Biochem. Cell Biol.* 95: 27–34.

Zhang, L., Wang, X., Zou, J., Liu, Y., and Wang, J. (2015a). Effects of an 11-nm DMSA-coated iron nanoparticle on the gene expression profile of two human cell lines, THP-1 and HepG2. *J Nanobiotechnol.* 13: 3.

Zhang, L.W., Yang, J., Barron, A.R., and Monteiro-Riviere, N.A. (2009). Endocytic mechanisms and toxicity of a functionalized fullerene in human cells. *Toxicol. Lett.* 191 (2–3): 149–157.

Zhang, X.-F., Park, J.-H., Choi, Y.-J., Kang, M.-H., Gurunathan, S., and Kim, J.-H. (2015b). Silver nanoparticles cause complications in pregnant mice. *Int. J. Nanomed.* 10: 7057–7071.

Zhao, X., Takabayashi, F., and Ibuki, Y. (2016a). Coexposure to silver nanoparticles and ultraviolet A synergistically enhances the phosphorylation of histone H2AX. *J. Photochem. Photobiol. B: Biol.* 162: 213–222.

Zhao, X., Toyooka, T., and Ibuki, Y. (2017). Silver nanoparticle-induced phosphorylation of histone H3 at serine 10 is due to dynamic changes in actin filaments and the activation of Aurora kinases. *Toxicol. Lett.* 276: 39–47.

Zhao, Y., Li, L., Min, L.-J., Zhu, L.-Q., Sun, Q.-Y., Zhang, H.-F., Liu, X.-Q., Zhang, W.-D., Ge, W., Wang, -J.-J., Liu, J.-C., and Hao, Z.-H. (2016b). Regulation of microRNAs, and the correlations of microRNAs and their targeted genes by zinc oxide nanoparticles in ovarian granulosa cells. *PLoS One* 11 (5): e0155865.

Zhou, W., Tian, D., He, J., Yan, X., Zhao, J., Yuan, X., and Peng, S. (2019). Prolonged exposure to carbon nanoparticles induced methylome remodeling and gene expression in zebrafish heart. *J. Appl. Toxicol.* 39 (2): 322–332.

Zou, Y., Li, Q., Jiang, L., Guo, C., Li, Y., Yu, Y., Li, Y., Duan, J., and Sun, Z. (2016). DNA hypermethylation of CREB3L1 and Bcl-2 associated with the mitochondrial-mediated apoptosis via PI3K/Akt pathway in human BEAS-2B cells exposure to silica nanoparticles. *PLoS One* 11 (6): e0158475.

3

Endocrine Disruptors: Genetic, Epigenetic, and Related Pathways

Rais A. Ansari[1,*], *Saleh Alfuraih*[1], *Kamran Shekh*[2], *Yadollah Omidi*[1], *Saleem Javed*[3], *and Saghir A. Shakil*[4,5,6]

[1] *Department of Pharmaceutical Sciences, College of Pharmacy, Nova Southeastern University, Fort Lauderdale, FL, USA*
[2] *Yordas Group, Hamilton, ON, Canada*
[3] *Department of Biochemistry, Aligarh Muslim University, Aligarh, India*
[4] *ToxInternational Inc, Hilliard, OH, USA*
[5] *Department of Biomedical and Biological Sciences, Aga Khan University, Karachi, Pakistan*
[6] *Institute of Environmental Science and Meteorology, University of the Philippines-Diliman, Quezon City, Philippines*
* *Corresponding author*

Introduction

To improve human health and increase agricultural productivity to feed the ever-increasing human population, humans have made chemicals and introduced them into an environment where they do not occur naturelly that. Some of these chemicals are used as pesticides and fertilizers in agriculture, plasticizers in the production of plastics, antimicrobials for the treatment of microbial infections, and firefighting chemicals as flame retardants. The initial intent for the development of these chemicals is their desirable persistence in the environment; they are used to achieve long-lasting beneficial effects. However, the persistence of these chemicals has proven to be the striking bane to health as several of these chemicals continue to accumulate in biological systems through the food chain, producing adverse health effects, both in humans and wildlife. Mostly, these chemicals are present in the environment in relatively small amounts and often transfer from small to large organisms in the food chain eventually reaching humans. Humans and wildlife are consistently exposed to these chemicals via food, water, personal care products, and sometimes through the air. Although these chemicals are used with the intended target(s)/goal(s) in mind, they are often drifted from their targeted use and have become ubiquitous in the environment. Many of these chemicals affect human and wildlife health via multiple pathways causing many health effects. One of these health effects is their ability to disrupt the hormonal signaling pathways vital for endocrine functions, growth, and differentiation. These chemicals disrupt hormonal signaling, which is part of the endocrine system, hence named endocrine-disrupting chemicals (EDCs). Even though the metabolic processes of human and wildlife is capable of biotransforming/metabolizing these persistent environmental chemicals, life as we know has not been adopted enough, either by gene duplication or increasing their activity, to handle the challenge of ever-increasing (both in numbers and amounts) synthetic EDCs. Additionally, in some cases, biotransformed/metabolized

Impact of Engineered Nanomaterials in Genomics and Epigenomics, First Edition. Edited by Saura C. Sahu.

product may be an EDC (Erguc et al. 2021). EDCs are both natural and man-made chemicals; the main emphasis of this chapter is man-made EDCs. A list of common man-made EDCs is presented in Tables 3.1 and 3.2.

It is difficult to pinpoint common feature(s) in the chemical structures of the vast number of compounds responsible for endocrine disruption (ED) as they act via different mechanisms to cause ED. However, some structural features may indicate the possibility of chemicals being endocrine disrupters, such as the presence of a phenol ring without halogens or sulfone functional group substitution next to hydroxyl groups (Kabir et al. 2015).

Table 3.1 A representative list of endocrine-disrupting chemicals (EDCs).

Category	Name
4-Nonyphenols	4-Nonylphenol, branched and linear; 4-nonylphenol, branched and linear, ethoxylated
4-*Tert* octyl phenols	4-(1,1,3,3-Tetramethylbutyl) phenol; 4-(1,1,3,3-tetramethyl butyl) phenol, ethoxylated
4-Heptylphenol, 4-*Tert*-pentylphenol	4-Heptylphenol, branched and linear; p-(1,1-dimethyl propyl) phenol
Phthalates	Bis(2-ethylhexyl) phthalate, DEHP; diisobutyl phthalate, DIBP; dibutyl phthalate, DBP; benzyl butyl phthalate, BBP
Benzophenones	Benzophenone-1; 2,4-dihydroxybenzophenone; benzophenone-2; 2,2′,4,4′-tetrahydroxybenzophenone; benzophenone-3; oxybenzone; 4,4′-dihydroxybenzophenone
3-BC, MBC, EHMC	3-Benzylidene camphor (3-BC); 1,7,7-trimethyl-3-(phenyl methylene) bicyclo[2.2.1]heptan-2-one 3-(4-methyl benzylidene) camphor; 1,7,7-trimethyl-3-[(4-methyl phenyl) methylene]bicyclo[2.2.1] heptan-2-one 2-ethylhexyl 4-methoxycinnamate
Bisphenols F and S	Bisphenol F, S
BHT and BHA	Butylated hydroxytoluene; tert-butylhydroxyanisole (BHA); *tert*butyl-4-methoxyphenol
Carbon disulfide	Carbon disulfide
Dithiocarbamates	Metam-sodium, zineb, ziram, thiram
MTBE	*Tert*-butyl methyl ether; MTBE
Parabens	Methylparaben, ethylparaben; propylparaben; propyl 4-hydroxybenzoate; butylparaben; butyl 4-hydroxybenzoate
Other phenol derivatives	4-nitrophenol; 2,4,6-tribromophenol; resorcinol
PCP, tebuconazole, and triclosan	Pentachlorophenol (PCP); tebuconazole; triclosan
Phthalates	Di-butylphthalate (DBP); butyl-benzyl-phthalate (BBP); diethyl phthalate (DEP); dihexyl phthalate (DHP); dicyclohexyl phthalate (DCHP); dioctyl phthalate (DOP); diisodecyl phthalate (DiDP); diundecyl phthalate (DuDP), branched and linear
Quadrosilan	2,6-*cis*Diphenylhexamethylcyclotetrasiloxane
Triphenyl phosphate	Triphenyl phosphate

Source: Adapted from Dodson et al. (2012), https://www.chemsafetypro.com/Topics/Restriction/UN_list_identified_endocrine_disrupting_chemicals_EDCs.html.

Table 3.2 Common EDCs encountered by humans.

Application	Common EDCs
Insecticides/pesticides	DDT; chlorpyriphos; atrazine; 2,4-D; glyphosate
Food contact materials	BPA; phthalates; phenol
Electronics and building materials	PCBs; brominated flame retardant
Medical tubing; personal care products	Phthalates
Antibacterial; surface removal of bacteria	Triclosan
Textile and clothing	Perfluorochemicals
Cosmetics; cleaners	Parabens; phthalates; glycol ethers
Antifoulant for painting the bottom of ships	Tributyltin
Surfactants	Nonylphenol
Contraceptives	Ethinyl estradiol

Source: Dodson et al. (2012), Gore et al. (2014), Kabir et al. (2015).

The harmful effects of EDCs are attributed to their ability to regulate transcription by the nuclear hormone receptor (NR) superfamily (Delfosse et al. 2015; Giulivo et al. 2016; Nebert 2017). EDCs can also mediate genomic and non-genomic effects via membrane-anchored NRs (Levin and Hammes 2016). NR can also affect G-protein coupled receptor-30 (GPCR30)- mediated effects (Prossnitz and Arterburn 2015). A graphical depiction of ED by genomic and non-genomic mechanisms is presented in Figure 3.1.

A. Flame Retardants and Insulators: Due to the nonflammability of polybrominated biphenyls, they are used in electronics and furniture such as sofas and mattresses. Polybrominated biphenyls have been linked to ED, especially, disruption of the thyroid function (Legler and Brouwer 2003). Proper thyroid function is important for the development of fetuses and children. Often these chemicals leach from the source materials and contaminate other substances and/or food. Similar compounds, such as polychlorinated biphenyls (PCBs), were used as transformer

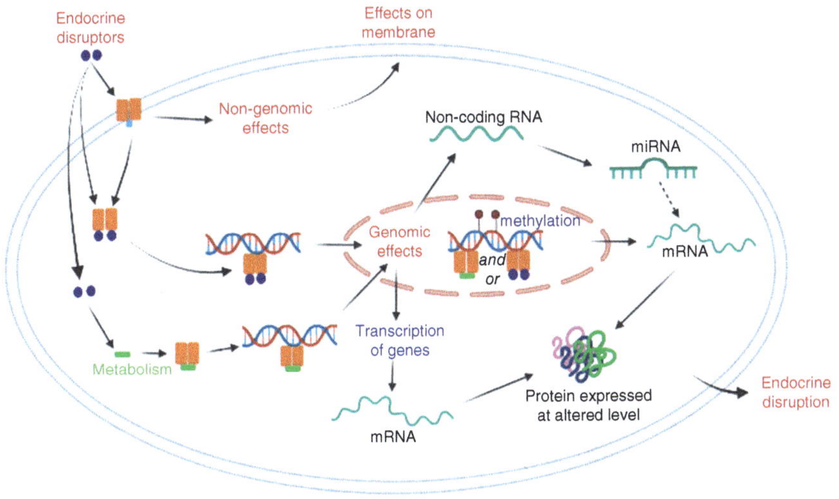

Figure 3.1 Mechanism of genomic and non-genomic endocrine disruption. Created using BioRender.com.

fluid due to their nonflammability and thermal stability; the use of PCBs was banned by the US EPA in 1976. Large amounts of PCBs are still present in the environment from their use in transformers and also stored in warehouses. The release of PCBs in the environment was usually from transformers leaking or catching fire and from industrial landfills where these transformers and containers used to store PCBs (with residual amounts remaining in them) were eventually disposed of. PCBs are ubiquitous in the environment with exposure to almost every living organism including humans (Jugan et al. 2010).

B. Phthalates and Bisphenol A: Phthalates and bisphenol A (BPA) chemicals are used in plastic to provide the desired pliability. Many foods and beverage containers, their plastic wraps, and tubing leach phthalates with time; they are also released during microwaving/warming foods in plastic containers. Now, many companies produce plastic containers and tubing without phthalates and sell them as "phthalate-free". Phthalates are known to interfere with the production of androgen. The eight most common phthalates are butyl benzyl phthalate (BBP), di-n-butyl phthalate (DBP), di(2-ethylhexyl) phthalate (DEHP), di-n-hexyl phthalate (DnHP), di-isodecyl phthalate (DiDP), di-isononyl phthalate (DiNP), diethyl phthalate (DEP), and di-n-octyl phthalate (DnOP) (Fisher 2004).

The Office of Environmental Health Hazard Assessment (OEHHA) of the California Environmental Protection Agency (Cal EPA) is the lead agency for the implementation of the Safe Drinking Water and Toxic Enforcement Act of 1986 (Proposition 65). OEHHA has developed safe harbor levels for chemicals that cause developmental/reproductive toxicity (maximum allowable dose levels or MADLs) and cancer (no significant risk level or NSRL). Proposition 65 safe harbor levels provide information that can be used to determine if product warnings are required for exposure to listed chemicals. OEHHA added BBP to the Proposition 65 list of chemicals in 2005 with MADL of 1200 μg/day (https://oehha.ca.gov/proposition-65/chemicals/butyl-benzyl-phthalate); DBP in 2005 with MADL of 8.7 μg/day (https://oehha.ca.gov/proposition-65/chemicals/di-n-butyl-phthalate-dbp); DEHP in 1988 with MADL of 410 μg/day and NSRL of 310 μg/day (https://oehha.ca.gov/chemicals/di2-ethylhexylphthalate); DnHP in 2005 with MADL of 2200 μg/day (https://oehha.ca.gov/chemicals/di-n-hexyl-phthalate-dnhp); DiDP in 2007 with MADL of 2200 μg/day (https://oehha.ca.gov/proposition-65/chemicals/di-isodecyl-phthalate-didp); and DiNP in 2013 with NSRL of 146 μg/day (https://oehha.ca.gov/chemicals/diisononyl-phthalate).

Bisphenol A, a class of synthetic phenolic compounds, is mainly used for the production of polycarbonate plastics and epoxy resins. It is used in many water bottles and plastic containers. Due to its widespread application, exposure to BPA in humans occurs ubiquitously (Ma et al. 2019). The use of BPA in children's feeding bottles is banned. Exposure to BPA is linked to reduced ova quality and viability; this observation was made in individuals seeking fertility treatments. Due to its presence in water bottles and plastic containers, the leaching of BPA is enhanced during the heating and microwaving of these products (Lim et al. 2009; Muncke 2011; Takao et al. 2002). Exposure to sunlight or acidic foods such as tomatoes and pineapple juice also increases the leaching of BPA into food products (NIEHS 2021). OEHHA added BPA to the Proposition 65 list of chemicals in 2020 with a MADL of 3 μg/day (https://oehha.ca.gov/proposition-65/chemicals/bisphenol-bpa).

C. Metals: Two important metals, lead (Pb) and cadmium (Cd), are linked to reproductive toxicity (Rafati Rahimzadeh et al. 2017; Vignesh et al. 2011). The lead-based paints are banned in the United States in 1978; however, Pb-based paints find their way into the United States from countries where their use is still allowed (EPA (US) 2022b). Pb is still used in the production of plastics as a softening and stabilizing agent (Turner and Filella 2021). The presence of Pb has been reported in 18% of the children's products manufactured in Russia and surrounding Central Asian republics (IPEN 2011), 15% of the products manufactured in Philippines (House of Representative, P. 2017), and up to 10%

products manufactured in certain cities in China (IPEN 2011). The release of Cd in the environment is also from the human usage of Cd in products, such as batteries, plastic stabilizers, alloys, and surface coating (Nokes and Weaver 2014). In addition to its nephrotoxicity, Cd is toxic to the reproductive system (Rafati Rahimzadeh et al. 2017).

Lead is a naturally occurring metal element that makes up about 14 ppm (~0.001%) of Earth's crust. Lead is generally found combined with other elements. Due to its ubiquitous nature, Pb is found in water (~5 μg/L in fresh and sea water), air (~0.1 μg/m3), and soil (~100 mg/kg) (Rieuwerts 2015). Lead is used in several industries and products due to its desirable properties. Humans can be exposed to Pb by ingesting food and/or water contaminated with Pb, eating food grown in areas with high soil and/or water Pb level, and can enter the body by breathing dust containing Pb or through the skin. The most dangerous form of occupational exposure to Pb is through inhalation of fine dust and fumes by workers in lead mining and smelting and battery manufacturing/disposal/recycling industries. Notably, lead is known to affect many organs and systems, including the brain, and kidney, and interfere with the proper functioning of enzymes. Lead seems to mimic calcium and crosses the blood-brain barrier causing central nervous system effects and impairing the learning process, especially in children. Lead can cause developmental/reproductive toxicity and is a possible carcinogenic agent of the kidney in humans (https://oehha.ca.gov/chemicals/lead-and-lead-compounds). In 1987, OEHHA added Pb to the list of chemicals known to cause cancer, and in 1992, to cause developmental/reproductive toxicity under Proposition 65 with the NSRL of 15 μg/day and MADL of 0.5 μg/day (https://oehha.ca.gov/chemicals/lead-and-lead-compounds).

Cadmium is a naturally occurring element that makes up about 0.1 ppm of Earth's crust. Cadmium is always associated with other minerals and is produced as a byproduct of mining, smelting, and refining other elements. Due to its ubiquitous nature, Cd is found in water (~1 μg/L in freshwater and ~50 ng/L in seawater), air (~5 ng/m3), and soil (~2 mg/kg) (Rieuwerts 2015). Rocks mined for phosphate fertilizers contain varying amounts of Cd, resulting in its presence in fertilizers and agricultural soils, and can be absorbed by plants. Humans can be exposed to Cd by drinking water containing Cd, by eating food grown in areas with high soil and/or water Cd levels, and can enter the body by breathing dust containing Cd or through the skin. Additionally, Cd exposure also occurs from cigarette smoking as tobacco leaves absorb/accumulate Cd from the soil/fertilizer (ATSDR 2011). The most dangerous form of occupational exposure to cadmium is the inhalation of fine dust and fumes or the ingestion of highly soluble cadmium compounds. Cadmium is known to affect many organs and systems, e.g., renal, respiratory, cardiovascular, immune, and endocrine systems, prostrate as well as developmental/reproductive processes. Cadmium is a known human carcinogen of the lung, prostate, and kidney (https://oehha.ca.gov/media/downloads/proposition-65/chemicals/cd-hid.pdf). In 1987, OEHHA added Cd to the list of chemicals known to cause cancer and, in 1997, to cause developmental/reproductive toxicity under Proposition 65 with NSRL of 0.05 μg/day and MADL of 4.1 μg/day (https://oehha.ca.gov/chemicals/cadmium).

D. Pesticides: Insecticides and Herbicides: Pesticides are used to control pests; they are designed either to disrupt the nervous or reproductive system of the pests. Even though pesticides are applied at extremely low rates (e.g., ~0.3–2 kg/hectare chlorpyrifos and ~0.7 kg/acre 2,4-D), the relative dose (e.g., /kg body weight) is magnitudes higher to pests, as they are small, resulting in intended toxicity (e.g., death). Notably, humans and wildlife might be exposed causing far less toxicity due to their relatively small doses. The mechanism of toxicity (e.g., neurotoxicity) of pesticides may be similar in the targeted pests, wildlife, and even humans. Additionally, pesticides may have certain undesirable effects such as ED in nontargeted aquatic (e.g., fish, amphibians, reptiles) and terrestrial animals and humans.

Dichlorodiphenyltrichloroethane (DDT) and its metabolites, such as dichlorodiphenyldichloroethane (DDE), possess an estrogenic potential and are best known for ED properties (OEHHA 2007). Due to the persistence of DDT in the environment and its ED property, US EPA banned DDT for most of its uses in the summer of 1972, exempting public health uses under some conditions (Perma.cc 2019). During the late 1970s, the EPA also began banning organochlorine pesticides (i.e., aldrin, dieldrin, chlordane, heptachlor, toxaphene, and mirex) that were chemically similar to DDT. Use of DDT under public health exemption continued, for example, in June 1979, the California Department of Health Services was permitted to use DDT to suppress flea vectors of bubonic plague. Moreover, the production (estimated at over 300 tons annually) of DDT continued in the United States for foreign markets until 1985 (ATSDR 2022). Agricultural use of DDT was banned in most developing countries in the 1970s and 1980s. The Stockholm Convention on Persistent Organic Pollutants, which took effect in 2004, put a global ban on several persistent organic pollutants, and restricted DDT uses in the vector control program. Despite the worldwide ban, their agricultural uses have been continued in India, North Korea, and possibly elsewhere (EPA (US) 2022a; UNEP 1999). As of 2013, an estimated 3,000–4,000 tons of DDT were produced for disease vector control, including 2,786 tons in India (van den Berg 2009). In South Africa, levels of DDT in the fat of chickens were reported to be hundreds of times higher than the level in the environment when DDT was used for the eradication of malaria as bioaccumulated through the food chain (Becher et al. 2013). Recent evidence points out that DDT could be linked to non-Hodgkin's lymphoma, breast cancer, premature birth, disruption of the menstrual cycle, and the problem with lactation, in addition to lower sperm counts among fertile adults globally (Michael et al. 2022).

Organophosphate pesticides (OP) were developed to ensure that when used as pesticides, they do not persist in the environment and do not accumulate in body fat like organochlorines. The OPs are degraded fast in the environment in the sun and rain. A few OPs, such as chlorpyrifos, bind tightly to soil and degrade slowly. Some of them are also detected in indoor air, dust, carpets, on children's toys in homes where products with chlorpyrifos have been used, and in agricultural products (ECHA 2010; Gurunathan et al. 1998). Chlorpyriphos exposure is linked to developmental delays and attention deficit disorder (ADHD) in children (CHADD 2019). EDCs, in addition to directly affecting the reproductive systems, also affect the endocrine axis resulting in developmental effects. The United States banned the residential use of chlorpyriphos in 2000, which resulted in a sharp decline in blood chlorpyriphos levels, reducing to half within two years in children living in New York (Earthjustice 2019). Other pesticides reported having ED properties include atrazine, a herbicide that disrupts the hypothalamus-pituitary axis, and glyphosate, another herbicide that is linked to obesity and behavioral and cognitive functions (Amato et al. 2021).

E. PFAS Chemicals: Presently, another group of compounds referred to as per-and polyfluoroalkyl substances (PFAS) have become ubiquitous in the environment with rising concerns about their biological activity toward disrupting endocrine function (Spanne 2022). These are a group of over 9,000 man-made chemicals that are being used as oil and water repellants on the surface coating of cookware, carpets, and textiles. PFAS are extremely stable chemicals and accumulate in the environment over time. Due to their stability and persistence in the environment, they are called forever chemicals. They can transfer from agricultural products to humans and can contaminate drinking water supplies (Giesy et al. 2010; Xu et al. 2021). The presence of PFAS has been detected near its manufacturing facilities, military bases (EWG 2019), and firefighting training facilities where PFAS foam is used to extinguish the fire (ITRC 2022). It also enters the food supply chain via food packaging materials and contaminated soils (Schreder and Dickman 2018). Dispersal of PFAS is also reported via air (Brusseau et al. 2019). CDC has observed that the US population has been exposed to PFAS (ATSDR 2020). In addition to mimicking fatty acids, they interfere with the hormonal system. A probable link between high cholesterol, thyroid disease, ulcerative colitis, testicular cancer, kidney cancer, and pregnancy-induced hypertension was observed near Parkersburg, WV, where people were exposed to perfluorooctanoic acid (PFOA), another form of PFAS (The PFAS Project Lab 2019).

Toxic Effects of EDCs on Wildlife and Humans

EDCs Effects on Wildlife

Wildlife is not exposed to a single but a mixture of EDCs. Exposure of wildlife with EDCs occurs via air, water, food, soil, and sediment to a plethora of chemical agents that persist in food and wildlife habitats. Wildlife is exposed to EDCs in the same manner as humans (Becher et al. 2013). Fish are often exposed to EDCs when their aquatic habitat receives EDCs in the runoff from treated areas like agricultural fields. Certain EDCs are soluble in water and fish can accumulate them easily in their tissues. Exposure to aquatic animals also occurs from the oil tanker, fuel extraction processes, and oil splits. Polar bears that feed largely on seals could be exposed to EDCs through the food chain; the highest levels of PFOS, PCBs, and organochlorine insecticides have been reported in polar bears (Becher et al. 2013). The chemicals that are found in wildlife are persistent organic pollutants (POPs), such as DDT, dieldrin, chlordane, PCBs, dioxins, and polybrominated diphenyl ethers (PBDEs). The list of EDCs that are commonly found in wildlife is presented in Table 3.3 (Gonsioroski et al. 2020).

Many studies have been conducted to pinpoint the effects of EDCs on wildlife. A high incidence of liver tumors was found in teleost fish that were exposed to EDCs through water (Rober et al. 1996). The exact cause of liver tumors was suspected to be due to exposure to PCBs, DDT, and polyaromatic hydrocarbons (PAHs). However, the role of EDCs could not be ruled out as contraceptives are routinely detected in fish muscles and plasma, and ever higher amounts of contraceptives are flushed into toilets with excreta and they are not completely removed during municipal wastewater treatment before releasing the water into rivers/seas (Lovett 2012).

Table 3.3 A list of EDCs commonly found in wildlife.

EDCs	Wildlife exposed
Human contraceptives (EE2, levonorgestrel)	Present in plasma and muscle of fish
Pharmaceuticals: antiepileptic-carbamazepine, antidepressants-fluoxetine, sertraline, venlafaxine, citalopram, norfluoxetine, diphenhydramine, diltiazem	Fish: liver and muscle
Triclosan	Aquatic organisms, such as algae, invertebrates, small fish, and dolphins
Benzotriazole, parabens, OPs	Body tissues of wildlife near sewage treatments

Effects During Development

Birthweight, preterm birth, infant mortality, and causes of future morbidity have been monitored since 1990 and reported to have significantly increased in the United States (Martin et al. 2010). Several studies attempted to link the birthweight with possible chemical exposure. However, the link between EDCs and birthweight is inconsistent and the link between exposure to POPs during pregnancy and decreased birthweight is not clear (Wigle et al. 2008). Low birthweight due to low levels of exposure to chemicals among the European birth cohort has been reported (Govarts et al. 2012). In this study, linkage to PCBs, but not DDT, to low birthweight was observed. Several studies have established an association between exposure to BPA, PBDE, and phthalates and low birthweight; however, others have found no such link. Based on these contrasting findings, an association between low birthweight and exposure to EDCs is not conclusive (Harley et al. 2011; Zhang et al. 2009). Similarly, the link between preterm birth and the role of gestational outcome among women is not conclusive as both preterm and full-term delivery is reported among women exposed to phthalates. Therefore, the role of preterm delivery due to the presence of EDCs cannot be concluded. One aspect of these chemicals, i.e., increased inflammation by activating circulating macrophages, can be linked as the cause of preterm delivery (Adibi et al. 2009; Cantonwine et al. 2010; Latini et al. 2003; Meeker et al. 2009; Wolff et al. 2008).

Delayed Effects

Often male reproductive tract abnormalities, such as cryptorchidism and hypospadias have been linked to EDCs (Main et al. 2010). Reduction in semen quality and reduced sperm counts, as a cause of infertility and increased incidence of testicular cancer, have been linked to EDCs exposure (Huyghe et al. 2003; Main et al. 2010; Swan et al. 2000). In a large study, maternal occupational exposure to antiandrogenic compounds, such as phthalates and hairspray, was linked to hypospadias (Ormond et al. 2009). In another study, prenatal exposure to pesticides was linked to hypospadias (Gaspari et al. 2011). The risk of cryptorchidism was also reported with the presence of PBDE in breast milk; however, the role of placental PBDE was not linked to cryptorchidism (Main et al. 2007). There are some studies linking various forms of developmental abnormalities due to the presence of EDCs and their exposure to developing children (Main et al. 2006; Suzuki et al. 2012; Swan 2008).

Due to the presence of confounding factors, such as exposure to toxic gases, particulate matter, heavy metals, and urban stress, establishing the link of EDCs with hypospadias is difficult to prove. More direct evidence came from the reduced size of phalluses of alligators grown in captivity because of the inability of males to transfer sperm during copulation due to the decreased size of their penises (Guillette et al. 1996). It was observed that a population of juvenile alligators inhabiting Lake Apopka possessed significantly smaller-sized penises (24% average decrease) with lower plasma concentrations of testosterone (70% lower concentrations) when compared with animals of similar size in Lake Woodruff (Guillette et al. 1996). While the reduction in penis size of juvenile alligators of Lake Apopka was not linked to testosterone, a higher level of the antiandrogenic DDT breakdown product p,p'-DDE was found/stored in their fat (Guillette et al. 1996).

Transgenerational Effects

Endocrine disruptors have been known to decrease the fertility of both men and women, thus making humans less fertile (Diamanti-Kandarakis et al. 2009; Gore et al. 2014). Endocrine disruptors have been found to affect the development and maturation of sperms. These chemicals affect steroidogenesis, thus the production of a hormone (Sidorkiewicz et al. 2017; Sweeney et al. 2015). EDCs affect the development of ovarian follicles influencing the female reproductive process (Hannon and Flaws 2015; Patel et al. 2015; Rattan et al. 2017). EDCs decrease the fertility of both sexes in animal models (Diamanti-Kandarakis et al. 2009; Gore et al. 2015). Besides the direct effect on the reproductive system, recently, it has been found that EDCs exposure produces multigenerational effects (Chen et al. 2015; Manikkam et al. 2012; Zhou et al. 2017). Both multigenerational and transgenerational effects are observed with EDCs exposure (Figure 3.2). When F0 males or nonpregnant F0 females are exposed to an EDC, while effects on F1 generation are observed, it is considered multigenerational through exposure in the germline. If the effects are observed in the F2 generation with no F1 generation exposure, it is considered a transgenerational effect (Skinner 2008, 2014b; Xin et al. 2015). Exposure of EDC to F0 (filial) generation during pregnancy results in offspring exposure that is F1 generation. As the germ cells were exposed in the F1 generation, the next generation (F2) was also exposed. Such an exposure during pregnancy is referred to as multigenerational. The subsequent generation (F3) is not directly exposed and the observed effects will be referred to as transgenerational (Skinner 2008, 2014b; Xin et al. 2015).

Identification of EDC: Methods

There are several methods of determining ED activities in biological systems (Table 3.4), from simple cell proliferation assays, such as of MCF-7 cells to determining estrogenic activity (Miyakoshi et al. 2009) or prostate cancer cell line, which is positive for androgen expression (LnCAP) to determine the androgenic activity of EDCs (Ait-Aissa et al. 2010; Ankley et al. 2009; Shah et al. 2008). In certain conditions, the production of thyroid hormones, their binding to thyroid receptors (TR), transport of constituents, such as iodine, etc. are also considered as part of detail ED activity determination (OECD 2013). Studies, such as ligand-binding (purified NR) with EDCs, gene reporter assays (3–4 response elements fused with a basal promoter driving the expression of reporter gene; luciferase or β-galactosidase with internal control of renilla luciferase) (Bagchi Bhattacharjee and Paul Khurana 2014; OECD 2013; Shanle and Xu 2011; Sun et al. 2016), induction of vitellogenin gene (xenoestrogen assay), or antigen-antibody interactions are also used (Hiramatsu et al. 2005).

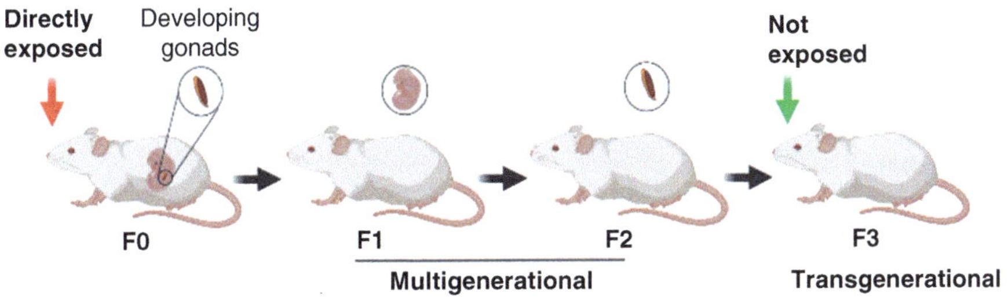

Figure 3.2 Schematic representation of multigenerational and transgenerational effects from prenatal exposure to EDCs.

Table 3.4 In vitro and in vivo based assays for the identification of EDCs with estrogenic activity.

Assay type	Source	References
In vitro		
Binding with ER	Uterine cytosol of rat	(EDSTAC 1998)
Cell proliferation assays • *E-SCREEN assay*	MCF7 proliferation	(Soto et al. 1995; Silva et al. 2007)
Receptor (ER) dimerization assay • *Fluorescence resonance energy transfer* • *Bioluminescence resonance energy transfer*		(Tamrazi et al. 2002; Liu et al. 2003; Powell et al. 2010; Powell and Xu 2008)
ER Transcriptional activation	hERα-Hela-9903 stably transfected cell line	(EDSTAC 1998)
	Firefly reporter (Luc) T47DKBLuc cell line	(Wilson et al. 2004)
	HELN-ERα and HELN-ERβ reporter cell lines	(Balaguer et al. 1999; Lemaire et al. 2006)
	Alkaline phosphate reporter cell line (HEK 293) with stable expression of ERα and ERβ	(Barkhem et al. 1998)
Metabolic assays	Steroidogenesis assay in H295R cells (adrenal cortex cells)	(EDSTAC 1998)
	Aromatase inhibition assay	(EDSTAC 1998)
	Steroidogenesis assay in JEG-3 cells	(Letcher et al. 1999)
	Steroidogenic enzyme expression assay in H295R cells	(Zhang et al. 2005)
In vivo		
Uterotropic assay in rodent	21–22 days administration of test compound	(Odum et al. 1997)
Pubertal female rat assay	20-day pubertal female assay/ Sprague-Dawley rats	(Kim et al. 2002)
Fish reproduction assay	Short-term reproductive assay	(OECD 2012)

Source: Adapted from Shanle and Xu (2011).

Genetic Pathways

Nuclear Receptor-Mediated Assays

Modular Structure of Nuclear Receptors

The nuclear receptors possess a common modular structure that is depicted in Figure 3.3 (Weikum et al. 2018). Except for SHP and DAX (orphan members of the NR), NR possesses six domains from A–E, and in some NRs, E domain may be absent. The NH_2-terminal domain varies in length and seems highly disorganized when the detailed structure of all 48 NRs is examined. The NH_2-terminal domain contains activator function-1 (AF-1) that can interact with coregulators in

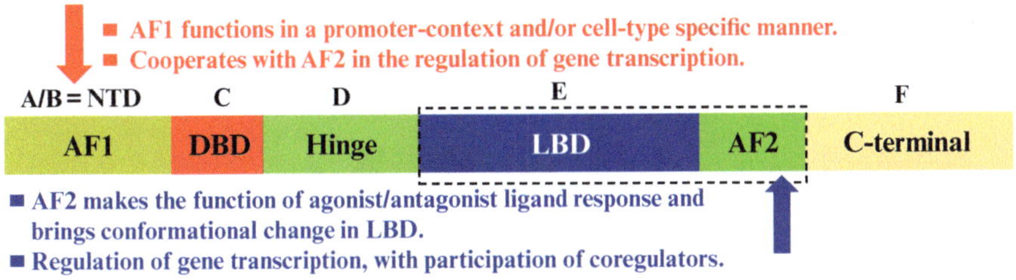

Figure 3.3 Changes in the modular structure of nuclear receptors.

a promoter-specific manner (Kumar and Thompson 2003). Variations in NH_2-terminal produce isoforms through alternative splicings such as multiple forms of the glucocorticoid receptor (GR), GRB, and GRA (Weikum et al. 2017) and TR (Sadow et al. 2003), while variation in the carboxylic end also produces isoforms such as multiple forms of ERβ (Zhao et al. 2008).

The NH_2-terminal variant structure of NRs is shown in Figure 3.4. Due to the size differences in the NH_2-terminal, the transcriptional activity of NR varies. The larger the size of the NH_2-terminal, the greater the transcriptional activity of NR that is exhibited (Weikum et al. 2018). This is the reason ERα possesses higher transcriptional activity compared with ERβ (Yasar et al. 2017). The ERβ2, ERβ4, and ERβ5 are all carboxyl-terminal variants lacking the AF-2 domain and cannot bind to estrogen, and cannot engage in transcriptional activation. However, ERβ2, ERβ4, and ERβ5 can homodimerize with ERβ or heterodimerize with ERα affecting the transcriptional activities of wild-type receptors (Tremblay et al. 1999).

The ERα variants, ERα46 and ERα36, are the two variants of ERα (Figure 3.5). The ERα46 arises as a result of alternate splicing, lacking the amino-terminal A/B domain, which is common in carcinogenesis. ERα46 acts as a competitive inhibitor of ERα. ERα36 contains 27 amino acids in the carboxyl-terminal instead of the regular 138 amino acids of ERα and arises due to alternate splicing; therefore, ERα36 lacks both AF-1 and AF-2 domains (Yasar et al. 2017). Palmitoylation of ERα36 results in plasma membrane localization and can participate in membrane-initiated signaling by estrogen. The interaction of ERα36 with ERα in relation to transcriptional activity is not well defined with estrogen and, therefore, the transcriptional activities of EDCs, especially of xenoestrogens, are speculative (Yasar et al. 2017).

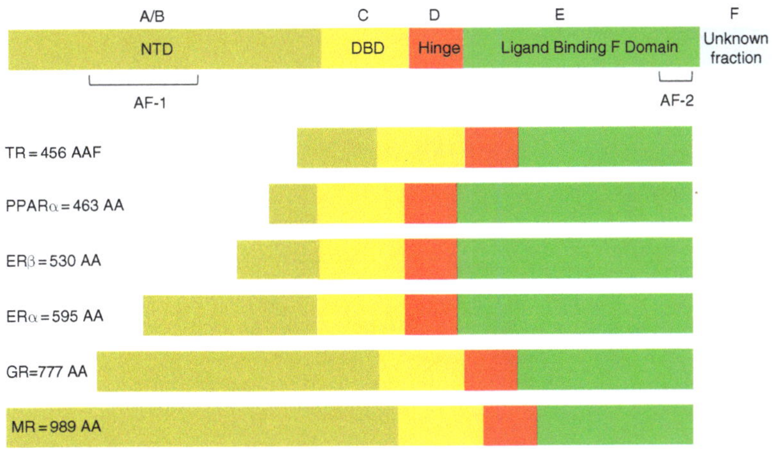

Figure 3.4 Modular structure of some nuclear receptors involved in endocrine disruption.

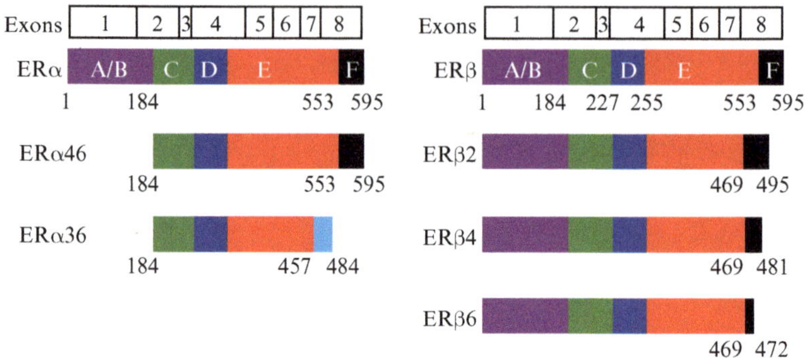

Figure 3.5 Schematic representation of the estrogen receptor isoforms. Source: Adapted from Muyan et al. (2017).

The EDCs engage NR and modulate their transcriptional activities. There is a total of 48 NR, of which 24 are identified with ligands while the remaining 24 receptors' ligands are not identified (Weikum et al. 2018). There are two types of receptors, type I steroid receptors that remain bonded with heat shock protein-90 and remain sequestered in cytoplasm while type II is nonsteroidal receptors that remain bound to the nuclear response element or hormone response element in the genomic DNA in the nucleus (Figure 3.6). The effects of EDCs on NR signaling is not the only ligand-dependent activity of NRs that can disrupt endocrine organs, but it is the best-understood mechanism.

A fraction of nuclear receptors remains bound with clathrin-coated segments of plasma membranes. Signaling from the plasma membrane part of NR is known to produce both genomic and non-genomic effects, especially from estrogen receptors, androgen receptors, and mineralocorticoid receptors (Levin and Hammes 2016; Trevino and Gorelick 2021). Extracellular steroid receptors, such as ER, AR, PRA, and PRB have been found in the plasma membrane of cancer cells. ERα, PRA, and PRB are found to be associated with the plasma membrane in breast cancer cells and the binding of agonists leads to rapid signaling via G-protein coupled receptors and cell proliferation (Levin 2011). The palmitoylation of cysteine residues of AR, PRB, ERα, and ERβ is responsible for membrane anchoring of the steroid receptors (Levin and Hammes 2016). The action of EDCs via membrane-bound ERs, PRs, and AR is not well defined. In addition to cytoplasmic and plasma membrane-bound receptors, ERα is also found in the endoplasmic reticulum (ER), where it is involved in calcium signaling and ER stress (Chami et al. 2008; Revankar et al. 2005). ERα and ERβ have also been found inside mitochondria (Chen et al. 2004). Below, we will describe each steroid signaling step and cite the effects of EDCs where available.

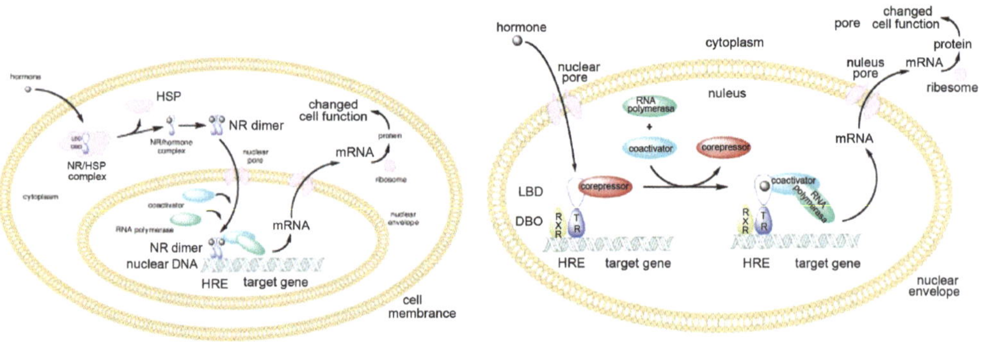

Figure 3.6 Modulation of EDCs with NR and their transcriptional activities.

Phosphorylation-Mediated Signaling Pathways of Nuclear Receptors and Other Transcription Factors: Link to EDC

Nuclear receptor phosphorylation occurs on serine and threonine residues by various kinases. Phosphorylation is associated with enhanced transcriptional activities of nuclear receptors more often than inhibition (TR and PPARγ) (Shao and Lazar 1999). The dephosphorylated form of carbohydrate response element binding protein-α (CREBPα) can bind to its response element and activate fatty acid synthase (FAS) and acetyl-CoA carboxylase (ACC) genes involved in fatty acid synthesis regulation. Estrogen phosphorylates the membrane-bound ERα's serine and threonine residues that activate protein kinase-A (PKA) and AMP-activated kinase (AMPK) causing phosphorylation of serine residues of CREBPα in adipocytes that shuttle out the phospho-CREBPα from nucleus to cytoplasm thus blunts the activation of FAS and ACC genes. An opposing action happens when insulin binds to insulin receptors and activates the protein phosphatase A2 (PP2A) that dephosphorylates phospho-CREBPα, thus shuttling the CREBPα to the nucleus activating FAS and ACC genes (Levin and Hammes 2016). The actions of EDCs in adipocytes via membrane-bound ERα in the regulation of CREBPα for FAS and ACC activation is not known. However, the action of diethylstilbestrol (DES) in the myometrial cell is known; DES activates ERα that activates PI3K-protein kinase B/AKT signaling. It phosphorylates Zeste homolog 2 (EZH2), a methyl transferase of histone H3, at lysine-27. The methylation of lysine is an epigenetic event. Phosphorylation of EZH2 decreases methyl transferase activity, thus decreasing gene expression, which also includes ER-responsive genes (Levin and Hammes 2016). Similarly, membrane-bound AR also exerts genomic effects in prostate cancer cells, independent of ligand-AR binding in the cytoplasm (Levin and Hammes 2016).

ER-Signaling Pathways

In addition to direct ligand–DNA interaction by ERα and ERβ, ERs can be tethered to other transcription factors or transactivate other transcription factors to exhibit the effects. ERα activation has been demonstrated to transactivate the Sp1 binding site (GC rich). Antiestrogen, ICI 182,780, and tamoxifen, activate the Sp1-reporter construct in the same fashion as estrogen. This effect was not observed when ERβ was expressed after transfection. EDCs, such as octylphenol, nonylphenol, bisphenol A, and many xenoestrogens activated Sp1-mediated transcriptional activation via E_2-ERα when tested in MCF7 cell line (Wu et al. 2008). ER activation is antagonistic to NF-κB-mediated transcription. Both ERs transrepression NF-κB-mediated transcription. The corresponding author of this chapter has investigated the effects of methoxychlor, o,p'-DDT, and dieldrin and has found that expression/activation of ERα caused transrepression of NF-κB reporter by E_2 and the aforementioned xenoestrogens/EDCs (Ansari and Gandy 2007). The activities of these xenoestrogens were less when ERβ was expressed (unpublished data). These tests were performed in Cos-7 cells that are devoid of significant ERs expression (Ansari and Gandy 2007). WAY-169916 is not a steroidal compound and does not activate ERs but inhibits NF-κB-mediated transcription when ERα and ERβ are expressed. Due to its inhibitory effects on NF-κB, the compound, Way-169916, possesses significant anti-inflammatory effects in the absence of significant effects of the classical ER activation such as uterotrophic effects. This is one of the examples where EDCs could have very specific effects in the presence of ERs when typical ER-signaling pathways are absent (Chadwick et al. 2005).

Xenoandrogens and Metabolic Syndrome

The risk of developing diabetes has been linked to the increased application of pesticides and their presence in the environment. Long-term exposure causes an increased risk of diabetes mellitus (Saldana et al. 2007; Starling et al. 2014). Cyclodiene group of insecticides, such as aldrin, dieldrin, and heptachlor (cage compound), are associated with an increased risk of diabetes (Montgomery et al. 2008). The herbicide 2,4,5-trichlorophenoxyacid (2,4,5-T) is used to contain dioxins as the contaminant. Dioxins not only cause interference with the reproductive process but also affect the glucose transport process (Enan et al. 1992; Kogevinas 2001). Organochlorine insecticides and dioxins exposure were shown to be associated with the risk of developing diabetes mellitus in humans (Lee et al. 2011; Pesatori et al. 1998; Remillard and Bunce 2002; Rylander et al. 2005). Pesticides and other commonly used compounds causing diabetes mellitus are listed in Table 3.5.

AR Signaling Pathways

Like ER, AR is a classical NR that participates in transcriptional activation by homodimerization. Human AR is located on an X-chromosome with a size of 2757 nucleotide cDNA and a protein size of 110 kDa size (van Laar et al. 1989). NRs ligand-binding domain contains 12 helixes from H1–H12 and AR does not contain an H2 helix which makes AR different from other NRs (Bohl et al. 2005; Kuiper et al. 1989).

Endocrine disturbance via AR signaling not only produces malfunction of reproductive processes but also causes metabolic disturbance. Organochlorine insecticides or their metabolites mimic androgen and affect androgens-AR-mediated transcription activity. Herbicide, 2,4,5-T, atrazine, and insecticide, aldrin, dieldrin, and heptachlor behave as anti-androgen in the *in vitro* assay (Fang et al. 2003). The presence of chlorine plays a more significant role in binding to AR than ER. Atrazine contains only one chlorine and fails to bind to AR in AR-competitive binding assay; however, it exhibits a positive EDC response in EDTSTAC (Stoker et al. 2000). Dieldrin and chlorinated insecticides act as an antiandrogen *in vitro* studies (Sakkiah et al. 2018). Both dieldrin, and other chlorinated insecticides reduce the intake and retention of androgen like 5-hydroxydihydrotestosterone

Table 3.5 A representative list of pesticides/commonly used compounds causing diabetes mellitus.

Compound	Type of diabetes	Mechanism	References
Aldrin	Diabetes mellitus	Long-term hyperglycemia after exposure	(Montgomery et al. 2008)
Dieldrin	Diabetes mellitus	Long-term hyperglycemia after exposure	(Montgomery et al. 2008; Starling et al. 2014)
Hepatchlor	Diabetes mellitus	Long-term hyperglycemia after exposure	(Montgomery et al. 2008)
Atrazine	Diabetes mellitus	Long-term hyperglycemia after exposure	(Montgomery et al. 2008; Saldana et al. 2007)
2,4,5-T	Diabetes mellitus	Long-term hyperglycemia after exposure	(Saldana et al. 2007; Starling et al. 2014)
Bisphenol A	Diabetes mellitus, type I	Long-term hyperglycemia after exposure	(Bodin et al. 2014)

Source: Adapted from Sakkiah et al. (2018).

Table 3.6 *In vitro* and *in vivo* based assays for the identification of EDCs with androgenic activity.

Assay	Source/description	References
AR binding assay	Rat prostate cytosol	(Zoeller et al. 2012)
H295 steroidogenesis assay	Human adrenal gland cells	(OECD 2011)
Hershberger bioassay in rats	Mechanistic *in vivo* screening assay for androgen agonists, antagonists, and 5α-reductase inhibitors	(Zoeller et al. 2012)
Pubertal male rat assay	Measures of androgen-dependent organ weights, age, and body weight at the time of preputial separation, testis histology, and serum testosterone levels	(Zoeller et al. 2012)

(5-DHT) (Sakkiah et al. 2018). The relative binding affinity of heptachlor (0.0133), aldrin (0.0096), and 2,4,5-T (0.0007) are quite low as compared with androgens binding with AR in the binding assay (Fang et al. 2003). Some of the commonly used assays to determine the androgenic activity of EDCs are presented in Table 3.6.

Mechanism of ED

Epigenetic Mechanism

The EDC exposure results in epigenetic changes without any genetic change, such as base mutation or base deletion/insertion. Epigenetic changes are different from genetic changes. It includes DNA methylation, and histone modification, such as acetylation and changes in noncoding RNA that are linked to gene regulation (Skinner 2016, 2014a).

Methylation and Gene Regulation

The CpG islands are found in promoters (Deaton and Bird 2011). Methylation of residue in CpG reduces the promoter activity of genes and causes repressed gene expression. Methyl transfer on C-residue in CpG occurs by methyltransferases (Schubeler 2015). Methylation of C-residue in CpG closes the chromatin and thus decreases the ability of transcription factors to bind to the promoter and transactivate (Phillips 2008). In contrast, the acetylation of histones opens the chromatin and allows the transcription factors to bind, thus transactivating the gene (Grunstein 1997). Hyperacetylation is linked to increased gene expression while hypoacetylation is linked to the repression of gene expression (Lee and Grant 2019).

Bisphenol A. Exposure of neonatal male rats to BPA for the first five days caused changes in gene expression of ERα and ERβ in testes in adults. It also caused an increase in DNA methyltransferase (DNMT3A and DNMT3B) (Doshi et al. 2011). Transplacental and translactational exposure of male rat pups to BPA causes hypermethylation of DNMT3A and ERα (El Henafy et al. 2020). Transplacental along with translactational exposure is longer compared with translactational exposure alone with higher dose-dependent effects (El Henafy et al. 2020).

Exposure to bisphenol A and its structural analogs, bisphenol F (BPF) and bisphenol S (BPS), causes methylation. Although both BPS and BPS are at least 10-fold less potent in estrogenic activity than BPA, both compounds increased the proliferation of ER + breast cancer cell line

(MCF-7 CV) (Kim et al. 2017). A 24-h BPA exposure induced ER-dependent DNMT1 transcription in MCD-7 cells. These effects were reversed by the inhibition of ER. Similarly, an increase in TET methylcytosine dioxygenase 2 (TET2) transcript was also observed with exposure to BPA (Awada et al. 2019).

Preimplantation exposure to BPA reduces steroidogenesis; as the markers of steroidogenesis, StAR, P450scc, Cyp17a1, and 3β-hydroxysteroid dehydrogenase (3β-HSD) are reduced. The effects occur due to the reduction in histone acetylation of the StAR promoter, which creates a closed chromatin state (Hong et al. 2016). Sertoli cells are nutritional support for the developing germ cells and exposure to environmental toxicants causes testicular abnormalities due to epigenetic changes (Sadler-Riggleman et al. 2019). BPA exposure to mouse Sertoli cells alters microRNA (miRNA) expression profile, corresponding gene expression modification, and reproductive patterns (Cho et al. 2010). Changes in spermatocyte proliferation and sperm motility in a murine model were observed after BPA treatment due to alterations in the methylation of MYBPH and PRKCD genes (Yin et al. 2016).

Bisphenol A alters the methylation patterns of genes if exposure occurs during development. Hypermethylation of estrogen receptor promoter of rat testis has been observed after neonatal exposure to BPA (Doshi et al. 2011). Estradiol and BPA developmental exposure increased the susceptibility of prostate cancer by epigenetically regulating phosphodiesterase type 4 variant 4 (Ho et al. 2006). Estradiol and BPA neonatal exposure altered promoter methylation and gene expression of nucleosomal binding protein 1(nsbp1) and hippocalcin-like protein 1 (hpcal1) genes (Tang et al. 2012). It also altered the transcriptional program of dnmt3a/b and mbd2/4 genes in rat prostate gland throughout life after neonatal exposure (Tang et al. 2012). In another study, the effects of BPA exposure were determined using human fetal liver. It was observed that BPA exposure caused hypermethylation of CpG island while hypomethylation of CG shores, shelves, and repetitive regions, such as long interspersed nuclear elements (LINEs) have been observed in aggressive cancers (Faulk et al. 2015).

Di(2-Ethylhexyl) Phthalate. Humans are exposed to DHEP, which possesses ED activity, due to its presence in a variety of consumer products. Some studies have demonstrated that prenatal exposure to DHEP might cause altered DNA methylation and produce effects on organ weight, and body weight in a sex-specific manner (Neier et al. 2019). Phthalate exposure has been linked as one of the causes of adiposity in Mexican children through adolescence due to altered methylation (Bowman et al. 2019). Effects on the sexual differentiation process have been observed after prenatal exposure of rats with BPA and DHEP (Abdel-Maksoud et al. 2015). In an attempt to decipher the genes affected by maternal exposure, the methylation of genes was followed in cord blood. It was observed that genes that were differentially methylated were *IRAK4* and *ESM1*, the inflammatory response genes, *BRCA1* and *LASP1*, cancer-associated genes and *CNPY1*, endocrine function, and male fertility genes, *IFT140, TESC*, and *PRDM8* (Solomon et al. 2017). DHEP is metabolized in the liver and the perinatal DEHP exposure causes DNA methylation in the liver of both juvenile and adult mice with differential DNA methylation of thousands of sex-specific genes in five-month-old mice observed in the liver even after ceasing exposure postweaning (PND21) (Liu et al. 2021).

Cadmium. Cadmium does not cross the placental–fetal barrier and is retained in the placenta, thereby increasing its concentration in the placenta (Rudge et al. 2009; Sakamoto et al. 2013). Higher levels of Cd in the placenta have been observed in mothers who delivered neonates with low birth weight (Llanos and Ronco 2009). The methylation of long interspersed nuclear element 1 (LINE1) by Cd exposure was studied in humans by Boeke et al. (2012). Due to a transposable element, it is methylated and silent in a normal context; Cd exposure periconceptually causes hypomethylation of LINE-1 in cord blood (Boeke et al. 2012). Since LINE-1 possesses enzymes for its replication and transposase activity, hypomethylation might increase the transposase activity, thereby causing adverse effects, such as an increase in blood pressure later in life (Barchitta et al. 2014; Bellavia et al. 2013). The study also reported contrasting findings where Cd intake was associated with LINE-1 hypermethylation in maternal blood in the first trimester. A similar study among Argentinian women reported hypomethylation of LINE-1 retrotransposon (Hossain et al. 2012). In another study, no significant difference in genome-wide methylation (approximately 450,000 sites) of CpG was observed in women from Bangladesh who were exposed to Cd only through food (Kippler et al. 2013). Markedly, gender-specific Cd-related methylations were observed: in boys, hypermethylation with increasing Cd concentration in cord blood was observed in 96% of the top CpGs, while CpGs among girls were 29% hypermethylated. Pathway analysis indicated only a gene of apoptosis among boys while embryonic and organ development-associated genes in girls. The maternal CpG methylation related to lower birth weight of neonates correlated with Cd-related fetal growth impairment (Vilahur et al. 2015). Animal studies involving maternal Cd exposure and offspring methylation pattern indicated a specific effect, where GRs in male offspring were hypermethylated due to increased expression of DNMTA3; in female offspring, increased expression of GRs was observed due to decreased activity of DNTM3A. The animal study is in line with the human study, in which hypermethylation in newborn males and hypomethylation in newborn females with Cd exposure were observed (Kippler et al. 2013). In a few *in vitro* studies, exposed cells with Cd and the effects on methylation were investigated. Mouse embryonic stem cells when exposed to $CdCl_2$ for 1 and 24 h, caused a significant reduction in monomethylation of lysine residues in histone proteins (Gadhia et al. 2012). Reduction in histone lysine methylation causes the opening of the chromatin and transcriptional activation. In another study, human embryonic lung fibroblasts were exposed to Cd for 2 months at concentrations of 1.2 or 1.5 μM/L and global hypermethylation was observed (Jiang et al. 2008). The hypermethylation was linked to the upregulation of methylation genes, DNMT1, DNMT3A, and DNMT3B (Jiang et al. 2008). In another study, Cd and nicotine exposure (cigarette smoking) resulted in the hypermethylation of an overlapping 30 genes. Analysis of methylation revealed that the apoptosis-related gene; PRR13 was differentially methylated in fetal DNA after Cd exposure, which was reconfirmed by methylation-specific tests (Sanders et al. 2014).

Arsenic. Arsenic is a non-mutagenic human carcinogen whose mechanism of carcinogenesis is not well established; it is also classified as an EDC. Arsenic was added to the list of chemicals known to cause cancer under Proposition 65 in 1987 with the NSRL of 0.06 μg/day for inhalation and 10 μg/day for all other routes (https://oehha.ca.gov/chemicals/arsenic). Several mechanisms such as mitochondrial involvement in carcinogenesis and epigenetic regulation of genes by methylation of specific genes for ED and carcinogenesis have been proposed. In a study, placental JEG3 trophoblasts were exposed to inorganic arsenic methylated and altered 12 GR-mediated genes that play a significant role in fetal and placental development (Meakin et al. 2019). Prenatal exposure to arsenic causes epigenetic imprinting, resulting in ED (Gangopadhyay et al. 2019). Detoxification of arsenic involves *S*-adenosyl methionine (SAM) which serves as a universal donor for methyl transferases (DNTMs).

It is proposed that arsenic metabolism involving methylation may deplete intracellular methyl stores leading to the hypomethylation of genes (Cheng et al. 2012). Both hypo- and hypermethylation of genes occur, such as increased histone methylation, H3K9me2, and decreased H3K27me3 (Salnikow and Zhitkovich 2008; Zhou et al. 2008). Arsenic exposure decreases DNMT activity by downregulating the DNMT gene itself (Reichard et al. 2007). Using human keratinocytes, inhibition of DNMT1 and DNMT3a was demonstrated at 0.2 μM arsenite concentration; depletion of SAM and lower global methylation were also observed (Reichard et al. 2007). Sodium arsenite treatment of pregnant mice at 85 ppm in drinking water during gestation days 8–18 caused hypomethylation of the GC-rich regions of hepatic DNA of newborn male mice (Xie et al. 2007). Treatment of rat liver cell line, TRL1215, with sodium arsenite at low concentrations (125–300 nM) for more than 18 weeks resulted in decreased SAM pool, global hypomethylation, and malignant transformation (Zhao et al. 1997). Later, using the microarray technique, Chen et al. (2001) identified the upregulation of various oncogenes, such as myc, ras, erbB2, and met. Arsenic exposure causes hypermethylation of tumor suppressor genes (Jensen et al. 2008); exposure of lung adenocarcinoma cell, A549, with sodium arsenite (0.08–2.0 μM) or sodium arsenate (30–300 μM) caused hypermethylation of p53 promoter (Mass and Wang 1997).

Humans exposed to arsenic and diagnosed with arsenicosis in Mexico were used for determining the methylation of CpG using peripheral lymphocytes. Compared with control/nonexposed subjects, arsenicosis individuals exhibited hypermethylation of 182 genes against 183 genes, 17 of which were tumor suppressor genes (Smeester et al. 2011). Besides affecting methylation, arsenic exposure increases the expression of c-jun, c-fos genes by histone H3, and increased phosphorylation and acetylation (Li et al. 2003). Arsenate is capable of altering chromatin structure thus affecting gene expression by inhibiting histone deacetylase, which increases the acetylation and opens the chromatin for gene expression. Increased global acetylation of HepG2 cells was observed with the treatment of arsenic at 7.5 μM (Ramirez et al. 2008). However, contradictory results were reported, i.e., lower histone acetylation in human uroepithelium when exposed to arsenic (Chu et al. 2011).

Arsenic exposure causes changes in miRNA expression (Marsit et al. 2006). Arsenic exposure to human lymphoblastoid cells at 2 μmol/L upregulated miRNA-22, miRNA34a, miRNA-221, and miRNA-222 while downregulating miRNA-210. The miRNAs were identified from a pool of 385 miRNA that was subjected to microarray analysis (Marsit et al. 2006).

In another study, the miRNA expression profile from the liver of Sprague-Dawley rats was investigated after 0.1, 1, 10, and 100 mg/L for 60 days. Multiple miRNA expressions were noted, either increased or decreased. The miRNAs that were identified were Nfe212-regulated antioxidant genes that included glutamate-cysteine ligase (GCL) catalytic subunit (GCLC) and modifier subunit (GCLM). These genes are involved in reduced glutathione (GSH) synthesis. The low-dose exposure increased the expression of GCLC and GCLM while high doses decreased the expression. The study concluded that arsenic exposure altered genome-wide miRNA expression (Ren et al. 2015).

Nickel. Nickel as metal finds its application in chemical catalysts, manufacturing of alkaline batteries, welding products, etc. Humans are exposed to Ni due to industrial activities, such as electroplating pigment production and the production of Ni-based compounds as intermediates for further use (Medicine 2004). Ni exposure is associated with human male reproductive abnormalities and is recognized as an endocrine disruptor. High blood Ni levels among welders have been linked to abnormal sperms (Danadevi et al. 2003; Hjollund et al. 1998). It causes disruption of steroidogenesis and thus spermatogenesis (Apostoli and Catalani 2011; Sun et al. 2011; Wirth and Mijal 2010). It may affect the secretion of luteinizing hormone (LH) that is involved in the synthesis of testosterone by Leydig cells and follicle-stimulating hormone (FSH) that maintains spermatogenesis in seminiferous tubule (Pizent et al. 2012). In a study, among workers exposed to Ni in Egypt, several sexual abnormalities, such as a decrease in libido, premature ejaculation, impotence, and infertility with changes in LH, FSH, and testosterone were observed (Beshir et al. 2016). Nickel was added to the list of chemicals known to cause cancer under proposition 65 in 1989 and is under consideration for listing to cause developmental/reproductive effects, no safe harbor levels are provided yet (https://oehha.ca.gov/chemicals/nickel-and-nickel-compounds).

Studies related to epigenetic changes linked to EDC are lacking; however, studies linking to epigenetic changes due to Ni exposure and carcinogenesis are available. Ni, as a divalent ion, condenses the chromatin, and condensed chromatin is not a good substrate during transcription (Ellen et al. 2009; Li et al. 2007). Ni exposure as Ni_3S_2 to wild-type C57BL/6 and p53± mice led to the formation of malignant histiocytomas and tumor development associated with methylation of the p16 promoter (Govindarajan et al. 2002). Nickel exposure silences the tumor suppressor gene Fhit (Kowara et al. 2004). Silencing of the DNA repair gene, 6-O-methylguanine DNA methyltransferase (MGMT), via epigenetic mechanism with hypermethylation and histone modifications including higher levels of H3K9me2 and lower levels of acetylation (H4ac and H3K9ac) has also been observed (Ji et al. 2008). Ni-based hypoacetylation is linked to the inhibition of histone acetyltransferase (HAT) activity with no effects on histone deacetylase (HDAC) activity (Kang et al. 2003).

Lead. There is conflicting evidence showing that the accumulation of Pb affects endocrine glands. Lead affects the hypothalamic-pituitary axis and reduces the production of thyroid-stimulating hormone (TSH), growth hormone (GH), follicle-stimulating hormone (FSH), and luteinizing hormone (LH), and thus affects corresponding receptor-based function (Doumouchtsis et al. 2009). The lower levels of GH could be due to the GH-releasing hormone (GHRH). In individuals with short-term Pb exposure, the levels of LH and FSH are higher than that of normal testosterone levels, while longer-term exposure lowers testosterone concentration (Doumouchtsis et al. 2009). In a study correlating Pb levels of the patella and tibia with cord blood methylation, it was found that LINE-1 and Alu1 methylation were inversely linked (Pilsner et al. 2009). Reduced LINE-1 DNA methylation was also found among older men correlating to Pb in bone (Wright et al. 2010).

Role of Noncoding RNAs

Most of the noncoding RNAs (ncRNA) are expressed at substantially lower levels compared with most mRNAs, pointing to their regulatory role. In addition to their role during development, the ncRNA responds to environmental stimuli and stresses (De Santa et al. 2010; Dinger et al. 2008; Faghihi et al. 2008; Kim et al. 2010; Li et al. 1999). Noncoding RNAs could be infrastructural (ribosomal, transfer, small nuclear, and small nucleolar RNAs) and regulatory RNAs (miRNAs, Piwi-interacting RNAs [piRNAs], small interfering RNAs [siRNAs], and long noncoding RNAs [lncRNAs]) (Ponting et al. 2009; Yao et al. 2019). miRNAs control the genes at posttranscriptional levels, i.e., regulating the levels of mRNAs. Notably, degradation and destabilization of mRNA are among the several mechanisms of miRNA-based gene regulation (Guo et al. 2010; Hutvagner and Zamore 2002). The effects of EDCs on cellular functions by genomic and non-genomic forms are summarized in Figure 3.1.

Transgenerational Inheritance of Epigenetics Induced by EDCs

Bisphenol A: Bisphenol A is the predominantly used bisphenol (Dianin 1891). Approximately 8 million tons of BPA are produced annually (Vom Saal et al. 2012). Humans are exposed directly and indirectly to BPA at varying levels. It is estimated that globally, humans are exposed to around 30 ng/kg body weight/per day BPA (Huang et al. 2017, 2012; Rezg et al. 2014). The major route of BPA exposure in humans is ingestion (Chen et al. 2016).

Certain miRNA genes possess estrogen response elements in the promoter, which could become a target for weak estrogenic endocrine disruptors, such as BPA. miRNA-21, miRNA-155, and miRNA-124 are miRNA genes that could be modulated by ERα and ERβ (Klinge 2009). Another miRNA, miRNA-224, is involved in oocyte maturation and was found dysregulated in rat granulosa after *in vivo* treatment with BPA (Lite et al. 2019). The miRNA-224 is involved in regulating enzymes for hormone synthesis, including aromatase (Lite et al. 2019). Studies related to oocyte development and BPA effects are not that extensive. Other studies involving other estrogen-responsive organs, such as the breast, indicating that BPA reduces the levels of miRNA-21, Let-7, and mIR-15b while increasing the levels of mIR-638, miRNA-663, and miRNA-1915 in breast cancer cells (Tilghman et al. 2012).

EPA has established a dose of 50 μg/kg/day BPA as the safe dose. *In utero* exposure to <50 and 50 μg/kg/day doses of BPA has been associated with increased DNA methylation. The increased DNA methylation results in the dysregulation of glucose levels via impaired insulin regulation (Newbold et al. 2007).

DHEP: In an attempt to decipher the potential effects of exposure to DEHP on changes in ovarian miRNA expression profile during mouse primordial follicle development, ovaries from newborn mice were cultured *in vitro* and exposed to 0, 10, and 100 μM DHEP for 72 h. Changes in miRNA were observed at 100 μM concentration. DHEP exposure decreased miRNA-19a-3p, whose target is phosphatase and tensin homolog (PTEN). Decreased miRNA-19a-3p will increase the levels of PTEN with implications for phosphorylation of AKT decreasing phospho-AKT levels. Decreased phospho-AKT will induce the expression of Bax and Bcl-2 (Zhang et al. 2019). Bax and Bcl-2 are two gene regulators of the mitochondrial apoptotic pathway (Linehan et al. 2003). The product of Bcl-2 suppresses the apoptotic signal and can contribute to cell growth or oncogenesis. Bax is a homolog of Bcl-2 and promotes apoptosis. Apoptosis stimuli are dependent on the Bcl-2, its homolog, and the Bax ratio. Excess Bcl-2 will protect cell survival but when Bax is in excess, the cell will be susceptible to apoptosis due to Bax homodimerization (Oltvai et al. 1993). The role of Bax vs. Bcl-2 derangement in the context of DHEP exposure and ED is not fully understood (Oltvai et al. 1993).

Anti-Thyroids

In addition to direct interaction with thyroid hormone receptors, EDCs disrupt thyroid gland function by acting on the hypothalamus-pituitary-thyroid gland axis, which is presented in Table 3.7. A pictorial illustration of the perturbation in thyroid hormones in through hypothalamus-pituitary-thyroid axis described for trifluralin, an herbicide, by Saghir et al. (2008) is shown in Figure 3.7. They affect receptors of thyroid releasing hormone, receptor of TSH, sodium/iodide symporter activity, thyroid hormone transport proteins (e.g., transthyretin, thyroid binding globulin, thyroid deiodinase activity), besides direct effects on TR, the nuclear receptor superfamily member (Fini et al. 2007; Kitamura et al. 2005). Polychlorinated biphenyls act both as agonists and antagonists at TR in various tissues depending upon the congeners. Hydroxylated metabolites bind TR as an agonist that activates thyroid response genes in the pituitary and liver via its thyroid response elements (Giera et al. 2011). PCBs can also antagonize the effects of triiodothyronine (T3) which binds to activate TR-mediated transcription (Giera et al. 2011). Studies in humans and animals demonstrated that PCB and PCDD can disrupt thyroid function at the concentrations found in humans (Ghassabian and Trasande 2018). Like PCBs and PCDDs, PBDEs can inhibit T3 binding to TR and disrupt TR-mediated transcription (Fini et al. 2007; Kitamura et al. 2005). Epidemiological studies have linked prenatal PBDE exposure with possible thyroid disruptive properties leading to several neuropsychological, cognitive, and motor function abnormalities among children (Chen et al. 2014; Ghassabian and Trasande 2018; Herbstman et al. 2010; Roze et al. 2009).

Bisphenol A mimics estrogen and thus produces health effects of neurodevelopment, uterine cancer, immune toxicity, neurotoxicity, and obesity (Shafei et al. 2018; Watkins et al. 2017). BPA binds with estrogen receptors, alpha, and beta (ERα and ERβ) (Acconcia et al. 2015; Grimaldi et al. 2019). BPA and analogs inhibit androgen receptor (AR) spliced and mutant forms, and

Table 3.7 List of thyroid EDCs and their targets.

Class of EDC	Mechanism of action	References
Phthalates	Receptor of thyroid releasing hormone (TRHR) in the hypothalamus and pituitary Receptor of thyroid-stimulating hormone (TSHR) in the thyroid gland Gene expression of T4 and T4 synthesis, transport, and metabolism	(Ghassabian and Trasande 2018; Ye et al. 2017; Oliveira et al. 2018; Chen et al. 2021)
BPA	Thyroid receptor (TR) in the pituitary and THR	(Oliveira et al. 2018)
PFAS	Thyroid hormone transport, thyroid gland deiodinase activity	(Coperchini et al. 2020; Wang et al. 2018; Ghassabian and Trasande 2018)
Polybrominated biphenyls and their ethers	Thyroid hormone transport, THRs, thyroid hormone metabolism, deiodinase activity of thyroid gland	(Oliveira et al. 2018; Vuong et al. 2018; Byrne et al. 2018)
Polychlorinated biphenyls and dibenzodioxins (PCDD)	Thyroid hormone transport, THRs	(Berlin et al. 2021; Turyk et al. 2007)
Sodium/iodine transporter (symporter) inhibitors	Iodine transport by the thyroid gland	(Wang et al. 2018)
Pesticides	Thyroid hormone transport, THRs	(Leemans et al. 2019)

pregnane-X-receptor (PXR) (Acconcia et al. 2015; Grimaldi et al. 2019). In addition, they can bind to aryl hydrocarbon receptors (AhRs) and peroxisome proliferator-activated receptors (PPAR) (Bishop-Bailey et al. 2000). BPA and analogs also bind to orphan ER-related receptor-gamma (Grimaldi et al. 2019). Bisphenol AF where one of the hydrogens of the methyl group is replaced with fluorine (BP-AF) is a full agonist of ERα but an antagonist of ERβ (Matsushima et al. 2010). BPA is defined as a weak estrogen-mimicking chemical but it could be involved in membrane-associated NR signaling (McLachlan et al. 1984). Data indicate that BPA is involved in ERα-mediated extranuclear signaling activating ERK/MAPK and AKT (Bolli et al. 2008; Marino et al. 2012). Therefore, BPA is involved in non-genomic effects via ER. In addition, BPA releases Ca++ from intracellular stores leading to exocytosis and cell motility (Wozniak et al. 2005). Since the action of BPA on ERβ are antagonistic, ERβ activation inhibits p38/MAPK (Bolli et al. 2010; Marino et al. 2012).

Triclosan [(5-chloro-2-(2,4-dichlorophenoxy)phenol)] is an antimicrobial agent that possesses a chlorinated phenolic structure, humans are exposed to it due to its presence in household products, such as dishwashing soap, food containers, and cutting boards that are listed as antibacterial, odor-fighting, or keepers of fresh food, etc. It functions via the activation of PXR and by the inhibition of diiodothyronine (T2) sulfotransferases. Triclosan structural similarity to TH suggests its possible effects on TH-mediated effects (Paul et al. 2007). The Second Korean National Environmental Health Survey (2012–14) showed a positive association between triclosan exposure and serum TSH concentrations in females with marginal significance. No significant link was observed between triclosan exposure and serum total T3 and T4 levels in both males and females, and TSH in males (Ha et al. 2019). Experimentally, it has been demonstrated

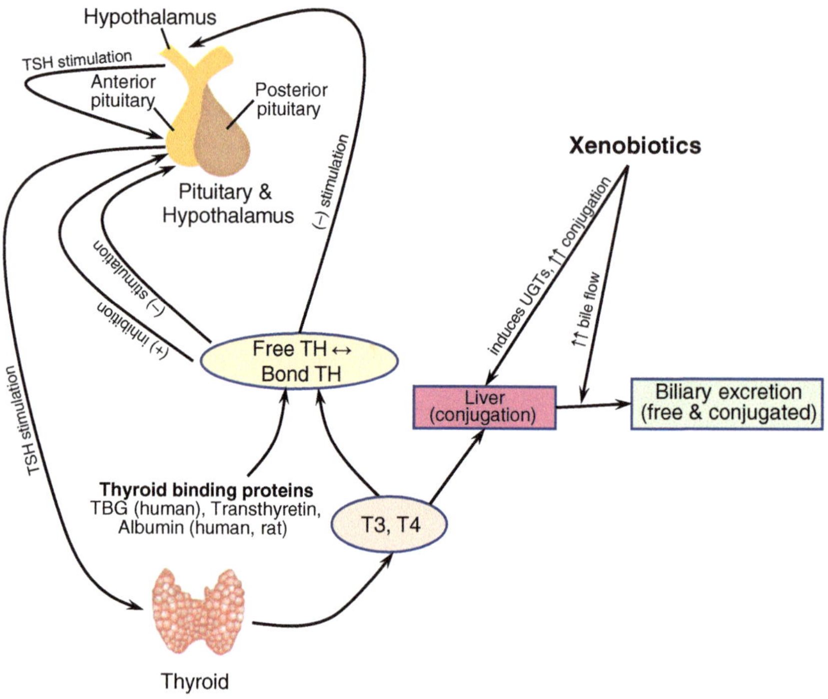

Figure 3.7 Possible xenobiotic-induced perturbation in thyroid hormones in the hypothalamic-pituitary-thyroid axis that can lead to thyroid effects. Abbreviations: TRH, thyrotropin releasing hormone; TSH, thyroid stimulating hormone; T3, triiodothyronine; T4, thyroxine; TH, thyroid hormones; TBG, thyroxine binding globulin; UGT, uridinediphosphate glucuronosyltransferase.

that triclosan does not affect androgen-dependent weight gains in male rats; however, it affects the levels of thyroid hormone in male juvenile rats (Zorrilla et al. 2009). Table 3.7 shows a list of EDCs targeting the thyroid.

Available literature links EDC exposure with alterations in thyroid function. As mentioned above, PFOA, a derivative of PFAS [perfluorohexane sulfonate (PFHxS), and 2-(N-methyl-perfluorooctane sulfonamide) acetate (MeFOSAA)], exposure to mothers alters the thyroid function in both pregnant mothers and male offsprings. A study, Project Viva, detected 6 PFAS and T4 hormones in pregnant women (732 cases) at 10 weeks and in male offspring (480 cases). Higher levels of four out of six PFAS were found in mothers with lower T4 levels, same PFAS were also found in their newborns. The TSH levels were not affected by PFAS in pregnant women except for those who were autoimmuned with thyroid peroxidase antibodies, which is a marker of thyroid disease (Preston et al. 2018; Woodmansee 2018).

Organotin

Genomic Signaling and Effects

Organotin compounds were developed in the twentieth century, and some of them possess fungicidal properties (Abdellah et al. 2009). Organotins are used as a catalyst for polyurethane foam and in silicone production. It is being used as a surface disinfectant in hospitals and veterinary hospitals. It is also used in laundering sanitizers and scintillation detectors of gamma and X-rays. As a pesticide, it is used to control fungi, mites, and mollusks. Due to its commercial use, production is increasing (Okoro et al. 2014).

The effect of organotin exposure on reproductive systems is exactly the opposite in humans to what is observed in gastropods. In gastropods, organotins, such as tributyl tin (TBT) and triphenyl tin (TPT), inhibit aromatase, at nanomolar levels, which converts androgen to estrogen and produces irreversible sexual abnormality referred to as "imposex", while in humans organotins increase the levels of estrogens (Bettin et al. 1996; Horiguchi et al. 1997; Matthiessen and Gibbs 2009; Nakanishi 2008). Using human choriocarcinoma cells, it was observed that the activities of aromatase and 17beta-hydroxysteroid dehydrogenase type I (17beta-HSD I) were increased (Nakanishi 2008). Organotins also increase the human chorionic gonadotropin (hCG) hormone in the range of 3–100 nM concentration. These observations point out that organotins are potent stimulators of estrogen and hCG and could potentially affect the pregnancy by affecting the placental estrogen and hCG synthesis (Nakanishi 2008). Organotins have been identified as the agonist of retinoid X receptor (RXR) and peroxisome proliferator-activated receptor (PPAR) gamma (Kanayama et al. 2005). Human aromatase is regulated by two promoters, and RXR and PPARγ regulate the activity of aromatase differently in the ovary and placenta. In human ovaries, organotins activate RXR and PPARγ, which downregulate the activity of NF-κB and activate the P II promoter. Thus RXR and PPARγ counteract the activity of NF-κB, downregulating the expression of aromatase (Nakanishi 2008). In the case of the placenta, the organotin-activated RXR binds to the promoter of exon 1.1, which contains the retinoid X response element (RXRE), and activates the transcription (Nakanishi 2008). Both RXR and PPARγ are members of the nuclear receptor superfamily, which is implicated in the genomic effects of EDCs.

Epigenetic Effects of Organotin

Studies have demonstrated that treatment of SH-SY5Y neuroblastoma *in vitro* with TBT causes increased methylation of the promoter of nuclear respiratory factor-1 (NRF-1) and increases the amount of methyl-CpG-binding protein 2 (MeCP2) resulting in decreased expression of NRF-1 and neurotoxicity (Hanaoka et al. 2018). Global hypomethylation in the liver of marble trout (*S. marmoratus*) was observed after 48 days of exposure to TBT and TPT (Wang et al. 2009). Organotin compounds can increase gene expression by increasing histone acetylation, which opens the closed state of DNA and allows transcriptional activation (Osada et al. 2005).

TCDD and Related Compounds

Role of AhR: AhR, which is known as a dioxin receptor due to binding with dioxin, belongs to the basic helix-loop-helix (bHLH) PER-ARNT-SIM (PAS) protein family that binds to DNA following heterodimerization with the nuclear translocator protein (ARNT) (Nebert 2017). AhR also possesses a modular structure like NR. The basic helix loop helix (bHLH) N-terminal domain is involved in DNA recognition. Two PAS domains (PAS-A and PAS-B) are present in tandem repeats which encompass 50 amino acids. The bHLH of PAS-A domains of AhR and ARNT dimerize to form a stable DNA–heteromer complex. The ligand binds to the PAS-B domain. The C-terminal domain functions as a transcriptional activation domain, like NRs (Denison et al. 2011; Kewley et al. 2004). AhR binds with many ligands that include tryptophan metabolites, man-made dietary indoles in addition to TCDD, and many xenobiotics, such as PAHs (Murray et al. 2014). The list of xenobiotics that binds to AhR is presented in Table 3.8.

Table 3.8 The list of xenobiotic ligands that binds to AhR.

Non-pharmaceuticals	Pharmaceuticals
Agonist	
A: Halogenated aromatic hydrocarbons	B: Pharmaceuticals
1. TCDD (2,3,7,8-tetrachlorodibenzo-p-dioxin)	1. Tranilast
2. Dibenzofurans	2. Leflutamide
3. Biphenyls	3. Omeprazole
4. Polyaromatic hydrocarbons	
5. 3-Methylcholanthrene	
6. Benzo[a]pyrne	
7. Benzanthracene	
8. Benzoflavone	
Antagonist	
1. StemRegenin 1	
2. GNF 351	
3. CH-223191	
4. 6,2′,4′-trimethoxyflavone	

Source: Adapted from Murray et al. (2014).

TCDD possesses a high-affinity binding to the aryl hydrocarbon receptor (AhR) (Baccarelli et al. 2004). Benzo(a)pyrene, which is the product of industrial activities, low-temperature burning of plastic, charred BBQ meat, and a component of cigarette smoke, also binds and activates AhR-mediated transcription (Birnbaum 1994; Buck Louis 2014). AhR is an orphan nuclear receptor and knockout of AhR causes reduced fertility, indicating the presence of endogenous compounds that might activate AhR (Denison and Nagy 2003; Seidel et al. 2001).

TCDD is produced by the incomplete combustion of fossil fuels, plastic, wood, and industrial waste. TCDD was a contaminant in 2,4-D and agent orange (2,4,5-T); agent orange was used as a defoliant during the Vietnam War (Bruner-Tran et al. 2017). TCDD is a class of compounds comprising dioxin or dioxin-like compounds, which include PCDDs (polychlorinated dibenzodioxins) PCDFs, (polychlorinated dibenzofurans), and PCBs (polychlorinated biphenyls).

TCDD and Genetic Response

TCDD and several polycyclic hydrocarbons are recognized as the ligand for AhR and activate transcription from AhR that resides in the cytoplasm in an inactive form and translocates to the nucleus after ligand-mediated activation (Poland et al. 1976). After dimerization with aryl hydrocarbon receptor nuclear translocator, the heterodimerized partner of hypoxia-inducible factor, which also binds to xenobiotic response element (5′-TNGCGTG-3′) to activate transcription and increase the expression of genes, such as cytochrome P450 (CYP), 1A1, CYP1B1, indoleamine 2,3-dioxygenase 1 (IDO1), tryptophan 2,3-dioxygenase 2 (TDO2), and aryl hydrocarbon receptor repressor (AhRR), which provides the feedback regulation of AhR (Matsushita et al. 1993; Reyes et al. 1992; Rothhammer and Quintana 2019). The modular structure of AhR is depicted in Figure

Figure 3.8 The modular structure of AhR.

3.8. AhR activation not only activates drug-metabolizing enzymes by its direct transcriptional activation but also functions in cross-talks with ER (Beischlag and Perdew 2005; Ohtake et al. 2003; Wormke et al. 2003) GR (Dvorak et al. 2008; Vrzal et al. 2009), PR (Kuil et al. 1998), AR (Safe 2005), and TR (Pocar et al. 2006) pathways and thus affects the transcriptional regulation by nuclear receptors.

TCDD-Mediated Epigenetic Response

TCDD decreases progesterone responsiveness due to epigenetic modification of the progesterone gene (*Pgr*) (Rattan et al. 2017). Experimentally when mice are exposed to TCDD, reduced expression of progesterone receptors in endometrial tissue occurs (Bruner-Tran et al. 2017). Similarly, altered expression of transforming growth factor-β2 occurs, which is important in the maintenance of pregnancy (Nayyar et al. 2007). Exposure to TCDD is linked to endometriosis among women and methylation of progesterone receptor has been observed in endometrial stromal cells of women undergoing endometriosis, which is an implication in terms of decidualization (Bruner-Tran et al. 2017). The endometrial cells isolated from disease-free women do not exhibit hypermethylation of PR, which could be hypermethylated by treatments with inflammatory cytokines, such as interleukin-1 and tumor necrosis factor-α (Bruner-Tran et al. 2017; Wu et al. 2007). Experimental and human studies demonstrate that loss of PR activity occurs due to heightened immune response after TCDD exposure (Bruner-Tran et al. 2017). An animal study demonstrated that exposure to TCCD could lead to transgenerational hypermethylation of PR (Bruner-Tran et al. 2017). Progesterone receptors (and GRs) transrepress NF-κB-mediated transcription and downregulate inflammatory response; that is why glucocorticoids find their application in asthma treatment, which is an inflammatory condition (McKay and Cidlowski 1998; van der Velden 1998). Reduced expression of PR will allow NF-κB-mediated transcription and produce an inflammatory response. As is the case with methylation, which could be transient and not passed on to the next generation, studies on methylation alterations may not be consistent. Both hypo- and hypermethylation of HOXA10 have been reported in women with endometriosis compared with non-diseased women (Andersson et al. 2014; Kim et al. 2007).

Conclusions

EDCs are a diverse group of compounds, added to the environment through both natural (e.g., aflatoxin) and anthropogenic sources. This chapter mostly describes the endocrine-disrupting effect of anthropogenic chemicals. Most of these chemicals are designed to resist degradation to be effective in their intended use (e.g., pesticides, fire-resistant, fabric coating). However, their persistence has proven to be a bane when it comes to ED effects. These chemicals at lower doses disrupt the endocrine function, including the reproduction of aquatic and terrestrial animals. The possibility of effects on humans also exists; however, being a larger species, the effects are not as pronounced as in aquatic and terrestrial animals. EDCs disrupt endocrine function at far below levels of exposure where other typical toxicity of the xenobiotics is not observed. The traditional scientific methods for assessing the human health impact of EDCs are currently inadequate.

Many of the EDCs do not have structural alerts. Notably, they span from pesticides to PCBs, PBDE, phthalates, dioxins, PFAS, BPA, metals (e.g., Pb, Ni, and As), and naturally occurring phytoestrogens (e.g., aflatoxin, genistein, daidzein in soy products). Most of the traditional toxicology methods are not appropriate to determine ED activity. Often, animal and resource-intensive use of molecular biological techniques are recommended as a rapid and less expensive alternative approach before embarking on the use of whole animal studies. Two of the molecular biological techniques, which can be used to determine ED activities of chemicals, are (i) nuclear receptor superfamily mediated transcriptional activation/interference, and (ii) transcriptional transrepression and gene regulation by AhR. Chemical exposure to the biological system usually causes DNA methylation, often CpG methylation, silencing transcription and gene expression. The binding of hormones to nuclear receptors can lead to transcriptional activation and 26S proteasomal-mediated degradation. Understanding the degradation of NR-EDC complex with 26S proteasome may further our understanding of the mechanism of action of EDCs. Collectively, this chapter described the effects of EDCs via nuclear receptor superfamily mediated transcriptional interference, transrepression of transcription, and epigenetics involving DNA methylation and micro-RNA-based mechanism using representative EDCs.

References

Abdellah, M.A., Hadjikakou, S.K., Hadjiliadis, N., Kubicki, M., Bakas, T., Kourkoumelis, N., Simos, Y.V., Karkabounas, S., Barsan, M.M., and Butler, I.S. (2009). Synthesis, characterization, and biological studies of organotin(IV) derivatives with o- or p-hydroxybenzoic acids. *Bioinorg. Chem. Appl.* 542979.

Abdel-Maksoud, F.M., Leasor, K.R., Butzen, K., Braden, T.D., and Akingbemi, B.T. (2015). Prenatal exposures of male rats to the environmental chemicals bisphenol A and Di(2-Ethylhexyl) phthalate impact the sexual differentiation process. *Endocrinology* 156: 4672–4683.

Acconcia, F., Pallottini, V., and Marino, M. (2015). Molecular mechanisms of action of BPA. *Dose Response* 13: 1559325815610582.

Adibi, J.J., Hauser, R., Williams, P.L., Whyatt, R.M., Calafat, A.M., Nelson, H., Herrick, R., and Swan, S.H. (2009). Maternal urinary metabolites of Di-(2-Ethylhexyl) phthalate in relation to the timing of labor in a US multicenter pregnancy cohort study. *Am. J. Epidemiol.* 169: 1015–1024.

Ait-Aissa, S., Laskowski, S., Laville, N., Porcher, J.M., and Brion, F. (2010). Anti-androgenic activities of environmental pesticides in the MDA-kb2 reporter cell line. *Toxicol. Vitro* 24: 1979–1985.

Amato, A.A., Wheeler, H.B., and Blumberg, B. (2021). Obesity and endocrine-disrupting chemicals. *Endocr. Connect.* 10: R87–R105.

Andersson, K.L., Bussani, C., Fambrini, M., Polverino, V., Taddei, G.L., Gemzell-Danielsson, K., and Scarselli, G. (2014). DNA methylation of HOXA10 in eutopic and ectopic endometrium. *Hum. Reprod.* 29: 1906–1911.

Ankley, G., Mihaich, E., Stahl, R., et al. (2009). Overview of a workshop on screening methods for detecting potential (anti-) estrogenic/androgenic chemicals in wildlife. *Environ. Toxicol. Chem.*

Ansari, R.A. and Gandy, J. (2007). Determining the transrepression activity of xenoestrogen on nuclear factor-kappa B in Cos-1 cells by estrogen receptor-alpha. *Int. J. Toxicol.* 26: 441–449.

Apostoli, P. and Catalani, S. (2011). Metal ions affecting reproduction and development. *Met. Ions Life Sci.* 8: 263–303.

ATSDR. (2011) *(Agency for Toxic Substances and Disease Registry): How Are People Exposed to Cadmium?* [Online]. https://www.atsdr.cdc.gov/csem/cadmium/How-Are-People-Exposed-to-Cadmium.html.

ATSDR. (2020). *(Agency for Toxic Substances and Disease Registry): What are the health effects of PFAS?* https://www.atsdr.cdc.gov/pfas/health-effects/index.html (accessed 11/29/2022).

ATSDR. (2022). *(Agency for Toxic Substances and Disease Registry): Toxicological Profile for DDT, DDE, and DDD* [Online]. https://wwwn.cdc.gov/TSP/ToxProfiles/ToxProfiles.aspx?id=81&tid=20.

Awada, Z., Nasr, R., Akika, R., Cahais, V., Cuenin, C., Zhivagui, M., Herceg, Z., Ghantous, A., and Zgheib, N.K. (2019). DNA methylome-wide alterations associated with estrogen receptor-dependent effects of bisphenols in breast cancer. *Clin. Epigenetics* 11: 138.

Baccarelli, A., Pesatori, A.C., Masten, S.A., Patterson, D.G., Jr., Needham, L.L., Mocarelli, P., Caporaso, N.E., Consonni, D., Grassman, J.A., Bertazzi, P.A., and Landi, M.T. (2004). Aryl-hydrocarbon receptor-dependent pathway and toxic effects of TCDD in humans: a population-based study in Seveso, Italy. *Toxicol. Lett.* 149: 287–293.

Bagchi Bhattacharjee, G. and Paul Khurana, S.M. (2014). In vitro reporter assays for screening of chemicals that disrupt androgen signaling. *J. Toxicol.* 2014: 701752.

Balaguer, P., Francois, F., Comunale, F., Fendt, H., Boussioux, A.M., Pons, M., Nicolas, J.C., and Casellas, C. (1999). Reporter cell lines to study the estrogenic effects of xenoestrogens. *Sci. Total Environ.* 233: 47–56.

Barchitta, M., Quattrocchi, A., Maugeri, A., Vinciguerra, M., and Agodi, A. (2014). LINE-1 hypomethylation in blood and tissue samples as an epigenetic marker for cancer risk: a systematic review and meta-analysis. *PLoS One* 9: e109478.

Barkhem, T., Carlsson, B., Nilsson, Y., Enmark, E., Gustafsson, J., and Nilsson, S. (1998). Differential response of estrogen receptor alpha and estrogen receptor beta to partial estrogen agonists/antagonists. *Mol. Pharmacol.* 54: 105–112.

Becher, G., Bergman, Å., Bjerregaard, P., Bornman, R., Brandt, I., Heindel, J. J., Iguchi, T., Jobling, S., Kidd, K. A., Kortenkamp, A., Muir, D. C. G., Ochieng, R., Skakkebæk, N. E., Toppari, J., Woodruff, T. J., & Zoeller, R. T. (2013). *State of the Science of Endocrine Disrupting Chemicals – 2012.* World Health Organization.

Beischlag, T.V. and Perdew, G.H. (2005). ER alpha-AHR-ARNT protein–protein interactions mediate estradiol-dependent transrepression of dioxin-inducible gene transcription. *J. Biol. Chem.* 280: 21607–21611.

Bellavia, A., Urch, B., Speck, M., Brook, R.D., Scott, J.A., Albetti, B., Behbod, B., North, M., Valeri, L., Bertazzi, P.A., Silverman, F., Gold, D., and Baccarelli, A.A. (2013). DNA hypomethylation, ambient particulate matter, and increased blood pressure: findings from controlled human exposure experiments. *J. Am. Heart Assoc.* 2: e000212.

Berlin, M., Barchel, D., Brik, A., Kohn, E., Livne, A., Keidar, R., Tovbin, J., Betser, M., Moskovich, M., Mandel, D., Lubetzky, R., Ovental, A., Factor-Litvak, P., Britzi, M., Ziv-Baran, T., Koren, R., Klieger, C., Berkovitch, M., Matok, I., and Marom, R. (2021). Maternal and newborn thyroid hormone, and the association with polychlorinated biphenyls (PCBs) burden: the EHF (environmental health fund) birth cohort. *Front. Pediatr.* 9: 705395.

Beshir, S., Ibrahim, K.S., Shaheen, W., and Shahy, E.M. (2016). Hormonal perturbations in occupationally exposed nickel workers. *Open Access Maced. J. Med. Sci.* 4: 307–311.

Bettin, C., Oehlmann, J., and Stroben, E. (1996). TBT-induced impo-sex in marine neogastropods is mediated by an increasing andro-gen level. *Helgol. Meeresunters* 50: 299–317.

Birnbaum, L.S. (1994). Evidence for the role of the Ah receptor in response to dioxin. *Prog. Clin. Biol. Res.* 387: 139–154.

Bishop-Bailey, D., Hla, T., and Warner, T.D. (2000). Bisphenol A diglycidyl ether (BADGE) is a PPARgamma agonist in an ECV304 cell line. *Br. J. Pharmacol.* 131: 651–654.

Bodin, J., Bolling, A.K., Becher, R., Kuper, F., Lovik, M., and Nygaard, U.C. (2014). Transmaternal bisphenol A exposure accelerates diabetes type 1 development in NOD mice. *Toxicol. Sci.* 137: 311–323.

Boeke, C.E., Baccarelli, A., Kleinman, K.P., Burris, H.H., Litonjua, A.A., Rifas-Shiman, S.L., Tarantini, L., and Gillman, M. (2012). Gestational intake of methyl donors and global LINE-1 DNA methylation in maternal and cord blood: prospective results from a folate-replete population. *Epigenetics* 7: 253–260.

Bohl, C.E., Gao, W., Miller, D.D., Bell, C.E., and Dalton, J.T. (2005). Structural basis for antagonism and resistance of bicalutamide in prostate cancer. *Proc. Natl. Acad. Sci. U.S.A.* 102: 6201–6206.

Bolli, A., Bulzomi, P., Galluzzo, P., Acconcia, F., and Marino, M. (2010). Bisphenol A impairs estradiol-induced protective effects against DLD-1 colon cancer cell growth. *IUBMB Life* 62: 684–687.

Bolli, A., Galluzzo, P., Ascenzi, P., Del Pozzo, G., Manco, I., Vietri, M.T., Mita, L., Altucci, L., Mita, D.G., and Marino, M. (2008). Laccase treatment impairs bisphenol A-induced cancer cell proliferation affecting estrogen receptor alpha-dependent rapid signals. *IUBMB Life* 60: 843–852.

Bowman, A., Peterson, K.E., Dolinoy, D.C., Meeker, J.D., Sanchez, B.N., Mercado-Garcia, A., Tellez-Rojo, M.M., and Goodrich, J.M. (2019). Phthalate exposures, DNA methylation and adiposity in Mexican children through adolescence. *Front. Public Health* 7: 162.

Bruner-Tran, K.L., Gnecco, J., Ding, T., Glore, D.R., Pensabene, V., and Osteen, K.G. (2017). Exposure to the environmental endocrine disruptor TCDD and human reproductive dysfunction: translating lessons from murine models. *Reprod. Toxicol.* 68: 59–71.

Brusseau, M.L., Yan, N., Van Glubt, S., Wang, Y., Chen, W., Lyu, Y., Dungan, B., Carroll, K.C., and Holguin, F.O. (2019). Comprehensive retention model for PFAS transport in subsurface systems. *Water Res.* 148: 41–50.

Buck Louis, G.M. (2014). Persistent environmental pollutants and couple fecundity: an overview. *Reproduction* 147: R97–R104.

Byrne, S.C., Miller, P., Seguinot-Medina, S., Waghiyi, V., Buck, C.L., von Hippel, F.A., and Carpenter, D.O. (2018). Associations between serum polybrominated diphenyl ethers and thyroid hormones in a cross-sectional study of a remote Alaska Native population. *Sci. Rep.* 8: 2198.

Cantonwine, D., Meeker, J.D., Hu, H., Sanchez, B.N., Lamadrid-Figueroa, H., Mercado-Garcia, A., Fortenberry, G.Z., Calafat, A.M., and Tellez-Rojo, M.M. (2010). Bisphenol A exposure in Mexico City and risk of prematurity: a pilot nested case control study. *Environ. Health* 9: 62.

CHADD. (2019). Children and Adults with Attention-Deficit/Hyperactivity Disorder (CHADD): Pesticides Could Play a Role for Some People with ADHD.

Chadwick, C.C., Chippari, S., Matelan, E., Borges-Marcucci, L., Eckert, A.M., Keith, J.C., Jr., Albert, L.M., Leathurby, Y., Harris, H.A., Bhat, R.A., Ashwell, M., Trybulski, E., Winneker, R.C., Adelman, S.J., Steffan, R.J., and Harnish, D.C. (2005). Identification of pathway-selective estrogen receptor ligands that inhibit NF-kappaB transcriptional activity. *Proc. Natl. Acad. Sci. U.S.A.* 102: 2543–2548.

Chami, M., Oules, B., Szabadkai, G., Tacine, R., Rizzuto, R., and Paterlini-Brechot, P. (2008). Role of SERCA1 truncated isoform in the proapoptotic calcium transfer from ER to mitochondria during ER stress. *Mol. Cell* 32: 641–651.

Chen, A., Yolton, K., Rauch, S.A., Webster, G.M., Hornung, R., Sjodin, A., Dietrich, K.N., and Lanphear, B.P. (2014). Prenatal polybrominated diphenyl ether exposures and neurodevelopment in U.S. children through 5 years of age: the HOME study. *Environ. Health Perspect.* 122: 856–862.

Chen, H., Liu, J., Merrick, B.A., and Waalkes, M.P. (2001). Genetic events associated with arsenic-induced malignant transformation: applications of cDNA microarray technology. *Mol. Carcinog.* 30: 79–87.

Chen, J., Wu, S., Wen, S., Shen, L., Peng, J., Yan, C., Cao, X., Zhou, Y., Long, C., Lin, T., He, D., Hua, Y., and Wei, G. (2015). The mechanism of environmental endocrine disruptors (DEHP) induces epigenetic transgenerational inheritance of cryptorchidism. *pLoS One* 10: e0126403.

Chen, J.Q., Delannoy, M., Cooke, C., and Yager, J.D. (2004). Mitochondrial localization of eRalpha and eRbeta in human MCF7 cells. *Am. J. Physiol. Endocrinol. Metab.* 286: E1011–22.

Chen, W.Y., Shen, Y.P., and Chen, S.C. (2016). Assessing bisphenol A (BPA) exposure risk from long-term dietary intakes in Taiwan. *Sci. Total Environ.* 543: 140–146.

Chen, Y., Zhang, W., Chen, J., Wang, N., Chen, C., Wang, Y., Wan, H., Chen, B., and Lu, Y. (2021). Association of phthalate exposure with thyroid function and thyroid homeostasis parameters in Type 2 diabetes. *J. Diabetes Res.* 2021: 4027380.

Cheng, T.F., Choudhuri, S., and Muldoon-Jacobs, K. (2012). Epigenetic targets of some toxicologically relevant metals: a review of the literature. *J. Appl. Toxicol.* 32: 643–653.

Cho, H., Kim, S.J., Park, H., Oh, M-J, Yu, S.Y., Park, C., et al. (2010). A relationship between miRNA and gene expression in the mouse Sertoli cell line after exposure to bisphenol A. *BioChip J.* 4: 75–81.

Chu, F., Ren, X., Chasse, A., Hickman, T., Zhang, L., Yuh, J., Smith, M.T., and Burlingame, A.L. (2011). Quantitative mass spectrometry reveals the epigenome as a target of arsenic. *Chem. Biol. Interact.* 192: 113–117.

Coperchini, F., Croce, L., Ricci, G., Magri, F., Rotondi, M., Imbriani, M., and Chiovato, L. (2020). Thyroid disrupting effects of old and new generation PFAS. *Front. Endocrinol. (Lausanne)* 11: 612320.

Danadevi, K., Rozati, R., Reddy, P.P., and Grover, P. (2003). Semen quality of Indian welders occupationally exposed to nickel and chromium. *Reprod. Toxicol.* 17: 451–456.

De Santa, F., Barozzi, I., Mietton, F., Ghisletti, S., Polletti, S., Tusi, B.K., Muller, H., Ragoussis, J., Wei, C.L., and Natoli, G. (2010). A large fraction of extragenic RNA pol II transcription sites overlap enhancers. *pLoS Biol.* 8: e1000384.

Deaton, A.M. and Bird, A. (2011). CpG islands and the regulation of transcription. *Genes Dev.* 25: 1010–1022.

Delfosse, V., Maire, A.L., Balaguer, P., and Bourguet, W. (2015). A structural perspective on nuclear receptors as targets of environmental compounds. *Acta Pharmacol. Sin.* 36: 88–101.

Denison, M.S. and Nagy, S.R. (2003). Activation of the aryl hydrocarbon receptor by structurally diverse exogenous and endogenous chemicals. *Annu. Rev. Pharmacol. Toxicol.* 43: 309–334.

Denison, M.S., Soshilov, A.A., He, G., DeGroot, D.E., and Zhao, B. (2011). Exactly the same but different: promiscuity and diversity in the molecular mechanisms of action of the aryl hydrocarbon (dioxin) receptor. *Toxicol. Sci.* 124: 1–22.

Diamanti-Kandarakis, E., Bourguignon, J.P., Giudice, L.C., Hauser, R., Prins, G.S., Soto, A.M., Zoeller, R.T., and Gore, A.C. (2009). Endocrine-disrupting chemicals: an Endocrine Society scientific statement. *Endocr. Rev.* 30: 293–342.

Dianin, A.P. (1891). Kondensation von ketonen mit phenolen (Übers. aus d. russ.). *Zhurnal Rus. Fiz. Obs. (J. Russ. Phys. Chem. Soc.)* 23: 488–517, 523–546, 601–611.

Dinger, M.E., Amaral, P.P., Mercer, T.R., Pang, K.C., Bruce, S.J., Gardiner, B.B., Askarian-Amiri, M.E., Ru, K., Solda, G., Simons, C., Sunkin, S.M., Crowe, M.L., Grimmond, S.M., Perkins, A.C., and Mattick, J.S. (2008). Long noncoding RNAs in mouse embryonic stem cell pluripotency and differentiation. *Genome Res.* 18: 1433–1445.

Dodson, R.E., Nishioka, M., Standley, L.J., Perovich, L.J., Brody, J.G., and Rudel, R.A. (2012). Endocrine disruptors and asthma-associated chemicals in consumer products. *Environ. Health Perspect.* 120: 935–943.

Doshi, T., Mehta, S.S., Dighe, V., Balasinor, N., and Vanage, G. (2011). Hypermethylation of estrogen receptor promoter region in adult testis of rats exposed neonatally to bisphenol A. *Toxicology* 289: 74–82.

Doumouchtsis, K.K., Doumouchtsis, S.K., Doumouchtsis, E.K., and Perrea, D.N. (2009). The effect of lead intoxication on endocrine functions. *J. Endocrinol. Invest.* 32: 175–183.

Dvorak, Z., Vrzal, R., Pavek, P., and Ulrichova, J. (2008). An evidence for regulatory cross-talk between aryl hydrocarbon receptor and glucocorticoid receptor in HepG2 cells. *Physiol. Res.* 57: 427–435.

Earthjustice. (2019). *New York Bans Brain-damaging Pesticide Chlorpyrifos* [Online]. https://earthjustice.org/news/press/2019/new-york-bans-brain-damaging-pesticide-chlorpyrifos.

ECHA. (2010). *Chlorpyrifos: General Fact Sheet* [Online]. http://npic.orst.edu/factsheets/chlorpgen.html.

EDSTAC. (1998). *Endocrine Disruptor Screening and Testing Advisory Committee (EDSTAC) Final Report*. Environmental Protection Agency.

El Henafy, H.M.A., Ibrahim, M.A., Abd El Aziz, S.A., and Gouda, E.M. (2020). Oxidative stress and DNA methylation in male rat pups provoked by the transplacental and translactational exposure to bisphenol A. *Environ. Sci. Pollut. Res. Int.* 27: 4513–4519.

Ellen, T.P., Kluz, T., Harder, M.E., Xiong, J., and Costa, M. (2009). Heterochromatinization as a potential mechanism of nickel-induced carcinogenesis. *Biochemistry* 48: 4626–4632.

Enan, E., Liu, P.C., and Matsumura, F. (1992). TCDD (2,3,7,8-tetrachlorodibenzo-P-dioxin) causes reduction in glucose uptake through glucose transporters on the plasma membrane of the guinea pig adipocyte. *J. Environ. Sci. Health B* 27: 495–510.

EPA (US). (2022a). DDT-- A Brief History and Status.

EPA (US). (2022b). Lead. *Protect Your Family from Sources of Lead*. https://www.epa.gov/lead/protect-your-family-sources-lead.

Erguc, E.I., Tascioglu-Aliyev, A., Entezari, B., and Gurer-Orhan, H. (2021). The role of biotransformation in the activity of endocrine disruptors. *Curr. Drug Metab.* 22: 628–644.

EWG. (2019). *Mapping PFAS Chemical Contamination at 206 U.S. Military Sites* [Online]. https://www.ewg.org/research/mapping-pfas-chemical-contamination-206-us-military-sites.

Faghihi, M.A., Modarresi, F., Khalil, A.M., Wood, D.E., Sahagan, B.G., Morgan, T.E., Finch, C.E., St Laurent, G., 3rd, Kenny, P.J., and Wahlestedt, C. (2008). Expression of a noncoding RNA is elevated in Alzheimer's disease and drives rapid feed-forward regulation of beta-secretase. *Nat. Med.* 14: 723–730.

Fang, H., Tong, W., Branham, W.S., Moland, C.L., Dial, S.L., Hong, H., Xie, Q., Perkins, R., Owens, W., and Sheehan, D.M. (2003). Study of 202 natural, synthetic, and environmental chemicals for binding to the androgen receptor. *Chem. Res. Toxicol.* 16: 1338–1358.

Faulk, C., Kim, J.H., Jones, T.R., McEachin, R.C., Nahar, M.S., Dolinoy, D.C., and Sartor, M.A. (2015). Bisphenol A-associated alterations in genome-wide DNA methylation and gene expression patterns reveal sequence-dependent and non-monotonic effects in human fetal liver. *Environmental Epigenetics*, 1(1): dvv006.

Fini, J.B., Le Mevel, S., Turque, N., Palmier, K., Zalko, D., Cravedi, J.P., and Demeneix, B.A. (2007). An in vivo multiwell-based fluorescent screen for monitoring vertebrate thyroid hormone disruption. *Environ. Sci. Technol.* 41: 5908–5914.

Fisher, J.S. (2004). Environmental anti-androgens and male reproductive health: focus on phthalates and testicular dysgenesis syndrome. *Reproduction* 127: 305–315.

Gadhia, S.R., Calabro, A.R., and Barile, F.A. (2012). Trace metals alter DNA repair and histone modification pathways concurrently in mouse embryonic stem cells. *Toxicol. Lett.* 212: 169–179.

Gangopadhyay, S., Sharma, V., Chauhan, A., and Srivastava, V. (2019). Potential facet for prenatal arsenic exposure paradigm: linking endocrine disruption and epigenetics. *Nucleus* 62 (2): 127–142.

Gaspari, L., Paris, F., Jandel, C., Kalfa, N., Orsini, M., Daures, J.P., and Sultan, C. (2011). Prenatal environmental risk factors for genital malformations in a population of 1442 French male newborns: a nested case-control study. *Hum. Reprod.* 26: 3155–3162.

Ghassabian, A. and Trasande, L. (2018). Disruption in thyroid signaling pathway: a mechanism for the effect of endocrine-disrupting chemicals on child neurodevelopment. *Front. Endocrinol. (Lausanne)* 9: 204.

Giera, S., Bansal, R., Ortiz-Toro, T.M., Taub, D.G., and Zoeller, R.T. (2011). Individual polychlorinated biphenyl (PCB) congeners produce tissue- and gene-specific effects on thyroid hormone signaling during development. *Endocrinology* 152: 2909–2919.

Giesy, J.P., Naile, J.E., Khim, J.S., Jones, P.D., and Newsted, J.L. (2010). Aquatic toxicology of perfluorinated chemicals. *Rev. Environ. Contam. Toxicol.* 202: 1–52.

Giulivo, M., Lopez de Alda, M., Capri, E., and Barcelo, D. (2016). Human exposure to endocrine disrupting compounds: their role in reproductive systems, metabolic syndrome and breast cancer. A review. *Environ. Res.* 151: 251–264.

Gonsioroski, A., Mourikes, V.E., and Flaws, J.A. (2020). Endocrine disruptors in water and their effects on the reproductive system. *Int. J. Mol. Sci.* 21.

Gore, A.C., Chappell, V.A., Fenton, S.E., Flaws, J.A., Nadal, A., Prins, G.S., Toppari, J., and Zoeller, R.T. (2015). EDC-2: the endocrine society's second scientific statement on endocrine-disrupting chemicals. *Endocr. Rev.* 36: E1–E150.

Gore, A.C., Crews, D., Doan, L.L., Merill, M.L., Patisaul, H., and Zota, A. (2014). Introduction to Endocrine Disrupting Chemicals (EDCs) – a Guide for Public Interest Organizations and Policy-makers. Endocrine Society (2014).

Govarts, E., Nieuwenhuijsen, M., Schoeters, G., Ballester, F., Bloemen, K., de Boer, M., Chevrier, C., Eggesbo, M., Guxens, M., Kramer, U., Legler, J., Martinez, D., Palkovicova, L., Patelarou, E., Ranft, U., Rautio, A., Petersen, M.S., Slama, R., Stigum, H., Toft, G., Trnovec, T., Vandentorren, S., Weihe, P., Kuperus, N.W., Wilhelm, M., Wittsiepe, J., and Bonde, J.P. Obelix & Enrieco. (2012). Birth weight and prenatal exposure to polychlorinated biphenyls (PCBs) and dichlorodiphenyldichloroethylene (DDE): a meta-analysis within 12 European Birth Cohorts. *Environ. Health Perspect.* 120: 162–170.

Govindarajan, B., Klafter, R., Miller, M.S., Mansur, C., Mizesko, M., Bai, X., LaMontagne, K., Jr., and Arbiser, J.L. (2002). Reactive oxygen-induced carcinogenesis causes hypermethylation of p16(Ink4a) and activation of MAP kinase. *Mol. Med.* 8: 1–8.

Grimaldi, M., Boulahtouf, A., Toporova, L., and Balaguer, P. (2019). Functional profiling of bisphenols for nuclear receptors. *Toxicology* 420: 39–45.

Grunstein, M. (1997). Histone acetylation in chromatin structure and transcription. *Nature* 389: 349–352.

Guillette, L.J., Jr., Pickford, D.B., Crain, D.A., Rooney, A.A., and Percival, H.F. (1996). Reduction in penis size and plasma testosterone concentrations in juvenile alligators living in a contaminated environment. *Gen. Comp. Endocrinol.* 101: 32–42.

Guo, H., Ingolia, N.T., Weissman, J.S., and Bartel, D.P. (2010). Mammalian microRNAs predominantly act to decrease target mRNA levels. *Nature* 466: 835–840.

Gurunathan, S., Robson, M., Freeman, N., Buckley, B., Roy, A., Meyer, R., Bukowski, J., and Lioy, P.J. (1998). Accumulation of chlorpyrifos on residential surfaces and toys accessible to children. *Environ. Health Perspect.* 106: 9–16.

Ha, N.-Y., Kim, D.H., and Ryu, J.Y. (2019). Relationship between triclosan exposure and thyroid hormones: the Second Korean National Environmental Health Survey (2012–2014). *Ann. Occup. Environ. Med.* 5: e22.

Hanaoka, S., Ishida, K., Tanaka, S., Sakamoto, S., Okuda, K., Sanoh, S., Ohta, S., and Kotake, Y. (2018). Tributyltin induces epigenetic changes and decreases the expression of nuclear respiratory factor-1. *Metallomics* 10: 337–345.

Hannon, P.R. and Flaws, J.A. (2015). The effects of phthalates on the ovary. *Front. Endocrinol. (Lausanne)* 6: 8.

Harley, K.G., Chevrier, J., Aguilar Schall, R., Sjodin, A., Bradman, A., and Eskenazi, B. (2011). Association of prenatal exposure to polybrominated diphenyl ethers and infant birth weight. *Am. J. Epidemiol.* 174: 885–892.

The Zero Waste Memoirs. (2022). *The Healthiest Water Bottles for you & the Planet!* [Online]. https://zerowastememoirs.com/healthiest-water-bottles.

Herbstman, J.B., Sjodin, A., Kurzon, M., Lederman, S.A., Jones, R.S., Rauh, V., Needham, L.L., Tang, D., Niedzwiecki, M., Wang, R.Y., and Perera, F. (2010). Prenatal exposure to PBDEs and neurodevelopment. *Environ. Health Perspect.* 118: 712–719.

Hiramatsu, N., Cheek, A.O., Sullivan, C.V., Matsubara, T., and Hara, A. (2005). Vitellogenesis and endocrine disruption. *Biochemistry and Molecular Biology of Fishes.* 6: 431–471.

Hjollund, N.H., Bonde, J.P., Jensen, T.K., Ernst, E., Henriksen, T.B., Kolstad, H.A., Giwercman, A., Skakkebaek, N.E., and Olsen, J. (1998). Semen quality and sex hormones with reference to metal welding. *Reprod. Toxicol.* 12: 91–95.

Ho, S.M., Tang, W.Y., Belmonte de Frausto, J., and Prins, G.S. (2006). Developmental exposure to estradiol and bisphenol A increases susceptibility to prostate carcinogenesis and epigenetically regulates phosphodiesterase type 4 variant 4. *Cancer Res.* 66: 5624–5632.

Hong, J., Chen, F., Wang, X., Bai, Y., Zhou, R., Li, Y., and Chen, L. (2016). Exposure of preimplantation embryos to low-dose bisphenol A impairs testes development and suppresses histone acetylation of StAR promoter to reduce production of testosterone in mice. *Mol. Cell. Endocrinol.* 427: 101–111.

Horiguchi, T., Shiraishi, H., Shimizu, M., and Morita, M. (1997). Effects of triphenyltin chloride and five other organotin compounds on the development of imposex in the rock shell, Thais clavigera. *Environ. Pollut.* 95: 85–91.

Hossain, M.B., Vahter, M., Concha, G., and Broberg, K. (2012). Low-level environmental cadmium exposure is associated with DNA hypomethylation in Argentinean women. *Environ. Health Perspect.* 120: 879–884.

House of Representative, P. (2017). November 2017 Regulatory Update. *Philippines Approves Acts to Regulate Safety Labeling and Certain Hazardous Chemicals in Children's Toys, School Supplies and Child Care Articles.*

Huang, R.P., Liu, Z.H., Yuan, S.F., Yin, H., Dang, Z., and Wu, P.X. (2017). Worldwide human daily intakes of bisphenol A (BPA) estimated from global urinary concentration data (2000-2016) and its risk analysis. *Environ. Pollut.* 230: 143–152.

Huang, Y.Q., Wong, C.K., Zheng, J.S., Bouwman, H., Barra, R., Wahlstrom, B., Neretin, L., and Wong, M.H. (2012). Bisphenol A (BPA) in China: a review of sources, environmental levels, and potential human health impacts. *Environ. Int.* 42: 91–99.

Hutvagner, G. and Zamore, P.D. (2002). A microRNA in a multiple-turnover RNAi enzyme complex. *Science* 297: 2056–2060.

Huyghe, E., Matsuda, T., and Thonneau, P. (2003). Increasing incidence of testicular cancer worldwide: a review. *J. Urol.* 170: 5–11.

IPEN. (2011). Children's product in China. *Toxic heavy metals found in children's products on the Chinese market.*

ITRC. (2022). *PFAS — Per- and Polyfluoroalkyl Substances* [Online]. https://pfas-1.itrcweb.org/3-firefighting-foams.

Jensen, T.J., Novak, P., Eblin, K.E., Gandolfi, A.J., and Futscher, B.W. (2008). Epigenetic remodeling during arsenical-induced malignant transformation. *Carcinogenesis* 29: 1500–1508.

Ji, W., Yang, L., Yu, L., Yuan, J., Hu, D., Zhang, W., Yang, J., Pang, Y., Li, W., Lu, J., Fu, J., Chen, J., Lin, Z., Chen, W., and Zhuang, Z. (2008). Epigenetic silencing of O6-methylguanine DNA methyltransferase gene in NiS-transformed cells. *Carcinogenesis* 29: 1267–1275.

Jiang, G., Xu, L., Song, S., Zhu, C., Wu, Q., Zhang, L., and Wu, L. (2008). Effects of long-term low-dose cadmium exposure on genomic DNA methylation in human embryo lung fibroblast cells. *Toxicology* 244: 49–55.

Jugan, M.L., Levi, Y., and Blondeau, J.P. (2010). Endocrine disruptors and thyroid hormone physiology. *Biochem. Pharmacol.* 79: 939–947.

Kabir, E.R., Rahman, M.S., and Rahman, I. (2015). A review on endocrine disruptors and their possible impacts on human health. *Environ. Toxicol. Pharmacol.* 40: 241–258.

Kanayama, T., Kobayashi, N., Mamiya, S., Nakanishi, T., and Nishikawa, J. (2005). Organotin compounds promote adipocyte differentiation as agonists of the peroxisome proliferator-activated receptor gamma/retinoid X receptor pathway. *Mol. Pharmacol.* 67: 766–774.

Kang, J., Zhang, Y., Chen, J., Chen, H., Lin, C., Wang, Q., and Ou, Y. (2003). Nickel-induced histone hypoacetylation: the role of reactive oxygen species. *Toxicol. Sci.* 74: 279–286.

Kewley, R.J., Whitelaw, M.L., and Chapman-Smith, A. (2004). The mammalian basic helix-loop-helix/PAS family of transcriptional regulators. *Int. J. Biochem. Cell Biol.* 36: 189–204.

Kim, H.S., Shin, J.H., Moon, H.J., Kim, T.S., Kang, I.H., Seok, J.H., Kim, I.Y., Park, K.L., and Han, S.Y. (2002). Evaluation of the 20-day pubertal female assay in Sprague-Dawley rats treated with DES, tamoxifen, testosterone, and flutamide. *Toxicol. Sci.* 67: 52–62.

Kim, J.J., Taylor, H.S., Lu, Z., Ladhani, O., Hastings, J.M., Jackson, K.S., Wu, Y., Guo, S.W., and Fazleabas, A.T. (2007). Altered expression of HOXA10 in endometriosis: potential role in decidualization. *Mol. Hum. Reprod.* 13: 323–332.

Kim, J.Y., Choi, H.G., Lee, H.M., Lee, G.A., Hwang, K.A., and Choi, K.C. (2017). Effects of bisphenol compounds on the growth and epithelial mesenchymal transition of MCF-7 CV human breast cancer cells. *J. Biomed. Res.* 31: 358–369.

Kim, T.K., Hemberg, M., Gray, J.M., Costa, A.M., Bear, D.M., Wu, J., Harmin, D.A., Laptewicz, M., Barbara-Haley, K., Kuersten, S., Markenscoff-Papadimitriou, E., Kuhl, D., Bito, H., Worley, P.F., Kreiman, G., and Greenberg, M.E. (2010). Widespread transcription at neuronal activity-regulated enhancers. *Nature* 465: 182–187.

Kippler, M., Engstrom, K., Mlakar, S.J., Bottai, M., Ahmed, S., Hossain, M.B., Raqib, R., Vahter, M., and Broberg, K. (2013). Sex-specific effects of early life cadmium exposure on DNA methylation and implications for birth weight. *Epigenetics* 8: 494–503.

Kitamura, S., Kato, T., Iida, M., Jinno, N., Suzuki, T., Ohta, S., Fujimoto, N., Hanada, H., Kashiwagi, K., and Kashiwagi, A. (2005). Anti-thyroid hormonal activity of tetrabromobisphenol A, a flame retardant, and related compounds: affinity to the mammalian thyroid hormone receptor, and effect on tadpole metamorphosis. *Life Sci.* 76: 1589–1601.

Klinge, C.M. (2009). Estrogen regulation of MicroRNA expression. *Curr. Genomics* 10: 169–183.

Kogevinas, M. (2001). Human health effects of dioxins: cancer, reproductive and endocrine system effects. *Hum. Reprod. Update* 7: 331–339.

Kowara, R., Salnikow, K., Diwan, B.A., Bare, R.M., Waalkes, M.P., and Kasprzak, K.S. (2004). Reduced Fhit protein expression in nickel-transformed mouse cells and in nickel-induced murine sarcomas. *Mol. Cell. Biochem.* 255: 195–202.

Kuil, C.W., Brouwer, A., van der Saag, P.T., and van der Burg, B. (1998). Interference between progesterone and dioxin signal transduction pathways. Different mechanisms are involved in repression by the progesterone receptor A and B isoforms. *J. Biol. Chem.* 273: 8829–8834.

Kuiper, G.G., Faber, P.W., van Rooij, H.C., van der Korput, J.A., Ris-Stalpers, C., Klaassen, P., Trapman, J., and Brinkmann, A.O. (1989). Structural organization of the human androgen receptor gene. *J.Mol. Endocrinol.* 2: R1–4.

Kumar, R. and Thompson, E.B. (2003). Transactivation functions of the N-terminal domains of nuclear hormone receptors: protein folding and coactivator interactions. *Mol. Endocrinol.* 17: 1–10.

Latini, G., De Felice, C., Presta, G., Del Vecchio, A., Paris, I., Ruggieri, F., and Mazzeo, P. (2003). In utero exposure to di-(2-ethylhexyl)phthalate and duration of human pregnancy. *Environ. Health Perspect.* 111: 1783–1785.

Lee, C.Y. and Grant, P.A. (2019). Role of histone acetylation and acetyltransfearses in gene regulation. In: *Toxicogenetics,* S.D. McCullough and D.C. Dolinoy (eds), 3–30. Academic Press.

Lee, D.H., Lind, P.M., Jacobs, D.R., Jr., Salihovic, S., van Bavel, B., and Lind, L. (2011). Polychlorinated biphenyls and organochlorine pesticides in plasma predict development of type 2 diabetes in the elderly: the prospective investigation of the vasculature in Uppsala Seniors (PIVUS) study. *Diabetes Care* 34: 1778–1784.

Leemans, M., Couderq, S., Demeneix, B., and Fini, J.B. (2019). Pesticides with potential thyroid hormone-disrupting effects: a review of recent data. *Front. Endocrinol. (Lausanne)* 10: 743.

Legler, J. and Brouwer, A. (2003). Are brominated flame retardants endocrine disruptors? *Environ. Int.* 29: 879–885.

Lemaire, G., Mnif, W., Mauvais, P., Balaguer, P., and Rahmani, R. (2006). Activation of alpha- and beta-estrogen receptors by persistent pesticides in reporter cell lines. *Life Sci.* 79: 1160–1169.

Letcher, R.J., van Holsteijn, I., Drenth, H.J., Norstrom, R.J., Bergman, A., Safe, S., Pieters, R., and van den Berg, M. (1999). Cytotoxicity and aromatase (CYP19) activity modulation by organochlorines in human placental JEG-3 and JAR choriocarcinoma cells. *Toxicol. Appl. Pharmacol.* 160: 10–20.

Levin, E.R. (2011). Minireview: extranuclear steroid receptors: roles in modulation of cell functions. *Mol. Endocrinol.* 25: 377–384.

Levin, E.R. and Hammes, S.R. (2016). Nuclear receptors outside the nucleus: extranuclear signalling by steroid receptors. *Nat. Rev. Mol. Cell Biol.* 17: 783–797.

Li, B., Carey, M., and Workman, J.L. (2007). The role of chromatin during transcription. *Cell* 128: 707–719.

Li, J., Gorospe, M., Barnes, J., and Liu, Y. (2003). Tumor promoter arsenite stimulates histone H3 phosphoacetylation of proto-oncogenes c-fos and c-jun chromatin in human diploid fibroblasts. *J. Biol. Chem.* 278: 13183–13191.

Li, T., Spearow, J., Rubin, C.M., and Schmid, C.W. (1999). Physiological stresses increase mouse short interspersed element (SINE) RNA expression in vivo. *Gene* 239: 367–372.

Lim, D.S., Kwack, S.J., Kim, K.B., Kim, H.S., and Lee, B.M. (2009). Potential risk of bisphenol A migration from polycarbonate containers after heating, boiling, and microwaving. *J. Toxicol. Environ. Health Part A* 72: 1285–1291.

Linehan, W.M., Walther, M.M., and Zbar, B. (2003). The genetic basis of cancer of the kidney. *J. Urol.* 170: 2163–2172.

Lite, C., Ahmed, S., Santosh, W., and Seetharaman, B. (2019). Prenatal exposure to bisphenol-A altered miRNA-224 and protein expression of aromatase in ovarian granulosa cells concomitant with elevated serum estradiol levels in F1 adult offspring. *J. Biochem. Mol. Toxicol.* 33: e22317.

Liu, J., Knappenberger, K.S., Kack, H., Andersson, G., Nilsson, E., Dartsch, C., and Scott, C.W. (2003). A homogeneous in vitro functional assay for estrogen receptors: coactivator recruitment. *Mol. Endocrinol.* 17: 346–355.

Liu, S., Wang, K., Svoboda, L.K., Rygiel, C.A., Neier, K., Jones, T.R., Cavalcante, R.G., Colacino, J.A., Dolinoy, D.C., and Sartor, M.A. (2021). Perinatal DEHP exposure induces sex- and tissue-specific DNA methylation changes in both juvenile and adult mice. *Environ. Epigenet.* 7: dvab004.

Llanos, M.N. and Ronco, A.M. (2009). Fetal growth restriction is related to placental levels of cadmium, lead and arsenic but not with antioxidant activities. *Reprod. Toxicol.* 27: 88–92.

Lovett, R.A. (2012). Human drugs make fish flounder. *Nature.* https://doi.org/10.1038/nature.2012.11843.

Ma, Y., Liu, H., Wu, J., Yuan, L., Wang, Y., Du, X., Wang, R., Marwa, P.W., Petlulu, P., Chen, X., and Zhang, H. (2019). The adverse health effects of bisphenol A and related toxicity mechanisms. *Environ. Res.* 176: 108575.

Main, K.M., Kiviranta, H., Virtanen, H.E., Sundqvist, E., Tuomisto, J.T., Tuomisto, J., Vartiainen, T., Skakkebaek, N.E., and Toppari, J. (2007). Flame retardants in placenta and breast milk and cryptorchidism in newborn boys. *Environ. Health Perspect.* 115: 1519–1526.

Main, K.M., Mortensen, G.K., Kaleva, M.M., Boisen, K.A., Damgaard, I.N., Chellakooty, M., Schmidt, I.M., Suomi, A.M., Virtanen, H.E., Petersen, D.V., Andersson, A.M., Toppari, J., and Skakkebaek,

N.E. (2006). Human breast milk contamination with phthalates and alterations of endogenous reproductive hormones in infants three months of age. *Environ. Health Perspect.* 114: 270–276.
Main, K.M., Skakkebaek, N.E., Virtanen, H.E., and Toppari, J. (2010). Genital anomalies in boys and the environment. *Best Pract. Res. Clin. Endocrinol. Metab.* 24: 279–289.
Manikkam, M., Guerrero-Bosagna, C., Tracey, R., Haque, M.M., and Skinner, M.K. (2012). Transgenerational actions of environmental compounds on reproductive disease and identification of epigenetic biomarkers of ancestral exposures. *PLoS One* 7: e31901.
Marino, M., Pellegrini, M., La Rosa, P., and Acconcia, F. (2012). Susceptibility of estrogen receptor rapid responses to xenoestrogens: physiological outcomes. *Steroids* 77: 910–917.
Marsit, C.J., Eddy, K., and Kelsey, K.T. (2006). MicroRNA responses to cellular stress. *Cancer Res.* 66: 10843–10848.
Martin, J.A., Hamilton, B.E., Sutton, P.D., Ventura, S.J., Mathews, T.J., and Osterman, M.J. (2010). Births: final data for 2008. *Natl. Vital. Stat. Rep.* 59 (1): 3–71.
Mass, M.J. and Wang, L. (1997). Arsenic alters cytosine methylation patterns of the promoter of the tumor suppressor gene p53 in human lung cells: a model for a mechanism of carcinogenesis. *Mutat. Res.* 386: 263–277.
Matsushima, A., Liu, X., Okada, H., Shimohigashi, M., and Shimohigashi, Y. (2010). Bisphenol AF is a full agonist for the estrogen receptor ERalpha but a highly specific antagonist for ERbeta. *Environ. Health Perspect.* 118: 1267–1272.
Matsushita, N., Sogawa, K., Ema, M., Yoshida, A., and Fujii-Kuriyama, Y. (1993). A factor binding to the xenobiotic responsive element (XRE) of P-4501A1 gene consists of at least two helix-loop-helix proteins, Ah receptor and Arnt. *J. Biol. Chem.* 268: 21002–21006.
Matthiessen, P. and Gibbs, P.E. (2009). Critical appraisal of the evidence for tributyltin-mediated endocrine disruption in mollusks. *Environ. Toxicol. Chem.* 17: 37–43.
McKay, L.I. and Cidlowski, J.A. (1998). Cross-talk between nuclear factor-kappa B and the steroid hormone receptors: mechanisms of mutual antagonism. *Mol. Endocrinol.* 12: 45–56.
McLachlan, J.A., Korach, K.S., Newbold, R.R., and Degen, G.H. (1984). Diethylstilbestrol and other estrogens in the environment. *Fundam. Appl. Toxicol.* 4: 686–691.
Meakin, C.J., Martin, E.M., Szilagyi, J.T., Nylander-French, L.A., and Fry, R.C. (2019). Inorganic arsenic as an endocrine disruptor: modulation of the glucocorticoid receptor pathway in placental cells via CpG methylation. *Chem. Res. Toxicol.* 32: 493–499.
Medicine, N.L.O. (2004). *Nickel* [Online]. National Center for Biotechnology Information. https://pubchem.ncbi.nlm.nih.gov/#query=Nickel.
Meeker, J.D., Hu, H., Cantonwine, D.E., Lamadrid-Figueroa, H., Calafat, A.M., Ettinger, A.S., Hernandez-Avila, M., Loch-Caruso, R., and Tellez-Rojo, M.M. (2009). Urinary phthalate metabolites in relation to preterm birth in Mexico City. *Environ. Health Perspect.* 117: 1587–1592.
Michael, C.A., Hofmann, J.N., Charles, F.L., Cynthia, J.H., Kathryn, H.B., Joseph, B., Dennis, W.B., Kent, T., Dale, P.S., Jane, A.H., Stella, K., Gabriella, A., Jay, H.L., Aaron, B., and Laura, E.B.F. (2022). Non-Hodgkin Lymphoma risk and insecticide, fungicide and fumigant use in the Agricultural Health Study.
Miyakoshi, T., Miyajima, K., Takekoshi, S., and Osamura, R.Y. (2009). The influence of endocrine disrupting chemicals on the proliferation of ERalpha knockdown-human breast cancer cell line MCF-7; new attempts by RNAi technology. *Acta Histochem. Cytochem.* 42: 23–28.
Montgomery, M.P., Kamel, F., Saldana, T.M., Alavanja, M.C., and Sandler, D.P. (2008). Incident diabetes and pesticide exposure among licensed pesticide applicators: Agricultural Health Study, 1993–2003. *Am. J. Epidemiol.* 167: 1235–1246.
Muncke, J. (2011). Endocrine disrupting chemicals and other substances of concern in food contact materials: an updated review of exposure, effect and risk assessment. *J. Steroid Biochem. Mol. Biol.* 127: 118–127.

Murray, I.A., Patterson, A.D., and Perdew, G.H. (2014). Aryl hydrocarbon receptor ligands in cancer: friend and foe. *Nat. Rev. Cancer* 14: 801–814.

Nakanishi, T. (2008). Endocrine disruption induced by organotin compounds; organotins function as a powerful agonist for nuclear receptors rather than an aromatase inhibitor. *J. Toxicol. Sci.* 33: 269–276.

Nayyar, T., Bruner-Tran, K.L., Piestrzeniewicz-Ulanska, D., and Osteen, K.G. (2007). Developmental exposure of mice to TCDD elicits a similar uterine phenotype in adult animals as observed in women with endometriosis. *Reprod. Toxicol.* 23: 326–336.

Nebert, D.W. (2017). Aryl hydrocarbon receptor (AHR): "pioneer member" of the basic-helix/loop/helix per-Arnt-sim (bHLH/PAS) family of "sensors" of foreign and endogenous signals. *Prog. Lipid Res.* 67: 38–57.

Neier, K., Cheatham, D., Bedrosian, L.D., and Dolinoy, D.C. (2019). Perinatal exposures to phthalates and phthalate mixtures result in sex-specific effects on body weight, organ weights and intracisternal A-particle (IAP) DNA methylation in weanling mice. *J. Dev. Orig. Health Dis.* 10: 176–187.

Newbold, R.R., Padilla-Banks, E., Snyder, R.J., and Jefferson, W.N. (2007). Perinatal exposure to environmental estrogens and the development of obesity. *Mol. Nutr. Food Res.* 51: 912–917.

NIEHS (National Institute of Environmental Sciences) (2021). *Your Environment, Your Health: Bisphenol A (BPA)*, https://www.niehs.nih.gov/health/topics/agents/sya-bpa/index.cfm (accessed 11/29/2022).

Nokes, C. and Weaver, L. (2014). Cadmium in Groundwater: review of regional council data.

Odum, J., Lefevre, P.A., Tittensor, S., Paton, D., Routledge, E.J., Beresford, N.A., Sumpter, J.P., and Ashby, J. (1997). The rodent uterotrophic assay: critical protocol features, studies with nonyl phenols, and comparison with a yeast estrogenicity assay. *Regul. Toxicol. Pharmacol.* 25: 176–188.

OECD. (2011). Test No. 456: H295R Steroidogenesis Assay.

OECD. (2012). Fish Short-Term Reproduction Assay (FSTRA) (OECD TG 229).

OECD. (2013). InVitro & Ex Vivo Assays for Identification of Modulators of Thyroid Hormone Signalling.

OEHHA (California Office of Environmental Health Hazard Assessment) (2007). DDE, https://oehha.ca.gov/media/downloads/crnr/dart0907078.pdf.

Ohtake, F., Takeyama, K., Matsumoto, T., Kitagawa, H., Yamamoto, Y., Nohara, K., Tohyama, C., Krust, A., Mimura, J., Chambon, P., Yanagisawa, J., Fujii-Kuriyama, Y., and Kato, S. (2003). Modulation of oestrogen receptor signalling by association with the activated dioxin receptor. *Nature* 423: 545–550.

Okoro, H.K., Fatoki, F.A., Adekola, B.J., and Snyman, R.G. (2014). Organotin compounds. In: *Encyclopedia of Toxicology*, 3e. Elsevier.

Oliveira, K.J., Chiamolera, M.I., Giannocco, G., Pazos-Moura, C.C., and Ortiga-Carvalho, T.M. (2018). Thyroid function disruptors: from nature to chemicals. *J. Mol. Endocrinol.* 62: R1–R19.

Oltvai, Z.N., Milliman, C.L., and Korsmeyer, S.J. (1993). Bcl-2 heterodimerizes in vivo with a conserved homolog, Bax, that accelerates programmed cell death. *Cell* 74: 609–619.

Ormond, G., Nieuwenhuijsen, M.J., Nelson, P., Toledano, M.B., Iszatt, N., Geneletti, S., and Elliott, P. (2009). Endocrine disruptors in the workplace, hair spray, folate supplementation, and risk of hypospadias: case-control study. *Environ. Health Perspect.* 117: 303–307.

Osada, S., Nishikawa, J., Nakanishi, T., Tanaka, K., and Nishihara, T. (2005). Some organotin compounds enhance histone acetyltransferase activity. *Toxicol. Lett.* 155: 329–335.

Patel, S., Zhou, C., Rattan, S., and Flaws, J.A. (2015). Effects of endocrine-disrupting chemicals on the ovary. *Biol. Reprod.* 93: 20.

Paul, K.B., Hedge, J.M., Devito, M.J., and Crofton, K.M. (2010). Short-term exposure to triclosan decreases thyroxine in vivo via upregulation of hepatic catabolism in Young Long-Evans rats. *Toxicol Sci.* 113(2): 367–379.

Perma.cc (2019). YU7E-H9L4: DDT regulatory history: a brief survey (to 1975). https://perma.cc/YU7E-H9L4.

Pesatori, A.C., Zocchetti, C., Guercilena, S., Consonni, D., Turrini, D., and Bertazzi, P.A. (1998). Dioxin exposure and non-malignant health effects: a mortality study. *Occup. Environ. Med.* 55: 126–131.

The PFAS Project Lab. (2019). *Social Science Environmental Health Research Institute: The PFAS Project Lab* [Online]. Northeastern University: Social Science Environmental Health Research Institute. https://pfasproject.com/parkersburg-west-virginia.

Phillips, T. (2008). The role of methylation in gene expression. *Nat. Educ.* 1: 116.

Pilsner, J.R., Hu, H., Ettinger, A., Sanchez, B.N., Wright, R.O., Cantonwine, D., Lazarus, A., Lamadrid-Figueroa, H., Mercado-Garcia, A., Tellez-Rojo, M.M., and Hernandez-Avila, M. (2009). Influence of prenatal lead exposure on genomic methylation of cord blood DNA. *Environ. Health Perspect.* 117: 1466–1471.

Pizent, A., Tariba, B., and Zivkovic, T. (2012). Reproductive toxicity of metals in men. *Arh. Hig. Rada Toksikol.* 63 (Suppl 1): 35–46.

Pocar, P., Klonisch, T., Brandsch, C., Eder, K., Frohlich, C., Hoang-Vu, C., and Hombach-Klonisch, S. (2006). AhR-agonist-induced transcriptional changes of genes involved in thyroid function in primary porcine thyrocytes. *Toxicol. Sci.* 89: 408–414.

Poland, A., Glover, E., and Kende, A.S. (1976). Stereospecific, high affinity binding of 2,3,7,8-tetrachlorodibenzo-p-dioxin by hepatic cytosol. Evidence that the binding species is receptor for induction of aryl hydrocarbon hydroxylase. *J. Biol. Chem.* 251: 4936–4946.

Ponting, C.P., Oliver, P.L., and Reik, W. (2009). Evolution and functions of long noncoding RNAs. *Cell* 136: 629–641.

Powell, E., Huang, S.X., Xu, Y., Rajski, S.R., Wang, Y., Peters, N., Guo, S., Xu, H.E., Hoffmann, F.M., Shen, B., and Xu, W. (2010). Identification and characterization of a novel estrogenic ligand actinopolymorphol A. *Biochem. Pharmacol.* 80: 1221–1229.

Powell, E. and Xu, W. (2008). Intermolecular interactions identify ligand-selective activity of estrogen receptor alpha/beta dimers. *Proc. Natl. Acad. Sci. U.S.A.* 105: 19012–19017.

Preston, E.V., Webster, T.F., Oken, E., Claus Henn, B., McClean, M.D., Rifas-Shiman, S.L., Pearce, E.N., Braverman, L.E., Calafat, A.M., Ye, X., and Sagiv, S.K. (2018). Maternal plasma per- and polyfluoroalkyl substance concentrations in early pregnancy and maternal and neonatal thyroid function in a prospective birth cohort: project viva (USA). *Environ. Health Perspect.* 126: 027013.

Prossnitz, E.R. and Arterburn, J.B. (2015). International union of basic and clinical pharmacology. XCVII. G Protein-coupled estrogen receptor and its pharmacologic modulators. *Pharmacol. Rev.* 67: 505–540.

Rafati Rahimzadeh, M., Rafati Rahimzadeh, M., Kazemi, S., and Moghadamnia, A.A. (2017). Cadmium toxicity and treatment: an update. *Caspian J. Intern. Med.* 8: 135–145.

Ramirez, T., Brocher, J., Stopper, H., and Hock, R. (2008). Sodium arsenite modulates histone acetylation, histone deacetylase activity and HMGN protein dynamics in human cells. *Chromosoma* 117: 147–157.

Rattan, S., Zhou, C., Chiang, C., Mahalingam, S., Brehm, E., and Flaws, J.A. (2017). Exposure to endocrine disruptors during adulthood: consequences for female fertility. *J. Endocrinol.* 233: R109–R129.

Reichard, J.F., Schnekenburger, M., and Puga, A. (2007). Long term low-dose arsenic exposure induces loss of DNA methylation. *Biochem. Biophys. Res. Commun.* 352: 188–192.

Remillard, R.B. and Bunce, N.J. (2002). Linking dioxins to diabetes: epidemiology and biologic plausibility. *Environ. Health Perspect.* 110: 853–858.

Ren, X., Gaile, D.P., Gong, Z., Qiu, W., Ge, Y., Zhang, C., Huang, C., Yan, H., Olson, J.R., Kavanagh, T.J., and Wu, H. (2015). Arsenic responsive microRNAs in vivo and their potential involvement in arsenic-induced oxidative stress. *Toxicol. Appl. Pharmacol.* 283: 198–209.

Revankar, C.M., Cimino, D.F., Sklar, L.A., Arterburn, J.B., and Prossnitz, E.R. (2005). A transmembrane intracellular estrogen receptor mediates rapid cell signaling. *Science* 307: 1625–1630.

Reyes, H., Reisz-Porszasz, S., and Hankinson, O. (1992). Identification of the Ah receptor nuclear translocator protein (Arnt) as a component of the DNA binding form of the Ah receptor. *Science* 256: 1193–1195.

Rezg, R., El-Fazaa, S., Gharbi, N., and Mornagui, B. (2014). Bisphenol A and human chronic diseases: current evidences, possible mechanisms, and future perspectives. *Environ. Int.* 64: 83–90.

Rieuwerts, J. (2015). *The Elements of Environmental Pollution*. London: Taylor & Francis Group.

Rober, J.K., George, P.D., Chris, D.R., Fenner-Crisp, P., Gray, L.E., Kaattari, S., Lucier, G., Michael, L., Michael, J.M., Maczka, C., Miller, R., Moore, J., Roslind, R., Geoffrey, S., Daniel, M.S., Thomas, S., and Hugh, A.T. (1996). Research needs for the risk assessment of health and environmental effects of endocrine disruptors: a report of the U.S. EPA-sponsored workshop. *Environ. Health Perspect.* 104: 715–740.

Rothhammer, V. and Quintana, F.J. (2019). The aryl hydrocarbon receptor: an environmental sensor integrating immune responses in health and disease. *Nat. Rev. Immunol.* 19: 184–197.

Roze, E., Meijer, L., Bakker, A., Van Braeckel, K.N., Sauer, P.J., and Bos, A.F. (2009). Prenatal exposure to organohalogens, including brominated flame retardants, influences motor, cognitive, and behavioral performance at school age. *Environ. Health Perspect.* 117: 1953–1958.

Rudge, C.V., Rollin, H.B., Nogueira, C.M., Thomassen, Y., Rudge, M.C., and Odland, J.O. (2009). The placenta as a barrier for toxic and essential elements in paired maternal and cord blood samples of South African delivering women. *J. Environ. Monit.* 11: 1322–1330.

Rylander, L., Rignell-Hydbom, A., and Hagmar, L. (2005). A cross-sectional study of the association between persistent organochlorine pollutants and diabetes. *Environ. Health* 4: 28.

Sadler-Riggleman, I., Klukovich, R., Nilsson, E., Beck, D., Xie, Y., Yan, W., and Skinner, M.K. (2019). Epigenetic transgenerational inheritance of testis pathology and Sertoli cell epimutations: generational origins of male infertility. *Environ. Epigenet.* 5: dvz013.

Sadow, P.M., Chassande, O., Koo, E.K., Gauthier, K., Samarut, J., Xu, J., O'Malley, B.W., and Weiss, R.E. (2003). Regulation of expression of thyroid hormone receptor isoforms and coactivators in liver and heart by thyroid hormone. *Mol. Cell. Endocrinol.* 203: 65–75.

Safe, S.H. (2005). *DTIC ADA436964: Inhibitory Ah Receptor-Androgen Receptor Crosstalk in Prostate Cancer. Defence Technical Information Center*, https://archive.org/details/DTIC_ADA436964.

Saghir S.A., Charles G.D., Kan L.H.L., Dryzga M.D., Brzak K.A., and Clark A.J. (2008). Mechanism of trifluralin-induced thyroid tumors in rats. *Toxicol Lett.* 180: 38–45.

Schreder, E. and Dickman J. (2018). *Take Out Toxics: PFAS Chemicals in Food Packaging*. https://48h57c2l31ua3c3fmq1ne58b-wpengine.netdna-ssl.com/wp-content/uploads/2019/05/Take-Out-Toxics-Full-Report.pdf

Sakamoto, M., Yasutake, A., Domingo, J.L., Chan, H.M., Kubota, M., and Murata, K. (2013). Relationships between trace element concentrations in chorionic tissue of placenta and umbilical cord tissue: potential use as indicators for prenatal exposure. *Environ. Int.* 60: 106–111.

Sakkiah, S., Guo, W., Pan, B., Kusko, R., Tong, W., and Hong, H. (2018). Computational prediction models for assessing endocrine disrupting potential of chemicals. *J. Environ. Sci. Health C: Environ. Carcinog. Ecotoxicol. Rev.* 36: 192–218.

Saldana, T.M., Basso, O., Hoppin, J.A., Baird, D.D., Knott, C., Blair, A., Alavanja, M.C., and Sandler, D.P. (2007). Pesticide exposure and self-reported gestational diabetes mellitus in the Agricultural Health Study. *Diabetes Care* 30: 529–534.

Salnikow, K. and Zhitkovich, A. (2008). Genetic and epigenetic mechanisms in metal carcinogenesis and cocarcinogenesis: nickel, arsenic, and chromium. *Chem. Res. Toxicol.* 21: 28–44.

Sanders, A.P., Smeester, L., Rojas, D., DeBussycher, T., Wu, M.C., Wright, F.A., Zhou, Y.H., Laine, J.E., Rager, J.E., Swamy, G.K., Ashley-Koch, A., Lynn Miranda, M., and Fry, R.C. (2014). Cadmium exposure and the epigenome: exposure-associated patterns of DNA methylation in leukocytes from mother–baby pairs. *Epigenetics* 9: 212–221.

Schubeler, D. (2015). Function and information content of DNA methylation. *Nature* 517: 321–326.

Seidel, S.D., Winters, G.M., Rogers, W.J., Ziccardi, M.H., Li, V., Keser, B., and Denison, M.S. (2001). Activation of the Ah receptor signaling pathway by prostaglandins. *J. Biochem. Mol. Toxicol.* 15: 187–196.

Shafei, A., Ramzy, M.M., Hegazy, A.I., Husseny, A.K., El-Hadary, U.G., Taha, M.M., and Mosa, A.A. (2018). The molecular mechanisms of action of the endocrine disrupting chemical bisphenol A in the development of cancer. *Gene* 647: 235–243.

Shah, S., Hess-Wilson, J.K., Webb, S., Daly, H., Godoy-Tundidor, S., Kim, J., Boldison, J., Daaka, Y., and Knudsen, K.E. (2008). 2,2-bis(4-chlorophenyl)-1,1-dichloroethylene stimulates androgen independence in prostate cancer cells through combinatorial activation of mutant androgen receptor and mitogen-activated protein kinase pathways. *Mol. Cancer Res.* 6: 1507–1520.

Shanle, E.K. and Xu, W. (2011). Endocrine disrupting chemicals targeting estrogen receptor signaling: identification and mechanisms of action. *Chem. Res. Toxicol.* 24: 6–19.

Shao, D. and Lazar, M.A. (1999). Modulating nuclear receptor function: may the phos be with you. *J. Clin. Invest.* 103: 1617–1618.

Sidorkiewicz, I., Zareba, K., Wolczynski, S., and Czerniecki, J. (2017). Endocrine-disrupting chemicals – Mechanisms of action on male reproductive system. *Toxicol. Ind. Health* 33: 601–609.

Silva, E., Scholze, M., and Kortenkamp, A. (2007). Activity of xenoestrogens at nanomolar concentrations in the E-Screen assay. *Environ. Health Perspect.* 115 (Suppl 1): 91–97.

Skinner, M.K. (2008). What is an epigenetic transgenerational phenotype? F3 or F2. *Reprod. Toxicol.* 25: 2–6.

Skinner, M.K. (2014a). Endocrine disruptor induction of epigenetic transgenerational inheritance of disease. *Mol. Cell. Endocrinol.* 398: 4–12.

Skinner, M.K. (2014b). Environmental stress and epigenetic transgenerational inheritance. *BMC Med.* 12: 153.

Skinner, M.K. (2016). Endocrine disruptors in 2015: epigenetic transgenerational inheritance. *Nat. Rev. Endocrinol.* 12: 68–70.

Smeester, L., Rager, J.E., Bailey, K.A., Guan, X., Smith, N., Garcia-Vargas, G., Del Razo, L.M., Drobna, Z., Kelkar, H., Styblo, M., and Fry, R.C. (2011). Epigenetic changes in individuals with arsenicosis. *Chem. Res. Toxicol.* 24: 165–167.

Solomon, O., Yousefi, P., Huen, K., Gunier, R.B., Escudero-Fung, M., Barcellos, L.F., Eskenazi, B., and Holland, N. (2017). Prenatal phthalate exposure and altered patterns of DNA methylation in cord blood. *Environ. Mol. Mutagen.* 58: 398–410.

Soto, A.M., Sonnenschein, C., Chung, K.L., Fernandez, M.F., Olea, N., and Serrano, F.O. (1995). The E-SCREEN assay as a tool to identify estrogens: an update on estrogenic environmental pollutants. *Environ. Health Perspect.* 103 (Suppl 7): 113–122.

Spanne, A. (2022). *What are PFAS* [Online]. https://www.ehn.org/what-are-pfas-2656619391/what-does-pfas-stand-for.

Starling, A.P., Umbach, D.M., Kamel, F., Long, S., Sandler, D.P., and Hoppin, J.A. (2014). Pesticide use and incident diabetes among wives of farmers in the Agricultural Health Study. *Occup. Environ. Med.* 71: 629–635.

Stoker, T.E., Laws, S.C., Guidici, D.L., and Cooper, R.L. (2000). The effect of atrazine on puberty in male Wistar rats: an evaluation in the protocol for the assessment of pubertal development and thyroid function. *Toxicol. Sci.* 58: 50–59.

Sun, S., Park, E.J., Choi, Y.H., Lee, H.S., Ahn, B.Y., and Dong, M.S. (2016). Development and pre-validation of an in vitro transactivation assay for detection of (anti)androgenic potential compounds using 22Rv1/MMTV cells. *Reprod. Toxicol.* 60: 156–166.

Sun, Y., Ou, Y., Cheng, M., Ruan, Y., and van der Hoorn, F.A. (2011). Binding of nickel to testicular glutamate-ammonia ligase inhibits its enzymatic activity. *Mol. Reprod. Dev.* 78: 104–115.

Suzuki, Y., Yoshinaga, J., Mizumoto, Y., Serizawa, S., and Shiraishi, H. (2012). Foetal exposure to phthalate esters and anogenital distance in male newborns. *Int. J. Androl.* 35: 236–244.

Swan, S.H. (2008). Environmental phthalate exposure in relation to reproductive outcomes and other health endpoints in humans. *Environ. Res.* 108: 177–184.

Swan, S.H., Elkin, E.P., and Fenster, L. (2000). The question of declining sperm density revisited: an analysis of 101 studies published 1934–1996. *Environ. Health Perspect.* 108: 961–966.

Sweeney, M.F., Hasan, N., Soto, A.M., and Sonnenschein, C. (2015). Environmental endocrine disruptors: effects on the human male reproductive system. *Rev. Endocr. Metab. Disord.* 16: 341–357.

Takao, Y., Lee, H.C., Kohra, S., and Arizono, K. (2002). Release of Bisphenol A from food can lining upon heating. *J. Health Sci.* 48: 331–334.

Tamrazi, A., Carlson, K.E., Daniels, J.R., Hurth, K.M., and Katzenellenbogen, J.A. (2002). Estrogen receptor dimerization: ligand binding regulates dimer affinity and dimer dissociation rate. *Mol. Endocrinol.* 16: 2706–2719.

Tang, W.Y., Morey, L.M., Cheung, Y.Y., Birch, L., Prins, G.S., and Ho, S.M. (2012). Neonatal exposure to estradiol/bisphenol A alters promoter methylation and expression of Nsbp1 and Hpcal1 genes and transcriptional programs of Dnmt3a/b and Mbd2/4 in the rat prostate gland throughout life. *Endocrinology* 153: 42–55.

Tilghman, S.L., Bratton, M.R., Segar, H.C., Martin, E.C., Rhodes, L.V., Li, M., McLachlan, J.A., Wiese, T.E., Nephew, K.P., and Burow, M.E. (2012). Endocrine disruptor regulation of microRNA expression in breast carcinoma cells. *PLoS One* 7: e32754.

Tremblay, G.B., Tremblay, A., Labrie, F., and Giguere, V. (1999). Dominant activity of activation function 1 (AF-1) and differential stoichiometric requirements for AF-1 and -2 in the estrogen receptor alpha-beta heterodimeric complex. *Mol. Cell. Biol.* 19: 1919–1927.

Treviño, L.S. and Gorelick, D.A. (2021). The interface of nuclear and membrane steroid signaling. *Endocrinology* 162.: bqab107.

Turner, A. and Filella, M. (2021). Lead in plastics – recycling of legacy material and appropriateness of current regulations. *J. Hazard. Mater.* 404: 124131.

Turyk, M.E., Anderson, H.A., and Persky, V.W. (2007). Relationships of thyroid hormones with polychlorinated biphenyls, dioxins, furans, and DDE in adults. *Environ. Health Perspect.* 115: 1197–1203.

UNEP. (1999). *(United Nations Environment Programme): The 12 Initial POPs under the Stockholm Convention* [Online]. http://chm.pops.int/TheConvention/ThePOPs/The12InitialPOPs/tabid/296/Default.aspx.

van den Berg, H. (2009). Global status of DDT and its alternatives for use in vector control to prevent disease. *Environ. Health Perspect.* 117: 1656–1663.

van der Velden, V.H. (1998). Glucocorticoids: mechanisms of action and anti-inflammatory potential in asthma. *Mediat. Inflamm.* 7: 229–237.

van Laar, J.H., Bolt-de Vries, J., Voorhorst-Ogink, M.M., and Brinkmann, A.O. (1989). The human androgen receptor is a 110 kDa protein. *Mol. Cell. Endocrinol.* 63: 39–44.

Vigeh, M., Smith, D.R., and Hsu, P.-C. (2011). How does lead induce male infertility. *Iran. J. Reprod. Med.* 9: 1–8.

Vilahur, N., Vahter, M., and Broberg, K. (2015). The epigenetic effects of prenatal cadmium exposure. *Curr. Environ. Health Rep.* 2: 195–203.

Vom Saal, F.S., Nagel, S.C., Coe, B.L., Angle, B.M., and Taylor, J.A. (2012). The estrogenic endocrine disrupting chemical bisphenol A (BPA) and obesity. *Mol. Cell. Endocrinol.* 354: 74–84.

Vrzal, R., Stejskalova, L., Monostory, K., Maurel, P., Bachleda, P., Pavek, P., and Dvorak, Z. (2009). Dexamethasone controls aryl hydrocarbon receptor (AhR)-mediated CYP1A1 and CYP1A2 expression and activity in primary cultures of human hepatocytes. *Chem. Biol. Interact.* 179: 288–296.

Vuong, A.M., Braun, J.M., Webster, G.M., Thomas Zoeller, R., Hoofnagle, A.N., Sjodin, A., Yolton, K., Lanphear, B.P., and Chen, A. (2018). Polybrominated diphenyl ether (PBDE) exposures and thyroid hormones in children at age 3 years. *Environ. Int.* 117: 339–347.

Wang, J., Hallinger, D.R., Murr, A.S., Buckalew, A.R., Simmons, S.O., Laws, S.C., and Stoker, T.E. (2018). High-throughput screening and quantitative chemical ranking for sodium-iodide symporter inhibitors in ToxCast Phase I Chemical Library. *Environ. Sci. Technol.* 52: 5417–5426.

Wang, Y., Wang, C., Zhang, J., Chen, Y., and Zuo, Z. (2009). DNA hypomethylation induced by tributyltin, triphenyltin, and a mixture of these in *Sebastiscus marmoratus* liver. *Aquat. Toxicol.* 95: 93–98.

Watkins, D.J., Sanchez, B.N., Tellez-Rojo, M.M., Lee, J.M., Mercado-Garcia, A., Blank-Goldenberg, C., Peterson, K.E., and Meeker, J.D. (2017). Phthalate and bisphenol A exposure during in utero windows of susceptibility in relation to reproductive hormones and pubertal development in girls. *Environ. Res.* 159: 143–151.

Weikum, E.R., Knuesel, M.T., Ortlund, E.A., and Yamamoto, K.R. (2017). Glucocorticoid receptor control of transcription: precision and plasticity via allostery. *Nat. Rev. Mol. Cell Biol.* 18: 159–174.

Weikum, E.R., Liu, X., and Ortlund, E.A. (2018). The nuclear receptor superfamily: a structural perspective. *Protein Sci.* 27: 1876–1892.

Wigle, D.T., Arbuckle, T.E., Turner, M.C., Berube, A., Yang, Q., Liu, S., and Krewski, D. (2008). Epidemiologic evidence of relationships between reproductive and child health outcomes and environmental chemical contaminants. *J. Toxicol. Environ. Health B: Crit. Rev.* 11: 373–517.

Wilson, V.S., Bobseine, K., and Gray, L.E., Jr. (2004). Development and characterization of a cell line that stably expresses an estrogen-responsive luciferase reporter for the detection of estrogen receptor agonist and antagonists. *Toxicol. Sci.* 81: 69–77.

Wirth, J.J. and Mijal, R.S. (2010). Adverse effects of low level heavy metal exposure on male reproductive function. *Syst. Biol. Reprod. Med.* 56: 147–167.

Wolff, M.S., Engel, S.M., Berkowitz, G.S., Ye, X., Silva, M.J., Zhu, C., Wetmur, J., and Calafat, A.M. (2008). Prenatal phenol and phthalate exposures and birth outcomes. *Environ. Health Perspect.* 116: 1092–1097.

Woodmansee, W.W. (2018). Thyroid and pregnancy: Endocrine disrupting chemicals during pregnancy and thyroid function in the mother and the baby. *Clinical Thyroidology for the Public*, 11(7): 5–6.

Wormke, M., Stoner, M., Saville, B., Walker, K., Abdelrahim, M., Burghardt, R., and Safe, S. (2003). The aryl hydrocarbon receptor mediates degradation of estrogen receptor alpha through activation of proteasomes. *Mol. Cell. Biol.* 23: 1843–1855.

Wozniak, A.L., Bulayeva, N.N., and Watson, C.S. (2005). Xenoestrogens at picomolar to nanomolar concentrations trigger membrane estrogen receptor-alpha-mediated Ca2+ fluxes and prolactin release in GH3/B6 pituitary tumor cells. *Environ. Health Perspect.* 113: 431–439.

Wright, R.O., Schwartz, J., Wright, R.J., Bollati, V., Tarantini, L., Park, S.K., Hu, H., Sparrow, D., Vokonas, P., and Baccarelli, A. (2010). Biomarkers of lead exposure and DNA methylation within retrotransposons. *Environ. Health Perspect.* 118: 790–795.

Wu, F., Khan, S., Wu, Q., Barhoumi, R., Burghardt, R., and Safe, S. (2008). Ligand structure-dependent activation of estrogen receptor alpha/Sp by estrogens and xenoestrogens. *J. Steroid Biochem. Mol. Biol.* 110: 104–115.

Wu, Y., Strawn, E., Basir, Z., Halverson, G., and Guo, S.W. (2007). Aberrant expression of deoxyribonucleic acid methyltransferases DNMT1, DNMT3A, and DNMT3B in women with endometriosis. *Fertil. Steril.* 87: 24–32.

Xie, Y., Liu, J., Benbrahim-Tallaa, L., Ward, J.M., Logsdon, D., Diwan, B.A., and Waalkes, M.P. (2007). Aberrant DNA methylation and gene expression in livers of newborn mice transplacentally exposed to a hepatocarcinogenic dose of inorganic arsenic. *Toxicology* 236: 7–15.

Xin, F., Susiarjo, M., and Bartolomei, M.S. (2015). Multigenerational and transgenerational effects of endocrine disrupting chemicals: a role for altered epigenetic regulation? *Semin. Cell Dev. Biol.* 43: 66–75.

Xu, B., Liu, S., Zhou, J.L., Zheng, C., Jin, W., Chen, B., Zhang, T., and Qui, W. (2021). PFAS and their substitutes in groundwater: occurrence, transformation and remediation. *J. Hazard. Mater.* 412: 125159.

Yao, R.W., Wang, Y., and Chen, L.L. (2019). Cellular functions of long noncoding RNAs. *Nat. Cell Biol.* 21: 542–551.

Yasar, P., Ayaz, G., User, S.D., Gupur, G., and Muyan, M. (2017). Molecular mechanism of estrogen–estrogen receptor signaling. *Reprod. Med. Biol.* 16: 4–20.

Ye, H., Ha, M., Yang, M., Yue, P., Xie, Z., and Liu, C. (2017). Di2-ethylhexyl phthalate disrupts thyroid hormone homeostasis through activating the Ras/Akt/TRHr pathway and inducing hepatic enzymes. *Sci. Rep.* 7: 40153.

Yin, L., Dai, Y., Jiang, X., Liu, Y., Chen, H., Han, F., Cao, J., and Liu, J. (2016). Role of DNA methylation in bisphenol A exposed mouse spermatocyte. *Environ. Toxicol. Pharmacol.* 48: 265–271.

Zhang, J.N., Zhang, R.Q., Liu, J.C., Li, L., Shen, W., and Sun, X.F. (2019). Di (2-ethylhexyl) phthalate exposure impairs the microRNAs expression profile during primordial follicle assembly. *Front. Endocrinol. (Lausanne)* 10: 877.

Zhang, X., Yu, R.M., Jones, P.D., Lam, G.K., Newsted, J.L., Gracia, T., Hecker, M., Hilscherova, K., Sanderson, T., Wu, R.S., and Giesy, J.P. (2005). Quantitative RT-PCR methods for evaluating toxicant-induced effects on steroidogenesis using the H295R cell line. *Environ. Sci. Technol.* 39: 2777–2785.

Zhang, Y., Lin, L., Cao, Y., Chen, B., Zheng, L., and Ge, R.S. (2009). Phthalate levels and low birth weight: a nested case-control study of Chinese newborns. *J. Pediatr.* 155: 500–504.

Zhao, C., Dahlman-Wright, K., and Gustafsson, J.A. (2008). Estrogen receptor beta: an overview and update. *Nucl. Recept. Signal* 6: e003.

Zhao, C.Q., Young, M.R., Diwan, B.A., Coogan, T.P., and Waalkes, M.P. (1997). Association of arsenic-induced malignant transformation with DNA hypomethylation and aberrant gene expression. *Proc. Natl. Acad. Sci. U.S.A.* 94: 10907–10912.

Zhou, C., Gao, L., and Flaws, J.A. (2017). Prenatal exposure to an environmentally relevant phthalate mixture disrupts reproduction in F1 female mice. *Toxicol. Appl. Pharmacol.* 318: 49–57.

Zhou, X., Sun, H., Ellen, T.P., Chen, H., and Costa, M. (2008). Arsenite alters global histone H3 methylation. *Carcinogenesis* 29: 1831–1836.

Zoeller, R.T., Brown, T.R., Doan, L.L., Gore, A.C., Skakkebaek, N.E., Soto, A.M., Woodruff, T.J., and Vom Saal, F.S. (2012). Endocrine-disrupting chemicals and public health protection: a statement of principles from The Endocrine Society. *Endocrinology* 153: 4097–4110.

Zorrilla, L.M., Gibson, E.K., Jeffay, S.C., Crofton, K.M., Setzer, W.R., Cooper, R.L., and Stoker, T.E. (2009). The effects of triclosan on puberty and thyroid hormones in male Wistar rats. *Toxicol. Sci.* 107: 56–64.

4

Nanoplastics in Agroecosystem and Phytotoxicity

An Evaluation of Cytogenotoxicity and Epigenetic Regulation

Piyoosh Kumar Babele[1,*] *and Ravi Kant Bhatia*[2]

[1] *College of Agriculture, Rani Lakshmi Bai Central Agricultural University, Jhansi, India*
[2] *Department of Biotechnology, Himachal Pradesh University, Shimla, India*
* *Corresponding author*

Introduction

Plastics manufacturing and application are expanding all the time. Since the first synthetic plastic (Bakelite) was produced in 1907, total plastic output has exceeded to 367 million metric tons in 2020 (Plastics Europe 2021). The term "plastics" now refers to the production volumes and current consumption patterns of five major commodity plastics: polypropylene, polyethylene, polyvinyl chloride, polystyrene (PS), and polyethylene terephthalate, which are produced by the use of additives (plasticizers) to modify the properties depending on their applications (Lebreton and Andrady 2019). Because of its robustness, chemical stability, persistence, and bioaccumulation character, most of the plastic is still present in the environment (Shams et al. 2021). Although plastics have societal benefits and a wide range of applications in every walk of life, their unrestricted use creates an emerging problem of plastic pollution in the environment and is becoming increasingly prominent (Kumari et al. 2022; Rhodes 2018). Generally, plastics in the global ecosystem are distributed between three fractions: (i) plastics in use, (ii) post-consumer managed and mismanaged plastic wastes, and (iii) urban/municipal litter. According to a new OECD report, only 9% of plastic waste is recycled, while 22% is mismanaged globally (Global Plastics Outlook 2022). Plastics used in packaging and other single-use plastics dominate mismanaged urban litter and municipal plastic waste, which is inadequately contained in open dumps and is therefore transportable via runoff and wind.

Mismanaged and urban litter plastics enter the environment and eventually in marine and terrestrial ecosystems, accidentally or otherwise. In the ecosystem, mismanaged plastic wastes will degrade/disintegrate into smaller and smaller pieces such as microplastics (MPs) and nanoplastics (NPs) (Kumari et al. 2022). Primary MPs (5 mm to 0.1 μm) and NPs (<0.1 μm) are produced intentionally (primary particles), such as fibers, beads, and pellets, which are discharges from personal care products, synthetic fabrics, and various other industries (Golwala et al. 2021; Yang et al. 2021). Secondary particles are unintentionally generated from larger plastic products (i.e. from the manufacturing of the plastic objects and/or degradation) under the natural weathering process, such as ultraviolet radiation (photoinduced degradation), decomposition, and disintegration through hydrolysis, rainwater, thermal, ozone-induced, mechanochemical, catalytic, and biological processes (Chamas et al. 2020; Golwala et al. 2021; Yang et al. 2021). Generally, when plastic particles break down, they gain new physical and chemical properties, increasing the risk of a toxic effect they could have on our environment, and pose a notable threat on the ecosystem (in particular to the biota) and, eventually, human health (Ya et al. 2021). MPs and NPs pollution is listed by the UN Environmental

Impact of Engineered Nanomaterials in Genomics and Epigenomics, First Edition. Edited by Saura C. Sahu.

Program (UNEP 2014) as one of the top 10 environmental issues due to their presence in aquatic and terrestrial ecosystems and thereafter harmful impacts. A large number of scientific journals mainly focus on the occurrence, detection, characterization, and toxicology of MP/NPs in marine and freshwater ecosystems and their transfer into the food web (reviewed in Benson et al. 2022; Bhuyan 2022; Stapleton 2021; Yuan et al. 2022). MP/NPs have been reported in numerous marine organisms, including fishes and other species that are used for human consumption (Bhuyan 2022; Yuan et al. 2022). *In vitro* experiments to assess the effect of NPs have been mostly performed on marine species (Bhuyan 2022; Yuan et al. 2022), *Caenorhabditis elegans* (Jewett et al. 2022; Yu et al. 2021), mouse models (Huang et al. 2022), and human cell lines (Ballesteros et al. 2020; Domenech et al. 2021; Shi et al. 2021); these cellular assessments showed that once absorbed, NPs may reach the nucleus where they exert toxic effects depending on the model species, human cell lines, as well as on the particle size distribution and monomeric composition/characteristics of plastics polymers. NPs pose acute cytotoxicity, chronic toxicity, (cardiovascular toxicity, hepatotoxicity, and neurotoxicity) (Yuan et al. 2022), carcinogenicity, genotoxicity (DNA damage) (Ballesteros et al. 2020), developmental toxicity and had an enhanced expression of inflammatory regulators that can alter the immune function (Ballesteros et al. 2020; López de las Hazas et al. 2021). Several other reports documented that ingestion of different MP/NPs dysregulates key molecular signaling pathways, and may induce key epigenetic changes that might be associated with the onset of several metabolic disorders including cancer (Barguilla et al. 2022; Kumar et al. 2022). NPs may imbalance the cell's redox homeostasis by dysregulating the production of reactive oxygen species (ROS) and reactive nitrogen species (RNS) excessive (beyond optimum levels) production of ROS and RNS leads to oxidative stress and thus alters the redox status of the cell (Huang et al. 2022; Ramakrishnan et al. 2022).

Terrestrial and agroecosystems are very prone to plastic pollution due to mismanaged plastic litter and the widespread use of plastic products in modern agricultural practices. Soils are one of the main reservoirs of terrestrial and agricultural plastics and are known to contain larger quantities of plastic particles (approximately 4–23 times higher) compared with the ocean and marine ecosystem (Pathan et al. 2020; Yang et al. 2021). Despite the potential entry of these emerging pollutants into agricultural systems and eventually crop plants (base of the terrestrial food web), they have largely been overlooked in ecotoxicological and nanotoxicological studies. Although NPs may share many of the properties of nanoparticles/engineered nanomaterials (ENMs), unlike ENMs that include intentional/desired formulations of metal, metal oxides, and other polymers, NPs are produced incidentally in ecosystems by the disintegration of larger plastic objects from the environmental processes. Moreover, NPs have been largely treated as an extension of MPs; however, size-dependent properties of NPs are fairly different from MPs with respect to their transport properties, interactions with soil parameters and biota, analytical challenges, bioavailability, potential toxicity, and leaching times for plasticizers and contaminants. Consequently, NPs should be considered as a unique class of environmental pollutants, distinct from both MPs and ENMs, henceforth new strategies and methods specific to NPs should be developed that are directly relevant to the ecotoxicological and nanotoxicological studies. A detailed investigation on the fate of NPs inside plant cells has not yet been carried out. The mechanisms by which NPs penetrate/internalized plant cells are not well established and how they induce cytogenotoxicity and trigger epigenetic changes are not understood. Further research is urgently required to enable the field of phytotoxicology to explore the impacts of NPs on agroecosystem and human health. The aim of this chapter is to summarize the screening of studies concerned mainly on plant–NPs interactions, adsorption, uptake, and transport mechanisms. In addition, the nanotoxicity of NPs in plants and crops will be discussed. Our main focus is to highlight the mechanism of cytotoxicity, genotoxicity, and epigenetic alterations in plant cells upon NPs exposure.

Fate and Behavior of NPs in Agroecosystem and Soil Environment

In terrestrial ecosystems, NP sources are classified as primary or secondary. MPs are mostly derived through the use of organic fertilizers, composts, and sewage sludge; the use of plastic mulches in agriculture; and flooding and irrigation with plastic-contaminated wastewater in agricultural operations (Keller et al. 2020). Secondary sources mostly include air degradation of big plastic goods employed by farmers in agricultural techniques, as well as landfill leachate (Figure 4.1). In the soil ecosystem, many biochemical processes transform the characteristics, such as size, shape, charge, surface coating, agglomeration rate, density of NPs, thereby affecting their biological fate, mobility, and bioavailability (Awet et al. 2018). The nano-scale dimension and large surface area of NPs make them possible to pass through the biological barriers and potentially enhance their bioactivity as well as toxicity (Awet et al. 2018). Detrimental effects of NPs on soil had already been investigated and it was reported that NPs can alter physical, chemical, and biological properties of soils by directly influencing bulk density, water-stable soil aggregates, water-holding capacity, and organic matter of soil (Pathan et al. 2020). Soil type and nature influence the migration of NPs, which in turn alter the soil properties by influencing the pH, soil structure, soil fertility and nutrient, water-stable aggregates, and soil microbial diversity (De Souza Machado et al. 2018, 2019; Zhao et al. 2021). Soil is one of the great scientific frontiers and the rhizosphere is the most dynamic section of that frontier where belowground biogeochemical processes occur. The rhizosphere is a dynamic region governed by complex interactions between plant roots, root-associated soil biota, and the soil stressors that drive important ecosystem processes (Cheng et al. 2019). It was reported that combined NPs and Cd stress trigger specific changes in root exudation of organic acids and bacterial community composition by the accumulation of NPs in the root of *Arabidopsis thaliana* (Yoon et al. 2021). This study also reveals a metabolic preference for aromatic compounds that drive the degradation of NPs in the rhizosphere and provide critical implications relevant to food security that NPs will contaminate crops as well, and in turn, transfer along the human food chain (Yoon et al. 2021).

Soil biota plays a vital role in soil ecosystem, and on plant growth and performance. Soil biota includes the micro-organisms (bacteria, fungi, archaea, and algae), soil animals (protozoa, nematodes, mites, springtails, spiders, insects, and earthworms), and plants. Soil biota and soil enzymes with high catalytic activities are closely connected with multiple soil biochemical processes; these enzymes act as an indicator for evaluating soil fertility and play an essential role in the regulation of soil nutrient cycling for nutrients such as C, N, and P (Awet et al. 2018; Liang et al. 2021). With the increase of NPs, these pollutant materials have now begun to demonstrate serious implications for key soil ecosystem functions, such as soil microbial activity and carbon (Rillig et al. 2021), and nitrogen (Iqbal et al. 2020; Seeley et al. 2020) nutrient cycling. Soil NPs significantly change the bacterial and fungal diversity (Lear et al. 2021) and affect the composition and abundance of microbes, and arthropod and nematode communities (Lin et al. 2020). It has been demonstrated that MPs exhibit significant effects on the activity of the urease, catalase activities, fluorescein diacetate hydrolase, and phenol oxidase, which can cause short-term changes in soil quality (Fei et al. 2020; Huang et al. 2019). In a study, the short-term effects of PS nanoparticles were investigated on the activity and functional diversity of soil enzymes. The results showed the reduction in activities of dehydrogenase, N-(leucine-aminopeptidase), P-(alkaline-phosphatase), and C-(β-glucosidase and cellobiohydrolase) enzymes (Awet et al. 2018). These changes in soil structure and functions may translate to plant and animal consequences and present potential concerns for food quality and safety, ultimately threatening human health (De Souza Machado et al. 2019). These anthropogenic adulterants are continuously polluting the terrestrial ecosystem and pose unfavorable effects on soil fertility, crop production, food security, and human health (Figure 4.1).

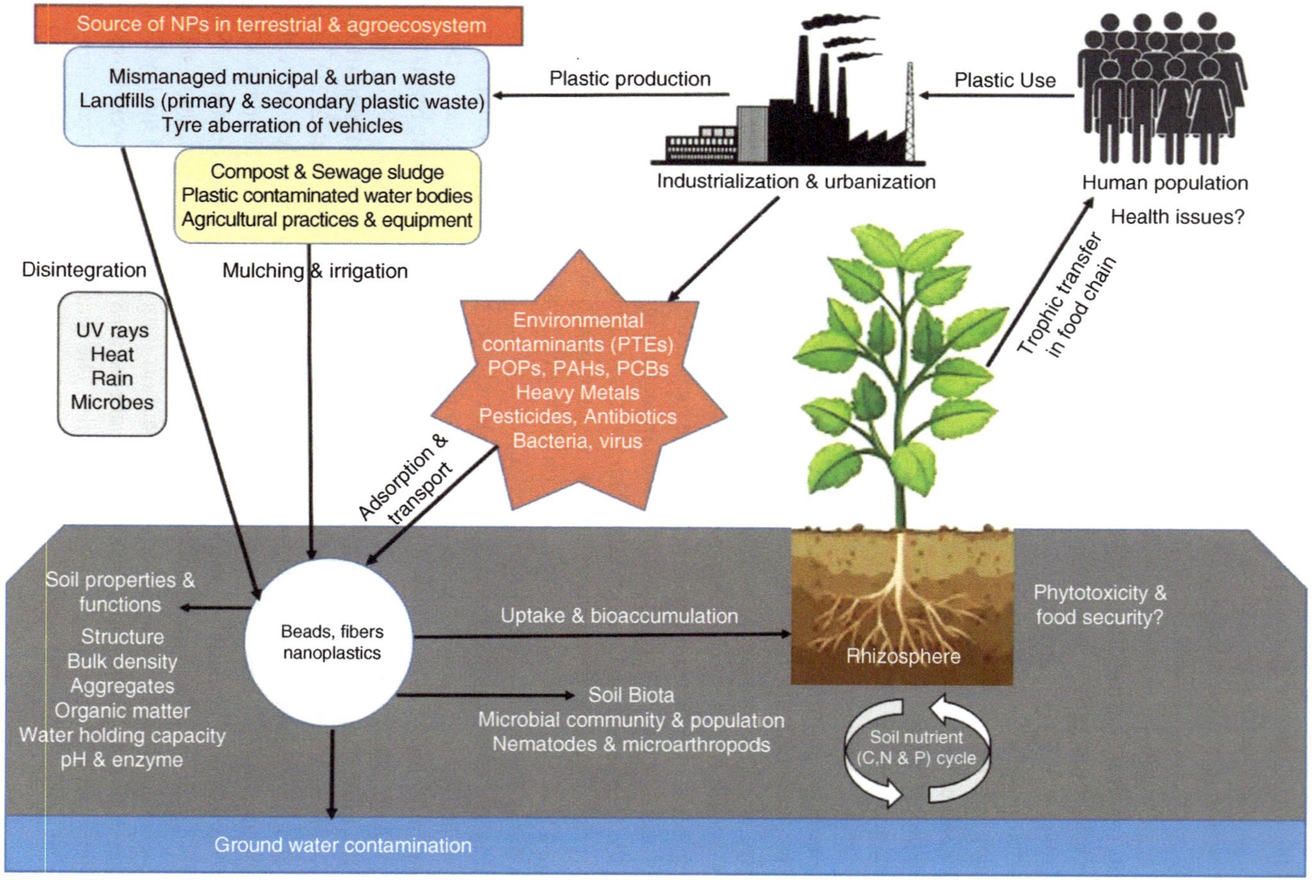

Figure 4.1 Sources and fate of nanoplastics (NPs) in the agroecosystem. Plastic waste from sewage sludge ends up in soil either directly by mulching and irrigation with contaminated water or indirectly by mismanaged plastic litter and landfills. These plastics disintegrate into NPs by several physical and biological processes in the environment. NPs interfere with soil particles and alter their physical and chemical properties as well as the soil biota functions and eventually soil's nutrient cycling.

Therefore, a better understanding of these NPs-driven processes are critical for maintaining the health of the planet and feeding the organisms that live on it.

Uptake and Accumulation of NPs in Plants

Plants and crops growing on plastic-contaminated soils are exposed to NPs directly. According to a new study, NPs may accumulate in plant tissues, which can have both direct ecological consequences as well as implications for agricultural sustainability and food safety. Researchers have identified several potential routes for ENMs to be absorbed at the root surface, followed by uptake and inter/intracellular mobility within plant tissue, while aerial exposure is absorbed by foliage, primarily via cuticles, hydathodes, stigma, stomata, and trichomes (Murali et al. 2022; Ruttkay-Nedecky et al. 2017), but the actual mode of ENMs and NPs absorption into plants is still unclear (Figure 4.2). In the soil environment, NPs are first accumulated in the rhizosphere, and in the rhizosphere, these particles are particularly adsorbed on root hairs, where they are transformed into other plant tissues (Chen et al. 2022). Although the uptake and accumulation of MP/NPs by edible plants and crops have recently been qualitatively investigated, the precise pathways and mechanisms remain ambiguous due to the lack of quantitative methods for NPs uptake in plants. Recently several methods were proposed, which utilize gas chromatography-mass spectrometry, inductively coupled plasma mass spectrometry, and provide new possibilities for screening NPs in plants (Li et al. 2021; Luo et al. 2022). A few recent studies suggested the uptake, translocation, and accumulation in plants. Li et al. (2020) analyze the uptake of different MPs by crop plants, such as wheat and lettuce; fluorescently labeled plastic beads were trapped in cells outside the root cap

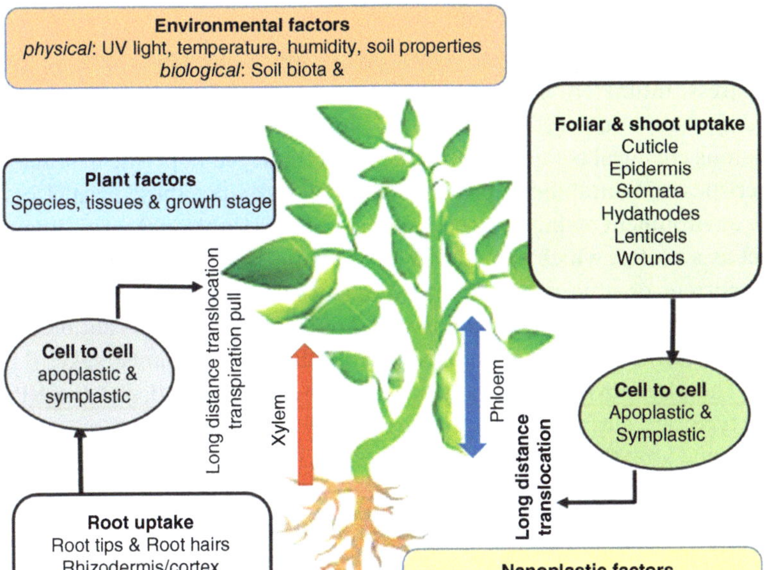

Figure 4.2 Major sites and uptake routes of nanoplastics entry into plant system and their translocation to different parts and factors influencing their adsorption, penetration, and biotransformation in plants.

mucilage and they penetrate the stele of both species using the crack-entry mode at sites of lateral root emergence, and these particles were subsequently transported from the roots to the shoots. The luminescence signals of 0.2-m plastic particles were found on the cell walls of the cortical tissue of the roots and were mostly seen in the vascular system, indicating that the plastic particles moved through the intercellular channels, through the apoplastic transport system and the transpirational pull was the main driving force of their movement (Li et al. 2020).

NPs can also move through small extracellular channels and reach the vasculature, which is responsible for water transport (Schwab et al. 2020). The water transport system that supports NPs can rapidly travel to the stem, leaves, and perhaps fruits. Confocal microscopy with 3D laser scanning clearly shows that NPs are absorbed from the root tips to the leaves, as validated by fluorescent signals and SEM. NPs (100 nm) may be delivered from root to shoot via the xylem route and then carried into the grain. Another route for plastics to enter the plant leaf is through the stomata via foliar spray and subsequently travel to the vasculature (Azeem et al. 2021). The uptake pathways of PS particles doped with the europium chelate Eu–β-diketonate (PS-Eu) have recently been qualitatively investigated in wheat and lettuce. Uptake and accumulation of PS-Eu NPs were detected mainly in the roots as confirmed by SEM, while transport to the shoots was limited (Luo et al. 2022). It might be concluded that NPs can accumulate in plants, depending on their size, shape, and surface charge (Sun et al. 2020). Due to conflicting findings on the uptake and accumulation of NPs on plants, it is difficult to draw any solid conclusion. In-depth research is needed to understand the processes of NPs absorption and transport in plants. The knowledge on the impacts of NPs in plant systems is critical for improving our understanding of their absorption and accumulation mechanisms in plants. Knowledge of NPs absorption and bioaccumulation mechanisms in plants is critical to better understand the mechanism of phyotoxicity.

NPs and Phytotoxicity

NPs cause phytotoxicity and impede plant development by inflicting direct mechanical damage as well as indirect oxidative stress, ultimately interfering with plant physiology and metabolism. When they are internalized in plant tissues/cells, they also deregulate genetic expression and epigenetic markers. The numerous chemical compounds (plasticizers) employed in plastic manufacturing to improve characteristics and durability represent additional risk since these plasticizers seep out in the rhizosphere environment owing to biochemical processes (Hahladakis et al. 2018). Moreover, NPs can also act as a vector, which can adsorb other potential environmental pollutants/contaminants, e.g., persistent organic pollutants, polycyclic aromatic hydrocarbons, polychlorinated biphenyls (PCBs), pesticides and herbicides, and pharmaceuticals, such as antibiotics (Lozano et al. 2021; Yu et al. 2019). The combination of NPs with other contaminants makes the joint effect more complex. The phytotoxicity of NPs depends on the characteristics of the NPs (exposure dose, size, shape, type, age, surface charge, and functionalization) (Sun et al. 2020) and of the plants (species, tissues, and growth stage) (Azeem et al. 2022).

Morphological and Physiological Responses

A few studies have reported the direct impacts of NPs on plant physiology and metabolism thus affecting growth and productivity. A report documented a direct impact on physiological processes, the study showed NPs type and dose-dependent effects on rice plants exposed to PS-NPs; there was a significant reduction in shoot biomass and root length (about 25%), with further losses in germination, root biomass, and plant height (Zhou et al. 2021). Similar to the type, the dose size of NPs also have distinct toxicity as reported root lengths of cucumber (*Cucumis sativus*) plants were reduced by roughly 12% at 100 nm and by approximately 8% at 300 nm after exposure to NPs, but no effects were seen at 500 or 700 nm (Z. Li et al. 2021). The unfavorable effect of NPs on seed germination and reduction in root length and biomass and plant height was related to plastic type, and the summary of these findings was reviewed by Azeem et al. (2022), Chen et al. (2022), and Mateos-Cárdenas et al. (2021). The photosynthetic rate is known to be a proxy for plant stress; NPs also affect plant photosynthesis, stomatal conductance, and transient transpiration, while effecting significant increases in intercellular CO_2 concentration and ultimately plant development as reflected by the reduction in leaf growth, chlorophyll, carotenoids, and other metabolites concentration (Chen et al. 2022; Y. Li et al. 2021; Mateos-Cárdenas et al. 2021). Foliar exposure to PS-NPs significantly decreased the dry weight, height, and leaf area, as well as pigment content (chlorophyll a, b, and carotenoid), which was considerably reduced in lettuce (*Lactuca sativa* L.) (Lian et al. 2021).

Biochemical and Metabolic Responses

Some recent studies show that NPs cause oxidative damage to plants and alter several typical biochemical indices of plants (Figure 4.3). Prior studies have looked at these processes and reviewed them elsewhere (Azeem et al. 2021; Chen et al. 2022; Yin et al. 2021). A study demonstrates the phytotoxicity of PS-NPs by foliar exposure by a significant increase in electrolyte leakage rate and a decrease in total antioxidant capacity (Lian et al. 2021). A study describes the response of rice (*Oryza sativa* L.) roots to NPs treatment significantly enhances the antioxidant enzyme activities reflecting the oxidative stress response of rice roots (Zhou et al. 2021). Lian et al. discovered that NPs (100 nm) increased superoxide dismutase SOD activity (23%) and catalase activity CAT (22%) in *C. sativus* (Lian et al. 2022). Several studies have looked at several typical biochemical and metabolic indices of plants, such as lignin, metabolites, and phytohormones, and these are possible avenues for future research (Campanale et al. 2022; Zhou et al. 2021). Zhou et al. (2021) reported that carbon metabolism was activated (e.g., increased carbon and soluble sugar contents), whereas jasmonic acid and lignin biosynthesis were inhibited in rice upon NPs exposure to roots (Zhou et al. 2021). Root exudates and mucilage are important barriers to plant pollutant buildup. Sun et al. (2020) found that amino-modified polystyrene NPs ($PS-NH_2$) stimulated root exudates, which disrupted PS stability and lowered plastic absorption in *Arabidopsis thaliana* (Sun et al. 2020). Recent research findings provide information on the specific changes in root exudation of organic acids and bacterial community composition in response to combined NPs and Cd stress revealing a metabolic preference for aromatic compounds that drive the degradation of NPs in the rhizosphere (Yoon et al. 2021). Metabolomics analysis revealed that all PS-NPs treatments altered the leaf metabolic profiles were profoundly modulated by PS-NPs mainly by regulating carbon, energy, and amino acid metabolism, and promoting amino acid biosynthesis, mitogen-activated protein kinase signaling pathway in plants, and plant hormone signal transduction in wheat leaves (Lian et al. 2022). This growing information shed new light on the phytotoxic mechanism and environmental implication behind the interaction of NPs and crop plants, and advances our understanding of the potential adverse effect induced by the presence of NPs in agricultural systems.

Can NPs Cause Cytogenotoxicity and Dysregulate Epigenetic Markers in Plants?

Pollution from nanoparticles has emerged as one of the world's most severe environmental challenges. The interaction of plastic particles with plants causes oxidative stress (ROS generation) and has a negative impact on photosynthesis, metabolism, gene expression, signaling, and other cellular processes (Chen et al. 2022), which could be classified as another type of environmental (abiotic) stress. We propose that ENMs and nanoplastics (NPs) have comparable properties. They are of identical size and form, and have equivalent surface qualities. Because of these parallels in characteristics, some of the advances made in studies over the past decade on the destiny and consequences of ENMs in the environment may be relevant to NPs (González-Morales et al. 2020; Karami Mehrian and De Lima 2016; Murali et al. 2022; Wang et al. 2013). Several studies demonstrated that ENMs (metals, metal oxides, carbon nanotubes, polymers, etc.) cause mechanical damage and accumulate inside the cells. Oxidative stress, cytotoxicity, genotoxicity, genomic instability, epigenetic changes, and inflammation are suggested as the possible mechanisms of ENMs-induced toxicity (Babele et al. 2018; Babele, Singh et al. 2019; Babele et al. 2021; Gedda et al. 2019; Kumar Babele 2019; Wang et al. 2013). Although several studies documented the presence of NPs in plant tissues/cells, the molecular mechanism of toxicity (cytotoxicity, genotoxicity, and epigenetic regulation) in plants and crops are currently very scarce. Some of the effects of NPs on the plant cells can likely be extrapolated using previous research on ENMs.

In response to any kind of environmental stress, cells regulated their gene/protein expression by a series of molecular mechanisms, which then interfere with several cellular processes of central dogma, such as mRNA transcription, translation, processing, transport, and cell death (Guarino et al. 2022). Several genetic and epigenetic regulatory mechanisms tightly work together to regulate gene expression (Figure 4.3). Deregulation in gene expression can result in a remodeling of the metabolic profile in plants, leading to the production, accumulation, or degradation of molecules to meet the requirements of the plants in response to environmental stress (Chang et al. 2020). Numerous studies have revealed that many epigenetic markers participate in environmental stress responses (Chang et al. 2020; Lämke and Bäurle 2017; Saeed et al. 2022). Epigenetic markers (heritable, but do not follow the known patterns of inheritance), mainly DNA methylation, histone modifications, histone variants, and some noncoding RNAs (ncRNA), influence the structure and accessibility of chromatin and thereby the biological function of the chromatin (Chang et al. 2020; Lämke and Bäurle 2017; Saeed et al. 2022). ROS and nitric oxide (NO) are hallmarks of stress responses; it was predicted that they are also pivotal in mediating chromatin dynamics during environmental responses (Lindermayr et al. 2020).

NPs Cause Cytogenotoxicity

NPs produce toxicity by accumulating ROS and inducing oxidative stress, as seen by considerable alterations in antioxidative enzymes, as previously described and discussed. ROS overproduction results in cytotoxicity and genotoxicity by causing lipid peroxidation, mitochondrial malfunction, DNA damage, and protein denaturation. In relevant research, oxidative damage to vascular plants is typically measured by measuring antioxidant enzyme activity and malondialdehyde levels (Faisal et al. 2016; Pignattelli et al. 2020; Zhou et al. 2021). ROS are produced at many organelle and cellular sites, including cell walls, endoplasmic reticulum, plasma membranes, apoplasts, peroxisomes, mitochondria, and chloroplasts (Azeem et al. 2021; Babele et al. 2018; Babele, Kumar et al. 2019; Dai et al. 2018; S. Li et al. 2021). ROS accumulation is highly related to plant growth and development, and stress defense, and NPs may stimulate a rise in ROS concentration, which induce oxidative damage, which is directly related to genotoxicity and cytotoxicity (Giorgetti et al. 2020). Lian et al. (2022) analyzed the transcriptome response of hydroponically grown wheat (*Triticum aestivum* L.) upon PS-NPs exposure and significantly altered gene expression patterns in a tissue-specific manner (Lian et al. 2022). The research investigates the mechanisms of action of PS-NPs on seeds of *Allium cepa*. PS-NPs were internalized in different cellular compartments evidenced by histological localization observed under a transmission electron microscope. Furthermore, cytological analysis

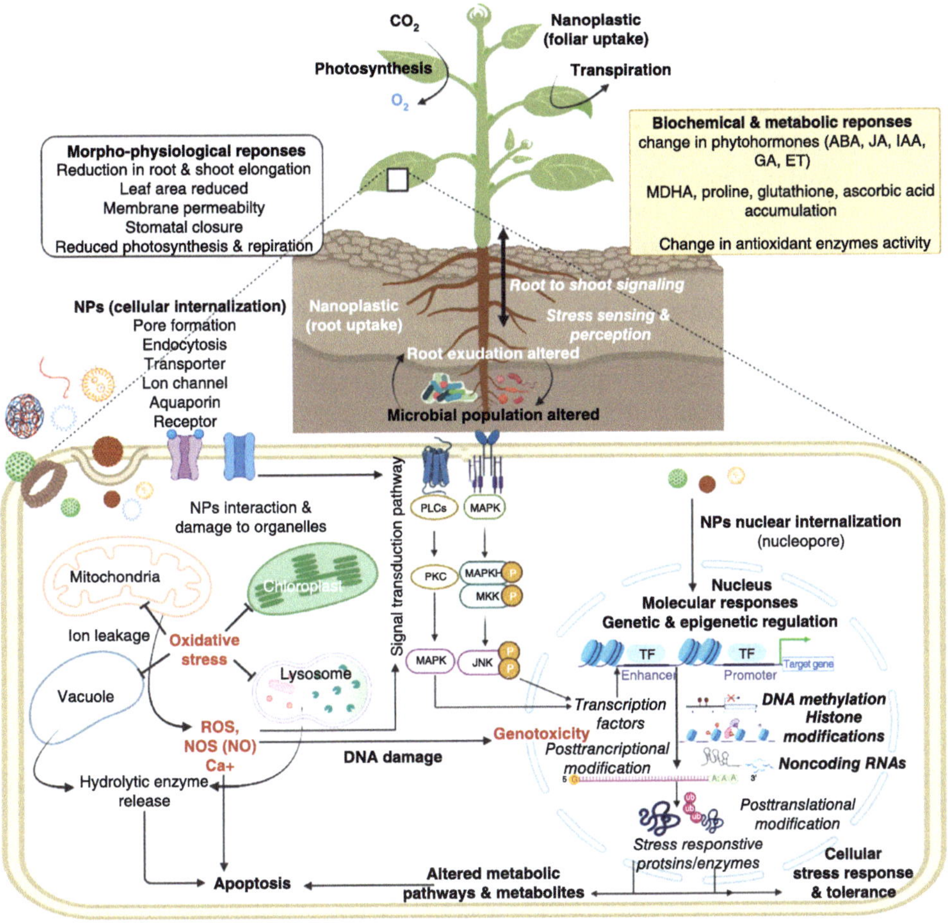

Figure 4.3 Schematic diagram showing the routes of cellular internalization of nanoplastics in plants and morpho-physiological, biochemical, and molecular responses. Upon internalization, they cause oxidative stress and damage to cellular organelles, which, in turn, induce the generation of ROS and RNS, ion leakage, and release of proteolytic enzymes; together, these cause cytogenotoxicity and activation of programmed cell death (apoptosis). Cells activate cell-signaling cascades and, genetic and epigenetic regulatory mechanisms to mitigate the toxic effects of NPs.

of root meristems indicated cytotoxicity (reduction of the mitotic index) and genotoxicity (induction of cytogenetic anomalies and micronuclei), starting from the lowest dose (Giorgetti et al. 2020).

NPs and Epigenetic Regulation

It is now very well established that NPs can cause mechanical damage and cross biological membranes and influence cellular signaling; however, the cellular and systemic toxicities associated with these emerging pollutant exposures and the role of epigenetic alterations in their toxicity remain relatively unexplored. A few review articles summarize current evidence of alterations in cytosine DNA methylation and histone modifications in response to nanomaterials and nanoparticles exposures *in vivo* and *in vitro* mainly on animal models and human cell lines (Gedda et al. 2019; Pogribna and Hammons 2021). A few pieces of research have focused on dissecting the mechanisms of genotoxicity and epigenetic responses in human cell lines (Ballesteros et al. 2020). A study provides comprehensive evidence on the impact of ZnO-NP on epigenetic modification; it was evident that ZnO-NPs may remodel the chromatin ultrastructure and transcription program by downregulating the histone deacetylases (HDA3) gene in tomato (Pejam et al. 2021). We believe these research approaches could be translated into plant cells to delineate the nanotoxicity of NPs in plants. It is believed that epigenetic alterations have the potential to induce prolonged changes in the genetic regulation and any failure in the detection of these changes could lead to unexpected and unpleasant effects in the metabolically active cells. The studies focusing on epigenetic regulation in plant systems are not conducted. Most of the epigenetic impacts of plastic particles have been evaluated using animal models (López de las Hazas et al. 2021). Yu et al. (2021) reported that NPs exposure reduced the expression of epigenesis-related genes met-2, set-2, and spr-5 genes in *C. elegans*. Studies demonstrate that NPs exposure can cause multi- and trans-generational reproduction defects that may be associated with germline toxicity and epigenetic regulation (Yu et al. 2021).

Conclusion and Perspectives

NPs pollution, which is one of the world's most significant environmental challenges, has been evolving, resulting in environmental stress with a detrimental impact on plant growth and development along with severe consequences for agricultural output and substantial economic losses. Nonetheless, further studies should be conducted to better understand the destiny of plastics in agroecosystems and their interaction with plants (particularly crops), and phytotoxicology studies are critically needed. NPs can modify soil physicochemical qualities and have a significant influence on rhizosphere biochemistry through influencing microbial community composition and diversity, as well as plant performance. The impact of different kinds, sizes, and dosages of NPs on plant interaction, translocation, and the likely repercussions of soil and foliar application on diverse plant species urgently need a sincere and serious analysis of their future implications on the agroecosystem and human health. Growing data suggest a tight relationship between ROS metabolism (cell redox state) and epigenetic control during plant development and environmental adaptation and any changes in ROS levels that have resulted in visible epigenetic alterations; however, further research is needed to determine how redox state and epigenetic changes influence gene expression. Future research must determine whether antioxidant system activation can deal with the rising ROS concentrations, as well as look at several typical biochemical indices of plants, such as metabolism and phytohormones alteration (metabolomics), genetic and epigenetic (transcriptomics and proteomics) markers, as possible avenues that may aid in risk assessments of plastic and food safety in agroecosystems. To better comprehend the interrelationships between plastic waste, ecosystems, and human civilization, an integrated framework, and policies are the demand of the present. There are many solutions to mitigate plastic pollution, such as by reducing the use of single-use plastics, implementing and abiding by legislation to curb plastic production, improving waste management as well as holding plastic producers responsible for the waste they generate, while always recycling plastic products, and promoting good practices as using biodegradable plastics. Collectively, the general public should be informed of MP/NPs, how they might pollute soil, and how to protect soil against particle pollution. Furthermore, governments and policymakers can enact the required laws and regulations to restrict plastic production and can also provide economic incentives to encourage manufacturers to adopt alternatives to plastic (e.g., bioplastics). We believe that with these simple practices and policies, we can expect a greener and a more sustainable future for our planet suitable for agricultural development and better human health.

References

Awet, T.T. et al. (2018). Effects of polystyrene nanoparticles on the microbiota and functional diversity of enzymes in soil. *Environ. Sci. Eur.* Springer Verlag, 30 (1): 1–10. doi: 10.1186/S12302-018-0140-6/FIGURES/3.

Azeem, I. et al. (2021). Uptake and accumulation of nano/microplastics in plants: a critical review. *Nanomaterials.* Multidisciplinary Digital Publishing Institute, 11 (11): 2935. doi: 10.3390/NANO11112935.

Azeem, I. et al. (2022). Microplastic and nanoplastic interactions with plant species: trends, meta-analysis, and perspectives. *Environ. Sci. Technol. Lett.* American Chemical Society. doi: 10.1021/ACS.ESTLETT.2C00107.

Babele, P.K. et al. (2018). Zinc oxide nanoparticles induce toxicity by affecting cell wall integrity pathway, mitochondrial function and lipid homeostasis in *Saccharomyces cerevisiae. Chemosphere.* Elsevier Ltd, 213: 65–75. doi: 10.1016/j.chemosphere.2018.09.028.

Babele, P.K., Kumar, J., and Chaturvedi, V. (2019). Proteomic de-regulation in cyanobacteria in response to abiotic stresses. *Front. Microbiol.* 10 (June). doi: 10.3389/fmicb.2019.01315.

Babele, P.K., Singh, A.K., and Srivastava, A. (2019). Bio-inspired silver nanoparticles impose metabolic and epigenetic toxicity to *Saccharomyces cerevisiae. Front. Pharmacol.* 10. doi: 10.3389/fphar.2019.01016.

Babele, P.K., Verma, M.K., and Bhatia, R.K. (2021). Carbon nanotubes: a review on risks assessment, mechanism of toxicity and future directives to prevent health implication. *Biocell.* Tech Science Press, 45 (2): 267. doi: 10.32604/BIOCELL.2021.013409.

Ballesteros, S. et al. (2020). Genotoxic and immunomodulatory effects in human white blood cells after ex vivo exposure to polystyrene nanoplastics. *Environ. Sci. Nano.* The Royal Society of Chemistry, 7 (11): 3431–3446. doi: 10.1039/D0EN00748J.

Barguilla, I. et al. (2022). Nanoplastics and arsenic co-exposures exacerbate oncogenic biomarkers under an in vitro long-term exposure scenario. *Int. J. Mol. Sci.* MDPI, 23 (6). doi: 10.3390/IJMS23062958/S1.

Benson, N.U. et al. (2022). Micro(nano)plastics prevalence, food web interactions, and toxicity assessment in aquatic organisms: a review. *Front. Mar. Sci.* Frontiers Media SA, 9: 291. doi: 10.3389/FMARS.2022.851281/BIBTEX.

Bhuyan, M.S. (2022). Effects of microplastics on fish and in human health. *Front. Environ. Sci.* Frontiers Media SA, 10: 250. doi: 10.3389/FENVS.2022.827289/BIBTEX.

Campanale, C. et al. (2022). Microplastics pollution in the terrestrial environments: poorly known diffuse sources and implications for plants. *Sci. Total Environ.* Elsevier B.V., 805. doi: 10.1016/J.SCITOTENV.2021.150431.

Chamas, A. et al. (2020). Degradation rates of plastics in the environment. *Cite This: ACS Sustain. Chem. Eng.* 8: 3511. doi: 10.1021/acssuschemeng.9b06635.

Chang, Y.N. et al. (2020). Epigenetic regulation in plant abiotic stress responses. *J. Integr. Plant Biol.* John Wiley & Sons, Ltd, 62 (5): 563–580. doi: 10.1111/JIPB.12901.

Chen, G. et al. (2022). Effects of micro(nano)plastics on higher plants and the rhizosphere environment. *Sci. Total Environ.* Elsevier, 807: 150841. doi: 10.1016/J.SCITOTENV.2021.150841.

Cheng, Y.T., Zhang, L., and Yang He, S. (2019). Review plant–microbe interactions facing environmental challenge. doi: 10.1016/j.chom.2019.07.009.

Dai, Y. et al. (2018). Interaction of CuO nanoparticles with plant cells: internalization, oxidative stress, electron transport chain disruption, and toxicogenomic responses. *Environ. Sci. Nano.* The Royal Society of Chemistry, 5 (10): 2269–2281. doi: 10.1039/C8EN00222C.

De Souza Machado, A.A. et al. (2018). Impacts of microplastics on the soil biophysical environment. *Environ. Sci. Technol.* American Chemical Society, 52 (17): 9656–9665. doi: 10.1021/ACS.EST.8B02212/SUPPL_FILE/ES8B02212_SI_002.PDF.

De Souza Machado, A.A. et al. (2019). Microplastics can change soil properties and affect plant performance. *Environ. Sci. Technol.* American Chemical Society, 53 (10): 6044–6052. doi: 10.1021/ACS.EST.9B01339/SUPPL_FILE/ES9B01339_SI_003.PDF.

Domenech, J. et al. (2021). Long-term effects of polystyrene nanoplastics in human intestinal Caco-2 cells. *Biomolecules* 11 (10). doi: 10.3390/BIOM11101442.

Faisal, M. et al. (2016). Cobalt oxide nanoparticles aggravate DNA damage and cell death in eggplant via mitochondrial swelling and NO signaling pathway. *Biol. Res.* BioMed Central Ltd., 49 (1): 1–13. doi: 10.1186/S40659-016-0080-9/FIGURES/8.

Fei, Y. et al. (2020). Response of soil enzyme activities and bacterial communities to the accumulation of microplastics in an acid cropped soil. *Sci. Total Environ.* Elsevier, 707: 135634. doi: 10.1016/J.SCITOTENV.2019.135634.

Gedda, M.R. et al. (2019). Epigenetic aspects of engineered nanomaterials: is the collateral damage inevitable? *Front. Bioeng. Biotechnol.* 7 (September). doi: 10.3389/fbioe.2019.00228.

Giorgetti, L. et al. (2020). Exploring the interaction between polystyrene nanoplastics and *Allium cepa* during germination: internalization in root cells, induction of toxicity and oxidative stress. *Plant Physiol. Biochem.* Elsevier Masson SAS, 149: 170–177. doi: 10.1016/J.PLAPHY.2020.02.014.

Global Plastics Outlook. (2022). Global Plastics Outlook. OECD. doi: 10.1787/DE747AEF-EN.

Golwala, H. et al. (2021). Solid waste: an overlooked source of microplastics to the environment. *Sci. Total Environ.* Elsevier, 769: 144581. doi: 10.1016/J.SCITOTENV.2020.144581.

González-Morales, S. et al. (2020). The ecology of nanomaterials in agroecosystems. In: *Nanomaterials for Agriculture and Forestry Applications*, 313–355. Elsevier. doi: 10.1016/B978-0-12-817852-2.00013-5.

Guarino, F. et al. (2022). An epigenetic alphabet of crop adaptation to climate change. *Front. Genet.* Frontiers Media S.A., 13: 266. doi: 10.3389/FGENE.2022.818727/BIBTEX.

Hahladakis, J.N. et al. (2018). An overview of chemical additives present in plastics: migration, release, fate and environmental impact during their use, disposal and recycling. *J. Hazard. Mater.* Elsevier B.V., 344: 179–199. doi: 10.1016/J.JHAZMAT.2017.10.014.

Huang, T. et al. (2022). Maternal exposure to polystyrene nanoplastics during gestation and lactation induces hepatic and testicular toxicity in male mouse offspring. *Food Chem. Toxicol.* Pergamon, 160: 112803. doi: 10.1016/J.FCT.2021.112803.

Huang, Y. et al. (2019). LDPE microplastic films alter microbial community composition and enzymatic activities in soil. *Environmental Pollution (Barking, Essex: 1987). Environ. Pollut.* 254 (Pt A). doi: 10.1016/J.ENVPOL.2019.112983.

Iqbal, S. et al. (2020). Unraveling consequences of soil micro- and nanoplastic pollution on soil–plant system: implications for nitrogen (N) cycling and soil microbial activity. *Chemosphere.* Pergamon, 260: 127578. doi: 10.1016/J.CHEMOSPHERE.2020.127578.

Jewett, E. et al. (2022). Microplastics and their impact on reproduction—can we learn from the *C. elegans* model? *Front. Toxicol.* Frontiers, 0: 25. doi: 10.3389/FTOX.2022.748912.

Karami Mehrian, S. and De Lima, R. (2016). Nanoparticles cyto and genotoxicity in plants: mechanisms and abnormalities. *Environ. Nanotechnol. Monit. Manag.* Elsevier, 6: 184–193. doi: 10.1016/J.ENMM.2016.08.003.

Keller, A.S., Jimenez-Martinez, J., and Mitrano, D.M. (2020). Transport of nano- and microplastic through unsaturated porous media from sewage sludge application. *Environ. Sci. Technol.* American Chemical Society, 54 (2): 911–920. doi: 10.1021/ACS.EST.9B06483/SUPPL_FILE/ES9B06483_SI_001.PDF.

Kumar Babele, P. (2019). Zinc oxide nanoparticles impose metabolic toxicity by de-regulating proteome and metabolome in *Saccharomyces cerevisiae*. *Toxicol. Rep.* 6. doi: 10.1016/j.toxrep.2018.12.001.

Kumar, R. et al. (2022). Micro(nano)plastics pollution and human health: how plastics can induce carcinogenesis to humans? *Chemosphere* 298. doi: 10.1016/J.CHEMOSPHERE.2022.134267.

Kumari, A. et al. (2022). Microplastic pollution: an emerging threat to terrestrial plants and insights into its remediation strategies. *Plants*. Multidisciplinary Digital Publishing Institute, 11 (3): 340. doi: 10.3390/PLANTS11030340.

Lämke, J. and Bäurle, I. (2017). Epigenetic and chromatin-based mechanisms in environmental stress adaptation and stress memory in plants. *Genome Biol.* BioMed Central, 18 (1): 1–11. doi: 10.1186/S13059-017-1263-6.

Lear, G. et al. (2021). Plastics and the microbiome: impacts and solutions. *Environ. Microbiome*. BioMed Central, 16 (1): 1–19. doi: 10.1186/S40793-020-00371-W.

Lebreton, L. and Andrady, A. (2019). Future scenarios of global plastic waste generation and disposal. *Palgrave Commun.* Palgrave, 5 (1): 1–11. doi: 10.1057/s41599-018-0212-7.

Li, C. et al. (2021). Quantification of nanoplastic uptake in cucumber plants by pyrolysis gas chromatography/mass spectrometry. *Environ. Sci. Technol. Lett.* American Chemical Society, 8 (8): 633–638. doi: 10.1021/ACS.ESTLETT.1C00369/SUPPL_FILE/EZ1C00369_SI_001.PDF.

Li, L. et al. (2020). Effective uptake of submicrometre plastics by crop plants via a crack-entry mode. *Nat. Sustain.* Nature Publishing Group, 3 (11): 929–937. doi: 10.1038/s41893-020-0567-9.

Li, S. et al. (2021). Polystyrene microplastics disturb the redox homeostasis, carbohydrate metabolism and phytohormone regulatory network in barley. *J. Hazard. Mater.* Elsevier B.V., 415. doi: 10.1016/J.JHAZMAT.2021.125614.

Li, Y. et al. (2021). Impacts of micro- and nanoplastics on photosynthesis activities of photoautotrophs: a mini-review. *Front. Microbiol.* Frontiers Media SA, 12: 3324. doi: 10.3389/FMICB.2021.773226/BIBTEX.

Li, Z. et al. (2021). The distribution and impact of polystyrene nanoplastics on cucumber plants. *Environ. Sci. Pollut. Res.* Springer Science and Business Media Deutschland GmbH, 28 (13): 16042–16053. doi: 10.1007/S11356-020-11702-2/FIGURES/8.

Lian, J. et al. (2021). Foliar-applied polystyrene nanoplastics (PS-NPs) reduce the growth and nutritional quality of lettuce (*Lactuca sativa* L.). *Environ. Pollut.* Elsevier Ltd, 280. doi: 10.1016/J.ENVPOL.2021.116978.

Lian, J. et al. (2022). Nanotoxicological effects and transcriptome mechanisms of wheat (*Triticum aestivum* L.) under stress of polystyrene nanoplastics. *J. Hazard. Mater.* Elsevier, 423: 127241. doi: 10.1016/J.JHAZMAT.2021.127241.

Liang, Y. et al. (2021). Effects of microplastic fibers on soil aggregation and enzyme activities are organic matter dependent. *Front. Environ. Sci.* Frontiers Media S.A., 9: 97. doi: 10.3389/FENVS.2021.650155/BIBTEX.

Lin, D. et al. (2020). Microplastics negatively affect soil fauna but stimulate microbial activity: insights from a field-based microplastic addition experiment. *Proc. R. Soc. B.* The Royal Society, 287 (1934). doi: 10.1098/RSPB.2020.1268.

Lindermayr, C. et al. (2020). Interactions between metabolism and chromatin in plant models. *Mol. Metab.* Elsevier, 38: 100951. doi: 10.1016/J.MOLMET.2020.01.015.

López de las Hazas, M.-C., Boughanem, H., and Dávalos, A. (2021). Untoward effects of micro- and nanoplastics: an expert review of their biological impact and epigenetic effects. *Advances in Nutrition (Bethesda, Md.). Adv. Nutr.* doi: 10.1093/ADVANCES/NMAB154.

Lozano, Y.M. et al. (2021). Microplastic shape, polymer type, and concentration affect soil properties and plant biomass. *Front. Plant Sci.* Frontiers Media S.A., 12: 169. doi: 10.3389/FPLS.2021.616645/BIBTEX.

Luo, Y. et al. (2022). Quantitative tracing of uptake and transport of submicrometre plastics in crop plants using lanthanide chelates as a dual-functional tracer. *Nat. Nanotechnol.* Nature Publishing Group, 17 (4): 424–431. doi: 10.1038/s41565-021-01063-3.

Mateos-Cárdenas, A. et al. (2021). Adsorption, uptake and toxicity of micro- and nanoplastics: effects on terrestrial plants and aquatic macrophytes. *Environ. Pollut.* Elsevier, 284: 117183. doi: 10.1016/J.ENVPOL.2021.117183.

Murali, M. et al. (2022). Fate, bioaccumulation and toxicity of engineered nanomaterials in plants: current challenges and future prospects. *Sci. Total Environ.* Elsevier, 811: 152249. doi: 10.1016/J.SCITOTENV.2021.152249.

Pathan, S.I. et al. (2020). Soil pollution from micro- and nanoplastic debris: a hidden and unknown biohazard. *Sustainability*. Multidisciplinary Digital Publishing Institute, 12 (18): 7255. doi: 10.3390/SU12187255.

Pejam, F. et al. (2021). Zinc oxide nanoparticles mediated substantial physiological and molecular changes in tomato. *PLoS ONE*. Public Library of Science, 16 (3). doi: 10.1371/JOURNAL.PONE.0248778.

Pignattelli, S., Broccoli, A., and Renzi, M. (2020). Physiological responses of garden cress (*L. sativum*) to different types of microplastics. *Sci. Total Environ.* Elsevier B.V., 727. doi: 10.1016/J.SCITOTENV.2020.138609.

Plastics Europe (2021). https://plasticseurope.org/wp-content/uploads/2021/12/Plastics-the-Facts-2021-web-final.pdf

Pogribna, M. and Hammons, G. (2021). Epigenetic effects of nanomaterials and nanoparticles. *J. Nanobiotechnology*. BioMed Central Ltd, 19 (1): 1–18. doi: 10.1186/S12951-020-00740-0/FIGURES/2.

Ramakrishnan, M. et al. (2022). Redox status of the plant cell determines epigenetic modifications under abiotic stress conditions and during developmental processes. *J. Adv. Res.* Elsevier. doi: 10.1016/J.JARE.2022.04.007.

Rhodes, C.J. (2018). Plastic pollution and potential solutions. *Sci. Prog.* NLM (Medline), 101 (3): 207–260. doi: 10.3184/003685018X15294876706211.

Rillig, M.C., Leifheit, E., and Lehmann, J. (2021). Microplastic effects on carbon cycling processes in soils. *PLOS Biol.* Public Library of Science, 19 (3): e3001130. doi: 10.1371/JOURNAL.PBIO.3001130.

Ruttkay-Nedecky, B. et al. (2017). Nanoparticles based on essential metals and their phytotoxicity. 15: 33. doi: 10.1186/s12951-017-0268-3.

Saeed, F., Chaudhry, U.K., Bakhsh, A., Raza, A., Saeed, Y., Bohra, A., and Varshney, R.K. (2022). Moving beyond DNA sequence to improve plant stress responses. Frontier in Genetics. 13:874648. doi: 10.3389/fgene.2022.874648

Schwab, F., Rothen-Rutishauser, B., and Petri-Fink, A. (2020). When plants and plastic interact. *Nat. Nanotechnol.* Nature Publishing Group, 15 (9): 729–730. doi: 10.1038/s41565-020-0762-x.

Seeley, M.E. et al. (2020). Microplastics affect sedimentary microbial communities and nitrogen cycling. *Nat. Commun.* Nature Publishing Group, 11 (1): 1–10. doi: 10.1038/s41467-020-16235-3.

Shams, M., Alam, I., and Mahbub, M.S. (2021). Plastic pollution during COVID-19: plastic waste directives and its long-term impact on the environment. *Environ. Adv.* Elsevier, 5: 100119. doi: 10.1016/J.ENVADV.2021.100119.

Shi, Q. et al. (2021). Toxicity in vitro reveals potential impacts of microplastics and nanoplastics on human health: a review. Taylor & Francis. doi: 10.1080/10643389.2021.1951528.

Stapleton, P.A. (2021). Microplastic and nanoplastic transfer, accumulation, and toxicity in humans. *Curr. Opin. Toxicol.* Elsevier, 28: 62–69. doi: 10.1016/J.COTOX.2021.10.001.

Sun, X.D. et al. (2020). Differentially charged nanoplastics demonstrate distinct accumulation in *Arabidopsis thaliana*. *Nat. Nanotechnol.* Nature Publishing Group, 15 (9): 755–760. doi: 10.1038/s41565-020-0707-4.

UNEP (2014). https://www.unep.org/resources/annual-report/unep-2014-annual-report

Wang, H. et al. (2013). Engineered nanoparticles may induce genotoxicity. *Environ. Sci. Technol.* American Chemical Society, 47 (23): 13212–13214. doi: 10.1021/ES404527D/ASSET/IMAGES/LARGE/ES-2013-04527D_0002.JPEG.

Ya, H. et al. (2021). Recent advances on ecological effects of microplastics on soil environment. *Sci. Total Environ.* Elsevier B.V., 798. doi: 10.1016/J.SCITOTENV.2021.149338.

Yang, L. et al. (2021). Microplastics in soil: a review on methods, occurrence, sources, and potential risk. *Sci. Total Environ.* Elsevier, 780: 146546. doi: 10.1016/J.SCITOTENV.2021.146546.

Yin, L. et al. (2021). Interactions between microplastics/nanoplastics and vascular plants. *Environ. Pollut.* Elsevier, 290: 117999. doi: 10.1016/J.ENVPOL.2021.117999.

Yoon, H. *et al* (2021). Fragmentation of nanoplastics driven by plant–microbe rhizosphere interaction during abiotic stress combination. *Environ. Sci. Nano.* The Royal Society of Chemistry, 8 (10): 2802–2810. doi: 10.1039/D1EN00230A.

Yu, C.W., Luk, T.C., and Liao, V.H.C. (2021). Long-term nanoplastics exposure results in multi and trans-generational reproduction decline associated with germline toxicity and epigenetic regulation in *Caenorhabditis elegans*. *J. Hazard. Mater.* 412. doi: 10.1016/J.JHAZMAT.2021.125173.

Yu, F. et al. (2019). Adsorption behavior of organic pollutants and metals on micro/nanoplastics in the aquatic environment. *Sci. Total Environ.* Elsevier B.V., 694. doi: 10.1016/J.SCITOTENV.2019.133643.

Yuan, Z., Nag, R., and Cummins, E. (2022). Human health concerns regarding microplastics in the aquatic environment – from marine to food systems. *Sci. Total Environ.* Elsevier, 823: 153730. doi: 10.1016/J.SCITOTENV.2022.153730.

Zhao, T., Lozano, Y.M., and Rillig, M.C. (2021). Microplastics increase soil pH and decrease microbial activities as a function of microplastic shape, polymer type, and exposure time. *Front. Environ. Sci.* Frontiers Media S.A., 9: 235. doi: 10.3389/FENVS.2021.675803/BIBTEX.

Zhou, C.Q. et al. (2021). Response of rice (*Oryza sativa* L.) roots to nanoplastic treatment at seedling stage. *J. Hazard. Mater.* Elsevier, 401: 123412. doi: 10.1016/J.JHAZMAT.2020.123412.

5

Metal Oxide Nanoparticles and Graphene-Based Nanomaterials

Genotoxic, Oxidative, and Epigenetic Effects

Delia Cavallo, Pieranna Chiarella, Anna Maria Fresegna, Aureliano Ciervo, Valentina Del Frate, and Cinzia Lucia Ursini

Department of Occupational and Environmental Medicine, Epidemiology and Hygiene, Italian Workers' Compensation Authority-INAIL, Monte Porzio Catone, Rome, Italy

Introduction

In 2022, for their exceptional properties (according to Nanotechnology Products Database (21/03/2022) https://product.statnano.com), NMs are present in 9,586 products from 2,874 companies in sectors as: electronics (19.9%), medicine (12.1%), construction (10.2%), cosmetics (9.5%), textile (8.6%), automotive (7.4%), environment (5.9%), food (3.8%), home appliance (3.6%), petroleum (3.1%), agriculture (2.4%), printing (1.8%), and others. Consequently, humans can be exposed to NMs intentionally or unintentionally.

Unintentional release of NMs in the environment includes a release due to the life cycle of products incorporating NPs, such as paints, cosmetics, and sunscreens (Sun et al. 2016; Wu et al. 2020a) and accidental spills or industrial liquid effluents, such as those emitted by textile industries during the washing of nanotextiles (Yetisen et al. 2016).

Several biological models have been used in ecotoxicity studies from bacteria to fishes; however, to evaluate the impact of NMs on human health, animals and several mammalian cell lines have been used to perform studies on the toxicity for different organs or systems, particularly the respiratory tract because inhalation represents one of the main routes of NMs exposure. Recently, in vitro models based on 3D structures with a single cell type or co-cultures of two or more cell types have been introduced and their application in toxicology could furnish more reliable data than standard two-dimensional models for genotoxicity evaluation in humans.

The genome is constantly exposed to genotoxic insult that can lead to DNA damage with potential consequences for health including carcinogenesis (Huang et al. 2019). Concerns have been raised about NM genotoxicity although the majority of the studies on NM toxicity have focused on cytotoxicity and used high concentrations. A recent review of all the studies published in the past 20 years, using "nanomaterials" and "toxicology" as keywords, found that among 3,500 studies, only 900 focused on genotoxic effects (Huang et al. 2019). Several studies have shown that NMs induce genotoxic effects, such as chromosomal fragmentation, DNA strand breaks, point mutations, oxidative DNA adducts, apoptosis, hypoxic responses, mitochondrial dysfunction, and epigenetic modifications, but the results are still inconsistent, inconclusive, or even contradictory, so it is difficult to make conclusions on the NM characteristics that cause genotoxicity (Huang et al. 2019).

Metal oxide nanoparticles (NPs) are among the most widely used NPs in applications as cosmetics, drug and medicine industry, detergents, agricultural systems, environment, antibacterial

Impact of Engineered Nanomaterials in Genomics and Epigenomics, First Edition. Edited by Saura C. Sahu.

agents, paints, and textiles (Attarilar et al. 2020). In particular, metal oxide NPs became also significant components in applied nanotechnology for trace gas sensors, batteries, magnetic storage media, solar cells, catalysis, energy conversion, architecture, food, antennas (including microstrip and patch-type optically transparent antennas) rectifers, optoelectronics, and electronics (Chavali and Nikolova 2019).

Graphene-based nanomaterials (GBNMs) for their unique electrical, thermal, optical, and mechanical properties are increasingly applied in several fields, such as many areas of industry, biomedical applications, electronics, energy, and sensors, raising concerns about possible human exposure and hazards in environmental and occupational settings. Several in vitro and in vivo studies demonstrated for GBNMs genotoxic, oxidative, epigenetic effects, and inflammatory response, but reliable data on their potential toxic effects on general population are still scarce.

This chapter focuses on the knowledge currently available about the possible genotoxic, oxidative, and epigenetic effects of metal oxides on human health, products containing NMs grown enormously in the past decades, particularly focusing on those more widely used, such as TiO_2, ZnO, Ag/AgO, CuO, Co_3O_4, and SiO_2 and on GBMNs. We describe the results of the most recent studies that evaluated the genotoxic, oxidative, and epigenetic effects in vivo and in vitro of such NMs. In addition, since workers employed in the production or handling of NMs are more exposed than the general population, the available biomonitoring studies on exposed workers are reported to clarify the potential health effects of such exposure.

Physicochemical Properties of NMs and Toxicity

Physical (including shape, type, size) and chemical (surface charge, solubility, dose, and chemistry) properties of NMs affect their capability to enter certain organs across specific pathways, to be accumulated in cell organelles, or to be transported to other organelles with an impact on their toxicity; therefore, the same material with different shapes and sizes can change the response of live tissue and identify whether a NP is toxic or not (Attarilar et al. 2020). The shape of NPs also plays an important role in the metabolism of their components in the body. Spherical NPs are phagocytosed and released at a faster rate than their high-aspect-ratio counterparts (Behzadi et al. 2017).

Properties of NPs like charge, crystal structure, surface area, dose, mass, and susceptibility of particular cell type determine their mechanism of toxicity (Manuja et al. 2021).

The surface charge influences cell take-up of NPs. The net positive and negative charges are associated with increased toxicity, while neutral surfaces should have biosafety (Sun et al. 2016). The charge of particle surface also depends on pH and influences the speed and orientation pathways of their cellular uptake. The binding areas on the surface of NPs increase the possibility of interacting with DNA, RNA, proteins, and lipids and influence the degree of cytotoxicity (Chusuei et al. 2013). Solubilizing media or solvents also influence the particle size, dispersion, and agglomeration of metal NPs, thus affecting the toxicity. Also crystal structure of the same NPs influences the toxic response.

Several studies show that the genotoxicity of NMs not only depends on dose and exposure duration, but also on their size, surface properties, chemical composition as well as on shape. Moreover, since aging processes may change the surface properties as well as chemical composition and size of the NMs, aging could influence the genotoxic potential of NMs, probably reducing it, but the few available studies are inconclusive (Kohl et al. 2020).

Mechanism of NM Genotoxicity

Genotoxicity of NMs could result from two main mechanisms: primary (direct or indirect) or secondary genotoxicity (Figure 5.1).

Primary direct genotoxicity is caused by direct interaction of NMs with the genome, requiring physical contact of NMs with DNA in the nucleus that could lead to DNA damage, such as DNA breaks and other DNA lesions, large DNA malformation, or chromosomal damage (Kohl et al. 2020). The primary indirect mechanism arises from NM-induced reactive oxygen species (ROS), or from toxic ions released from dissolution of NMs and production of intracellular ROS via the Fenton-type reaction (Doak and Dusinska 2017; OECD 2014). In particular, Sharifi et al. (2012) found that CuO, ZnO, NiO, and Cr_2O_3 NPs release metal ions into the culture medium inducing oxidative stress through the formation of ROS and the generation of OOH• and OH• radicals from H_2O_2 under the Fenton reaction. ROS are able to interact with DNA and RNA inducing genotoxicity. For some NMs, both mechanisms can occur simultaneously after NM exposure. Oxidative stress is considered a key mechanism in primary indirect genotoxicity. Free radicals may interact with cellular biomolecules including DNA. ROS may reach the DNA causing purine or pyrimidine oxidation lesions and strand breaks. If the damage is not repaired, it can also result in gene mutations and/or larger chromosomal damage. NMs can promote DNA damage also via other molecules that either have the capacity to interact with DNA or interfere with DNA replication and cell division. NMs can interact, for example, with protein kinases responsible for the regulation of cell cycle events, such as DNA replication and cell division (Magdolenova et al. 2015). Gene expression and epigenetics represent new endpoints in the study of NM genotoxicity. NM exposure has been reported to induce the deregulation of genes involved in several biological processes, including the DNA damage response and DNA repair, cell cycle progression, oxidative stress, and inflammatory

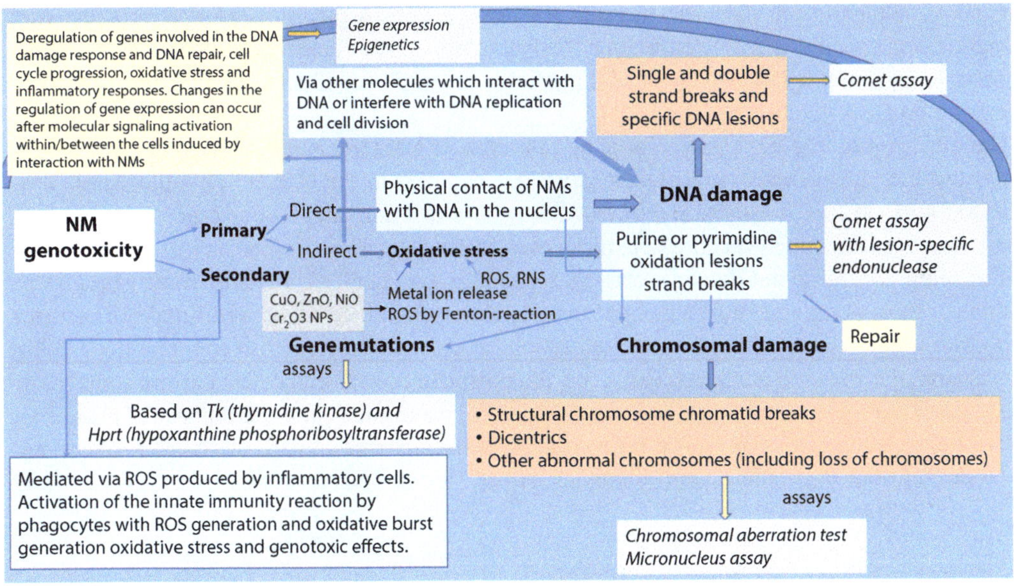

Figure 5.1 Mechanisms of nanomaterial genotoxicity and related biomarkers.

responses (Rossner et al. 2020; Smolkova et al. 2015). Many of these processes have been shown to lead to indirect genotoxicity (Smolkova et al. 2017). Changes in the regulation of gene expression can occur in response to molecular signaling activated within or between the cells after interaction with chemicals and/or NMs. The induction of genotoxicity by direct or indirect mechanism depends on the mechanism of NM uptake and on their capability to enter the nucleus. The smallest NMs (with a size of only a few nanometers) could penetrate the nucleus via nuclear pores. NMs can enter the nucleus also during mitosis when the nuclear membrane is dissolved. Once in the nucleus, NMs can directly interact with DNA, or with chromosomal structures inducing direct genotoxicity.

Secondary genotoxicity is induced by ROS produced by inflammatory cells stimulated by NMs and it is considered one of the main mechanisms of genotoxicity of NMs. Inhaled NMs can activate phagocytes to induce oxidative burst and inflammation that represents the initial defense mechanism against foreign materials, involved in clearance. If the clearance of NMs by macrophages fails, it can cause a chronic immune cell response and inflammation leading to genotoxic effects (Kohl et al. 2020).

Genotoxicity biomarkers include a battery of assays, generally evaluated in prokaryotic systems, cell culture models in vitro and in vivo, able to detect DNA damage, gene mutations, and chromosomal damage used as sensitive genotoxicity endpoints.

Detection of DNA damage (single- and double-strand breaks and specific DNA lesions) belongs to the group of so-called indicative tests that are also used in hazard assessment of NMs (Di Bucchianico et al. 2017; Moreno-Villanueva et al. 2011; Watson et al. 2014). DNA strand breaks can be evaluated using the comet assay, which is the method of choice for measuring damage in cellular DNA (Collins et al. 2017). This assay (also known as single-cell gel electrophoresis) is a simple, robust, reliable, and user-friendly method, widely used for many years for genotoxicity testing of chemicals, and it is the most commonly applied method to test genotoxicity of NMs (Sramkova et al. 2019). The comet assay is able to detect from about hundred up to several thousand breaks per cell, allowing the detection also of weak genotoxic agents. A modification of the comet assay, using a lesion-specific endonuclease after the lysis step, allows the detection of damaged bases. Formamidopyrimidine DNA glycosylase (Fpg) has been employed to measure the effects of oxidative stress on DNA (Collins 2017).

The most commonly used assays for gene mutation of NMs are based on thymidine kinase or hypoxanthine phosphoribosyltransferase genes as described in the OECD Test Guidelines (TGs) (OECD TG 2014; OECD 490 2015; OECD 476 1997), with small adaptations especially regarding NM characterization and exposure (Dusinska et al. 2019).

Regarding chromosomal damage, clastogenicity and aneugenicity are important genotoxicity endpoints that identify agents causing structural chromosome or chromatid breaks, dicentrics, and other abnormal chromosomes, including loss of chromosomes. These could be detected by chromosomal aberration (CA) test. However, the most common test to detect chromosomal damage is the in vitro micronucleus (MN) assay that detects micronuclei in the cytoplasm of interphase cells. The micronucleus assay detects both structural chromosome damage (clastogenic effect) and numerical chromosome alterations (aneugenic effect) (Kirsch-Volders et al. 2011).

Epigenetic Effects of Nanomaterials

Nowadays, epigenetics is considered an essential component affecting human beings, where the modifications induced by the surrounding exposure may be translated into effects on the genome and on health. In brief, these novel epigenetic effects are the following: (i) DNA methylation on

CpG islands by DNA protein methyltransferase; (ii) chemical modifications on histones; (iii) remodeling of chromatin conformation; (iv) nucleosomes positioning; (v) synthesis and regulation of noncoding microRNA.

In general, two important epigenetic events on chromatin may occur: (i) gene hypomethylation on the CpG sequences involving RNA transcription and activation of gene and (ii) gene hypermethylation on the CpG sequences determining gene silencing with consequent suppression of the RNA messenger (Figure 5.2).

CpG islands are known as DNA methylation in promoter regions known to regulate gene expression through transcriptional silencing of the corresponding gene. From a chemical point of view, DNA addition of a methyl group (CH_3) by S-adenosylmethionine on Carbon 5 of the cytosine preceding guanine occurs in the area defined as the "CpG island." To simplify, the DNA methylation and modifications generally end up with the silencing of specific genes as well as with noncoding, repetitive, and non-transposable intronic sequences (Moore et al. 2013). It has been widely recognized that the epigenetic pattern is not fixed and can be still inherited through cell generations (Lacal and Ventura 2018). Pathological epigenetic changes are now considered additional mechanisms of gene alterations and chromosomal aberrations, interfering with gene expression. However, it seems that the epigenetic effects on humans depend not only on the personal individual exposure encountered during life, but also on the response of the subject who is related to the genetic variability inherited from ancestors. These events can impact the body so profoundly as to permanently alter the epigenetic profile (Figure 5.3).

The epigenetic modifications may affect the gene–environmental interactions, highlighting them as potential molecular targets after exposure to novel substances such as NMs. It has also been acknowledged that environmental stress factors, including the environmental and toxic chemical exposure, may result in epigenetic effects on the ecosystem. The major concern of scientists is the transgenerational epigenetic inheritance where the associated changes in the phenotype persist after the noxious substance is removed (Bicho et al. 2020; Vandegehuchte and Janssen 2014).

Some authors have already emphasized that the potential epigenetic properties of NMs could not only affect the exposed individuals but also their offspring (Stoccoro et al. 2013). Nanomaterials can induce epigenetic changes on DNA, RNA, and protein levels. The epigenetics of gene expression patterns can be regulated through the modification of DNA, histones, and small and long noncoding RNAs. Epigenetic events can regulate and modify the gene expression after NMs exposure. DNA methylation controls the mRNA synthesis blocking the binding of transcription factors to the gene promoter. The histone tails modifications make gene promoters accessible to transcription factors, promoting the binding of DNA to the nucleosomes and regulating gene expression activating gene transcription. Also the microRNAs (miRNA) are recruited binding to the complementary sequence of the target mRNA, inducing degradation or block of translation. Many other

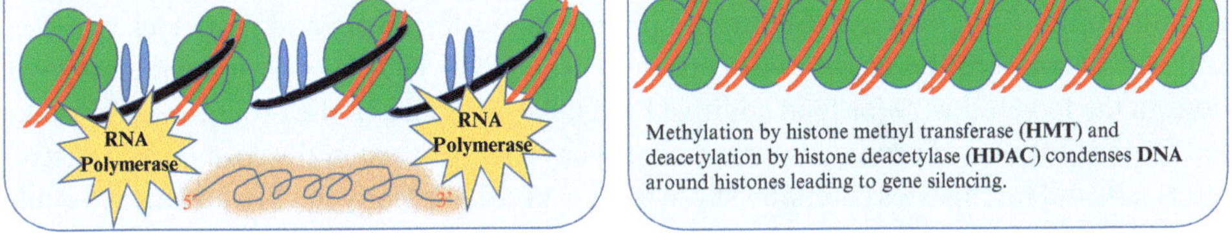

Figure 5.2 Main epigenetic events on chromatin.

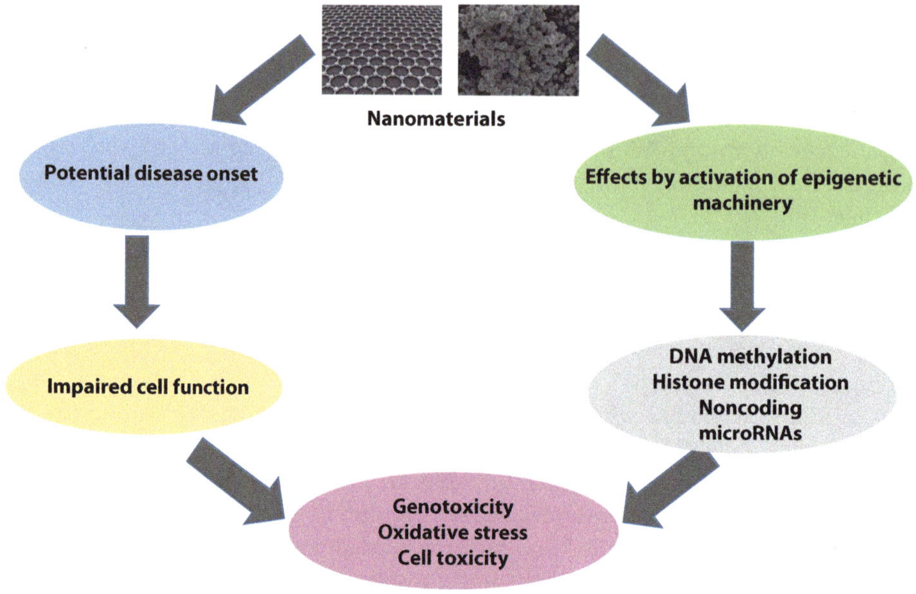

Figure 5.3 Potential NMs mediated effect on humans. *Source:* AlexanderAlUS / Wikimedia commons / CC BY SA-3.0

factors can bind to chromatin-modifying proteins and recruit their catalytic activity to specific sites in the genome. Disregulation of post-translational modifications can result in aberrant gene expression, which causes disease progression. However, such uncertainty in the relation of the physicochemical properties of NM toxicity, the novelty and complexity of the dose–response relationships, resulting in difficult data extrapolation and the measurement of NM exposure levels emerge like critical issues in the establishment of traditional risk assessment. An appropriate risk evaluation should furnish adequate estimates of NM exposure and will be a guideline for appropriate management of strategies to ensure human safety and protection (Gedda et al. 2019).

Studies on Genotoxic and Oxidative Effects of Metal Oxides and Graphene-Based Nanomaterials

Titanium Dioxide NPs

Titanium dioxide (TiO_2) NPs are used in many fields, such as environmental protection and building engineering, medicine, agriculture, and the food and cosmetic industry. With regard to its catalytic properties, TiO_2 NPs are used in self-cleaning materials for building, water and sewage treatment, gas combustion, as an antibacterial material for decontamination, as well as a catalyst in organic synthesis. Their biomedical applications include pharmaceuticals and medical devices. In the agriculture industry, they are used in the production of fertilizers and pesticides that can significantly affect soil fertility, growth of plants, and crop yield. TiO_2 NPs have a wide range of applications in the food industry (as food additive E171); they are used in the processing and packing of food for the purposes of product improvement. They are also used in the cosmetics industry, pharmaceuticals, and toothpastes (Baranowska-Wójcik et al. 2020). Many in vitro and in vivo studies on TiO_2 NP genotoxicity were available (Wani and Shadab 2020) and are reported in Table 5.1.

Table 5.1 In vitro and in vivo studies on genotoxic and oxidative effects of metal oxide nanomaterials.

Nanomaterial	Model (in vitro/in vivo)	Exposure conditions	Methods	Findings	Reference
TiO_2 NPs	A549 cell line BEAS-2B cell line	1,5,10,20,40 μg/ml 2 and 24 h	Fpg-comet assay	Direct and oxidative DNA damage	Fresegna et al. (2021)
TiO_2 NPs	Human–hamster hybrid (A_L) cells	10 μg/ml of fresh (0-day) or aged (60-day) TiO_2 NPs 72 h	Immunofluorescence staining for γ-H2AX detection Intracellular ROS level detection	The smaller TiO_2 NPs induce genotoxicity by increasing the production of mitochondrial ROS	Wang et al. (2019)
TiO_2 NPs	HUVECs	1, 5, and 25 μg/ml 4 and 24 h	Alkaline comet assay; CBMN assay ROS test	Increase in DNA damage and micronucleus frequency. Increased levels of intracellular ROS	Liao et al. (2019b)
TiO_2 NPs	Human sperm cells of 125 men	Co-exposed to 1 μg/L of TiO_2NPs and 10 μM of $CdCl_2$ 15, 30, 45, and 90 min	Comet assay TUNEL test DCF assay	Significant loss of sperm DNA integrity. Increased threshold of sperm DNA fragmentation Oxidative stress	Santonastaso et al. (2019)
TiO_2 NPs	PNT1A cells A375 cell line SH-SY5Y cell line	100 μg/ml 6 h	Fluorescence microscopy comet test	Release of higher amounts of ROS DNA fragmentation	Skubalova et al. (2019)
TiO_2 NPs	Murine RAW 264.7 cells —————— Swiss-Albino mice	10–100 μg/ml ----------------------- 200 and 500 mg/kg 90 days	Comet assay MN assay CA assay	Increased comet assay parameters (tail length, % DNA in tail, tail moment and Olive moment) at the higher doses of TiO_2 NPs Formation of MN and chromosomal breakage	Chakrabarti et al. (2019)

(Continued)

Table 5.1 (Continued)

Nanomaterial	Model (in vitro/in vivo)	Exposure conditions	Methods	Findings	Reference
TiO_2 NPs	C57BL/6 J mice	Intratracheal instillation 1, 3, 28, 90, or 180 days post-exposure	Comet test	DNA strand breaks could be considered as chance findings	Danielsen et al. (2020)
TiO_2 NPs	Swiss-Albino male mice	Gavage 28 days	Chromosomal aberrations assay Comet assay	Rutile was genotoxic and clastogenic	Manivannan et al. (2020)
TiO_2 NPs	Male Swiss-Albino mice (18–25 g)	Oral administration 50,250,500 mg/kg body weight 5 days	Biochemical assays in hepatic tissue Chromosome aberrations analysis from bone marrow cells	Dose-dependent increase in the percentage of chromosomal aberrations Increase in ROS	Ali et al. (2019)
TiO_2 NPs	B6C3F1 male mice	Intraperitoneal injection 50 mg/kg 4 h	hOGG1 and EndoIII comet assay	Standard comet assay highlighted DNA damage only in liver. Enzyme-modified comet assay detected oxidative DNA damage in both liver and lungs	Li et al. (2017)
ZnO NPs	Human nasal mucosa	0.1–20 μg/ml	Comet assay	Genotoxic effects	Hackenberg et al. (2017)
ZnO NPs	MDCK cell line	12–123 μM 24 h	Cytokinesis-block micronucleus assay Alkaline comet assay	Genotoxic effects accompanied by a reduction of GST and CAT activity	Kononenko et al. (2017)
ZnO NPs	CHL/IU cell line	24, 48, and 72 h	MN assay chromosomal aberration test, chemiluminescence	Chromosome aberrations Oxidative stress Genotoxicity	Yanagisawa et al. (2018)
ZnO NPs	BMSCs	5, 10, and 15 μg/ml 24 h	Comet assay Fpg-comet assay Caspase-3 gene expression	Significant DNA damage Oxidative stress Genotoxicity	Moratin et al. (2018)
ZnO NPs	MRC5 fibroblasts	0,1,10,25,50,75 and 100 μg/ml 24, 48, and 72 h	Comet assay DCFDA by FACS	ROS production Oxidative DNA damage	Ng et al. (2017)

ZnO NPs	Human peripheral blood lymphocytes	1,12.5,25,50,100,250,500,1000 and 2000 ppm 72 h	MN and CA assays	Genotoxicity	Akbaba and Türkez (2018)
ZnO NPs	Male *Gallus gallus domesticus* chicks	2 intraperitoneal applications (one per day) 0.245 and 245.26 mg/kg 2 days	Optical microscopy by using immersion lenses	Formation of different types of erythrocyte nuclear abnormalities	de Andrade Vieira et al. (2019)
ZnO NPs	Swiss mice	Intraperitoneal administration 150 mg/kg 10 days	MN assay SOD and CAT activities, reduced GSH and TBARs activities	Cytogenotoxicity Oxidative stress	Fadoju et al. (2019)
ZnO NPs	Swiss mice	Oral administration 300 and 2000 mg/kg 2 days at 24 h apart	CA and MN tests Randomly amplified polymorphic DNA assay	Genotoxicity at 2000 mg/kg	Srivastav et al. (2017)
ZnO NPs	Zebrafish	1,5,10,20,50,100,200 mg/l 30 days	Comet assay MN test	Genotoxicity at 200 mg/l Oxidative stress	Du et al. (2016)
ZnO NPs	Female C57BL/6J BomTac mice 7 weeks of age	Single intratracheal instillation 0.2,0.7,2 μg/mouse 1, 3, and 28 days	BAL, lung and liver analysis Comet assay	BAL fluid: increased DNA strand breaks at low dose at 28 days post-exposure. Lung tissue increased levels of DNA strand breaks at day 28 at medium-dose groups. No increases in DNA strand break in liver tissue	Hadrup et al. (2019)
ZnO NPs	Hens	25,50,100 mg/kg 24 weeks	Effects of ZnO NPs on plasma metabolome	ZnO NPs altered liver metabolism enzymes	Zhang et al. (2018)
ZnO NPs	Murine Ehrlich ascites carcinoma (ESC) bearing mice	Oral 50,300,500 mg/kg 7 days	Histopathological examination of liver, kidney, muscle and ESC Oxidative stress biomarkers; comet assay; P53 and caspase-3 immunohistochemical examinations	Antitumor activity at 500 mg/kg	El-Shorbagy et al. (2019)

(*Continued*)

Table 5.1 (Continued)

Nanomaterial	Model (in vitro/in vivo)	Exposure conditions	Methods	Findings	Reference
ZnO NPs	Female Swiss-Albino rats	Oral 100,200,300 mg/kg 28 days	Histopathological analysis of rat's liver. Expression of Bax and P53 genes	Anticancer at 300 mg/kg	Rahimi Kalateh Shah Mohammad et al. (2019)
Ag NPs	BEAS-2B cell line RAW 264.7 cell line	5–25 μg/ml 4 and 24 h	Comet and MN assays 8-oxo-dG and 8-oxo-dA measurement by LC–MS/MS	BEAS-2B more sensitive to the genotoxic insult respect to the RAW 264.7. No significant increase in chromosomal and DNA damage was induced by AgNPs in RAW 264.7	Andreoli et al. (2021)
Ag PTs (green synthesized) and Ag NPs	CHO-K1 HEK293 B16F10 HeLa MCF-7 Hep G2 cell lines —————— Male Swiss-Albino Male C57BL/6J mice lines	1,2,4 μg/ml 24 h ----------------------- Intraperitoneal injection 5 mg/kg/bw Alternate day over 10 days (5 doses)	Histopathological examination TUNEL assay Flow cytometry (DCFDA) Immunohistochemistry study CAs assay Mitotic Index test MN assay	Ag PTs cause less in vitro effects (ROS, cell cycle alterations, apoptosis) than those by Ag NPs Genotoxicity: Ag PTs are less toxic than Ag NPs Tumor regression: A regression of the melanoma tumor through a p53-mediated apoptotic pathway in mice treated with Ag PTs	Veeragoni et al. (2022)
Ag NPs	THP-1 macrophages	5,10,25 μg/ml 24 h	CM-H2DCFH-DA assay	Small decrease in ROS at the highest concentration	Pavlin et al. (2022)
Ag NPs	RAW 264.7 cell line	0.36 and 3.6 μg/ml 4 and 24 h	ROS levels analysis Confocal microscopy	Initial increase of ROS levels and DNA repair foci (4 h-exposure) followed by cell death by apoptosis and loss of cell viability (24 h-exposure)	Glinski et al. (2021)

Ag NPs	C57BL/6 mESCs	5.0 μg/ml 24 h	Flow cytometry analysis	Dose-dependent increase in oxidative stress and apoptosis	Gao et al. (2017)
Ag NPs	N9 cell line N27 cell line	50 μg/mL 24 h	ELISA Griess reaction DCFH-DA assay	Significant anti-inflammatory effects, reducing lipopolysaccharide-stimulated ROS, nitric oxide, and TNFα production	Gonzalez-Carter et al. (2017)
Ag NPs	Zebrafish ovarian follicle cells	30 μg/ml 2 h	Fluorescent microscopy with Hoechst 33342 Gene expression analysis	Concentration-dependent increase in the apoptosis of ovarian follicle cells and a germinal vesicle breakdown	Chen et al. (2017)
Ag NPs	Balb/C mice	Intratracheal instillation 5 mg/kg 1 and 7 days	Tissue homogenates analysis ELISA	Biodistribution of AgNPs in lung, brain, and liver Oxidative stress markers; significant increase of GSH at 1 day after exposure in tissues. No significant changes observed at 7 days after exposure. DNA damage: the concentration of 8-OH-dG significantly increased at 1 day after instillation. No effect on 8-OH-dG observed 7 days after instillation	Ferdous et al. (2021)
Ag NPs	Female Sprague-Dawley rats	Intravenous administration 5 mg/kg 24 h	Histopathological examination Bone marrow chromosome aberration test	Accumulation of Ag NPs in the immune system Significantly increased chromosome breakage and polyploidy cell rates	Wen et al. (2017)
Ag NPs	Adult male Wistar rats	Oral administration 5 and 50 mg/kg 79 days	qRT-PCR. Assay of DNA fragmentation. ELISA. Markers of oxidative stress and antioxidant parameters. Histopathological examination	Cell death, disturbed neurotransmitter and cytokine production, and ROS generation in brain	Yousef et al. (2019)
Nano-sized and fine CuO particles	BEAS-2B cell line	0.95–190 μg/mL CuO NPs and 1.9–475 μg/mL CuO fine particles 3, 6, and 24 h	Comet assay MN assay	Increase in DNA damage Increase in micronucleated cells Stronger genotoxicity of nanosized than fine CuO particles	Siivola et al. (2020)

(*Continued*)

Table 5.1 (Continued)

Nanomaterial	Model (in vitro/in vivo)	Exposure conditions	Methods	Findings	Reference
Cu/CuO NPs	WI-38 cell line A549 cell line	201.26 μg/ml for WI-38 209.70 μg/ml for A549 cells 72 h	Comet assay	Oxidative stress DNA damage	Fahmy et al. (2020)
CuO NPs and CuO MPs	BEAS-2B cell line	1–50 μg/ml 24 h	Gene expression analyses via high-throughput RT-qPCR Flow cytometry analysis	Oxidative stress Transcriptional changes of genes related to copper overload DNA damage	Strauch et al. (2017)
CuO NPs	A549 cell line	5–15 μg/ml	Comet assay Flow cytometry	Increase in micronuclei DNA damage Oxidative stress	Akhtar et al. (2016)
CuO NPs	HBEC cell line A549 cell line	CuO NPs aerosol at 5 ml/min flow rate regulated by six mass flow controllers 2 or 4 h	Carboxy-H2DCFDA assay	Oxidative stress	Jing et al. (2015)
CuO NPs	Male Sprague Dawley rats (100–120 g)	1/40, 1/20, and 1/10 times of LD_{50} (2075 mg/kg) 28 days by gavage	Oxidative stress and nitrosative stress analysis in renal tissue homogenate RT-PCR and Western blot analysis for CYP450 isoforms The activities of CYP 1 A1, 2C11, 2D6, 2E1, and 3 A1 in vitro. Protein expression of the nuclear receptors. Signaling pathway analysis	Increase in oxidative injury Increase in NF-kB, p38, and ERK1/2 inflammatory markers in the kidney Change in the activity of cytochrome P450 enzymes	Xu et al. (2018)
CuO NPs	Male Sprague-Dawley rats mice	Oral gavage 100,200 and 400 mg/kg/day 7 days	Liver tissue analysis Measurements of oxidative stress and signaling pathway	Oxidative injury via upregulation of inflammatory mediators iNOS and MAPKs	Tang et al. (2018)

CuO NPs	C57BL/6 mice A549 cell line BEAS-2B cell line	Instillation into the lungs following dispersion in cell culture medium 1, 2.5, 5, and 10 mg/kg body weight	Histological analysis and immunohistochemistry TUNEL staining; determination of hydroxyproline content; BALF collection and quantification of TGF-β1; realtime-PCR; MTT assay; annexin V and propidium iodine (PI) staining; measurement of intracellular ROS generation; Western blot analysis	Significant increase in inflammatory markers Significant reduction of SOD activity Oxidative stress and apoptosis	Lai et al. (2018)
CuO NPs	Male Wister albino rats of 8 weeks old	Via oral gavage 5 and 50 mg/kg/day 14 days	Antioxidant enzyme activities analysis	Oxidative stress	Anreddy et al. (2018)
Co_3O_4 NPs	BEAS-2B cell line A549 cell line	1,5,10, 20,40 μg/ml 2 and 24 h	Fpg-comet assay	Direct DNA damage and significant oxidative-inflammatory effects	Cavallo et al. (2015)
Co_3O_4 NPs	BEAS-2B cell line A549 cell line	0.1–100 μg/ml 24 h	Comet assay Annexin V-FITC-PI apoptosis assay ELISA	Genotoxic and apoptotic effects on A549 cells Oxidative effects in all cell types	Abudayyak et al. (2017)
Co, CoO, and Co_3O_4 NPs	A549 cell line	20 and 40 microg/ml Co NPs 40 and 60 microg/ml CoO and Co_3O_4 NPs 3 and 24 h	Comet assay DCFH-DA assay	Induction of DNA strand breaks and oxidative damage by Co and CoO NPs with higher reactivity for Co. ROS generating capacity of the different NPs	Cappellini et al. (2018)

(Continued)

Table 5.1 (Continued)

Nanomaterial	Model (in vitro/in vivo)	Exposure conditions	Methods	Findings	Reference
Co_3O_4 NPs	V79 cell line	20–100 μg/ml 24 h	PI dye exclusion assay; LDH release assay; MTT assay; mitochondrial membrane potential ($\Delta\psi m$); ROS generation (flow cytometric method)	ROS-mediated genotoxic potential	Lugun et al. (2022)
Co_3O_4 NPs	Rats	Lung instillation	Total protein, LDH and cytokine measurement	Immune-inflammatory response resulting in lung interstitial inflammation	Cho et al. (2012)
CoO NPs	Mice	Subcutaneous injection 100, 200, 500, and 1000 μg/kg body weight	Serum LDH, SGOT, and creatinine. Cell viability; TNF-a and IL-10 by ELISA; apoptotic proteins caspase 8, caspase 3, p38, p53 and anti-apoptotic BCL-2 XL, pAKT by ELISA	ROS induction Increased pro-inflammatory response	Chattopadhyay et al. (2015)
Co_3O_4 NPs	56 patients with metal-on-metal hip resurfacing containing Co_3O_4 NPs	Followed prospectively over a ten-year period	Cobalt and chromium levels determined by atomic absorption spectrometry with electrothermal atomisation (ETA-AAS)	Release cobalt and chromium ions in the serum of patients Potential risk of carcinogenicity	deSouza et al. (2010)
Amorphous silica NPs	BEAS-2B cell line	5 μg/ml 40 passages (about 18 weeks)	Genome-wide transcriptional analysis Microarray and bioinformatics analysis	DNA breakage Chromosomal damage Multinucleation	Guo et al. (2017)
SiNPs	HUVECs	1, 5, 25 μg/ml	Comet assay MN test ROS test	DNA damage Increase intercellular ROS	Zhou et al. (2019)

SiO_2 NPs	CHL cell line	4 and 24 h	Ames test CA test	Genotoxicity	Lee et al. (2020)
	Rats	Via oral gavage 4 days	Comet assay MN test		
Amorphous silica NM200 and NM203	BEAS-2B cell line	0.1,1,10,100 μg/ml 24 and 48 h	MN assay Comet assay	Significant increase of DNA damage for NM203	Zijno et al. (2020)
Several types of differently sized SiO_2 and TiO_2	Human intestinal cells	physiologically relevant doses 48 h	Comet assay MN flow cytometry assay	Not cause genotoxicity	Franz et al. (2020)
SiO_2 NPs	BALB/c mice	Oral administration 0,25,50,100,200 mg/kg 84 days	Whole-genome analysis	Expression of genes associated with ROS production and ER stress	Hu et al. (2019)

In Vitro Studies

Several studies suggest genotoxic effects of TiO_2 NPs via induction of ROS production leading to DNA damage, but they appear to be independent of the size and form of the said NPs. Moreover, due to their redox properties, TiO_2 NPs can alter the DNA repair proteins inactivating them (Fenoglio et al. 2009).

TiO_2 occurs in three crystalline forms: the rutile crystal has a prismatic shape, the anatase octahedral, and the brookite an orthorhombic shape. Different crystal structures of TiO_2 induce different toxicity responses as indicated by Fresegna et al. (2021) in human bronchial and alveolar cells. The study showed that anatase TiO_2 NPs induced slight oxidative effects only in alveolar cell line and rutile TiO_2 NPs induced genotoxic/oxidative effects in both cell lines. Wang et al. (2019) found that three types of TiO_2 NPs in different sizes (anatase 5 nm, anatase 15 nm, and a mixture of rutile and anatase <100 nm) prepared fresh or aged (day 60) induced toxicity in an aging time-independent and size-dependent manner in human–hamster hybrid cells (A_L). The smaller TiO_2 NPs induced more genotoxicity by increasing the production of mitochondrial ROS. Different sizes of anatase (10–100 nm) induce an increase in DNA damage and micronucleus frequency by comet assay and cytokinesis-block micronucleus assay respectively, in human umbilical vein endothelial cells (HUVECs) (Liao et al. 2019b). Testing low concentrations (1 and 10 μg/L) of a mixture of anatase and rutile in human sperm cells, Santonastaso et al. (2019) found a statistically significant loss of sperm DNA integrity after 30 min of exposure, measured by the comet assay and an increased threshold of sperm DNA fragmentation by TUNEL test. Using the comet test, Skubalova et al. (2019) found that a mixture of 63% anatase and 37% brookite TiO_2 NPs causes a release of ROS, resulting in DNA fragmentation, in three different human cell lines (PNT1A normal prostatic epithelial immortalized cells, A375 human melanoma cell line, and SH-SY5Y human neuroblastoma cell line). Chakrabarti et al. (2019) found increased DNA damage using the comet assay in murine RAW 264.7 cells.

In Vivo Studies

Several in vivo studies show that TiO_2 NPs accumulate in the lungs, alimentary tract, liver, heart, spleen, kidneys, and cardiac muscle (Baranowska-Wojcik et al. 2020). Li et al. (2017) used standard and enzyme-modified (hOGG1 and EndoIII) comet assay in B6C3F1 male mice exposed to anatase NPs via intraperitoneal injection. Standard comet assay highlighted DNA damage only in liver, despite a measured accumulation in liver and lung; enzyme-modified comet assay detected oxidative DNA damage (adducts) in both liver and lungs. Chakrabarti et al. (2019) have also conducted experiments in vivo, exposing Swiss-Albino mice for 90 days to TiO_2 NPs. They found an increased formation of DNA breaks at higher doses, including micronuclei, and chromosomal breakage. Ali et al. (2019) found a dose-dependent increase of chromosomal aberrations percentage (inversely proportional to the size of the particles) in mice exposed by oral administration to increasing doses of TiO_2 NPs (sized 21 and 80 nm). Danielsen et al. (2020) exposed C57BL/6 J mice by intratracheal instillation to four anatase TiO_2 NPs, differing in shape and size. They found only a few statistically significant and not-dose-dependent increases of DNA strand breaks, assessed by the comet test. Manivannan et al. (2020) studied the genotoxicity of rutile NPs (25 nm) in Swiss-Albino male mice exposed to 0.2, 0.4, and 0.8 mg/kg body weight, by gavage over a period of 28 days. The chromosomal aberrations assay showed genotoxic, clastogenic, and cytotoxic effects and the comet assay showed DNA damage in liver, spleen, and thymus cells.

Zinc Oxide NPs

Zinc oxide (ZnO) NPs represent the most commonly used NPs among different NMs applied in personal care products, sensors, antibacterial creams, and biomedical applications. Once ZnO NPs enter the cell, they tend to release their metallic ions while they get dissolved in the cytoplasm.

Several studies revealed that the interaction of these metal ions is the main cause of genotoxicity through inactivation of the metal-protein involved in the DNA repair system and in other basic cellular processes (Shukla et al. 2021). The results of several in vitro and in vivo studies are reported in Table 5.1.

In Vitro Studies

Several in vitro studies demonstrated the capability of ZnO NPs to induce genotoxicity in different cell types, such as nasal mucosa cells, V79 cells, rat liver, kidney and lung tissues, human skin fibroblast cells and keratinocytes, rat neuronal cells, and mouse testicular cells (Mortezaee et al. 2019). Among the most recent studies published after 2015, Hackenberg et al. (2017) used the comet assay on human nasal mucosa cells exposed to 0.1–20 μg/ml of TiO_2 and ZnO NPs, alone and in combination. They found genotoxic effects only for ZnO NPs at the tested concentrations. Kononenko et al. (2017), using the cytokinesis-block micronucleus assay and alkaline comet assay, found that ZnO NPs induce genotoxic effects, accompanied by a reduction of GST and CAT activity on MDCK cells. This study demonstrates that the size of ZnO particles has an important role in inducing genotoxicity and that dissolved Zn ions from ZnO NPs in cell culture medium is not the only cause of cellular effect. Yanagisawa et al. (2018) found dose-dependent significant increase of micronuclei frequency and chromosome aberrations, particularly structural aberrations, in Chinese hamster lung (CHL/IU) cells exposed to ZnO NPs for 24, 48, and 72 hours. The study suggests that ZnO NPs evoke genotoxicity through superoxide radical-induced oxidative stress derived from mitochondrial damage in CHL/IU cells. S9 mixture appears to contribute to a further increase in genotoxicity through the generation of superoxide radicals by the metabolic activation of ZnO NPs. Moratin et al. (2018) found that ZnO induced significant DNA damage on human mesenchymal bone marrow stem cells already at the lowest used concentration (5 μg/ml) evaluated by the comet assay, and the fpg-enzyme allowed to determine the influence of oxidative stress on the mechanisms of nanoparticle-induced genotoxicity. Ng et al. (2017) demonstrated using the comet assay that ZnO NP exposure causes ROS production, leading to an accumulation of 8-OHdG followed by oxidative DNA damage in human lung fibroblasts. Akbaba and Türkez (2018) evaluated using the micronucleus and chromosome aberration techniques on human peripheral blood lymphocytes, the genotoxicity of ZnO, aluminum oxide (Al_2O_3), and b-tricalcium phosphate (b-TCP) ($Ca_3(PO_4)_2$) NPs and found genotoxic effects only for ZnO.

In Vivo Studies

Several in vivo studies showed immunotoxic, genotoxic, and oxidative effects of exposure to ZnO NPs (Keerthana and Kumar 2020).

In particular, de Andrade Vieira et al. (2019) reported different types of erythrocyte nuclear abnormalities in male *Gallus gallus domesticus* chicks exposed to ZnO NPs. Fadoju et al. (2019) and Srivastav et al. (2017) have found genotoxicity of ZnO NPs in Swiss mice, and Du et al. (2016) reported oxidative stress in zebrafish. Pulmonary acute phase response and inflammation was demonstrated by Hadrup et al. (2019) in mice exposed to low doses of ZnO NPs administered by intratracheal instillation. In hens exposed to increasing concentrations of ZnO NPs for 24 weeks, in liver, Zhang et al. (2018) found changes in plasma metabolites involved in antioxidative mechanisms, energy metabolism, and lipid metabolism. It is important to note that several studies have shown antitumoral effects of ZnO NPs under in vivo conditions. El-Shorbagy et al. (2019), on Ehrlich solid carcinoma (ESC) bearing mice, found antitumor activity at the highest concentration (500 mg/kg). Similarly, Rahimi Kalateh Shah Mohammad et al. (2019) found an anticancer effect at 300 mg/kg after 28 days of exposure on female Swiss-Albino rats. In conclusion, given the conflicting results of in vivo and in vitro effects, the safety of these particles should be further investigated (Keerthana and Kumar 2020).

Silver and Silver Oxide NPs

Silver (Ag) NPs have gained much attention in recent years owing to their antimicrobial and anticancer properties, in addition to their use in wound repair and bone healing, as vaccine adjuvant, antidiabetic agent, and in biosensors (Xu et al. 2020). Like most NMs, the potential toxicity of Ag NPs depends on the routes of administration and the properties or characteristics of the NPs, such as the size, shape, and concentration. In particular, the mechanisms of action of Ag NPs mainly involve the release of silver ions (Ag^+), the generation of ROS, the induction of oxidative stress resulting in DNA damage, and the destruction of membrane structure (Shukla et al. 2021; Xu et al. 2020). Regarding the release of Ag ions, Wang et al. (2015) found that the cytotoxicity of AgNPs is largely due to the chemical transformation of particulate silver from elemental silver Ag^0 to Ag^+ ions and Ag-O-, then Ag-S- species. The results of some in vitro and in vivo studies are reported in Table 5.1.

In Vitro Studies

The potential of Ag NPs to induce genes associated with cell cycle progression, chromosomal damage, cell cycle arrest, and cell death was reported in human BEAS-2B cells, umbilical vein endothelial cells, and hepatocellular liver carcinoma cells at various concentrations (Liao et al. 2019a). Gonzalez-Carter et al. (2017) reported a dose-dependent increase in nitric oxide (NO) and TNF-α values in mouse microglia N9 cell line and N27 neuronal cells exposed to 50 μg/mL of AgNPs. In mouse embryonic stem cells, Gao et al. (2017) found a dose-dependent increase in oxidative stress and apoptosis after 24 hours of exposure to AgNPs. Some genotoxic effect of 20-nm AgNPs (evaluated by comet and micronucleus assays) on BEAS-2B was found by Andreoli et al. (2021). Glinski et al. (2021), exposing RAW 264.7 to AgNPs, found an initial increase of ROS levels and DNA repair foci (4-hour exposure) followed by cell death by apoptosis, and loss of cell viability (24-hour exposure). On exposing THP-1 macrophages to different types of NPs including silver, Pavlin et al. (2022) saw a small decrease in ROS at the highest concentration.

Many studies concern the effects of Ag NPs obtained by biosynthesis from plant extracts. Veeragoni et al. (2022) found that biosynthesized (green) Ag NPs (employing the crude extract of *Padina tetrastromatica* (Ag-PTs)) cause less in vitro effects (ROS, cell cycle alterations, apoptosis) than those by chemically synthesized (Ag-NPs) . This could be due to the presence of phytochemicals (aldehydes and ketones) that retain the Ag ions that are mainly responsible for the genotoxic action.

In Vivo Studies

Several studies have been performed to evaluate the in vivo toxicity of Ag NPs. Some studies showed no induction of adverse effects while other studies reported adverse effects varying from a minimal inflammatory response to the presence of inflammatory lesions in the lungs (Ferdous and Nemmar 2020). In a multi-organ study, Wen et al. (2017) found an accumulation of AgNPs in the immune system of Sprague-Dawley rats, following an acute intravenous dose of Ag NPs. They also observed significantly increased chromosome breakage and polyploidy cell rates. Yousef et al. (2019) found cell death, disturbed neurotransmitter and cytokine production, and ROS generation in rat brain after oral administration of Ag NPs (5 and 50 mg/kg). More recently, in a study by Ferdous et al. (2021), markers of inflammation (TNF-α and IL-6), DNA damage, and apoptosis were significantly increased in lung, brain, and liver of mice exposed by tracheal instillation to AgNPs. They also found a significant increase of reduced glutathione, nitric oxide, and 8-isoprostane in most of the organs investigated. Finally, as well as in vitro, many in vivo studies concern the antitumor activity of green synthesized silver NPs.

In the aforementioned study, Veeragoni et al. (2022) also evaluated the genotoxic effect and tumor regression induced in mice using chemically and green synthesized AgNPs. The results show that concerning genotoxicity, green synthesized particles are less toxic than their chemical counterparts. Regarding tumor regression, they observed a regression of the melanoma tumor through a p53-mediated apoptotic pathway in mice treated with green synthesized particles.

Copper and Copper Oxide NPs

Copper and copper oxide (Cu and CuO) NPs can be used for antimicrobial bioactive coatings, as catalysts, semiconductors, chemical sensors, supercapacitors, near-infrared filters, magnetic storage media, and energy conversion devices, potentially increasing the possibility of occupational exposure to fumes and aerosols containing CuO NPs (Siivola et al. 2020). Several studies demonstrated high cytotoxicity of CuO NPs; however, its potential genotoxicity received less attention. CuO NPs exposure of the living systems results in ROS generation, oxidative stress, inflammation, cytotoxicity, genotoxicity, and immunotoxicity (Naz et al. 2020). The results of recent in vitro and in vivo studies are reported in Table 5.1.

In Vitro Studies

Lung epithelial cells treated with CuO NPs (9.2 nm size) at various concentrations showed a reduction in cell cytotoxicity and an increased level of oxidative stress (dose-dependent) in a study performed by Jing et al. (2015). Akhtar et al. (2016) demonstrated that CuO NPs induced a dose-dependent increase in micronuclei and DNA damage evaluated using the comet assay in A549 cells. Morever, the CuO NPs induced dose-dependent oxidative stress by the induction of ROS and lipid peroxidation along with glutathione depletion. A comparison of cytotoxicity, cellular uptake, and intracellular copper accumulation induced by the exposure to CuO NPs with those induced by CuO MPs (microparticles) and water soluble $CuCl_2$ were investigated by Strauch et al. (2017) in human bronchial epithelial cell line BEAS-2B. They also compared the results found on BEAS-2B with those obtained in adenocarcinoma cell line A549 previously performed. According to these authors, the high cytotoxicity and effects on gene expression by CuO NPs can be induced by the intracellular copper ion release, with subsequent copper accumulation in the cytoplasm and the nucleus. GSH depletion and activation of Nrf2 in HeLa S3 cells confirmed oxidative stress induction, mainly restricted to CuO NPs. Modulation of gene expression by CuO NPs appeared to be primarily oxidative stress-related and was more pronounced in redox-sensitive BEAS-2B cells. Siivola et al. (2020) using the comet and micronucleus assays evaluated genotoxicity and cellular uptake of nano-sized and fine CuO particles in human bronchial epithelial (BEAS-2B) cells . They found that fine- and nano-sized CuO particles induced a dose-dependent increase in DNA damage, with nano-sized CuO damaged DNA at lower doses and at higher levels than fine CuO. CuO NPs also induced a dose-dependent increase in micronucleated cells at cytotoxic doses. BEAS-2B cells internalized more nano-sized CuO particles in respect to fine CuO and this different behavior could explain the stronger genotoxic effect and more intracellular dissolution of NPs compared with the fine particles. Fahmy et al. (2020) found in normal WI-38 and cancer A549 lung cells that Cu/CuO NPs with size <20 nm induced cytotoxicity and genotoxicity to the IC50 value, corresponding to 201.26 μg/ml for WI-38 and to 209.70 μg/ml for A549 cells. Moreover, this study demonstrated that Cu/CuO NPs induced oxidative stress damage in addition to DNA damage in both cell types.

In Vivo Studies

Xu et al. (2018) showed that the administration of Cu NPs to rats lead to a dose-dependent increase in oxidative injury and an increase in NF-kB, p38, and ERK1/2 inflammatory markers in the kidney. This was also associated with a change in the activity of cytochrome P450 enzymes. Cu NPs can also induce oxidative injury via the upregulation of inflammatory mediators, iNOS and MAPKs as shown by Tang et al. (2018) in rat livers. In an in vivo study, Lai et al. (2018) evaluated the late effect of lung exposure to CuO NPs in C57BL/6 mice. CuO NPs were instilled into the lungs following dispersion in cell culture medium. The results showed a significant increase in inflammatory markers and a significant reduction of superoxide dismutase (SOD) activity, which leads to oxidative stress and apoptosis (Lai et al. 2018). Anreddy (2018) demonstrated that oral exposure of CuO NPs in rats causes significant toxicity to the liver and it might be due to oxidative stress.

Cobalt Oxide Nanoparticles

Cobalt oxide (mainly Co_3O_4) NPs are increasingly used in biomedical applications, electronics, and energy storage. In particular, Co_3O_4 NPs are widely used for their magnetic properties, in biomedicine, for drug delivery, hyperthermic cancer treatment, and as new contrast agents for medical imaging, such as magnetic resonance imaging. The high reactivity and characteristics that make cobalt-based NPs so attractive also raise concerns about the potential health effects. The results of several in vitro and in vivo studies are reported in Table 5.1.

In Vitro Studies

Several in vitro studies report for cobalt oxide NPs DNA damage, oxidative stress, inflammatory response (Alarifi et al. 2013; Alinovi et al. 2015; Chattopadhyay et al. 2015), and downregulation of the transcription of genes involved in immune response (Verstraelen et al. 2014). In the study by Cavallo et al. (2015), Co_3O_4 NPs induced moderate cytotoxicity, direct DNA damage only at the highest concentration, and significant oxidative-inflammatory effects at low concentrations on normal bronchial (BEAS-2B) while higher genotoxic-oxidative potential was shown by tumoral lung (A549) cells. Also a study by Abudayyak, Gurkaynak, and Özhan (2017) on different human cell lines showed genotoxic and apoptotic effects only for A549 cells, while oxidative effects were found in all cell types. Cappellini et al. (2018) in a study using three different Co-based NPs (Co, CoO, and Co_3O_4) on different cell lines showed cellular uptake of all NPs in lung cells and induction of DNA strand breaks and oxidative damage (with the comet assay) by Co and CoO NPs with higher reactivity for Co. ROS-mediated genotoxic potential of Co_3O_4 NPs on CHL fibroblast cell line (V79) was recently demonstrated, also at noncytotoxic concentrations, by Lugun et al. (2022).

In Vivo Studies

Immune-inflammatory response resulting in lung interstitial inflammation was demonstrated in vivo in rats exposed to Co_3O_4 NPs by lung instillation (Cho et al. 2012). ROS induction and increased pro-inflammatory response were found by Chattopadhyay et al. (2015) in mice exposed to CoO NPs. Moreover, in orthopedics, the introduction of metal-on-metal total disc replacement of degenerated vertebrae containing CoCr NPs, which seems to release cobalt and chromium ions in the serum of patients (deSouza et al. 2010), raises concerns about the potential risk of carcinogenicity.

Silicon Dioxide NPs

SiO_2 NPs are currently used in food, cosmetic industry, medicine, cancer therapy, enzyme immobilization, and in dentistry as an abrasive agent. They are among the most widely commercialized manufactured NMs. Silicon dioxide exists as crystalline, amorphous, and fused. Amorphous silica has been considered less toxic than crystalline silica, but its toxicity is still disputed. Various in vitro and in vivo studies on both forms of silica are available for their genotoxicity and oxidative effects and are reported in Table 5.1.

In Vitro Studies

Amorphous silica NPs have been reported to induce cell apoptosis, autophagy, DNA breakage, chromosomal damage, multinucleation, and activation of key molecular events involved in carcinogenesis, such as malignant transformation via p53 signaling (Guo et al. 2017). A study that exposed HUVECs to four sizes of Si NPs observed the capability of these NPs to induce DNA damage evaluated by the comet assay and micronucleus test, negatively correlated with the size of the NPs, and showed an increase in the intercellular ROS (Zhou et al. 2019). Some studies reported by Yazdimamaghani et al. (2019) found similar genotoxic effects mediated by ROS induction for amorphous silica nanoparticles (SNPs).

Lee et al. (2020) studied the genotoxic potential of SiO_2NPs (18 nm) using four standard genotoxicity tests: the Ames test, in vitro CA test in CHL fibroblasts, in vivo micronucleus test in rats (via oral gavage) and additional in vitro comet assay in CHL fibroblasts (for 4- and 24-hour exposure). They found positive response only by comet assay, which is the most sensitive test for the evaluation of genotoxicity of NPs. Also Zijno et al. (2020) evaluated the in vitro genotoxic potential of two forms of amorphous silica (NM200 and NM203) produced by different synthesis methods (precipitated and pyrogenic, respectively), with different morphology (more complex and branched structure for NM203 than the NM200). In particular, they exposed human bronchial BEAS-2B cells to these NMs and using the micronucleus and comet assays, found a statistically significant increase of DNA damage for NM203 at 100 μg/ml only by using the fpg-comet assay. Otherwise Franz et al. (2020) demonstrate using the comet assay and MN flow cytometry that physiologically relevant doses of several types of SiO_2 and TiO_2 NPs from different sources did not cause genotoxicity in human intestinal cells.

In Vivo Studies

Adverse effects were mainly observed in acutely exposed animals, while no significant effects were found in chronically exposed ones with route of administration and physicochemical properties of silica NPs influencing the toxicokinetics (Murugadoss et al. 2017). Despite the many available studies on acute toxicity, there are inadequate investigations on long-term genotoxicity and gene expression profile by amorphous silica NPs (Yazdimamaghani et al. 2019). Brandão et al. (2021) investigated the in vivo pulmonary toxicity of four variants of amorphous SiO_2 after intratracheal instillation in rats and short-term inhalation study for SiO_2.7, using TiO_2 NM105 as a benchmark NM. In the instillation study, exposure to SiO_2.7 or SiO_2.40 markedly increased oxidative DNA lesions in rat lung cells of the exposure group at every tested dose. They observed a significant increase in DNA strand breaks of the lung cells exposed to 0.5 mg/m^3 of SiO_2-7 or 50 mg/m^3 of TiO_2NM105 in both groups. The detected gene expression changes suggest that oxidative stress and/or inflammation pathways are likely implicated in the induction of (oxidative) DNA damage. An overview on the toxicity of synthetic amorphous silica (SAS) and on the assessment of the associated risk of the oral exposure has been published by Brand et al. (2021), focusing mainly on the hepatotoxic effects as liver seems to be a critical target organ. ROS generation in serum and liver was studied weekly by Hu et al. (2019) within the 90-day study period in which mice were exposed to 4 different SAS (including 2 types of NPs). Pyrogenic SAS NPs affected serum and liver biomarkers indicative of liver ROS after seven weeks of exposure. This was combined with the altered expression of genes associated with ROS production and endoplasmic reticulum (ER) stress in liver (Hu et al. 2019). As SAS is present in many everyday products, humans will be exposed to SAS on a daily basis over a prolonged period and, taking into account all the available studies, effects on liver cannot be excluded, so further studies on the potential effects resulting from chronic, oral exposure to food-grade SAS are warranted (Brand et al. 2021).

Graphene-Based NMs

Graphene-based nanomaterials (GBNMs) include graphene (G), graphene oxide (GO) with its reduced form (rGO), and graphene quantum dots (GQDs). GBNMs have a two-dimensional structure constituted by a single or a few layers of hexagonal rings of carbon atoms in a lattice honeycomb arrangement (Ristic et al. 2021; Zhang et al. 2017). GBNMs are promising materials, with interesting physical properties: excellent electrical and thermal conductivity, large specific surface area, low toxicity, and good biocompatibility (Zhang et al. 2017). Their production and use are increasing day by day in many applications, such as energy storage, environmental protection, and industrial manufacturing (Wu et al. 2021a). In addition, they are also widely used in nano-biotechnology and in medicine, for example, in sensitive biosensors, drug delivery systems, near-infrared light imaging techniques, and cancer photothermal therapy (Banerjee 2018; de Melo-Diogo et al. 2018; Zhang et al. 2017). Given the speed of design/production and the huge amount of new graphene-based products (Guo et al. 2022; Kong et al. 2019; Xiao et al. 2016), the challenge for the scientific community is to be involved in the exploration of the potential adverse effects on human health and environment, even before these substances are placed on the market.

In Vitro Studies

GBNMs have been tested (most of all GO) in different cell lines in vitro. In most cases, genotoxic and oxidative damage is detected (ROS increase). The results are heterogeneous and, in some cases, conflicting. In Table 5.2, some recent studies are reported.

May et al. (2022) found no genotoxic effects by measuring DNA strand breaks on Jurkat E6-I lymphoblastoid cell line exposed to several concentrations of different NMs, including GO. Creutzenberg et al. (2022) have performed a series of experiments on different cell lines (primary rat alveolar macrophages, MRC-5 lung fibroblasts, normal rat alveolar macrophage cell line NR8383, and RAW 264.7 cells). They exposed the cells for 24 hours to seven different GBNs (two single-layer graphene [SLG], two graphene nanoplatelets [GNP], one carboxyl graphene, one GO, one graphite oxide). The SLG samples were found to be the most genotoxic, while none of the GBNs tested induced oxidative damage. While investigating neurotoxicity and biological adverse effects of commercial GO nanosheets in cultured human SH-SY5Y neuroblastomal cells, Xiaoli et al. (2021) found that the ultrasonication in medium changed oxidation status and surface reactivity of GO, causing lipid peroxidation and then cell membrane damage. Once internalized, GO causes a disruption of mitochondrial homeostasis, and this stressful condition could lead, as a result, to cellular apoptosis.

Zhang et al. (2021) produced rGO starting from GO by gamma irradiation. Then they exposed rat myocardial cells (H9C2) in vitro to GO and rGO, and measured ROS (for in vivo, see below). They found that the total ROS content increases with the increasing radiation dose and is mainly

Table 5.2 In vitro and in vivo studies on genotoxic and oxidative effects of graphene-based nanomaterials.

Nanomaterial	Model (in vitro/in vivo)	Exposure conditions	Methods	Findings	Reference
GO	Jurkat E6-I lymphoblastoid cell line	0,1.56,3.125,6.25,12.5,25, 50,100 μg/ml 3 and 24 h	Comet assay Fluorometric analysis of DNA unwinding (FADU) assay	No genotoxic effect	May et al. (2022)
GO	Human SH-SY5Y neuroblastoma cell line	40,60,80 μg/ml 6 and 24 h	DCFH-DA dye assay Confocal microscopy	Lipid peroxidation Disruption of mitochondrial homeostasis	Xiaoli et al. (2021)
Two SLGs, two GNPs, CG, GO, and graphite oxide	Primary rat alveolar macrophages (AM) MRC-5 lung fibroblasts NR8383 cell line RAW 264.7 cells ------------------------ 5 Female Wistar rats	25 μg/cm^2 24 h ------------------------ 28-day nose-only inhalation study 0.2,0.8,3.2 mg/m^3 Day 29 post-exposure	hOGG1-modified alkaline comet assay	SLG samples were most genotoxic No GNPs induced oxidative damage ------------------------ GNP had almost no lung toxicity/ pro-inflammatory potential, in contrast to a moderate effect found with SLG.	Creutzenberg et al. (2022)
GO and rGO	Rat myocardial cells (H9C2) ------------------------ 40 healthy C57BL/6 male mice (8 weeks old)	50 mg/mL 24 h ------------------------ Injection via tail vein 4 mg/kg 14 days	Total ROS assay kit based on DCFH-DA staining Myocardial tissues analysis Lipid peroxidation indicators	Cardiotoxicity by mediating lipid peroxidation, oxidative stress and mitochondrial dysfunction	Zhang et al. (2021)
GO and rGO	Human RPE cells (ARPE-19)	10,50,100,200 μg/ml 6, 12, 24, 48, and 72 h	ROS assay Alkaline comet assay	ROS-dependent DNA damage	Ou et al. (2021)

(Continued)

Table 5.2 (Continued)

Nanomaterial	Model (in vitro/in vivo)	Exposure conditions	Methods	Findings	Reference
GP	Human THP-1 cell line	0–60 μg/ml 40 h	GSH assay Cytokinesis-block micronucleus test	No induction of oxidative stress A significant dose-dependent increase in DNA damage	Malkova et al. (2021)
GO Nanosheets (pristine, reduced, and hydrated)	THP-1 cells BEAS-2B cells ----------------------- Eight-week-old male C57BL/6 mice	0–200 μg/ml 24 or 48 h ----------------------- Oropharyngeal aspiration 2 mg/kg 40 h	Lipid peroxidation assay BALF analysis RBC hemolysis assay	Acute lung inflammation Lipid peroxidation	Li et al. (2018)
GQDs	Gastric tissues of mice	Ulcer induced by swim stress for 4 h 2 and 10 mg/kg body weight	TUNEL assay	Effective ROS scavenger by suppressing excessive oxidative stress	Choudhary et al. (2022)
GNPs (Pristine and with different functional groups)	Six-week-old female Wistar rats	Intratracheal instillation 0.3, 1, 3 mg/rat 24 h, 1 week, and 4 weeks	Cell-free DCFH-DA assay BALF analysis	High dose-dependent potential for the generation of ROS Pulmonary inflammogenicity	Lee et al. (2017)

GO: graphene oxide; SLG: single-layer graphene; GNP: graphene nanoplatelets; CG: carboxyl graphene; rGO: reduced graphene oxide; GP: graphene platelets; GQDs: graphene quantum dots.

generated in mitochondria. While exposing human retinal pigment epithelium cells (ARPE-19), Ou et al. (2021) evaluated the genotoxicity and ROS production induced by GO and different rGOs (with different oxygen contents, obtained by Na citrate). Both materials induce ROS-dependent DNA damage, which is higher with rGO. The authors suppose that this depends on the C–O bonds, which in the GO are saturated (C–O) and in the rGO, unsaturated (C = O).

Malkova et al. (2021) evaluated the genotoxic effect and oxidative stress induced by two types of graphene platelets (GP) in the human acute monocytic leukemia cell line (THP-1). They found no evidence of the induction of oxidative stress, and a significant dose-dependent increase in DNA damage may be caused, according to the authors, by the direct interaction of GP with DNA during cell division or with the cytoskeleton and mitotic apparatus. Li et al. (2018) have used five types of GO nanosheets (one pristine, two rGO, and two hydrated GO [hGO]) to expose macrophage THP-1 and epithelial BEAS-2B cells (and murine lung in vivo by oropharyngeal aspiration), to demonstrate that the surface oxidation state and carbon radical content play an important role in the induction of GO toxicity. They observed that hGO (with the highest carbon radical density) was responsible for cell death in mammalian cells via lipid peroxidation of the surface membrane. In the same cells, rGO has a lesser effect on viability and higher uptake; results have been confirmed by in vivo studies.

In Vivo Studies

GBNMs have also been tested in various organisms. Here we report some recent in vivo studies (summarized in Table 5.2). Lee et al. (2017) while testing six different types of graphene nanoparticles (GNPs), both pristine and functionalized (GNP_{dot}, GNP_{COOH}, GNP_{O2}, GNP_{N2}, GNP_{F}, GNP_{NH2}) by intratracheal instillation in Wistar rats, found a very high dose-dependent potential for the generation of ROS. In addition, they found a dose-dependent inflammation at 24 h after instillation, statistically significant only in positively charged GNPs (GNP_{N2} and GNP_{NH2}). Also in vivo, Zhang et al. (2021) (for in vitro, see above) found significant increase in ROS production and mitochondrial damage. They evaluated lipid peroxidation in heart tissues collected from C57BL/6 male mice and exposed (injected via tail vein) to GO and rGO for 14 days. While exposing Balb/c mice to GQDs and evaluating for lipid peroxidation and protein oxidation among other tests, Choudhary et al. (2022) found that GQDs are effective ROS scavengers, as they reduce the inflammatory burden by suppressing excessive oxidative stress.

Studies on Epigenetic Effects of Metal Oxides and Graphene-Based Nanomaterials

Several papers have been published on the up- and downregulation of epigenetic factors contributing to clarify specific epigenetic mechanisms. Here we have focused on selected studies, including metal oxide NPs, i.e. SiO_2, ZnO, TiO_2, Ag NPs, and GO (Table 5.3). A collection of significant in vitro and in vivo studies implying the use of NPs and NMs exposure has been available in the review of Pogribna and Hammons (2021). Here selected groups of experiments have been chosen to demonstrate the epigenetic effects on several tissue culture models, including in vitro, in vivo animal and human studies. The exposure to NPs on such models may help clarify novel and possible unknown epigenetic effects on humans and animals.

In Table 5.3, several in vitro and in vivo studies performed on mouse, human cell lines, and exposed workers were reported. The most common epigenetic modifications have been categorized as (i) changes in methylation; (ii) changes in histone modifications; (iii) changes in

Table 5.3 In vitro and in vivo studies on epigenetic effects of nanomaterials.

Nanomaterial	Model (in vitro/in vivo)	Exposure conditions	Methods	Findings	Reference
TiO_2 NPs	Human MRC5 lung fibroblasts	0.5 and 4 μg/ml 24 and 48 h	Enzyme-linked immunosorbent assay–based Imprint® Methylated DNA Quantification Kit	Global DNA hypomethylation decrease in total DNA methyltransferase activity	Patil et al. (2016)
Silica- or citrate-coated TiO_2 NPs	A549 cell line	40 μg/cm^2 48 and 72 h	ELISA method able to detect LINE-1 methylation degree	Global DNA hypomethylation Demethylation of LINE1 repetitive elements	Stoccoro et al. (2017)
SiO_2 NPs	WBC of 31 nano-SiO_2 handling workers and 43 controls		High-performance liquid chromatography	Global DNA hypomethylation	Liou et al. (2017)
Nano-composite materials	Venous blood samples of 20 long-term Nanocomposite exposed workers and 20 controls		Infinium MethylationEPIC BeadChips	Changes in CpG methylation Hypomethylated CpG sites	Rossnerova et al. (2020)
Crystalline silica particle Min-U-Sil^R 5	Bhas 42 cell line	15 and 25 μg/cm^2 48 h	ELISA Western blot analysis	~80% decrease in global DNA methylation Increase in the levels of DNMT3a and DNMA3b	Seidel et al. (2017)
Nano-SiO_2 and micro-SiO_2 particles	Human HaCaT cells	2.5, 5 and 10 μg/ml (nano-SiO_2) 10 μg/ml (micro-SiO_2) 24 h	Real-time Q-PCR and western blotting Methylation-specific PCR and Bisulfite sequencing assay	Promoter hypermethylation Decreased expression of PARP1 protein	Gong et al. (2012)
Nano-SiO_2 and micro-SiO_2 particles	Human HaCaT cells	2.5, 5 and 10 μg/ml (nano-SiO_2) 10 μg/ml (micro-SiO_2) 24 h	Western blotting RT-PCR, Immuno-fluorescence assay, Flow cytometric analysis	Global hypoacetylation Decrease in DNMT1, DNMT3 and MBDs	Gong et al. (2010)
Ag NPs	Beas-2B cell line	1 μg/ml 3 and 6 weeks	RNA-sequencing analysis Genome-wide DNA methylation analysis Transcriptomics analysis	Several genes differentially expressed Marginal effects on DNA methylation Induction of EMT and cell transformation	Gliga et al. (2018)

Table 5.3 (Continued)

Nanomaterial	Model (in vitro/ in vivo)	Exposure conditions	Methods	Findings	Reference
Ag NPs	A549 cell line	0,10,25,50,100 and 200 μg/ml 24, 48, and 72 h	Western blot analysis	Deacetylation of histone H3 tails and elevation of total histone H3 Phosphorylation of p53	Blanco et al. (2017)
Ag NPs	HT22 mouse hippocampal neuronal cell line	1–20 μg/ml 48 h	EpiQuik™ Assay	Increase in DNMT1, DNMT3a, and DNMT3b	Mytych et al. (2017)
ZnO NPs	HEK293 cells	25 and 50 μg/ml 48 h	MethylFash Quantification Kits qPCR	Global DNA (5-mC) hypomethylation, increase in DNA hydroxymethylation (5-hmC) associated with high expression of TET1 and TET2 genes	Choudhury et al. (2017)
CuO NPs	Eight-week-old BALB/c male mice	Intratracheal instillation 2.5 mg/kg 24 h	BALF and lung tissue analysis Analysis expression of DNA methylation machinery	Global DNA hypermethylation reduced expression of DNMT1, DNMT3a, DNMT3b and Tet1	Lu et al. (2016)
GO	Mice	Instillation in the caudal mouse vein	RNA and DNA methylation sequencing analysis	Differential distribution of GO in the liver leading to compromised hepatic functions Dysregulation of key epigenetic signaling pathways regulating mRNA transcription and DNA methylation with changes in liver functional zonation	Wu et al. (2020b)
rGO	BEAS-2B cells	1 and 10 μg/mL	Illumina's MethylationEPIC arrays and bisulphite pyrosequencing	Absence of DNA methylation changes	Pérez et al. (2020)
SLGO, FLGO Other GNPs	BEAS-2B cells	10 and 50 mg/L	Global DNA methylation assay	Dose-dependent global DNA hypermethylation by single- and few-layered SLGO and FLGO and global hypomethylation by other GNPs	Chatterjee et al. (2016)

noncoding miRNAs; (iv) involvement of DNA methyltransferase (DNMT1, DNMT3A, and DNMT3B); (v) CpG hypermethylation; (vi) CpG hypomethylation. All these effects may influence the genome with such epigenetic factors contributing to modify human health and disease.

Metal Oxide Nanomaterials

Stoccoro et al. (2017) investigated in the A549 bronchial cell line the cytotoxic, genotoxic, and epigenetic effects of uncoated anatase TiO_2 NPs and TiO_2 NPs coated with silica and sodium citrate. The results indicated that TiO_2 NPs are able to induce DNA breaks and also oxidative damage to DNA in A549 cells. Global DNA methylation in treated cells has also been observed to detect correlation with micronuclei formation. Only after 72 hours of exposure, a significant demethylating effect was induced by the silica- and the citrate-coated NPs, the latest inducing the strongest effects. Titanium dioxide (TiO_2) and zinc oxide (ZnO) NPs are good candidates for applications in consumer products. However, there are some specific epigenetic effects that need to be investigated further to understand the potential dangers in humans and in animals. For instance, SiO_2 NPs have been shown to decrease the mRNA expression of PARP-1, increase the DNA methylation levels of PARP-1 promoter, decrease in global DNA methylation and the related methyltransferase, Dnmt1, Dnmt3A, and MBD2 (Patil et al. 2016). In a further study, human embryonic kidney cells (HEK-293) have been chosen as model system to assess the effects of ZnO NPs toxicity (Choudhury et al. 2017). The authors observed global reduction in 5-methylcytosine and increase in 5-hydroxymethylcytosine. A significant increase in the expression of (TET)-methylcytosine dioxygenase genes but not in the expression of DNA-methyltransferases (DNMTs) have been detected. Based on these findings, the authors suggested that ZnO NPs induce abundant increase in ROS promoting structural and functional anomalies in HEK-293 cells. In particular ZnO-NP-induced ROS may activate global hypomethylation in cells leading to expression of TET-enzymes, avoiding the DNMT interference. Since the global DNA demethylation is considered the hallmark of the majority of malignant tumors, once these features have been acquired, it is likely the epigenetic modifications could be propagated to the future progeny (D'Urso and Brickner 2014). Therefore, the following studies can be used as a platform for the assessment of epigenomic toxicity of ZnO NPs in humans in light of their use in commercial products (Choudhury et al. 2017). Further interesting experimental results regarding the nanotoxicity affecting epigenetic endpoints have been recently reported (Musolino et al. 2022). Here the understanding of the effects of NPs on the epigenome requires further epigenetic studies not only to correlate the epigenome with the transcriptional status of the cells, but also to investigate the mechanisms by which NPs might cause epigenetic lesions. Although the authors believe that intersecting epigenetics and nanotechnology will push both fields forward, the assays currently used to define nanotoxicity are still not able to evaluate the latent toxicity. The researchers' conclusion is that the complexity of epigenetics and the heritability of epigenetic biomarkers should make more progress, even though NPs are emerging as potential tools for designing epigenetic drugs against severe diseases such as cancer.

Graphene-Based Nanomaterials

Fewer studies have reported the epigenetic effects of GO. Wu et al. (2020b) analyzed the differential distribution of GO in liver lobules in mice. GO instilled in the caudal mouse vein accumulated in the mouse liver. A differential distribution of GO in the liver occurred, leading to compromised hepatic functions. The RNA and DNA methylation sequencing analysis found out that GO-induced changes in liver functional zonation could be ascribed to the dysregulation of key epigenetic signaling pathways regulating mRNA transcription and DNA methylation, by TET-dependent signaling. Taken together, these results underlined the genetic and epigenetic mechanisms in GO-induced liver zonation alterations.

In a further study, GO, inhibited the growth and promoted the differentiation of glioblastoma stem-like cells (GSCs) (Wang et al. 2020). Researchers found that GO affected cells by inducing cycle arrest and epigenetic regulation. The results suggested that GO could promote differentiation and reduce malignancy in GSCs through epigenetic mechanisms, demonstrating that GO might be a potent anti-glioblastoma multiforme tumor agent that could be useful for future clinical applications (Wang et al. 2020). In contrast, Pérez et al. (2020) did not find significant epigenetic mechanisms in rGO. These scientists synthesized rGO through thermal exfoliation/reduction of graphite oxide, and characterized the resulting material. By Illumina's MethylationEPIC arrays and bisulphite pyrosequencing, they analyzed the genome-wide and global DNA methylation dynamics associated with the medium-term exposure of human lung epithelial cells to rGO at concentrations of 1 and 10 μg/mL. The absence of DNA methylation changes caused by rGO at low doses with medium-term exposures showed no genome-wide or global DNA methylation changes associated with both conditions. The authors concluded that medium-term rGO exposure does not have significant effects on the DNA methylation patterns in human lung epithelial cells (Pérez et al. 2020).

Chatterjee and collaborators (Chatterjee et al. 2016) reported dose-dependent increase in global DNA methylation in human bronchial epithelial cells exposed to single- and few-layered GO, while other graphene family NMs induced hypomethylation. In general, alterations induced by GO in DNA methylation can disrupt several cell functions, such as permeability of the cell membrane and reduction of the density of cytoskeletal protein fibers.

Zhao et al. (2016) designed experiments on the nematode worm model *Caenorhabditis elegans*. The investigators reported the presence of GO toxicity in the reproductive organs and reduction of gonad development. Furthermore, specific micro-RNAs (miRNA), such as mir-2210, mir-360, mir-4810, mir-4805, mir-259, mir-82, and mir-4807 were found upregulated, suppressing the DNA damage-signaling pathways. GO-induced toxicity was enhanced by mutation of mir-360 and decreased by its overexpression, indicating a relevant role for mir-360 in GO toxicity (Zhao et al. 2016). The authors reported that basic studies in animal models are necessary for understanding the epigenetic effects, since long time evaluations of epigenetic changes caused by GO exposure have not been performed.

Studies on Workers – Genotoxic and Oxidative Effects of Occupational Exposure to Metal Oxides Nanoparticles, SiO_2 NPs, and Graphene-Based Nanomaterials

Epidemiological studies are still needed to evaluate the potential health effects of NMs. Workers are more exposed to NMs than general population because of their involvement in research and development, production, use, recycling, and disposal of these materials or products containing them. However, it is very difficult to perform biomonitoring studies of workers handling only one kind of NM, so most of the fewer available studies on their genotoxic and oxidative effects, took into consideration workers employed in different plants that produce various types of NMs. Table 5.4 presents the available studies on genotoxic and oxidative effects of occupational exposure to metal oxide NPs, SiO_2 NPs, and graphene-based NMs.

One of the first available studies on the potential toxicity of exposure to NMs enrolled 158 nanomaterial-handling workers and 104 nonexposed workers from 14 manufacturing plants in Taiwan (Liao et al. 2014). They investigated pulmonary and cardiovascular disease markers, inflammation

Table 5.4 Studies on genotoxic and oxidative effects of occupational exposure to metal oxide nanoparticles, SiO_2 NPs and graphene-based nanomaterials.

Nanomaterial	Populations and study design	Biological matrices	Methods	Findings	References
CNTs, TiO_2, SiO_2, nanoresins, nanosilver, nanogold, nanoclay, nanoalumina metal oxides	Exposed workers ($n = 158$) and controls ($n = 104$) After a 6-month follow-up period: exposed workers ($n = 124$) and controls ($n = 77$) Exposure assessment by control banding	Blood, urine, EBC	Antioxidant enzyme activity (SOD and GPX), inflammation and oxidative damage markers (8-idroxydeoxyguanosine, N^7-methyl guanosine and isoprostane) Comet assay	No significant differences between exposed workers and controls for the changes in oxidative damage markers No significant differences in changes between exposed and controls for genotoxicity markers Antioxidant enzymes associated with NM handling after 6 months	Liao et al. (2014)
Nanosilver, SiO_2, TiO_2, CNT, nanoresins, nanogold, nanoclay, nanoalumin, and metal oxides	Exposed workers ($n = 206$) and controls ($n = 108$) Follow-up no less than 2 times in 4 years Exposure assessment by control banding	Blood, urine, EBC	Antioxidant enzyme activity (SOD and GPX), inflammation and oxidative damage markers (8-idroxydeoxyguanosine, N^7-methyl guanosine and isoprostane) Comet assay MN assay	Lack of significant change of oxidative damages and genotoxic markers	Wu et al. (2019)
Metal oxide nanomaterials (NMs)	Workers handling nano-TiO_2 ($n = 26$), nano-SiO_2 ($n = 31$), ITO ($n = 30$) and controls ($n = 43$) Exposure assessment by control banding	Urine, leukocytes, EBC	Oxidative stress biomarkers (8-idroxydeoxyguanosine, isoprostane) Methylation assay	The exposure to metal oxide NMs may lead to global methylation, DNA oxidative damage and lipid peroxidation	Liou et al. (2017)
NanoTiO_2	Exposed workers ($n = 34$) and controls ($n = 45$) Exposure assessment: detailed workplace aerosol measurements	EBC, urine	Malondialdehyde, 4-hydroxy-trans-hexenal, 4-hydroxy-trans-nonenal, 8-isoprostaglandin F2α and aldehydes C6–C12	Higher values of all markers in the EBC of exposed workers with dose-dependent association with the exposure Induction of oxidative stress of TiO_2 NP exposure	Pelclova et al. (2017)

NanoTiO_2	Exposed workers ($n = 36$) and controls ($n = 45$) working with TiO_2 for at least 6 months (2012 and 2013) Exposure assessment: detailed workplace aerosol measurements	EBC	Quantitative analysis of Ti in EBC by ICP-MS Analysis of markers of oxidative (8-OHdG, 8-OHG, 5-OHMeU, o-Tyr, 3-ClTyr, 3-NOTyr) stress after solid-phase extraction by LC-ESI-MS/MS	Titanium in the EBC significantly higher in production workers vs research workers and unexposed controls. Dose-dependent biological effect of occupational exposure to TiO2 nanoparticles	Pelclova et al. (2016)
Graphene or SiO_2 NPs	Exposed workers ($n = 12$) and controls ($n = 11$) Exposure assessment: detailed workplace measurements and personal monitoring	Exfoliated buccal cells, lymphocytes, serum, urine, EBC	BMCyt assay Fpg-comet assay (lymphocytes) Urinary oxidized DNA bases (8-hydroxy-2′-deoxyguanosine, 8-oxo-7,8-dihydroguanosine and 8-oxo-7,8-dihydroguanine) Oxidative stress biomarkers in EBC (8-isoprostane, malondialdehyde and 4-hydroxy-2-nonhenal), FeNo, and cytokine (IL-6, IL-8 and TNFα) by ELISA	Slight increase of MN frequency vs controls for both the groups of workers (by BMCyt assay) Direct DNA damage higher in the group of silica workers and oxidative DNA damage higher in graphene workers (by Fpg-comet assay) Lack of statistically significant increases of biomarkers of oxidative stress in NM workers groups vs controls in EBC and urine	Ursini et al. (2021)
Nanocomposite materials	Researchers exposed during processing nanocomposite materials ($n = 21$) and controls (21) Long-term (mean 17.8 years) Short-term (pre-shift vs. post-shift) Exposure assessment: detailed workplace aerosol measurements	Whole venous blood	MN assay using human pan centromeric probes	Lack of effects after long-term NP exposure on the total frequencies of micronuclei Increased levels of chromosomal breaks in the exposed group of short-term NP exposure	Rossnerova et al. (2019)
Nanocomposite materials	*2016*: Workers long-term exposed (20) and controls ($n = 20$) *2017*: Exposed workers ($n = 20$) and controls ($n = 20$) Pre-shift and post-shift samples in 2016 and 2017 Exposure assessment: detailed workplace aerosol measurements	Leukocytes	Alkaline version of the enzyme (fpg and ENDO III) modified comet assay	Increase of single- and double-strand breaks in DNA (DNA-SB) and oxidized bases Higher genotoxic risk vs controls Higher susceptibility to NP exposure in males	Novotna et al. (2020)

(*Continued*)

Table 5.4 (Continued)

Nanomaterial	Populations and study design	Biological matrices	Methods	Findings	References
Nanocomposites	*2017*: Exposed workers ($n = 20$) and controls ($n = 20$) *2018*: EBC, plasma and urine of workers ($n = 21$) and controls ($n = 18$) Pre-shift and post-shift samples in 2017 and 2018 Exposure assessment: detailed workplace aerosol measurements	EBC, plasma and urine	Oxidative stress biomarkers (MDA, C6-12, 8-isoprostane, 8-OHdG, 8-OHG, 5-OHMeU, o-Tyr and 3-NOTyr) after solid-phase extraction using LC-ESI-MS/MS	In both years, pre- and post-shift elevations of oxidative stress markers in EBC and plasma were higher than in urine of the workers handling nanoparticles	Pelclova et al. (2020)
Copier-emitted nanoparticles (titanium dioxide, iron oxide, and fumed silica and several metals)	*Acute exposure study* Samples of 9 volunteers. Acute exposure study: 5–6 h/day in a copy center. *Chronic exposure study* Samples of 6 full-time copier operators ($n = 6$) and controls ($n = 11$) Exposure assessment: detailed workplace aerosol measurements including monitoring for O_3, CO, CO_2, and total VOCs	Urine	Urinary oxidative DNA damage markers (8-hydroxydeoxyguanosine, 8-OHdG; 8-hydroxyguanosine, 8-OHG; 5-hydroxymethyl uracil 5-OHMeU), lipid peroxidation markers (8-isoprostane; 4-hydroxynonenal, HNE), and protein oxidation biomarkers (o-tyrosine, 3-chlorotyrosine, and 3-nitrotyrosine) analyzed by LC-ESI-MS/MS	NPs from photocopiers induce systemic oxidative stress by damaging DNA, RNA, and lipids	Zhang et al. (2022)

and oxidative stress markers, antioxidant enzymes, and genotoxicity markers. The workers were mainly exposed to CNTs, SiO_2, TiO_2, and nanoAg, and the study showed that antioxidant enzymes and cardiovascular markers were significantly associated with nanomaterial-handling during the six-month follow-up period. The same research group published the results of a longitudinal follow-up (in an interval of four years) of health effects among workers handling engineered NMs from the same plants in Taiwan (Wu et al. 2019). Oxidative damage, antioxidant enzyme activity, DNA damage, and genotoxicity were evaluated by comet and micronucleus assays. They found no evidence of potential adverse health effects among the enrolled workers except for the increase of antioxidant enzymes. From the same 14 nanomaterial manufacturing units in Taiwan, Liou et al. (2017) selected workers handling nano-TiO_2 ($n = 26$), nano-SiO_2 ($n = 31$), and indium tin oxide (ITO) ($n = 30$) and 43 controls to perform the first study to assess global methylation, oxidative DNA damage, and lipid peroxidation in workers with occupational exposure to metal oxide NMs. They evaluated oxidative stress measuring urinary and white blood cell (WBC) 8-hydroxydeoxyguanosine (8-OHdG), and exhaled breath condensate (EBC) 8-isoprostane. As an epigenetic alteration, they measured WBC global methylation. Urinary 8-OHdG and EBC 8-isoprostane were higher in exposed workers than in the controls, whereas they found higher WBC 8-OHdG and lower global methylation only in ITO-handling workers. Significant positive correlations were found between WBC and urinary 8-OHdG. In addition, they demonstrated a significant negative correlation between WBC 8-OHdG and global methylation. The study suggests that exposure to metal oxide NMs may induce global hypomethylation, DNA oxidative damage, and lipid peroxidation. Pelclova et al. (2017) studied workers employed in the production of TiO_2 NPs evaluating in the EBC and urine malondialdehyde, 4-hydroxy-trans-hexenal, 4-hydroxy-trans-nonenal, 8-isoprostaglandin F2α and aldehydes C_6–C_{12}. They found higher values of all the studied markers in the EBC of exposed workers vs controls with dose-dependent exposure association. In earlier studies, the same laboratory documented an increase in markers of inflammation, as well as DNA and protein oxidative damage, in EBC of workers exposed to nanoTiO_2 (Pelclova et al. 2016). Altogether, they demonstrated the induction of oxidative stress of TiO_2 NP exposure. Our laboratory (Ursini et al. 2021) recently published a pilot study that aimed to identify a panel of sensitive biomarkers and suitable biological matrices to evaluate genotoxic and oxidative effects induced in workers involved in the production process of graphene or SiO_2 NPs. We monitored workers using the Buccal Micronucleus Cytome (BMCyt) assay, fpg-comet test (lymphocytes), oxidized DNA bases 8-oxoGua, 8-oxoGuo, and 8-oxodGuo measurements (urine), analysis of oxidative stress biomarkers in EBC, FENO measurement, and cytokines release detection (serum). This explorative study, performed on 12 workers vs 11 controls, demonstrates that BMCyt and fpg-comet assays are the most sensitive biomarkers of early, still reparable, genotoxic and oxidative effects. In fact, using the buccal micronucleus assay, we detected genotoxic effects in both the exposed groups in terms of higher percentage vs controls of MN positive subjects (those subjects with micronucleated cells frequency exceeding a fixed cut-off value of 1.5‰). Using the fpg-comet assay, we found DNA damage only in the group of silica workers and oxidative damage in graphene workers. However, these results need to be confirmed on a higher number of subjects. Rossnerova et al. (2019) biomonitored a group of researchers exposed to nanocomposites and did not find, after long-term NP exposure, effects on the total frequencies of micronuclei (MN) by the MN assay with human pan centromeric probes used to distinguish between centromere positive (CEN+) and centromere negative (CEN−) MN and to detect clastogenic and aneugenic effects. They observed a shift in the types of aberrations toward the increased levels of chromosomal breaks in the exposed group, suggesting a clastogenic effect NP exposure. Novotna et al. (2020) evaluated the DNA damage in the leukocytes of workers who are occupationally exposed long term to NPs. In particular, the enrolled workers were involved in the welding of metal surfaces, smelting of mixtures containing nanoadditives, and machining of the finished nanocomposites. The chronic effects of NPs on DNA (pre-shift samples) and the acute effects of exposure during the shift (the difference between pre- and post-shift samples) were studied. By the enzyme-modified (Fpg and endonuclease III) comet assay, they found an increase of single- and double-strand breaks in DNA (DNA-SB) and oxidized bases. Acute exposure induced a further increase of DNA-SB. The welding and smelting of nanocomposites represented a higher genotoxic risk than the milling and grinding of nanocomposite surfaces. They also found a higher susceptibility to NP exposure in males. Pelclova et al. (2020) published the results of a three-year follow-up of a 2016 study, on nanocomposite plant workers, in which only EBC had been examined. The follow-up study analyzed eight markers of oxidative stress in EBC, plasma, and urine, in both pre- and post-shift samples in 2017 and 2018. Nanocomposite mass concentration during machining was higher in 2017 than in 2018. In both years, they found pre-and post-shift elevations in EBC and plasma to be higher than in urine biomarkers. In this study, EBC and plasma appear to be the ideal fluids for the biomonitoring of oxidative stress arising from engineered NMs. Zhang et al. (2022) evaluated oxidative stress induced by acute and chronic exposures to copier-emitted NPs using urinary oxidative stress biomarkers of DNA damage (8-hydroxydeoxyguanosine, 8-OHdG; 8 hydroxyguanosine, 8-OHG; 5-hydroxymethyl uracil 5-OHMeU), lipid peroxidation (8-isoprostane; 4-hydroxynonenal, HNE), and protein oxidation biomarkers (o-tyrosine, 3-chlorotyrosine, and 3-nitrotyrosine). For acute exposure, they asked nine young, healthy, and nonsmoking volunteers to spend 5–6 hours/day in a copy center, whereas for chronic exposure, they enrolled six full-time copier operators, whose primary job was printing and photocopying. By the physicochemical characterization of toners and airborne nano-scale fraction, they found engineered NPs, including titanium dioxide, iron oxide, and fumed silica as well as several metals. The chemical composition of airborne nano-scale fraction reflected toner chemistry. They confirm that NPs from photocopiers induce systemic oxidative stress, leading to DNA (8-OHdG), RNA (8-OHG), and lipid (HNE, 8-isoprostane) oxidation.

Conclusions

Most of the studies performed until now have been done using standard test assays that showed genotoxic and oxidative effects. However, the application of NPs in all sectors cannot avoid the possible exposure to such materials, particularly in the occupational field. Therefore, there is an urgent need to understand whether the exposure to such materials represents a hazard for humans and their environment. A preventive approach represents the best way to reduce as much as possible the potential toxic effects of these materials, particularly for workers.

References

Abudayyak, M., Gurkaynak, T.A., and Özhan, G. (2017). In vitro evaluation of cobalt oxide nanoparticle-induced toxicity. *Toxicol. Ind. Health* 33 (8): 646–654. doi: 10.1177/0748233717706633.

Akbaba, G.B. and Türkez, H. (2018). Investigation of the genotoxicity of aluminum oxide, β-tricalcium phosphate, and zinc oxide nanoparticles in vitro. *Int. J. Toxicol.* 37 (3): 216–222. doi: 10.1177/1091581818775709.

Akhtar, M.J., Kumar, S., Alhadlaq, H.A., Alrokayan, S.A., Abu-Salah, K.M., and Ahamed, M. (2016). Dose-dependent genotoxicity of copper oxide nanoparticles stimulated by reactive oxygen species in human lung epithelial cells. *Toxicol. Ind. Health* 32 (5): 809–821. doi: 10.1177/0748233713511512.

Alarifi, S., Ali, D., Al Omar Suliman, Y., Ahamed, M., Siddiqui, M.A., and Al-Khedhairy, A.A. (2013). Oxidative stress contributes to cobalt oxide nanoparticles-induced cytotoxicity and DNA damage in human hepatocarcinoma cells. *Int. J. Nanomed.* 8: 189–199. doi: 10.2147/ijn.S37924.

Ali, S.A., Rizk, M.Z., Hamed, M.A., Aboul-Ela, E.I., El-Rigal, N.S., Aly, H.F., and Abdel-Hamid, A.Z. (2019). Assessment of titanium dioxide nanoparticles toxicity via oral exposure in mice: effect of dose and particle size. *Biomarkers* 24 (5): 492–498. doi: 10.1080/1354750x.2019.1620336.

Alinovi, R., Goldoni, M., Pinelli, S., Campanini, M., Aliatis, I., Bersani, D., Lottici, P.P., Iavicoli, S., Petyx, M., Mozzoni, P., and Mutti, A. (2015). Oxidative and pro-inflammatory effects of cobalt and titanium oxide nanoparticles on aortic and venous endothelial cells. *Toxicol. Vitro* 29 (3): 426–437. doi: 10.1016/j.tiv.2014.12.007.

Andreoli, C., Prota, V., De Angelis, I., Facchini, E., Zijno, A., Meccia, E., Barletta, B., Butteroni, C., Corinti, S., Chatgilialoglu, C., Krokidis, M.G., Masi, A., Condello, M., Meschini, S., Di Felice, G., and Barone, F. (2021). A harmonized and standardized in vitro approach produces reliable results on silver nanoparticles toxicity in different cell lines. *J. Appl. Toxicol.* 41 (12): 1980–1997. doi: 10.1002/jat.4178.

Anreddy, R.N.R. (2018). Copper oxide nanoparticles induces oxidative stress and liver toxicity in rats following oral exposure. *Toxicol. Rep.* 5: 903–904. doi: 10.1016/j.toxrep.2018.08.022.

Attarilar, S., Yang, J., Ebrahimi, M., Wang, Q., Liu, J., Tang, Y., and Yang, J. (2020). The toxicity phenomenon and the related occurrence in metal and metal oxide nanoparticles: a brief review from the biomedical perspective. *Front. Bioeng. Biotechnol.* 8: 822. doi: 10.3389/fbioe.2020.00822.

Banerjee, A.N. (2018). Graphene and its derivatives as biomedical materials: future prospects and challenges. *Interface Focus* 8 (3): 20170056. doi: 10.1098/rsfs.2017.0056.

Baranowska-Wojcik, E., Szwajgier, D., Oleszczuk, P., and Winiarska-Mieczan, A. (2020). Effects of titanium dioxide nanoparticles exposure on human health – a review. *Biol. Trace Elem. Res.* 193 (1): 118–129. doi: 10.1007/s12011-019-01706-6.

Behzadi, S., Serpooshan, V., Tao, W., Hamaly, M.A., Alkawareek, M.Y., Dreaden, E.C., Brown, D., Alkilany, A.M., Farokhzad, O.C., and Mahmoudi, M. (2017). Cellular uptake of nanoparticles: journey inside the cell. *Chem. Soc. Rev.* 46 (14): 4218–4244. doi: 10.1039/c6cs00636a.

Bicho, R.C., Roelofs, D., Mariën, J., Scott-Fordsmand, J.J., and Amorim, M.J.B. (2020). Epigenetic effects of (nano)materials in environmental species – Cu case study in *Enchytraeus crypticus*. *Environ. Int.* 136: 105447. doi: 10.1016/j.envint.2019.105447.

Blanco, J., Lafuente, D., Gómez, M., García, T., Domingo, J.L., and Sánchez, D.J. (2017). Polyvinyl pyrrolidone-coated silver nanoparticles in a human lung cancer cells: time- and dose-dependent influence over p53 and caspase-3 protein expression and epigenetic effects. *Arch. Toxicol.* 91 (2): 651–666. doi: 10.1007/s00204-016-1773-0.

Brand, W., van Kesteren, P.C.E., Peters, R.J.B., and Oomen, A.G. (2021). Issues currently complicating the risk assessment of synthetic amorphous silica (SAS) nanoparticles after oral exposure. *Nanotoxicology* 15 (7): 905–933. doi: 10.1080/17435390.2021.1931724.

Brandão, F., Costa, C., Bessa, M.J., Dumortier, E., Debacq-Chainiaux, F., Hubaux, R., Salmon, M., Laloy, J., Stan, M.S., Hermenean, A., Gharbia, S., Dinischiotu, A., Bannuscher, A., Hellack, B., Haase, A., Fraga, S., and Teixeira, J.P. (2021). Genotoxicity and gene expression in the rat lung tissue following instillation and inhalation of different variants of amorphous silica nanomaterials (aSiO(2) NM). *Nanomaterials (Basel)* 11 (6). doi: 10.3390/nano11061502.

Cappellini, F., Hedberg, Y., McCarrick, S., Hedberg, J., Derr, R., Hendriks, G., Odnevall Wallinder, I., and Karlsson, H.L. (2018). Mechanistic insight into reactivity and (geno)toxicity of well-characterized nanoparticles of cobalt metal and oxides. *Nanotoxicology* 12 (6): 602–620. doi: 10.1080/17435390.2018.1470694.

Cavallo, D., Ciervo, A., Fresegna, A.M., Maiello, R., Tassone, P., Buresti, G., Casciardi, S., Iavicoli, S., and Ursini, C.L. (2015). Investigation on cobalt-oxide nanoparticles cyto-genotoxicity and inflammatory response in two types of respiratory cells. *J. Appl. Toxicol.* 35 (10): 1102–1113. doi: 10.1002/jat.3133.

Chakrabarti, S., Goyary, D., Karmakar, S., and Chattopadhyay, P. (2019). Exploration of cytotoxic and genotoxic endpoints following sub-chronic oral exposure to titanium dioxide nanoparticles. *Toxicol. Ind. Health* 35 (9): 577–592. doi: 10.1177/0748233719879611.

Chatterjee, N., Yang, J., and Choi, J. (2016). Differential genotoxic and epigenotoxic effects of graphene family nanomaterials (GFNs) in human bronchial epithelial cells. *Mutat. Res. Genet. Toxicol. Environ. Mutagen.* 798-799: 1–10. doi: 10.1016/j.mrgentox.2016.01.006.

Chattopadhyay, S., Dash, S.K., Tripathy, S., Das, B., Mandal, D., Pramanik, P., and Roy, S. (2015). Toxicity of cobalt oxide nanoparticles to normal cells; an in vitro and in vivo study. *Chem. Biol. Interact.* 226: 58–71. doi: 10.1016/j.cbi.2014.11.016.

Chavali, M.S. and Nikolova, M.P. (2019). Metal oxide nanoparticles and their applications in nanotechnology. *SN Appl. Sci.* 1 (6): 607. doi: 10.1007/s42452-019-0592-3.

Chen, S.X., Yang, X.Z., Deng, Y., Huang, J., Li, Y., Sun, Q., Yu, C.P., Zhu, Y., and Hong, W.S. (2017). Silver nanoparticles induce oocyte maturation in zebrafish (*Danio rerio*). *Chemosphere* 170: 51–60. doi: 10.1016/j.chemosphere.2016.12.016.

Cho, W.S., Duffin, R., Bradley, M., Megson, I.L., Macnee, W., Howie, S.E., and Donaldson, K. (2012). NiO and Co3O4 nanoparticles induce lung DTH-like responses and alveolar lipoproteinosis. *Eur. Respir. J.* 39 (3): 546–557. doi: 10.1183/09031936.00047111.

Choudhary, P., Biswas, S., Kandoth, N., Tayde, D., Chatterjee, A., Chattopadhyay, S., Das, A., Swarnakar, S., and Pramanik, S.K. (2022). Graphene quantum dots alleviate ROS-mediated gastric damage. *iScience* 25 (4): 104062. doi: 10.1016/j.isci.2022.104062.

Choudhury, S.R., Ordaz, J., Lo, C.L., Damayanti, N.P., Zhou, F., and Irudayaraj, J. (2017). From the cover: zinc oxide nanoparticles-induced reactive oxygen species promotes multimodal cyto- and epigenetic toxicity. *Toxicol. Sci.* 156 (1): 261–274. doi: 10.1093/toxsci/kfw252.

Chusuei, C.C., Wu, C.H., Mallavarapu, S., Hou, F.Y., Hsu, C.M., Winiarz, J.G., Aronstam, R.S., and Huang, Y.W. (2013). Cytotoxicity in the age of nano: the role of fourth period transition metal oxide nanoparticle physicochemical properties. *Chem. Biol. Interact.* 206 (2): 319–326. doi: 10.1016/j.cbi.2013.09.020.

Collins, A., El Yamani, N., and Dusinska, M. (2017). Sensitive detection of DNA oxidation damage induced by nanomaterials. *Free Radic. Biol. Med.* 107: 69–76. doi: 10.1016/j.freeradbiomed.2017.02.001.

Collins, A.R. (2017). The use of bacterial repair endonucleases in the comet assay. In: *Drug Safety Evaluation: Methods and Protocols* (ed. J.-C. Gautier), 173–184. New York: Springer New York.

Creutzenberg, O., Oliveira, H., Farcal, L., Schaudien, D., Mendes, A., Menezes, A.C., Tischler, T., Burla, S., and Ziemann, C. (2022). PLATOX: integrated in vitro/in vivo approach for screening of adverse lung effects of graphene-related 2D nanomaterials. *Nanomaterials (Basel)* 12 (8). doi: 10.3390/nano12081254.

D'Urso, A. and Brickner, J.H. (2014). Mechanisms of epigenetic memory. *Trends Genet.* 30 (6): 230–236. doi: 10.1016/j.tig.2014.04.004.

Danielsen, P.H., Knudsen, K.B., Strancar, J., Umek, P., Koklic, T., Garvas, M., Vanhala, E., Savukoski, S., Ding, Y., Madsen, A.M., Jacobsen, N.R., Weydahl, I.K., Berthing, T., Poulsen, S.S., Schmid, O., Wolff, H., and Vogel, U. (2020). Effects of physicochemical properties of TiO2 nanomaterials for pulmonary inflammation, acute phase response and alveolar proteinosis in intratracheally exposed mice. *Toxicol. Appl. Pharmacol.* 386: 114830. doi: 10.1016/j.taap.2019.114830.

de Andrade Vieira, J.E., de Oliveira Ferreira, R., Marcel Dos Reis Sampaio, D., Pereira da Costa Araújo, A., and Malafaia, G. (2019). An insight on the mutagenicity and cytotoxicity of zinc oxide nanoparticles in *Gallus gallus domesticus* (Phasianidae). *Chemosphere* 231: 10–19. doi: 10.1016/j.chemosphere.2019.05.111.

de Melo-Diogo, D., Lima-Sousa, R., Alves, C.G., Costa, E.C., Louro, R.O., and Correia, I.J. (2018). Functionalization of graphene family nanomaterials for application in cancer therapy. *Colloids Surf. B: Biointerfaces* 171: 260–275. doi: 10.1016/j.colsurfb.2018.07.030.

deSouza, R.M., Parsons, N.R., Oni, T., Dalton, P., Costa, M., and Krikler, S. (2010). Metal ion levels following resurfacing arthroplasty of the hip: serial results over a ten-year period. *J. Bone Joint Surg.* 92 (12): 1642–1647. doi: 10.1302/0301-620x.92b12.24654.

Di Bucchianico, S., Cappellini, F., Le Bihanic, F., Zhang, Y., Dreij, K., and Karlsson, H.L. (2017). Genotoxicity of TiO2 nanoparticles assessed by mini-gel comet assay and micronucleus scoring with flow cytometry. *Mutagenesis* 32 (1): 127–137. doi: 10.1093/mutage/gew030.

Doak, S.H. and Dusinska, M. (2017). NanoGenotoxicology: present and the future. *Mutagenesis* 32 (1): 1–4. doi: 10.1093/mutage/gew066.

Du, J., Wang, S., You, H., Jiang, R., Zhuang, C., and Zhang, X. (2016). Developmental toxicity and DNA damage to zebrafish induced by perfluorooctane sulfonate in the presence of ZnO nanoparticles. *Environ. Toxicol.* 31 (3): 360–371. doi: 10.1002/tox.22050.

Dusinska, M., Mariussen, E., Rundén-Pran, E., Hudecova, A.M., Elje, E., Kazimirova, A., El Yamani, N., Dommershausen, N., Tharmann, J., Fieblinger, D., Herzberg, F., Luch, A., and Haase, A. (2019). In vitro approaches for assessing the genotoxicity of nanomaterials. *Methods Mol. Biol.* 1894: 83–122. doi: 10.1007/978-1-4939-8916-4_6.

El-Shorbagy, H.M., Eissa, S.M., Sabet, S., and El-Ghor, A.A. (2019). Apoptosis and oxidative stress as relevant mechanisms of antitumor activity and genotoxicity of ZnO-NPs alone and in combination with N-acetyl cysteine in tumor-bearing mice. *Int. J. Nanomed.* 14: 3911–3928. doi: 10.2147/ijn.S204757.

Fadoju, O., Ogunsuyi, O., Akanni, O., Alabi, O., Alimba, C., Adaramoye, O., Cambier, S., Eswara, S., Gutleb, A.C., and Bakare, A. (2019). Evaluation of cytogenotoxicity and oxidative stress parameters in male Swiss mice co-exposed to titanium dioxide and zinc oxide nanoparticles. *Environ. Toxicol. Pharmacol.* 70: 103204. doi: 10.1016/j.etap.2019.103204.

Fahmy, H.M., Ebrahim, N.M., and Gaber, M.H. (2020). In-vitro evaluation of copper/copper oxide nanoparticles cytotoxicity and genotoxicity in normal and cancer lung cell lines. *J. Trace Elem. Med. Biol.* 60: 126481. doi: 10.1016/j.jtemb.2020.126481.

Fenoglio, I., Greco, G., Livraghi, S., and Fubini, B. (2009). Non-UV-induced radical reactions at the surface of TiO2 nanoparticles that may trigger toxic responses. *Chemistry* 15 (18): 4614–4621. doi: 10.1002/chem.200802542.

Ferdous, Z., Al-Salam, S., Yuvaraju, P., Ali, B.H., and Nemmar, A. (2021). Remote effects and biodistribution of pulmonary instilled silver nanoparticles in mice. *NanoImpact.* 22: 100310. doi: 10.1016/j.impact.2021.100310.

Ferdous, Z. and Nemmar, A. (2020). Health impact of silver nanoparticles: a review of the biodistribution and toxicity following various routes of exposure. *Int. J. Mol. Sci.* 21 (7). doi: 10.3390/ijms21072375.

Franz, P., Bürkle, A., Wick, P., and Hirsch, C. (2020). Exploring flow cytometry-based micronucleus scoring for reliable nanomaterial genotoxicity assessment. *Chem. Res. Toxicol.* 33 (10): 2538–2549. doi: 10.1021/acs.chemrestox.0c00071.

Fresegna, A.M., Ursini, C.L., Ciervo, A., Maiello, R., Casciardi, S., Iavicoli, S., and Cavallo, D. (2021). Assessment of the influence of crystalline form on cyto-genotoxic and inflammatory effects induced by TiO(2) nanoparticles on human bronchial and alveolar cells. *Nanomaterials (Basel)* 11 (1). doi: 10.3390/nano11010253.

Gao, X., Topping, V.D., Keltner, Z., Sprando, R.L., and Yourick, J.J. (2017). Toxicity of nano- and ionic silver to embryonic stem cells: a comparative toxicogenomic study. *J. Nanobiotechnol.* 15 (1): 31. doi: 10.1186/s12951-017-0265-6.

Gedda, M.R., Babele, P.K., Zahra, K., and Madhukar, P. (2019). Epigenetic aspects of engineered nanomaterials: is the collateral damage inevitable? *Front. Bioeng. Biotechnol.* 7: 228. doi: 10.3389/fbioe.2019.00228.

Gliga, A.R., Di Bucchianico, S., Lindvall, J., Fadeel, B., and Karlsson, H.L. (2018). RNA-sequencing reveals long-term effects of silver nanoparticles on human lung cells. *Sci. Rep.* 8 (1): 6668. doi: 10.1038/s41598-018-25085-5.

Glinski, A., Lima de Souza, T., Zablocki da Luz, J., Bezerra Junior, A.G., Camargo de Oliveira, C., de Oliveira Ribeiro, C.A., and Filipak Neto, F. (2021). Toxicological effects of silver nanoparticles and cadmium chloride in macrophage cell line (RAW 264.7): an in vitro approach. *J. Trace Elem. Med. Biol.* 68: 126854. doi: 10.1016/j.jtemb.2021.126854.

Gong, C., Tao, G., Yang, L., Liu, J., Liu, Q., Li, W., and Zhuang, Z. (2012). Methylation of PARP-1 promoter involved in the regulation of nano-SiO2-induced decrease of PARP-1 mRNA expression. *Toxicol. Lett.* 209 (3): 264–269. doi: 10.1016/j.toxlet.2012.01.007.

Gong, C., Tao, G., Yang, L., Liu, J., Liu, Q., and Zhuang, Z. (2010). SiO(2) nanoparticles induce global genomic hypomethylation in HaCaT cells. *Biochem. Biophys. Res. Commun.* 397 (3): 397–400. doi: 10.1016/j.bbrc.2010.05.076.

Gonzalez-Carter, D.A., Leo, B.F., Ruenraroengsak, P., Chen, S., Goode, A.E., Theodorou, I.G., Chung, K.F., Carzaniga, R., Shaffer, M.S., Dexter, D.T., Ryan, M.P., and Porter, A.E. (2017). Silver nanoparticles reduce brain inflammation and related neurotoxicity through induction of H(2) S-synthesizing enzymes. *Sci. Rep.* 7: 42871. doi: 10.1038/srep42871.

Guo, C., Wang, J., Yang, M., Li, Y., Cui, S., Zhou, X., Li, Y., and Sun, Z. (2017). Amorphous silica nanoparticles induce malignant transformation and tumorigenesis of human lung epithelial cells via P53 signaling. *Nanotoxicology* 11 (9–10): 1176–1194. doi: 10.1080/17435390.2017.1403658.

Guo, Q., Yang, Y., Zhao, L., Chen, J., Duan, G., Yang, Z., and Zhou, R. (2022). Graphene oxide toxicity in W(1118) flies. *Sci. Total Environ.* 805: 150302. doi: 10.1016/j.scitotenv.2021.150302.

Hackenberg, S., Scherzed, A., Zapp, A., Radeloff, K., Ginzkey, C., Gehrke, T., Ickrath, P., and Kleinsasser, N. (2017). Genotoxic effects of zinc oxide nanoparticles in nasal mucosa cells are antagonized by titanium dioxide nanoparticles. *Mutat. Res. Genet. Toxicol. Environ. Mutagen.* 816-817: 32–37. doi: 10.1016/j.mrgentox.2017.02.005.

Hadrup, N., Rahmani, F., Jacobsen, N.R., Saber, A.T., Jackson, P., Bengtson, S., Williams, A., Wallin, H., Halappanavar, S., and Vogel, U. (2019). Acute phase response and inflammation following pulmonary exposure to low doses of zinc oxide nanoparticles in mice. *Nanotoxicology* 13 (9): 1275–1292. doi: 10.1080/17435390.2019.1654004.

Hu, H., Fan, X., Guo, Q., Wei, X., Yang, D., Zhang, B., Liu, J., Wu, Q., Oh, Y., Feng, Y., Chen, K., Hou, L., and Gu, N. (2019). Silicon dioxide nanoparticles induce insulin resistance through endoplasmic reticulum stress and generation of reactive oxygen species. *Part Fibre Toxicol.* 16 (1): 41. doi: 10.1186/s12989-019-0327-z.

Huang, R., Zhou, Y., Hu, S., and Zhou, P.-K. (2019). Targeting and non-targeting effects of nanomaterials on DNA: challenges and perspectives. *Rev. Environ. Sci. Bio/Technol.* 18 (4): 617–634. doi: 10.1007/s11157-019-09510-3.

Jing, X., Park, J.H., Peters, T.M., and Thorne, P.S. (2015). Toxicity of copper oxide nanoparticles in lung epithelial cells exposed at the air–liquid interface compared with in vivo assessment. *Toxicol. Vitro* 29 (3): 502–511. doi: 10.1016/j.tiv.2014.12.023.

Keerthana, S. and Kumar, A. (2020). Potential risks and benefits of zinc oxide nanoparticles: a systematic review. *Crit. Rev. Toxicol.* 50 (1): 47–71. doi: 10.1080/10408444.2020.1726282.

Kirsch-Volders, M., Plas, G., Elhajouji, A., Lukamowicz, M., Gonzalez, L., Vande Loock, K., and Decordier, I. (2011). The in vitro MN assay in 2011: origin and fate, biological significance, protocols, high throughput methodologies and toxicological relevance. *Arch. Toxicol.* 85 (8): 873–899. doi: 10.1007/s00204-011-0691-4.

Kohl, Y., Rundén-Pran, E., Mariussen, E., Hesler, M., El Yamani, N., Longhin, E.M., and Dusinska, M. (2020). Genotoxicity of nanomaterials: advanced in vitro models and high throughput methods for human hazard assessment – a review. *Nanomaterials (Basel)* 10 (10). doi: 10.3390/nano10101911.

Kong, W., Kum, H., Bae, S.H., Shim, J., Kim, H., Kong, L., Meng, Y., Wang, K., Kim, C., and Kim, J. (2019). Path towards graphene commercialization from lab to market. *Nat. Nanotechnol.* 14 (10): 927–938. doi: 10.1038/s41565-019-0555-2.

Kononenko, V., Repar, N., Marušič, N., Drašler, B., Romih, T., Hočevar, S., and Drobne, D. (2017). Comparative in vitro genotoxicity study of ZnO nanoparticles, ZnO macroparticles and ZnCl(2) to MDCK kidney cells: size matters. *Toxicol. Vitro* 40: 256–263. doi: 10.1016/j.tiv.2017.01.015.

Lacal, I. and Ventura, R. (2018). Epigenetic inheritance: concepts, mechanisms and perspectives. *Front. Mol. Neurosci.* 11: 292. doi: 10.3389/fnmol.2018.00292.

Lai, X., Zhao, H., Zhang, Y., Guo, K., Xu, Y., Chen, S., and Zhang, J. (2018). Intranasal delivery of copper oxide nanoparticles induces pulmonary toxicity and fibrosis in C57BL/6 mice. *Sci. Rep.* 8 (1): 4499. doi: 10.1038/s41598-018-22556-7.

Lee, G.H., Kim, Y.S., Kwon, E., Yun, J.W., and Kang, B.C. (2020). Toxicologic evaluation for amorphous silica nanoparticles: genotoxic and non-genotoxic tumor-promoting potential. *Pharmaceutics* 12 (9). doi: 10.3390/pharmaceutics12090826.

Lee, J.K., Jeong, A.Y., Bae, J., Seok, J.H., Yang, J.Y., Roh, H.S., Jeong, J., Han, Y., Jeong, J., and Cho, W.S. (2017). The role of surface functionalization on the pulmonary inflammogenicity and translocation into mediastinal lymph nodes of graphene nanoplatelets in rats. *Arch. Toxicol.* 91 (2): 667–676. doi: 10.1007/s00204-016-1706-y.

Li, R., Guiney, L.M., Chang, C.H., Mansukhani, N.D., Ji, Z., Wang, X., Liao, Y.P., Jiang, W., Sun, B., Hersam, M.C., Nel, A.E., and Xia, T. (2018). The surface oxidation of graphene oxide determines membrane damage, lipid peroxidation, and cytotoxicity in macrophages in a pulmonary toxicity model. *ACS Nano* 12 (2): 1390–1402. doi: 10.1021/acsnano.7b07737.

Li, Y., Yan, J., Ding, W., Chen, Y., Pack, L.M., and Chen, T. (2017). Genotoxicity and gene expression analyses of liver and lung tissues of mice treated with titanium dioxide nanoparticles. *Mutagenesis* 32 (1): 33–46. doi: 10.1093/mutage/gew065.

Liao, C., Li, Y., and Tjong, S.C. (2019a). Bactericidal and cytotoxic properties of silver nanoparticles. *Int. J. Mol. Sci.* 20 (2). doi: 10.3390/ijms20020449.

Liao, F., Chen, L., Liu, Y., Zhao, D., Peng, W., Wang, W., and Feng, S. (2019b). The size-dependent genotoxic potentials of titanium dioxide nanoparticles to endothelial cells. *Environ. Toxicol.* 34 (11): 1199–1207. doi: 10.1002/tox.22821.

Liao, H.Y., Chung, Y.T., Lai, C.H., Wang, S.L., Chiang, H.C., Li, L.A., Tsou, T.C., Li, W.F., Lee, H.L., Wu, W.T., Lin, M.H., Hsu, J.H., Ho, J.J., Chen, C.J., Shih, T.S., Lin, C.C., and Liou, S.H. (2014). Six-month follow-up study of health markers of nanomaterials among workers handling engineered nanomaterials. *Nanotoxicology* 8 (Suppl 1): 100–110. doi: 10.3109/17435390.2013.858793.

Liou, S.H., Wu, W.T., Liao, H.Y., Chen, C.Y., Tsai, C.Y., Jung, W.T., and Lee, H.L. (2017). Global DNA methylation and oxidative stress biomarkers in workers exposed to metal oxide nanoparticles. *J. Hazard. Mater.* 331: 329–335. doi: 10.1016/j.jhazmat.2017.02.042.

Lu, X., Miousse, I.R., Pirela, S.V., Moore, J.K., Melnyk, S., Koturbash, I., and Demokritou, P. (2016). In vivo epigenetic effects induced by engineered nanomaterials: a case study of copper oxide and laser printer-emitted engineered nanoparticles. *Nanotoxicology* 10 (5): 629–639. doi: 10.3109/17435390.2015.1108473.

Lugun, O., Singh, J., Thakur, R.S., and Pandey, A.K. (2022). Cobalt oxide (Co3O4) nanoparticles induced genotoxicity in Chinese hamster lung fibroblast (V79) cells through modulation of reactive oxygen species. *Mutagenesis* 37 (1): 44–59. doi: 10.1093/mutage/geac005.

Magdolenova, Z., Drlickova, M., Henjum, K., Rundén-Pran, E., Tulinska, J., Bilanicova, D., Pojana, G., Kazimirova, A., Barancokova, M., Kuricova, M., Liskova, A., Staruchova, M., Ciampor, F., Vavra, I., Lorenzo, Y., Collins, A., Rinna, A., Fjellsbø, L., Volkovova, K., Marcomini, A., Amiry-Mog haddam, M., and Dusinska, M. (2015). Coating-dependent induction of cytotoxicity and genotoxicity of iron oxide nanoparticles. *Nanotoxicology* 9 (Suppl 1): 44–56. doi: 10.3109/17435390.2013.847505.

Malkova, A., Svadlakova, T., Singh, A., Kolackova, M., Vankova, R., Borsky, P., Holmannova, D., Karas, A., Borska, L., and Fiala, Z. (2021). In vitro assessment of the genotoxic potential of pristine graphene platelets. *Nanomaterials (Basel)* 11 (9). doi: 10.3390/nano11092210.

Manivannan, J., Banerjee, R., and Mukherjee, A. (2020). Genotoxicity analysis of rutile titanium dioxide nanoparticles in mice after 28 days of repeated oral administration. *The Nucleus* 63 (1): 17–24. doi: 10.1007/s13237-019-00277-0.

Manuja, A., Kumar, B., Kumar, R., Chhabra, D., Ghosh, M., Manuja, M., Brar, B., Pal, Y., Tripathi, B.N., and Prasad, M. (2021). Metal/metal oxide nanoparticles: toxicity concerns associated with their physical state and remediation for biomedical applications. *Toxicol. Rep.* 8: 1970–1978. doi: 10.1016/j.toxrep.2021.11.020.

May, S., Hirsch, C., Rippl, A., Burkle, A., and Wick, P. (2022). Assessing genotoxicity of ten different engineered nanomaterials by the novel semi-automated FADU assay and the alkaline comet assay. *Nanomaterials (Basel)* 12 (2). doi: 10.3390/nano12020220.

Moore, L.D., Le, T., and Fan, G. (2013). DNA methylation and its basic function. *Neuropsychopharmacology* 38 (1): 23–38. doi: 10.1038/npp.2012.112.

Moratin, H., Scherzad, A., Gehrke, T., Ickrath, P., Radeloff, K., Kleinsasser, N., and Hackenberg, S. (2018). Toxicological characterization of ZnO nanoparticles in malignant and non-malignant cells. *Environ. Mol. Mutagen.* 59 (3): 247–259. doi: 10.1002/em.22156.

Moreno-Villanueva, M., Eltze, T., Dressler, D., Bernhardt, J., Hirsch, C., Wick, P., von Scheven, G., Lex, K., and Bürkle, A. (2011). The automated FADU-assay, a potential high-throughput in vitro method for early screening of DNA breakage. *Altex* 28 (4): 295–303. doi: 10.14573/altex.2011.4.295.

Mortezaee, K., Najafi, M., Samadian, H., Barabadi, H., Azarnezhad, A., and Ahmadi, A. (2019). Redox interactions and genotoxicity of metal-based nanoparticles: a comprehensive review. *Chem. Biol. Interact.* 312: 108814. doi: 10.1016/j.cbi.2019.108814.

Murugadoss, S., Lison, D., Godderis, L., Van Den Brule, S., Mast, J., Brassinne, F., Sebaihi, N., and Hoet, P.H. (2017). Toxicology of silica nanoparticles: an update. *Arch. Toxicol.* 91 (9): 2967–3010. doi: 10.1007/s00204-017-1993-y.

Musolino, E., Pagiatakis, C., Serio, S., Borgese, M., Gamberoni, F., Gornati, R., Bernardini, G., and Papait, R. (2022). The Yin and Yang of epigenetics in the field of nanoparticles. *Nanoscale Adv.* 4 (4): 979–994. doi: 10.1039/D1NA00682G.

Mytych, J., Zebrowski, J., Lewinska, A., and Wnuk, M. (2017). Prolonged effects of silver nanoparticles on p53/p21 pathway-mediated proliferation, DNA damage response, and methylation parameters in HT22 hippocampal neuronal cells. *Mol. Neurobiol.* 54 (2): 1285–1300. doi: 10.1007/s12035-016-9688-6.

Naz, S., Gul, A., and Zia, M. (2020). Toxicity of copper oxide nanoparticles: a review study. *IET Nanobiotechnol.* 14 (1): 1–13. doi: 10.1049/iet-nbt.2019.0176.

Ng, C.T., Yong, L.Q., Hande, M.P., Ong, C.N., Yu, L.E., Bay, B.H., and Baeg, G.H. (2017). Zinc oxide nanoparticles exhibit cytotoxicity and genotoxicity through oxidative stress responses in human lung fibroblasts and *Drosophila melanogaster*. *Int. J. Nanomed.* 12: 1621–1637. doi: 10.2147/ijn.S124403.

Novotna, B., Pelclova, D., Rossnerova, A., Zdimal, V., Ondracek, J., Lischkova, L., Vlckova, S., Fenclova, Z., Klusackova, P., Zavodna, T., Topinka, J., Komarc, M., Dvorackova, S., and Rossner, P. (2020). The genotoxic effects in the leukocytes of workers handling nanocomposite materials. *Mutagenesis* 35 (4): 331–340. doi: 10.1093/mutage/geaa016.

OECD. (1997). *Test No. 476: In vitro Mammalian Cell Gene Mutation Test*. Paris, France: OECD Publishing for the Testing Chemicals Section.

OECD. (2014). *Genotoxicity of Manufactured Nanomaterials: Report of the OECD Expert Meeting*, 472, 1–37. Amsterdam, The Netherlands: Elsevier.

OECD. (2015). *Test No. 490: In Vitro Mammalian Cell Gene Mutation Tests Using the Thymidine Kinase Gene*. Paris, France: OECD.

Ou, L., Lv, X., Wu, Z., Xia, W., Huang, Y., Chen, L., Sun, W., Qi, Y., Yang, M., and Qi, L. (2021). Oxygen content-related DNA damage of graphene oxide on human retinal pigment epithelium cells. *J. Mater. Sci. Mater. Med.* 32 (2): 20. doi: 10.1007/s10856-021-06491-0.

Patil, N.A., Gade, W.N., and Deobagkar, D.D. (2016). Epigenetic modulation upon exposure of lung fibroblasts to TiO(2) and ZnO nanoparticles: alterations in DNA methylation. *Int. J. Nanomed.* 11: 4509–4519. doi: 10.2147/ijn.S110390.

Pavlin, M., Lojk, J., Strojan, K., Hafner-Bratkovič, I., Jerala, R., Leonardi, A., Križaj, I., Drnovšek, N., Novak, S., Veranič, P., and Bregar, V.B. (2022). The relevance of physico-chemical properties and protein corona for evaluation of nanoparticles immunotoxicity-in vitro correlation analysis on THP-1 macrophages. *Int. J. Mol. Sci.* 23 (11). doi: 10.3390/ijms23116197.

Pelclova, D., Zdimal, V., Fenclova, Z., Vlckova, S., Turci, F., Corazzari, I., Kacer, P., Schwarz, J., Zikova, N., Makes, O., Syslova, K., Komarc, M., Belacek, J., Navratil, T., Machajova, M., and Zakharov, S. (2016). Markers of oxidative damage of nucleic acids and proteins among workers exposed to TiO2 (nano) particles. *Occup. Environ. Med.* 73 (2): 110–118. doi: 10.1136/oemed-2015-103161.

Pelclova, D., Zdimal, V., Kacer, P., Zikova, N., Komarc, M., Fenclova, Z., Vlckova, S., Schwarz, J., Makeš, O., Syslova, K., Navratil, T., Turci, F., Corazzari, I., Zakharov, S., and Bello, D. (2017). Markers of lipid oxidative damage in the exhaled breath condensate of nano TiO(2) production workers. *Nanotoxicology* 11 (1): 52–63. doi: 10.1080/17435390.2016.1262921.

Pelclova, D., Zdimal, V., Komarc, M., Schwarz, J., Ondracek, J., Ondrackova, L., Kostejn, M., Vlckova, S., Fenclova, Z., Dvorackova, S., Lischkova, L., Klusackova, P., Kolesnikova, V., Rossnerova, A., and Navratil, T. (2020). Three-year study of markers of oxidative stress in exhaled breath condensate in workers producing nanocomposites, extended by plasma and urine analysis in last two years. *Nanomaterials (Basel)* 10 (12). doi: 10.3390/nano10122440.

Pérez, R.F., Soto Fernández, A.Y., Bousquets Muñoz, P., Sierra, M.I., Tejedor, J.R., Morales-Sánchez, P., Valdés, A.F., Santamaría, R., Blanco, C., Torrecillas, R., Fraga, M.F., and Fernández, A.F. (2020). No genome-wide DNA methylation changes found associated with medium-term reduced graphene oxide exposure in human lung epithelial cells. *Epigenetics* 15 (3): 283–293. doi: 10.1080/15592294.2019.1666650.

Pogribna, M. and Hammons, G. (2021). Epigenetic effects of nanomaterials and nanoparticles. *J. Nanobiotechnol.* 19 (1): 2. doi: 10.1186/s12951-020-00740-0.

Rahimi Kalateh Shah Mohammad, G., Seyedi, S.M.R., Karimi, E., and Homayouni-Tabrizi, M. (2019). The cytotoxic properties of zinc oxide nanoparticles on the rat liver and spleen, and its anticancer impacts on human liver cancer cell lines. *J. Biochem. Mol. Toxicol.* 33 (7): e22324. doi: 10.1002/jbt.22324.

Ristic, B., Harhaji-Trajkovic, L., Bosnjak, M., Dakic, I., Mijatovic, S., and Trajkovic, V. (2021). Modulation of cancer cell autophagic responses by graphene-based nanomaterials: molecular mechanisms and therapeutic implications. *Cancers* 13 (16). doi: 10.3390/cancers13164145.

Rossner, P., Jr., Vrbova, K., Rossnerova, A., Zavodna, T., Milcova, A., Klema, J., Vecera, Z., Mikuska, P., Coufalik, P., Capka, L., Krumal, K., Docekal, B., Holan, V., Machala, M., and Topinka, J. (2020). Gene expression and epigenetic changes in mice following inhalation of copper(II) oxide nanoparticles. *Nanomaterials (Basel)* 10 (3). doi: 10.3390/nano10030550.

Rossnerova, A., Honkova, K., Pelclova, D., Zdimal, V., Hubacek, J.A., Chvojkova, I., Vrbova, K., Rossner, P., Jr., Topinka, J., Vlckova, S., Fenclova, Z., Lischkova, L., Klusackova, P., Schwarz, J., Ondracek, J., Ondrackova, L., Kostejn, M., Klema, J., and Dvorackova, S. (2020). DNA methylation profiles in a group of workers occupationally exposed to nanoparticles. *Int. J. Mol. Sci.* 21 (7). doi: 10.3390/ijms21072420.

Rossnerova, A., Pelclova, D., Zdimal, V., Rossner, P., Elzeinova, F., Vrbova, K., Topinka, J., Schwarz, J., Ondracek, J., Kostejn, M., Komarc, M., Vlckova, S., Fenclova, Z., and Dvorackova, S. (2019). The repeated cytogenetic analysis of subjects occupationally exposed to nanoparticles: a pilot study. *Mutagenesis* 34 (3): 253–263. doi: 10.1093/mutage/gez016.

Santonastaso, M., Mottola, F., Colacurci, N., Iovine, C., Pacifico, S., Cammarota, M., Cesaroni, F., and Rocco, L. (2019). In vitro genotoxic effects of titanium dioxide nanoparticles (n-TiO(2)) in human sperm cells. *Mol. Reprod. Dev.* 86 (10): 1369–1377. doi: 10.1002/mrd.23134.

Seidel, C., Kirsch, A., Fontana, C., Visvikis, A., Remy, A., Gaté, L., Darne, C., and Guichard, Y. (2017). Epigenetic changes in the early stage of silica-induced cell transformation. *Nanotoxicology* 11 (7): 923–935. doi: 10.1080/17435390.2017.1382599.

Sharifi, S., Behzadi, S., Laurent, S., Forrest, M.L., Stroeve, P., and Mahmoudi, M. (2012). Toxicity of nanomaterials. *Chem. Soc. Rev.* 41 (6): 2323–2343. doi: 10.1039/c1cs15188f.

Shukla, R.K., Badiye, A., Vajpayee, K., and Kapoor, N. (2021). Genotoxic potential of nanoparticles: structural and functional modifications in DNA. *Front. Genet.* 12: 728250. doi: 10.3389/fgene.2021.728250.

Siivola, K.M., Suhonen, S., Hartikainen, M., Catalán, J., and Norppa, H. (2020). Genotoxicity and cellular uptake of nanosized and fine copper oxide particles in human bronchial epithelial cells in vitro. *Mutat. Res. Genet. Toxicol. Environ. Mutagen.* 856–857: 503217. doi: 10.1016/j.mrgentox.2020.503217.

Skubalova, Z., Michalkova, H., Michalek, P., Strmiska, V., Guran, R., Merlos Rodrigo, M.A., Castkova, K., Hynek, D., Pekarik, V., Zitka, O., Adam, V., and Heger, Z. (2019). Prevalent anatase crystalline phase increases the cytotoxicity of biphasic titanium dioxide nanoparticles in mammalian cells. *Colloids Surf. B: Biointerfaces* 182: 110391. doi: 10.1016/j.colsurfb.2019.110391.

Smolkova, B., Dusinska, M., and Gabelova, A. (2017). Nanomedicine and epigenome. Possible health risks. *Food Chem. Toxicol.* 109 (Pt 1): 780–796. doi: 10.1016/j.fct.2017.07.020.

Smolkova, B., El Yamani, N., Collins, A.R., Gutleb, A.C., and Dusinska, M. (2015). Nanoparticles in food. Epigenetic changes induced by nanomaterials and possible impact on health. *Food Chem. Toxicol.* 77: 64–73. doi: 10.1016/j.fct.2014.12.015.

Sramkova, M., Kozics, K., Masanova, V., Uhnakova, I., Razga, F., Nemethova, V., Mazancova, P., Kapka-Skrzypczak, L., Kruszewski, M., Novotova, M., Puntes, V.F., and Gabelova, A. (2019). Kidney nanotoxicity studied in human renal proximal tubule epithelial cell line TH1. *Mutat. Res. Genet. Toxicol. Environ. Mutagen.* 845: 403017. doi: 10.1016/j.mrgentox.2019.01.012.

Srivastav, A.K., Kumar, A., Prakash, J., Singh, D., Jagdale, P., Shankar, J., and Kumar, M. (2017). Genotoxicity evaluation of zinc oxide nanoparticles in Swiss mice after oral administration using chromosomal aberration, micronuclei, semen analysis, and RAPD profile. *Toxicol. Ind. Health* 33 (11): 821–834. doi: 10.1177/0748233717717842.

Stoccoro, A., Di Bucchianico, S., Coppedè, F., Ponti, J., Uboldi, C., Blosi, M., Delpivo, C., Ortelli, S., Costa, A.L., and Migliore, L. (2017). Multiple endpoints to evaluate pristine and remediated titanium dioxide nanoparticles genotoxicity in lung epithelial A549 cells. *Toxicol. Lett.* 276: 48–61. doi: 10.1016/j.toxlet.2017.05.016.

Stoccoro, A., Karlsson, H.L., Coppedè, F., and Migliore, L. (2013). Epigenetic effects of nano-sized materials. *Toxicology* 313 (1): 3–14. doi: 10.1016/j.tox.2012.12.002.

Strauch, B.M., Niemand, R.K., Winkelbeiner, N.L., and Hartwig, A. (2017). Comparison between micro- and nanosized copper oxide and water-soluble copper chloride: interrelationship between intracellular copper concentrations, oxidative stress and DNA damage response in human lung cells. *Part Fibre Toxicol.* 14 (1): 28. doi: 10.1186/s12989-017-0209-1.

Sun, T.Y., Bornhöft, N.A., Hungerbühler, K., and Nowack, B. (2016). Dynamic probabilistic modeling of environmental emissions of engineered nanomaterials. *Environ. Sci. Technol.* 50 (9): 4701–4711. doi: 10.1021/acs.est.5b05828.

Tang, H., Xu, M., Shi, F., Ye, G., Lv, C., Luo, J., Zhao, L., and Li, Y. (2018). Effects and mechanism of nano-copper exposure on hepatic cytochrome P450 enzymes in rats. *Int. J. Mol. Sci.* 19 (7). doi: 10.3390/ijms19072140.

Ursini, C.L., Fresegna, A.M., Ciervo, A., Maiello, R., Del Frate, V., Folesani, G., Galetti, M., Poli, D., Buresti, G., Di Cristo, L., Sabella, S., Iavicoli, S., and Cavallo, D. (2021). Occupational exposure to graphene and silica nanoparticles. Part II: pilot study to identify a panel of sensitive biomarkers of genotoxic, oxidative and inflammatory effects on suitable biological matrices. *Nanotoxicology* 15 (2): 223–237. doi: 10.1080/17435390.2020.1850903.

Vandegehuchte, M.B. and Janssen, C.R. (2014). Epigenetics in an ecotoxicological context. *Mutat. Res. Genet. Toxicol. Environ. Mutagen.* 764-765: 36–45. doi: 10.1016/j.mrgentox.2013.08.008.

Veeragoni, D., Deshpande, S., Rachamalla, H.K., Ande, A., Misra, S., and Mutheneni, S.R. (2022). In vitro and in vivo anticancer and genotoxicity profiles of green synthesized and chemically synthesized silver nanoparticles. *ACS Appl. Bio. Mater.* 5 (5): 2324–2339. doi: 10.1021/acsabm.2c00149.

Verstraelen, S., Remy, S., Casals, E., De Boever, P., Witters, H., Gatti, A., Puntes, V., and Nelissen, I. (2014). Gene expression profiles reveal distinct immunological responses of cobalt and cerium dioxide nanoparticles in two in vitro lung epithelial cell models. *Toxicol. Lett.* 228 (3): 157–169. doi: 10.1016/j.toxlet.2014.05.006.

Wang, J., Wang, J., Liu, Y., Nie, Y., Si, B., Wang, T., Waqas, A., Zhao, G., Wang, M., and Xu, A. (2019). Aging-independent and size-dependent genotoxic response induced by titanium dioxide nanoparticles in mammalian cells. *J. Environ. Sci. (China)* 85: 94–106. doi: 10.1016/j.jes.2019.04.024.

Wang, L., Zhang, T., Li, P., Huang, W., Tang, J., Wang, P., Liu, J., Yuan, Q., Bai, R., Li, B., Zhang, K., Zhao, Y., and Chen, C. (2015). Use of synchrotron radiation–analytical techniques to reveal chemical origin of silver-nanoparticle cytotoxicity. *ACS Nano* 9 (6): 6532–6547. doi: 10.1021/acsnano.5b02483.

Wang, X., Zhou, W., Li, X., Ren, J., Ji, G., Du, J., Tian, W., Liu, Q., and Hao, A. (2020). Graphene oxide suppresses the growth and malignancy of glioblastoma stem cell-like spheroids via epigenetic mechanisms. *J. Transl. Med.* 18 (1): 200. doi: 10.1186/s12967-020-02359-z.

Wani, M.R. and Shadab, G. (2020). Titanium dioxide nanoparticle genotoxicity: a review of recent in vivo and in vitro studies. *Toxicol. Ind. Health* 36 (7): 514–530. doi: 10.1177/0748233720936835.

Watson, C., Ge, J., Cohen, J., Pyrgiotakis, G., Engelward, B.P., and Demokritou, P. (2014). High-throughput screening platform for engineered nanoparticle-mediated genotoxicity using CometChip technology. *ACS Nano* 8 (3): 2118–2133. doi: 10.1021/nn404871p.

Wen, H., Dan, M., Yang, Y., Lyu, J., Shao, A., Cheng, X., Chen, L., and Xu, L. (2017). Acute toxicity and genotoxicity of silver nanoparticle in rats. *PLoS One* 12 (9): e0185554. doi: 10.1371/journal.pone.0185554.

Wu, F., Seib, M., Mauel, S., Klinzing, S., and Hicks, A.L. (2020a). A citizen science approach estimating titanium dioxide released from personal care products. *PLoS One* 15 (7): e0235988. doi: 10.1371/journal.pone.0235988.

Wu, K., Zhou, Q., and Ouyang, S. (2021a). Direct and indirect genotoxicity of graphene family nanomaterials on DNA-A review. *Nanomaterials (Basel)* 11 (11). doi: 10.3390/nano11112889.

Wu, W.T., Li, L.A., Tsou, T.C., Wang, S.L., Lee, H.L., Shih, T.S., and Liou, S.H. (2019). Longitudinal follow-up of health effects among workers handling engineered nanomaterials: a panel study. *Environ. Health: Glob. Access Sci. Source.* 18 (1): 107. doi: 10.1186/s12940-019-0542-y.

Wu, Y., Feng, W., Liu, R., Xia, T., and Liu, S. (2020b). Graphene oxide causes disordered zonation due to differential intralobular localization in the liver. *ACS Nano* 14 (1): 877–890. doi: 10.1021/acsnano.9b08127.

Xiao, X., Li, Y., and Liu, Z. (2016). Graphene commercialization. *Nat. Mater.* 15 (7): 697–698. doi: 10.1038/nmat4665.

Xiaoli, F., Yaqing, Z., Ruhui, L., Xuan, L., Aijie, C., Yanli, Z., Chen, H., Lili, C., and Longquan, S. (2021). Graphene oxide disrupted mitochondrial homeostasis through inducing intracellular redox deviation and autophagy-lysosomal network dysfunction in SH-SY5Y cells. *J. Hazard. Mater.* 416: 126158. doi: 10.1016/j.jhazmat.2021.126158.

Xu, L., Wang, Y.Y., Huang, J., Chen, C.Y., Wang, Z.X., and Xie, H. (2020). Silver nanoparticles: synthesis, medical applications and biosafety. *Theranostics* 10 (20): 8996–9031. doi: 10.7150/thno.45413.

Xu, M., Tang, H., Zhou, X., Chen, H., Dong, Q., Zhang, Y., Ye, G., Shi, F., Lv, C., Jing, B., He, C., Zhao, L., and Li, Y. (2018). Effects and mechanisms of sub-chronic exposure to copper nanoparticles on renal cytochrome P450 enzymes in rats. *Environ. Toxicol. Pharmacol.* 63: 135–146. doi: 10.1016/j.etap.2018.08.004.

Yanagisawa, H., Seki, Y., Yogosawa, S., Takumi, S., Shimizu, H., and Suka, M. (2018). Potential role of mitochondrial damage and S9 mixture including metabolic enzymes in ZnO nanoparticles-induced oxidative stress and genotoxicity in Chinese hamster lung (CHL/IU) cells. *Mutat. Res. Genet. Toxicol. Environ. Mutagen.* 834: 25–34. doi: 10.1016/j.mrgentox.2018.07.003.

Yazdimamaghani, M., Moos, P.J., Dobrovolskaia, M.A., and Ghandehari, H. (2019). Genotoxicity of amorphous silica nanoparticles: status and prospects. *Nanomedicine: Nanotechnol. Biol. Med.* 16: 106–125. doi: 10.1016/j.nano.2018.11.013.

Yetisen, A.K., Qu, H., Manbachi, A., Butt, H., Dokmeci, M.R., Hinestroza, J.P., Skorobogatiy, M., Khademhosseini, A., and Yun, S.H. (2016). Nanotechnology in textiles. *ACS Nano* 10 (3): 3042–3068. doi: 10.1021/acsnano.5b08176.

Yousef, M.I., Abuzreda, A.A., and Kamel, M.A.E.L.N. (2019). Neurotoxicity and inflammation induced by individual and combined exposure to iron oxide nanoparticles and silver nanoparticles. *J. Taibah Univ. Sci.* 13 (1): 570–578. doi: 10.1080/16583655.2019.1602351.

Zhang, J., Cao, H.Y., Wang, J.Q., Wu, G.D., and Wang, L. (2021). Graphene oxide and reduced graphene oxide exhibit cardiotoxicity through the regulation of lipid peroxidation, oxidative stress, and mitochondrial dysfunction. *Front. Cell Dev. Biol.* 9: 616888. doi: 10.3389/fcell.2021.616888.

Zhang, Q., Wu, Z., Li, N., Pu, Y., Wang, B., Zhang, T., and Tao, J. (2017). Advanced review of graphene-based nanomaterials in drug delivery systems: synthesis, modification, toxicity and application. *Mater. Sci. Eng. C: Mater. Biol. Appl.* 77: 1363–1375. doi: 10.1016/j.msec.2017.03.196.

Zhang, W., Zhao, Y., Li, F., Li, L., Feng, Y., Min, L., Ma, D., Yu, S., Liu, J., Zhang, H., Shi, T., Li, F., and Shen, W. (2018). Zinc oxide nanoparticle caused plasma metabolomic perturbations correlate with hepatic steatosis. *Front. Pharmacol.* 9: 57. doi: 10.3389/fphar.2018.00057.

Zhang, Y., Bello, A., Ryan, D.K., Demokritou, P., and Bello, D. (2022). Elevated urinary biomarkers of oxidative damage in photocopier operators following acute and chronic exposures. *Nanomaterials (Basel)* 12 (4). doi: 10.3390/nano12040715.

Zhao, Y., Wu, Q., and Wang, D. (2016). An epigenetic signal encoded protection mechanism is activated by graphene oxide to inhibit its induced reproductive toxicity in *Caenorhabditis elegans*. *Biomaterials* 79: 15–24. doi: 10.1016/j.biomaterials.2015.11.052.

Zhou, F., Liao, F., Chen, L., Liu, Y., Wang, W., and Feng, S. (2019). The size-dependent genotoxicity and oxidative stress of silica nanoparticles on endothelial cells. *Environ. Sci. Pollut. Res. Int.* 26 (2): 1911–1920. doi: 10.1007/s11356-018-3695-2.

Zijno, A., Cavallo, D., Di Felice, G., Ponti, J., Barletta, B., Butteroni, C., Corinti, S., De Berardis, B., Palamides, J., Ursini, C.L., Fresegna, A.M., Ciervo, A., Maiello, R., and Barone, F. (2020). Use of a common European approach for nanomaterials' testing to support regulation: a case study on titanium and silicon dioxide representative nanomaterials. *J. Appl. Toxicol.* 40 (11): 1511–1525. doi: 10.1002/jat.4002.

6

Epigenotoxicity of Titanium Dioxide Nanoparticles

*Carlos Wells, Marta Pogribna, Beverly Lyn-Cook, and George Hammons**

Division of Biochemical Toxicology, National Center for Toxicological Research, U.S. Food & Drug Administration, Jefferson, Arkansas USA
** Corresponding author*

Introduction

There has been a tremendous growth in the use of nanoparticles and nanomaterials in the past few years with applications in various fields, such as medicine, pharmacy, agriculture, engineering, and household products (Khan et al. 2019). Nanoparticles are small materials ranging in size from 1 to 100 nm. With a decrease in particle size, the ratio of surface areas to volume of nanomaterials increases, forming the basis for the differences in their physicochemical properties when compared with their bulk material (Salata 2004). Given our increased exposure to nanomaterials, their potential biological effects have assumed increased importance and elevated concerns of the potential health risks to humans. Extensive studies have been conducted to access the toxic effects of nanoparticles using both *in vitro* and *in vivo* model systems (Hou et al. 2019; Song et al. 2016). One of the primary causes of nanotoxicity is the excessive reactive oxygen species (ROS) production through the nanoparticles' interaction with biological molecules, such as proteins, lipids, and DNA, resulting in oxidative stress (Fu et al. 2014), lipid peroxidation (Pacurari et al. 2008), and inflammation (Valko et al. 2007).

Among the various nanomaterials, titanium dioxide nanoparticles (TiO_2 NPs) are one of the most abundant and widely used in the world (Shabbir et al. 2021; Zhang et al. 2015). There are three crystalline structures: anatase, rutile, and brookite. Anatase is the most common type and rutile is the most stable form, while brookite is the rarest (Fresegna et al. 2021). TiO_2 NPs exist in various nanostructures, such as nanoparticles, nanotubes, nanorods and nanowires, nanofilms, nanosheets, and nano-coatings. Approximately four million tons of this nanomaterial are consumed annually worldwide (Ortlieb 2010). In addition, TiO_2 accounts for 70% of the total production volume of pigments worldwide (Baan et al. 2006) and is in the top five nanoparticles used in consumer products (Shukla et al. 2011). TiO_2 NPs have a variety of applications and can be used in, for example, paints, coatings, plastics, papers, inks, medicines, pharmaceuticals, ceramic, and metallurgy (Kaida et al. 2004; Wang et al. 2007a; Wolf et al. 2003).

TiO_2 NPs are also used in the food industry and personal care products, including toothpastes, lip balms, shampoos, deodorants, and sunscreens (Heringa et al. 2016). In the United States, TiO_2 NPs can be used in food if its content does not exceed 1% of the total weight of the product containing TiO_2 (US Food and Drug Administration 2005). TiO_2 NPs are added to many foodstuffs, including cheeses and sauces, skimmed milk, ice cream, and confectionery products, e.g., as coating on sweets and chewing gum (Bachler et al. 2015; Chen et al. 2013; Dudefoi et al. 2017; Peters et al.

Impact of Engineered Nanomaterials in Genomics and Epigenomics, First Edition. Edited by Saura C. Sahu.

2014). Its content in sweets, and in particular candy, chewing gum, chocolate, and white-coated products, compared with other products, is very high, reaching 2.5 mg titanium (Ti)/g of food. Food-grade TiO_2 consists of approximately 40% of nanoparticles (<100 nm) and 60% of microparticles (>100 nm) (Dorier et al. 2017; Proquin et al. 2017; Weir et al. 2012). In the United States, the estimated average exposure of children below 10 years of age is 1–2 mg TiO_2/kg body weight (bw)/day and above 10 years old, 0.2–0.7 mg TiO_2/kg bw/day. In the United Kingdom, the estimated exposure is on average 2–3 mg TiO_2/kg bw/day for children younger than 10 years and 1 mg TiO_2/kg bw/day for children over 10 years of age (Weir et al. 2012). Another study performed in the Netherlands showed an intake of TiO_2 in the same range as in the United States (Rompelberg et al. 2016). This study estimated the mean average exposure to be 0.67 mg TiO_2/kg bw/day between 2 and 6 years old, 0.17 mg TiO_2/kg bw/day between 7 and 69 years old, and 0.06 mg TiO_2/kg bw/day above 70 years old. In addition, TiO_2 has long been used as a component for articulating prosthetic implants, especially for the hip and knee (Jacobs et al. 1991; Sul 2010). These implants occasionally fail due to degradation of the materials in the implant or a chronic inflammatory response to the implant material (Patri et al. 2009; Shi et al. 2013).

The wide variety of applications of TiO_2 NPs has been leading to reasonable considerations for the potential harm of these nanoparticles to human health and the environment. Humans are exposed to these nanoparticles through dermal, oral, or respiratory routes (Shakeel et al. 2016; Wang et al. 2007a). Traditionally, TiO_2 NPs have been considered as poorly soluble, low toxicity particles (Participants IRSIW 2000). However, this view was challenged after lung tumors developed in rats after two years of exposure to high concentrations of fine TiO_2 particles (Lee et al. 1985). The International Agency for Research on Cancer (IARC) has classified TiO_2 as a Group 2B carcinogen (possibly carcinogenic to humans) (IARC 2010). Most recently, food-grade TiO_2 will be banned as a food additive in the EU after the European Food Safety Authority (EFSA) ruled that TiO_2 is no longer safe when used as a food additive (EFSA 2016). EFSA concluded that it did not have sufficient data to calculate a safe daily intake. An acceptable daily intake could not be established.

Although controversial results have been found when studying the toxicity of TiO_2 NPs, perhaps due to different particle sizes, doses, culture media, or test methods (Fresegna et al. 2021), increasing evidence has shown that TiO_2 NPs exert a variety of adverse health effects, including liver function damage, nephrotoxicity, and pulmonary toxicity, suggesting considerable risk to human health (Chen et al. 2006; Duan et al. 2010; Rashid et al. 2012; Wang et al. 2007b). Many of the *in vitro* and *in vivo* studies reporting significant toxicity of TiO_2 NPs showed loss of cell viability, oxidative stress, inflammatory response, genotoxicity, and apoptotic cell death (Iavicoli et al. 2011; Shi et al. 2013; Zhang et al. 2015). In print plant manufacturing units, workers were found to have symptoms such as shortness of breath because they were exposed to polyacrylate nanoparticles coupled with TiO_2 NPs without any protective procedures (Song et al. 2009). Other clinical signs of TiO_2 toxicity may include rashes on the face, hands, and forearms (Lu et al. 2008). Available data on TiO_2 NPs toxicity to humans remain, however, limited.

Most studies on the toxicity mechanisms of nanomaterials have focused on the effects of nanomaterials on cells at the transcriptome and proteome levels. However, there is increasing evidence demonstrating that exposure to nanoparticles can induce epigenetic alterations in the target tissues and cells even at sub-cytotoxic, low doses (Hou et al. 2019; Pogribna and Hammons 2021; Stoccoro et al. 2013; Wong et al. 2017). Epigenetics is defined as “a stably heritable phenotype resulting from changes in a chromosome without alterations in the DNA sequence” (Felsenfeld 2014). Epigenetic modifications mainly include DNA methylation, histone tail modifications (PTM), and noncoding RNAs (ncRNA), particularly microRNAs (miRNAs) and long noncoding RNAs (lncRNA) expression (Portela and Esteller 2010; Wei et al. 2017).

The epigenome contributes to regulating gene expression and critical cellular functions, from gene imprinting to gene reprogramming, by switching genes "on" and "off". Epigenetics constitutes an important link between genotype and phenotype and plays a critical role in the regulation of numerous cellular processes (Chung and Herceg 2020). It is recognized that epigenetic regulatory mechanisms can change dynamically in response to environmental stressors and that altered epigenetic mechanisms can play an essential role in the development and progression of human diseases, including cancer and neurological, autoimmune, cardiovascular, and metabolic diseases (Feil and Fraga 2012; Ferrari, Carugno et al. 2019; Nilsson and Skinner 2015; Portela and Esteller 2010). Consequentially, nanoparticles can influence our epigenome by targeting epigenetic regulatory molecules and changing the gene expression pattern of the exposed cells, which can ultimately contribute to the development or progression of diseases (Lim et al. 2017; Sierra et al. 2016; Wong et al. 2017; Yu et al. 2020).

Studies demonstrating the effects of TiO_2 NPs on DNA methylation, histone modifications, and miRNA expression in *in vitro* cellular systems and *in vivo* animal models are described in this review. The epigenotoxicity of TiO_2 NPs is a critical contributor to the potential harm of this nanomaterial to human health. Integration of epigenetics, as a necessary and beneficial component in the risk assessment of TiO_2 NPs, is discussed with the application of adverse outcome pathways (AOPs) as a potentially useful approach. Evidence establishing the cellular uptake and tissue biodistribution of TiO_2 NPs in various systems is included. Additionally, epidemiological studies have shown the potential harm associated with exposure to TiO_2 NPs in human populations.

Cellular Uptake and Biodistribution

The main physiological routes of NP uptake are inhalation, dermal absorption, or ingestion. The extent of NP exposure is affected by their characteristics, influencing absorption, metabolism, distribution, and excretion (Frohlich and Roblegg 2016). While inhalation may be the primary route for TiO_2 exposure in the workplace, in nonoccupational settings, titanium concentrations in air are very low, in the range of 0.01–0.1 $\mu g/m^3$, resulting in a maximum exposure through inhalation of only 1 μg Ti/day. Oral exposure, e.g., from food, toothpaste, and medicines, seems to be the most important source for TiO_2 in nonoccupational settings (Rompelberg et al. 2016). To understand the toxicity of nanoparticles, there is the need to measure accurately how they are distributed in the body after exposure and how they interact with pertinent organs and cells (Marquis et al. 2009). Ideally, biodistribution studies should be performed *in vivo*, but *in vitro* models are often needed for ethical and practical reasons. Safety assessment of nanoparticles requires techniques that are suitable to quantify tissue and cellular uptake of nanoparticles (Hsiao et al. 2016). Several analytical techniques have been used in demonstrating uptake (internalization) of TiO_2 NPs in an array of *in vitro* cellular systems representing various tissue types, including lung and colon among others. Transmission electron microscopy (TEM) analysis is the most common method. Substantial evidence has been established for TiO_2 NPs uptake in human- and animal-derived cells. Examples of these studies are presented.

In human lung epithelial BEAS-2B cells exposed to 20 $\mu g/mL$ TiO_2 NPs for 2 weeks, cellular uptake measurement by TEM showed high internalization (Sierra et al. 2017). This cellular uptake of TiO_2 NPs was reported to be dose- and time-dependent and the NPs and their nanoaggregates were mainly confined to vacuoles, although they were also present on the surface of the nuclear membrane. In another study, the uptake of a panel of TiO_2 NPs (pristine, citrate-coated, silica-coated, and TiO_2 Aeroxide® P25, as benchmark material) was compared in human alveolar epithelial type-II-like A549 cells (Stoccoro et al. 2017). After cells were exposed to 20 μg/cm^2 of TiO_2 NPs for 48 h, TEM analysis showed mostly single particle uptake for P25, while for pristine, citrate- and silica-coated TiO_2 NPs, mainly NP agglomerates were found. TiO_2 NPs were found both outside and inside cells where they were mostly localized in lysosomes and early and late endosomes. P25 uptake was qualitatively and quantitatively lower than pristine, citrate- and silica-coated samples.

The cellular uptake, accumulation, and localization of TiO_2 NPs in rat (C6) and human (U373) glial cells were analyzed using time-lapse microscopy (TLM) and TEM (Huerta-Garcıa et al. 2015). TiO_2 NPs were taken up by both cell types and internalized at very short times after exposure. Cells were treated with 5 μg/cm^2 of TiO_2 NPs. No specific localization of TiO_2 NPs into particular organelles was found; in contrast, they were primarily localized into large vesicles in the cytoplasm. TiO_2 NPs internalization was also evaluated in embryonic H9c2 rat cardiomyoblasts (Huerta-Garcia et al. 2019). Cells were treated with TiO_2 NPs (anatase, <25 nm), 20 μg/cm^2 for 5–30 min, 1–6 h, or 1–7 days and analyzed by TEM and TLM. TiO_2 NPs were rapidly internalized. Small aggregates were observed inside cells after 30 min of exposure, whereas large aggregates formed after 1 h. The size of internalized aggregates was dependent on exposure time. In AGS cells (a human stomach epithelial cell line) exposed to food-grade TiO_2 (about 150 nm, a single anatase phase), TEM images confirmed that food-grade TiO_2 localized in the perinuclear region of the AGS cells 24 h after treatment (40 μg/mL) (Han et al. 2021).

Other analytical techniques have also been employed. Uptake analysis performed with human HTR-8/SVneo (HTR) cells (human trophoblastic cells) treated with 1, 10, and 100 μg/mL TiO_2 NPs (anatase:rutile [80:20], 21 nm) demonstrated dose-dependent uptake using flow cytometry (Mao et al. 2018). TEM analysis conducted to observe the location of TiO_2 NPs showed that the nanoparticles accumulated in the cytoplasm. Cellular uptake of three forms of TiO_2 NPs – anatase (NM-102; ~21 nm), rutile (NM-104; ~26 nm), and anatase–rutile (NM-105; anatase:rutile [88.2:11.8]; ~21 nm) – was compared in the human bronchial epithelial cell line 16-HBE (Ghosh et al. 2017). Cells were treated with 3.125, 12.5, and 25 μg/ml TiO_2 NPs. Flow cytometry analysis indicated a dose-dependent increase in the TiO_2 NPs uptake/interaction with the cells. The uptake of TiO_2 NPs was the highest for anatase, which could be due to the higher surface area of the particle.

Using a fluorescence approach for visualization, fluorescein isothiocyanate conjugated TiO_2 NPs (AEROXIDE P25) (125 and 1250 μM) was shown to be internalized in the cytoplasm of human buccal epithelial cells (TR146) by confocal microscopy (Tay et al. 2014). In MCF-7 cells treated with 0.12 μM TiO_2 NPs, cellular uptake was demonstrated with confocal microscopy using an in situ posttreatment labeling approach with optical fluorescence detection (Brown et al. 2018). Cellular uptake of TiO_2 NPs (28 nm) has also been shown in human colorectal adenocarcinoma HT29 cells (Schneider et al. 2017). Cells were treated with 2–10 μg/mL TiO_2 NPs and analyzed by inductively coupled plasma mass spectrometry (ICP-MS). Although the measured intake increased with concentration, only a small amount was up-taken.

Hyperspectral Raman imaging in combination with multivariate classification to quantify the distribution of nanoparticles was employed in lung epithelial A549 cells exposed to TiO_2 NPs (anatase:rutile [80:20]) for 4 or 48 hours (Ahlinder et al. 2013). Up to 40% of spectra from the cell nucleus showed Raman bands associated with nanoparticles. In another study, enhanced dark-field microscopy combined with hyperspectral imaging was utilized to visualize the cellular uptake and binding of nanoparticles in NL20 lung cells (Pogribna et al. 2020). Optical images of TiO_2 NPs-treated cells and changes in the spectral profile of the TiO_2 nanoparticles allowed visualization of the nanoparticles in various organelles in the cell including the nucleus. Other studies have also reported nanoparticles to be localized inside cell nuclei (Hackenberg et al. 2011; Singh et al. 2007).

The biodistribution of TiO_2 NPs after exposure by several routes in various tissues in *in vivo* animal models has been examined. *In vivo* testing revealed that after inhalation or oral exposure, TiO_2 NPs may accumulate in different internal organs (Shabbir et al. 2021). In one study, various TiO_2 NPs, which differ with respect to primary particle size, crystalline form, and hydrophobicity, were investigated in rats up to 90 days postexposure after oral and intravenous administration of a single or five repeated doses (Geraets et al. 2014). For the oral study, dose levels were 6.8–8.5 mg TiO_2/kg bw/day for the single dose treatment and a cumulative dose range of 34.1–42.4 mg TiO_2/kg bw for male rats and 54.5–59.9 mg TiO_2/kg bw for rats. Analysis of the liver, spleen, and mesenteric lymph nodes indicated that some minor absorption occurs in the gastrointestinal tract, but to a very limited extent. Both after single and repeated intravenous (IV) exposure, titanium rapidly distributed from the systemic circulation to the liver, spleen, kidney, lung, heart, brain, thymus, and reproductive organs. Liver, spleen, and lungs were identified as primary target tissues for TiO_2 NMs. Highest levels were observed in the liver, but redistribution from the liver to the spleen was observed over the 90-day post-exposure period. Elimination was very slow with only small amounts of TiO_2 NPs eliminated in up to 90 days post exposure. Limited uptake in combination with slow elimination may result in potential tissue accumulation in long-term exposures. Only minor differences in kinetic profile were observed among the various particles, although in an earlier study in mice, the size of nanoparticles affected the accumulation of TiO_2 NPs in different organs (Wang et al. 2007c). Larger particles (80 nm) were mainly accumulated in the liver, whereas smaller particles (25 nm) were found in the spleen and, to a lesser extent, in the kidneys and lungs. Animals were exposed to a single high oral dose (5,000 mg/kg bw). However, much more knowledge is needed on the differences in kinetics related to physicochemical characteristics such as size.

A similar kinetic profile was observed in rats exposed to TiO_2 NPs in two earlier studies (Fabian et al. 2008; van Ravenzwaay et al. 2009). In these studies, a single IV administration of (5 mg TiO_2 NPs /kg bw) resulted in the highest levels (in descending order) in liver, spleen, lung, and kidney. The effects of TiO_2 NPs (anatase; 6–7 nm) exposure in the spleen have also been examined in mice dosed by intraperitoneal injection for 45 consecutive days (Li et al. 2010a). The results demonstrated that TiO_2 NPs accumulated in the mouse spleen, leading to congestion and lymph nodule proliferation of spleen tissue, and splenocyte apoptosis. ICP-MS was used to analyze the titanium concentration in the samples.

In a 13-week subchronic oral toxicity study, the absorption, distribution, excretion, and toxicity of TiO_2 NPs (anatase:rutile [70:30]; 20–30 nm; doses up to 1,042 mg/kg bw/day) were evaluated after repeated daily oral administration in rats (Cho et al. 2013). There were no clear dose-related increases in titanium levels in liver, spleen, kidney, and brain, indicating very low systemic bioavailability. Titanium levels were analyzed using ICP-MS. It was suggested that at high dose levels, absorption of TiO_2 may be reduced due to agglomeration/aggregation of the particles in the gastrointestinal tract. Another study also reported that oral administration of 5 mg/kg bw of TiO_2 NPs as a bolus to rats did not lead to any significant uptake of TiO_2 (measured as titanium) from the gastrointestinal tract to blood, urine, or to various organs at any of the time intervals studied over a 96-hour post-administration period (MacNicoll et al. 2015).

In contrast, other studies have reported adverse hazardous effects following oral administration of TiO_2 NPs in rats and mice. In a study evaluating the reproductive and endocrine effects of TiO_2 NPs (anatase, < 25 nm), titanium levels in the thyroid, ovaries, and spleen after short-term exposure were determined (Tassinari et al. 2014). Rats were orally exposed for 5 days to 0, 1, or 2 mg TiO_2 NPs/kg bw/day. Single particle-ICP-MS analysis revealed a significant increase in titanium levels in spleen at the high-dose group; both TiO_2 NPs and their aggregates were present. In another study, ovarian injury and gene-expressed characteristics in female mice induced by intragastric administration of

TiO_2 NPs (5–6 nm; 10 mg/kg) for 90 consecutive days were investigated (Gao et al. 2012). The results indicated that TiO_2 NPs can accumulate in the ovary and result in ovarian damage, cause an imbalance of mineral element distribution and sex hormones, and decrease fertility or the pregnancy rate and oxidative stress in mice. ICP-MS was used to analyze the titanium. Similarly, when testicular damage and alterations in gene expression profiles in male mice induced by intragastric administration of 2.5, 5, and 10 mg/kg bw of TiO_2 NPs for 90 consecutive days were examined, TiO_2 NPs were observed to cross the blood-testis barrier to reach the testis and accumulate therein, which, in turn, resulted in testicular lesions, sperm malformations, and alterations in serum sex hormone levels (Gao et al. 2013). With increased TiO_2 NPs doses, the net increase in body weight, testicular, and testes masses were significantly reduced, and titanium accumulation was greatly increased.

In a study in which mice were exposed to 10 mg/kg bw TiO_2 NPs (5–6 nm) for 90 days, this long-term exposure to TiO_2 NPs resulted in titanium accumulation in the liver and TiO_2 NP aggregation in hepatocyte nuclei, an inflammatory response, hepatocyte apoptosis, and liver dysfunction (Cui et al. 2012). The amount of titanium accumulated in the mouse liver was 2440 ng/g tissue for a 90-day exposure. Information for oral absorption and the elimination from tissues following longer time periods is limited. The organ-tissue distribution and clearance of Degussa P25 TiO_2 NPs were determined after IV administration to rats (0.95 mg/kg bw) using an inductively coupled plasma sector field mass spectrometer (Shinohara et al. 2014). Liver and spleen TiO_2 burden was significantly higher in the treatment group and did not decrease up to 30 days after administration, while TiO_2 burden in the lung, kidney, heart, and blood decreased over time. The titanium content in feces and urine was unchanged.

TiO_2 NPs have been found to also accumulate in the intestine in rats and migrate to other organs (Jani et al. 1994). Analysis of rats dosed with 500 nm TiO_2 particles daily for 10 days showed systemic absorption and distribution of titanium (for example, blood, liver, and lung): 6.5% of an oral dose was estimated to be absorbed. In another study, the subacute toxicity of instilled TiO_2 NPs was examined in mice intratracheally instilled with TiO_2 NPs, once per week for 4 consecutive weeks, at a total dose of 13.2 mg/kg bw (Li et al. 2010b). At 28 days post-instillation, the results showed that instilled TiO_2 NPs could induce lung damage and change the permeability of the alveolar-capillary barrier. With the ability to have access to blood circulation and reach extrapulmonary tissues, TiO_2 NPs could lead to injury at additional tissues, such as liver and kidney.

In a recent study, E171 (food-grade TiO_2) was observed to accumulate in the stomach wall of rats administrated with 1,000 mg/kg bw of E171 for 90 days (Han et al. 2021). Also, to explore the main routes of excretion, Ti concentrations were measured (by TEM) in the colon, kidney, and spleen harvested at necropsy. The Ti concentration increased only in the colon in both sexes. TEM images revealed that E171 accumulated in the cytosol and nuclei of various cells comprising the colon tissue and forms lamella-like structures.

Data on human oral absorption of TiO_2 are limited. Biokinetic/gastrointestinal uptake studies were implemented with human volunteers following oral exposures to TiO_2 particles of various sizes (Jones et al. 2015). In these studies, nine volunteers received a five mg/kg bw single oral dose of TiO_2 dispersed in water. Three successive studies using the same volunteers were carried out using particle sizes of 15 nm (nanoscale), 100 nm (nanoscale), and <5,000 nm (pigment grade) TiO_2 particles. Doses were administered at least four weeks apart. The extent of absorption, excretion, and translocation of TiO_2 particles was assessed by the determination of titanium in appropriate fluids (blood, serum, urine, or culture medium) using ICP-MS. In addition, full blood count and liver function tests were performed. Very little TiO_2 absorption from the gastrointestinal tract was reported for any of the particle sizes tested. High baseline values may have prevented observations on the appearance of titanium in the circulation following ingestion of TiO_2.

Other human volunteer studies, however, showed elevated blood Ti levels after ingestion of TiO_2. In an early human volunteer study, blood Ti levels before and after oral administration of TiO_2 (160–380 nm in gelatin capsules; dose 23 or 46 mg) were assessed (Bockmann et al. 2000). A typical persorption curve with three distinct maxima was observed. Measurable increases in total titanium levels were shown to occur in the blood stream. However, an expected doubling of blood levels after doubling of dose was not demonstrated, indicating that absorption may be biologically limited. In a more recent study to replicate the observations in the earlier study and to provide additional evidence for particle uptake, seven human volunteers with normal intestinal permeability were orally administered 100 mg pharmaceutical/food-grade TiO_2 (Pele et al. 2015). Blood samples were collected from 0.5 to 10 hours post ingestion and analyzed for the presence of reflectant bodies (particles) by dark field microscopy and for total Ti by ICP-MS. Blood film analyses showed a bimodal pattern of particle uptake, with early absorption of particles (two hours) with a peak maximum at six hours following ingestion. The results of the study showed that a fraction of pharmaceutical/food-grade TiO_2 is absorbed systemically by humans following ingestion. Although penetration of NPs into the human body appears to be low, long-term exposure can cause accumulation of NPs due to their limited excretion rate. It was noted that commonly used toxicity screening is based on testing rather high-dose short exposures. In humans, long-term low-dose exposures occur more often, and so accumulation of minor changes rather than acute toxicity is probable. Currently, knowledge is lacking for the biological effects of combined, long-term NP exposure.

Two studies have now provided evidence of TiO_2 absorption in humans by the characterization and quantification of both total Ti and TiO_2 particles in postmortem human tissue samples. In the first study, livers and spleens from 15 postmortem samples were analyzed by inductively coupled plasma high-resolution mass spectrometry (ICP-HRMS) detection methods and scanning electron microscopy with energy dispersive X-ray spectrometry (Heringa et al. 2018). The analyses showed that approximately all TiO_2 is present as particles in human liver and spleen, with sizes ranging 85–550 and 85–720 nm, respectively, with ≥24% of nano-size (<100 nm). It was noted that the levels detected were below the doses regarded as safe in animals, but half were above the dose that is deemed safe for liver damage in humans when considering several commonly applied uncertainty factors. The second study, conducted with an additional set of postmortem tissue samples from 15 deceased persons, showed the presence of total Ti, and particulate TiO_2 in the liver, spleen, kidney, and the intestinal tissues jejunum and ileum (Peters et al. 2020). In general, the particulate TiO_2 explained 80% of the total Ti concentration, indicating that most titanium in these organ tissues was particulate material. The detected particles comprised primary particles, aggregates, and agglomerates, and were in the range of 50–500 nm with a mode in the range of 100–160 nm. About 17% of the detected TiO_2 particles had a size <100 nm.

DNA Methylation and TiO_2 Nanoparticles

DNA methylation is a major epigenetic mechanism that involves the addition of a methyl group to the fifth position of the cytosine residue of CpG dinucleotides resulting in 5'-methylated cytosine (5-mC) (Bird 2002; Moore et al. 2013). Methylation/demethylation of CpG sites is important for the maintenance of cell-specific or tissue-specific gene expression. DNA methylation plays a role in embryogenesis, imprinting, X chromosome inactivation, and maintaining genome integrity. Global hypomethylation has been associated with increased chromosome instability and oncogenesis and is commonly accompanied by gene-specific hypermethylation of CpG islands. Methylation of cytosines within CpG sequences, and their subsequent interaction with methyl-CpG binding proteins, may induce chromatin conformational modifications and inhibit the access of the transcriptional machinery to gene promoter regions, thus altering gene expression levels (Bogdanovic and Veenstra 2009). Other binding proteins may be involved, such as UHRF1, which interacts with DNMT1 and is implicated in maintenance DNA methylation (Lui et al. 2013). Promoter hypermethylation is commonly associated with gene silencing (Martín-Subero and Esteller 2011). DNA methylation is established and maintained by DNA methyltransferases (DNMT1, DNMT3a, and DNMT3b), which catalyze the transfer of a methyl group from S-adenosyl-l-methionine to the cytosine of a CpG dinucleotide (Bestor 2000; Okano et al. 1999; Pradhan et al. 1999; Robertson et al. 1999). In particular, DNMT1 is the maintenance methyltransferase that specifically works on hemimethylated DNA strands, whereas *de novo* methylation, the establishment of new methylation, is carried out by DNMT3a and DNMT3b. Demethylation, in contrast, can occur either passively or through active mechanisms in which 10–11 translocator proteins (TETs), 5-hydroxymethylcytosine formation (5-hmC), and thymidine glycosidases are all implicated (Bochtler et al. 2017; Yang et al. 2020). The active process of DNA demethylation involves modification of the 5-mC base by either oxidation or deamination followed by replacement of the modified base.

It is now well established that aberrations in DNA methylation play a critical role in cancer (Baylin and Herman 2000; Esteller 2008; Suzuki and Bird 2008). Several types of aberration in the DNA methylation machinery occur in cancer, including hypermethylation of tumor-suppressor genes, aberrant expression of DNMTs, as well as global hypomethylation. Additionally, aberrant DNA methylation is associated with complex nonmalignant diseases, such as type 2 diabetes, major depression, schizophrenia, hypertension, autoimmune, and cardiovascular diseases (Baccarelli et al. 2010; Nilsson and Skinner 2015; Schwarz 2010). Assessing the effects of nanoparticles on DNA methylation is, thereby, a necessary part of the risk assessment of the toxicological potential of these materials, given the critical role of alterations in DNA methylation in the development and progression of numerous pathological states and diseases. TiO_2 NPs have been shown to affect both global DNA methylation levels and methylation at specific gene promoters as well as expression of enzymes/proteins involved in the regulation of DNA methylation, such as methyl-CpG binding proteins and DNMTs (Table 6.1).

In an early study, the effect of TiO_2 NPs (anatase) on the HaCaT human keratinocyte cell line was examined (Tucci et al. 2013). The results showed a reduced S-adenosyl homocysteine concentration following treatment of HaCaT cells with 50 and 100 μg/mL TiO_2 NPs, as well as a decreased level of homocysteine. The resulting dysfunction of the methionine cycle and methionine deficiency suggested an impairment in DNA methylation reactions of the cells after TiO_2 NPs exposure. Other significant biochemical changes were indicative of oxidative stress and depressed energy metabolism. In another study, DNA methylation-based epigenetic changes upon exposure to several concentrations of TiO_2 NPs were evaluated in the lung fibroblast (MRC5) cell line (Patil et al. 2016). Cells were exposed to TiO_2 NPs for 24 and 48 hours at concentrations of 0.5 and 4 μg/mL. An enzyme-linked immunosorbent assay-based immunochemical assay revealed a dose-related decrease in global DNA methylation and DNA methyltransferase activity. A direct correlation between the concentration of NPs, global methylation levels, and expression levels of DNMT1, 3A, and 3B genes upon exposure was found. It was noted that global hypomethylation occurs early in tumorigenesis and predisposes cells to genomic instability and further genetic changes. Also, clonogenic assaying showed a dose-response in the inhibition of growth and proliferation of TiO_2 NPs-treated cells.

Results from a study examining the effect of different crystal phases of TiO_2 NPs (anatase, rutile, and anatase:rutile mixture; 20–26 nm) on global DNA methylation in human bronchial epithelial

Table 6.1 Summary of DNA methylation changes induced by TiO_2 NPs.

Study type	TiO_2 properties/cell or animal model/ experimental conditions	Epigenetic changes	Associated events	References
In vitro	TiO_2 NPs (anatase), human keratinocyte HaCaT cells, 50 and 100 μg mL^{-1}	Reduced SAH concentrations; decreased homocysteine level	Biochemical indictors of oxidative stress and depressed energy metabolism	(Tucci et al. 2013)
In vitro	TiO_2 NPs, lung fibroblast (MRC5) cells, 500 ng mL^{-1} and 4 μg mL^{-1} for 24 and 48 h	Dose-related decrease in global DNA methylation and methyltransferase activity; concentration of NPs correlated with global methylation and Dnmt1, 3A, and 3B gene levels	Dose-response inhibited growth and proliferation of treated cells	(Patil et al. 2016)
In vitro	TiO_2 NPs (anatase, rutile, anatase, and rutile mixture, 20–26 nm), human bronchial epithelial (16HBE) cells, 3.25 and 25 μg mL^{-1} for 24 h	Global DNA hypomethylation		(Ghosh et al. 2017)
In vitro	TiO_2 NPs (anatase, 25 nm), neuronal (PC12) cells, 2.5–80 μg mL^{-1}	Global DNA methylation decreased; DNMT1, DNMT3a, DMNT3b upregulated		(Song et al. 2017)
In vitro	TiO_2 NPs (EVONIK, 21 nm), human macrophages (THP-1), murine macrophages (RAW264.7), and small airway epithelial (SAEC) cells, 500 ng mL^{-1} and 30 μg mL^{-1} for 24 h	Reactivation of transposable elements: ORF1, ORF2, SINE B1, and SINE B2 in RAW264.7 cells; increased expression of *Tet2*	Inflammatory response and oxidative stress	(Lu et al. 2016)
In vitro	TiO_2 NPs (uncoated, silica- or citrate-coated, P25), human lung epithelial (A549) cells, 40 μg cm^{-2} for 48 or 72 h	Induced demethylation of LINE1 sequences by all tested materials at 72 h	Induced genotoxicity (chromosomal damage and oxidative DNA damage)	(Stoccoro et al. 2017)
In vitro	TiO_2 NPs (nanotube morphology 25 nm, anatase morphology 60 nm), human bronchial epithelial (16HBE) and non-small cell lung cancer (A549) cells, 100 ng to 100 μg mL^{-1} for 48 h	Global DNA hypomethylation; altered expression levels of DMNT3b, MBD2, *Tet1*, *Tet2*, and *Tet3*	Induced time and concentration-dependent decrease in cell viability; increase in cell death and intracellular ROS levels	(Ma et al. 2017)

(Continued)

Table 6.1 (Continued)

Study type	TiO_2 properties/cell or animal model/ experimental conditions	Epigenetic changes	Associated events	References
In vitro	TiO_2 NPs (anatase:rutile [86:14], 21 nm), bronchial epithelial (BEAS-2B) and human lung epithelial (A549) cells, 1–100 μg mL^{-1} for 4, 24, or 48 h	Gene-specific DNA methylation alterations, increased methylation levels of *APE1*, *POLD3*, *MRE11A*, and *PMS2*	Downregulation of genes encoding DNA repair proteins correlated with increased methylation of DNA repair gene promoters; downregulation of NRF2 and BRCA1; induced oxidatively generated DNA damage and impaired DNA repair activities	(Biola-Clier et al. 2017)
In vitro	TiO_2 NPs (Aeroxide P25), colorectal (CaCo2), liver (HepG2), lung (NL20), and skin (A431) cells, 100 mg mL^{-1} for 24 or 72 h	Decrease in global methylation in CaCo2, HepG2, and A431 cells; increased promoter methylation in *CDKN1A*, *DNAJC15*, *GADD45A*, *GDF15*, *INSIG1*, *SCARA3*, *TP53*, and *BNIP3* genes; aberrant expression of DNMT1, DNMT3a, DNMT3b, MBD2, and UHRF1	(Altered promoter methylation of these genes associated with diseases)	(Pogribna et al. 2020)
In vitro	TiO_2 NPs (anatase, 22.1 nm), human lung epithelial (A549) cells, 6.25–100 μg mL^{-1} for 24 h	High level of poly(ADP-ribose) polymerase 1 (PARP-1) methylation	Dose-dependent cell viability decrease; increase of lactate dehydrogenase activity (cell membrane damage); increase of ROS generation	(Bai et al. 2015)
In vitro	TiO_2 NPs (anatase, 3974 nm), human peripheral blood mononuclear cells, 25–100 μg mL^{-1} for 4 h	Dose-dependent global DNA hypomethylation		(Malakootian et al. 2021)
In vivo	TiO2 NPs (anatase and rutile, 21 nm), adult male albino rats (liver), 100 mg $kgbw^{-1}$ for 6 weeks through an orogastric tube	Global DNA hypomethylation	Oxidative stress, increased blood levels of malonaldehyde, reduced glutathione levels	(El Dine et al. 2018)

Table 6.1 (Continued)

Study type	TiO_2 properties/cell or animal model/ experimental conditions	Epigenetic changes	Associated events	References
In vivo	TiO_2 NPs (anatase, 25 nm), 5-week (young) and 10-week adult (old) NIH mice (lung), (20 mg $kgbw^{-1}$) day^{-1} for 30 days with intranasal instillation	Young group: decreased global methylation and hydroxymethylation; altered expression levels of *DNMT1*, *DNMT3b*, *DNMT3a*, *Tet1*, and *Tet3*; altered methylation in *TNFα* and *Thy1*	Enriched cancer pathway, as demonstrated by KEGG pathway analysis	(Ma et al. 2019)
Cohort study	TiO_2 plant workers (urine)	Global hypomethylation	Elevated levels of oxidative biomarkers, 8-oxoGuo, 8-oxodGuo, and $3NO_2Tyr$	Buonaurio et al. 2020

16-HBE cells found significant hypomethylation at 3.25 and 25 μg/mL (24 h) for the various forms using LC–MS/MS analysis (Ghosh et al. 2017). However, no apparent effect of the crystal phase was observed. From cyto-genotoxicity assaying, 3.25 μg/mL was designated as a sub-cyto-genotoxic concentration, suggesting the need for considering proper precaution and epigenetic endpoints in the safety assessment of TiO_2 NPs in various consumer products. Another study explored whether a sublethal concentration of TiO_2 NPs (anatase; 25 nm) could affect global DNA methylation levels in PC12, a neuronal cell line, and characterized the relationship between DNA methylation mechanism and TiO_2 NPs treatment (Song et al. 2017). The concentrations were 2.5–80 μg/mL. The global DNA methylation levels in TiO_2 NPs-treated cells decreased at all concentrations. The lowest levels were observed at 2.5 μg/mL. The relative gene expression of DNMT1, DNMT3a, and DNMT3b was significantly upregulated after cells were incubated with different TiO_2 NPs concentrations.

The epigenetic effects of TiO_2 NPs were assessed in a panel of cells relevant to inhalation exposures (human and murine macrophages, THP-1 and RAW264.7, respectively, and human small airway epithelial cells [SAEC]) (Lu et al. 2016). The cells were exposed to TiO_2 (from EVONIK; 21 nm) at two doses (0.5 and 30 μg/mL) for 24 h. Although there were no significant differences identified in the levels of global methylation (5-mC and 5-hmC) in response to TiO_2 NPs, exposure to a lower dose of TiO_2 resulted in significant and profound reactivation of transposable elements (TEs) L1 ORFs–ORF1 and ORF2 as well as SINE B1 and SINE B2 in RAW264.7 and exposure to a high dose of TiO_2 resulted in increased expression of *Tet2* in these cells, demonstrating modest epigenetic alterations. Additionally, exposure to TiO_2 NPs resulted in inflammatory response and oxidative stress. In investigating the epigenetic properties of several forms (uncoated, coated with silica or citrate, and the benchmark P25) of TiO_2 NPs in the human lung epithelial cell line A549, the effect on global methylation was found to be time-dependent in cells treated with 40 $\mu g/cm^2$ TiO_2 NPs (a concentration noted to be within the exposure range representative of realistic human exposure) (Stoccoro et al. 2017). After 48 hours of exposure, TiO_2 NPs did not affect the methylation degree of LINE-1 sequences (LINE-1 methylation analysis is noted to be frequently used as a surrogate of global DNA methylation levels and associated with genome instability). On the other hand, after a 72-hour exposure, all the NPs tested induced a demethylation effect. It was noted that

LINE-1 hypomethylation is associated with elevated risk of several disease states. TiO_2 NPs were shown to induce genotoxicity (chromosomal damage and oxidative DNA damage) in this cellular system, suggesting that genotoxic and epigenotoxic mechanisms could work in association with the induction of adverse health effects and are able to benefit from each other.

The effects of TiO_2 NPs (nanotube morphology, 25 nm; anatase morphology, 60 nm) on DNA methylation were tested in two cell lines relevant to inhalation exposure, namely human bronchial epithelial cell line 16HBE and human non-small cell lung cancer A549 cells, with treatment concentrations ranging from 0.1 to 100 μg/mL for 48 hours (Ma et al. 2017). The TiO_2 NPs resulted in decreased global DNA methylation (hypomethylation) at all concentrations in both cell lines, suggesting that these NPs induce cellular epigenomic toxicity. However, in 16HBE cells, the global DNA methylation levels decreased only after exposure to TiO_2-N25 at 100 μg/mL. Additionally, altered expression levels of methylation-related genes and proteins (DNMT3b, MBD2, TET1, TET2, TET3) were observed in both cell lines posttreatment with TiO_2 NPs. The TiO_2 NPs also induced time- and concentration-dependent decrease in cell viability in both A549 and 16HBE cells, as well as a rise in cell death and increased intracellular ROS levels. It was noted that these results found that the alterations in global DNA methylation were associated with the cytotoxic effects of TiO_2 NPs and may further contribute to the higher risk of respiratory diseases.

Altered DNA methylation was investigated in human noncancerous bronchial epithelial BEAS-2B cells and epithelial alveolar cancer A549 cells exposed to TiO_2 NPs (Biola-Clier et al. 2017). In cells exposed to 1–100 μg/mL TiO_2 NPs (anatase/rutile [86:14]; 21 nm) for 4, 24, or 48 hours, no change in the overall DNA methylation level by 5-Me-dC quantification was detected. Alterations in gene-specific DNA methylation were, however, observed. Using an array technology, the methylation profile of 20 DNA repair gene promoters in cells exposed to TiO_2 NPs was determined. Exposure of A549 cells to 100 μg/mL of TiO_2 NPs for 4 or 24 hours led to an increase in methylation levels of *APE1*, *POLD3*, *MRE11AI*, and *PMS2*. It was noted that the concentration may be high and unrealistic as an acute dose; however, due to increased intracellular accumulation from chronic occupational exposure, it was suggested that this exposure model could nevertheless be informative from a risk-assessment perspective. The downregulation of the genes encoding DNA repair proteins correlated with increased methylation of DNA repair gene promoters and downregulation of NRF2 and BRCA1. Results also showed that TiO_2 NPs induced significant oxidative DNA damage and impaired DNA repair activities.

In a comprehensive study, the effect of TiO_2 NPs exposure on DNA methylation and expression levels of DNA methyltransferases, MBD2, and UHRF was investigated in a panel of cells relevant to portals of human exposure (Caco-2 [colorectal], HepG2 [liver], NL20 [lung], and A-431 [skin]) (Pogribna et al. 2020). Cells were treated with 100 mg/mL TiO_2 (Aeroxide P25) for 24 or 72 hours. Array technology was used to evaluate the degree of promoter methylation across a defined panel of genes. The signature panel profiled genes indicative of cellular stress responses and toxicity. Decrease in global DNA methylation in cell lines Caco-2, HepG2, and A-431 exposed to TiO_2 NPs was shown. Across the four cell lines, eight genes (*CDKN1A*, *DNAJC15*, *GADD45A*, *GDF15*, *INSIG1*, *SCARA3*, *TP53*, and *BNIP3*) were identified in which promoters were methylated after exposure. Several genes were affected in multiple cell lines and the promoter methylation status of *CDKN1A* and *SCARA3*, for example, was altered in all four cell lines. It was noted that for each of the eight genes, altered promoter methylation is associated with disease. The results also revealed aberrant expression of epigenetic regulatory genes involved in DNA methylation (DNMT1, DNMT3a, DNMT3b, MBD2, and UHRF1) in TiO_2 NPs exposed cells, which was cell-type dependent.

Human alveolar epithelial cells A549 treated with anatase TiO_2 NPs (anatase; 22.1 nm) for 24 h (6.25–100 μg/mL) resulted in a high level of poly(ADP-ribose) polymerase 1 (PARP-1) methylation

(Bai et al. 2015). PARP-1, a highly conserved DNA-binding protein, is involved in many molecular and cellular processes, which include DNA repair, proliferation, and chromatin modification (Chevanne et al. 2007). PARP-1 was also found in a complex with the DNA methyltransferase DNMT1, the histone H3K9 methyltransferase G9a, and the histone ubiquitin ligase Np95, indicative of a link between poly(ADP-ribosyl)ation and the epigenome (Sharif et al. 2007) and was described as a fundamental constituent of the transcription machinery that interacts with and modulates the activities of several transcription factors (Kraus 2008). The results showed that TiO_2 NPs also decreased the cell viability in a dose-dependent manner, in accordance with the increase of lactate dehydrogenase activity, which indicated membrane damage of cells, and significantly increased the generation of ROS after exposure for 24 h. Using the methyltransferase inhibitor 5-aza-2′-deoxycytidine and the ROS scavenger α-lipoic acid, evidence was provided suggesting that there is a cross talk between oxidative stress and PARP-1 methylation in TiO_2 NPs-induced A549 cells toxicity.

DNA methylation effects of TiO_2 NPs have also been demonstrated in *in vivo* studies. The epigenetic changes of the genome (epigenome), in the form of DNA methylation in liver tissue samples, were evaluated in adult male albino rats orally administered TiO_2 NPs (mixed rutile and anatase, 21 nm) (El Dine et al. 2018). The TiO_2 NPs suspension (100 mg/kg of body weight) was administered through an orogastic tube daily for 6 weeks. Exposure to TiO_2 NPs caused a decrease in global DNA methylation (hypomethylation) in liver tissue samples. This exposure also provoked oxidative stress, increased blood levels of malonaldehyde, and decreased reduced glutathione levels. In another study, the comparative effects of TiO_2 NPs (anatase; 25 nm) on DNA methylation were investigated in lung of 5-week-old (young) and 10-week-old (adult) (old) NIH mice following nasal inhalation (intranasal instillation) of TiO_2 NPs at a dose of 20 mg/kg (bw)/day for 30 days (Ma et al. 2019). Decreased global methylation and hydroxymethylation were only found in the younger group. Expression levels of *Dnmt1*, *Dnmt3b*, *Dnmt3a*, *Tet1*, and *Tet3* were also changed. Altered methylation in the promoter of inflammatory gene *TNF-α* and tissue fibrosis gene *Thy-1* were also observed, which were found to be associated with the inflammatory response and fibration. In addition, KEGG pathway analysis showed that the most enriched pathways in the young mice included "pathway in cancer." These alterations in global DNA methylation and hydroxymethylation were associated with the toxic effects of TiO_2 NPs and may further contribute to the higher risk of respiratory diseases.

The effect of TiO_2 NPs (anatase; 39–74 nm) on DNA methylation in human peripheral blood mononuclear cells (PBMCs) at concentrations of 25–100 μg/mL in a 4-hour exposure was investigated by measuring the percentage of 5-mC alterations using an ELISA assay (Malakootian et al. 2021). According to the results of the MTT test, 100 μg/mL was found to be maximum nontoxic concentration. The percentage of 5-mC in treated PBMCs revealed that TiO_2 NPs resulted in DNA hypomethylation in PBMCs. The effect was dose dependent.

Histone Modifications and TiO_2 Nanoparticles

Histones are fundamental components of the transcriptional regulatory machinery. H2A, H2B, H3, and H4 histones assemble as an octamer and wrap DNA (Blackledge and Klose 2010). Posttranslational modifications, mostly at a specific position within amino-terminal- and carboxy-terminal-histone tails (Bannister and Kouzarides 2011), lead to changes in chromatin structure, thereby modifying the accessibility of DNA to transcription factors, either directly by changing the charge on histone proteins, or indirectly by employing proteins that bind to specific modifications

(Kouzarides 2007). These posttranslational modifications function as a major component of gene expression regulation, thereby controlling several processes, including apoptosis, DNA replication, and DNA damage response (Millar and Grunstein 2006). At least 12 posttranslational modifications (e.g., methylation, acetylation, sumoylation, ubiquitination, and phosphorylation) have been identified in over 60 different residues on the histone proteins (Kouzarides 2007; Lachner et al. 2003). Depending on the posttranslation modifications to the histone, the level of chromatin condensation varies, and chromatin exist as euchromatin (less condensed and contains transcriptionally active genes) or heterochromatin (well condensed and present in centromere and telomere regions of chromosomes) (Grewal and Elgin 2007). (Increased acetylation results in open chromatin configuration and transcription activation, while methylation of histones can result either in repression or activation of transcription [Rice and Allis 2001]). Histone modifications are mediated by a complex of histone-modifying enzymes. The enzymes involved in "writing" and "erasing" these reversible marks include histone acetyltransferases, histone deacetylases, histone methyltransferases, histone demethylases, histone kinases, histone phosphatases, histone ubiquitinases, histone deubiquitinases, and others (Allis et al. 2007; Chervona and Costa 2012). Their aberrant functions cause misregulation of chromatin structure and activity with impact on access of transcription factors to DNA and gene transcription.

Dysregulation of histone modifications and histone-modifying enzyme expression has been implicated in the pathogenesis of human diseases, including a variety of cancers (Kurdistani 2011; Roberti et al. 2019), cardiovascular diseases (Abi Kahalil 2014; Loernzen et al. 2012), neurological diseases (Jakovcevski and Akbarian 2012; Hwang et al. 2017; Sharma et al. 2019), and autoimmune diseases (Deng et al. 2019). Given the potential role of this epigenetic mechanism in disease etiology, including histone modification effects can be an important part of the risk assessment of the toxicological potential of nanoparticles.

Although less studied than the two other major epigenetic mechanisms, several studies have evaluated the impact of nanoparticles on histone modifications and histone-modifying enzyme expression (Hou et al. 2019; Pogribna and Hammons 2021; Stoccoro et al. 2013; Wong et al. 2017). These studies have shown that nanoparticle exposure can induce changes in histone modification patterns, as well as in the expression or activity of histone-modifying enzymes that are known for their roles in causing changes in the pattern of histone modifications (Table 6.2). Notably, these changes can differ according to the type of nanoparticles and cells used. Although only a few studies have been reported, several have involved TiO_2 NPs.

In one study, experiments measured changes in gene expression following an acute exposure to TiO_2 NPs and changes inherited by progeny cells (Jayaram et al. 2019). Treatment of human A549 lung epithelial carcinoma cells with TiO_2 NPs (anatase:rutile [80:20]; 21 nm) at a concentration of 800 μg/mL led to significant (>2,000 differentially expressed genes) changes in gene expression following a 24-hour incubation. A survey of the epigenetic modifiers in the differentially expressed genes at Generation 1 indicated that histone deacetylases HDAC9, HDAC10, and histone acetylases HAT1, KAT2A, and ESCO2 were all altered, suggesting that TiO_2 NPs exposure leads to marked changes in the histone acetylation pattern, which could result in activation and silencing of their specific target genes. Gene ontology analysis indicated that other altered genes were enriched in intercellular signaling transduction, membrane docking, and endogenous stimuli, while silenced genes are enriched in signal release, regulation of secretion, and exocytosis. Genes that were altered in both Generation 1 and progeny cells (Generation 10) have been shown to be epigenetically regulated, providing support that TiO_2 NPs exposure can regulate epigenetic modification. Accompanying the changes in gene expression was the production of intracellular ROS, specifically superoxide, along with changes in oxidative stress-related genes.

Table 6.2 Summary of changes in histone modifications and histone-modifying enzymes induced by TiO_2.

Study type	TiO_2 properties/cell or animal model/ experimental conditions	Epigenetic changes	Associated events	References
In vitro	TiO_2 NPs (anatase:rutile [80:20], 21 nm), human lung epithelial (A549) cells, 800 μg mL^{-1} for 24 h	>2000 differentially expressed genes; altered expression levels of HDAC9, HDAC10, HAT1, KAT2A, and ESCO2	Other altered genes were enriched in intracellular signaling transduction, membrane docking, and endogenous stimuli; silenced genes were enriched in signal release, secretion regulation, and exocytosis; production of ROS, changes in oxidative stress genes	(Jayaram et al. 2019)
In vivo	TiO_2 NPs (anatase:rutile [80:20], 21 nm), pregnant Sprague Dawley rats, Aerosols (10 ± 0.5 mg m^3) for 78 days	ChIP bioinformatic analysis with H3K4Me3 and H3K27Me3 found altered canonical pathways involved in growth regulation and cell cycle/ apoptosis signaling and those involved in cancer and immunity	Increases in H3K27me3 at promoter regions for infection capacity and growth signaling with kidney and liver disfunction loci indicate epigenetic regulation that could alter an organism's susceptibility to disease and potential predisposition to future insult	(Stapleton et al. 2018)
In vitro	TiO_2 NPs, alveolar macrophage-like (THP-1), human lung epithelial (A549), and human pulmonary microvascular endothelial (HPMECST-1.6R) cells, 200 and 800 μg mL^{-1} for 24 h	Phosphorylation of H2AX histone (γH2AX) in THP-1 macrophages and HPMEC-ST1.6R cells	DNA damage in THP-1 macrophages and HPMEC-ST1.6R cells; DNA repair pathways activated in THP-1 macrophages but not in HPMEC-ST1.6R cells; ROS generation was correlated with exposure concentration of NPs in all cell lines	(Hanot-Roy et al. 2016)
In vitro	TiO_2 NPs (anatase, 15 nm); human dermal fibroblasts, 1, 3, and 10 μg mL^{-1} for 24 h	Increased phosphorylation of H2AX	Inhibited overall rate of DNA synthesis and replicon initiation events frequency in DNA combed fibers	(Prasad et al. 2013)
In vitro	TiO_2 NPs (anatase:rutile: amorphous [7385:1417:213], 22 nm), human skin fibroblasts (BJ), 10 – 500 μg mL^{-1}	Increased H2AX phosphorylation (500 μg mL^{-1} did not substantially increase the level of γH2AX)	Dose-dependent cytotoxicity and genotoxic effects (genotoxic effects could be detected at much lower concentrations of NPs)	(Setyawati et al. 2013)
In vitro	TiO_2 NPs (anatase, 5 nm and <5 μm), Human lung epithelial (A549) cells, 250 μg – 1 mg mL^{-1} of both particles and 1 – 100 μg mL^{-1} of TiO_2 NPs	**γ**H2AX detected after treatment with both TiO_2 nano and microparticles (small quantity (from 1 μg mL^{-1}) of TiO_2 NPs could generate **γ**H2AX; in microparticles, similar generation was detected from over 75 μgmL^{-1})		(Toyooka et al. 2012)

Alterations in the F1 generation of Sprague-Dawley rats that result from maternal TiO_2 NPs inhalation during gestation were identified using chromatin immunoprecipitation (ChIP) and transcriptomic analyses as well as pathway analysis (Stapleton et al. 2018). Pregnant dams were exposed to TiO_2 NPs (anatase:rutile [80:20]; 21 nm) aerosols (10 ± 0.5 mg/m^3) for 7–8 days and 20 fetal hearts were isolated. ChIP analysis was conducted with the marks histone 3 lysine 4 trimethylation (H3K4me3) and histone 3 lysine 27 trimethylation (H3K27me3). From a ChIP bioinformatic analysis, one of the most prominent pathways for H3K4me3 affected by TiO_2 NPs maternal exposure involved infectious disease. In general, the canonical pathways altered after exposure involve the regulation of growth and cell cycle/apoptosis signaling. For the promoter regions associated with H3K27me3, the majority of signaling changes involve cancer and immunity. Increases in H3K4me3 at promoter regions for infection capacity and growth signaling as well as loci involving kidney and liver dysfunction suggest epigenetic regulation that could significantly alter an organism's susceptibility to disease and potential predisposition to future insult. Comparative analysis to examine how changes between the H3K4me3 mark and RNA transcript data aligned illustrated the common pathways involving both inflammation and organismal growth signaling. For toxicological functions, the molecular profile for cardiac dysfunction was significantly decreased compared with the controls, while kidney dysfunction was increased. The comparative analysis suggested that maternal TiO_2 NPs exposure can cause significant changes to how the progeny develop, changing the epigenetic landscape, which can directly affect transcript abundance. The two most prominently impacted mechanisms were inflammatory signaling and cardiac–renal–hepatic pathology/toxicity.

There have been several reports demonstrating the phosphorylation of the histone family member H2AX (γH2AX) resulting from exposure to TiO_2 NPs. The effects (genotoxic potential) of TiO_2 NPs (P25) in three model cell lines: alveolar macrophage-like THP-1 cells, alveolar epithelial A549 cells, and human pulmonary microvascular endothelial cells (HPMEC-ST1.6R cells) were evaluated to further understand the toxicological behavior of TiO_2 NPs and the underlying mechanisms (Hanot-Roy et al. 2016). Various endpoints were assessed after a 24-hour exposure to TiO_2 NPs. In the cell lines, DNA damage was persistent. The three cell lines exhibited different profiles of activated signaling in response to DNA damage. In differentiated THP-1 macrophages, phosphorylation of H2AX was observed, at concentrations of 200 and 800 μg/ml, with downstream activation of DNA repair pathway. In HPMEC-ST1.6R cells, phosphorylation of H2AX was observed, but no DNA repair pathway activation. Otherwise, P53 and checkpoint protein ChK1 phosphorylation occurred at 200 and 800 μg/mL, correlating with a cell cycle arrest. Activation of signaling pathways related to DNA damage was also not observed in A549 cell line 24 hours posttreatment. The generation of ROS was correlated with the exposure concentration of nanoparticles in all three cell lines. ROS generation suggested their role in inducing oxidative stress leading to DNA damage in most treated cells.

The effects of TiO_2 NPs on the DNA damage response and DNA replication dynamics were also investigated in human dermal fibroblasts, specifically, examining the relationship between TiO_2 NPs and the DNA damage response pathways regulated by ATM/Chk2 and ATR/Chk1 (Prasad et al. 2013). Cells were exposed to non-cytotoxic concentrations (i.e., 1, 3, and 10 μg/mL or 0.285, 0.857, and 2.85 μg/cm^2) of TiO_2 NPs (anatase; 15 nm) for 24 h. The results showed increased phosphorylation of H2AX, which can be used as a marker of DNA double strand breaks. ATM and Chk2 were also increased after exposure. In addition, TiO_2 NPs inhibited the overall rate of DNA synthesis and frequency of replicon initiation events in DNA combed fibers. In another study, the potential cytotoxicity and genotoxicity of TiO_2 NPs (anatase:rutile:amorphous [73–85:14–17:2–13]; 22 nm) was evaluated in a human skin fibroblast cell line (BJ) (Setyawati et al. 2013). Genotoxic influence was ascertained by profiling DNA damage via detection of γH2AX expression. About 10 μg/mL of TiO_2 NPs treatment resulted in increased phosphorylation of H2AX; however, increasing TiO_2 NPs concentration (500 μg/mL) did not substantially increase the level of γH2AX. Dose-dependent cytotoxicity and genotoxic effects of TiO_2 NP were observed; however, the genotoxic effect could be detected at a much lower concentration.

The genotoxicity of two different sized TiO_2 particles (anatase; 5 nm and <5000 nm) in human lung adenocarcinoma epithelial A549 cells based on the phosphorylation of histone H2AX (γ-H2AX) was examined (Toyooka et al. 2012). Cells were treated with high doses (250–1000 μg/mL) or lower doses (1–100 μg/mL) of both TiO_2 particles. γ-H2AX foci distributed in the nucleus were detected after treatment with both TiO_2 nano- and microparticles; however, the degree of γ-H2AX generation was different. In nanoparticles, a small amount (from 1 μg/mL) of TiO_2 could generate γ-H2AX. In microparticles, a similar generation of γ-H2AX was detected at concentrations >75 μg/ml.

MicroRNAs and TiO_2 Nanoparticles

miRNAs belong to a class of single-stranded, small noncoding RNAs (~22 nucleotides) that act as translational repressors and regulate diverse genes involved in vital cellular processes, including cellular proliferation, cell death, development, and differentiation (Ambros 2004; Bartel 2004). The 5′-end of mature miRNA consists of a "seed region", containing 2–8 nucleotides that determine the mRNA target. Based on the complementarity of the "seed region" and their mRNA target, the miRNA either cleaves and degrades the mRNA target or inhibits protein translation. High complementarity results in mRNA cleavage through a RNA interference (RNAi) mechanism. In contrast, partial complementarity, which is commonly seen in mammalian cells, results in translational inhibition (Bartel 2009). While each miRNA is predicted to have numerous mRNA targets, a single mRNA can also be the target of multiple miRNAs (Miranda et al. 2006; Shyamasundar et al. 2015).

The role of miRNAs in the RNA interference process is to regulate gene expression by interfering with mRNA processes, affecting mRNA stability, targeting mRNA for degradation, or both; in rare cases, it can also increase gene transcription (Mathers et al. 2010). miRNAs can also be involved in epigenetic regulation through establishing DNA methylation and may influence chromatin structure by regulating histone modifiers (Yao et al. 2019). Their capacity for dynamic interaction with DNA, RNA, and proteins confers on miRNAs the ability to mediate in the various epigenetic processes through the cell's response to diverse external or internal stimuli (Vrijens et al. 2015). Eukaryotic cells use RNA interference to silence transposable elements after transcription, protecting the cell against genomic damage. More than 1,050 miRNAs have been identified in human cells with impact on almost all genetic pathways, through their effects on transcription factors, receptors, and transporters (Esquela-Kerscher and Slack 2006; Griffiths-Jones et al. 2008). The aberrant expression of miRNAs has been linked to various human diseases, including cancer, Alzheimer's disease, and heart disease (Montgomery and van Rooij 2010; Olive et al. 2010; Provost 2010; Raitoharju et al. 2011; Wang et al. 2017). Increasing evidence show that nanoparticles are capable of inducing changes in microRNA expression (Gedda et al. 2019; Yu et al. 2020). Therefore, understanding the epigenotoxicity of these nanomaterials associated with miRNA expression is critical in assessing their potential for harm. Most studies have been conducted in cellular model systems (Table 6.3).

An early study was performed on the effects of surface-coated TiO_2 NPs on the pulmonary system using DNA and miRNA expression profiling (Halappanavar et al. 2011). Female C57BL/6BomTac mice were exposed for 1 hour daily to 42.4 (SEM) mg surface-coated TiO_2 NP/m^3 for 11 consecutive

Table 6.3 Summary of changes in microRNA expression induced by TiO_2.

Study type	TiO_2 properties/ cell or animal model/ experimental conditions	Epigenetic changes	Associated events	References
In vivo	TiO_2 NPs (surface-coated), C57BL/6BomTac mice (pulmonary system), 42.4 ± 2.9 mg m^3, 1 h daily for 11 days by inhalation	Altered miR1, miR449a, miR135b, and miR21 expression (several altered miRNAs implicated in inflammation and immune response with others linked to a cancer phenotype)	Changes in gene expression associated with acute phase inflammation and immune response; increased levels of serum amyloid A3, several CXC and CC motif chemokines and cytokine tumor necrosis factor genes	(Halappanavar et al. 2011)
In vitro	TiO_2 NPs (20 nm), human lung epithelial (BEAS-2B) cells, 1.34 μg cm^2 at 3 and 6 weeks	Altered expression in 33 miRNAs related to inflammation, cellular stress, or carcinogenesis processes, including miR21, miR148b, miR154, miR200a, miR200b, miR218, and miR939; a subset including miR23a, miR25, miR96, miR210, and miR502, which were noted to act as oncogenes, were found sufficient to identify exposure effects	BEAS-2B cells exhibited an oncogenic phenotype after treatment	(Ballesteros et al. 2021)
In vitro	TiO_2 NPs (25 nm), human colon cancer (HCT116) and normal colon (NCM4603) cells, 30 μg mL^{-1} for 24 h	34 altered microRNAs at 3 μg mL^{-1} and 24 at 30 μg mL^{-1}, including miRNAs 378b and 378g (in both cell lines)	Upregulated genes involved in metabolism, human disease, and environmental information processes; increased ROS production in both cell lines	(Li et al. 2020)
In vitro	TiO_2 NPs (21 nm), mouse RAW264.7 cells, 100 μg mL^{-1} for 24 h	21 miRNAs expression altered, pathways primarily affected were involved in cell communication and signal transduction	Upregulation of miR29b3p reinforced apoptosis of cells	(Xu et al. 2020)
In vitro	TiO_2 NPs (P25, 21 nm), mouse RAW264.7 cells, 100 μg mL^{-1} for 24 h	140 miRNAs changed, 131 upregulated and 9 downregulated (miR27a5p and miR1823p were downregulated, miR27b3p, miR181d5p, and miR350 were upregulated)	P13KAkt, p53, MAPK, and Wnt signaling pathways along with apoptosis affected; apoptosis rate increased in miR350 mimic transfected RAW264.7 cells	(Sui et al. 2018)

Table 6.3 (Continued)

Study type	TiO_2 properties/ cell or animal model/ experimental conditions	Epigenetic changes	Associated events	References
In vitro	TiO_2 NPs (anatase, <25 nm), human monocytic (THP-1) cells, 10 $\mu g\ mL^{-1}$ for 8 h	Altered expression identified in 12 miRNAs with miR-155-5p determined as the best candidate for functional validation	Severe DNA damage; elevated expression of DNA damage repairing genes; miR1555p mimic transfection influenced the expression of *ATM*, *APAF1*, and *RAD51*	(Hu and Palić 2020)
In vitro	TiO_2 NPs, human lung epithelial (BEAS-2B) cells, 3–100 $\mu g\ mL^{-1}$ for 1 day and 1–30 $\mu g\ mL^{-1}$ for 3 days	Alterations in both mRNA and miRNA expression (pathway analysis identified genes involved in cell cycle regulation, inflammatory response, acute phase response, apoptosis, NRF2-mediated oxidative stress response, and P13K/AKT signaling pathways); miRNA 17–92 cluster (induced) and let-7 miRNA family (suppressed) were altered	(The miRNA 17–92 cluster, an oncogenic microRNA cluster, and let-7 miRNA family, tumor-suppressor miRNA, are involved in lung cancer formation)	(Thai et al. 2015)
In vitro	TiO_2 NPs (anatase:rutile [90:10], 30–40 nm), differentiated THP-1 cells, 100 $\mu g\ mL^{-1}$ for 6 or 24 h	Co-regulated miRNA cluster identified that regulates the expression of genes involved in cell cycle regulation, immune response, and response to NPs	7 key miRNAs were proposed as possible biomarkers of a potential broad range of metal-based NPs	(Ndika et al. 2019)
In vitro	TiO_2 NPs (anatase:rutile [80:20], 21 nm), human HTR8/ SVneo (HTR) cells, 1, 10, and 100 $\mu g\ mL^{-1}$	Dose-dependent increases in miR1013p and miR965p	Autophagy was observed and cell migration ability severely impaired; miRNA upregulation affected the expression of mTOR and Bcl-2 in regulation of autophagy	(Mao et al. 2018)
In vitro	TiO_2 NPs, human lung epithelial (A549) cells	Downregulation of autophagy-related miR21 and miR30a	Autophagy was induced (miR21 is involved in all cancer-related processes, including tumorigenesis, progression, and metastasis, and its expression is associated with clinical pathological factors and prognosis)	(Alinovi et al. 2017)

(Continued)

Table 6.3 (Continued)

Study type	TiO_2 properties/ cell or animal model/ experimental conditions	Epigenetic changes	Associated events	References
In vitro	TiO_2 NPs (near-spherical, 22 nm), human bronchial epithelial (BEAS-2B) cells, 6.25, 12.5, and 25 $\mu g\ mL^{-1}$ for 24 h	Downregulation of miR34a in a dose-dependent manner	Induction of autophagy; increased expression of B cell lymphoma/leukemia2 (Bcl-2); overexpressed miR34a enhanced TiO_2 NP-induced autophagy and cell death through targeted downregulation of Bcl-2, as demonstrated using overexpression of miR34a by lentivirus vector transfection	(Bai et al. 2016)
In vitro	E171 (food-grade TiO_2), human stomach epithelial (AGS) cells, 40 $\mu g\ mL^{-1}$ for 24 h	Immune response-related miRNAs were the most strongly affected, miRNA 3908 being most downregulated	Expression of ER stress (calnexin, IRE1a, Bip, PERK, and Ero1a), pyroptosis (caspase1, IL18, and NALP3), autophagy (p62 and LAMP1), and iron metabolism related (ferritin heavy chain) proteins enhanced, accompanied by the peroxisomal (catalase) and mitochondrial antioxidant (SOD2)	(Han et al. 2021)
In vivo	TiO_2 NPs (P25 powder, anatase:rutile [80:20], 21 nm), wild type and miRNA378a knockout mice, aerodynamic diameter of 182 ± 1.70 nm, 11.09 mg m^3 for 4 h	Knockout, and knockdown to a lesser extent, of miRNA378a improved cardiac and mitochondrial function	Altering global bioenergetic and respiration capacity	(Hathaway et al. 2019)

days by inhalation and were sacrificed 5 days following the previous exposure. The pulmonary response to TiO_2 NPs was characterized using DNA microarrays and pathway-specific PCR arrays and related to data on pulmonary inflammation from bronchial lavages. Inhalation of TiO_2 NPs resulted in changes in the expression of genes associated with the acute phase, inflammation, and immune response, with concomitant changes in several miRNAs. Altered microRNAs included miR-1, miR-449a, miR-135b, and miR-21. It was noted that several of the altered miRNAs are implicated in inflammation and immune response and others are linked to a cancer phenotype. Increased levels of mRNA for acute phase markers, including serum amyloid A-1 and serum amyloid A-3, several C-X-C and C-C motif chemokines, and cytokine tumor necrosis factor genes, occurred in parallel with the induced changes in specific miRNAs.

The role of miRNA expression in connecting TiO_2 NPs exposure with impaired cardiac function was investigated by manipulating miRNA to attenuate the effects of the nanoparticle exposure (Hathaway et al. 2019). Wild type and miRNA-378a knockout mice were exposed to TiO_2 NPs (P25 powder; anatase:rutile [80:20]; 21 nm) with an aerodynamic diameter of 182 ± 1.70 nm and a mass concentration of 11.09 mg/m^3 for 4 hours. Cardiac function, utilizing a Vevo 2100 Imaging System, electron transport chain complex activities, and mitochondrial respiration assessed cardiac and mitochondrial function. Immunoblotting and qPCR examined molecular targets of miRNA-378a. Knockout, and knockdown to a lesser extent, of miRNA-378a improved cardiac and mitochondrial function by altering global bioenergetic and respiration capacity following TiO_2 NPs inhalation exposure.

Other studies have been conducted using *in vitro* systems. The expression levels of the miRNAs were evaluated in BEAS-2B cells following exposure to TiO_2 NPs (20 nm) (10 μg/mL, equivalent to 1.34 μg/cm^2) for 3 and 6 weeks ("long-term") (Ballesteros et al. 2021). Altered expression changes were shown in a battery of 33 miRNAs related to inflammation, cellular stress, and the carcinogenesis processes. Included were miR-21, miR-148b, miR-154, miR-200a, miR-200b, miR-218, and miR-939. These miRNA-target genes are implicated in cell cycle arrest, signal transduction, and adherent junctions. It was also noted that miR-21 targets some common tumor-suppressor genes, such as PTEN or PDCD4, and has been related to poor prognosis in many types of cancers. A subset of five miRNAs (miR-23a, miR-25, miR-96, miR-210, and miR-502), which were noted to act as oncogenes, was found to be sufficient to identify the effects induced by TiO_2 NPs exposure. BEAS-2B cells also exhibited an oncogenic phenotype according to the results obtained in a soft-agar assay, providing evidence that long-term, low-dose exposure to TiO_2 NPs induces miRNA expression changes directly associated with an oncogenic phenotype.

miRNA profiling induced by TiO_2 NPs (25 nm) was analyzed in human colon cancer (HCT116) and normal colon (NCM460) cell lines (Li et al. 2020). Comparative analysis identified 34 and 24 miRNAs that were significantly altered in the TiO_2 NPs-treated cells at concentrations of 3 μg/mL and 30 μg/mL, respectively, for 24 hours. Functional classification demonstrated that a large proportion of genes involved in metabolism, human disease, and environmental information process were significantly upregulated by TiO_2 NPs. Bioinformatic analysis suggested that miRNA 378 might be an early indicator of cellular response to exogenous stimuli with apoptotic signals. Furthermore, TiO_2 NPs significantly altered the expression of miRNA 378b and 378g in HCT116 and NCM460 cell lines; miRNA 378 has biological functions that regulate a variety of tumor cells, including cell proliferation, migration and invasion, and drug resistance. TiO_2 NPs also increased ROS production in the two types of colon cell lines.

Alterations in the expression profiles of miRNAs were identified in mouse RAW264.7 cells exposed to TiO_2 NPs (21 nm) (100 μg/mL) for 24 hours (Xu et al. 2020). Among the miRNA expression profile, 21 miRNAs were found to be affected by TiO_2. Bioinformatic analyses showed that the pathways regulated by differentially expressed miRNAs were mainly involved in cell communication and signal transduction. By investigating the effect of TiO_2 exposure on immune-related signaling pathways, including T cell receptors, B cell receptors, and natural killer cell-mediated cytotoxicity, 12 miRNAs were found to be involved in these three pathways. miR-29b-3p was selected for further study because the analysis suggested that miR-29b-3p functioned both in the T and B cell receptor signaling pathways. Dual-luciferase reporter gene assays showed that the upregulation of miR-29b-3p reinforced the apoptosis of cells. The results supported the involvement of immune regulation by miRNAs in the epigenotoxicity of TiO_2 NPs. In another study in mouse RAW264.7 cells, the role of miRNA in regulating the cytotoxicity of TiO_2 NPs was investigated (Sui et al. 2018). Cells were treated with 100 μg/mL TiO_2 NPs (P25; 21 nm) for 24 hours (for miRNA analysis) and screened for cytotoxicity-related miRNAs. The differentially expressed miRNAs were detected using Illumina HiSeq™ 2000 sequencing. The expression of 140 miRNAs was significantly changed in response to TiO_2 NPs, of which 131 miRNAs were upregulated and 9 miRNAs were downregulated. DIANA-mirPath software was applied to perform differential expression miRNAs involved in the signaling pathway analysis. PI3K-Akt signaling pathway, p53 signaling pathway, MAPK signaling pathway, Wnt signaling pathway, and apoptosis were identified as being involved in TiO_2 NPs exposure. qRT-PCR verification showed that miR-27a-5p and miR-182-3p were downregulated in the TiO_2 NPs exposed group, while miR-27b-3p, miR-181d-5p, and miR-350 were upregulated. Bioinformatic analysis results showed miR-350 was mainly involved in apoptosis-related signal transduction pathways. In the MiR-350 mimic (transfected into RAW264.7 cells), the apoptosis rate of cells was significantly increased. The results of biological function and target gene regulation studies showed miR-350 may promote RAW264.7 cell apoptosis through the negative regulation of the *phosphatidylinositol 3-kinase-R3* gene; phosphatidylinositol 3-kinase signaling is implicated in cell proliferation metabolic control, migration, and differentiation.

The role of miRNAs in DNA damage induced by *in vitro* exposure to TiO_2 NPs was examined in the human monocytic cell line (THP-1) (Hu and Palic 2020). Cells were incubated with fluorescent dye Alizarin Red S-labeled TiO_2 NPs (anatase; <25 nm) at 10 μg/mL for 8 hours. Altered expression was identified in 12 miRNAs, with miR-155-5p being determined as the best candidate for functional validation. Severe DNA damage was also triggered by TiO_2 NPs, and expression of DNA damage repairing genes was elevated post nano-TiO_2 exposure. Transfection of miR-155-5p mimic influenced the expression of miR-155-5p-related genes responsible for DNA damage (*ATM*, *APAF-1*, and *RAD51*) post TiO_2 NPs exposure, suggesting significant involvement of miR-155-5p in gene repair mechanisms triggered by adverse effects of TiO_2 NPs in monocytes. In another study, human lung epithelial BEAS-2B cells were treated with 3–100 μg/ml TiO_2 NPs *in vitro* for 1 day and 1–30 μg/mL of TiO_2 NPs 3 days before mRNA and miRNA expression analyses, respectively (Thai et al. 2015). The analysis was conducted using Affymetrix mRNA or miRNA microarrays. Although the tested doses were not cytotoxic, there were alterations in both mRNA and miRNA expression. Pathway analysis of differentially expressed genes and miRNAs identified genes involved in cell cycle regulation, inflammatory response, acute phase response, apoptosis, NRF2-mediated oxidative stress response, and PI3K/AKT signaling pathways. The miRNA 17-92 cluster, an oncogenic microRNA cluster, and the let-7miRNA family, tumor-suppressor miRNA, that are involved in lung cancer formation were altered by TiO_2 NPs treatment, with the miR 17–92 cluster being induced and the let-7 family being suppressed. The observed changes demonstrated that *in vitro* treatment in human lung cells can cause diverse but coordinated pathway alterations associated with changes in *in vivo* response to tumorigens.

The cross talk between nanoparticle exposure, mRNA, and miRNA expression was investigated by integrating gene and miRNA expression profiles that employed RNA sequencing together with a quantitative strategy to identify miRNAs (Ndika et al. 2019). Differentiated THP-1 cells were exposed to TiO_2 NPs (rutile/anatase [90:10]; 30–40 nm); 100 μg/mL (60 μg/cm^2), for 6 or 24 hours. Exposure was normalized to induce a maximum 15% cellular cytotoxicity. TiO_2 NPs triggered clearly distinct changes on the miR transcriptome. Via canonical correlation and pathway enrichment analyses, a co-regulated miR-mRNA cluster was identified. Biological process and pathway enrichment analyses functionally implicated the co-regulated miR-mRNA cluster in the cellular response of differentiated THP-1 cells to TiO_2 NPs exposure, indicating that THP-1 miRNAs/isomiRs regulate the expression of genes involved in cell cycle regulation, immune response, and response to these nanoparticles. The study also used this integrative miRNA/mRNA profiling to analyze several other metal-based nanoparticles in an effort to develop predictive classification of hazardous nanomaterials based on shared features as an approach in risk assessment. As a result, seven key miRNAs were proposed as possible biomarkers of potentially a broad range of metal-based nanoparticles.

The role of miRNAs in the regulation of autophagy has been explored. Autophagy is recognized as an emerging toxicity mechanism triggered by nanomaterials. Toxicological consequences of autophagy are regarded as a potential mechanism of cell death (Stern et al. 2012; Yu et al. 2014; Zabirnyk et al. 2007). Dysregulation of autophagy has been implicated in numerous human diseases (Mizushima et al. 2008; Yang and Klionsky 2010). In one study, human HTR-8/SVneo (HTR) cells were exposed to TiO_2 NPs (anatase:rutile [80:20]; 21 nm) (Mao et al. 2018). Dose-dependent increases in miR-101-3p and miR-96-5p were observed after treatment with different concentrations (1, 10, and 100 μg/mL). Autophagy was observed and cell migration ability was severely impaired. Further analysis showed that the upregulation of miRNAs affected the expression of mTOR and Bcl-2 in regulating autophagy.

In another study, the effect of TiO_2 NPs exposure on the expression of autophagy-related miRNAs (in the absence of cytotoxic effects) was assessed in human lung carcinoma A549 cells (Alinovi et al. 2017). A significant and persistent downregulation of these autophagy-related miRNAs (miR-21 and miR-30a) in TiO_2 NPs-treated cultures was observed, although this response occurred at different times and to a different extent. It was noted that miR-21 is involved in all cancer-related processes including tumorigenesis, progression, and metastasis, and its expression is associated with clinical–pathological factors and prognosis. TiO_2 NPs also caused autophagy. The impact of miR34a on Bcl-2 mRNA and protein expression and its molecular actions in the regulation of TiO_2 NPs-induced autophagy *in vitro* with human bronchial epithelial normal BEAS-2B cells was also examined (Bai et al. 2016). In cells exposed to suspensions of TiO_2 NPs (near-spherical; 22 nm) (6.25, 12.5, and 25 μg/mL) for 24 hours, downregulation of miR34a and increased the expression of B cell lymphoma/leukemia-2 (Bcl-2) was observed in a dose-dependent manner. Activation of TiO_2 NPs-induced autophagy was also found. Overexpressed miR34a was shown to enhance TiO_2 NP-induced autophagy and cell death through targeted downregulation of Bcl-2 in BEAS-2B cells in experiments performed using overexpression of miR34a by lentivirus vector transfection.

The effect on the gene profile in AGS cells, a human stomach epithelial cell line, incubated with E171 (food-grade TiO_2) (40 μg/mL) for 24 hours was evaluated (Han et al. 2021). A gene profile analysis showed that immune response-related miRNAs were most strongly affected by E171 exposure. The expression of miRNA 3908 was the most downregulated; miRNA 3908 is suggested to inhibit cancer progression by inducing apoptosis. Also, the expression of the ER stress- (calnexin, IRE-1a, Bip, PERK, and Ero-1a), pyroptosis- (caspase-1, IL-18, and NALP3), autophagy- (p62 and LAMP-1), and iron metabolism-related (ferritin heavy chain) proteins was enhanced in the E171-treated cells accompanied by the peroxisomal (catalase) and mitochondrial antioxidant (SOD-2).

Risk Assessment

Exposure to TiO_2 NPs has been demonstrated in various populations. These epidemiological studies have been generally cross-sectional in nature and have shown limited evidence of malignant or nonmalignant health effects (Schulte et al. 2019). It has been noted that limitations in many of these studies, including inconsistencies in exposure assessments and the healthy worker survivor effect, may explain the lack of association between cancers and exposure to TiO_2 as considered in previous evaluations (Guseva et al. 2020). The potential harm of TiO_2 NPs exposure to human health has, however, been reported.

In a cross-sectional study conducted in a manufacturing plant in eastern China, cardiopulmonary effects among workers exposed to TiO_2 NPs were explored (Zhao et al. 2018). The majority of TiO_2 particles were rod-shaped or irregularly spheroidal with a purity rate of 96%; the main crystal structure was rutile. Total mass concentration of particles was 3.17 mg/m^3. The mass concentration of nanoparticles was 1.22 mg/m^3 accounting for 39% of the total mass. Evidence was found for adverse cardiopulmonary effects of exposure to TiO_2 NPs. Lung damage markers (SP-D [surfactant protein D] and pulmonary function), cardiovascular disease markers (VCAM-1, ICAM-1, LDL, and TC), oxidative stress markers (SOD and MDA), and inflammation markers (IL-8, IL-6, IL-1b, TNF-a, and IL-10) were associated with exposure to TiO_2 NPs. Among these markers, SP-D showed a time (dose)–response pattern within exposed workers. In another investigation, leukotriene levels were significantly elevated in the exhaled breath condensate of exposed workers relative to controls and well correlated with workplace titanium concentrations (Pelclova et al. 2016b). A significant increase in markers of oxidative stress damage of nucleic acids and proteins in the EBC was evident in the more exposed production workers, involved in micronation, calcination, and other TiO_2-ENM production activities, than in the lower exposed groups of research workers and controls (Pelclova et al. 2016a).

Untargeted metabolomics was used to analyze the effects of exposure to TiO_2 NPs on urine metabolite profiles of workers in a cross-sectional study conducted in a TiO_2 NP manufacturing plant in Jinan, China (Chen et al. 2021). Scanning electron microscopy (SEM) observations showed that most of the TiO_2 NPs were rod-shaped or irregular spherical. The size distribution of the particles (Feret diameter) was 194 ± 108 nm. The purity of TiO_2 NPs was about 96% and the main crystal structure was rutile. Metabolomic analysis found 1,760 differentially expressed metabolites in the TiO_2 NP exposed group, of which 60 differential metabolites were simultaneously confirmed by one-dimensional and multi-dimensional statistical analysis. Pathway enrichment analysis further found that the metabolic pathway of long chain acyl-CoA dehydrogenase deficiency (Lcad) was significantly enriched. These metabolites had a biological basis associated with adverse health effects.

Studies have also assessed the effect of TiO_2 NPs exposure on epigenetic markers. Epigenetic assessment has the potential to provide great insight into the biological changes associated with exposures and disease thus bridging genetics or environment and disease. Global methylation, oxidative DNA damage, and lipid peroxidation were assessed in workers with occupational exposure to metal oxide nanomaterials (Liou et al. 2017). TiO_2-exposed workers showed increased oxidative damage (measured by urinary 8-OHdG and EBC 8-isoprostane) but did not show hypomethylation. It was suggested that this may be because exposure to TiO_2 NPs among workers in Taiwan was not sufficiently high to induce epigenetic alterations. In a more recent study, urinary biomarkers in workers of a plant producing TiO_2 pigment powder, having 0.25 μm average particle size and an ultrafine fraction, were evaluated (Buonaurio et al. 2020). TiO_2 NPs exposure resulted in global hypomethylation; urinary concentration of 5-meC was lower in the exposed subjects than in those not exposed. Elevated levels for the oxidative stress biomarkers 8-oxoGuo, 8-oxodGuo, and $3NO_2$Tyr were shown following exposure. Several types of environmental exposures have been demonstrated to induce epigenetic alterations (Ferrari, Pavanello et al. 2019), including other nanomaterials (Rossnerova et al. 2020). The incorporation of epigenetics in this field can provide new measurement tools for biomarker-based analyses of exposures or disease in epidemiology. Epigenetics may represent a mechanistic link between environmental exposures, or genetics, and many common diseases, or can simply provide a quantitative biomarker for exposure or disease for areas of epidemiology currently lacking such measures. Human epidemiological data for nanomaterial exposure remain, however, limited, and epigenetic endpoints are insufficient in themselves as indicators of an adverse response without a connection with phenotype.

The history of risk assessment for engineered nanomaterials has a fairly short time span but during this time, various approaches have been utilized (Schulte et al. 2018). Recently, the concept of AOPs has been proposed as a framework to organize and communicate the existing knowledge on the toxicity mechanisms and outcomes across levels of biological organization, using cause-and-effect relationships to connect molecular initiating events (MIE) to adverse outcomes (AO) (Ankley et al. 2010). In an AOP, a MIE (exposure to a toxicant) is shown to induce a series of key events (KEs) (e.g., DNA methylation change, transcriptional changes, sequence changes, cell proliferation), which, in turn, are shown to result in an AO (e.g., cancer, a disease state). Although an AOP is itself a collection of correlative evidence, it provides a framework for organizing data across multiple disciplines in a biologically plausible and evidence-based manner. The AOP approach enables the integration of findings from epidemiology studies, which are generally observational associations between an exposure and adverse health outcomes, and *in vitro* and *in vivo* studies (Murugadoss et al. 2021). The totality of this evidence can be used to establish a causative link between the exposure and the adverse outcome. As such, the AOP framework provides systematic knowledge about key toxic mechanisms, thus being very effective at characterizing the individual biological impact and toxicological potential of substances and significantly improving the prediction of adverse effects. The AOP framework has emerged as a robust approach for anchoring mechanistic understanding to potential health effects in humans. From a risk assessment perspective, the AOP framework provides a structured framework to evaluate existing information available for a substance of interest; a way to identify data gaps and efficiently generate missing information to increase confidence in decision-making and assessments of risk; and a framework to apply an iterative approach until sufficient information is gathered for decision-making (Ede et al. 2020).

There have been several reports illustrating the application of the AOP approach for nanotoxicology. This approach offers an alternative to traditional animal-based testing in nanosafety assessment since the traditional approach may not be feasible to generate health hazard information for hundreds of nanomaterials and their property derivatives that have found commercial application and provide human health risk assessment (HHRA) in a timely manner (Halappanavar et al. 2020). Current efforts within nanotoxicology have particularly focused on case study-based and data mining approaches, which aim at developing new AOPs or refining existing AOPs based on one or several model stressors, which are then generalized to other stressors.

In one report, the AOP framework was used to rationally combine mechanistic knowledge relating to both nanomaterial- and chemically induced liver toxicity (Gerloff et al 2017). A well-advanced AOP for (chemically induced) liver-fibrosis was used to describe similarities and differences to mechanisms known for nanomaterial-derived toxicity, focused on metal oxide nanomaterials, such as TiO_2 or SiO_2. Hepatitis was, however, defined as the AO for the nanomaterial AOP since it was not possible to describe a full nanomaterial-relevant AOP leading to liver fibrosis based on nanomaterial toxicology literature alone. Also, the development of liver inflammation (hepatitis) due to systemic exposure to metal oxides is a well-documented phenomenon and a known intermediate event leading to the development of fibrosis. In this putative AOP, the proposed (M)IE or initial KE is lysosomal damage. KEs in the AOP for metal oxide nanomaterials include ROS formation, mitochondria damage, hepatocyte injury/apoptosis, inflammatory cytokine upregulation, and infiltration of inflammatory cells. The results demonstrated that the mechanistic knowledge captured in AOPs that have been developed to describe chemically induced toxicity could be utilized to fill knowledge gaps related to the toxicity of poorly soluble nanomaterials. The major differences in the toxicodynamics of nanomaterials and chemicals were noted to lie in the initial upstream events, and, in particular, the (M)IE. However, chemically and nanomaterial-induced toxicological processes share downstream events that lead to a particular AO.

Additionally, a recent study was conducted to extract and integrate relevant information from the literature/database to build a strategy to test the potential of a nanomaterial to induce an initiating KE leading to AO through causally linked KEs (Murugadoss et al. 2021). The approach was based on using existing AOPs to predict if a given nanomaterial has the potential to induce a MIE leading to an AO through a series of KEs. Initially, potential MIEs reported for nanomaterials in the literature were identified. A search in the AOP-Wiki using the identified MIEs (as keywords) to find associated AOPs that can be applied to nanomaterials and could be verified using *in vitro* and in silico approaches through testing the involved KEs identified five potential MIEs for NMs, including DNA damage/methylation. Instead of MIE, the term "initial key event (initial KE)" was used because not all identified events occur at the molecular level. The AOP-Wiki search also revealed potential AOPs linked to the initially identified MIEs, including 19 AOPs associated with "DNA damage, oxidative DNA damage, DNA strand breaks, and DNA methylation".

Specifically, an AOP for TiO_2 has been proposed, which leads to lung tumors in rats after prolonged inhalation (Braakhuis et al. 2021). Based on scientific peer-reviewed literature, the ability of TiO_2 to induce each KE in the AOPs for inhalation leading to tumor formation was evaluated. The rationale behind the suggested AOP was that, upon chronic inhalation exposure to TiO_2 at relatively high concentrations, the particles can accumulate in the lungs. In case the accumulation of particles overwhelms the clearance capacity of the lung, impaired clearance is induced (the IE). Proposed KEs are ROS generated (KE1), induced oxidative stress (KE2), continuous recruitment of neutrophils and persistent inflammation (KE3), induced persistent epithelial injury (KE4), DNA damage of the epithelial cells (KE5), regenerative cell proliferation (KE6), and hyperplasia (KE7), which ultimately lead to lung tumors (the AO). Similar to the inhalation route, it was suggested that TiO_2 might induce or promote colon tumors via inflammation and ROS production. In case of excess ROS production, oxidative stress might be induced, leading to tissue damage. The suggested AOP for TiO_2 leading to intestinal tumors was very similar to that for lung tumors, although compared with the inhalation route, it was noted that less information was available about each of the KEs for oral exposure to TiO_2.

In another report, a putative AOP for lung cancer associated with pulmonary deposition and retention of poorly soluble nanoparticles, covering aspects of both primary and secondary genotoxicity is described with TiO_2 NPs as one of the representative stressors (Nymark et al. 2021). The putative AOP extended on the proposed AOP operative for the assessment of TiO_2 carcinogenicity in rats, which is described above (Braakhuis et al. 2021), and opens the possibility to build iteratively further on the KE descriptions. The MIE, KEs, and AO in the suggested AOP were aligned with relevant existing KEs in the AOP-Wiki, as well as selected evidence for their association with exposure to nanoparticles. Also, in investigating a variety of endpoints to assess the toxicity of a variety of TiO_2 NPs to induce potentially adverse health effects in an *in vitro* human respiratory epithelial cell model, several of the signaling pathways affected by these NPs correlated well with the AOPs that have been observed in *in vivo* animal treatments; these include inflammation, stress-related pathways, fibrosis, and cancer-related pathways (Thai et al. 2019).

An AOP following the ingestion of food-grade TiO_2 has been postulated, resulting from the evaluation of recent toxicological studies on food-grade TiO_2 and TiO_2 NPs in *ex vivo*, *in vitro*, and *in vivo* experiments along the gastrointestinal route (Bischoff et al. 2021). The molecular mechanisms potentially leading to the initiation of E171-induced adverse health effects were presumed to be closely related to the potential of TiO_2 to induce ROS formation and to promote inflammation. Mechanistic studies *in vitro* and *in vivo*, mostly performed with TiO_2 nanoparticles, suggested that the absorption of TiO_2/E171 can induce the formation of ROS and could lead to a misbalance of the oxidative–antioxidative system. Two postulated initiating events (IE) and several KEs related to adverse effects after TiO_2 ingestion were identified. KEs included ROS generation, oxidative stress, persistent inflammation, persistent epithelial injury, increased cell proliferation, and DNA damage, resulting in preneoplastic lesions and, ultimately, intestinal adenomas/carcinomas (AO).

A recent study was conducted to assess whether oral exposure to TiO_2 in animals can potentially induce effects in the liver and intestine at concentrations relevant for humans (Brand et al. 2020). Postulated AOPs describing chains of events leading to the induction of steatosis, edema, and fibrosis in the liver, and the induction of intestinal tumors were used to structure the available information from animal studies and to assess whether the associated KEs and the AOs are likely to occur due to TiO_2 exposure, and at which dose. The suggested AOP for liver effects was a compilation of two existing AOPs, amended with information from recent research. In addition, the available information on organ concentrations of Ti (resulting from exposure to TiO_2) from oral animal studies was compared with recently reported concentrations found in human postmortem organs. The overview obtained with the AOP approach indicated that TiO_2 can trigger KEs in the liver and intestine, including ROS generation, induction of oxidative stress and inflammation. It was noted that TiO_2 seems to be able to exert these early effects in animal studies at titanium liver concentrations that are only a factor of 30 and 6 times higher than the median and highest liver concentration found in humans, respectively, supporting the potential of adverse effects on the liver in humans as a result of oral TiO_2 exposure.

The integration of epigenetic data into the risk assessment process can be beneficial and aid in the development of a new paradigm of toxicity-pathway-based risk assessment (National Research Council 2007, 2010). Epigenetic data have the potential to inform both mechanisms and modes of action and in combination with genome data may identify novel modes of action. Measures of epigenetic alterations can also contribute to dose-response analyses employed in HHRA frameworks directed toward regulating environmental pollutants and consumer chemicals (Herceg et al. 2013; Kuppusamy et al. 2015; Ray et al. 2014). The AOP framework has the potential to demonstrate a causal relationship between chemical exposures, epigenome, and adverse health outcomes, recognizing that the links between epigenetic modifications and health outcomes are often correlative and do not distinguish between cause-and-effect or common-cause relationships (Angrish et al. 2018). Included in this approach would be collecting and organizing existing scientific evidence into a framework that establishes evidence-based biological plausibility.

Arsenic, for example, has been used as a case to design an epigenetics-based AOP framework structuring arsenic-induced cancer by DNA methylation modification (van der Ven et al. 2017). In developing the AOP, overlap between epigenetic targets that have been associated with arsenic exposure and arsenic exposure-associated disease in humans as well as targets derived from *in vitro* cell culture models were identified. The approach combined the advantages of arsenic exposure assessment, association with apical outcomes, and human relevance from epidemiology with support for cause and effect. Moreover, it included the possibility of direct testing from *in vitro* models. Broadly, after initial induction of a modified DNA methylation profile, processes (molecular networks) involving gene expression may be changed and subsequently lead to altered cell functions, which in time may promote carcinogenesis. Specifically, in this putative AOP, exposure to arsenic leads to an unknown MIE, followed by KEs of decreased DNA methylation, decreased p53 expression, increased DNA damage, increased mutation, increased tumor formation, and AO cancer. Previously an AOP was also proposed for cadmium-induced epigenetic effects (Ruiter et al.

2016). The AOP postulates that cadmium–thiol interactions may lead to depletion of the methyl donor *S*-adenosyl-methionine, resulting in methylome alterations, and may, additionally, result in oxidative stress, which may lead to DNA oxidation, and subsequently altered DNA methyltransferase activity. In this way, DNA methylation may be affected at a critical developmental stage, causing the observed apical phenotypes.

As discussed above, there is increasing evidence that TiO_2 NPs can induce epigenetic alterations, and molecular and epidemiologic results link TiO_2 NPs exposure to potential harm to human health. It seems reasonable that epigenetics can potentially be used to inform AO pathways in human safety and risk assessment for this nanomaterial. In many of these studies demonstrating the epigenetic alterations, other toxic events were also observed (Figure 6.1). As examples, generation of ROS, inflammatory response and oxidative stress, oxidative DNA damage, genotoxicity (chromosomal damage), increased lipid peroxidation, and tissue fibrosis were associated with altered DNA methylation. Changes in histone modifications and histone-modifying enzymes expression were associated with the production of ROS, DNA damage, and changes in oxidative stress-related genes. Expression of miRNAs was also associated with ROS production and with cardiac and mitochondrial dysfunction, autophagy, or an oncogenic phenotype. Most these associated events overlap with KEs proposed in the AOPs for nanomaterials, including TiO_2 NPs. Several of the affected miRNAs were noted to have biological functions that are involved in inflammatory or immune response and cancer-related processes, including tumorigenesis, progression, and metastasis. Also, bioinformatic analysis in several of the studies showed that differentially expressed miRNAs are involved in various disease-related pathways, including the PI3K-Akt, p53, MAPK, and Wnt signaling pathways. There is also an increasing body of evidence that gene expression changes are central to AOPs (Groh et al. 2015). Changes in gene expression can be induced epigenetically. Organizing epigenetic information for TiO_2 NPs into AOP frameworks may yield a clear, transparent, scientific, and evidence-based approach to linking causally epigenome toxicity

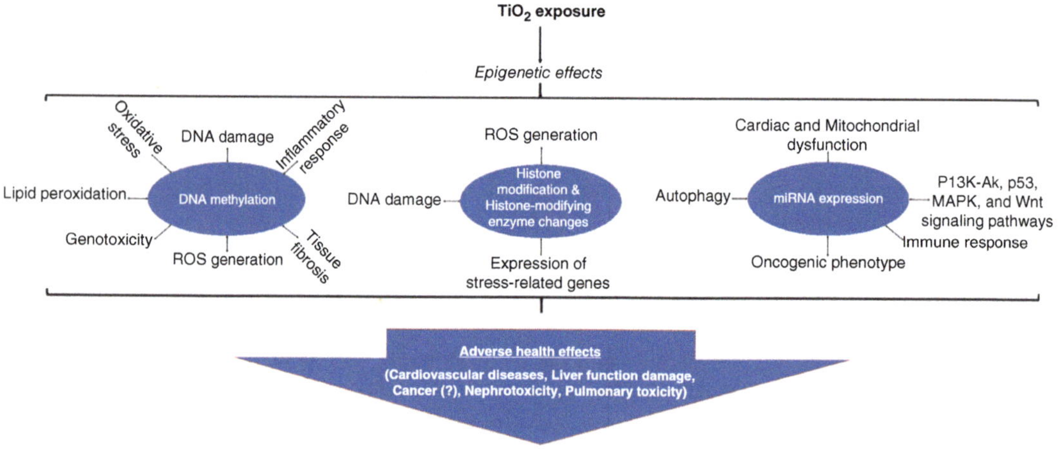

Figure 6.1 TiO_2 exposure induces alterations in DNA methylation, histone modifications and histone-modifying enzymes, and miRNA expression. These epigenetic changes have been shown to be associated with other toxic events, including generation of ROS, inflammatory response, oxidative stress, oxidative DNA damage, genotoxicity (chromosomal damage), increased lipid peroxidation, tissue fibrosis, changes in oxidative stress-related genes, cardiac and mitochondrial dysfunction, autophagy, an oncogenic phenotype, and activation of various disease-related signaling pathways (PI3K-Akt, p53, MAPK, and Wnt). Integration of epigenetic events with other toxic events may offer pathways to adverse health effects resulting from exposure to TiO_2 nanomaterials.

to otherwise correlative human health effects. It is possible that these epigenetic data can be layered with other diverse data and be used to inform events within the AOP framework so that there is sufficient evidence to conclude clear relationships between exposure, epigenetic modification, and adverse outcome (Angrish et al. 2018).

Challenges abound in linking epigenetic effects with outcomes in risk assessment that support the use of the AOP framework to organize epigenetic data into an inference-based biologically plausible framework (Chung and Herceg 2020; Miousse et al. 2015). These obstacles include understanding the complex range of epigenetic modifications and their combinatorial effects, the large number of environmental agents to be tested, and the lack of data that quantitatively evaluate the epigenetic effects of environmental agent exposure. Epigenetic mechanisms act in concert with their effect on chromatin accessibility and gene regulation; a single end-point measure is often insufficient to gain a thorough understanding of the epigenotoxicity of an agent. There is also the current uncertainty as to what constitutes a normal, adaptive response to an exogenous stimulus and what is an adverse, disease-associated alteration. Specific epigenetic end points that can be utilized in safety assessment as well as assays and model systems for assessment of these parameters need to be further identified and evaluated for consistent implementation. Another challenge is the fact that the health outcomes associated with exposure and epigenetic alterations can occur years or generations after the initial exposure.

Despite these significant challenges, the remarkable advances in epigenetics have provided important insights into the epigenetic mechanisms underlying diseases. Technological developments, including novel methodologies and approaches for evaluation of existing epigenetic endpoints, as well as establishment of new epigenetic parameters, promise to increase significantly the robustness of these methods to allow effective application in safety assessment (Dusinska et al. 2017; Miousse et al. 2015). These technological advances will enable the identification of cell-type-specific epigenetic profiles and will allow the distinction between adverse effects and adaptive response. Accumulation of a sufficient amount of reproducible data on epigenetic changes induced by nanomaterials and convincing evidence for their association with disease risk will permit the selection of the most representative model systems, end points to be evaluated, and techniques to be employed for reliable testing of epigenotoxicity of nanomaterials.

Conclusion

Accumulating evidence establishes that TiO_2 NPs induce alterations in DNA methylation, histone modifications, and miRNA expression. Data are available in *in vitro* cellular systems, *in vivo* animal models, and, to a limited extent, in humans. TiO_2 NPs have also been shown to exert a variety of adverse health effects. Assessment of the risk of TiO_2 NPs to human health is, therefore, critical in developing guidance for the safe use of this nanomaterial. Given the role of epigenetics in disease etiology and progression, understanding the epigenotoxicity of TiO_2 NPs will be required in the assessment. In general, epigenetic effect research broadens nanomaterial toxicity evaluation. Useful specific biomarkers that reflect nanomaterial toxicity could be uncovered and critical mechanistic insight can be gained from an increased understanding of the epigenetic biology of nanomaterials. Existing evidence demonstrates an eligibility of epigenetic end points for their incorporation into nanomaterial risk and safety assessment. Additionally, combining epigenetic analyses with other toxicological approaches to assess the safety of nanomaterials would be beneficial in providing a more comprehensive evaluation method.

Disclaimer

This chapter reflects the views of the authors and not necessarily those of the U.S. Food and Drug Administration.

References

Abi Kahalil, C. (2014). The emerging role of epigenetic in cardiovascular disease. *Ther. Adv. Chronic Dis.* 5: 178–187.

Ahlinder, L., Ekstrand-Hamarstrom, B., Geladi, P., and Osterlund, L. (2013). Large uptake of titania and iron oxide nanoparticles in the nucleus of lung epithelial cells as measured by Raman imaging and multivariate classification. *Biophys. J.* 105 (2): 310–319.

Alinovi, R., Pinelli, S., Ravanetti, F., Galetti, M., Pelosi, G., De Palma, G., Apostoli, P., Cacchioli, A., Mutti, A., and Mozzoni, P. (2017). Titanium dioxide aggregating nanoparticles induce autophagy and under-expression of microRNA 21 and 30a in A549 cell line: a comparative study with cobalt(II, III) oxide nanoparticles. *Toxicol. In Vitro* 42: 76–85.

Allis, C.D., Berger, S.L., Cote, J., Dent, S., Jenuwien, T., Kouzarides, T. et al. (2007). New nomenclature for chromatin-modifying enzymes. *Cell* 131: 633–636.

Ambros, V. (2004). The functions of animal microRNAs. *Nature* 431: 350–355.

Angrish, M.M., Allard, P., McCullough, S.D., Druwe, I.L., Helbling, C.L., Hines, E., and Chorley, B.N. (2018). Epigenetic applications in adverse outcome pathways and environmental risk evaluation. *Environ. Health Perspect.* 126 (4): 045001-1-12.

Ankley, G.T., Bennett, R.S., Erickson, R.J., Hoff, D.J., Hornung, M.W., Johnson, R.D., Mount, D.R., Nichols, J.W., Russom, C.L., Schmieder, P.K., Serrrano, J.A., Tietge, J.E., and Villeneuve, D.L. (2010). Adverse outcome pathways: a conceptual framework to support ecotoxicology research and risk assessment. *Environ. Toxicol. Chem.* 29 (3): 730–741.

Baan, R., Straif, K., Grosse, Y., Secretan, B., El Ghissassi, F., and Cogliano, V. (2006). Carcinogenicity of carbon black, titanium dioxide, and talc. *Lancet Oncol.* 7: 295–296.

Baccarelli, A., Rienstra, M., and Benjamin, E.J. (2010). Cardiovascular epigenetics: basic concepts and results from animal and human studies. *Circ. Cardiovasc. Genet.* 3: 567–573.

Bachler, G., von Goetz, N., and Hungerbuhler, K. (2015). Using physiologically based pharmacokinetic (PBPK) modeling for dietary risk assessment of titanium dioxide (TiO_2) nanoparticles. *Nanotoxicology* 9: 373–380.

Bai, W., Chen, Y., and Gao, A.. (2015). Cross talk between poly(ADP-ribose) polymerase 1 methylation and oxidative stress involved in the toxic effect of anatase titanium dioxide nanoparticles. *Int. J. Nanomed.* 10: 5561–5569.

Bai, W., Chen, Y., Sun, P., and Gao, A. (2016). Downregulation of B-cell lymphoma/leukemia-2 by overexpressed microRNA 34a enhanced titanium dioxide nanoparticle-induced autophagy in BEAS-2B cells. *Int. J. Nanomed.* 11: 1959–1971.

Ballesteros, S., Vales, G., Velázquez, A., Pastor, S., Alaraby, M., Marcos, R., and Hernández, A. (2021). MicroRNAs as a suitable biomarker to detect the effects of long-term exposures to nanomaterials. Studies on TiO2 NP and MWCNT. *Nanomaterials* 11: 3458.

Bannister, A.J. and Kouzarides, T. (2011). Regulation of chromatin by histone modifications. *Cell Res.* 21: 381–395.

Bartel, D.P. (2004). MicroRNAs: genomics, biogenesis, mechanism, and function. *Cell* 116 (2): 281–297.

Bartel, D.P. (2009). MicroRNAs: target recognition and regulatory functions. *Cell* 136 (2): 215–233.

Baylin, S.B. and Herman, J.G. (2000). DNA hypermethylation in tumorigenesis: epigenetics joins genetics. *Trends Genet.* 16: 168–174.

Bestor, T.H. (2000). The DNA methyltransferases of mammals. *Hum. Mol. Genet.* 9: 2395–2402.

Biola-Clier, M., Beal, D., Caillat, S., Libert, S., Armand, L., Herlin-Boime, N., Sauvaigo, S., Douki, T., and Carriere, M. (2017). Comparison of the DNA damage response in BEAS-2B and A549 cells exposed to titanium dioxide nanoparticles. *Mutagenesis* 32: 161–172.

Bird, A. (2002). DNA methylation patterns and epigenetic memory. *Genes. Dev.* 16: 6–21.

Bischoff, N.S., de Kok, T.M., Sijm, D.T.H.M., van Breda, S.G., Briedé, J.J., Castenmiller, J.J.M., Opperhuizen, A., Chirino, Y.I., Dirven, H., Gott, D., Houdeau, E., Oomen, A.G., Poulsen, M., Rogler, G., and van Loveren, H. (2021). Possible adverse effects of food additive E171 (titanium dioxide) related to particle specific human toxicity, including the immune system. *Int. J. Mol. Sci.* 22 (1): 207.

Blackledge, N.P. and Klose, R.J. (2010). Histone lysine methylation: an epigenetic modification? *Epigenomics* 2 (1): 151–161.

Bochtler, M., Kolano, A., and Xu, G.L. (2017). DNA demethylation pathways: additional players and regulators. *BioEssays* 39: 1–13.

Böckmann, J., Lahl, H., Eckert, T., and Unterhalt, B. (2000). Blood levels of titanium before and after oral administration of titanium dioxide. *Pharmazie* 55 (2): 140–143.

Bogdanovic, O. and Veenstra, G.J. (2009). DNA methylation and methyl-CpG binding proteins: developmental requirements and function. *Chromasoma* 188: 549–565.

Braakhuis, H.M., Gosens, I., Heringa, M.B., Oomen, A.G., Vandebriel, R., Groenewold, M., and Cassee, F.R. (2021). Mechanism of action of TiO2: recommendations to reduce uncertainties related to carcinogenic potential. *Annu. Rev. Pharmacol. Toxicol.* 61: 203–223.

Brand, W., Peters, R.J.B., Braakhuis, H.M., Maślankiewicz, L., and Oomen, A.G. (2020). Possible effects of titanium dioxide particles on human liver, intestinal tissue, spleen and kidney after oral exposure. *Nanotoxicology* 14 (7): 985–1007.

Brown, K., Thurn, T., Xin, L., Liu, W., Bazak, R., Chen, S., Lai, B., Vogt, S., Jacobsen, C., Paunesku, T., and Woloschak, G.E. (2018). Intracellular in situ labelling of TiO_2 nanoparticles for fluorescence microscopy detection. *Nano. Rs.* 11 (1): 1–20.

Buonaurio, F., Astolfi, M.L., Canepari, S., Di Basilio, M., Gibilras, R., Mecchia, M., Papacchini, M., Paci, E., Pigini, D., and Tranfo, G. (2020). Urinary oxidative stress biomarkers in workers of a titanium dioxide based pigment production plant. *Int. J. Environ. Res. Public Health* 17 (23): 9085.

Chen, H.W., Su, S.F., Chien, C.T., Lin, W.H., Yu, S.L., Chou, C.C., Chen, J.J., and Yang, P.C. (2006). Titanium dioxide nanoparticles induce emphysema-like lung injury in mice. *FASEB J.* 20: 2393–2395.

Chen, X.X., Cheng, B., Yang, Y.X., Cao, A., Liu, J.H., Du, L.J., Liu, Y., Zhao, Y., and Wang, H. (2013). Characterization and preliminary toxicity assay of nano-titanium dioxide additive in sugar-coated chewing gum. *Small* 9: 1765–1774.

Chen, Z., Han, S., Zhang, J., Zheng, P., Liu, X., Zhang, Y., and Jia, G. (2021). Exploring urine biomarkers of early health effects for occupational exposure to titanium dioxide nanoparticles using metabolomics. *Nanoscale* 13: 4122–4132.

Chervona, Y. and Costa, M. (2012). Histone modifications and cancer: biomarkers of prognosis? *Am. J. Cancer Res.* 2: 589–597.

Chevanne, M., Calia, C., Zampieri, M., Cecchinelli, B., Caldini, R., Monti, D., Bucci, L., Franceschi, C., and Caiafa, P. (2007). Oxidative DNA damage repair and parp 1 and parp 2 expression in Epstein-Barr virus-immortalized B lymphocyte cells from young subjects, old subjects, and centenarians. *Rejuvenation Res.* 10 (2): 191–204.

Cho, W.S., Kang, B.C., Lee, J.K., Jeong, J., Che, J.H., and Seok, S.H. (2013). Comparative absorption, distribution, and excretion of titanium dioxide and zinc oxide nanoparticles after repeated oral administration. *Part. Fibre Toxicol.* 10: 9.

Chung, F.F. and Herceg, Z. (2020). The promises and challenges of toxico-epigenomics: environmental chemicals and their impacts on the epigenome. *Environ. Health Perspect.* 128 (1): 15001.

Cui, Y., Liu, H., Ze, Y., Zengli, Z., Cheng, Z., Cheng, J., Hu, R., Gao, G., Wang, L., Tang, M., and Hong, F. (2012). Gene expression in liver injury caused by long-term exposure to titanium dioxide nanoparticles in mice. *Toxicol. Sci.* 1287: 171–185.

Deng, Q., Luo, Y., Chang, C., Zwergel, C., Artico, M., Taurone, S., Ralli, M., Greco, A., and Mai, A. (2019). The emerging role of epigenetics in human autoimmune disorders. *Clin. Epigenetics* 11: 34.

Dorier, M., Beal, D., Marie-Desvergne, C., Dubosson, M., Barreau, F., Houdeau, E., Herlin-Boime, N., and Carriere, M. (2017). Continuous in vitro exposure of intestinal epithelial cells to E171 food additive causes oxidative stress, inducing oxidation of DNA bases but no endoplasmic reticulum stress. *Nanotoxicology* 1: 54.

Duan, Y., Liu, J., Ma, L., Li, N., Liu, H., Wang, J., Zheng, L., Liu, C., Wang, X., Zhao, X., Yan, J., Wang, S., Wang, H., Zhang, X., and Hong, F. (2010). Toxicological characteristics of nanoparticulate anatase titanium dioxide in mice. *Biomaterials* 32: 894–899.

Dudefoi, W., Moniz, K., Allen-Vercoe, E., Ropers, M.-H., and Virginia, K. (2017). Impact of food grade and nano-TiO2 particles on a human intestinal community. *Food Chem. Toxicol.* 106: 242–249.

Dusinska, M., Tulinska, J., El Yamani, N., Kuricova, M., Liskova, A., Rollerova, E., Rundén-Pran, E., and Smolkova, B. (2017). Immunotoxicity, genotoxicity and epigenetic toxicity of nanomaterials: New strategies for toxicity testing?. *Food Chem. Toxicol.* 109 (Pt 1): 797–811.

Ede, J.D., Lobaskin, V., Vogel, U., Lynch, I., Halappanavar, S., Doak, S.H., Roberts, M.G., and Shatkin, J.A. (2020). Translating scientific advances in the AOP framework to decision making for nanomaterials. *Nanomaterials* 10 (6): 1229.

EFSA (2016). Re-evaluation of titanium dioxide (E 171) as a food additive. *EFSA J.* 14 (9).

El Dine, F.M.M.B., Hussein, H.A.S.M., and Ghazala, R.A. (2018). Evaluation of epigenetic changes of liver tissue induced by oral administration of titanium dioxide nanoparticles and possible protective role of *Nigella sativa* oil, in adult male albino rats. *Nanomed. J.* 5: 192–198.

Esquela-Kerscher, A. and Slack, F.J. (2006). Oncomirs – microRNAs with a role in cancer. *Nat. Rev. Cancer* 6: 259–269.

Esteller, M. (2008). Epigenetics in cancer. *N. Engl. J. Med.* 358: 1148–1159.

Fabian, E., Landsiedel, R., Ma-Hock, L., Wiench, K., Wohlleben, W., and van Ravenzwaay, B. (2008). Tissue distribution and toxicity of intravenously administered titanium dioxide nanoparticles in rats. *Arch. Toxicol.* 82: 151–157.

Feil, R. and Fraga, M.F. (2012). Epigenetics and the environment: emerging patterns and implications. *Nat. Rev. Genet.* 13 (2): 97–109.

Felsenfeld, G. (2014). A brief history of epigenetics. *Cold Spring Harb. Perspect. Biol.* 1: 6.

Ferrari, L., Carugno, M., and Bollati, V. (2019a). Particulate matter exposure shapes DNA methylation through the lifespan. *Clin. Epigenetics* 11: 129.

Ferrari, L., Pavanello, S., and Bollati, V. (2019b). Molecular and epigenetic markers as promising tools to quantify the effect of occupational exposures and the risk of developing non-communicable diseases. *Med. Lav.* 110 (3): 168–190.

Fresegna, A.M., Ursini, C.J., Ciervo, A., Maiello, R., Casciardi, S., Iavicoli, S., and Cavallo, D. (2021). Assessment of the influence of crystalline form on cyto-genotoxic and inflammatory effects induced by TiO2 nanoparticles on human bronchial and alveolar cells. *Nanomaterials* 11: 253.

Frohlich, E. and Roblegg, E. (2016). Oral uptake of nanoparticles: human relevance and the role of in vitro systems. *Arch. Toxicol.* 90: 2297–2314.

Fu, P.P., Xia, Q., Hwang, H.M., Ray, P.C., and Yu, H. (2014). Mechanisms of nanotoxicity: generation of reactive oxygen species. *J. Food Drug Anal.* 22: 64–75.

Gao, G., Ze, Y., Li, B., Zhao, X., Zhang, T., Sheng, L., Hu, R., Gui, S., Sang, Z., Sun, Q., Cheng, J., Cheng, Z., Wang, L., Tang, M., and Hong, F. (2012). Ovarian dysfunction and gene-expressed characteristics of female mice caused by long-term exposure to titanium dioxide nanoparticles. *J. Hazard. Mater.* 243: 19–27.

Gao, G., Ze, Y., Zhao, X., Zhang, T., Sang, Z., Zheng, L., Ze, X., et al. (2013). Titanium dioxide nanoparticle-induced testicular damage, spermatogenesis suppression, and gene expression alterations in male mice. *J. Hazard. Mater.* 258–259: 133–143.

Gedda, M.R., Babele, P.K., Zahra, K., and Madhukar, P. (2019). Epigenetic aspects of engineered nanomaterials: is the collateral damage inevitable? *Front. Bioeng. Biotechnol.* 7: 228.

Geraets, L., Oomen, A.G., Krystek, P., Jacobsen, N.R., Håkan, W., Laurentie, M., Verharen, H.W., Brandon, E.F.A., and de Jong, W.H. (2014). Tissue distribution and elimination after oral and intravenous administration of different titanium dioxide nanoparticles in rats. *Part. Fibre Toxicol.* 11: 30.

Gerloff, K., Landesmann, B., Worth, A., Munn, S., Palosaari, T., and Whelan, M. (2017). The Adverse Outcome Pathway approach in nanotoxicology. *Comput. Toxicol.* 1: 3–11.

Ghosh, M., Öner, D., Duca, R.-C., Cokic, S.M., Seys, S., Kerkhofs, S., Van Landuyt, K., Hoet, P., and Godderis, L. (2017). Cyto-genotoxic and DNA methylation changes induced by different crystal phases of TiO2-np in bronchial epithelial (16-HBE) cells. *Mutat. Res.* 796: 1–12.

Grewal, S.I. and Elgin, S.C. (2007). Transcription and RNA interference in the formation of heterochromatin. *Nature* 447 (7143): 399–406.

Griffiths-Jones, S., Saini, H.K., van Dongen, S., and Enright, A.J. (2008). miRBase: tools for microRNA genomics. *Nucleic Acids Res.* 36: D154–D158.

Groh, K.J., Carvalho, R.N., Chipman, J.K., Denslow, N.D., Halder, M., Murphy, C.A., Roelofs, D., Rolaki, A., Schirmer, K., and Watanabe, K.H. (2015). Development and application of the adverse outcome pathway framework for understanding and predicting chronic toxicity: I. Challenges and research needs in ecotoxicology. *Chemosphere* 120: 764–777.

Guseva, C.I., Fraize-Frontier, S., Michel, C., and Charles, S. (2020). Weight of epidemiological evidence for titanium dioxide risk assessment: current state and further needs. *J. Expo. Sci. Environ. Epidemiol.* 30 (3): 430–435.

Hackenberg, S., Friehs, G., and Kleinsasser, N. (2011). Nanosized titanium dioxide particles do not induce DNA damage in human peripheral blood lymphocytes. *Environ. Mol. Mutagen.* 52: 264–268.

Halappanavar, S., Jackson, P., Williams, A., Jensen, K.A., Hougaard, K.S., Vogel, U., Yauk, C.L., and Wallin, H. (2011). Pulmonary response to surface-coated nanotitanium dioxide particles includes induction of acute phase response genes, inflammatory cascades, and changes in microRNAs: a toxicogenomic study. *Environ. Mol. Mutagen.* 252 (6): 425–439.

Halappanavar, S., van den Brule, S., Nymark, P., Gate, L., Seidel, C., Valentino, S., Zhernovkov, V., Danielsen, P., De Vizcaya-Ruiz, A., Wolff, H., Stoeger, T., Boyadziev, A., Poulsen, S., Sørli, J.B., and Vogel, U. (2020). Adverse outcome pathways as a tool for the design of testing strategies to support the safety assessment of emerging advanced materials at the nanoscale. *Part. Fibre Toxicol.* 17.

Han, H.-Y., Yang, M.-J., Yoon, C., Lee, G.-H., Kim, D.-W., Kim, T.-W., Kwak, M., Heo, M.B., Lee, T.G., Kim, S., Oh, J.-H., Lim, H.-J., Oh, I., Yoon, S., and Park, E.-J. (2021). Toxicity of orally administered food-grade titanium dioxide nanoparticles. *J. Appl. Toxicol.* 41 (7): 1127–1147.

Hanot-Roy, M., Tubeuf, E., Guilbert, A., Bado-Nilles, A., Vigneron, P., Trouiller, B., Braun, A., and Lacroix, G. (2016). Oxidative stress pathways involved in cytotoxicity and genotoxicity of titanium dioxide (TiO2) nanoparticles on cells constitutive of alveolo-capillary barrier in vitro. *Toxicol. In Vitro* 33: 125–135.

Hathaway, Q.A., Durr, A.J., Shepherd, D.L., Pinti, M.V., Brandebura, A.N., Nichols, C.E., Kunovac, A., Goldsmith, W.T., Friend, S.A., Abukabda, A.B., Fink, G.K., Nurkiewicz, T.R., and Hollander, L.M. (2019). miRNA-378a as a key regulator of cardiovascular health following engineered nanomaterial inhalation exposure. *Nanotoxicology* 13 (5): 644–663.

Herceg, Z., Lambert, M.P., van Veldhoven, K., Demetriou, C., Vineis, P., Smith, M.T., Straif, K., and Wild, C.P. (2013). Towards incorporating epigenetic mechanisms into carcinogen identification and evaluation. *Carcinogenesis* 34: 1955–1967.

Heringa, M.B., Geraets, L., van Eijkeren, J.C., Vandebriel, R.J., de Jong, W.H., and Oomen, A.G. (2016). Risk assessment of titanium dioxide nanoparticles via oral exposure, including toxicokinetic considerations. *Nanotoxicology* 10 (10): 1515–1525.

Heringa, M.B., Peters, R.J.B., Bleys, R.L.A.W., van der Lee, M.K., Tromp, P.C., van Kesteren, P.C.E., van Eijkeren, J.C.H., Undas, A.K., Oomen, A.G., and Bouwmeester, H. (2018). Detection of titanium particles in human liver and spleen and possible health implications. *Part. Fibre Toxicol.* 15 (1): 15.

Hou, J., Wang, L., Wang, C., Zhang, S., Liu, H., Li, S., and Wang, X. (2019). Toxicity and mechanisms of action of titanium dioxide nanoparticles in living organisms. *J. Environ. Sci.* 75: 40–53.

Hsiao, I.-L., Bierkandt, F.S., Reichardt, P., Luch, A., Huang, Y.-J., Jakubowski, N., Tentschert, J., and Haase, A. (2016). Quantification and visualization of cellular uptake of TiO_2 and Ag nanoparticles: comparison of different ICP-MS techniques. *J. Nanobiotechol.* 14: 50.

Hu, M. and Palić, D. (2020). Role of MicroRNAs in regulation of DNA damage in monocytes exposed to polystyrene and TiO_2 nanoparticles. *Toxicol. Rep.* 7: 743–751.

Huerta-García, E., del Pilar Ramos-Godinez, M., López-Saavedra, A., Alfaro-Moreno, E., Gómez-Crisóstomo, N.P., Colín-Val, Z., Sánchez-Barrera, H., and López-Marure, R. (2019). Internalization of titanium dioxide nanoparticles is mediated by actin-dependent reorganization and clathrin- and dynamin-mediated endocytosis in H9c2 rat cardiomyoblasts. *Chem. Res. Toxicol.* 32: 578–588.

Huerta-Garcıa, E., Ma´rquez-Ramı´rez, S.G., del Pilar Ramos-Godinez, M., Lopez-Saavedra, A., Herrera, L.A., Parra, A., Alfaro-Moreno, E., Gomez, E.O., and Lopez-Marure, R. (2015). Internalization of titanium dioxide nanoparticles by glial cells is given at short times and is mainly mediated by actin reorganization-dependent endocytosis. *Neurotoxicology* 51: 27–37.

Hwang, J.Y., Aromolaran, K.A., and Zukin, R.S. (2017). The emerging field of epigenetics in neurodegeneration and neuroprotection. *Nat. Rev. Neurosci.* 18: 347–361.

IARC (2010). IARC monographs on the evaluation of carcinogenic risks to humans: carbon black, titanium dioxide, and talc. *Lyon Fr. World Health Organ. Int. Agency Res. Cancer* 93.

Iavicoli, I., Eeso, V., Fontana, L., and Bergamaschi, A. (2011). Toxicological effects of titanium dioxide nanoparticles: a review of in vitro mammalian studies. *Eur. Rev. Med. Pharmacol. Sci.* 15: 481–508.

Jacobs, J.J., Skipor, A.K., Black, J., Urban, R., and Galante, J.O. (1991). Release and excretion of metal in patients who have a total hip-replacement component made of titanium-base alloy. *J. Bone Joint Surg. Am.* 73: 1475–1486.

Jakovcevski, M. and Akbarian, S. (2012). Epigenetic mechanisms in neurological disease. *Nat. Med.* 18: 1194–1202.

Jani, P.U., McCarthy, D.E., and Florence, A.T. (1994). Titanium dioxide (rutile) particle uptake from the rat GI tract and translocation to systemic organs after oral administration. *Int. J. Pharm.* 105: 157–168.

Jayaram, D.T., Kumar, A., Kippner, L.E., Ho, P.-Y., Kemp, M.L., Fan, Y., and Payne, C.K. (2019). TiO_2 nanoparticles generate superoxide and alter gene expression in human lung cells. *RSC Adv.* 9: 25039–25047.

Jones, K., Morton, J., Smith, I., Jurkschat, K., Harding, A.H., and Evans, G. (2015). Human in vivo and in vitro studies on gastrointestinal absorption of titanium dioxide nanoparticles. *Toxicol. Lett.* 233 (2): 95–101.

Kaida, T., Kobayashi, K., Adachi, M., and Suzuki, F. (2004). Optical characteristics of titanium oxide interference film and the film laminated with oxides and their applications for cosmetics. *J. Cosmet. Sci.* 55: 219–220.

Khan, I., Saeed, K., and Khan, I. (2019). Nanoparticles: properties, applications and toxicities. *Arab. J. Chem.* 12: 908–931.

Kouzarides, T. (2007). Chromatin modifications and their function. *Cell* 128: 693–678.

Kraus, W.L. (2008). Transcriptional control by PARP-1: chromatin modulation, enhancer-binding, coregulation, and insulation. *Curr. Opin. Cell Biol.* 20 (3): 294–302.

Kuppusamy, S.P., Kaiser, J.P., and Wesselkamper, S.C. (2015). Epigenetic regulation in environmental chemical carcinogenesis and its applicability in human health risk assessment. *Int. J. Toxicol.* 34: 384–392.

Kurdistani, S.K. (2011). Histone modifications in cancer biology and prognosis. *Prog. Drug Res.* 67: 91–106.

Lachner, M., O'sullivan, R.J., and Jenuwein, T. (2003). An epigenetic roadmap for histone lysine methylation. *J. Cell Sci.* 116 (Pt. 11): 2117–2124.

Lee, K.P., Trochimowicz, H.J., and Reinhardt, C.F. (1985). Pulmonary response of rats exposed to titanium dioxide (TiO2) by inhalation for two years. *Toxicol. Appl. Pharmacol.* 79: 179–192.

Li, N., Duan, Y., Hong, M., Zheng, L., Fei, M., Zhao, X., Wang, J., Cui, Y., Liu, H., Cai, J., Gong, S., Wang, H., and Hong, F. (2010a). Spleen injury and apoptotic pathway in mice caused by titanium dioxide nanoparticles. *Toxicol. Lett.* 195 (2–3): 161–168.

Li, W., Jia, M.X., Deng, J., Wang, J.H., Zuberi, Z., Yang, S., Ba, J., and Chen, Z. (2020). MicroRNA response and toxicity of potential pathways in human colon cancer cells exposed to titanium dioxide nanoparticles. *Cancers (Basel)* 12 (5): 1236.

Li, Y., Li, J., Yin, J., Li, W., Kang, C., Huang, Q., and Li, Q. (2010b). Systematic influence induced by 3 nm titanium dioxide following intratracheal instillation of mice. *J. Nanosci. Nanotechnol.* 10 (12): 8544–8549.

Lim, J.P., Baeg, G.H., Srinivasan, D.K., Dheen, S.T., and Bay, B.H. (2017). Potential adverse effects of engineered nanomaterials commonly used in food on the miRNome. *Food Chem. Toxicol.* 109: 771–779.

Liou, S.H., Wu, W.T., Liao, H.Y., Chen, C.-Y., Tsai, C.-Y., Jung, W.T., and Lee, H.-L. (2017). Global DNA methylation and oxidative stress biomarkers in workers exposed to metal oxide nanoparticles. *J. Hazard. Mater.* 331: 329–335.

Loernzen, J.M., Martino, F., and Thum, T. (2012). Epigenetic modifications in cardiovascular diseases. *Basic Res. Cardiol.* 107: 245.

Lu, N., Zhu, Z., Zhao, X., Tao, R., Yang, X., and Gao, Z. (2008). Nano titanium dioxide photocatalytic protein tyrosine nitration: a potential hazard of TiO2 on skin. *Biochem. Biophys. Res. Commun.* 370 (4): 675–680.

Lu, X., Miousse, I.R., Pirela, S.V., Melnyk, S., Koturbash, I., and Demokritou, P. (2016). Short-term exposure to engineered nanomaterials affects cellular epigenome. *Nanotoxicology* 10 (2): 140–150.

Lui, X., Gao, Q., Li, P., Zhao, Q., Zhang, J., Li, J., Koseki, H., and Wong, J. (2013). UHRF1 targets DNMT1 for DNA methylation through cooperative binding of hemi-methylated DNA and methylated H3K9. *Nat. Commun.* 4: 1563.

Ma, Y., Guo, Y., Wu, S., Lv, Z., Zhang, Q., and Ke, Y. (2017). Titanium dioxide nanoparticles induce size-dependent cytotoxicity and genomic DNA hypomethylation in human respiratory cells. *RSC Adv.* 7: 23560–23572.

Ma, Y., Guo, Y., Ye, H., Huang, K., Lv, Z., and Ke, Y. (2019). Different effects of titanium dioxide nanoparticles instillation in young and adult mice on DNA methylation related with lung inflammation and fibrosis. *Ecotoxicol. Environ. Saf.* 176: 1–10.

MacNicoll, A., Kelly, M., Aksoy, H., Kramer, E., Bouwmeester, H., and Chaudhry, Q. (2015). A study of the uptake and biodistribution of nano-titanium dioxide using in vitro and in vivo models of oral intake. *J. Nanopart. Res.* 17: 66.

Malakootian, M., Nasir, A., Osornio-Vargas, A.R., and Faraji, M. (2021). Effect of titanium dioxide nanoparticles on DNA methylation of human peripheral blood mononuclear cells. *Toxicol. Res. (Camb)* 10 (5): 1045–1051.

Mao, Z., Yao, M., Li, Y., Fu, Z., Li, S., Tang, Q., Han, X., and Xia, Y. (2018). miR-96-5p and miR-101-3p as potential intervention targets to rescue TiO_2 NP-induced autophagy and migration impairment of human trophoblastic cells. *Biomater. Sci.* 6: 3273–3283.

Marquis, B.J., Love, S.A., and Haynes, C.L. (2009). Analytical methods to assess nanoparticle toxicity. *Analyst (Lond.)* 134: 425–439.

Martín-Subero, J.I. and Esteller, M. (2011). Profiling epigenetic alterations in disease. *Adv. Exp. Med. Biol.* 711: 162–177.

Mathers, J.C., Strathdee, G., and Relton, C.L. (2010). Induction of epigenetic alterations by dietary and other environmental factors. *Adv. Gene.* 71: 3–39.

Millar, C.B. and Grunstein, M. (2006). Genome-wide patterns of histone modifications in yeast. *Nat. Rev. Mol. Cell Biol.* 7 (9): 657–666.

Miousse, I.R., Currie, R., Datta, K., Ellinger-Ziegelbauer, H., French, J.E., Harrill, A.H., Koturbash, I., Lawton, M., Mann, D., Meehan, R.R., Moggs, J.G., O'Lone, R., Rasoulpour, R.J., Reijo Pera, R.A., and Thompson, K. (2015). Importance of investigating epigenetic alterations for industry and regulators: an appraisal of current efforts by the Health and Environmental Sciences Institute. *Toxicology* 335: 11–19.

Miranda, K.C., Huynh, T., Tay, Y., Ang, Y.S., Tam, W.L., Thomson, A.M. et al. (2006). A pattern-based method for the identification of MicroRNA binding sites and their corresponding heteroduplexes. *Cell* 126 (6): 1203–1217.

Mizushima, N., Levine, B., Cuervo, A.M., and Klionsky, D.J. (2008). Autophagy fights disease through cellular self-digestion. *Nature* 451 (7182): 1069–1075.

Montgomery, R.L. and van Rooij, E. (2010). MicroRNA regulation as a therapeutic strategy for cardiovascular disease. *Curr. Drug Targets* 11: 936–942.

Moore, L.D., Le, T., and Fan, G. (2013). DNA methylation and its basic function. *Neuropsychopharmacology* 38: 23–38.

Murugadoss, S., Vrček, I.V., Pem, B., Jagiello, K., Judzinska, B., Sosnowska, A., Martens, M., Willighagen, E.L., Puzyn, T., Dusinska, M., Cimpan, M.R., Fessard, V., and Hoet, P.H. (2021). A strategy towards the generation of testable adverse outcome pathways for nanomaterials. *ALTEX* 38: 580–594.

National Research Council (NRC) (2007). *Toxicity Testing in the 21st Century: A Vision and a Strategy.* Washington, DC: National Academies Press.

National Research Council (NRC) (2010). *Toxicity-Pathway-based Risk Assessment: Preparing for a Paradigm Change.* Washington, DC: National Academies Press.

Ndika, J., Seemab, U., Poon, W.L., Fortino, V., El-Nezami, H., Karisola, P., and Alenius, H. (2019). Silver, titanium dioxide, and zinc oxide nanoparticles trigger miRNA/isomiR expression changes in THP-1 cells that are proportional to their health hazard potential. *Nanotoxicology* 13 (10): 1380–1395.

Nilsson, E.E. and Skinner, M.K. (2015). Environmentally induced epigenetic transgenerational inheritance of disease susceptibility. *Transl. Res.* 165: 12–17.

Nymark, P., Karlsson, H.L., Halappanavar, S., and Vogel, U. (2021). Adverse outcome pathway development for assessment of lung carcinogenicity by nanoparticles. *Front. Toxicol.* 3: 653386.

Okano, M., Bell, D.W., Haber, D.A., and Li, E. (1999). DNA methyltransferases Dnmt3a and Dnmt3b are essential for de novo methylation and mammalian development. *Cell* 99: 247–257.

Olive, V., Jiang, I., and He, H. (2010). mir-17–92, a cluster of miRNAs in the midst of the cancer network. *Int. J. Biochem. Cell Biol.* 42: 1348–1354.

Ortlieb, M. (2010). White giant or white dwarf?: particle size distribution measurements of TiO2. *GIT Lab. J. Europe* 14: 42–43.

Pacurari, M., Yin, X.J., Zhao, J., Ding, M., Leonard, S.S., Schwegler-Berry, D., Ducatman, B.S., Sbarra, D., Hoover, M.D., Castranova, V., and Vallyathan, V. (2008). Raw single-wall carbon nanotubes induce oxidative stress an active MAPKs, AP-1, NF-kappaB, and Akt in normal and malignant human mesothelial cells. *Environ. Health Perspect.* 116: 1211–1217.

Participants IRSIW (2000). The relevance of the rat lung response to particle overload for human risk assessment: a workshop consensus report. *Inhal. Toxicol.* 12: 1–17.

Patil, N.A., Gade, W.N., and Deobagkar, D.D. (2016). Epigenetic modulation upon exposure of lung fibroblasts to TiO2 and ZnO nanoparticles: alterations in DNA methylation. *Int. J. Nanomed.* 11: 4509–4519.

Patri, A., Umbreit, T., Zheng, J., Nagashima, K., Goering, P., Francke-Carroll, S., Gordon, E., Weaver, J., Miller, T., and Sadrieh, N. (2009). Energy dispersive X-ray analysis of titanium dioxide nanoparticle distribution after intravenous and subcutaneous injection in mice. *J. Appl. Toxicol.* 29: 662–672.

Pelclova, D., Zdimal, V., Fenclova, Z., Vlckova, S., Turci, F., Corazzari, I., Kacer, P., Schwarz, J., Zikova, N., Makes, O., Syslova, K., Komarc, M., Belacek, J., Navratil, T., Machajova, M., and Zakharov, S. (2016a). Markers of oxidative damage of nucleic acids and proteins among workers exposed to TiO_2 (nano) particles. *Occup. Environ. Med.* 73 (2): 110–118.

Pelclova, D., Zdimal, V., Kacer, P., Fenclova, Z., Vlckova, S., Komarc, M., Navratil, T., Schwarz, J., Zikova, N., Makes, O., Syslova, K., Belacek, J., and Zakharov, S. (2016b). Leukotrienes in exhaled breath condensate and fractional exhaled nitric oxide in workers exposed to TiO_2 nanoparticles. *J. Breath Res.* 10 (3): 036004.

Pele, L.C., Thoree, V., Bruggraber, S.F., Koller, D., Thompson, R.P., Lomer, M.C., and Powell, J.J. (2015). Pharmaceutical/food grade titanium dioxide particles are absorbed into the bloodstream of human volunteers. *Part. Fibre Toxicol.* 12 (1): 26.

Peters, R.J., van Bemmel, G., Herrera-Rivera, Z., Helsper, H.P., Marvin, H.J., Weigel, S., Tromp, P.C., Oomen, A.G., Rietveld, A.G., and Bouwmeester, H. (2014). Characterization of titanium dioxide nanoparticles in food products: analytical methods to define nanoparticles. *J. Agric. Food Chem.* 62: 6285–6293.

Peters, R.J.B., Oomenb, A.G., van Bemmel, G., van Vliet, L., Undas, A.K., Munniks, S., Bleys, R.L.A.W., Tromp, P.C., Brand, W., and van der Lee, M. (2020). Silicon dioxide and titanium dioxide particles found in human tissues. *Nanotoxicology* 14 (3): 420–443.

Pogribna, M. and Hammons, G. (2021). Epigenetic effects on nanomaterials and nanoparticles. *J. Nanobiotechnology* 19: 1–18.

Pogribna, M., Koonce, N.A., Mathew, A., Word, B., Patri, A.K., Lyn-Cook, B., and Hammons, G. (2020). Effect of titanium dioxide nanoparticles on DNA methylation in multiple human cell lines. *Nanotoxicology* 14 (4): 534–553.

Portela, A. and Esteller, M. (2010). Epigenetic modifications and human disease. *Nat. Biotechnol.* 28: 1057–1068.

Pradhan, S., Bacolla, A., Wells, R.D., and Roberts, R.J. (1999). Recombinant human DNA (cytosine-5) methyltransferase. I. Expression, purification, and comparison of de novo and maintenance methylation. *J. Biol. Chem.* 274: 33002–33010.

Prasad, R.Y., Chastain, P.D., Nikolaishvili-Feinberg, N., Smeester, L., Kaufmann, W.K., and Fry, R.C. (2013). Titanium dioxide nanoparticles activate the ATM-Chk2 DNA damage response in human dermal fibroblasts. *Nanotoxicology* 7 (6): 1111–1119.

Proquin, H., Rodriguez-Ibarra, C., Moonen, C.G., Urrutia Ortega, I.M., Briede, J.J., de Kok, T.M., van Loveren, H., and Chirino, Y.I. (2017). Titanium dioxide food additive (E171) induces ROS formation and genotoxicity: contribution of micro and nanosized fractions. *Mutagenesis* 32 (1): 139–149.

Provost, P. (2010). MicroRNAs as a molecular basis for mental retardation, Alzheimer's and prion diseases. *Brain Res.* 1338: 58–66.

Raitoharju, E., Lyytikäinen, L.P., Levula, M., Oksala, N., Mennander, A., Tarkka, M., Klopp, N., Illig, T., Kahonen, M., Karhunenn, P.J., Laaksonen, R., and Lehtimaki, T. (2011). MiR-21, miR-210, miR-34a, and miR-146a/b are up-regulated in human atherosclerotic plaques in the Tampere Vascular Study. *Atherosclerosis* 219: 211–217.

Rashid, M.M., Travcer, P.F., and Tomsic, B. (2012). Influence of titanium dioxide nanoparticles on human health and the environment. *Namomaterials* 11: 2354.

Ray, P.D., Yosim, A., and Fry, R.C. (2014). Incorporating epigenetic data into the risk assessment process for the toxic metals arsenic, cadmium, chromium, lead, and mercury: strategies and challenges. *Front. Genet.* 5: 201.

Rice, J.C. and Allis, C.D. (2001). Histone methylation versus histone acetylation: new insights into epigenetic regulation. *Curr. Opin. Cell Biol.* 13: 263–273.

Roberti, A., Valdes, A.F., Torrecilla, R., Fraga, M.F., and Fernandez, A.F. (2019). Epigenetics in cancer therapy and nanomedicine. *Clin. Epigenetics* 11 (1): 81.

Robertson, K.D., Uzvolgyi, E., Liang, G., Talmadge, C., Sumegi, J., Gonzales, F.A., and Jones, P.A. (1999). The human DNA methyltransferases (DNMTs) 1, 3a, and 3b: coordinate mRNA expression in normal tissues and overexpression in tumors. *Nucleic Acids Res.* 27: 2292–2298.

Rompelberg, C., Heringa, M.B., van Donkersgoed, G., Drijvers, J., Roos, A., Westenbrink, S., Peters, R., van Bemmel, G., Brand, W., and Oomen, A.G. (2016). Oral intake of added titanium dioxide and its nanofraction from food products, food supplements and toothpaste by the Dutch population. *Nanotoxicology* 10 (10): 1404–1414.

Rossnerova, A., Honkova, K., Pelclova, D., Zdimal, V., Hubacek, J.A., Chvojkova, I., Vrbova, K., Rossner, P., Jr., Topinka, J., Vlckova, S., Fenclova, Z., Lischkova, L., Klusackova, P., Schwarz, J., Ondracek, J., Ondrackova, L., Kostejn, M., Klema, J., and Dvorackova, S. (2020). DNA methylation profiles in a group of workers occupationally exposed to nanoparticles. *Int. J. Mol. Sci.* 21 (7): 2420.

Ruiter, S., Sippel, J., Bouwmeester, M.C., Lommelaars, T., Beekhof, P., Hodemaekers, H.M., Bakker, F., van den Brandhof, E.J., Pennings, J.L., and van der Ven, L.T. (2016). Programmed effects in neurobehavior and antioxidative physiology in zebrafish embryonically exposed to cadmium: observations and hypothesized adverse outcome pathway framework. *Int. J. Mol. Sci.* 17 (11): 1830.

Salata, O.V. (2004). Applications of nanoparticles in biology and medicine. *J. Nanobiotechnology* 2 (1): 3.

Schneider, T., Westermann, M., and Glei, M. (2017). In vitro uptake and toxicity studies on metal nanoparticles and metal oxide nanoparticles in human HT29 cells. *Arch. Toxicol. 91: 3517–3527.*

Schulte, P.A., Kuempel, E.D., and Drew, N.M. (2018). Characterizing risk assessments for the development of occupational exposure limits for engineered nanomaterials. *Regul. Toxicol. Pharmacol.* 95: 207–219.

Schulte, P.A., Leso, V., Niang, M., and Iavicoli, I. (2019). Current state of knowledge on the health effects of engineered nanomaterials in workers: a systematic review of human studies and epidemiological investigations. *Scand J. Work Environ. Health* 45 (3): 217–238.

Schwarz, D.A. (2010). Epigenetics and environmental lung disease. *Proc. Am. Thorac. Soc.* 7: 123–125.

Setyawati, M.I., Khoo, P.K.S., Eng, B.H., Xiong, S., Zhao, X., Das, G.K., Tan, T.T.-Y., Loo, J.S.C., Leong, D.T., and Ng, K.W. (2013). Cytotoxic and genotoxic characterization of titanium dioxide, gadolinium oxide, and poly(lactic-co-glycolic acid) nanoparticles in human fibroblasts. *J. Biomed. Mater. Res. Part A* 101 (3): 633–640.

Shabbir, S., Kulyar, M.F.-E.-A., Bhutta, Z.A., Boruah, P., and Asif, M. (2021). Toxicological consequences of titanium dioxide nanoparticles (TiO2NPs) and their jeopardy to human population. *BioNanoScience* 11: 621–632.

Shakeel, M., Jabeen, F., Shabbir, S., Asghar, M.S., Khan, M.S., and Chaudhry, A.S. (2016). Toxicity of nano-titanium dioxide (TiO2-NP) through various routes of exposure: a review. *Biol. Trace Elem. Res.* 172: 1–36.

Sharif, J., Takebayashi, K., Muto, M., Takebayashi, S.-I., Suetake, I., Iwamatsu, A., Endo, T.A., Shinga, J., Mizutani-Koseki, Y., Toyoda, T., Okamura, K., Tajima, S., Mitsuya, K., and Okano, M. (2007). The SRA protein Np95 mediates epigenetic inheritance by recruiting Dnmt1 to methylated DNA. *Nature* 450 (7171): U908–U925.

Sharma, S., Sarathal, K.C., and Taliyan, R. (2019). Epigenetics in neurodegenerative diseases: the role of histone deacetylases. *CNS Neurol. Disord. Drug Targets* 18: 11–18.

Shi, H., Magaye, V., Casltranove, V., and Zhao, J. (2013). Titanium dioxide nanoparticles: a review of current toxicological data. *Part. Fibre Toxicol.* 10: 15.

Shinohara, N., Danno, N., Ichinose, T., Sasaki, T., Fukui, H., Honda, K., and Gamo, M. (2014). Tissue distribution and clearance of intravenously administered titanium dioxide (TiO(2)) nanoparticles. *Nanotoxicology* 8: 132–141.

Shukla, R.K., Sharma, V., Pandey, A.K., Singh, S., Sultana, S., and Dhawan, A. (2011). ROS-mediated genotoxicity induced by titanium dioxide nanoparticles in human epidermal cells. *Toxicol. In Vitro* 25: 231–241.

Shyamasundar, S., Ng, C.T., Yung, L.Y., Dheen, S.T., and Bay, B.H. (2015). Epigenetic mechanisms in nanomaterial-induced toxicity. *Epigenomic* 7 (3): 395–411.

Sierra, M.I., Rubio, L., Bayón, G.F., Cobo, I., Menendez, P., Morales, P., Mangas, C., Urdinguio, R.G., Lopez, V., Valdes, A., Vales, G., Marcos, R., Torrecillas, R., Fernández, A.F., and Fraga, M.F. (2017). DNA methylation changes in human lung epithelia cells exposed to multi-walled carbon nanotubes. *Nanotoxicology* 11 (7): 857–870.

Sierra, M.I., Valdés, A., Fernández, A.F., Torrecillas, R., and Fraga, M.F. (2016). The effect of exposure to nanoparticles and nanomaterials on the mammalian epigenome. *Int. J. Nanomed.* 11: 6297–6306.

Singh, S., Shi, T., and Schins, R.P. (2007). Endocytosis, oxidative stress and IL-8 expression in human lung epithelial cells upon treatment with fine and ultrafine TiO2: role of the specific surface area and of surface methylation of the particles. *Toxicol. Appl. Pharmacol.* 222: 141–151.

Song, B., Zhou, T., Liu, J., and Shao, L.Q. (2017). Measuring global DNA methylation to assess neurotoxicity of titanium dioxide nanoparticles. *Sci. Adv. Mater.* 9 (6): 1051–1056.

Song, B., Zhou, T., Yang, W., Liu, J., and Shao, L. (2016). Contribution of oxidative stress to TiO2 nanoparticle-induced toxicity. *Environ. Toxicol. Pharmacol.* 48: 130–140.

Song, Y., Li, X., and Du, X. (2009). Exposure to nanoparticles is related to pleural effusion, pulmonary fibrosis and granuloma. *Eur. Respir. J.* 34 (3): 559–567.

Stapleton, P.A., Hathaway, Q.A., Nichols, C.E., Abukabda, A.B., Pinti, M.V., Shepherd, D.L., McBride, C.R., Yi, J., Castranova, V.C., Hollander, J.M., and Nurkiewicz, T.R. (2018). Maternal engineered nanomaterial inhalation during gestation alters the fetal transcriptome. *Part. Fibre Toxicol.* 15: 3.

Stern, S.T., Adiseshaiah, P.P., and Crist, R.M. (2012). Autophagy and lysosomal dysfunction as emerging mechanisms of nanomaterial toxicity. *Part. Fibre Toxicol.* 9: 20.

Stoccoro, A., Di Bucchianico, S., Coppede, F., Ponti, J., Uboldi, C., Blosi, M., Delpivo, C., Ortelli, S., Costa, A.L., and Migliore, L. (2017). Multiple endpoints to evaluate pristine and remediated titanium dioxide nanoparticles genotoxicity in lung epithelial A549 cells. *Toxicol. Lett.* 276: 48–61.

Stoccoro, A., Karlsson, H.L., Coppede, F., and Migliore, L. (2013). Epigenetic effects of nanosized materials. *Toxicology* 313: 3–14.

Sui, J., Fu, Y., Zhang, Y., Ma, S., Yin, L., Pu, Y., and Liang, G. (2018). Molecular mechanism for miR-350 in regulating of titanium dioxide nanoparticles in macrophage RAW264.7 cells. *Chem. Biol. Interact.* 280 (25): 77–85.

Sul, Y.T. (2010). Electrochemical growth behavior, surface properties, and enhanced in vivo bone response of TiO2 nanotubes on microstructured surfaces of blasted, screw-shaped titanium implants. *Int. J. Nanomed.* 5: 87–100.

Suzuki, M.M. and Bird, A. (2008). DNA methylation landscape: provocative insights from epigenomics. *Nat. Rev. Genes.* 9: 465–472.

Tassinari, R., Cubadda, F., Moracci, G., Aureli, F., D'Amato, M., Valeri, M., De Berardis, B., Raggi, A., Mantovani, A., Passeri, D., Rossi, M., and Maranghi, F. (2014). Oral, short-term exposure to titanium dioxide nanoparticles in Sprague-Dawley rat: focus on reproductive and endocrine systems and spleen. *Nanotoxicology* 8: 654–662.

Tay, C.Y., Fang, W., Setyawati, M.I., Chia, S.L., Tan, K.S., Hong, C.H.L., and Leong, D.T. (2014). Nano-hydroxyapatite and nano-titanium dioxide exhibit different subcellular distribution and apoptotic profile in human oral epithelium. *ACS Appl. Mater. Interfaces* 6: 6248–6256.

Thai, S.F., Jones, C.P., Nelson, G.B., Vallanat, B., Killius, M., Crooks, J.L., Ward, W.O., Blackman, C.F., and Ross, J.A. (2019). Differential effects of nano TiO_2 and CeO_2 on normal human lung epithelial cells in vitro. *J. Nanosci. Nanotechnol.* 19 (11): 6907–6923.

Thai, S.-F., Wallace, K.A., Jones, C.P., Carlton, P., Ren, H., Prasad, R.Y., Ward, W.O., Kohan, M.J., and Blackman, C.F. (2015). Signaling pathways and microRNA changes in nano-TiO2 treated human lung epithelial (BEAS-2B) cells. *J. Nanosci. Nanotechnol.* 15 (1): 492–503.

Toyooka, T., Amano, T., and Ibuki, Y. (2012). Titanium dioxide particles phosphorylate histone H2AX independent of ROS production. *Mutat. Res. Genet. Toxicol. Environ. Mutagen.* 742 (1–2): 84–91.

Tucci, P., Porta, G., Agostini, M., Dinsdale, D., Iavicoli, I., Cain, K., Finazzi-Agró, A., Melino, G., and Willis, A. (2013). Metabolic effects of TiO2 nanoparticles, a common component of sunscreens and cosmetics, on human keratinocytes. *Cell Death Dis.* 4 (3): e549.

US Food and Drug Administration (2005). Titanium dioxide. USFDA, Washington, DC. Code of Federal Regulations, Title 21, Section 73.575. http://www.accessdata.fda.gov/scripts/cdrh/cfdocs/cfCFR/CFRSearch.cfm?fr0.2573.575.

Valko, M., Leibfritz, D., Moncol, J., Cronin, M.T., Mazur, M., and Telser, J. (2007). Free radicals and antioxidants in normal physiological functions and human disease. *Int. J. Biochem. Cell Biol.* 39: 44–84.

van der Ven, L.T.M., Jelinek, J., Hodemaekers, H.M., Zwart, E.P., Ruiter, S., van den Brandhof, E.-J., Issa, J.-P.J., Pennings, J.L.A., and Luijten, M. (2017). An adverse outcome pathway analysis employing DNA methylation effects in arsenic-exposed zebrafish embryos supports a role of epigenetic events in arsenic-induced chronic disease. *Appl. Vitro Toxicol.* 3 (4): 312–324.

van Ravenzwaay, B., Landsiedel, R., Fabian, E., Burkhardt, S., Strauss, V., and Ma-Hock, L. (2009). Comparing fate and effects of three particles of different surface properties: nano-TiO(2), pigmentary TiO(2) and quartz. *Toxicol. Lett.* 186: 152–159.

Vrijens, K., Bollati, V., and Nawrot, T.S. (2015). MicroRNAs as potential signatures of environmental exposure or effect: a systematic review. *Environ. Health Perspect.* 123 (5): 300–441.

Wang, F., Ma, Y., Wang, H., and Qin, H. (2017). Reciprocal regulation between microRNAs and epigenetic machinery in colorectal cancer. *Oncol. Lett.* 13 (3): 1048–1057.

Wang, J.X., Li, Y.F., Zhou, G.Q., Li, B., Jiao, F., Chen, C.Y., Gao, Y.X., Zhao, Y.L., and Chai, Z.F. (2007a). Influence of intranasal instilled titanium dioxide nanoparticles on monoaminergic neurotransmitters of female mice at different exposure time. *Zhonghau Yu Fang Yi Xue Za Zhi* 41: 91–95.

Wang, J.J., Sanderson, B.J., and Wang, H. (2007b). Cyto- and genotoxicity of ultrafine TiO2 particles in cultured human lymphoblastoid cells. *Mutat. Res.* 628: 99–106.

Wang, J.B., Zhou, G., Chen, C., Yu, H., Wang, T., Ma, Y., Jia, G., Gao, Y., Li, B., Sun, J., Li, Y., Jial, F., Zhao, Y., and Chai, Z. (2007c). Acute toxicity and biodistribution of different sized titanium dioxide particles in mice after oral administration. *Toxicol. Lett.* 168: 176–185.

Wei, J.W., Huang, K., Yang, C., and Kang, C.S. (2017). Non-coding RNAs as regulators in epigenetics (Review). *Oncol. Rep.* 37 (1): 3–9.

Weir, A., Westerhoff, P., Fabricius, L., Hristovski, K., and von Goetz, N. (2012). Titanium dioxidenanoparticles in food and personal care products. *Environ. Sci. Technol.* 46: 2242–2250.

Wolf, R., Matz, H., Orion, E., and Lipozencic, J.. (2003). Sunscreens–the ultimate cosmetic. *Acta Dermatovenerol. Croat.* 11: 158–162.

Wong, B.S.E., Hu, Q., and Baeg, G.H. (2017). Epigenetic modulations in nanoparticle-mediated toxicity. *Food Chem. Toxicol.* 109 (Pt 1): 746–752.

Xu, S., Sui, J., Fu, Y., Wu, W., Liu, T., Yang, S., and Liang, G. (2020). Titanium dioxide nanoparticles induced the apoptosis of RAW264.7 macrophages through miR-29b-3p/NFAT5 pathway. *Environ. Sci. Pollut. Res.* 27: 26153–26162.

Yang, J., Bashkenova, N., Zang, R., Huang, X., and Wang, J. (2020). The roles of TET family proteins in development and stem cells. *Development* 147 (2): dev183129.

Yang, Z.F. and Klionsky, D.J. (2010). Mammalian autophagy: core molecular machinery and signaling regulation. *Curr. Opin. Cell Biol.* 22 (2): 124–131.

Yao, Q., Chen, Y., and Zhou, X. (2019). The roles of microRNAs in epigenetic regulation. *Curr. Opin. Chem. Biol.* 51: 11–17.

Yu, J., Loh, X.J., Luo, Y., Ge, S., Fan, X., and Ruan, J. (2020). Insights into the epigenetic effects of nanomaterials on cells. *Biomater. Sci.* 8: 763.

Yu, Y.B., Duan, J.C., Yu, Y., Li, Y., Liu, X., Zhou, X., Ho, K.-F., Tian, L., and Sun, Z. (2014). Silica nanoparticles induce autophagy and autophagic cell death in HepG2 cells triggered by reactive oxygen species. *J. Hazard. Mater.* 270: 176–186.

Zabirnyk, O., Yezhelyev, M., and Seleverstov, O. (2007). Nanoparticles as a novel class of autophagy activators. *Autophagy* 3 (3): 278–281.

Zhang, X., Li, W., and Yang, Z. (2015). Toxicology of nanosized titanium dioxide: an update. *Arch. Toxicol.* 89: 2207–2217.

Zhao, L., Zhu, Y., Chen, Z., Xu, H., Zhou, J., Tang, S., Xu, Z., Kong, F., Li, X., Zhang, Y., Li, X., Zhang, J., and Jia, G. (2018). Cardiopulmonary effects induced by occupational exposure to titanium dioxide nanoparticles. *Nanotoxicology* 12 (2): 169–184.

7

Toxicogenomics of Multi-Walled Carbon Nanotubes

*Pius Joseph**

Molecular Carcinogenesis Laboratory, Toxicology and Molecular Biology Branch, Health Effects Laboratory Division, National Institute for Occupational Safety and Health (NIOSH), 1095 Willowdale Road, Morgantown, WV 26505, USA
** Corresponding author*

Introduction

The International Standards Organization (ISO) defines nanomaterial as "material with any external dimension in the nanoscale or having internal structure or surface structure in the nanoscale" (ISO 2015). Nanoscale is defined as the size ranging approximately from 1 to 100 nm (ISO (International Standards Organization) 2015). Materials having at least one dimension in the nanoscale are considered by the US National Nanotechnology Initiative as nanomaterials (Yokel and MacPhail 2011). Nanomaterials or nanoparticles can be either naturally occurring or manufactured. Nanomaterials that are manufactured for specific purposes are commonly referred to as engineered nanomaterials (ENMs). ENMs are considered as advanced materials that are specifically engineered with unique and desirable properties intended for superior performance compared with conventional materials. ENMs can be generated by bottom-up processes, such as physical and chemical vapor deposition, liquid phase synthesis, and self-assembly, starting with chemical elements, such as carbon, metals, metal oxides, polymers, and biological molecules (Cao 2004). They can also be generated through top-down approaches by breaking down bulk materials to the nanoscale. ENMs exhibit many desirable properties that include, but are not limited to, their nanoscale size, light weight, and increased surface area, stiffness, tensile strength, and stability. Additionally, ENMs may exhibit superior electromagnetic, catalytic, thermal, and pharmacokinetic properties. The unique and desirable properties of ENMs have facilitated their use either directly or in products enabled by them. ENMs have found numerous commercial, industrial, and medical applications (De Volder et al. 2013; Eatemadi et al. 2014; Stoner et al. 2014). These include, but are not limited to, their use in cosmetics, personal care products, drug delivery, electronics, automobiles, and household and home improvement products. The widespread use of ENMs in products presents numerous opportunities for their exposure among workers who are engaged in the manufacture of such products and consumers who use them. Many of the same desirable properties of ENMs that enable their use in commerce, industries, and medicine have also raised serious concern from a human health and safety standpoint. For example, due to their light weight, ENMs can be easily aerosolized with potential for inhalation exposure among workers who are engaged in the manufacture and use of ENMs or ENM-containing products. Similarly, the size of ENMs in the nanoscale may enable their entry into deep anatomical areas in the lungs where interaction with cellular components may result in biological activity that may be quantitatively and/or qualitatively distinct from their large-sized counterparts of similar chemical composition (Castranova

Impact of Engineered Nanomaterials in Genomics and Epigenomics, First Edition. Edited by Saura C. Sahu.

2011). ENMs may enter the circulatory system from the lungs and translocate to other organs to result in systemic toxicity (Simeonova and Erdely 2009). Protection from adverse health effects in workers and consumers who are likely to be exposed to ENMs is a priority for various government and nongovernment agencies and organizations. This requires a clear understanding of the toxicity potential of ENMs and the mechanisms underlying the toxicity that are integral components in the assessment and management of the potential health risks associated with human exposure to ENMs. Current knowledge with respect to the lung toxicity caused by multi-walled carbon nanotubes (MWCNT), perhaps the most studied ENM, is presented in this chapter with an emphasis on the application of transcriptomics in investigating ENM-induced lung toxicity.

MWCNTs

MWCNTs are carbon-based nanomaterials consisting of multiple concentric layers of graphene sheets composed of carbon rings rolled into cylindrical structures. Typically, the diameter of a MWCNT is in the nanoscale and the length can range from a few nanometers to several millimeters or more. MWCNTs, due to their many desirable features, such as light weight, increased surface area, tensile strength, and stiffness and enhanced durability, have found multiple applications in commerce, industry, and medicine (Eatemadi et al. 2014; Stark et al. 2015; Stoner et al. 2014). Unfortunately, the same desirable characteristics that have facilitated the use of MWCNTs in commercial, industrial, and medical applications have raised concerns with respect to their biological activity that can result in adverse health effects. Some MWCNTs, because of their resemblance to asbestos with respect to the thin and needle-like shape, rigidity, and biopersistence, have raised concerns about the potential for cancer and mesothelioma (Donaldson et al. 2010; Fukushima et al. 2018; Kasai et al. 2016). Significant variations in size, shape, length, thickness of the wall, aspect ratio, presence of metal contaminants and their levels, and surface chemistry exist among the various preparations of MWCNTs, all of which are known to significantly influence the potential of MWCNTs for differing degrees of toxicity, and these are known to exist among the various preparations of MWCNTs (Johnston et al. 2010; Fraser et al. 2020, 2021).

MWCNTs, primarily due to their lightweight, can be easily aerosolized, and the airborne particles may present a significant risk for inhalation exposure to result in pulmonary and systemic toxicity in humans. In an exposure assessment study (Erdely et al. 2013), the mass concentration of elemental carbon present in the personal breathing zones (PBZ) of workers employed in eight US-based facilities manufacturing MWCNTs was determined as an index of their occupational exposure to MWCNTs. The inhalable elemental carbon mass concentration in PBZ samples collected exhibited an arithmetic mean of 10.6 $\mu g/m^3$ (geometric mass 4.21 $\mu g/m^3$), suggesting the presence of airborne MWCNT particles and their potential exposure among workers employed in the manufacture of MWCNTs. In another study conducted by Han et al. (2008), the total mass of all particulates in full shift air samples collected from a research facility handling MWCNTs was determined. The total mass of all the particulates in the PBZ samples collected from the facility ranged from non-detectable concentrations to 331.7 $\mu g/m^3$, further supporting the presence of airborne MWCNT particles in workplaces with potential human exposure. In a cross-sectional study, Dahm et al. (2018) measured exposure to carbon nanotubes and fibers among 108 participants enrolled from 12 facilities across the United States. Mean exposure to elemental carbon (EC) at the respirable and inhalable size aerosol fractions collected from the facilities was 1 and 6.22 $\mu g/m^3$, respectively. Additionally, ENM particles were detected in 18% and 70% of the sputum and hand/dermal samples, respectively, collected from the study participants, confirming exposures. Similar findings supporting the presence of airborne MWCNT particles in workplaces with potential human exposure have been reported in other studies (Dahm et al. 2012, 2013, 2015; Hedmer et al. 2014; Fatkhutdinova et al. 2016; Kuijpers et al. 2016; Shvedova et al. 2016; Lee et al. 2015; Vlaanderen et al. 2017). Currently, to protect workers from the potential adverse health effects associated with occupational exposure to the MWCNTs, NIOSH recommends that exposure to the ENM should be kept below an 8-hour time-weighted average of 1 $\mu g/m^3$ of respirable EC (NIOSH 2013).

There is limited epidemiological evidence with respect to the potential of MWCNT to cause adverse health effects in humans. The association between occupational exposure to MWCNTs and effects on lung health and the immune system was assessed in a cross-sectional study consisting of MWCNT-exposed workers and matching, unexposed control population (Vlaanderen et al. 2017). Lung function and fractional exhaled nitric oxide (FENO) were assessed to evaluate lung health whereas complete blood cell counts and serum levels of 51 immune markers and three pneumoproteins were determined to assess the effect of exposure to MWCNTs on the immune system. Regression analysis of the data obtained demonstrated significant correlation between exposure to MWCNT and increase in the serum levels of specific immune markers, such as the C-C motif ligand, basic fibroblast growth factor, and soluble interleukin-1 (IL-1) receptor II. There were also significant differences in the blood cell counts and FENO between the MWCNT-exposed and control populations. However, no MWCNT exposure-associated difference in lung function was detected. These results, therefore, indicated the early effects of occupational exposure to MWCNT on lung health and the immune system. Significant associations between MWCNT exposure and sputum levels of interleukins IL-1β, IL-6, tumor necrosis factor-α (TNF-α), and Krebs von den Lungen-6 were reported by Fatkhutdinova et al. (2016) in a cross-sectional study that included MWCNT-exposed workers and an unexposed, control population. Genome-wide differential expression analysis revealed changes in the expression profiles of mRNA and noncoding RNA in the blood samples obtained from a population that consisted of workers ($n = 8$) who were exposed to MWCNT for at least 6 months and for nonexposed controls ($n = 7$) (Shvedova et al. 2016). In a Korean population consisting of MWCNT-exposed workers and matching control subjects, Lee et al. (2015) determined lung function, hematology, and blood chemistry parameters. Additionally, markers of oxidative stress were determined in the exhaled breath condensate (EBC) collected from the subjects. No MWCNT exposure-related changes were detected in lung function, hematology, and blood chemistry parameters among the subjects. However, significant increases in markers of oxidative stress, such as malondialdehyde, n-hexanal, and 4-hydroxy-2-hexenal, were detected in the EBC of the MWCNT-exposed subjects that correlated with their blood molybdenum concentration, suggesting that the oxidative stress markers in EBC and blood molybdenum concentration may be useful biomarkers of MWCNT exposure.

Despite the availability of limited epidemiological data demonstrating the potential human health effects associated with occupational exposure to MWCNT, substantial evidence with respect to the pulmonary toxicity of MWCNT has been obtained from the results of studies conducted by employing animal models of human lung toxicity. These studies have demonstrated the induction of inflammation, oxidative stress, granuloma formation, fibrosis, tissue injury, genotoxicity, mesothelioma, tumorigenesis, and changes in the gene expression profile in the lungs in response to pulmonary exposure to MWCNTs as described in detail in the following sections. A vast majority of the *in vivo* studies that investigated the pulmonary toxicity caused by MWCNTs were conducted by employing either a mouse or rat as the animal model. Furthermore, in both these species, intratracheal instillation, pharyngeal aspiration, and inhalation were employed as exposure techniques to investigate pulmonary toxicity.

Lung Injury

Lung injury resulting from pulmonary exposure to MWCNTs has been determined either by assaying the lactate dehydrogenase (LDH) activity in the bronchial alveolar lavage (BAL) fluid or by conducting a histopathological analysis of the lungs. A significant elevation in the BAL level of LDH activity has been detected in the lungs in response to pulmonary exposure to toxic agents (Sager et al. 2020; Sellamuthu et al. 2011a), which has been considered as an indicator of lung toxicity. Significantly elevated BAL LDH activity in the lungs of MWCNT-exposed rats and mice has been reported (Aiso et al. 2010; Kasai et al. 2015; Porter et al. 2010; Sager et al. 2022), suggesting their potential for lung injury.

Lung histopathological changes, compared with appropriate control lungs, are most widely accepted as indicators of lung injury, resulting from pulmonary exposure to MWCNTs. Histological changes, suggestive of potential lung injury in MWCNT-exposed lungs, have been reported in most murine studies that investigated the pulmonary toxicity of MWCNTs. Despite the differences in the severity of the lung injury, depending on the dose and duration of the MWCNT exposure and the post-exposure time-period when histopathological analysis was performed (see details below), histopathological changes indicating lung injury have been detected in almost all toxicity studies reported so far.

Pharyngeal aspiration of MWCNT-7, one form of MWCNT, manufactured by Mitsui & Company (Tokyo, Japan) in mice at doses ranging from 10 to 80 μg and analysis of the lung sections obtained at post-exposure time intervals of 1, 7, 28, and 56 days revealed histopathological changes, including inflammation, fibrosis, and cellular proliferation (Porter et al. 2010). The MWCNT-7-induced pulmonary inflammation in mice was detected mainly in the interstitium but sometimes extending into the lumen at the bronchoalveolar junction, resulting in airway obstruction. Hypertrophy and hyperplasia of the bronchiolar and alveolar epithelial cells were also induced by MWCNT-7 in mice. In addition, granulomatous and fibrotic changes were detected in the lungs of MWCNT-7-administered mice. Penetration of the pleura by MWCNT particles and pleural inflammation, especially at the 56-day post-exposure time interval in mice administered the highest dose of 80 μg MWCNT-7, were also detected. Similar histological changes, indicative of lung injury, were also detected in mice that were exposed to MWCNT-7 by whole-body inhalation (Fraser et al. 2021; Mercer et al. 2013; Porter et al. 2013).

Whole-body inhalation exposure of rats to MWCNT-7 (1.25, 2.5, 5, or 10 mg/m^3, 6 hours/day, 3 days) resulted in dose-dependent histological changes in lungs that mainly consisted of the accumulation of inflammatory cells (alveolar macrophages [AMs] and neutrophils), thickening of the alveolar wall, and granuloma formation (Sager et al. 2022). Similar, dose-dependent histopathological changes were detected in male and female rats exposed by whole-body inhalation to MWCNT-7 (0.2, 1, and 5 mg/m^3 for 6 h/day, 5 days/week, for 13 weeks) (Kasai et al. 2015). Dose-dependent infiltration of inflammatory cells, type II cell hyperplasia, granuloma formation, and fibrosis with alveolar wall thickening were detected in the lungs of rats administered MWCNT-7 by intratracheal instillation (Aiso et al. 2010).

Inflammation

Induction of inflammation, in response to the presence of a toxic material in an organ or tissue, is characterized by the recruitment of leukocytes to the site where the toxic material is present and the release of inflammatory mediators, such as pro-inflammatory cytokines/chemokines. Induction of inflammation is generally considered as a protective response to fight against the potential harmful effects of toxic materials. However, excessive inflammatory response may result in tissue injury. The MWCNT entering the lungs are phagocytosed, mainly by the AMs, for their elimination from the lungs. However, due to their elongated, fibrous structure and poor solubility, MWCNTs are not efficiently phagocytosed and eliminated, resulting in their prolonged persistence in the lungs. For example, the retention halftimes, an index of biopersistence, of MWCNTs were 151, 350, 318, and 375 days, respectively, following a 13-week, sub-chronic inhalation exposure of rats to MWCNTs at 0.1, 0.4, 1.5, and 6 mg/m^3 (Pauluhn 2010). The deposition of MWCNT particles in the alveoli and their interaction with the AMs result in the activation of the macrophages and the release of signaling molecules, including pro-inflammatory cytokines/chemokines. The pro-inflammatory cytokines/chemokines released by the AMs function as chemoattractants and facilitate the recruitment of additional inflammatory phagocytes, mainly polymorphonuclear neutrophils (PMNs), from blood to the lungs. This, in turn, results in the release of additional pro-inflammatory cytokines/chemokines culminating in the induction of excess inflammation. Induction of persistent or chronic inflammation is characteristic to MWCNTs due to their relatively long half-time and biopersistence and has been recognized as a major mechanism underlying the MWCNT-induced lung toxicity (Ellinger-Ziegelbauer and Pauluhn 2009; Ma-Hock et al. 2009; Porter et al. 2010, 2013; Hamilton et al. 2013; Poulsen et al. 2015, 2016; Dong and Ma 2016a; Kasai et al. 2015; Fujita et al. 2016; Gate et al. 2019; Sager et al. 2022).

Regardless of the technique employed to administer MWCNTs (intratracheal instillation, oropharyngeal aspiration, and nose-only or whole-body inhalation exposure), pulmonary exposure of mice and rats to the ENM, in a majority of the studies reported so far, resulted in a dose-dependent pulmonary inflammatory response. The number of PMNs present in the BAL samples was determined as the index of lung inflammation at 1, 7, 28, and 56 days, following a single oropharyngeal aspiration of 0, 10, 20, 40, or 80 μg MWCNT-7 in 7-week-old male C57BL/6 mice (Porter et al. 2010). The MWCNT-7 dose-dependent increase in the number of PMNs detected in the BAL of the mice peaked at the 7-day post-exposure period and returned to the control level by the 56-day post-exposure period with the exception of the group of mice that received the highest dose of 80 μg MWCNT-7. In a recent dose–response study conducted in our laboratory (Sager et al. 2022), Fisher 344 rats were exposed by whole-body inhalation to air (control) or an aerosol containing MWCNT-7 particles, ranging in concentrations from 1.25 to 10 mg/m^3 for 6 hours/day for 3 days. Analysis of the bronchoalveolar lavage (BAL) cells at the end of the exposure period revealed significant and MWCNT-7 dose-dependent increases in the number of total BAL cells, AMs, and PMNs in the exposed rats, compared with the controls. The induction of lung inflammation in MWCNT-7-exposed rats, revealed by the increase in the number of BAL cells, was further supported by the significant and MWCNT-7 dose-dependent elevations in BAL levels of multiple cytokines (IL-1β, IL-10, IL-13, IL-16, IL-18), monocyte chemoattractant protein-1 (MCP-1), macrophage inflammatory protein-2 (MIP-2), and TNF-α (Sager et al. 2022). The inflammatory response to pulmonary exposure to a long and thick and a short and thin form of MWCNTs was determined in the lungs of Sprague-Dawley rats (Gate et al. 2019). The rats were exposed to two forms of MWCNT either by a single intratracheal instillation (180 and 540 μg) or by nose-only inhalation (0.5 and 1.5 mg/m^3, 2 × 3 hours/day, 5 days/week for 4 weeks). The number of PMNs infiltered into the lungs was determined at 1 and 28 days following the intratracheal or 3, 30, 90, and 180 days following inhalation exposure, as an index of MWCNT-induced lung inflammation. The dose-dependent inflammatory response detected in MWCNT-exposed rat lungs was similar, irrespective of the type of MWCNT or the method of administration, when the deposited dose was considered. This, in agreement with the findings of Poulsen et al. (2016), suggested that the deposited surface area of

MWCNT is an excellent predictor of pulmonary neutrophil infiltration, a hallmark of lung inflammation in response to pulmonary exposure to toxic agents, including ENMs. Induction of lung inflammation, a major response to the pulmonary exposure to MWCNTs, has also been demonstrated in several other studies that employed mouse (Dong and Ma 2016a; Hamilton et al. 2013; Porter et al. 2013; Poulsen et al. 2015) or rat (Ellinger-Ziegelbauer and Pauluhn 2009; Fujita et al. 2016; Kasai et al. 2015; Ma-Hock et al. 2009) as the lung toxicity model.

Oxidative Stress

Under normal physiological conditions, reactive oxygen species (ROS), such as superoxide anion, hydrogen peroxide, and hydroxyl radical – the by-products of cellular respiration and metabolic activities – play an important role in maintaining tissue homeostasis. The biological system is equipped with protective mechanisms capable of detoxifying toxic ROS. However, excessive generation of ROS may exceed the capability of the biological system to detoxify them. This may result in the accumulation of toxic ROS in tissues and organs that, in turn, may lead to oxidative stress and damage to cellular components, including macromolecules, such as lipids, DNA, and proteins. The oxidative stress resulting from excessive generation and tissue accumulation of ROS has been identified as a contributing factor in several chronic and potentially fatal diseases including cancer (Hayes et al. 2020).

Induction of oxidative stress has been identified as a major mechanism underlying the toxicity of inhaled particles such as crystalline silica (Sager et al. 2020) and asbestos (Pietrofesa et al. 2016) with sizes larger than those of ENMs. Multiple mechanisms have been identified to contribute to particle-induced generation of reactive oxidants and the resulting oxidative stress. The particles may possess reactive oxygen molecules on their surface, which, following their entry into the lungs, may directly result in the induction of oxidative stress (Vallyathan et al. 1988). Some of the inhaled particles may contain trace metals, such as iron as contaminants that may generate ROS through Fenton reaction, resulting in the induction of oxidative stress (Diabate et al. 2002; Voelkel et al. 2003). Another mechanism involved in the particle-induced generation of ROS is through their interaction with lung phagocytes, mainly AMs. The AMs phagocytose the inhaled particles deposited in the alveoli to eliminate them from the lungs, preventing their accumulation and the resulting toxicity. However, due to physical characteristics such as their elongated structure and insolubility, MWCNTs are not efficiently engulfed and removed by AMs. This may result in frustrated phagocytosis with eventual damage to the AMs and the release of the engulfed particles along with reactive and toxic intermediates, including ROS, potentially resulting in oxidative stress and tissue injury. Metal contaminants present in some preparations of MWCNTs have been found to be responsible for the induction of oxidative stress (Pulskamp et al. 2007). Some preparations of MWCNT-7, on the other hand, do not contain adequate quantities of metal contaminants to induce oxidative stress (Porter et al. 2010). Furthermore, the inability of CNTs to directly generate ROS in a cell-free system has been demonstrated (Fenoglio et al. 2006; Porter et al. 2010). In the absence of metal contaminants, the direct interaction of MWCNTs with AMs has been suggested as the mechanism responsible for the generation of ROS in the lungs (Sager et al. 2022).

In the study reported by Sager et al. (2022), rats were exposed by whole-body inhalation to air or an aerosol that contained increasing concentrations of MWCNT-7 for three days. The lung phagocytes obtained, following euthanasia, from the control and MWCNT-7-exposed rats were used in a chemiluminescence assay to determine the generation of reactive oxidants either by the entire population of the phagocytes or the AMs alone present in the lungs. The results obtained from the study (Sager et al. 2022) demonstrated a significant ($p < 0.05$) and dose-dependent generation of reactive oxidants in response to inhalation exposure to MWCNT-7 in rats. Similar to these results, a dose-dependent generation of reactive oxidants and their potential role in the lung toxicity induced in response to pulmonary exposure to MWCNT-7 have been reported by Kato et al. (2013).

The importance of reactive oxidant generation and the resulting oxidative stress in lung toxicity in response to exposure to MWCNT-7 has been demonstrated by Dong and Ma (2016b) by employing a knock-out mouse model that does not express the nuclear factor erythroid-2-related factor 2 (Nrf2) gene and matching wild-type mice. The Nrf2 is a redox-sensitive transcription factor that regulates the expression of key genes involved in protecting the biological system through the detoxification of toxic reactive oxidants (Kensler et al. 2007). The Nrf2 knock-out ($Nrf2^{-/-}$) and wild-type ($Nrf2^{+/+}$) mice were exposed to MWCNT-7 at doses of 0, 5, 20, and 40 μg by oropharyngeal aspiration and the AMs obtained from the mice were analyzed for the presence of ROS at post-exposure time intervals of 1, 3, 7, and 14 days. The number of ROS-producing AMs and the intensity of ROS staining in them, in response to the pharyngeal aspiration of MWCNT-7, were significantly higher in $Nrf2^{-/-}$ mice compared with $Nrf2^{+/+}$ mice. Similar differences were noticed in the parameters employed to determine oxidative damage to DNA and lipids between $Nrf2^{-/-}$ and $Nrf2^{+/+}$ mice, in response to their exposure to MWCNT-7. Collectively, these results demonstrated the induction of oxidative stress in response to pulmonary exposure to MWCNT-7 and the role of Nrf2 in regulating the MWCNT-7-induced oxidative stress.

Fibrosis

The high aspect ratio, rigidity, insolubility, and fibrous structure of MWCNT, like that of asbestos, have raised concerns about their potential, if inhaled, to cause lung fibrosis (Donaldson et al. 2006). There is compelling evidence in the literature suggesting the fibrogenic potential of MWCNT-7 with significant resemblance to that of a foreign substance-induced fibrotic response that is initiated with an acute phase response followed by chronic, interstitial fibrosis with or without granuloma formation (Mercer et al. 2013; Porter et al. 2010, 2013). In a dose–response and time-course study reported by Dong et al. (2015), MWCNT-7 particles were administered to C57BL/6J mice by pharyngeal aspiration at concentrations of 0, 5, 20, and 40 μg. The lungs obtained from the control and MWCNT-7 exposed mice were analyzed for histological changes, indicative of fibrosis, at post-exposure time intervals of 1, 3, 7, and 14 days. Rapid and dose-dependent appearance of collagen fibers in the lungs of the MWCNT-7 exposed mice were detected by Masson's trichrome and Picrosirius staining. Fibrotic foci, especially near MWCNT-7 deposited in the lungs, were also detected. The rapid appearance of lung fibrosis was detected in mouse lungs as early as one day following the MWCNT-7 exposure and persisted throughout the observation period. A similar lung fibrotic response was noticed by Porter et al. (2010) in another mouse study in which the animals were exposed to MWCNT-7 by whole-body inhalation (10 mg/m^3, 5 hours/day) for 2, 4, 8, or 12 days and their lungs were assessed at one day post exposure for histological changes indicative of fibrosis. Similar to the mouse studies, pulmonary exposure of rats to MWCNT-7 also resulted in lung fibrosis. In a dose–response study (Sager et al. 2022) in which rats were exposed by whole-body inhalation to MWCNT-7 at concentrations of 1.25, 2.5, 5, and 10 mg/m^3, 6 hours/day for 3 days, thickening of the alveolar wall and presence of collagen fibers, indicative of lung fibrosis, were detected at the end of the exposure period. In another rat whole-body inhalation study (Kasai et al. 2015), male and female rats were exposed to aerosols containing MWCNT-7 particles at concentrations of 0, 0.2, 1, and 5 mg/m^3 for 13 weeks. Increased accumulation of collagen fibers, in association with granulomatous changes, were detected in the lungs, which further supported the induction of lung fibrosis by MWCNT-7 in rats. Lung fibrosis was also detected in rats in response to intratracheal instillation of MWCNT particles (Muller et al. 2005).

A close relationship between the induction of lung inflammation and fibrosis in response to pulmonary exposure to MWCNT-7 has been reported in many studies (Dong and Ma 2016a; Dong et al. 2015; Porter et al. 2010). For example, the appearance of collagen fibers, a hallmark of fibrosis, is most abundantly detected in granulomas appearing in MWCNT-7 exposed lungs (Porter et al. 2013; Sager et al. 2022). By employing a mouse model for allergic asthma, Ryman-Rasmussen et al. (2009) have demonstrated the role of preexisting inflammation in MWCNT-induced lung fibrosis. Control and ovalbumin-sensitized mice were exposed to air or MWCNT (100 mg/m^3) by inhalation for 6 hours; at 1 and 14 days following the termination of the exposures, lung injury, inflammation, and fibrosis were determined. At 14 days post exposure, lung fibrosis was detectable only in mice that received both ovalbumin and MWCNT. Ovalbumin sensitization or MWCNT exposure alone did not result in lung fibrosis suggesting that the existence of inflammation is a requirement for MWCNT-induced lung fibrosis. It is generally accepted that a biphasic pathologic development of lung fibrosis that includes transition from an initial acute inflammatory response to interstitial fibrosis is characteristic of MWCNT-induced lung fibrosis.

Further insight into the mechanisms underlying MWCNT-7-induced lung fibrosis is presented in a series of papers published by Dong and Ma (2016b, 2017a, 2017b). Transgenic mice models were employed in their studies to demonstrate the role of specific genes and the mechanisms involved in MWCNT-7-induced lung fibrosis. In one of the studies (Dong and Ma 2016b), MWCNT-7 was administered in $Nrf2^{+/+}$ and $Nrf2^{-/-}$ mice by pharyngeal aspiration at doses of 0, 5, 20, and 40 μg, and the resulting lung toxicity, including induction of oxidative stress, inflammation, and fibrosis, was detected on 1, 3, 7, and 14 days following the exposure. Infiltration of phagocytes, especially PMNs, and the appearance of inflammatory and fibrotic foci were detected in all mice administered MWCNT-7. Similarly, generation of fluorescence due to the ROS-mediated oxidation of the fluorophore, dihydroethidium, was detected in the AMs obtained from the lungs of MWCNT-7-exposed mice. However, lung inflammation, fibrosis, and oxidative stress, in response to MWCNT-7 exposure, were more in MWCNT-7-administered $Nrf2^{-/-}$ mice compared with $Nrf2^{+/+}$ mice. These results demonstrated the protective role of the Nrf2 gene in MWCNT-7-induced lung toxicity. Furthermore, the findings suggested the interplay between Nrf2 and oxidative stress, inflammation, and fibrosis, determining the pathological outcomes of MWCNT-7 exposure.

The role of the gene, tissue inhibitor of metalloproteinase 1 (Timp1), in MWCNT-7-induced lung fibrosis and the underlying mechanisms were investigated by Dong and Ma (2017b). Timp1 wild-type ($Timp1^{+/+}$) and knock-out ($Timp1^{-/-}$) mice were exposed to 0 or 40 μg MWCNT-7 by pharyngeal aspiration and at post-exposure time periods of 1, 3, 7, and 14 days, and the lungs obtained from the euthanized mice were analyzed for fibrotic changes. Increased lung deposition of matrix proteins, collagen I, and fibronectin was detected in MWCNT-7-exposed mice, compared with the vehicle-only-administered mice, regardless of their genotype. Similarly, recruitment, activation, and proliferation of fibroblasts were higher in MWCNT-7-exposed mouse lungs. Differentiation of fibroblasts to myofibroblasts, an important step in fibrosis development (Barnes and Gorin 2011), was also higher in MWCNT-7-exposed mouse lungs. All fibrosis-related cellular events viz. recruitment, activation, and proliferation of fibroblasts, and their differentiation to myofibroblasts, taking place in response to exposure to MWCNT-7 were more in Timp1 wild-type mice compared with knock-out mice. These results, therefore, suggested the important role played by the Timp1 gene in MWCNT-7-induced lung fibrosis. Furthermore, it was demonstrated that the modulation of the MWCNT-7-induced pulmonary fibrotic response by Timp1 was mediated through the Timp1/CD63/integrin β1 axis and extracellular-signal-regulated kinase signaling.

Osteopontin (OPN; secreted phosphoprotein 1 or SPP1) is a gene that has been found overexpressed in the blood of workers (Fatkhutdinova et al. 2016) and the lungs of rats (Sager et al. 2022) in response to MWCNT-7 exposure. Osteopontin functions both as a cytokine and an extracellular matrix (ECM) protein and has been implicated in functions relevant to fibrosis, such as infiltration of inflammatory cells and tissue remodeling (Rittling and Singh 2015). In addition, OPN is highly overexpressed during wound healing (Liaw et al. 1998) and significant induction of OPN has been reported in animal models of fibrosis (Dong et al. 2016; Sabo-Attwood et al. 2011). The role of OPN in MWCNT-7-induced lung fibrosis was investigated by Dong and Ma (2017a) by employing OPN knock-out ($OPN^{-/-}$) and wild-type ($OPN^{+/+}$) mice. A single dose of 40 μg MWCNT-7 that is known to result in lung fibrosis in mouse was administered in $OPN^{-/-}$ and $OPN^{+/+}$ mice by pharyngeal aspiration. The lungs obtained from the vehicle and MWCNT-7-administered mice were analyzed for fibrosis at 1, 3, 7, 14, 28, and 56 days following exposure. Both OPN mRNA and protein were significantly overexpressed, at all post-exposure time intervals analyzed, in MWCNT-7-administered $OPN^{+/+}$ mice. Fibrotic lesions (appearance of fibrotic foci and detection of collagen fibers by Masson's trichrome staining) assessed in the lungs of MWCNT-7-exposed $OPN^{+/+}$ mice exhibited a biphasic response consisting of an early rapid phase that peaked on day 7 and resolved by day 14. The early phase was followed by a chronic phase, fully developed on day 28 and lasted until day 56. Compared with MWCNT-7-exposed $OPN^{+/+}$ mice, the lung fibrotic response detected in the $OPN^{-/-}$ counterparts was much milder. Expressions of ECM marker proteins and transformation of fibroblasts to myofibroblasts were much higher in the lungs of MWCNT-7-exposed $OPN^{+/+}$ mice, compared with the corresponding $OPN^{-/-}$ mice. The profibrotic role of OPN in the MWCNT-7-exposed mouse lungs was mediated through the activation of Smad-dependent TGF-β1 signaling.

Mesothelioma

The similarity of MWCNT to asbestos with regard to features, such as high aspect ratio, long and needle-like shape, rigidity, and biopersistence, have raised serious concerns regarding their potential to cause mesothelioma (Donaldson et al. 2006). Mesothelioma is a toxicological response unique to fibrous particles (Craighead et al. 1982). According to the fiber-pathogenicity paradigm (Donaldson and Tran 2004), exposure to biopersistent fibers with a diameter less than 3 μm and length more than 10–20 μm is likely to result in mesothelioma. Intraperitoneal administration of MWCNT in mice (Takagi et al. 2008, 2012) and rats (Rittinghausen et al. 2014) resulted in the appearance of mesothelioma. Similar results were obtained following intrascrotal administration of MWCNT in rats (Sakamoto et al. 2009). Donaldson et al. (2010) hypothesized that long fibrous particles deposited in the lungs are retained within the parietal pleura and initiate pathological events, such as the induction of inflammation and granuloma formation resulting in mesothelioma. This was based on the earlier demonstration that intraperitoneal administration of MWCNT resulted in asbestos-like pathogenicity in mice (Poland et al. 2008). Additional evidence for the length of the fibrous particles being the main determinant in the induction of mesothelioma is derived from the observation that pulmonary deposition of long but not short crocidolite particles resulted in pleural fibrosis (Adamson et al. 1993) and proliferative responses (Adamson et al. 1994) in mice. Similarly, intraperitoneal administration of longer fiber particles, compared with shorter particles, resulted in greater inflammatory and granuloma-generating response in mice (Donaldson et al. 1989; Goodglick and Kane 1990). It was suggested that the failure of the AMs to completely enclose or engulf the longer fibrous particles resulting in frustrated phagocytosis is likely responsible for their accumulation and initiation of pathological conditions in mesothelial cells culminating in the development of mesothelioma (Poland et al. 2008).

Experimental evidence for the retention of long MWCNT particles in the parietal pleura as a mechanism responsible for the induction of mesothelioma, resulting from exposure to MWCNT was obtained from a study conducted by Murphy et al. (2011). In the study, an intrapleural instillation method was employed to administer particles of different lengths directly into the pleura of mice through the chest wall. The particles viz. long and short MWCNT, long and short asbestos fibers, and nanoparticle carbon black, were administered in different groups of female C57BL/6 mice. At post-exposure time periods consisting of 24 hours, 7 days, 4 weeks, 12 weeks, and 24 weeks, the mice were euthanized, and pleural lavage was performed to assess pleural inflammatory response. Simultaneously, the parietal pleura of the mice was assessed for histological changes and retention of the administered particles. Pleural inflammation, characterized by the infiltration of granulocytes, was prominent in mice administered long, fibrous particles. Parietal pleura showed histological changes consistent with the inflammatory response detected in mice administered the particles. Cellular proliferation in the mesothelium of the parietal pleura was also detected only in mice administered long, fibrous particles. Furthermore, retention of the long, fibrous particles was detected in the pleural space of mice. The shorter particles were found deposited in the lymph nodes of mice, suggesting their elimination from the pleural space. Collectively, the results provided evidence for the length of the MWCNT particles as the major determinant of their potential to cause mesothelioma in the mouse model employed in the study. While the shorter MWCNT particles administered in the pleural cavity egressed through the stomata, the longer particles failed to be cleared from the pleura through the stomata. This resulted in the retention of longer MWCNT particles in the pleura of mice, leading to inflammatory and granulomatous responses characteristic of mesothelioma.

Lung Cancer

To date, there are no epidemiological data to support that human exposure to MWCNT results in lung cancer. Nevertheless, the International Agency for Research on Cancer (IARC) has classified MWCNT-7, one form of MWCNT that was employed in multiple animal toxicity studies (Porter et al. 2010; Dong and Ma 2016a, 2016b, 2017a, 2016b; Dong et al. 2015, 2016; Kasai et al. 2015; Umeda et al. 2013; Sager et al. 2022), as a Group 2B carcinogen (possibly carcinogenic to human) (IARC 2017). The classification of MWCNT-7 as a Group 2B carcinogen is based on the positive results obtained from mouse and rat carcinogenicity studies. In the mouse carcinogenicity study, Sargent et al. (2014) administered either the vehicle or a single intraperitoneal dose (10 μg/g body weight) of the tumor initiator, methylcholanthrene, in mice, which was followed by their whole-body inhalation exposure to air or MWCNT-7 (Mitsui and Company, Tokyo, Japan) at a concentration of 5 mg/m^3 for 5 hours/day for 15 days. At 17 months following the MWCNT-7 exposure, incidence of tumor formation was determined in the lungs obtained from the mice that received the vehicle (control), MCA alone, MWCNT-7 alone, or MCA plus MWCNT-7. The tumor incidence in the lungs of the mice that received either MCA or MWCNT-7 alone was not significantly different compared with the mice in the control group. On the other hand, a significantly higher incidence of tumorigenesis was detected in the lungs of the mice that received MCA plus MWCNT-7, compared with those received either MCA or MWCNT-7 alone. These results, therefore, suggested that MWCNT-7 is a tumor promoter in a mouse two-stage carcinogenesis model.

The lung carcinogenicity of MWCNT-7 was investigated in both sexes of the F344 rats by Kasai et al. (2016). The rats were chronically exposed by inhalation to air or aerosols containing MWCNT-7 at concentrations of 0.02, 0.2, or 2 mg/m^3, 6 hours/day, 5 days/week for 104 weeks and the incidence of tumorigenesis was recorded. A significant and MWCNT-7 dose-dependent increase in the incidence of bronchioloalveolar carcinoma, adenosquamous carcinoma, adenocarcinoma, and squamous cell carcinoma was detected in both the sexes of the rats. However, the incidence of lung tumorigenesis was more in male rats, suggesting a gender difference in the tumorigenic potential of MWCNT-7. For example, a statistically significant increase in lung tumor formation was detected in male and not in female rats that belonged to the 0.2 mg/m^3 exposure group. No tumor formation was detected in any organs or tissues other than the lungs despite the translocation of the administered MWCNT-7 particles into the pleural and peritoneal cavities in rats. The results of the rat carcinogenesis study by Kasai et al. (2016) suggested that MWCNT-7 is a complete carcinogen in rats. The results may also suggest that the dose and/or the duration of MWCNT-7 exposure employed in the mouse two-stage carcinogenesis study conducted by Sargent et al. (2014) was not sufficient to cause lung tumor formation in mice that were exposed to MWCNT-7 alone. The decreased sensitivity of mice, compared with rats, to develop lung tumors following pulmonary exposure to particles (Mauderly 1997) may also account for the absence of lung tumors detected in mice exposed to MWCNT-7 alone.

Contradictory results, with respect to the lung tumorigenic potential of another form of MWCNT, designated as MWCNT-N (Nikkiso, Tokyo, Japan), have been reported in rats. In a study conducted by Suzui et al. (2016), rats were intratracheally administered MWCNT-N at doses of 0 or 125 μg/rat, 8 times over a period of 2 weeks. Compared with the control rats, the incidence of lung tumors, at 109 weeks following MWCNT-N administration, was significantly higher in the exposed rats. In contrast to these findings, no lung tumors were detected in rats at 2 years following the intratracheal administration of MWCNT-N at doses of 0.67 or 3.3 mg/kg body weight (Nakanishi et al. 2015).

Genotoxicity

Genotoxicity has attracted much attention as a major mechanism responsible for the carcinogenicity of toxic agents. In a study reported by Kato et al. (2013), the genotoxicity potential of MWCNT-7 was demonstrated in *in vitro* cell culture and *in vivo* mouse models by employing a battery of tests. Exposure of human lung carcinoma A549 cells or Chinese hamster ovary AA8 cells to MWCNT-7 resulted in a dose-dependent increase in the frequency of the sister chromatid exchange or micronuclei, respectively, suggesting the *in vitro* genotoxicity potential of MWCNT-7. These results were further supported by the positive results obtained in mice in response to their exposure to MWCNT-7. Results of the comet assay revealed a significant increase in the incidence of DNA damage in the lungs of ICR mice intratracheally instilled with a single dose (0.05 or 0.2 mg/mouse) of MWCNT-7. A similar increase in the mutation frequency of the *gpt* gene, further supporting the genotoxicity of MWCNT-7, was detected in the lungs of the *gpt* delta transgenic mouse. Furthermore, dose-dependent increases in the formation of 8-oxo-7,8-dihydro-2′-deoxyguanosine and heptanone etheno-deoxyribonucleosides detected in the lungs of the MWCNT-7-exposed mice, prompted the authors to suggest the involvement of oxidative stress in the genotoxicity resulting from the exposure to MWCNT-7. Similar evidence of genotoxicity has been reported based on the observation of DNA strand breaks in mouse lungs in response to their pulmonary exposure to MWCNT (Catalan et al. 2016).

Further evidence for the genotoxicity of MWCNT was obtained from the results of a mouse study conducted by Poulsen et al. (2016). The genotoxicity of 10 commercial MWCNT samples that differ in their morphology and physical and chemical characteristics was determined in C57BL/6J mice. MWCNTs used in the study were either pristine or surface modified and exhibited differences in their morphology, chemical composition, surface area, and functionalization status. Various strengths of MWCNTs were intratracheally administered in mouse lungs at doses of 0, 6, 18, or 54 μg/mouse. At post-exposure time intervals of 1, 28, or 92 days, DNA strand breaks – indicators of genotoxicity – were determined by comet assay in the lungs of the control and MWCNT-administered mice. Furthermore, the authors, by multiple regression analysis of the data, determined the relationship, if any, between the physicochemical properties of the MWCNT samples and their genotoxicity potential. The authors found that pulmonary exposure to MWCNTs in mice resulted in lung genotoxicity and the Brunauer–Emmett–Teller (BET) surface area of MWCNT particles that correlates well and inversely with the diameter was a good predictor of MWCNT-induced genotoxicity. A lower BET surface area or a corresponding larger diameter of the MWCNT particles was associated with increased genotoxicity in mouse lungs. Similar findings were also reported by Fraser et al. (2020).

Whether exposure to MWCNT-7 results in changes in the chromosome number, a key genotoxic event potentially responsible for carcinogenesis, was investigated by employing immortalized human bronchial epithelial cells (BEAS-2B) and primary small airway epithelial cells (SAEC) (Siegrist et al. 2014). Exposure of the cultured BEAS-2B and SAEC cells to occupationally relevant doses of MWCNT-7 resulted in changes in chromosome number, suggesting induction of aneuploidy as a potential genotoxic mechanism involved in MWCNT-7-induced carcinogenesis. The chromosomal aberrations induced by MWCNT-7 exposure were attributed to the disruption of centrosomes and interference with the mitotic spindle apparatus in association with the cell cycle block. Furthermore, the genetically damaged, aneuploid BEAS-2B cells exhibited increased colony formation, a characteristic of the transformed phenotype, further supporting the induction of aneuploidy as a potential mechanism underlying the neoplastic potential of MWCNT-7.

Contrary to the above reports, Ema et al. (2013) did not find any evidence to support the genotoxicity of MWCNTs in rats. Male Crl/CD(SD) rats were intratracheally administered MWCNTs as a single dose (0.2 or 1 mg/kg body weight) or repeated doses of 0.04 or 0.2 mg/kg body weight, once a week for 5 weeks. The control and MWCNT-administered rats were euthanized (3 or 24 hours following a single instillation or 3 hours following the previous instillation in the repeatedly administered groups) and the DNA obtained from their lungs was analyzed using the comet assay for DNA damage. The findings of the study showed that MWCNT samples employed in the study, under exposure conditions, did not result in genotoxicity in rats. It is worth mentioning that significant lung inflammation was detected in MWCNT-administered rats, under the same exposure conditions, compared with the corresponding control rats. Similar to the findings of Ema et al. (2013), the absence of *gpt* gene mutation, an indicator of genotoxicity, has been reported in the lungs and bone marrow of F344 *gpt* delta rats that were intratracheally administered a single dose of MWCNT (Horibata et al. 2017).

Toxicogenomics of ENMs

Over the past 30 years, the toxicology field has undergone a major transformation. This was facilitated mainly by the realization that determination of global changes in cellular molecules, collectively referred to as the "-ome," may provide broader information regarding the toxicity of xenobiotics than the traditional approaches employed in toxicity assessment, such as the histological, biochemical, and physiological changes indicative of toxicity. Specifically, the "-ome" refers to the whole array of molecules, such as DNA, RNA, proteins, and lipids, present in the cells and metabolites generated from xenobiotics. This resulted in the emergence of a new branch in toxicology, referred to as toxicogenomics. The US National Research Council (NRC) defines toxicogenomics as "combin(ing) toxicology with information-dense genomic technologies to integrate toxicant-specific alterations in gene, protein, and metabolite expression patterns with phenotypic response of cells, tissues, and organisms" (NRC 2007). Toxicogenomics facilitates the interrogation of cellular response to a xenobiotic exposure at the global, molecular level (Alexander-Dann et al. 2018). The advances in molecular biology, computer science, mathematics, bioinformatics, and the detection and analysis of metabolites played significant roles in the emergence and growth of the various branches in toxicogenomics such as genomics (study of the genome or DNA), transcriptomics (study of the transcriptome or transcripts), proteomics (study of the proteome or proteins), lipidomics (study of the lipidome or lipids), and metabolomics or metabonomics (study of the metabolome or metabolites or small molecules). Transcriptomics, by far, is more advanced than the other branches in toxicogenomics and has played a prominent role in studying the toxicity of xenobiotics including that of ENMs. Transcriptome profiling and bioinformatic analysis of the data obtained with crystalline silica and acetaminophen have facilitated the identification of sensitive biomarkers for the detection of target organ toxicity (Bushel et al. 2007; Sellamuthu et al. 2011a) as well as determining the molecular mechanisms underlying the toxicity (Sellamuthu et al. 2011b).

Transcriptomics – Technical Aspects

Transcriptomics, in the context of toxicogenomics, consists of two important steps. The first step involves the detection and identification of the transcripts that are differentially expressed in response to exposure to a toxic agent. Subsequently, during the second step, the functional relevance or significance of the differentially expressed genes (DEGs) in the ensuing biological response or toxicity is determined. Various techniques are currently available to determine the differential gene expression (DGE) profile in biological systems in response to their exposure to a toxic agent. The selection of a specific technique employed in DGE profiling depends on the coverage or the number of transcripts that the investigator(s) is(are) interested in determining. Real-time polymerase chain reaction (RT-PCR) is the method of choice when the objective is to determine the profile of a limited number of specific DEGs. On the other hand, if the objective is to determine the expression profile of all or majority of the transcripts present in the transcriptome, high-throughput techniques, such as microarray or next generation sequence (NGS) or, more specifically, RNA sequencing (RNA-seq) analysis, are preferred (Joseph 2017).

Among the three popular techniques currently employed to determine the DGE profiling, RT-PCR is the least complex one. It involves reverse transcription of the high-quality RNA isolated from the control and toxicant exposed biological samples to synthesize cDNAs followed by the DNA polymerase-catalyzed amplification, detection, and quantification of the cDNAs of interest that are annealed to specific oligonucleotide primers. The expression of the DEGs determined by RT-PCR is typically normalized to that of one or more housekeeping genes to obtain more reliable and acceptable data. Finally, the fold changes in expression of the transcript(s) between the toxicant exposed and the appropriate control sample is calculated to determine the DGE profile (Joseph 2017).

The microarray technique consists of sequence-specific hybridization of the transcripts (targets) prepared from a biological sample with hundreds or thousands of the oligonucleotides or cDNAs (probes) arrayed on the microarray platform. The various steps involved in microarray analysis of the DGE profile are the reverse transcription of mRNAs to synthesize cDNAs, their labeling with fluorescent dye(s) to prepare targets, sequence-specific hybridization of the labeled targets with oligonucleotides or cDNAs arrayed on the microarray, washing to remove nonspecific binding of targets with probes and the free labels potentially contributing to background signal, scanning the arrays to detect the fluorescence intensity associated with the hybridized target–probe complex, comparison of the signal intensity with appropriate control and xenobiotic-exposed samples to detect and quantify the DGE profile (Joseph 2017). Since its inception in the early 1990s (Bumgarner 2013), microarray technology has been employed in studying the toxicity of xenobiotics affecting virtually every organ or system in the body in a variety of species.

RNA-sequence (RNA-seq) analysis is the latest advancement in profiling the whole transcriptome for DGE. Since RNA-seq is based on *de novo* sequence detection, the technique can detect virtually all transcripts, including novel or previously uncharacterized transcripts that are present in the transcriptome. RNA-seq consists of multiple steps, such as the preparation of sequencing libraries from the control and xenobiotic-exposed biological samples, sequencing the libraries, and quantification of the transcripts to determine DGE between the control and xenobiotic-exposed samples (Figure 7.1). Since RNA molecules such as rRNA and globin mRNA that are present in very high abundance in biological samples are known to result in their overrepresentation during RNA-seq, complicating the detection of less abundant transcripts and their differential expression, it is important to selectively remove those abundant RNA species prior to or during library preparation. The next step in RNA-seq analysis is the reverse transcription of RNA to synthesize cDNA that are ligated to adapters that serve as barcodes to identify the cDNAs prepared from a sample. The adapter-ligated cDNA library samples are subsequently subjected to high-throughput sequencing, typically to a read depth of 10–30 million reads per sample, and the data is further analyzed to determine the differences in the gene expression profile between the toxicant exposed and appropriately matched control samples.

Analysis of the raw data generated by the techniques employed in DGE profiling is a critical step involved in the identification and quantification of DEGs. Data analysis can be simple or very

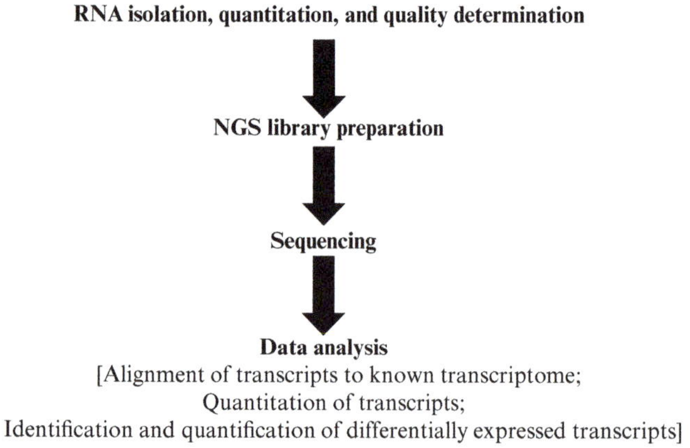

Figure 7.1 A schematic representation of the steps involved in differential gene expression profiling by RNA sequence analysis.

complex depending on the technique that is employed to generate gene expression data. Among the three currently employed techniques, data analysis is least complicated in the case of QRT-PCR and consists of normalization of the expression of gene(s) of interest (target gene[s]) to that of one or more housekeeping genes and calculating the fold changes in gene expression between the xenobiotic-exposed and matched control samples. Data analysis in the case of microarray is not as simple as QRT-PCR and typically consists of normalization of the hybridization signal intensities and calculating the fold changes in the expression of all of the detected genes (tens of thousands in the majority of cases) represented on the microarray between the sample sets. Data analysis in the case of RNA-seq, compared with QRT-PCR and microarray, is much more complex and requires significant expertise and computational resources (Stark et al. 2019). The first step in RNA-seq data analysis is alignment or mapping the sequence reads generated to a known transcriptome. Various alignment tools, such as TopHat (Kim et al. 2013), STAR (Dobin et al. 2013), or HISAT (Kim et al. 2015), are available and routinely employed. These tools facilitate the spliced alignment of the cDNA sequences generated by RNA-seq with the exon boundaries of the genes. The next step in data analysis is assignment of the sequences to genes or transcripts to determine their counts or abundance. Some of the common quantification tools employed in the RNA-seq analysis are RSEM (Li and Dewey 2011), CuffLinks (Trapnell et al. 2012), MMSeq (Turro et al. 2011), and HTSeq (Anders et al. 2015). Various factors, such as read depth, expression patterns of genes, GC content, and technical biases, are known to contribute to differences in the quantification of transcripts (Risso et al. 2011; Wagner et al. 2012) and, therefore, the data needs to be normalized and filtered to improve data quality. Quantile or median normalization (Bourgon et al. 2010) is commonly employed for this purpose. The final step in RNA-seq data analysis is differential expression modeling, determining the transcripts that may have changed in expression and quantifying the changes between the sample sets. Some of the tools commonly employed in this step of the RNA-seq analysis are edgeR (Robinson et al. 2010), DESeq2 (Love et al. 2014), and limma + voom (Law et al. 2014). Each of the steps involved in RNA-seq data analysis may significantly influence the DGE data output and, therefore, great care should be employed while analyzing the RNA-seq data to obtain biologically or toxicologically meaningful gene expression data.

Despite the challenges, especially those associated with data analysis, RNA-seq has several advantages compared with the microarray analysis in high-throughput DGE studies. Compared with microarray, RNA-seq is more sensitive and, therefore, capable of detecting transcripts that are present in very low abundance (Wang et al. 2014). Since microarray is based on the hybridization between the labeled transcripts (probes) expressed in the biological sample and the targets that are arrayed on the microarray platform, unknown or novel transcripts are not detected by the microarray technique. RNA-seq, on the other hand, involves *de novo* sequencing of all of the transcripts present in the biological specimen employed and, therefore, can detect novel transcripts and RNA species other than mRNA, such as miRNA, long noncoding RNA (lncRNA), and small nuclear RNA (snRNA) (Stark et al. 2019). RNA-seq, unlike microarray, can also detect splice variants of the transcripts. Despite being the newcomer in the field, RNA-seq is being increasingly utilized lately in DGE analysis in toxicogenomics (Joseph et al. 2021; Sager et al. 2020).

Toxicogenomics of MWCNTs – Animal Studies

The application of whole transcriptome profiling and bioinformatic analysis of the gene expression data to understand the molecular mechanisms of MWCNT-induced lung toxicity was recently investigated in our laboratory by employing a rat inhalation exposure and lung toxicity model

(Sager et al. 2022). Global gene expression profiles in the lung samples obtained from rats exposed to an aerosol containing increasing concentrations of MWCNT-7 and the corresponding air only exposed to controls was analyzed by NGS analysis. The number of the significantly differentially expressed genes (SDEGs) (fold change >1.5 and false discovery rate [$p < 0.05$]) detected in rat lungs (Sager et al. 2022), like the alterations in the toxicity parameters (Sager et al. 2022), exhibited a response to the cumulative dose of MWCNT-7 inhaled by the rats. Similarly, the fold changes in expressions of many of the SDEGs that are known to play a significant role in the pulmonary toxicity of MWCNT-7 were also dependent on the dose of the MWNCT-7 inhaled by the rats. The SDEGs identified in rat lungs by NGS analysis were used as input in the bioinformatic analysis to identify the biological functions and canonical pathways that were significantly enriched in response to inhalation exposure of the rats to MWCNT-7 and the resulting lung toxicity. The enrichment in the biological function and canonical pathway categories, similar to the alterations in the lung toxicity parameters, exhibited a dose response to the MWCNT-7 inhaled by the rats (Sager et al. 2022). Thus, the findings of the transcriptome analysis and those of lung toxicity determination suggested that the gene expression changes detected in rat lungs, in response to their exposure to MWCNT-7, were relevant to the lung toxicity induced by the inhaled MWCNT-7 in the rats.

Inhaled MWCNT particles, upon deposition in the lungs, are engulfed by phagocytes, mostly AMs, for their elimination from the lungs. The significant increase in the number of AMs detected in the MWCNT-7 exposed rat lungs (Sager et al. 2022) may be considered an adaptive response due to the increased demand for the AMs to facilitate phagocytosis and thus detoxification of the toxic MWCNT-7 particles. This view is further supported by the significant and MWCNT-7 dose-dependent enrichment in the functional and canonical pathway categories viz. recruitment of phagocytes, activation of macrophages, and phagosome formation as well as differential expressions of the genes involved in those categories (Sager et al. 2022). The interaction between the toxic particles deposited in the lungs and the AMs may result in the activation of the macrophages and the ensuing release of various signaling molecules. Activation of AMs by inhaled MWCNT-7 was supported by the dose-dependent increase in the transcript, *RETNLA1*, a marker for macrophage activation (Nair et al. 2009), in the MWCNT-7 exposed rat lungs (Sager et al. 2022). Cytokines/chemokines are prominent among the signaling molecules released by the activated AMs (Becker et al. 1991). The transcripts for several cytokines/chemokines viz. *CCl2*, *CCl7*, *CCl9*, *CCL22*, *CXCl2*, *CXCl3*, *CXCl6*, and *CXCL10* were significantly overexpressed in MWCNT-7-exposed rat lungs. The cytokines/chemokines released by the macrophages are involved in the recruitment of additional macrophages to the lungs, activation of the macrophages, and phagocytosis of toxic particles by the AMs and, therefore, are important determinants in deciding the fate of the particles with respect to their potential to induce lung toxicity.

The involvement of inflammation in lung toxicity and the potential for adverse health effects resulting from the exposure to MWCNT is well established (Dong and Ma 2016a; Porter et al. 2010; Poulsen et al. 2016). Results of the bioinformatic analysis of the gene expression data, in addition to supporting the induction of inflammation in MWCNT-7-exposed rat lungs, provided insights into the mechanisms underlying the induction of inflammation. Recruitment of phagocytes, primarily PMNs, from blood to the lungs is one of the first responses to pulmonary exposure to MWCNTs. The interaction of the MWCNT-7 particles deposited in the lungs with the AMs and/or the lung epithelium results in the release of pro-inflammatory signaling molecules. The recruitment of PMNs from the circulatory system to the lungs is a complex, multistep process that involves their tethering, activation, adhesion to the endothelium, rolling, and transmembrane release (Mizgerd 2002). The entire process is controlled by signaling molecules, including cytokines/chemokines that belong to the CC and CXC families (Mizgerd 2002). The overexpression of the transcripts for pro-inflammatory cytokines/chemokines in the lungs and pulmonary infiltration of the PMNs exhibited a response to the cumulative MWCNT-7 dose to which the rats were exposed, suggesting a possible relationship among overexpression of cytokines/chemokines, pulmonary infiltration of PMNs, and the lung injury detected in MWCNT-7 exposed rats (Sager et al. 2022).

Results of the bioinformatic analysis of the SDEGs, in addition to supporting the involvement of inflammation in MWCNT-7-induced lung toxicity, provided information regarding the potential molecular mechanisms underlying the toxicity. This included the enrichment of several inflammation-related IPA biological functions such as inflammatory response and inflammation of airways and canonical pathways, such as acute phase response signaling, LXR/RXR activation, complement system, and TREM1 signaling as well as the significant increases in the expression levels of several genes belonging to those categories in MWCNT-7-exposed rat lungs (Sager et al. 2022). The acute phase response is a rapid inflammatory response to defend tissues/organs against invading microorganisms (Perez 2019) and injury resulting from exposure to toxic agents (Saber et al. 2014). However, enhanced activation of the acute phase response and excessive release and accumulation of acute phase response signaling molecules may result in tissue damage and toxicity. Serum amyloid is a marker for acute phase response (Sack 2018) and a significant overexpression of the *Serum amyloid A like 1 (Saal1)* transcript was detected in MWCNT-7-exposed rat lungs. Similarly, the transcript for another gene, *Orosomucoid 1 (ORM1)*, involved in the acute phase response and inflammation (Alfadda et al. 2012; Ligresti et al. 2012) was highly overexpressed in MWCNT-7-exposed rat lungs, which further supported the involvement of acute phase response in the lung inflammation and toxicity induced by MWCNT-7, similar to that resulting from exposure to other particles that are toxic to the lungs (Joseph et al. 2021; Sager et al. 2020). Like the acute phase response, activation of the complement system is another important mechanism involved in the immune and inflammatory responses to tissue injury by toxic particles (Pandya and Wilkes 2014). Complement proteins, regulators of the complement system, are synthesized by lung epithelial cells (Strunk et al. 1988; Varsano et al. 2000) and fibroblasts (Volanakis 1995) in response to the release of pro-inflammatory cytokines (Huber-Lang et al. 2002). The complement proteins, in turn, may function as chemoattractants facilitating the infiltration of PMNs into the lungs to result in the induction of inflammation, such as that was detected in MWCNT-7 exposed rat lungs (Sager et al. 2022). The transcripts for many of the genes involved in complement response, viz. *Complement C1s (C1s), Complement 3 (C3), Complement 6 (C6), Complement factor B (CFB), Complement C1q A chain (C1QA), Complement C1q B chain (C1QB), Complement C1qC chain (C1QC), Integrin subunit alpha M (ITGAM), Integrin subunit alpha X (ITGAX)*, and *Integrin subunit beta 8 (ITGβ8)* were overexpressed in rat lungs and their overexpression was dependent on the cumulative MWCNT-7 dose to which the rats were exposed. Taken together, these findings suggested the activation of the complement system and its potential involvement in the MWCNT7-induced lung inflammation and injury detected in the rats. Triggering receptor expressed on myeloid cells 1 (TREM1) is a receptor that belongs to the immunoglobulin family of cell surface receptors. The association between activated TREM1 and its partners, the toll-like receptors (TLR), facilitates the release of cytokines, such as MCP-1, MIP-2, and TNF-α (Bleharski et al. 2003; Ornatowska et al. 2007), resulting in a pro-inflammatory response. The significant overexpression of the transcripts for *TREM1* and several *TLRs* as well as the elevated BALF levels of cytokines (MCP-1, MIP-2, and TNF-α) detected in the lungs are suggestive of the potential involvement of the TREM1 signaling pathway in the induction of lung inflammation and injury that resulted from inhalation exposure of rats to MWCNT-7.

Generation of reactive oxidants and the resulting oxidative stress, similar to the induction of inflammation, is another major mechanism involved in the toxicity and pathology associated with the exposure to MWCNT (Srivastava et al. 2011). The gene expression data and findings of the bioinformatic analysis supported the MWCNT-7-induced oxidative stress in rat lungs (Sager et al. 2022). Nrf2 is a redox-sensitive transcription factor that regulates the expression of key genes involved in protecting the biological system through the detoxification of toxic reactive oxidants (Kensler et al. 2007). MWCNT-7-induced oxidative stress, pulmonary inflammation, and fibrosis were significantly higher in the Nrf2 knock-out mouse, compared with the wild-type mouse (Dong and Ma 2016a), suggesting a protective role for Nrf2 in the deleterious effects associated with MWCNT-7 exposure. Genes that are members of the canonical pathway, Nrf2-mediated oxidative stress, were significantly overexpressed in the lungs of MWCNT-7-exposed rats. For example, *Superoxide dismutase 2 (SOD2)* – an Nrf2-regulated gene that was significantly overexpressed in the lungs of the MWCNT-7-exposed rats is primarily responsible for the generation of hydrogen peroxide, a toxic ROS, by dismutation of the superoxide anion. The *NADPH oxidase organizer 1 (NOXO1)* gene involved in the generation of toxic superoxide anion (Katsuyama et al. 2012) was also highly overexpressed in MWCNT-7-exposed rat lungs. Other genes involved in oxidative stress response such as *Lactoperoxidase (LPO)* (Sharma et al. 2013), *Hemeoxygenase 1 (HMOX1)* (Nakashima et al. 2018), and *Lipocalin 2 (LCN2)* (Roudkenar et al. 2007), also exhibited a significant and dose-dependent overexpression in MWCNT-7-exposed rat lungs. The absence of efficient detoxification of the ROS generated may facilitate their accumulation in lungs, resulting in oxidative stress and injury, including excessive inflammation and fibrosis, as previously reported in the case of the Nrf2 knock-out mouse, in response to MWCNT-7 exposure (Dong and Ma 2016b).

Pulmonary exposure of rats and mice to MWCNT is known to cause lung fibrosis (Porter et al. 2010; Dong et al. 2015; Kasai et al. 2015, 2016). Bioinformatic analysis of the gene expression data obtained from the lungs of MWCNT-7-exposed rats also provided information relevant to the molecular mechanisms involved in fibrosis. The *Serpine family E member 1 (SERPINE1)* gene plays a pivotal role in lung diseases through its involvement in chronic inflammation (Tiwari et al. 2016), tissue remodeling (Chen et al. 2021), and fibrosis (Ghosh and Vaughan 2012). Furthermore, *SERPINE1* gene polymorphism is a risk factor for respiratory diseases (Chen et al. 2021) and inhibitors of SERPINE1 protein have therapeutic application in treating upper respiratory diseases (Huang et al. 2012). Considering the role of the *SERPINE1* gene in respiratory diseases, the significant overexpression of its transcript detected in the lungs of rats should be of functional significance to MWCNT-7-induced lung fibrosis. Other genes whose expressions were significantly higher in MWCNT-7-exposed rat lungs, compared with the controls, and potentially involved in MWCNT-7-induced fibrosis were *secreted phosphoprotein 1 (SPP1), TIMP metallopeptidase inhibitor 1 (TIMP1), chitinase, acidic (CHIA), adenosine A1 receptor (ADORA1), pentraxin 3 (PTX3),* and *matrix metallopeptidase 12 (MMP12).* Osteopontin, the protein product of the SPP1 gene, through the activation of the TGF-β signaling pathway, facilitates the activation and differentiation of fibroblasts to myofibroblasts – a critical step involved in fibrosis (Dong and Ma 2017a). A similar role for TIMP1 in MWCNT-7-induced lung fibrosis through the activation and proliferation of fibroblasts has been identified (Dong and Ma 2017b). Extracellular matrix (ECM) formation, a critical event in fibrosis, is a complex process that involves the synthesis and degradation of ECM proteins, most notably collagen. The significant overexpression of *ADORA1, CHIA, MMP12,* and *PTX3* in MWCNT-7 exposed rat lungs, their established role in the maintenance of the ECM (Churg and Wright 2005; Cronstein 2011; Lee et al. 2012; Pilling et al. 2015), and the increase in collagen detected in the lungs should be considered as evidence for their potential involvement in the MWCNT-7-induced lung fibrosis in rats.

The IARC, based on the results of animal studies, has classified one form of the MWCNT, MWCNT-7, the same material used in our rat lung toxicity study, as a Group 2B human carcinogen

(IARC (International Agency for Research on Cancer) 2017). Whereas MWCNT-7 was identified as a tumor promoter in a mouse model (Sargent et al. 2014), prolonged inhalation exposure to MWCNT-7 alone resulted in significant increases in pre-neoplastic and neoplastic lesions in rat lungs, suggesting that MWCNT-7 is a complete carcinogen (Kasai et al. 2016). Despite the identification of MWCNT-7 as a carcinogen, the mechanisms underlying its carcinogenicity were not well understood. The absence of positive results in the Ames test (Ema et al. 2012) suggested that the carcinogenicity of MWCNT-7 may not be due to its role as a direct mutagen. On the other hand, changes in chromosome number were detected in cultured cells in response to exposure to MWCNT-7 (Asakura et al. 2010; Siegrist et al. 2014), attributing the induction of aneuploidy as a potential mechanism involved in the MWCNT-7-induced carcinogenesis. In fact, in immortalized human airway epithelial BEAS-2B cells, the chromosomal aberrations induced by MWCNT-7 exposure were attributed to the disruption of the centrosomes and interference with the mitotic spindle apparatus in association with cell cycle block (Siegrist et al. 2014). Furthermore, the genetically damaged, aneuploid BEAS-2B cells exhibited increased colony formation, a feature characteristic of the transformed phenotype, further supporting the induction of aneuploidy as a potential mechanism underlying the neoplastic potential of MWCNT-7.

The gene expression data obtained in our rat inhalation exposure and toxicity study (Sager et al. 2022) further supported the previously reported role of MWCNT-7 as a carcinogen (Kasai et al. 2016) and suggested potential mechanisms underlying its carcinogenicity. The biological function category, lung cancer, was significantly enriched in response to MWCNT-7 exposure in rat lungs (Sager et al. 2022) similar to that reported by Guo et al. (2012) in mice exposed to MWCNT-7 by pharyngeal aspiration. Similarly, cell cycle control of chromosomal replication and kinetochore metaphase signaling pathway were two of the significantly enriched canonical pathway categories in MWCNT-7 exposed rat lungs (Sager et al. 2022). The enrichment of these IPA categories and the overexpression of the genes involved in those categories, similar to the results of the lung toxicity assessment parameters employed in the study, were dependent on the MWCNT-7 dose to which the rats were exposed, suggesting the potential involvement of the overexpressed genes in mediating MWCNT-7-induced lung toxicity/carcinogenesis.

Faithful replication of chromosomal DNA is essential for the stable propagation of genetic information during cell division. This requires that chromosomal DNA replication, a very complex process, regulated by the expression of multiple genes, takes place only once per cell division. The genes that regulate the chromosomal DNA replication during cell division/cycle belong to three major classes – origin recognition complex (ORC), minichromosome maintenance (MCM), and cell division cycle (CDC), each regulating distinct and critical events involved in the faithful replication of chromosomal DNA (Dabral et al. 2019; Hizume et al. 2013; Perl et al. 2019). Overexpression of the ORC, MCM, and CDC genes resulting in excessive production of the encoded proteins and modulation of their functions may result in abnormal chromosomal DNA replication, leading to cell transformation and cancer (Issac et al. 2019; Semple and Duncker 2004; Tomita et al. 2011). Members of the ORC, MCM, and CDC gene families were significantly overexpressed in the lungs of MWCNT-7-exposed rats suggesting their potential involvement in lung cancer due to exposure to MWCNT-7 (Sager et al. 2022).

Another major event in cell division, relevant to carcinogenesis, is the replication of chromosomes taking place during metaphase and their subsequent segregation, equally, during anaphase to result in daughter cells with the same number of chromosomes as that of the parental cell. The kinetochore-spindle assembly formation plays a critical role in the equal segregation of the sister chromatids to the daughter cells. The entire process is controlled by proteins that belong to the KMN network consisting of Knl1, Mis12, and Ndc80 complexes (Varma and Salmon 2012).

Transcripts for the genes whose protein products play important roles in the kinetochore-spindle assembly formation, through their participation in the KMN network, were found significantly overexpressed in MWCNT-7-exposed rat lungs (Sager et al. 2022). The established role of KMN genes in oncogenesis (Bai et al. 2019; Gao et al. 2022), the observation that MWCNT-7-induced aneuploidy resulted in the acquisition of an oncogenic phenotype in human airway epithelial cells (Siegrist et al. 2014), and significant overexpression of the KMN network gene transcripts in MWCNT-7-exposed rat lungs may suggest that deregulation of expression of the genes involved in kinetochore metaphase signaling pathway is a potential mechanism underlying MWCNT-7-induced cancer.

Similar to the studies conducted in our laboratory, several other investigators have also employed gene expression profiling to determine the mechanisms underlying MWCNT-induced lung injury as well as to identify pulmonary disease outcomes of exposure to ENMs. Concordant changes in DNA methylation and lung global gene expression profiles were detected in mice at 56 days following administration of a single dose of MWCNT-7 by pharyngeal aspiration (Scala et al. 2021). Functional analysis of the gene expression data obtained from mice identified pathways involved in muscle contraction, immune system/inflammation, and extracellular matrix formation as the most affected ones and their potential involvement in MWCNT-induced lung toxicity. Poulsen et al. (2015) demonstrated that pulmonary transcriptional responses indicative of lung fibrosis varied for carbon nanotubes with differences in physical and chemical properties. In a mouse toxicogenomics study (Labib et al. 2016), an adverse outcome pathway framework approach was employed to identify key biological events linking MWCNT exposure to lung fibrosis. The investigators administered MWCNT with different physical and chemical properties in mouse lungs and major pathways perturbed by the exposures were identified by global gene expression profiling and bioinformatic analysis of the resulting data. The results obtained from the study showed that similar biological pathways were perturbed by different MWCNT particles in spite of the differences in their physical and chemical properties. Based on these results, the authors concluded that transcriptomic data can be employed as an effective mechanism-based approach to derive acceptable levels of exposure to nanomaterials in situations where epidemiological data are not available.

Toxicogenomics of MWCNT – Human Studies

Compared with the large number of animal studies, there have been only fewer human studies that investigated the potential toxicity and the health effects associated with exposure to MWCNT. In a study reported by Shvedova et al. (2016), the changes in the peripheral blood transcriptome of workers, in response to their exposure to MWCNT in a manufacturing facility, were investigated. The study population consisted of workers ($n = 8$) who had direct contact with an aerosol containing MWCNT for at least six months. A population consisting of professionals and/or technical staff ($n = 7$) from the same manufacturing facility with no known history of exposure to MWCNT served as the control. The detection of MWCNT particles in the PBZ of the workers engaged in manufacturing MWCNT confirmed their occupational exposure to the ENM. Global expression profiles of the long noncoding RNA (ncRNA), mRNA, and microRNA (miRNA) were determined in the peripheral blood samples collected from the control and MWCNT-exposed workers by microarray (lncRNA and mRNA) or NGS (miRNA) analysis. The resulting data were analyzed to determine the differences in the expression profiles of the RNA molecules in the blood samples obtained from the control and the MWCNT-exposed subjects. Significant differential expression (fold change ≥/± 1.5 and $p < 0.05$) of 977 ncRNAs, 785 mRNAs, and 11 miRNAs were detected between the blood samples obtained from the MWCNT-exposed and the nonexposed workers. The authors further reported that the changes in expressions of the ncRNAs and mRNAs were prominent in three workers who had higher MWCNT exposure compared with the remaining five workers with relatively lower levels of MWCNT exposure. Bioinformatic analysis of the expression data was performed to determine the potential disease outcomes associated with MWCNT exposure as well as the molecular mechanisms potentially involved in the diseases. Many of the dysregulated mRNAs and miRNAs detected in the blood samples obtained from the MWCNT-exposed workers were mostly associated with pulmonary inflammation and fibrosis, in agreement with the results obtained from several rodent toxicity studies (Ellinger-Ziegelbauer and Pauluhn 2009; Labib et al. 2016; Dong et al. 2016; Sager et al. 2022). The findings of the bioinformatic analysis of the transcriptomics data also suggested the potential for MWCNT exposure to result in cardiovascular and carcinogenic outcomes.

The finding that the blood transcriptome responds to pulmonary exposure to MWCNT in mice (Snyder-Talkington et al. 2015) and humans (Shvedova et al. 2016) may have significant application in the prevention of adverse health effects potentially resulting from exposure to MWCNT. It is known that gene expression changes taking place in the blood, in response to exposure to toxic agents, are relevant to target organ toxicity (Bushel et al. 2007; Sellamuthu et al. 2011a, 2012). Furthermore, the superior sensitivity of gene expression changes as indicators of toxicity compared with biochemical and histological changes taking place in the target organs has been established (Umbright et al. 2010). Blood gene expression signatures have been employed as highly sensitive surrogate biomarkers for the detection of early toxicity resulting from exposure to acetaminophen (Bushel et al. 2007) and crystalline silica (Sellamuthu et al. 2011a). Since small quantities of the blood required to conduct transcriptome profiling can be obtained by a minimally invasive procedure, blood gene expression signatures can be developed and employed to monitor human exposure to MWCNT. Additionally, understanding the molecular mechanisms involved in the toxicity, through bioinformatic analysis of the transcriptomics data, may provide the opportunity to develop therapeutics targeting the genes and/or the encoded proteins. Therefore, transcriptomic studies have the potential to prevent the development of life-threatening health effects that are likely associated with exposure to MWCNT.

Disclaimer:

The findings and conclusions in this report are those of the author and do not necessarily represent the official position of the National Institute for Occupational Safety and Health, Centers for Disease Control and Prevention.

References

Adamson, I.Y., Bakowska, J., and Bowden, D.H. (1993). Mesothelial cell proliferation after instillation of long or short asbestos fibers into mouse lung. *Am. J. Pathol.* 142 (4): 1209–1216.

Adamson, I.Y., Bakowska, J., and Bowden, D.H. (1994). Mesothelial cell proliferation: a nonspecific response to lung injury associated with fibrosis. *Am. J. Respir. Cell Mol. Biol.* 10 (3): 253–258.

Aiso, S., Yamazaki, K., Umeda, Y., et al. (2010). Pulmonary toxicity of intratracheally instilled multiwall carbon nanotubes in male Fischer 344 rats. *Ind. Health* 48 (6): 783–795.

Alexander-Dann, B., Pruteanu, L.L., Oerton, E., et al. (2018). Developments in toxicogenomics: understanding and predicting compound-induced toxicity from gene expression data. *Mol. Omics* 14 (4): 218–236.

Alfadda, A.A., Fatma, S., Chishti, M.A., et al. (2012). Orosomucoid serum concentrations and fat depot-specific mRNA and protein expression in humans. *Mol. Cells* 33 (1): 35–41.

Anders, S., Pyl, P.T., and Huber, W. (2015). HTSeq – a Python framework to work with high-throughput sequencing data. *Bioinformatics* 31 (2): 166–169.

Asakura, M., Sasaki, T., Sugiyama, T., et al. (2010). Genotoxicity and cytotoxicity of multi-wall carbon nanotubes in cultured Chinese hamster lung cells in comparison with chrysotile A fibers. *J. Occup. Health* 52 (3): 155–166.

Bai, T., Zhao, Y., Liu, Y., Cai, B., Dong, N., and Li, B. (2019). Effect of KNL1 on the proliferation and apoptosis of colorectal cancer cells. *Technol. Cancer Res. Treat.* 18: 1533033819858668.

Barnes, J.L. and Gorin, Y. (2011). Myofibroblast differentiation during fibrosis: role of NAD(P)H oxidases. *Kidney Int.* 79 (9): 944–956.

Becker, S., Quay, J., and Soukup, J. (1991). Cytokine (tumor necrosis factor, IL-6, and IL-8) production by respiratory syncytial virus-infected human alveolar macrophages. *J. Immunol.* 147 (12): 4307–4312.

Bleharski, J.R., Kiessler, V., Buonsanti, C., et al. (2003). A role for triggering receptor expressed on myeloid cells-1 in host defense during the early-induced and adaptive phases of the immune response. *J. Immunol.* 170 (7): 3812–3818.

Bourgon, R., Gentleman, R., and Huber, W. (2010). Independent filtering increases detection power for high-throughput experiments. *Proc. Natl. Acad. Sci. U.S.A.* 107 (21): 9546–9551.

Bumgarner, R. (2013). Overview of DNA microarrays: types, applications, and their future. *Curr. Protoc. Mol. Biol.* Chapter 22:Unit 22 21.

Bushel, P.R., Heinloth, A.N., Li, J., et al. (2007). Blood gene expression signatures predict exposure levels. *Proc. Natl. Acad. Sci. U.S.A.* 104 (46): 18211–18216.

Cao, G. (2004). *Nanostructures and Nanomaterials. Synthesis, Properties & Applications*. London: Imperial College Press.

Castranova, V. (2011). Overview of current toxicological knowledge of engineered nanoparticles. *J. Occup. Environ. Med.* 53 (6 Suppl): S14–17.

Catalan, J., Siivola, K.M., Nymark, P., et al. (2016). In vitro and in vivo genotoxic effects of straight versus tangled multi-walled carbon nanotubes. *Nanotoxicology* 10 (6): 794–806.

Chen, T.Y., Zhou, M., Lin, M.Q., et al. (2021). Research progress on the SERPINE1 protein and chronic inflammatory diseases of the upper respiratory tract: a literature review. *Int. Arch. Allergy Immunol.* 182 (11): 1097–1102.

Churg, A. and Wright, J.L. (2005). Proteases and emphysema. *Curr. Opin. Pulm. Med.* 11 (2): 153–159.

Craighead, J.E., Abraham, J.L., Churg, A., et al. (1982). The pathology of asbestos-associated diseases of the lungs and pleural cavities: diagnostic criteria and proposed grading schema. Report of the Pneumoconiosis Committee of the College of American Pathologists and the National Institute for Occupational Safety and Health. *Arch. Pathol. Lab. Med.* 106 (11): 544–596.

Cronstein, B.N. (2011). Adenosine receptors and fibrosis: a translational review. *F1000 Biol. Rep.* 3: 21.

Dabral, P., Uppal, T., Rossetto, C.C., and Verma, S.C. (2019). Minichromosome maintenance proteins cooperate with LANA during the G1/S phase of the cell cycle to support viral DNA replication. *J. Virol.* 93: 7.

Dahm, M.M., Evans, D.E., Schubauer-Berigan, M.K., Birch, M.E., and Deddens, J.A. (2013). Occupational exposure assessment in carbon nanotube and nanofiber primary and secondary manufacturers: mobile direct-reading sampling. *Ann. Occup. Hyg.* 57 (3): 328–344.

Dahm, M.M., Evans, D.E., Schubauer-Berigan, M.K., Birch, M.E., and Fernback, J.E. (2012). Occupational exposure assessment in carbon nanotube and nanofiber primary and secondary manufacturers. *Ann. Occup. Hyg.* 56 (5): 542–556.

Dahm, M.M., Schubauer-Berigan, M.K., Evans, D.E., et al. (2018). Exposure assessments for a cross-sectional epidemiologic study of US carbon nanotube and nanofiber workers. *Int. J. Hyg. Environ. Health* 221 (3): 429–440.

Dahm, M.M., Schubauer-Berigan, M.K., Evans, D.E., Birch, M.E., Fernback, J.E., and Deddens, J.A. (2015). Carbon nanotube and nanofiber exposure assessments: an analysis of 14 site visits. *Ann. Occup. Hyg.* 59 (6): 705–723.

De Volder, M.F., Tawfick, S.H., Baughman, R.H., and Hart, A.J. (2013). Carbon nanotubes: present and future commercial applications. *Science* 339 (6119): 535–539.

Diabate, S., Mulhopt, S., Paur, H.R., Wottrich, R., and Krug, H.F. (2002). In vitro effects of incinerator fly ash on pulmonary macrophages and epithelial cells. *Int. J. Hyg. Environ. Health* 204 (5–6): 323–326.

Dobin, A., Davis, C.A., Schlesinger, F., et al. (2013). STAR: ultrafast universal RNA-seq aligner. *Bioinformatics* 29 (1): 15–21.

Donaldson, K., Aitken, R., Tran, L., et al. (2006). Carbon nanotubes: a review of their properties in relation to pulmonary toxicology and workplace safety. *Toxicol. Sci.* 92 (1): 5–22.

Donaldson, K., Brown, G.M., Brown, D.M., Bolton, R.E., and Davis, J.M. (1989). Inflammation generating potential of long and short fibre amosite asbestos samples. *Br. J. Ind. Med.* 46 (4): 271–276.

Donaldson, K., Murphy, F.A., Duffin, R., and Poland, C.A. (2010). Asbestos, carbon nanotubes and the pleural mesothelium: a review of the hypothesis regarding the role of long fibre retention in the parietal pleura, inflammation and mesothelioma. *Part Fibre Toxicol.* 7: 5.

Donaldson, K. and Tran, C.L. (2004). An introduction to the short-term toxicology of respirable industrial fibres. *Mutat. Res.* 553 (1–2): 5–9.

Dong, J. and Ma, Q. (2016a). In vivo activation of a T helper 2-driven innate immune response in lung fibrosis induced by multi-walled carbon nanotubes. *Arch. Toxicol.* 90 (9): 2231–2248.

Dong, J. and Ma, Q. (2016b). Suppression of basal and carbon nanotube-induced oxidative stress, inflammation and fibrosis in mouse lungs by Nrf2. *Nanotoxicology* 10 (6): 699–709.

Dong, J. and Ma, Q. (2017a). Osteopontin enhances multi-walled carbon nanotube-triggered lung fibrosis by promoting TGF-beta1 activation and myofibroblast differentiation. *Part Fibre Toxicol.* 14 (1): 18.

Dong, J. and Ma, Q. (2017b). TIMP1 promotes multi-walled carbon nanotube-induced lung fibrosis by stimulating fibroblast activation and proliferation. *Nanotoxicology* 11 (1): 41–51.

Dong, J., Porter, D.W., Batteli, L.A., Wolfarth, M.G., Richardson, D.L., and Ma, Q. (2015). Pathologic and molecular profiling of rapid-onset fibrosis and inflammation induced by multi-walled carbon nanotubes. *Arch. Toxicol.* 89 (4): 621–633.

Dong, J., Yu, X., Porter, D.W., Battelli, L.A., Kashon, M.L., and Ma, Q. (2016). Common and distinct mechanisms of induced pulmonary fibrosis by particulate and soluble chemical fibrogenic agents. *Arch. Toxicol.* 90 (2): 385–402.

Eatemadi, A., Daraee, H., Karimkhanloo, H., et al. (2014). Carbon nanotubes: properties, synthesis, purification, and medical applications. *Nanoscale Res. Lett.* 9 (1): 393.

Ellinger-Ziegelbauer, H. and Pauluhn, J. (2009). Pulmonary toxicity of multi-walled carbon nanotubes (Baytubes) relative to alpha-quartz following a single 6h inhalation exposure of rats and a 3 months post-exposure period. *Toxicology* 266 (1–3): 16–29.

Ema, M., Imamura, T., Suzuki, H., Kobayashi, N., Naya, M., and Nakanishi, J. (2012). Evaluation of genotoxicity of multi-walled carbon nanotubes in a battery of in vitro and in vivo assays. *Regul. Toxicol. Pharmacol.* 63 (2): 188–195.

Ema, M., Masumori, S., Kobayashi, N., et al. (2013). In vivo comet assay of multi-walled carbon nanotubes using lung cells of rats intratracheally instilled. *J. Appl. Toxicol.* 33 (10): 1053–1060.

Erdely, A., Dahm, M., Chen, B.T., et al. (2013). Carbon nanotube dosimetry: from workplace exposure assessment to inhalation toxicology. *Part Fibre Toxicol.* 10 (1): 53.

Fatkhutdinova, L.M., Khaliullin, T.O., Vasil'yeva, O.L., et al. (2016). Fibrosis biomarkers in workers exposed to MWCNT. *Toxicol. Appl. Pharmacol.* 299: 125–131.

Fenoglio, I., Tomatis, M., Lison, D., et al. (2006). Reactivity of carbon nanotubes: free radical generation or scavenging activity? *Free Radic. Biol. Med.* 40 (7): 1227–1233.

Fraser, K., Hubbs, A., Yanamala, N., et al. (2021). Histopathology of the broad class of carbon nanotubes and nanofibers used or produced in U.S. facilities in a murine model. *Part Fibre Toxicol.* 18 (1): 47.

Fraser, K., Kodali, V., Yanamala, N., et al. (2020). Physicochemical characterization and genotoxicity of the broad class of carbon nanotubes and nanofibers used or produced in U.S. facilities. *Part Fibre Toxicol.* 17 (1): 62.

Fujita, K., Fukuda, M., Endoh, S., et al. (2016). Pulmonary and pleural inflammation after intratracheal instillation of short single-walled and multi-walled carbon nanotubes. *Toxicol. Lett.* 257: 23–37.

Fukushima, S., Kasai, T., Umeda, Y., Ohnishi, M., Sasaki, T., and Matsumoto, M. (2018). Carcinogenicity of multi-walled carbon nanotubes: challenging issue on hazard assessment. *J. Occup. Health* 60 (1): 10–30.

Gao, H., Pan, Q.Y., Wang, Y.J., and Chen, Q.F. (2022). Impact of KMN network genes on progression and prognosis of non-small cell lung cancer. *Anticancer Drugs* 33 (1): e398–e408.

Gate, L., Knudsen, K.B., Seidel, C., et al. (2019). *Toxicol. Appl. Pharmacol.* 375: 17–31.

Ghosh, A.K. and Vaughan, D.E. (2012). PAI-1 in tissue fibrosis. *J. Cell. Physiol.* 227 (2): 493–507.

Goodglick, L.A. and Kane, A.B. (1990). Cytotoxicity of long and short crocidolite asbestos fibers in vitro and in vivo. *Cancer Res.* 50 (16): 5153–5163.

Guo, F., Ma, N., Horibe, Y., Kawanishi, S., Murata, M., and Hiraku, Y. (2012). Nitrative DNA damage induced by multi-walled carbon nanotube via endocytosis in human lung epithelial cells. *Toxicol. Appl. Pharmacol.* 260 (2): 183–192.

Hamilton, R.F., Jr., Wu, Z., Mitra, S., Shaw, P.K., and Holian, A. (2013). Effect of MWCNT size, carboxylation, and purification on in vitro and in vivo toxicity, inflammation and lung pathology. *Part Fibre Toxicol.* 10 (1): 57.

Han, J.H., Lee, E.J., Lee, J.H., et al. (2008). Monitoring multiwalled carbon nanotube exposure in carbon nanotube research facility. *Inhal. Toxicol.* 20 (8): 741–749.

Hayes, J.D., Dinkova-Kostova, A.T., and Tew, K.D. (2020). Oxidative stress in cancer. *Cancer Cell* 38 (2): 167–197.

Hedmer, M., Isaxon, C., Nilsson, P.T., et al. (2014). Exposure and emission measurements during production, purification, and functionalization of arc-discharge-produced multi-walled carbon nanotubes. *Ann. Occup. Hyg.* 58 (3): 355–379.

Hizume, K., Yagura, M., and Araki, H. (2013). Concerted interaction between origin recognition complex (ORC), nucleosomes and replication origin DNA ensures stable ORC-origin binding. *Genes. Cells* 18 (9): 764–779.

Horibata, K., Ukai, A., Ogata, A., et al. (2017). Absence of in vivo mutagenicity of multi-walled carbon nanotubes in single intratracheal instillation study using F344 gpt delta rats. *Genes. Environ.* 39: 4.

Huang, W.T., Vayalil, P.K., Miyata, T., Hagood, J., and Liu, R.M. (2012). Therapeutic value of small molecule inhibitor to plasminogen activator inhibitor-1 for lung fibrosis. *Am. J. Respir. Cell Mol. Biol.* 46 (1): 87–95.

Huber-Lang, M., Younkin, E.M., Sarma, J.V., et al. (2002). Generation of C5a by phagocytic cells. *Am. J. Pathol.* 161 (5): 1849–1859.

IARC (International Agency for Research on Cancer). (2017). *Some Nanomaterials and Some Fibers.* Lyon, France: International Agency for Research on Cancer.

ISO (International Standards Organization). (2015). Nanotechnologies - Vocabulary - Part 2: nano-objects. ISO/TS 80004-2:2015(en). https://www.iso.org/obp/ui/#iso:std:iso:ts:80004:-2:ed-1:v1:en. (Accessed 7 January 2022).

Issac, M.S.M., Yousef, E., Tahir, M.R., and Gaboury, L.A. (2019). MCM2, MCM4, and MCM6 in breast cancer: clinical utility in diagnosis and prognosis. *Neoplasia* 21 (10): 1015–1035.

Johnston, H.J., Hutchison, G.R., Christensen, F.M., et al. (2010). A critical review of the biological mechanisms underlying the in vivo and in vitro toxicity of carbon nanotubes: the contribution of physico-chemical characteristics. *Nanotoxicology* 4 (2): 207–246.

Joseph, P. (2017). Transcriptomics in toxicology. *Food Chem. Toxicol.* 109 (Pt 1): 650–662.

Joseph, P., Umbright, C.M., Roberts, J.R., et al. (2021). Lung toxicity and gene expression changes in response to whole-body inhalation exposure to cellulose nanocrystal in rats. *Inhal. Toxicol.* 33 (2): 66–80.

Kasai, T., Umeda, Y., Ohnishi, M., et al. (2015). Thirteen-week study of toxicity of fiber-like multi-walled carbon nanotubes with whole-body inhalation exposure in rats. *Nanotoxicology* 9 (4): 413–422.

Kasai, T., Umeda, Y., Ohnishi, M., et al. (2016). Lung carcinogenicity of inhaled multi-walled carbon nanotube in rats. *Part Fibre Toxicol.* 13 (1): 53.

Kato, T., Totsuka, Y., Ishino, K., et al. (2013). Genotoxicity of multi-walled carbon nanotubes in both in vitro and in vivo assay systems. *Nanotoxicology* 7 (4): 452–461.

Katsuyama, M., Matsuno, K., and Yabe-Nishimura, C. (2012). Physiological roles of NOX/NADPH oxidase, the superoxide-generating enzyme. *J. Clin. Biochem. Nutr.* 50 (1): 9–22.

Kensler, T.W., Wakabayashi, N., and Biswal, S. (2007). Cell survival responses to environmental stresses via the Keap1-Nrf2-ARE pathway. *Annu. Rev. Pharmacol. Toxicol.* 47: 89–116.

Kim, D., Pertea, G., Trapnell, C., Pimentel, H., Kelley, R., and Salzberg, S.L. (2013). TopHat2: accurate alignment of transcriptomes in the presence of insertions, deletions and gene fusions. *Genome Biol.* 14 (4): R36.

Kim, J.E., Lee, S., Lee, A.Y., Seo, H.W., Chae, C., and Cho, M.H. (2015). Intratracheal exposure to multi-walled carbon nanotubes induces a nonalcoholic steatohepatitis-like phenotype in C57BL/6J mice. *Nanotoxicology* 9 (5): 613–623.

Kuijpers, E., Bekker, C., Fransman, W., et al. (2016). Occupational exposure to multi-walled carbon nanotubes during commercial production synthesis and handling. *Ann. Occup. Hyg.* 60 (3): 305–317.

Labib, S., Williams, A., Yauk, C.L., et al. (2016). Nano-risk science: application of toxicogenomics in an adverse outcome pathway framework for risk assessment of multi-walled carbon nanotubes. *Part Fibre Toxicol.* 13: 15.

Law, C.W., Chen, Y., Shi, W., and Smyth, G.K. (2014). Voom: precision weights unlock linear model analysis tools for RNA-seq read counts. *Genome Biol.* 15 (2): R29.

Lee, J.K., Sayers, B.C., Chun, K.S., et al. (2012). Multi-walled carbon nanotubes induce COX-2 and iNOS expression via MAP kinase-dependent and -independent mechanisms in mouse RAW264.7 macrophages. *Part Fibre Toxicol.* 9: 14.

Lee, J.S., Choi, Y.C., Shin, J.H., et al. (2015). Health surveillance study of workers who manufacture multi-walled carbon nanotubes. *Nanotoxicology* 9 (6): 802–811.

Li, B. and Dewey, C.N. (2011). RSEM: accurate transcript quantification from RNA-Seq data with or without a reference genome. *BMC Bioinform.* 12: 323.

Liaw, L., Birk, D.E., Ballas, C.B., Whitsitt, J.S., Davidson, J.M., and Hogan, B.L. (1998). Altered wound healing in mice lacking a functional osteopontin gene (spp1). *J. Clin. Invest.* 101 (7): 1468–1478.

Ligresti, G., Aplin, A.C., Dunn, B.E., Morishita, A., and Nicosia, R.F. (2012). The acute phase reactant orosomucoid-1 is a bimodal regulator of angiogenesis with time- and context-dependent inhibitory and stimulatory properties. *PLoS One* 7 (8): e41387.

Love, M.I., Huber, W., and Anders, S. (2014). Moderated estimation of fold change and dispersion for RNA-seq data with DESeq2. *Genome Biol.* 15 (12): 550.

Ma-Hock, L., Treumann, S., Strauss, V., et al. (2009). Inhalation toxicity of multiwall carbon nanotubes in rats exposed for 3 months. *Toxicol. Sci.* 112 (2): 468–481.

Mauderly, J.L. (1997). Relevance of particle-induced rat lung tumors for assessing lung carcinogenic hazard and human lung cancer risk. *Environ. Health Perspect.* 105 (Suppl 5): 1337–1346.

Mercer, R.R., Scabilloni, J.F., Hubbs, A.F., et al. (2013). Distribution and fibrotic response following inhalation exposure to multi-walled carbon nanotubes. *Part Fibre Toxicol.* 10: 33.

Mizgerd, J.P. (2002). Molecular mechanisms of neutrophil recruitment elicited by bacteria in the lungs. *Semin. Immunol.* 14 (2): 123–132.

Muller, J., Huaux, F., Moreau, N., et al. (2005). Respiratory toxicity of multi-wall carbon nanotubes. *Toxicol. Appl. Pharmacol.* 207 (3): 221–231.

Murphy, F.A., Poland, C.A., Duffin, R., et al. (2011). Length-dependent retention of carbon nanotubes in the pleural space of mice initiates sustained inflammation and progressive fibrosis on the parietal pleura. *Am. J. Pathol.* 178 (6): 2587–2600.

Nair, M.G., Du, Y., Perrigoue, J.G., et al. (2009). Alternatively activated macrophage-derived RELM-{alpha} is a negative regulator of type 2 inflammation in the lung. *J. Exp. Med.* 206 (4): 937–952.

Nakanishi, J., Morimoto, Y., Ogura, I., et al. (2015). Risk assessment of the carbon nanotube group. *Risk Anal.* 35 (10): 1940–1956.

Nakashima, K., Sato, T., Shigemori, S., Shimosato, T., Shinkai, M., and Kaneko, T. (2018). Regulatory role of heme oxygenase-1 in silica-induced lung injury. *Respir. Res.* 19 (1): 144.

NIOSH (National Institute for Occupational Safety and Health). (2013). Current intelligence Bulletin 65: occupational exposure to carbon nanotubes and nanofibers. Cincinnati, Ohio: US Department of Health and Human Services Centers for Disease Control and Prevention, National Institute for Occupational Safety and Health. Publication No. 2013–145. 2013.

NRC (National Research Council). (2007). *Toxicity Testing in the 21st Century: A Vision and a Strategy*. Washington, DC: The National Academies Press.

Ornatowska, M., Azim, A.C., Wang, X., et al. (2007). Functional genomics of silencing TREM-1 on TLR4 signaling in macrophages. *Am. J. Physiol. Lung. Cell Mol. Physiol.* 293 (6): L1377–1384.

Pandya, P.H. and Wilkes, D.S. (2014). Complement system in lung disease. *Am. J. Respir. Cell Mol. Biol.* 51 (4): 467–473.

Pauluhn, J. (2010). Subchronic 13-week inhalation exposure of rats to multiwalled carbon nanotubes: toxic effects are determined by density of agglomerate structures, not fibrillar structures. *Toxicol. Sci.* 113 (1): 226–242.

Perez, L. (2019). Acute phase protein response to viral infection and vaccination. *Arch. Biochem. Biophys.* 671: 196–202.

Perl, A.L., O'Connor, C.M., Fa, P., et al. (2019). Protein phosphatase 2A controls ongoing DNA replication by binding to and regulating cell division cycle 45 (CDC45). *J. Biol. Chem.* 294 (45): 17043–17059.

Pietrofesa, R.A., Velalopoulou, A., Albelda, S.M., and Christofidou-Solomidou, M. (2016). Asbestos induces oxidative stress and activation of Nrf2 signaling in murine macrophages: chemopreventive role of the synthetic lignan secoisolariciresinol diglucoside (LGM2605). *Int. J. Mol. Sci.* 17 (3): 322.

Pilling, D., Cox, N., Vakil, V., Verbeek, J.S., and Gomer, R.H. (2015). The long pentraxin PTX3 promotes fibrocyte differentiation. *PLoS One* 10 (3): e0119709.

Poland, C.A., Duffin, R., Kinloch, I., et al. (2008). Carbon nanotubes introduced into the abdominal cavity of mice show asbestos-like pathogenicity in a pilot study. *Nat. Nanotechnol.* 3 (7): 423–428.

Porter, D.W., Hubbs, A.F., Chen, B.T., et al. (2013). Acute pulmonary dose-responses to inhaled multi-walled carbon nanotubes. *Nanotoxicology* 7 (7): 1179–1194.

Porter, D.W., Hubbs, A.F., Mercer, R.R., et al. (2010). Mouse pulmonary dose- and time course-responses induced by exposure to multi-walled carbon nanotubes. *Toxicology* 269 (2–3): 136–147.

Poulsen, S.S., Jackson, P., Kling, K., et al. (2016). Multi-walled carbon nanotube physicochemical properties predict pulmonary inflammation and genotoxicity. *Nanotoxicology* 10 (9): 1263–1275.

Poulsen, S.S., Saber, A.T., Williams, A., et al. (2015). MWCNT of different physicochemical properties cause similar inflammatory responses, but differences in transcriptional and histological markers of fibrosis in mouse lungs. *Toxicol. Appl. Pharmacol.* 284 (1): 16–32.

Pulskamp, K., Diabate, S., and Krug, H.F. (2007). Carbon nanotubes show no sign of acute toxicity but induce intracellular reactive oxygen species in dependence on contaminants. *Toxicol. Lett.* 168 (1): 58–74.

Risso, D., Schwartz, K., Sherlock, G., and Dudoit, S. (2011). GC-content normalization for RNA-Seq data. *BMC Bioinform.* 12: 480.

Rittinghausen, S., Hackbarth, A., Creutzenberg, O., et al. (2014). The carcinogenic effect of various multi-walled carbon nanotubes (MWCNT) after intraperitoneal injection in rats. *Part Fibre Toxicol.* 11: 59.

Rittling, S.R. and Singh, R. (2015). Osteopontin in immune-mediated diseases. *J. Dent. Res.* 94 (12): 1638–1645.

Robinson, M.D., McCarthy, D.J., and Smyth, G.K. (2010). edgeR: a Bioconductor package for differential expression analysis of digital gene expression data. *Bioinformatics* 26 (1): 139–140.

Roudkenar, M.H., Kuwahara, Y., Baba, T., et al. (2007). Oxidative stress induced lipocalin two gene expression: addressing its expression under the harmful conditions. *J. Radiat. Res.* 48 (1): 39–44.

Ryman-Rasmussen, J.P., Tewksbury, E.W., Moss, O.R., Cesta, M.F., Wong, B.A., and Bonner, J.C. (2009). Inhaled multiwalled carbon nanotubes potentiate airway fibrosis in murine allergic asthma. *Am. J. Respir. Cell Mol. Biol.* 40 (3): 349–358.

Saber, A.T., Jacobsen, N.R., Jackson, P., et al. (2014). Particle-induced pulmonary acute phase response may be the causal link between particle inhalation and cardiovascular disease. *Wiley Interdiscip. Rev. Nanomed. Nanobiotechnol.* 6 (6): 517–531.

Sabo-Attwood, T., Ramos-Nino, M.E., Eugenia-Ariza, M., et al. (2011). Osteopontin modulates inflammation, mucin production, and gene expression signatures after inhalation of asbestos in a murine model of fibrosis. *Am. J. Pathol.* 178 (5): 1975–1985.

Sack, G.H., Jr. (2018). Serum amyloid A – a review. *Mol. Med.* 24 (1): 46.

Sager, T.M., Umbright, C.M., Mustafa, G.M., et al. (2020). Tobacco smoke exposure exacerbated crystalline silica-induced lung toxicity in rats. *Toxicol. Sci.* 178 (2): 375–390.

Sager, T.M., Umbright, C.M., Mustafa, G.M., et al. (2022). Pulmonary toxicity and gene expression changes in response to whole-body inhalation exposure to multi-walled carbon nanotubes in rats. *Inhal. Toxicol.* 34 (7–8): 200–218.

Sakamoto, Y., Nakae, D., Fukumori, N., et al. (2009). Induction of mesothelioma by a single intrascrotal administration of multi-wall carbon nanotube in intact male Fischer 344 rats. *J. Toxicol. Sci.* 34 (1): 65–76.

Sargent, L.M., Porter, D.W., Staska, L.M., et al. (2014). Promotion of lung adenocarcinoma following inhalation exposure to multi-walled carbon nanotubes. *Part Fibre Toxicol.* 11: 3.

Scala, G., Delaval, M.N., Mukherjee, S.P., Federico, A., Khaliullin, T.O., Yanamala, N., Fatkhutdinova, L.M., Kisin, E.R., Greco, D., Fadel, B., and Shvedova, A. (2021). Multi-walled carbon nanotubes elicit concordant changes in DNA methylation and gene expression following long-term pulmonary exposure in mice. *Carbon* 178: 563–572.

Sellamuthu, R., Umbright, C., Li, S., Kashon, M., and Joseph, P. (2011b). Mechanisms of crystalline silica-induced pulmonary toxicity revealed by global gene expression profiling. *Inhal. Toxicol.* 23 (14): 927–937.

Sellamuthu, R., Umbright, C., Roberts, J.R., et al. (2011a). Blood gene expression profiling detects silica exposure and toxicity. *Toxicol. Sci.* 122 (2): 253–264.

Sellamuthu, R., Umbright, C., Roberts, J.R., et al. (2012). Transcriptomics analysis of lungs and peripheral blood of crystalline silica-exposed rats. *Inhal. Toxicol.* 24 (9): 570–579.

Semple, J.W. and Duncker, B.P. (2004). ORC-associated replication factors as biomarkers for cancer. *Biotechnol. Adv.* 22 (8): 621–631.

Sharma, S., Singh, A.K., Kaushik, S., et al. (2013). Lactoperoxidase: structural insights into the function, ligand binding and inhibition. *Int. J. Biochem. Mol. Biol.* 4 (3): 108–128.

Shvedova, A.A., Yanamala, N., Kisin, E.R., Khailullin, T.O., Birch, M.E., and Fatkhutdinova, L.M. (2016). Integrated analysis of dysregulated ncRNA and mRNA expression profiles in humans exposed to carbon nanotubes. *PLoS One* 11 (3): e0150628.

Siegrist, K.J., Reynolds, S.H., Kashon, M.L., et al. (2014). Genotoxicity of multi-walled carbon nanotubes at occupationally relevant doses. *Part Fibre Toxicol.* 11: 6.

Simeonova, P.P. and Erdely, A. (2009). Engineered nanoparticle respiratory exposure and potential risks for cardiovascular toxicity: predictive tests and biomarkers. *Inhal. Toxicol.* 21 (Suppl 1): 68–73.

Snyder-Talkington, B.N., Dong, C., Zhao, X., et al. (2015). Multi-walled carbon nanotube-induced gene expression in vitro: concordance with in vivo studies. *Toxicology* 328: 66–74.

Srivastava, R.K., Pant, A.B., Kashyap, M.P., et al. (2011). Multi-walled carbon nanotubes induce oxidative stress and apoptosis in human lung cancer cell line-A549. *Nanotoxicology* 5 (2): 195–207.

Stark, R., Grzelak, M., and Hadfield, J. (2019). RNA sequencing: the teenage years. *Nat. Rev. Genet.* 20 (11): 631–656.

Stark, W.J., Stoessel, P.R., Wohlleben, W., and Hafner, A. (2015). Industrial applications of nanoparticles. *Chem. Soc. Rev.* 44 (16): 5793–5805.

Stoner, B.R., Brown, B., and Glass, J.T. (2014). Selected topics on the synthesis, properties and applications of multiwalled carbon nanotubes. *Diam. Relat. Mater.* 42: 49–57.

Strunk, R.C., Eidlen, D.M., and Mason, R.J. (1988). Pulmonary alveolar type II epithelial cells synthesize and secrete proteins of the classical and alternative complement pathways. *J. Clin. Invest.* 81 (5): 1419–1426.

Suzui, M., Futakuchi, M., Fukamachi, K., et al. (2016). Multiwalled carbon nanotubes intratracheally instilled into the rat lung induce development of pleural malignant mesothelioma and lung tumors. *Cancer Sci.* 107 (7): 924–935.

Takagi, A., Hirose, A., Futakuchi, M., Tsuda, H., and Kanno, J. (2012). Dose-dependent mesothelioma induction by intraperitoneal administration of multi-wall carbon nanotubes in p53 heterozygous mice. *Cancer Sci.* 103 (8): 1440–1444.

Takagi, A., Hirose, A., Nishimura, T., et al. (2008). Induction of mesothelioma in p53± mouse by intraperitoneal application of multi-wall carbon nanotube. *J. Toxicol. Sci.* 33 (1): 105–116.

Tiwari, N., Marudamuthu, A.S., Tsukasaki, Y., Ikebe, M., Fu, J., and Shetty, S. (2016). p53- and PAI-1-mediated induction of C-X-C chemokines and CXCR2: importance in pulmonary inflammation due to cigarette smoke exposure. *Am. J. Physiol. Lung. Cell Mol. Physiol.* 310 (6): L496–506.

Tomita, Y., Imai, K., Senju, S., et al. (2011). A novel tumor-associated antigen, cell division cycle 45-like can induce cytotoxic T-lymphocytes reactive to tumor cells. *Cancer Sci.* 102 (4): 697–705.

Trapnell, C., Roberts, A., Goff, L., et al. (2012). Differential gene and transcript expression analysis of RNA-seq experiments with TopHat and Cufflinks. *Nat. Protoc.* 7 (3): 562–578.

Turro, E., Su, S.Y., Goncalves, A., Coin, L.J., Richardson, S., and Lewin, A. (2011). Haplotype and isoform specific expression estimation using multi-mapping RNA-seq reads. *Genome Biol.* 12 (2): R13.

Umbright, C., Sellamuthu, R., Li, S., Kashon, M., Luster, M., and Joseph, P. (2010). Blood gene expression markers to detect and distinguish target organ toxicity. *Mol. Cell. Biochem.* 335 (1–2): 223–234.

Umeda, Y., Kasai, T., Saito, M., et al. (2013). Two-week toxicity of multi-walled carbon nanotubes by whole-body inhalation exposure in rats. *J. Toxicol. Pathol.* 26 (2): 131–140.

Vallyathan, V., Shi, X.L., Dalal, N.S., Irr, W., and Castranova, V. (1988). Generation of free radicals from freshly fractured silica dust. Potential role in acute silica-induced lung injury. *Am. Rev. Respir. Dis.* 138 (5): 1213–1219.

Varma, D. and Salmon, E.D. (2012). The KMN protein network – chief conductors of the kinetochore orchestra. *J. Cell. Sci.* 125 (Pt 24): 5927–5936.

Varsano, S., Kaminsky, M., Kaiser, M., and Rashkovsky, L. (2000). Generation of complement C3 and expression of cell membrane complement inhibitory proteins by human bronchial epithelium cell line. *Thorax* 55 (5): 364–369.

Vlaanderen, J., Pronk, A., Rothman, N., et al. (2017). A cross-sectional study of changes in markers of immunological effects and lung health due to exposure to multi-walled carbon nanotubes. *Nanotoxicology* 11 (3): 395–404.

Voelkel, K., Krug, H.F., and Diabate, S. (2003). Formation of reactive oxygen species in rat epithelial cells upon stimulation with fly ash. *J. Biosci.* 28 (1): 51–55.

Volanakis, J.E. (1995). Transcriptional regulation of complement genes. *Annu. Rev. Immunol.* 13: 277–305.

Wagner, G.P., Kin, K., and Lynch, V.J. (2012). Measurement of mRNA abundance using RNA-seq data: RPKM measure is inconsistent among samples. *Theory Biosci.* 131 (4): 281–285.

Wang, L., Stueckle, T.A., Mishra, A., et al. (2014). Neoplastic-like transformation effect of single-walled and multi-walled carbon nanotubes compared to asbestos on human lung small airway epithelial cells. *Nanotoxicology* 8 (5): 485–507.

Yokel, R.A. and MacPhail, R.C. (2011). Engineered nanomaterials: exposures, hazards, and risk prevention. *J. Occup. Med. Toxicol.* 6: 7.

8

Nano-Engineering in Traumatic Brain Injury

Najlaa Al-Thani[1], *Mohammad Z. Haider*[2], *Maryam Al-Mansoob*[3], *Stuti Patel*[4], *Salma M.S. Ahmad*[3], *Firas Kobeissy*[5,*], *and Abdullah Shaito*[6,*]

[1] *Research and Development Department, Barzan Holdings, 7178 Doha, Qatar*
[2] *Department of Basic Medical Sciences, College of Medicine, QU Health, Qatar University, Doha, Qatar*
[3] *Department of Biological and Environmental Sciences, College of Arts and Sciences, Qatar University, Doha, Qatar*
[4] *Department of Biology, University of Florida, Gainesville, Florida, USA*
[5] *Program for Neurotrauma, Neuroproteomics & Biomarkers Research, Departments of Emergency Medicine, Psychiatry, Neuroscience and Chemistry, University of Florida, Gainesville, FL 32611, Florida, USA*
[6] *Biomedical Research Center, College of Medicine, and Department of Biomedical Sciences at College of Health Sciences, Qatar University, Doha, Qatar*
* *Corresponding authors*

Introduction

Traumatic brain injury (TBI) is one of the prime causes of mortality and long-term disability in the world. Those who survive TBI face major socioeconomic consequences including job losses and divorces [1]. The global incidence of TBI is estimated to be 939 per 10,000, with falls, automobile accidents, wars, and sports being the most common cause [2–5]. TBI is more frequent in low- and middle-income nations, with youth and adolescents bearing the brunt of the burden [2–6]. TBI-induced changes can last for months or years after the injury and have a major impact on patients' quality of life. Seizures, sleep difficulties, neurodegenerative illnesses, neuroendocrine dysregulation, and psychological issues are all possible subsequent pathological symptoms caused by a single TBI or repeated assaults [7–10]. TBI severity is determined by the primary injury, which is the most important prognostic factor [11]. TBI cases are categorized as mild, moderate, or severe depending on the degree of the damage. An impact with a penetrating object is the most serious example, such as a gunshot that penetrates the skull. At least 75% of all TBIs recorded in the United States are categorized as minor, mild, or concussions, covering the spectrum of mild TBI (mTBI) [12]. TBI has two phases: (i) the primary phase, which is caused by an external mechanical assault, such as acceleration, deceleration, or rotational forces, and (ii) the secondary phase: a set of pathological events that begin after the initial mechanical impact. The second phase may occur minutes or days after the primary phase and encompass a variety of driving events, including oxidative stress, excitotoxicity, mitochondrial dysfunction, inflammation, edema, and neuronal cell death, all leading to further cerebral damage [13, 14].

When a physical force is applied to the brain, the white matter deforms, resulting in diffuse axonal damage and release of calcium ions (Ca^{2+}) from intracellular stores [15, 16]. Excessive release of excitatory neurotransmitters such as glutamate follows causing an influx of Ca^{2+} into postsynaptic terminals, depolarizing them and causing hypermetabolism, which then progresses to a metabolic depression, which can last several days [17, 18]. Mitochondria are key in the course of TBI pathology where an increase in Ca^{2+} concentration causes excessive Ca^{2+} absorption by mitochondria, resulting in mitochondrial membrane permeabilization, mitochondrial dysfunction, and an enhanced state of oxidative stress illustrated by an increase in reactive oxygen species (ROS) generation [19]. Oxidative stress contributes to the worsening of other TBI-related pathogenic pathways, such as cytoskeletal damage *via* calpain activation and neuroinflammation *via*

Impact of Engineered Nanomaterials in Genomics and Epigenomics, First Edition. Edited by Saura C. Sahu.

glial cell activation [20, 21]. After a brain injury, oxidative stress caused by ROS and free radical generation changes the crucial architecture of tight junction proteins at the blood-brain barrier (BBB) (one of the most important components of a healthy brain, acting as a barrier between the central nervous system and the rest of the body [22]), resulting in increased paracellular leakage [23]. Additionally, astrocytes promote BBB rupture following brain damage by activating paracellular channels, physically disrupting astrocyte-endothelial junctions, and digesting BBB matrix proteins, [24]. TBI-induced disruption of the BBB may be exploited to pass therapeutics through the damaged BBB; however, the magnitude and time frame of BBB disruption varies and is incompletely understood.

TBI treatment has only made gradual advances, and long-term damage following TBI remains a substantial health-care concern. At present, there are no Food and Drug Administration -approved drugs for TBI. The current standard of care for moderate-to-severe TBI includes ventilation and oxygenation measures, fluid management, hypothermia stimulation, intracranial pressure control, cerebral perfusion pressure, blood pressure management, nutrition and glucose level management, and surgery [25]. The complicated pathology that follows a brain injury restricts the possible diagnostic and treatment options. Furthermore, the BBB, which primarily regulates access of materials into the brain, poses as a barrier to the entry of small and macromolecular therapies into the brain [26]. As a result, diagnosis and therapy by systemic or local delivery of medications are mostly ineffective [27]. Mostly, current treatment strategies are limited by two main hurdles: (i) poor delivery and retention of therapeutics, restricting therapeutic thresholds; (ii) off-target toxicity induced by treatments that target biochemical derangement receptors rather than the biochemical derangements themselves, resulting in a loss of function in off-target cells [28].

In recent years, novel strategies are being developed to repair brain tissue injuries. The use of nanoparticles (NPs) has had a major share of research and investigation among these new strategies. NPs range from 1 to 100 nm in size and are typically used for pharmaceutical purposes [29]. The disruption of BBB in TBI allows the delivery of therapeutics *via* NPs, owing to their large surface area, nanoscale size, and unique physical and chemical characteristics. Due to these characteristics, drugs, proteins, vaccines, biological macromolecules (DNA, RNA, carbohydrates, etc.), and other hydrophilic/hydrophobic substances can be potentially delivered into damaged tissues *via* NPs [30].

Nanoparticles in the Treatment of TBI

NPs are a promising therapeutic strategy in the treatment of TBI due to their small size, allowing them to exploit the presence of BBB leaks in patients of TBI [30]. NPs are created using a range of materials that can provide them with unique physiochemical and biological properties [29]. NP synthesis, mechanism of action, and the utilized materials are all relevant matters to consider when employing NPs in TBI therapy. NPs as therapeutic delivery agents face challenges, such as bioavailability, toxicity load, and proinflammatory activity, and these also have to be considered when choosing NPs for therapy. A summary of NP synthesis, materials used, and mechanisms of action is shown in Figure 8.1.

Synthesis of Nanoparticles

NPs are constructed from several materials and consist of varying sizes (1–100 nm) that are a key factor in determining their usability [31]. In the initial stage of their synthesis, NPs are generated

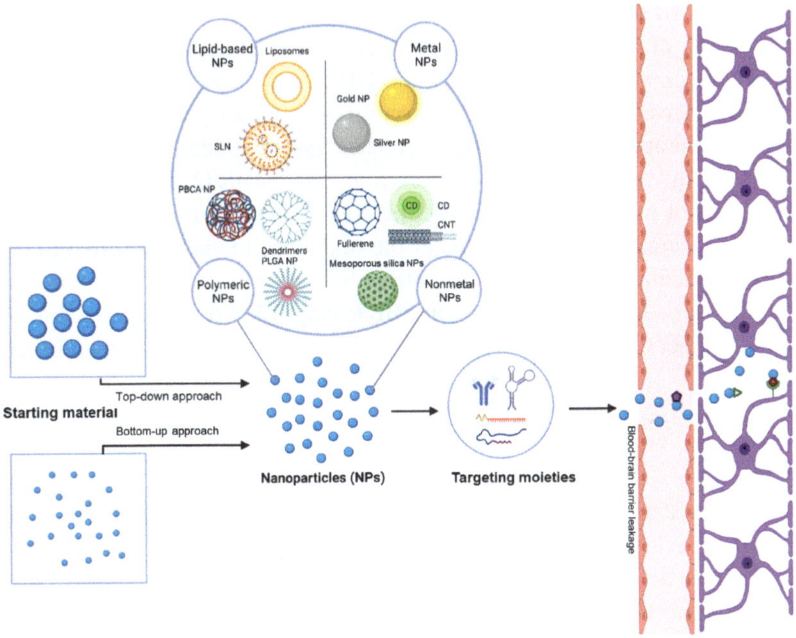

Figure 8.1 Nanoparticle synthesis, materials used, and utilization in TBI. PBCA: poly-butyl cyanoacrylate; PLGA: poly (lactic-co-glycolic acid); CD: carbon dots; CNT: carbon nanotubules.

through two main methods: a top-down approach and a bottom-up approach (Figure 8.1) [32]. The top-down approach creates NPs starting from larger particle-sized material that undergoes a deconstructive process leading to the generation of smaller units, creating NPs of the desired size. These deconstructive processes include mechanical milling, chemical etching, sputtering, laser ablation, and electro-explosion [33]. The bottom-up approach is simply the inverse, where smaller starting material undergoes sedimentation or reduction processes that allow the smaller particles to join together, creating NPs of the desired size. The sedimentation or reduction processes include spinning, template support synthesis, plasma or flame spraying synthesis, laser pyrolysis, chemical vapor-phase deposition, and atomic or molecular condensation [33]. As part of the bottom-up approach, biological vectors such as fungi, bacteria, and plants, by utilizing their biological production or degradation pathways for the synthesis of NPs, can be utilized to produce NPs. In this regard, plants and fungi are favored due to their large-scale production capability and eco-friendly attributes [33–35].

Mechanisms of Action of Nanoparticles in TBI

The two main methods of transport of NPs in the treatment of TBI are active transport and passive transport [36, 37]. Active transport requires energy expenditure as a result of receptor binding between NPs and surface receptor on target cells whereas passive transport does not require energy expenditure as NPs diffusion across a concentration gradient is made easier by the leaky BBB [36]. In addition, it is suspected that passive transport of NPs operates in an enhanced permeability and retention (EPR) effect by which molecules of certain sizes (typically liposomes, nanoparticles, and macromolecular drugs) tend to accumulate in a tumor mass or damaged tissue much more than their accumulation in normal tissues [38, 39]. Studies found that the extent of passive uptake of NPs is dependent on the severity of the disease [40], the degree of BBB breakdown [41], and the activation level of glial cells [42], as well as the size of NPs [31]. However, attachment of targeting moieties to the surface of the NPs will enhance the transport of passively transported NPs [39, 43]. These targeting moieties include aptamers, peptides, antibodies, glycopeptides, and a range of moieties that allow the NPs to bind more readily to the surface of target cells [31, 44–47]. Various studies have started the design of more optimal therapeutic NPs for the treatment of TBI. The newly designed NPs aimed to decrease oxidative stress in the brain and manage ischemia brain damage [43, 48], normalize mitochondrial dysfunction [49], reduce chronic inflammation in the brain [42, 50, 51], clear toxin overload in the brain due to BBB leakage [52], repair damaged neuronal membranes [53], and protect neurons from further damage to axonal conduction and connections [54].

Materials Used for the Synthesis of NPs in TBI Treatment

Materials used in the production of nanoparticles dictate the type of therapeutic cargo it can carry, including small molecule drugs, siRNA, and peptides cargoes [55]. The physicochemical properties of NPs such as their small particle size (<100 nm) and approximately neutral charge are critical to ensure optimal delivery and cellular uptake [56]. The types of NPs used in the treatment of TBI can be classified into four classes of materials: lipid-based NPs, metal NPs, nonmetal NPs, and polymeric NPs [29].

Lipid-Based NPs

Lipid-based NPs, such as liposomes and solid lipid nanoparticles (SLNs), both comprise vesicular lipid bilayers made of amphiphilic lipids [57]. Generally, lipid-based NPs are characteristically spherical and have a diameter of 10–100 nm [57]. Liposomes contain an aqueous center, allowing them to store and transport hydrophilic cargo, such as therapeutic drugs [58], whereas SLNs contain a solid hydrophobic lipid core allowing for small molecule delivery [59]. Altering the surface properties of SLNs with the addition of targeting ligands could improve delivery to the brain and limit reticuloendothelial system uptake, increasing the circulation and bioavailability of SLNs [60]. In the treatment of TBI, lipid-based NPs have been employed to treat cerebral ischemia [61], brain tumors [62], and delivering drugs across the BBB [58]. In addition, modified SLNs were utilized to improve drug delivery across the BBB through the adsorption of SLNs to blood capillaries of the brain [63, 64].

Metal NPs

Metal NPs, mainly gold (Au) and silver (Ag), typically of sizes below 200 nm [65], are initially formed from their elemental metallic atoms before undergoing surface modification to aid in peptide delivery [55]. Sizes of metal NPs used in biological applications should be uniform to maintain consistency of dose delivery [66]. In the case of Au and Ag NPs, they also possess localized surface plasmon resonance (LSPR) properties, leading to special optoelectrical characteristics and a broad absorption band in the visible light region of the electromagnetic spectrum [67]. Optoelectrical properties allow metal NPs to be used as a means of detection in biological tissues [68] or in select laser therapies of cancerous tissue by converting absorbed light into localized heat [69]. Ag NPs also have the added benefit of acting as antimicrobial agents [70]. In TBI, Au and Ag NPs have been researched as potential therapeutic delivery tools. Studies using metal NPs found clear delivery of NPs across the BBB into target regions [71], despite the lack of binding specificity of these

NPs, regardless of the addition of targeting moieties. Another challenge is the accumulation of Au and Ag NPs in healthy endothelial cells of the brain, leading to proinflammatory effects and cytotoxicity, mainly due to their nonbiodegradable nature [72, 73].

Nonmetal NPs

Nonmetal NPs are formed from a range of nonmetallic materials that provide NPs with select qualities. Carbon NPs, such as fullerenes, carbon nanotubules (CNT), and carbon dots (CDs), have created a noteworthy commercial interest due to their versatility and wide range of applications [74–76]. Carbon NPs gain their distinctive semi-conductive abilities from the sp^2 hybridization of each carbon atom, forming pentagonal and hexagonal carbon units [29]. In fullerenes, the carbon units are arranged in a ball-like structure and, depending on the number of carbon atoms, their sizes range from 7.114 nm (C_{60}) to 7.648 nm (C_{70}) [74]. When forming CNTs, single, double, or multiple layers of carbon NPs are wrapped into a cylindrical, tubular structure with a diameter of 1–2 nm [77]. CDs are the most recent carbon NP technology. CDs comprise five hexagonal carbon units surrounding one pentagonal carbon unit, creating a single carbon disc or "dot." CD sizes are typically less than 10 nm, making them particularly useful and capable in the treatment of TBI due to their exceptionally small size [76]. Moreover, additional moieties can be added to improve the function of CDs, such as specific functional groups that provide them with useful features, such as photoluminescence [78] or electroluminescence [79]. In addition, CDs can be used as carriers of small molecule therapeutics that can cross the BBB and deliver TBI treatments [80]. Other nonmetal NPs exist, such as novel ceramic NPs that include mesoporous silica (MS), amorphous, polycrystalline, dense, or hollow ceramic NPs [81]. MS NPs have an average diameter of less than 50 nm [82] and due to their porous nature, have a large surface area that can incorporate drugs and can be modified to attach targeting ligands to their surface [83]. Furthermore, MS NPs are ideal for transporting drugs that are sensitive to changes in pH as they are not affected by fluctuations in pH and can protect biological cargo, such as enzymes, against denaturation caused by pH and temperature changes [82]. In contrast to metal NPs, MS NPs are biodegradable, which increases their suitability in biological applications [84]. Importantly, MS NPs have been used as drug carriers in the treatment of TBI [84, 85].

Polymeric NPs

Polymeric NPs, also known as polymer NPs, encompass all NPs made of organic-based matter [29]. Polymeric NPs come in both spherical and capsular shapes, requiring different modes of the transported cargo [86]. Spherical types, similar to SLNs, consist of a matrix of different polymeric particles with a solid hydrophobic core, transporting small molecular cargoes within the NP [87], or small molecules adsorbed to their outer surface layer [88, 89]. On the other hand, capsular types, similar to liposomes, encapsulate cargo within an aqueous or oil-based core [90]. The first polymeric NP used to deliver therapeutic drug treatments across the BBB consisted of poly-butyl cyanoacrylate (PBCA) polymers [91]. Poly(lactic-co-glycolic acid) (PLGA) NPs are other polymeric NPs shown to be beneficial in TBI treatment. PLGA NPs, containing cerebrolysin as the therapeutic peptide, of varying sizes and different surface proteins were tested on control and brain-injured rats and showed mitigation of brain pathology after TBI [52]. Dendrimers are another class of polymeric NPs which was investigated in TBI treatment. Dendrimers have a symmetrical branch-like morphology around a central core that forms spheroidal structures in water [92]. In TBI, dendrimers were used to deliver therapeutic drugs targeting glial cells through the BBB leaks. This therapy reduced neuroinflammation of the highly active and inflamed glial cells present in TBI [93].

Limitations of the Use of NPs in TBI Therapy

Although the use of NPs in the treatment of TBI is promising, certain risks and challenges are involved, depending on the nature and material of the NPs used. The fundamental complications with the use of any type of NPs in the treatment of TBI include the replicable production of NPs especially in regard to the size, shape, complexity, and the production of monodisperse NPs, crucial factors impacting the effectiveness of therapeutic NPs [94, 95]. In addition, NPs are costly to produce especially compared with more conventional modes of drug delivery, although, more NPs are produced per cycle of production slightly offsetting NP production cost [55]. When using specific types of NPs in TBI therapy, specific challenges arise depending on the type and materials of NPs being used. These challenges include: toxicity load, scale-up, and safety of large-scale particle production processes. Metal NPs pose a significant toxicity risk. Ag NPs can disturb the mitochondrial respiratory chain, leading to increased production of ROS and interruption of ATP synthesis, which may result in heightened DNA damage [70]. In addition, strong interactions between Ag NPs and the cerebral microvasculature have been observed, leading to the onset of proinflammatory pathways, resulting in further BBB disturbance, neuronal degeneration, and astrocyte reactivity [73]. PLGA NPs pose another complication due to their lack of biocompatibility with brain tissue, resulting in proinflammatory responses [96, 97]. Indeed, activated astrocytes were observed by immunohistochemical staining in PLGA NPs-treated TBI brains; however, this may be due to mechanical trauma [97, 98]. In addition, cargoes encapsulated within PLGA NPs may become unstable due to their adsorption to the polymer matrix or their acidification by surrounding acids due to the disintegration of PLGA polymers [99]. In this regard, proton scavengers were proposed as a solution to stabilize cargo within PLGA NPs. Proton scavengers are basic amines, like 1–8-bis-(dimethylamino)naphthalene, which can be added as excipients to NPs therapy [100]. Proton scavengers can prevent the inactivation of cargo peptides *via* the acylation reaction of their amine groups with the ester bonds of PLGA polymers. In addition, they can reduce the amalgamation interaction of the cargo peptide while PEGylation with polyethylene glycol (PEG) prior to encapsulation, inducing reduced immunogenicity of the therapy [101, 102].

Conclusion

To conclude, TBI has led to the death and disability of millions of patients worldwide and remains without an efficient treatment. Basic research has served in the understanding of the mechanisms behind TBI incidence and the onset of chronic neurological damage, and helped develop novel therapeutic strategies, such as utilization of NPs. Usage of NPs in the treatment of TBI is primarily due to their small size (1–100 nm) and distinct and specialized materials (lipid-based, metal, non-metal, and polymeric), which allow NPs to facilitate the repair of cellular damage in TBI patients by targeting various cellular pathways. However, NPs are not hassle-free when applied in TBI therapy, and more efforts should be exerted to solve complications arising from NPs' low bioavailability, toxicity load, and proinflammatory activity in the brain. Nevertheless, NPs remain a remarkable treatment approach for TBI. The use of NPs in TBI has also shed light on the need to explore novel drug-based therapies and delivery methods, which will ultimately help in the improvement and development of future TBI therapies.

References

Norup, A., Kruse, M., Soendergaard, P.L., Rasmussen, K.W., and Biering-Sørensen, F. (2020). Socioeconomic consequences of traumatic brain injury: A Danish nationwide register-based study. *J. Neurotrauma* 37: 2694–2702.

Dewan, M.C., Rattani, A., Gupta, S., Baticulon, R.E., Hung, Y.-C., Punchak, M., Agrawal, A., Adeleye, A.O., Shrime, M.G., and Rubiano, A.M. (2018). Estimating the global incidence of traumatic brain injury. *J. Neurosurg.* 130: 1080–1097.

Peeters, W., van den Brande, R., Polinder, S., Brazinova, A., Steyerberg, E.W., Lingsma, H.F., and Maas, A.I. (2015). Epidemiology of traumatic brain injury in Europe. *Acta Neurochir.* 157: 1683–1696.

Phillips, S. and Woessner, D. (2015). Sports-related traumatic brain injury. *Prim. Care: Clin. Off. Pract.* 42: 243–248.

Wojcik, B.E., Stein, C.R., Bagg, K., Humphrey, R.J., and Orosco, J. (2010). Traumatic brain injury hospitalizations of US army soldiers deployed to Afghanistan and Iraq. *Am. J. Prev. Med.* 38: S108–S116.

Harrison, J.E., Berry, J.G., and Jamieson, L.M. (2012). Head and traumatic brain injuries among Australian youth and young adults July 2000–June 2006. *Brain Injury* 26: 996–1004.

Annegers, J.F., Hauser, W.A., Coan, S.P., and Rocca, W.A. (1998). A population-based study of seizures after traumatic brain injuries. *N. Engl. J. Med.* 338: 20–24.

Alvin, M., Jr, Sureyya, D., Allen, R.W., and James, L.G. (1984). Psychosocial functioning at 1 month after head injury. *Neurosurg.* 14: 393–399.

DeKosky, S.T., Blennow, K., Ikonomovic, M.D., and Gandy, S. (2013). Acute and chronic traumatic encephalopathies: Pathogenesis and biomarkers. *Nat. Rev. Neurol.* 9: 192–200.

Vespa, P.M. (2013). Hormonal dysfunction in neurocritical patients. *Curr. Opin. Crit. Care* 19: 107–112.

Haidar, M.A., Shakkour, Z., Barsa, C., Tabet, M., Mekhjian, S., Darwish, H., Goli, M., Shear, D., Pandya, J.D., and Mechref, Y. (2022). Mitoquinone helps combat the neurological, cognitive, and molecular consequences of open head traumatic brain injury at chronic time point. *Biomedicines* 10: 250.

Disease Control and Prevention. (2003). *National center for injury prevention and control: report to congress on mild traumatic brain injury in the United States: steps to prevent a serious public health problem.* Atlanta, GA: Centers for Disease Control and Prevention.

McKee, A.C. and Daneshvar, D.H. (2015). The neuropathology of traumatic brain injury. *Handb. Clin. Neurol.* 127: 45–66.

Kaur, P. and Sharma, S. (2018). Recent advances in pathophysiology of traumatic brain injury. *Curr. Neuropharmacol.* 16: 1224–1238.

Smith, D.H. and Meaney, D.F. (2000). Axonal damage in traumatic brain injury. *The Neuroscientist* 6: 483–495.

Stirling, D.P., Cummins, K., Wayne Chen, S., and Stys, P. (2014). Axoplasmic reticulum Ca2+ release causes secondary degeneration of spinal axons. *Ann. Neurol.* 75: 220–229.

Arciniegas, D.B. and Silver, J.M. (2006). Pharmacotherapy of posttraumatic cognitive impairments. *Behav. Neurol.* 17: 25–42.

Weil, Z.M., Gaier, K.R., and Karelina, K. (2014). Injury timing alters metabolic, inflammatory and functional outcomes following repeated mild traumatic brain injury. *Neurobiol. Dis.* 70: 108–116.

Fehily, B. and Fitzgerald, M. (2017). Repeated mild traumatic brain injury: Potential mechanisms of damage. *Cell Transplant.* 26: 1131–1155.

Mustafa, A.G., Wang, J.A., Carrico, K.M., and Hall, E.D. (2011). Pharmacological inhibition of lipid peroxidation attenuates calpain-mediated cytoskeletal degradation after traumatic brain injury. *J. Neurochem.* 117: 579–588.

Abdul-Muneer, P., Chandra, N., and Haorah, J. (2015). Interactions of oxidative stress and neurovascular inflammation in the pathogenesis of traumatic brain injury. *Mol. Neurobiol.* 51: 966–979.

Alves, J.L. (2014). Blood–brain barrier and traumatic brain injury. *J. Neurosci. Res.* 92: 141–147.

Shlosberg, D., Benifla, M., Kaufer, D., and Friedman, A. (2010). Blood–brain barrier breakdown as a therapeutic target in traumatic brain injury. *Nat. Rev. Neurol.* 6: 393–403.

Chen, Y. and Swanson, R.A. (2003). Astrocytes and brain injury. *J. Cereb. Blood Flow Metab.* 23: 137–149.

Al-Haj, N., Issa, H., Zein, O.E., Ibeh, S., Reslan, M.A., Yehya, Y., Kobeissy, F., Zibara, K., Eid, A.H., and Shaito, A. (2022). Phytochemicals as micronutrients: what is their therapeutic promise in the management of traumatic brain injury? In: *Role of Micronutrients in Brain Health* W. Mohamed, T. Yamashita, eds), 245–276. Springer.

Xiong, B., Wang, Y., Chen, Y., Xing, S., Liao, Q., Chen, Y., Li, Q., Li, W., and Sun, H. (2021). Strategies for structural modification of small molecules to improve blood–brain barrier penetration: A recent perspective. *J. Med. Chem.* 64: 13152–13173.

Silva, G.A. (2005). Nanotechnology approaches for the regeneration and neuroprotection of the central nervous system. *Surg. Neurol.* 63: 301–306.

Kochanek, P.M., Bramlett, H.M., Dixon, C.E., Dietrich, W.D., Mondello, S., Wang, K.K., Hayes, R.L., Lafrenaye, A., Povlishock, J.T., and Tortella, F.C. (2018). Operation brain trauma therapy: 2016 update. *Mil. Med.* 183: 303–312.

Khan, I., Saeed, K., and Khan, I. (2019). Nanoparticles: Properties, applications and toxicities. *Arab. J. Chem.* 12: 908–931.

Mitchell, M.J., Billingsley, M.M., Haley, R.M., Wechsler, M.E., Peppas, N.A., and Langer, R. (2021). Engineering precision nanoparticles for drug delivery. *Nat. Rev. Drug Discov.* 20: 101–124.

Cruz, L.J., Stammes, M.A., Que, I., van Beek, E.R., Knol-Blankevoort, V.T., Snoeks, T.J., Chan, A., Kaijzel, E.L., and Löwik, C.W. (2016). Effect of PLGA NP size on efficiency to target traumatic brain injury. *J. Control. Release* 223: 31–41.

Wang, Y. and Xia, Y. (2004). Bottom-up and top-down approaches to the synthesis of monodispersed spherical colloids of low melting-point metals. *Nano Lett.* 4: 2047–2050.

Iravani, S. (2011). Green synthesis of metal nanoparticles using plants. *Green Chem.* 13: 2638–2650.

Ahmed, S. and Ikram, S. (2016). Biosynthesis of gold nanoparticles: A green approach. *J. Photochem. Photobiol. B: Biol.* 161: 141–153.

Parveen, K., Banse, V., and Ledwani, L. Green synthesis of nanoparticles: Their advantages and disadvantages. In Proceedings of the AIP conference proceedings, 2016; p. 020048.

Bhaskar, S., Tian, F., Stoeger, T., Kreyling, W., de la Fuente, J.M., Grazú, V., Borm, P., Estrada, G., Ntziachristos, V., and Razansky, D. (2010). Multifunctional nanocarriers for diagnostics, drug delivery and targeted treatment across blood–brain barrier: Perspectives on tracking and neuroimaging. *Part. Fibre Toxicol.* 7: 1–25.

Mikitsh, J.L. and Chacko, A.-M. (2014). Pathways for small molecule delivery to the central nervous system across the blood–brain barrier. *Perspect. Medicin. Chem.* 6: PMC. S13384.

Dunkan, R. (2001). Tumor targeting by enhanced permeability and retention (EPR) effect. *Ann. Oncol.* 9: 39.

Kwon, E.J., Skalak, M., Lo Bu, R., and Bhatia, S.N. (2016). Neuron-targeted nanoparticle for siRNA delivery to traumatic brain injuries. *ACS Nano* 10: 7926–7933.

Bharadwaj, V.N., Lifshitz, J., Adelson, P.D., Kodibagkar, V.D., and Stabenfeldt, S.E. (2016). Temporal assessment of nanoparticle accumulation after experimental brain injury: Effect of particle size. *Sci. Rep.* 6: 1–12.

Bharadwaj, V.N., Rowe, R.K., Harrison, J., Wu, C., Anderson, T.R., Lifshitz, J., Adelson, P.D., Kodibagkar, V.D., and Stabenfeldt, S.E. (2018). Blood–brain barrier disruption dictates nanoparticle accumulation following experimental brain injury. *Nanomed.: Nanotechnol. Biol. Med.* 14: 2155–2166.

Alnasser, Y., Kambhampati, S.P., Nance, E., Rajbhandari, L., Shrestha, S., Venkatesan, A., Kannan, R.M., and Kannan, S. (2018). Preferential and increased uptake of hydroxyl-terminated PAMAM dendrimers by activated microglia in rabbit brain mixed glial culture. *Molecules* 23: 1025.

Singhal, A., Morris, V., Labhasetwar, V., and Ghorpade, A. (2013). Nanoparticle-mediated catalase delivery protects human neurons from oxidative stress. *Cell Death & Disease* 4: e903–e903.

Karatas, H., Aktas, Y., Gursoy-Ozdemir, Y., Bodur, E., Yemisci, M., Caban, S., Vural, A., Pinarbasli, O., Capan, Y., and Fernandez-Megia, E. (2009). A nanomedicine transports a peptide caspase-3 inhibitor across the blood–brain barrier and provides neuroprotection. *J. Neurosci.* 29: 13761–13769.

Zhao, H., Bao, X.-J., Wang, R.-Z., Li, G.-L., Gao, J., Ma, S.-H., Wei, J.-J., Feng, M., Zhao, Y.-J., and Ma, W.-B. (2011). Postacute ischemia vascular endothelial growth factor transfer by transferrin-targeted liposomes attenuates ischemic brain injury after experimental stroke in rats. *Human Gene Therapy* 22: 207–215.

Xu, K., Wang, H., Liu, L., Xu, W., Sheng, J., Fan, W., Yang, -Y.-Y., and Li, L. (2011). Efficacy of CG3R6TAT nanoparticles self-assembled from a novel antimicrobial peptide for the treatment of *Candida albicans* meningitis in rabbits. *Chemotherapy* 57: 417–425.

Tang, X., Liang, Y., Zhu, Y., Xie, C., Yao, A., Chen, L., Jiang, Q., Liu, T., Wang, X., and Qian, Y. (2015). Anti-transferrin receptor-modified amphotericin B-loaded PLA–PEG nanoparticles cure Candidal meningitis and reduce drug toxicity. *Int. J. Nanomedicine* 10: 6227.

Reddy, M.K. and Labhasetwar, V. (2009). Nanoparticle-mediated delivery of superoxide dismutase to the brain: An effective strategy to reduce ischemia-reperfusion injury. *FASEB J.* 23: 1384–1395.

Sharma, A., Liaw, K., Sharma, R., Zhang, Z., Kannan, S., and Kannan, R.M. (2018). Targeting mitochondrial dysfunction and oxidative stress in activated microglia using dendrimer-based therapeutics. *Theranostics* 8: 5529.

Nance, E., Porambo, M., Zhang, F., Mishra, M.K., Buelow, M., Getzenberg, R., Johnston, M., Kannan, R.M., Fatemi, A., and Kannan, S. (2015). Systemic dendrimer-drug treatment of ischemia-induced neonatal white matter injury. *J. Controlled Release* 214: 112–120.

Jiang, Y., Fay, J.M., Poon, C.D., Vinod, N., Zhao, Y., Bullock, K., Qin, S., Manickam, D.S., Yi, X., and Banks, W.A. (2018). Nanoformulation of brain-derived neurotrophic factor with target receptor-triggered-release in the central nervous system. *Adv. Funct. Mater.* 28: 1703982.

Ruozi, B., Belletti, D., Sharma, H.S., Sharma, A., Muresanu, D.F., Mössler, H., Forni, F., Vandelli, M.A., and Tosi, G. (2015). PLGA nanoparticles loaded cerebrolysin: Studies on their preparation and investigation of the effect of storage and serum stability with reference to traumatic brain injury. *Mol. Neurobiol.* 52: 899–912.

Ping, X., Jiang, K., Lee, S.-Y., Cheng, J.-X., and Jin, X. (2014). PEG-PDLLA micelle treatment improves axonal function of the corpus callosum following traumatic brain injury. *J. Neurotrauma* 31: 1172–1179.

Gaudin, A., Yemisci, M., Eroglu, H., Lepetre-Mouelhi, S., Turkoglu, O.F., Dönmez-Demir, B., Caban, S., Sargon, M.F., Garcia-Argote, S., and Pieters, G. (2014). Squalenoyl adenosine nanoparticles provide neuroprotection after stroke and spinal cord injury. *Nat. Nanotechnol.* 9: 1054–1062.

Bony, B.A. and Kievit, F.M. (2019). A role for nanoparticles in treating traumatic brain injury. *Pharmaceutics* 11: 473.

Laurent, S., Forge, D., Port, M., Roch, A., Robic, C., Vander Elst, L., and Muller, R.N. (2008). Magnetic iron oxide nanoparticles: Synthesis, stabilization, vectorization, physicochemical characterizations, and biological applications. *Chem. Rev.* 108: 2064–2110.

Mashaghi, S., Jadidi, T., Koenderink, G., and Mashaghi, A. (2013). Lipid nanotechnology. *Int. J. Mol. Sci.* 14: 4242–4282.

Lai, F., Fadda, A.M., and Sinico, C. (2013). Liposomes for brain delivery. *Expert Opin. Drug Deliv.* 10: 1003–1022.

Pardeshi, C., Rajput, P., Belgamwar, V., Tekade, A., Patil, G., Chaudhary, K., and Sonje, A. (2012). Solid lipid based nanocarriers: An overview/Nanonosači na bazi čvrstih lipida: Pregled. *Acta Pharm.* 62: 433–472.

Blasi, P., Giovagnoli, S., Schoubben, A., Ricci, M., and Rossi, C. (2007). Solid lipid nanoparticles for targeted brain drug delivery. *Adv. Drug Deliv. Rev.* 59: 454–477.

Ishii, T., Asai, T., Oyama, D., Agato, Y., Yasuda, N., Fukuta, T., Shimizu, K., Minamino, T., and Oku, N. (2013). Treatment of cerebral ischemia-reperfusion injury with PEGylated liposomes encapsulating FK506. *FASEB J.* 27: 1362–1370.

Orthmann, A., Zeisig, R., Süss, R., Lorenz, D., Lemm, M., and Fichtner, I. (2012). Treatment of experimental brain metastasis with MTO-liposomes: Impact of fluidity and LRP-targeting on the therapeutic result. *Pharm. Res.* 29: 1949–1959.

Wang, J.-X., Sun, X., and Zhang, Z.-R. (2002). Enhanced brain targeting by synthesis of 3′, 5′-dioctanoyl-5-fluoro-2′-deoxyuridine and incorporation into solid lipid nanoparticles. *Eur. J. Pharm. Biopharm.* 54: 285–290.

Martins, S., Tho, I., Reimold, I., Fricker, G., Souto, E., Ferreira, D., and Brandl, M. (2012). Brain delivery of camptothecin by means of solid lipid nanoparticles: Formulation design, in vitro and in vivo studies. *Int. J. Pharm.* 439: 49–62.

Villalobos-Noriega, J.M.A., Rodríguez-León, E., Rodríguez-Beas, C., Larios-Rodríguez, E., Plascencia-Jatomea, M., Martínez-Higuera, A., Acuña-Campa, H., García-Galaz, A., Mora-Monroy, R., Alvarez-Cirerol, F.J. et al. (2021). Au@Ag Core@Shell nanoparticles synthesized with *Rumex hymenosepalus* as antimicrobial agent. *Nanoscale Res. Lett.* 16. doi: 10.1186/s11671-021-03572-5.

Johnston, S.T., Faria, M., and Crampin, E.J. (2018). An analytical approach for quantifying the influence of nanoparticle polydispersity on cellular delivered dose. *J. R. Soc. Interface* 15: 20180364. doi: 10.1098/rsif.2018.0364.

Dreaden, E.C., Alkilany, A.M., Huang, X., Murphy, C.J., and El-Sayed, M.A. (2012). The golden age: Gold nanoparticles for biomedicine. *Chem. Soc. Rev.* 41: 2740–2779. doi: 10.1039/c1cs15237h.

Khlebtsov, N.G. and Dykman, L.A. (2010). Optical properties and biomedical applications of plasmonic nanoparticles. *J. Quant. Spectrosc. Radiat. Transf.* 111: 1–35.

Lim, -Z.-Z.J., Li, J.-E.J., Ng, C.-T., Yung, L.-Y.L., and Bay, B.-H. (2011). Gold nanoparticles in cancer therapy. *Acta Pharmacol. Sin.* 32: 983–990.

AshaRani, P., Low Kah Mun, G., Hande, M.P., and Valiyaveettil, S. (2009). Cytotoxicity and genotoxicity of silver nanoparticles in human cells. *ACS Nano* 3: 279–290.

Cheng, Y., Dai, Q., Morshed, R.A., Fan, X., Wegscheid, M.L., Wainwright, D.A., Han, Y., Zhang, L., Auffinger, B., and Tobias, A.L. (2014). Blood–brain barrier permeable gold nanoparticles: an efficient delivery platform for enhanced malignant glioma therapy and imaging. *Small* 10: 5137–5150.

Tee, J.K., Setyawati, M.I., Peng, F., Leong, D.T., and Ho, H.K. (2019). Angiopoietin-1 accelerates restoration of endothelial cell barrier integrity from nanoparticle-induced leakiness. *Nanotoxicology* 13: 682–700.

Trickler, W.J., Lantz, S.M., Murdock, R.C., Schrand, A.M., Robinson, B.L., Newport, G.D., Schlager, J.J., Oldenburg, S.J., Paule, M.G., and Slikker, W., Jr. (2010). Silver nanoparticle induced blood–brain barrier inflammation and increased permeability in primary rat brain microvessel endothelial cells. *Toxicol. Sci.* 118: 160–170.

Astefanei, A., Núñez, O., and Galceran, M.T. (2015). Characterisation and determination of fullerenes: A critical review. *Anal. Chim. Acta* 882: 1–21.

Aqel, A., Abou El-Nour, K.M., Ammar, R.A., and Al-Warthan, A. (2012). Carbon nanotubes, science and technology part (I) structure, synthesis and characterisation. *Arab. J. Chem.* 5: 1–23.

Wang, X., Feng, Y., Dong, P., and Huang, J. (2019). A mini review on carbon quantum dots: Preparation, properties, and electrocatalytic application. *Front. Chem.* 7: 671.

Ibrahim, K.S. (2013). Carbon nanotubes – properties and applications: A review. *Carbon Lett.* 14: 131–144.

Zhang, Q., Sun, X., Ruan, H., Yin, K., and Li, H. (2017). Production of yellow-emitting carbon quantum dots from fullerene carbon soot. *Sci. China Mat.* 60: 141–150.

Zhang, X., Zhang, Y., Wang, Y., Kalytchuk, S., Kershaw, S.V., Wang, Y., Wang, P., Zhang, T., Zhao, Y., and Zhang, H. (2013). Color-switchable electroluminescence of carbon dot light-emitting diodes. *ACS Nano* 7: 11234–11241.

Li, S., Amat, D., Peng, Z., Vanni, S., Raskin, S., De Angulo, G., Othman, A.M., Graham, R.M., and Leblanc, R.M. (2016). Transferrin conjugated nontoxic carbon dots for doxorubicin delivery to target pediatric brain tumor cells. *Nanoscale* 8: 16662–16669.

Sigmund, W., Yuh, J., Park, H., Maneeratana, V., Pyrgiotakis, G., Daga, A., Taylor, J., and Nino, J.C. (2006). Processing and structure relationships in electrospinning of ceramic fiber systems. *J. Am. Ceram. Soc.* 89: 395–407.

Carvalho, A., Fernandes, A.R., and Baptista, P.V. (2019). Nanoparticles as delivery systems in cancer therapy: focus on gold nanoparticles and drugs. In: *Applications of Targeted Nano Drugs and Delivery Systems (S. Mohapatra, S. Ranjan, N. Dasgupta, R. Mishra, S. Thomas, eds.)*, 257–295. Elsevier.

Kleinstreuer, C., Childress, E., and Kennedy, A. (2013). Chapter 10 - Targeted drug delivery: multifunctional nanoparticles and direct micro-drug delivery to tumors. In: *Transport in Biological Media* (ed. S.M. Becker and A.V. Kuznetsov), 391–416. Boston: Elsevier. https://www.sciencedirect.com/science/article/pii/B9780124158245000102.

Mann, A.P., Scodeller, P., Hussain, S., Joo, J., Kwon, E., Braun, G.B., Mölder, T., She, Z.-G., Kotamraju, V.R., Ranscht, B. et al. (2016). A peptide for targeted, systemic delivery of imaging and therapeutic compounds into acute brain injuries. *Nat. Commun.* 7: 11980. doi: 10.1038/ncomms11980.

Mo, J., He, L., Ma, B., and Chen, T. (2016). Tailoring particle size of mesoporous silica nanosystem to antagonize glioblastoma and overcome blood–brain barrier. *ACS Appl. Mater. Interfaces* 8: 6811–6825.

Mansha, M., Khan, I., Ullah, N., and Qurashi, A. (2017). Synthesis, characterization and visible-light-driven photoelectrochemical hydrogen evolution reaction of carbazole-containing conjugated polymers. *Int. J. Hydrog. Energy* 42: 10952–10961.

Khalil, I., Burns, A., Radecka, I., Kowalczuk, M., Khalaf, T., Adamus, G., Johnston, B., and Khechara, M. (2017). Bacterial-derived polymer poly-y-glutamic acid (y-PGA)-based micro/nanoparticles as a delivery system for antimicrobials and other biomedical applications. *Int. J. Mol. Sci.* 18: 313. doi: 10.3390/ijms18020313.

Abd Ellah, N.H. and Abouelmagd, S.A. (2017). Surface functionalization of polymeric nanoparticles for tumor drug delivery: Approaches and challenges. *Expert Opin. Drug Deliv.* 14: 201–214.

Abouelmagd, S.A., Meng, F., Kim, B.-K., Hyun, H., and Yeo, Y. (2016). Tannic acid-mediated surface functionalization of polymeric nanoparticles. *ACS Biomater. Sci. Eng.* 2: 2294–2303.

Rao, J.P. and Geckeler, K.E. (2011). Polymer nanoparticles: Preparation techniques and size-control parameters. *Prog. Polym. Sci.* 36: 887–913.

Kreuter, J., Alyautdin, R.N., Kharkevich, D.A., and Ivanov, A.A. (1995). Passage of peptides through the blood–brain barrier with colloidal polymer particles (nanoparticles). *Brain Res.* 674: 171–174.

Gillies, E.R. and Frechet, J.M. (2005). Dendrimers and dendritic polymers in drug delivery. *Drug Discov. Today* 10: 35–43.

Nance, E., Zhang, F., Mishra, M.K., Zhang, Z., Kambhampati, S.P., Kannan, R.M., and Kannan, S. (2016). Nanoscale effects in dendrimer-mediated targeting of neuroinflammation. *Biomaterials* 101: 96–107.

Ali, A., Zafar, H., Zia, M., Ul Haq, I., Phull, A.R., Ali, J.S., and Hussain, A. (2016). Synthesis, characterization, applications, and challenges of iron oxide nanoparticles. *Nanotechnol. Sci. Appl.* 9: 49–67. doi: 10.2147/nsa.s99986.

Lee, J.-H., Huh, Y.-M., Jun, Y.-W., Seo, J.-W., Jang, J.-T., Song, H.-T., Kim, S., Cho, E.-J., Yoon, H.-G., and Suh, J.-S. (2007). Artificially engineered magnetic nanoparticles for ultra-sensitive molecular imaging. *Nat. Med.* 13: 95–99.

Fournier, E., Passirani, C., Montero-Menei, C., and Benoit, J. (2003). Biocompatibility of implantable synthetic polymeric drug carriers: Focus on brain biocompatibility. *Biomaterials* 24: 3311–3331.

Lampe, K.J., Kern, D.S., Mahoney, M.J., and Bjugstad, K.B. (2011). The administration of BDNF and GDNF to the brain via PLGA microparticles patterned within a degradable PEG-based hydrogel: Protein distribution and the glial response. *J. Biomed. Mater. Res. Part A* 96: 595–607.

Emerich, D.F., Tracy, M.A., Ward, K.L., Figueiredo, M., Qian, R., Henschel, C., and Bartus, R.T. (1999). Biocompatibility of poly (DL-lactide-co-glycolide) microspheres implanted into the brain. *Cell Transplant.* 8: 47–58.

Ng, S.Y. and Lee, A.Y.W. (2019). Traumatic brain injuries: Pathophysiology and potential therapeutic targets. *Front. Cell. Neurosci.* 13: 528.

Houchin, M., Neuenswander, S., and Topp, E. (2007). Effect of excipients on PLGA film degradation and the stability of an incorporated peptide. *J.Control. Release* 117: 413–420.

Na, D.H. and DeLuca, P.P. (2005). PEGylation of octreotide: I. Separation of positional isomers and stability against acylation by poly (D, L-lactide-co-glycolide). *Pharm. Res.* 22: 736–742.

Domb, A.J., Turovsky, L., and Nudelman, R. (1994). Chemical interactions between drugs containing reactive amines with hydrolyzable insoluble biopolymers in aqueous solutions. *Pharm. Res.* 11: 865–868.

9

Application of Nanoemulsions in Food Industries

Recent Progress, Challenges, and Opportunities

*Ramesh Chaudhari, Vishva Patel, and Ashutosh Kumar**

Biological and Life Sciences, School of Arts and Sciences, Ahmedabad University, Central Campus, Navrangpura, Ahmedabad, Gujarat, India
Corresponding author

Introduction

Recent advances in nanotechnology have increased its usage in various new fields, among which the food industry is growing very rapidly. Nanomaterials are superior and indispensable in many fields of human activity due to their size-dependent characteristics (Chaudhari et al. 2021). There are several nano-enabled consumer options across engineering, medicine, agriculture, food industries, and biotechnology (Kumar et al. 2017). The concept of functional food has been introduced more than a decade ago. Still, the persistent challenges were increasing production and maintaining the nutritional value for health and its safety (Singh et al. 2017a). Nanostructured materials offer unique physicochemical features that allow for the development of novel, high-performance materials of high impact on food production, packaging, and storage (Ranjan et al. 2014). The major challenge for the food industry was to maintain the stability of such food with active ingredients and nutritional value (Meghani et al. 2020). Many scientist and researcher are working in the nano food sectors to improve the quality of the food or to increase the longevity of the nutritional value of food (Dhawan et al. 2018). The introduction of nanotechnology brought some solutions to this field. With the current scenario and demand of the food industry, one of the possible solutions is the use of nanoemulsions (Sharma et al. 2017).

Nanoemulsions are a metastable dispersion resulting from isotropic, thermodynamic stable components being immiscible with each other (Mansuri et al. 2022). A simple example is oil and water. Nanoemulsions are "fluid colloidal systems in which liquid droplets and/or liquid crystals are disseminated in a liquid," according to the International Union of Pure and Applied Chemistry (Salem and Ezzat 2018b). Various products have been developed to execute this system, like surfactants, emulsifiers, etc., to build a more stable component that can hold it for a more extended period. Primarily, there are three types of emulsions and all three differ in size based on stability, viscosity, appearance, solubility, hydrophilic–lipophilic balance (HLB) value, surface area, etc. (Fryd and Mason 2012).

Current developments in nanoemulsions have opened various applications in fields like pharmaceuticals, food, cosmetics, etc., allowing them to be used from small- to large-scale industries (Figure 9.1). Exceptionally controlled nanoemulsions can fill in as a superb model framework for exploring fundamental logical inquiries concerning soft matter. These enhancements demonstrate the significant advantages that nanoemulsions have over their microscale emulsion competitor (Fryd and Mason 2012). Improvements in formulation, characterization, and emulsification can

Impact of Engineered Nanomaterials in Genomics and Epigenomics, First Edition. Edited by Saura C. Sahu.

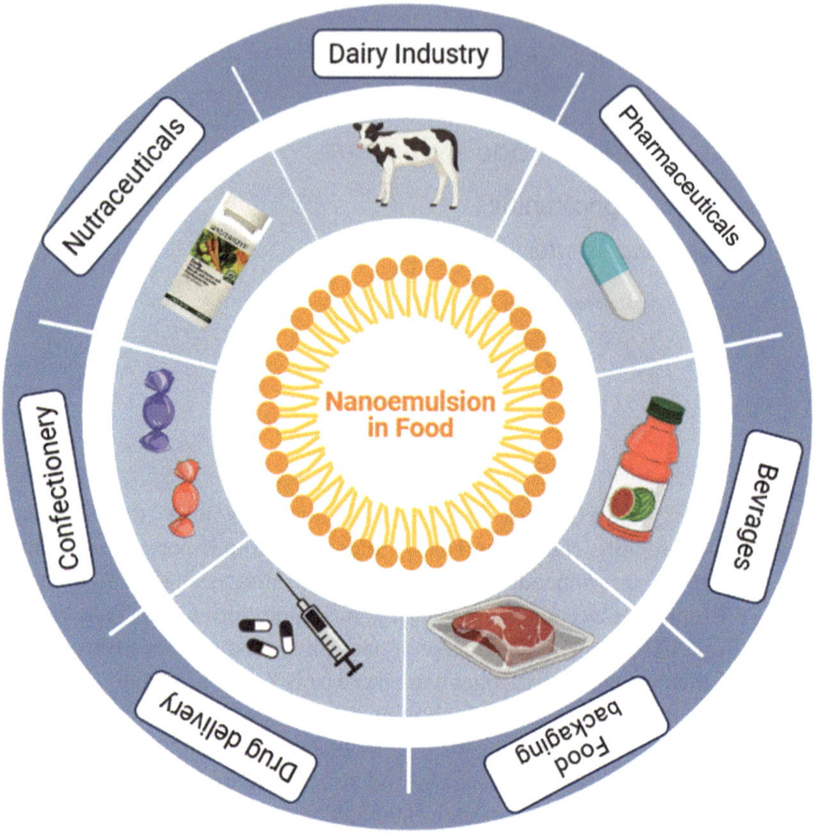

Figure 9.1 Applications of nanoemulsions in various industries ranging from a small- to a large scale.

ascend to the groundbreaking formation of nanoemulsions that are well controlled and have desirable morphological and physiological properties (Sugumar et al. 2015). Nanoemulsions' transport, mechanical, and optical capabilities can be manipulated and modified in conjunction with attractive materials, including coatings, biomolecules, and chemical additives (which can be joined simultaneously). These advancements allow for a superior nanoemulsion structure with more appealing layouts and features (Fryd and Mason 2012).

Nanoemulsions are gaining high importance because of their high kinetic stability. The degree of attraction force operating between drops decreases as droplet size increases, yet the degree of steric repugnance is unaffected by molecule size growth, making nanoemulsions immune to bead coagulation (Gupta et al. 2019). The small droplet size also prevents any fragile flocculation (small clusters) of the drops, resulting in a dispersed structure with no detachment.

Cardinal characteristics, like no coalescence, creaming, flocculation, and sedimentation, give a unique build-up to the nanoemulsion, often described as "approaching thermodynamic stability" (Gupta et al. 2019). The use of nanoemulsion is immense. However, it still has not reached a certain level where people prefer having it in their food. Various nanoemulsions have been tested for potential extraordinary ways to carry different functional foods (Assadpour and Mahdi Jafari 2019). Lipids, fats, essential vitamins, minerals, etc. are incorporated with the initiation of using biodegradable components in nanoemulsion; the side effects or minor hazardous problems are mainly minimized, which is responsible for the major function of nanoemulsions in the food business (Martínez-Ballesta et al. 2018).

Nanoemulsions are of various kinds, but the ones used in food industries are often referred to as food-grade nanoemulsions. They are synthesized using an emulsifier, ripening retarder, weighting agent, or texture modifier, all of which serve a specific role in the food industry, such as synthesizing nanoemulsion or increasing the viscosity, etc. (Martínez-Ballesta et al. 2018). The role of each component used in nanoemulsion is based on the application of the nanoemulsion (Espitia et al. 2019). For example, suppose the nanoemulsion needs to be of high viscosity. To synthesize a nanoemulsion of high viscosity, texture modifiers such as whey protein or gelatin or sugars, or polysaccharides are used, whereas, for the synthesis of balanced nanoemulsions for specific delivery target, lipophilic materials, such as ester gums, are used (Gurpreet and Singh 2018).

The focus of this chapter is to discuss the various applications of nanoemulsions in the food industry, as well as the regulations and safety issues related to this emerging technology. The synthesis process of nanoemulsions and the techniques used to develop nanoemulsions has also been discussed briefly.

Components of Nanoemulsions

A typical nanoemulsion contains an oil phase, water phase, and emulsifier as the main components (Figure 9.2) (Porras et al. 2008), and the physicochemical properties of these constituents determine the formation as well as the stabilization of the nanoemulsion. The major application in commercial products is oil-in-water nanoemulsion as they are very effective in encapsulating lipophilic content, e.g., Vitamin D, Vitamin E, Diclofenac, Paclitaxel (anticancer drug), etc. (Meghani et al. 2018). Oil-in-water nanoemulsions provides sustained release and can be converted into stealth nanoemulsion by PEG coating that makes preferable choice for drug delivery (Anarjan and Tan 2013; McClements and Rao 2011; Singh et al. 2017b).

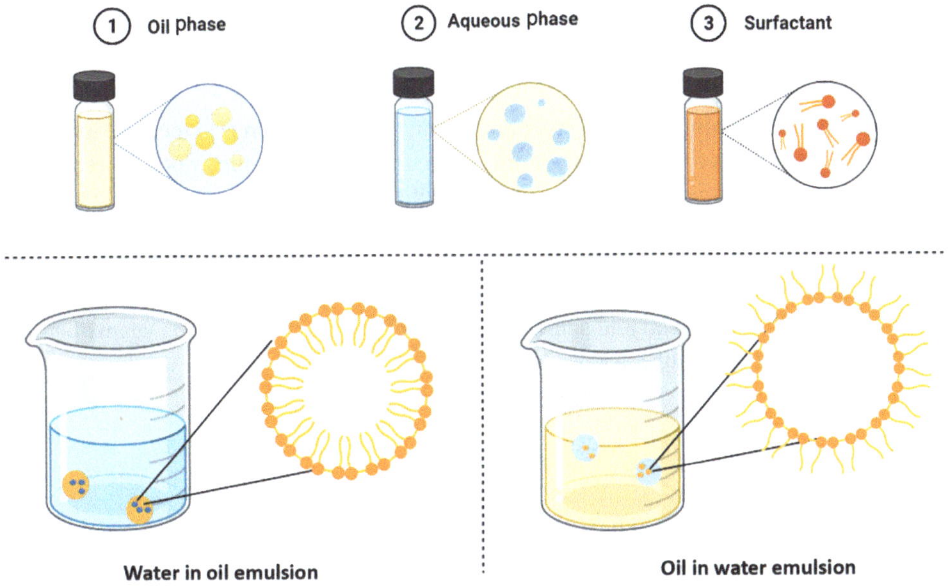

Figure 9.2 Ingredients and components of emulsion.

Oil Phase

Oil phase is useful in the production of food-grade nanoemulsions. Oil holds a vital property as it is considered the foundation of nanoemulsion. Various types of oils are found in the herbal sector, providing us ample opportunities to blend the nanoemulsion. Due to its vivacious property, individuals should precisely understand why and which oil to use and for what characteristics. Various characteristics, such as viscosity, water solubility, density, and polarity, influence the properties of nanoemulsions (Salem and Ezzat 2018a). Because of their nutritional advantages and low cost, monoacylglycerols, free fatty acids, waxes, minerals, and other substances are often used in nanoemulsions. Triacylglycerol from soybean, maize, flaxseed, olive, and fish is also commonly used (McClements and Rao 2011). A range of bioactive lipophilic chemicals are utilized to substitute antibiotics or flavoring agents in the food industry. These chemicals are utilized in the agricultural and pharmaceutical industries as hydrophobic pesticides and medicines. Nanoemulsions can be utilized as a colloidal delivery agent for these chemicals (Kong et al. 2011; Sonneville-Aubrun et al. 2004). Because of their unique qualities, such as a small droplet diameter (20–200 nm), better stability, excellent optical clarity, and improved bioavailability, nanoemulsions may be helpful (Meghani et al. 2018; Tadros et al. 2004).

Aqueous Phase

The selection of aqueous phase is crucial in terms of nanoemulsion stability and storage time. Maintaining the aqueous phase's pH, salt, and other current additive concentrations is beneficial; otherwise, they can impact the continuous phase's viscosity and oil droplet coating, reducing particle aggregation and shelf life of a nanoemulsion (Dasgupta et al. 2019a).

Polar molecules like carbohydrates, acids, minerals, water, etc., are used in the aqueous phase to prepare food-grade nanoemulsions. Water acts as a supporting tool with the oil as it blends the emulsion in a particular way (Polychniatou and Tzia 2014). Size and shape are the characteristics that help in building a proper emulsion. A brief crude emulsion can be made by mixing oil and water, which will split into two distinct phases after standing due to the coalescence of the scattered globules. Emulgents or emulsifying agents can impart stability to such systems (Jaiswal et al. 2015).

Stabilizers

The major objective to add stabilizers in the preparation of a nanoemulsion is to increase its stability for a long time. Stabilizers are commonly found in the form of emulsifiers, ripening retarders, and weighing agents (McClements 2015). Various emulsifiers have also been added to the emulsion, which aids in increasing the kinetic energy; as such, the products are amphiphilic proteins and phospholipids. Ripening retarders like mineral oil, long-chain triglyceride, and ester gum help stabilize the nanoemulsion by inhibiting the process of the Ostwald ripening (McClements and Rao 2011). Brominated vegetable oil and sucrose acetate isobutyrate like weighing agents help in balancing the densities of the emulsion (McClements 2011).

Approaches for Nanoemulsion Production

Nanoemulsion formulation strategies have been classified into high- and low-energy approaches. Effect on the stability and the droplet size is primarily affected by this approach (Sutradhar and Amin 2013). In order to disrupt oil and water phases, strong disruptive forces produced by mechanical devices are undertaken, which ultimately allow for the formation of oil droplets. Instantaneous production of the tiny droplets during the change in the environmental conditions results from the low-energy approach (Salem and Ezzat 2018a). The different mechanisms used in the formation of nanoemulsions is presented in Figure 9.3.

High-Energy Approaches

The particle size generated by high-energy techniques is governed by two competing processes within the homogenizer, namely droplet disruption and droplet coalescence. Disruptive forces generated by mechanical devices can produce tiny droplets that are required to form nanoemulsions. The devices encompass microfluidizers, ultrasonic devices, and high-pressure valve homogenizers.

High-energy approaches can produce food-grade nanoemulsions as they can be used with various emulsifiers and oil types. Only if the homogenization conditions are properly tuned, nanoemulsions can be created using triacylglycerol oil, essential oil, and flavor oil as the oil phase and surfactants, proteins, phospholipids, and polysaccharides as emulsifiers. Production of the tiny droplets can quickly be done simply by having an oil phase of a low interfacial tension or viscosity compared with having a high interfacial tension or high viscosity.

High-Pressure Valve Homogenizer

In the food business, the high-pressure valve homogenizer is the most frequently adopted method of generating traditional emulsions with tiny droplet sizes. They are more successful in reducing the size of droplets than emulsions made from two different liquids. A machine is made up of a piston pump and a narrow gap with a working pressure of up to 150 MPa. Excessive turbulence causes droplets to break down in the valve gap, causing the flow to become elongational and then turbulent. This method allows for scalable nanoemulsion creation, which is advantageous for

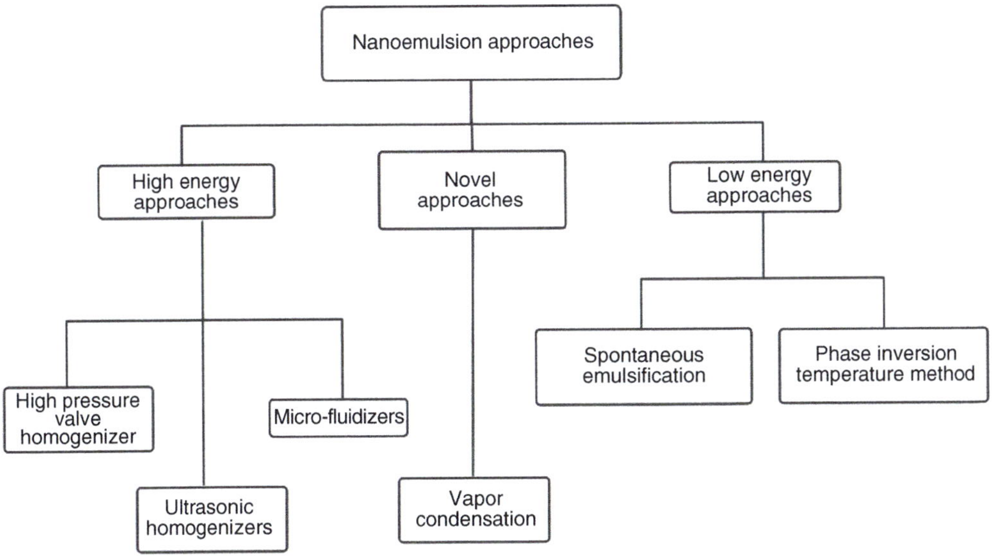

Figure 9.3 Classification of various approaches and techniques employed for the formation of nanoemulsion.

large-scale manufacturing. This method might be valuable in the food industry (Lee et al. 2014; Ranjan et al. 2016).

Microfluidizers

In a higher version of high-pressure homogenizers, microfluidizers are specially used for formulating emulsions. A narrow orifice is used, which helps in forcing a premix emulsion to undergo droplet disruption. Microfluidizers are mainly used to produce emulsion-based products in the pharmaceutical industry, nutraceutical emulsions and flavored emulsions in the beverage and food industry, and homogenized milk in the food industry (Ocampo-Salinas et al. 2016). Microfluidizers are more suitable for the pharma industry rather than the food industry. The food industry requires a high production rate that microfluidizers fail to provide (Lee et al. 2014).

Ultrasonic Homogenizers

Ultrasonic homogenizers convert electrical energy into acoustic energy and form vocal waves, which shear the droplet into fine droplets. This method helps to decrease droplet size just by using high-intensity waves. According to the frequency level, this method has two subgroups. One is a high-power ultrasound, which works on a low frequency and high intensity, and the second is a low-power ultrasound that works on a very high frequency. Ultrasonic methods can help sterilize the environment due to its high temperature and pressure (Ahari and Nasiri 2021).

To produce powerful, disruptive forces, high-intensity ultrasonic waves are employed. These forces are required to break down water and oil droplets into microscopic particles (McClements and Rao 2011). Sonicator probes incorporating piezoelectric quartz crystals provide the energy input. An oscillating electric voltage causes this to change. However, sonication may cause protein denaturation, polysaccharide depolymerization, or lipid oxidation in homogenization, which is a worry with this approach (McClements and Rao 2011).

Low-Energy Approaches

In the generation of emulsions utilizing low-energy techniques, spontaneous synthesis of oil droplets happens when either the content or the microenvironment is changed. For producing nanoemulsions based on low-energy procedures, spontaneous emulsification and phase inversion temperature methods are utilized. This method is more successful, but the kinds of oil and emulsifying agents that may be employed are more limited. Proteins and polysaccharides, for example, cannot be employed as emulsifiers in the preparation of nanoemulsions (McClements and Rao 2011).

Spontaneous Emulsification

Nanoemulsion formation occurs spontaneously when the organic and the aqueous phases are mixed. The strategies for spontaneous emulsification include solute diffusion between two phases, surface tension gradients, condensation or dispersion mechanisms, and interfacial turbulence (McClements and Rao 2011) The droplet size can be controlled by maintaining the mixing conditions and the composition of the two initial phases. No studies have shown where protein and polysaccharides have been used for spontaneous emulsification due to the physicochemical limitations (Shaker et al. 2019). Oil droplets need to be formed and immediately bud off from the interface.

Phase Inversion Temperature Method

The above-stated method can form varying temperature–time of different quantities of oil, water, and surfactant nanoemulsions. The breaking up of microemulsions leads to nanoemulsion formation, which leads to a more kinetically stable and irreversible formulation. The transformation of water-in-oil nanoemulsion to oil-in-water nanoemulsion or vice versa through a bicontinuous microemulsion phase is also used in this method. This process helps prevent the degradation of the encapsulated drug during processing, low-energy consumption, and an easy industrial scale-up (Salem and Ezzat 2018a).

Novel Approach for the Production of Nanoemulsion

High-energy methods involve breaking down a large droplet into finer nanodroplets; large droplets affect the adsorption rate on the droplet. As a result, it is challenging to maintain nanodroplets before they fuse. Sometimes low-energy methods (microfluidics) cannot produce nanoemulsions and are susceptible to blockage. Novel approaches are meant to eliminate the drawbacks of traditional methods and provide scalable and stable formations of nanoemulsion with low energy requirements (Kang et al. 2018).

Synthesizing Pickering Nanoemulsion by Vapor Condensation

Water vapor condensation on the composition of oil and organic surfactant can form nanoemulsions. It is a single-step process with minimal energy requirement. Kinetics of condensation increase the synthesis rate of nanoparticles, which accelerates nanoemulsion production. Water-in-oil nanoemulsions are created by condensing water vapor on a subcooled oil–surfactant solution. At the oil–air interface, water droplets form and diffuse freely inside the oil. This happens as a result of the oil–surfactant concentration impacting the oil-on-water spreading dynamics (Kang et al. 2018). The size of the nanoemulsions created using the condensation method is 100 nm. As an oil bath is placed in a humid environment with a high surfactant concentration, water condensation forms on the soil surface when the temperature falls below the dew point, resulting in the formation of nanoemulsions. Oil coats the surface of water molecules, preventing droplet diffusion. Nanoparticles in these oils prevent the droplets from coalescing. Even at low nanoparticle concentrations, this approach produces a precisely tailored nanoemulsion (Aswathanarayan and Vittal 2019).

Applications of Food-Grade Nanoemulsions

Food-grade nanoemulsion differs from other nanoemulsions by the components and the method for synthesis. Most of the food-grade nanoemulsions reported are known to be synthesized using a microfluidizer; one of the best examples of this technology is lemongrass essential oil. The benefits of this technology are predicted to have a significant influence on the nanotechnology business (Neethirajan and Jayas 2011; Roco 2001). Nanotechnology has already had a substantial influence on several sectors of food and related businesses, as evidenced by multiple papers, patent filings, and assessments (Neethirajan and Jayas 2011). Below are some of the most important nanoemulsion uses (Figure 9.4).

Encapsulation of Lipophilic Functional Food

Nanoencapsulation is a technique for packing substances using multiple approaches to provide product functionality that includes a controlled release of the main ingredient. The bioavailability of lipophilic substances encapsulated within them can be increased with the help of nanoemulsions.

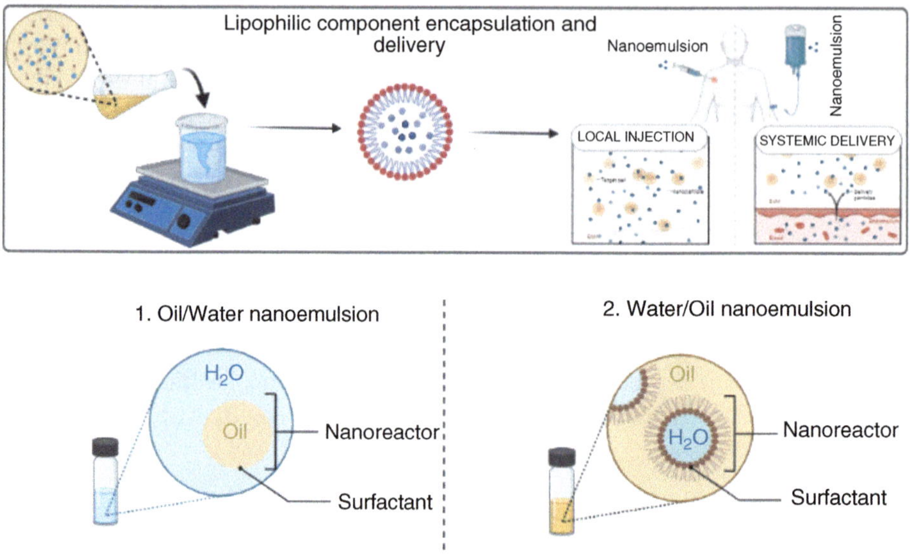

Figure 9.4 Formulation of oil-in-water nanoemulsion for local and systemic delivery of lipophilic content.

The use of nanoemulsions in food, medicines, and beverages has grown as a result of their benefits over traditional emulsions in a variety of applications. One of the major applications of nanoemulsion is that it leaves a pungent smell and taste in food by incorporating small molecules in the core (McClements and Rao 2011). Similarly, vitamins, proteins, antioxidants, lipids, carbohydrates, and other compounds have been encapsulated in order to improve their usefulness and stability. Researchers have also created a proprietary process that allows bioactive and active chemicals in nutraceutical goods to be nano-encapsulated. After their constituents have been supplied, these nanocapsules tend to break down and are absorbed like regular meals. Researchers looked at food additive encapsulation and controlled release (Yu et al. 2010).

Octenyl succinic anhydride-ε-polylysine can either be used as a surfactant or emulsifier, thereby acting as bifunctional molecules in the encapsulation of drugs or nutraceuticals or as antimicrobial agents (Sekhon 2010; Yu et al. 2009). Water solubility and absorption rate are typically improved via encapsulation. Before nanoemulsions are created, lipophilic components are combined in the oil phase and entrapped during the creation of oil-in-water nanoemulsions. The most prevalent compounds enclosed in industrial applications are lipids, proteins, and carbohydrates. Food components that are soluble in organic solvents, such as fatty acids, phospholipids, carotenoids, and oil-soluble vitamins, have benefited from encapsulation. They cannot be easily solved in food items due to their limited solubility in water, rendering them sensitive to oxidation. Due to encapsulation, an effective barrier for oxidation can be formed. The dietary fibers consist of carbohydrates. These fibers differ in the quantity of monosaccharides as well as vary in distribution and type. These fibers are encapsulated and serve an important function in cancer prevention, cholesterol-lowering, and constipation prevention, among other things. Fibers can be packed in the capsules through encapsulation without affecting the quality of the product, such as changing flavor, texture, etc.

Natural polymers like proteins and polysaccharides are used for the lipophilic components (de Vos et al. 2010). They possess many advantages, such as low toxicity, biocompatibility, and biodegradability. Encapsulation of lipophilic components, including bioactive lipids, such as fatty acids, conjugated linoleic acid, and butyric acid, and carotenoids, such as b-carotene, lycopene, lutein, proteins (insulin), and peptides, has aided in the creation of functional foods and improved food safety. Oil-soluble vitamins, such as A, D, E, and K, as well as flavors (essential oils) and nutraceuticals (co-enzyme Q), have all aided the food sector (Dima et al. 2015). Nanoemulsions may absorb taste and preserve it against temperature, oxidation, and enzymatic processes, as well as being stable across a wide pH range, extending food shelf life. Omega-3 supplementation may also show protective effects against diseases like cancer, asthma, myocardial infarction, etc.

One of the most successful examples of encapsulation is probiotics. Live mixtures of bacterial species are generally known as probiotics and can be incorporated into the food in yogurt, cheese, pudding, etc. Nanoencapsulation might potentially be used to produce specialized receptors. These nano-encapsulated customized probiotic bacterial cultures may also be able to operate as de novo vaccines due to their capacity to regulate immune responses (Vidhyalakshmi et al. 2009).

Expansion of the Functional Food Sector for the Application of Edible Coatings with Lipophilic Bioactive Substances

Consumers have grown more health-conscious in recent years, and the demand for healthier and safer foods has risen as a result. Growth of bacteria and environmental exposure are caused by a variety of issues such as peeling, ripping, or slicing. The inclusion of an edible coating extends the food's shelf life. These coatings include active ingredients that are good for the food product. Foods with edible coatings have a longer shelf life and have a higher nutritional content. Antimicrobials, antioxidants, anti-browning agents, texture enhancers, flavoring agents, and nutraceuticals are some examples (Acevedo-Fani et al. 2017). They also aid in the preservation of the product's color, wetness, and texture, as well as being biodegradable and nontoxic to the environment. Water-soluble substances such as vitamins (B and C), proteins, and pre-/probiotics may be easily combined with a biopolymer (polysaccharide and proteins) substrate to generate edible coatings (Dasgupta et al. 2019b). In the manufacture of edible coatings, the integration of lipophilic components, such as essential oils or fat-soluble chemicals posed a significant problem. As a result, extra processes to stabilize this combination were necessary. Other techniques were necessary to stabilize the mixture since the integration of lipophilic ingredients, such as essential oils or fat-soluble chemicals, constituted a significant problem in edible coating manufacture.

The organoleptic features of food items were vulnerable when coated with an active substance, and the optical and sensory qualities were affected. External influences and other dietary compositions also had an impact on the active components. For example, edible coatings containing essential or marine oils were prone to oxidation, lowering overall customer acceptance. Nanoemulsion was an advancement where the encapsulation of lipophilic ingredients increased activities like solubility, stability, and biological activity. Polysaccharides, such as cellulose derivatives, starch, and chitin, proteins, such as gelatin, maize zein, and wheat gluten, and lipids, such as oils, waxes, and resins, are among the coating materials. Sorbitol is sometimes used to improve the mechanical qualities of a product.

Bioactive films are enriched with essential oils, which reduces microbial activity. For example, alginate is used as a bioactive film, comprising garlic essential oil (Sánchez-González et al. 2011), which helps reduce microorganisms, such as *Escherichia coli*, *Staphylococcus aureus*, etc. Oregano oil and cinnamon oil encapsulated in polymers like chitosan can be used to get rid of *Escherichia coli* and *Listeria monocytogenes*. Rosemary oil coated with whey protein can be used to reduce *Salmonella enteritidis* (Chouhan et al. 2017). The diffusion rate of an antimicrobial agent can be slowed by maintaining high concentrations of the active ingredient, which increases the risks of contamination over a longer length of time.

Fruit coatings play an important role in keeping the color and firmness of fruits by minimizing water loss and lowering respiration rates (Sánchez-González et al. 2011). According to research, commercially coated oranges altered with *Lippia scaberrima* oil resulted in considerable weight loss (du Plooy et al. 2009). Basil gum containing oregano essential oil was used for freshly cut apricots. As a result, antioxidant and antimicrobial properties were seen, which prevented spoilage and increased shelf life (Hashemi et al. 2017). Bioactive coatings are also used for the process of meat and fish preservation in order to inhibit microbial growth. In meat and seafood, a decrease in lipid oxidation is also seen in edible coatings enriched with essential oil (Ojagh et al. 2010). Thus, this emerging technology helps in improving the physical stability of a product, thereby enhancing and bringing out the nutritional value of food products.

Invasion of Nanotechnology and Emulsion in Food Ingredients and Additives

Nanotechnology approaches are expected to yield their application more in the food industry. Many natural meals containing nanoscale components have been consumed for many years, and the structure of these foods determines their qualities. Food items containing nanostructures are being created with the goal of improving flavor, texture, and consistency. For example, the dairy industry produces emulsions (butter), foams (ice cream and whipped cream), gel networks (yogurt), complicated liquid (milk), and solid plastic using three fundamental micro- and nano-sized structures made up of whey proteins, fat globules, and casein micelles (cheese) (Aguilera and Stanley 1999). Low-fat nanostructured mayonnaise spreads and ice creams that are as "creamy" as their full-fat counterparts are now available to consumers. The nanoscale component enhances the color, texture, and taste of food (Morris et al. 2011). SiO_2 (Uboldi et al. 2012) and amorphous silica (Oberdörster et al. 2005) nanoparticles are used as a food additive, and TiO_2 nanoparticles are used as a coloring doughnut powdered sugar coating (Gerloff et al. 2009).

Nanotechnology may be used at any point in the food supply chain, from farm to fork. A cow's udder was used as an example of a nanodevice for manufacturing, assembling, and distributing proteins and fat into an aqueous phase. Later on, they become the building components for diverse protein architectures. Food undergoes microstructural changes as a result of this homogenization and fine milling process (Sekhon 2010). Even the addition of alginate nanoparticles containing zinc salt and iron to ice cream is being investigated (Sharifzadeh Baei 2013). Astaxanthin-enriched milk was created by mixing astaxanthin nanoemulsions with skimmed milk, which was then tested for cellular absorption and nano dispersion stability (Anarjan and Tan 2013). The nutritional content and functional quality of yogurt were improved by adding nanoemulsions with -3 polyunsaturated fatty acids or rice bran oil (Ganesan et al. 2014). To improve the digestible characteristics of food and natural extract, nanoemulsion is used as a suitable form. Carotene is an orange-red pigment that may be found in carrots and other bright vegetables (Yin et al. 2009). It is a member of the carotene family and the best-known provitamin A carotenoid (Yuan et al. 2008). Flavored nanoemulsion of β-carotene with improved digestibility has been applied (Qian et al. 2012). Indo-Pacific king mackerel steaks are processed using a nanoemulsion made with sunflower oil. The product's shelf life was extended to 48 hours, and no microbiological growth was detected after 12 hours (Joe et al. 2012).

Companies like Nestle and Unilever have applied nanotechnological applications to produce excellent and stable quality food products. Unilever has hinted at a low-fat ice cream with a fat percentage of 1% without sacrificing taste or quality. Nestle patented a water-in-oil emulsion product that promised to defrost frozen foods uniformly in the microwave (Durán and Marcato 2013). Polysorbates and other micelle-forming chemicals were added to the emulsion they created. Emulsifying agents like whey protein and casein micelles were used for entrapment, delivery of sensitive nutraceuticals, and protection. Perhaps nanotechnology has existed and prevailed for a long time, contributing to food science for ages (Silva et al. 2012).

Purple Rice Bran Oil Nanoemulsion Fortification of Frozen Yogurt

Purple rice oil is known for its high value of antioxidant properties, making it suitable for various health issues. It lowers cholesterol, is good for the skin, is rich in vitamins, and has anti-inflammatory properties. The surface qualities of ice cream are combined with the sour flavor of yogurt in frozen yogurt (Cruz et al. 2009). The desert market is touching its peak with the advent of various amounts of consumers. Due to this, there is an increasing need to better the product by inculcating various health supplements into it, making it healthier rather than having traditional desserts (Hagenmaier 1998). With the upsurge in consumers' awareness for more improved and healthier food, the demand for more organic and nutritional food has increased, which has pressured the manufacturers to produce products with healthier and more edible food, leading scientists to open doors to various oil, seed, and other edible components. Purple rice oil (PRBO) is high in nutraceutical substances, such as gamma oryzanol, tocopherols, and tocotrienols, all of which have been associated with improving health in humans by lowering blood cholesterol and providing antioxidant efficacy (Alfaro Sanabria 2012). Nanoemulsion drastically increases the product's nutritional profile by increasing the antioxidant material. Also, fortification may help to amplify the frozen yogurt market (Jang and Xu 2009). The food industry is gaining much attention in nanoemulsions due to its health benefits or integration of more health-related supplements, functional performances, and stability in food processing (Anarjan and Tan 2013).

Its method is combining ingredients to form a natural stirred yogurt with stabilizers/emulsifiers and sugar, then freezing the mixture in an ice cream freezer (Tamine and Robinson 2007). The oil was extracted from rice bran, and purple rice was formulated as it contributed to providing a reasonable amount of bioactive compounds. Oil is extracted by seeing its essential constituents and after which analyzing them. Frozen yogurt nanoemulsion with purple rice bran oil, frozen yogurt with emulsion components without purple rice bran oil, and plain frozen yogurt were all tested (Mahrous and Abd-El-Salam 2014). A nanoemulsion containing purple rice bran oil (NPRBO) was synthesized by ultrasonication and ultra-shearing. Firstly, the sample with nanoemulsion was found to be more stable than the control emulsion (Xu et al. 2001). The size of NPRBO was 199.2 nm, and the size was smaller than that of a control emulsion produced by sonication only (Alfaro Sanabria 2012). This study implied how, technically, purple bran rice oil carries significant importance by creating a valuable product in the dairy industry and providing insight into the resulting frozen yogurt.

Formation of Various Phytosomes and Using Them for Delivery in Herbal Products Without Resorting to Pharmacological Adjuvants

The advent of Ayurveda has created ample opportunities and opened various doors for undoubted success. It has been very promising how ancient people utilized it and nurtured it many years ago. With the help of modern science and research, we can take it to the next level. These were very versatile, being used in almost every field, starting from healing capacity to the overall betterment of the body. It was all reliable without any side effects, cost issues, or any need for other resources.

A unique drug-delivery system is a new method of delivering drugs that overcomes the limitations of previous methods. Significant progress has been made in the field of innovative drug-delivery systems (NDDS), particularly for herbal medications, in recent years (Afrin et al. 2018). NDDS looks for various phytochemicals to develop polymeric nanoparticles, liposomes, transferosome, pharmacosomes, nanocapsules, etc., which provide a large number of advantages for herbal drugs, such as improved tissue macrophages distribution, enhancement of bioavailability and solubility, sustained delivery, protection from vital toxicity, and protection from chemical and physical degradation. So, here is where phytosomes come into the picture (Sarangi and Padhi 2018). Phytosomes are defined as the composite of a natural product. The reaction of the stoic metric amount of phospholipids and phytoconstituents in an ideal solvent produces phospholipids, and the phospholipid–substrate interaction is due to the creation of hydrogen bonds between the polar heads of phospholipids (as revealed in the spectroscopic data) (Jain et al. 2010; Udapurkar et al. 2016).

Perhaps, phytosomes and liposomes are pretty similar but have some peculiar differences in phytosomes. By forming chemical connections with a hydrophilic phospholipid head, the phytoconstituent becomes an integral component of the membrane. The active herbal element is dissolved in the layers of the membrane, which are surrounded by phospholipids, in the liposome (Amin and Bhat 2012). As a result, phytosomes have strong absorbent capacity and bioavailability, and phytosomes are superior to liposomes in topical formulations (Karimi et al. 2015). This significance is an essential door to the importance of phytosomes research.

During the drug-delivery system, due to the affinity to lymph, the emulsion is distributed as an *in vivo* targeted method. Being prevented from direct touch with the body and packaged in the inner phase, the drug can be supported for a long time. After the designated drugs are made into the emulsion, the oil droplets collect inside the macrophages leading to a significantly high concentration of drugs in the spleen, kidney, and liver. In addition, the emulsion's size will impact its target distribution. By incorporating the herbal medicine into the emulsion, the hydrolyzed materials will be more stable, the medications' stimulation to tissues will be reduced, and drug penetrability to the skin will be improved. Herbal drugs like Bruceajavanica oil, Camptothecin, Zeodoary oil, and Coixenolide oil are formulated into the emulsion.

Primarily, the herbal nanoemulsion delivery system has the following advantages:

- An herbal formulated nanoemulsion delivers directly to the desired site of action.
- It improves and enhances the pharmacokinetics and solubility of the drugs.
- It provides transit of the drug through biological barriers and increases its bioavailability.
- Without destroying the surrounding environment, it takes the drug directly to the desired site of action.

Food Packaging

Nowadays, due to the busy lifestyle of people, focus is now diverted to the need for ready-to-eat foods. In addition, hygiene serves as the main purpose for sustainability and the ideal food quality. Various problems, like peeling, tearing of food, or slicing, are the reasons for microbial growth and exposure to the environment. Films and coatings are the two types of active packaging technologies available. Films are thin, preformed sheets of substantial strength that can be used for packaging or separate product layers. However, coatings are films generated when a product has been coated and then dried through a specific procedure, resulting in the production of the coating (Ahari and Naeimabadi 2021). Several studies propose the use of oil–water-based nanoemulsion in food packaging. There are two ways to induce nanoemulsion. The first is to directly insert into films and coatings or enclosed in a polymeric carrier before being put into films and coatings. To induce nanoemulsion into a film, we first have to break down the nanoemulsion in a polymeric solution by applying high shear stress. The use of nanoemulsion in food packaging proved to be effective (Ahari and Naeimabadi 2021).

The edible coating increases the nutritive and functional value when added with anti-browning agents, antioxidants, texture enhancers, flavoring agents, and antimicrobial or nutraceuticals. (Atarés and Chiralt 2016; Souza and Fernando 2016). Protein, pre-/probiotics, and water-soluble ingredients like vitamins can be integrated with the biopolymer matrix to form an efficient edible coating (Dasgupta et al. 2016). The assimilation of lipophilic substances like fat-soluble compounds or essential oils poses a big challenge in edible coating preparation. So, other procedures are required to stabilize this mixture.

The organoleptic features of food products were responsive when coated with active substances, and the optical and sensory aspects were altered. Eça et al. (2014) found that the active components were also responsible for changes caused by external variables and other dietary compositions. To address this problem, researchers concentrated their efforts on developing active edible coatings (Eça et al. 2014). The incorporation of susceptible (lipophilic) active components boosts the solubility, stability, and biological activity of nanoemulsions (Weiss et al. 2006). The dispersion phase, continuous phase, and emulsifiers are the three main components of edible oil–water nanoemulsion coatings (stabilizers or surface-active compounds). When a polysaccharide-based coating lacks surface-active characteristics to stabilize nano-sized oil droplets, emulsifiers are necessary.

Use in Confectionary

The confectionary industry is increasing tremendously, thereby having a market for different products, like toffee, bakery, biscuit products, and chocolates. This industry is classified as sugar and flour confectionary. The flour confectionary produces bakery products and biscuits, while the sugar confectionary is indulged in the production of toffees, candies, chocolates, and chewing gums. Although nanoemulsion flour confectionary has several uses, its use in the sugar sector may be limited to some extent (Dasgupta et al. 2019b). Nonfat particles suspended in a continuous fat phase are cocoa, sugar, and milk solids (cocoa butter). Various emulsifiers were utilized in the production of chocolate to prevent blooming, which can lead to oxidation and other unpleasant sensory changes. The emulsifier was also employed to increase the flow parameter of the chocolate for economic and operational reasons. The desired texture of the product can also be enhanced and achieved. Uniform air dispersion and foam stability were the two critical points considered and were expected to be applied in nanoemulsions. Cakes and sandwich biscuits represent a large market for nanoemulsions.

Nanoemulsions are also predicted to be used in flour and sugar confectionary goods that require better foam stability and consistent air dispersion. Cream is used in bakery items such as cakes and sandwich biscuits, creating a huge market for nanoemulsions (Dasgupta et al. 2019b). The addition of fat-in-sugar syrup emulsions to whipped montelimars also helps. The fat is dispersed in a combination of dextrose and sugar, together with proteins and other taste components, in the chewy confectionary items (Dasgupta and Ranjan 2018). Emulsifiers also act as a plasticizer, assisting in the dispersion of fat for optimal quality (Hobson et al. 2018).

Comparison of Nanoemulsion from Conventional Methods

Before using nanoemulsion, scientists used numerous approaches to obtain many materials used for encapsulation and coatings. However, all this approach subsided when nanoemulsion appeared. The drawbacks that these conventional methods faced were primarily the following:

1) The constituents might be destroyed when they come into contact with the acidic pH of the stomach (Abbas et al. 2013).

2) There can be a problem when minimum effective concentration cannot be achieved, as there are cases when components are metabolized in the liver before they outreach into the systemic circulation.
3) Polyphenols have poor absorbance issues as it leads to the outline of the lipid membrane's passage and eventually leads to lesser bioavailability (Kesarwani and Gupta 2013).
4) Issues can also be raised when there are chances of missing a dose that states poor patient compliance (Anal et al. 2013).
5) Variation in the body's blood levels can happen due to undermedication or overmedication.

On having observed this vital point and having had a severe need to overcome this, scientists were inclined to have a better aspect in the food industry by resolving this fallback and to create a better society by inculcating nanoemulsion into the picture.

Problems and Probable Solutions of Nanoemulsions

Ample research has been published on the current trends in nanoemulsions, their advantages, and other methods, deducing answers for predicting the future scope of the nanoemulsion. However, many drawbacks still prevail for nanoemulsions. Firstly, limited studies for the nanoemulsion view authentic health benefits in foods except for standard terms like increased bioavailability and increased nutrition. Here, the focus is not made to mislead people in the sense of the usefulness of nanoemulsions, but there is still a lot more to be explored in the actual usage of nanoemulsions in food products. Secondly, one of the most critical downsides considered is the cost of formulating the nanoemulsions. Due to the excessive use of instruments, multiple experiments are conducted, and the whole process leads to more extensive use of money for the production. Besides production, the preparation and application of the packaging and fortifying of foods also require more energy input and heavy instruments investment. These all lead to less work done toward the production part rather than the research part in the nanoemulsions. More work and research need to be carried out to mitigate the problems.

Probable solutions comprise designing the nanoemulsions while taking care of the cost and the nature of materials used. The cost can be reduced if bulk production can be considered. Every time a material is produced in a large proportion, if the supply–demand ratio is equal, it is known to be beneficial. However, the demand for the nanoemulsion would also have to increase to balance the cost. The cost of the nanoemulsion also increases as more time and energy are consumed in the formation and characterization. Suppose a steady protocol is formed universally to form any food-grade nanoemulsion. In that case, time will be cut down for the formation, allowing an advantage in the process followed after it.

While forming a nanoemulsion, there is no ideal protocol to be followed. Since the process is highly dependent on the type of the oil, nanoemulsion, and coating, it varies majorly to a large extent. Safety of the nanoemulsion is considered one of the problems faced currently. The materials used in the formation are not yet considered feasible and hence unsuitable in the food industry like synthetic oils, synthetic surfactants, organic solvents, or synthetic polymers. If the product is designed as food-grade, the components selected should be natural and not harmful at any concentration. Usage of food-grade components like triglyceride oils, polysaccharides, proteins, and flavor oils must be utilized to formulate food nanoemulsions. These ingredients are accepted legally, economically viable, natural, and not synthetic.

Research should focus on giving out nanoemulsions with better stability as they are considered metastable to a large extent. Theoretically, they are unstable, but they provide stability to the product in the long run. The focus should be to ameliorate this aspect as this holds importance in the production sector. It is vital to improve the encapsulated components' bioactivity for scaled-up production. Additional studies should emphasize the risks associated with and biological events in using nanoemulsion-based delivery systems in food products and packaging submissions to safeguard customers' safety. Nowadays, consumers are well versed in understanding which component is used in a particular product. This being considered, the relationship of nanoemulsion with carbohydrates, macromolecules, and fibers should be evaluated and assessed, delivering comprehensive data for validation and better consideration of the product to consumers.

Future Trends and Challenges

Nanoemulsion gel, a novel collection of soft materials displaying a strong and large elastic behavior even at a low dispersed phase volume fraction, has recently been discovered. Nanoemulsion gels have the potential to be employed as food components. New goods and procedures aimed at customizing and personalizing items have a bright future ahead of them. The first step will be to increase food safety and quality. Nanotechnology is providing new opportunities for the food business across the world. However, introducing nanoemulsions or nanoparticles into the food chain can lead to the accumulation of hazardous pollutants in food, which can be harmful to human health. Nanotechnology may be useful on a tiny scale, but the ramifications are huge (Nickols-Richardson 2007).

From farm to fork in the food supply chain, new and future developments have exceptional features. It is essential to maintain leadership in the food business and to cooperate with nanotechnology in the future. Food packaging resources may be used to increase nutritional content, lengthen shelf life, enhance food safety, prepare customers to select how food is destroyed or defiled, heal packaging tears, and continually release chemicals to extend the shelf life of the food in the container (Thiruvengadam et al. 2018). Because of the high compatibility of nanoemulsion structures, products such as fortified food and a wide range of drinks might play an important role in the near future. In order to give appropriate advantages, vitamin delivery techniques for food fortification must meet some essential features. A reasonable limit for food fortification should be maintained in terms of vitamin supply. Nanocarriers, which resemble nanoemulsions in structure, are used to transport pharmaceuticals and biomolecules, such as proteins, antibody fragments, DNA fragments, and RNA fragments, as the base for cancer biomarkers (Patel et al. 2019). Climate change, urbanization, sustainable use of natural resources, and environmental challenges such as runoff and pesticide and fertilizer accumulation are all difficulties that agriculturists face across the world. An alarming increase in demand to feed a population of 6–9 billion people by 2050 exacerbates these problems (Dasgupta et al. 2015). If taken care, nanoemulsions can help set a new positive path in this field, ameliorating the situation of food safety.

Regulations and Safety Aspects

Consumption of food exhibiting essential oil nanoemulsion has a vital point, which contributes to the regulations and safety aspect. Food and Drug Administration (FDA) and other European Food Safety Authority in Europe (EFSA) considered the natural antimicrobial nanoemulsion as a part of flavoring and additives. Some oils and extracts are considered safe for use, whereas there are restricted components in the food categories that FDA and EFSA [86] presented in this approach.

Furthermore, given the constraints of long-term nanotechnology development, the safety of nano-products has become critical to define. Various international authorities are already working on bringing nanoparticles safety regulation to human health and the environment, both individually and together (Gupta et al. 2017).

The regulatory authorities should be addressed through a protocol if any compounds have not been approved. Long- and short-term health effects and other safety aspects should be considered based on the identification, and the amount of product consumed (Tajkarimi et al. 2010). The risks associated with reactive molecules should be continuously monitored, like those constituting essential oils. For example, carvacrol is included among the registered flavorings and additives by the FDA, yet it is known to show a toxic effect (Donsi and Ferrari 2016). The regulatory issues concern ingredients and the use of nanotechnology in the case of nanoemulsions. World Health Organization and the FDA have also listed safety usage levels for the ingredients used in food-grade nanoemulsions.

Based on the development of regulations of nano-food, there are lists of recommendations:

1) Criteria such as physical and chemical properties, size range, measurement methods, and safety concerns should be considered.
2) Depending on the types of the product, the upper particle size limit for ultrafine food and nano-food may loosen to a value above 100 nm.
3) For correct categorization, management, and analysis of nano-food items, many categories such as liquid, powder, aerosol, suspension, emulsion, and liposome are used.
4) Food nanotechnology should be subjected to ethical research.
5) A safety evaluation or nanotoxicity study should be necessary for chemically changed nanoscale materials (Chau et al. 2007).

In the current state of knowledge, no accurate inclusion or exclusion criteria for nano-specific risk evaluation are known, and for the unique features of manufactured nanomaterials, no exact inclusion or exclusion criteria are known (Bundschuh et al. 2018). The biological destiny of a nanoemulsion particle is determined by the composition of the core and shell materials. When nanoemulsion is exposed to diverse physiological circumstances, the droplets in the nanoemulsion may alter as they transit through the gastrointestinal system (Karthik et al. 2017). The composition, size distribution, physical condition, and electrical charge of the nanoemulsion might all change. Understanding the release and absorption of encapsulated components of a nanoemulsion that is controlled by the nanoemulsion's characteristics is also crucial. For this, there should be a precaution taken before using nanoemulsion, which determines the stability and flexibility of using it inside or outside the body.

Conclusion

Nanoscale structures have grown in prominence in the food industry during the previous decade. As a result, the focus of research and activity in this subject has shifted dramatically. Nanoemulsion is the most desired form in technical applications in the food business due to qualities, such as large surface area, optical transparency, adjustable rheology, and thermodynamic stability. Various preparation methods, such as high-energy and low-energy approaches including microfluidizers, ultrasonic homogenization, spontaneous emulsification, and phase inversion methods, are available, as detailed in earlier sections. Encapsulation, food packaging, the food business, and food safety have all produced promising results and applications. In terms of manufacturing, cost, and characterization of the resulting nanoemulsion and the food systems to which it will be applied, nanoemulsion application to food systems still offers product safety and acceptability difficulties that must be addressed. Nanoemulsion, on the other hand, has the potential to outperform traditional emulsion in the future since it can be employed in a variety of food industries. The food sector is also looking for ways to incorporate lipophilic functional chemicals into new food items. In addition, nanoemulsion gel for basic chemicals and food items will be introduced. Nanoemulsion has the potential to transform the current food system by improving food nutritional quality and fostering a safe and healthy food culture.

Conflict of Interest

Authors have no conflict of interest.

Acknowledgments

Financial assistance by the Gujarat Institute for Chemical Technology (GICT) for establishing a facility for environmental risk assessment of chemicals and nanomaterials is gratefully acknowledged.

References

Abbas, S., Hayat, K., Karangwa, E., Bashari, M., and Zhang, X. (2013). An overview of ultrasound-assisted food-grade nanoemulsions. *Food Eng. Rev.* 5: 139–157.

Acevedo-Fani, A., Soliva-Fortuny, R., and Martín-Belloso, O. (2017). Nanoemulsions as edible coatings. *Curr. Opin. Food Sci.* 15: 43–49.

Afrin, S., Jahan, I., Hasan, A.N., and Deepa, K.N. (2018). Novel approaches of herbal drug delivery. *J. Pharm. Res. Int.* 21(5): 1–11.

Aguilera, J.M. and Stanley, D.W. (1999). *Microstructural Principles of Food Processing and Engineering*. Springer Science & Business Media.

Ahari, H. and Naeimabadi, M. (2021). Employing nanoemulsions in food packaging: shelf life enhancement. *Food Eng. Rev.* 13: 858–883.

Ahari, H. and Nasiri, M. (2021). Ultrasonic technique for production of nanoemulsions for food packaging purposes: a review study. *Coatings* 11: 847.

Alfaro Sanabria, L.A. (2012). Development of a frozen yogurt fortified with a nanoemulsion containing purple rice bran oil.

Amin, T. and Bhat, S.V. (2012). A review on phytosome technology as a novel approach to improve the bioavailability of nutraceuticals. *Int. J. Adv. Res. Technol.* 1: 1–5.

Anal, A., Bhowmik, D., and Gopinath, H. (2013). Overview on controlled release dosage forms. *Int. J. Pharma. Sci.* 3: 258–269.

Anarjan, N. and Tan, C.P. (2013). Chemical stability of astaxanthin nanodispersions in orange juice and skimmed milk as model food systems. *Food Chem.* 139: 527–531.

Assadpour, E. and Mahdi Jafari, S. (2019). A systematic review on nanoencapsulation of food bioactive ingredients and nutraceuticals by various nanocarriers. *Crit. Rev. Food Sci. Nutr.* 59: 3129–3151.

Aswathanarayan, J.B. and Vittal, R.R. (2019). Nanoemulsions and their potential applications in food industry. *Front. Sustain. Food Syst.* 95.

Atarés, L. and Chiralt, A. (2016). Essential oils as additives in biodegradable films and coatings for active food packaging. *Trends Food Sci. Technol* 48: 51–62.

Bundschuh, M., Filser, J., Lüderwald, S., McKee, M.S., Metreveli, G., Schaumann, G.E., Schulz, R., and Wagner, S. (2018). Nanoparticles in the environment: where do we come from, where do we go to? *Environ. Sci. Eur.* 30: 6.

Chau, C.-F., Wu, S.-H., and Yen, G.-C. (2007). The development of regulations for food nanotechnology. *Trends Food Sci. Technol.* 18: 269–280.

Chaudhari, R., Patel, P., Meghani, N., Nasra, S., and Kumar, A. (2021). Fabrication of methotrexate-loaded gold nanoconjugates and its enhanced anticancer activity in breast cancer. *3 Biotech* 11: 175.

Chouhan, S., Sharma, K., and Guleria, S. (2017). Antimicrobial activity of some essential oils—present status and future perspectives. *Medicines* 4: 58.

Cruz, A.G., Antunes, A.E., Sousa, A.L.O., Faria, J.A., and Saad, S.M. (2009). Ice-cream as a probiotic food carrier. *Food Res. Int.* 42: 1233–1239.

Dasgupta, N. and Ranjan, S. (2018). *An Introduction to Food Grade Nanoemulsions*. Springer.

Dasgupta, N., Ranjan, S., and Gandhi, M. (2019a). Nanoemulsion ingredients and components. *Environ. Chem. Lett.* 17: 917–928.

Dasgupta, N., Ranjan, S., and Gandhi, M. (2019b). Nanoemulsions in food: market demand. *Environ. Chem. Lett.* 17: 1003–1009.

Dasgupta, N., Ranjan, S., Mundekkad, D., Ramalingam, C., Shanker, R., and Kumar, A. (2015). Nanotechnology in agro-food: from field to plate. *Food Res. Int.* 69: 381–400.

Dasgupta, N., Ranjan, S., Mundra, S., Ramalingam, C., and Kumar, A. (2016). Fabrication of food grade vitamin E nanoemulsion by low energy approach, characterization and its application. *Int. J. Food Prop.* 19: 700–708.

de Vos, P., Faas, M.M., Spasojevic, M., and Sikkema, J. (2010). Encapsulation for preservation of functionality and targeted delivery of bioactive food components. *Int. Dairy J.* 20: 292–302.

Dhawan, A., Singh, S., Kumar, A., and Shanker, R. (2018). *Nanobiotechnology: Human Health and the Environment*. CRC Press.

Dima, Ş., Dima, C., and Iordăchescu, G. (2015). Encapsulation of functional lipophilic food and drug biocomponents. *Food Eng. Rev.* 7: 417–438.

Donsi, F. and Ferrari, G. (2016). Essential oil nanoemulsions as antimicrobial agents in food. *J. Biotechnol.* 233: 106–120.

du Plooy, W., Regnier, T., and Combrinck, S. (2009). Essential oil amended coatings as alternatives to synthetic fungicides in citrus postharvest management. *Postharvest Biol. Technol.* 53: 117–122.

Durán, N. and Marcato, P.D. (2013). Nanobiotechnology perspectives. Role of nanotechnology in the food industry: a review. *Int J. Food Sci. Technol.* 48: 1127–1134.

Eça, K.S., Sartori, T., and Menegalli, F.C. (2014). Films and edible coatings containing antioxidants – a review. *Braz. J. Food Technol.* 17: 98–112.

Espitia, P.J., Fuenmayor, C.A., and Otoni, C.G. (2019). Nanoemulsions: synthesis, characterization, and application in bio-based active food packaging. *Compr. Rev. Food Sci. Food Saf.* 18: 264–285.

Fryd, M.M. and Mason, T.G. (2012). Advanced nanoemulsions. *Annu. Rev. Phys. Chem.* 63: 493–518.

Ganesan, B., Brothersen, C., and McMahon, D.J. (2014). Fortification of foods with omega-3 polyunsaturated fatty acids. *Crit. Rev. Food Sci. Nutr.* 54: 98–114.

Gerloff, K., Albrecht, C., Boots, A.W., Förster, I., and Schins, R.P. (2009). Cytotoxicity and oxidative DNA damage by nanoparticles in human intestinal Caco-2 cells. *Nanotoxicology* 3: 355–364.

Gupta, G.S., Shanker, R., Dhawan, A., and Kumar, A. (2017). Impact of nanomaterials on the aquatic food chain. In: *Nanoscience in Food and Agriculture 5* (ed. S. Ranjan, N. Dasgupta and E. Lichtfouse). Springer.

Gupta, P.K., Bhandari, N., Shah, H.N., Khanchandani, V., Keerthana, R., Nagarajan, V., and Hiremath, L. (2019). An update on nanoemulsions using nanosized liquid in liquid colloidal systems. In: *Nanoemulsions-Properties, Fabrications and Applications* (ed. K.S. Koh and V.L. Wong). IntechOpen.

Gurpreet, K. and Singh, S. (2018). Review of nanoemulsion formulation and characterization techniques. *Indian J. Pharm. Sci.* 80: 781–789.

Hagenmaier, R.D. (1998). Wax microemulsion formulations used as fruit coatings. *Proc. Fla. State Hort. Soc.* 111: 251–254.

Hashemi, S.M.B., Khaneghah, A.M., Ghahfarrokhi, M.G., and Eş, I. (2017). Basil-seed gum containing *Origanum vulgare* subsp. viride essential oil as edible coating for fresh cut apricots. *Postharvest Biol. Technol.* 125: 26–34.

Hobson, J.J., Edwards, S., Slater, R.A., Martin, P., Owen, A., and Rannard, S.P. (2018). Branched copolymer-stabilised nanoemulsions as new candidate oral drug delivery systems. *RSC Adv.* 8: 12984–12991.

Jain, N., Gupta, B.P., Thakur, N., Jain, R., Banweer, J., Jain, D.K., and Jain, S. (2010). Phytosome: a novel drug delivery system for herbal medicine. *Int. J. Pharm. Sci. Drug Res.* 2: 224–228.

Jaiswal, M., Dudhe, R., and Sharma, P. (2015). Nanoemulsion: an advanced mode of drug delivery system. *3 Biotech* 5: 123–127.

Jang, S. and Xu, Z. (2009). Lipophilic and hydrophilic antioxidants and their antioxidant activities in purple rice bran. *J. Agric Food Chem.* 57: 858–862.

Joe, M.M., Chauhan, P.S., Bradeeba, K., Shagol, C., Sivakumaar, P.K., and Sa, T. (2012). Influence of sunflower oil based nanoemulsion (AUSN-4) on the shelf life and quality of Indo-Pacific king mackerel (*Scomberomorus guttatus*) steaks stored at 20 C. *Food Control* 23: 564–570.

Kang, D.J., Bararnia, H., and Anand, S. (2018). Synthesizing pickering nanoemulsions by vapor condensation. *ACS Appl. Mater. Interfaces* 10: 21746–21754.

Karimi, N., Ghanbarzadeh, B., Hamishehkar, H., Keivani, F., Pezeshki, A., and Gholian, M.M. (2015). Phytosome and liposome: the beneficial encapsulation systems in drug delivery and food application. *Appl. Food Biotechnol.* 2: 17–27.

Karthik, P., Ezhilarasi, P., and Anandharamakrishnan, C. (2017). Challenges associated in stability of food grade nanoemulsions. *Crit. Rev. Food Sci. Nutr.* 57: 1435–1450.

Kesarwani, K. and Gupta, R. (2013). Bioavailability enhancers of herbal origin: an overview. *Asian Pac. J. Trop. Biomed* 3: 253–266.

Kong, M., Chen, X.G., Kweon, D.K., and Park, H.J. (2011). Investigations on skin permeation of hyaluronic acid based nanoemulsion as transdermal carrier. *Carbohydr. Polym.* 86: 837–843.

Kumar, A., Singh, S., Shanker, R., and Dhawan, A. (2017). *Nanotoxicology: Challenges for Biologists.* The Royal Society of Chemistry.

Lee, L., Hancocks, R., Noble, I., and Norton, I.T. (2014). Production of water-in-oil nanoemulsions using high pressure homogenisation: a study on droplet break-up. *J. Food Eng.* 131: 33–37.

Mahrous, H. and Abd-El-Salam, R. (2014). Production of a functional frozen yoghurt fortified with omega-3 and vitamin E. *Am. J. Food Nutr.* 2: 77–84.

Mansuri, A., Chaudhari, R., Nasra, S., Meghani, N., Ranjan, S., and Kumar, A. (2022). Development of food-grade antimicrobials of fenugreek oil nanoemulsion-bioactivity and toxicity analysis. *Environ. Sci. Pollut. Res. Int. 2022 (26).*

Martínez-Ballesta, M., Gil-Izquierdo, Á., García-Viguera, C., and Domínguez-Perles, R. (2018). Nanoparticles and controlled delivery for bioactive compounds: outlining challenges for new "smart-foods" for health. *Foods* 7: 72.

McClements, D.J. (2011). Edible nanoemulsions: fabrication, properties, and functional performance. *Soft Matter* 7: 2297–2316.

McClements, D.J. (2015). Nanoscale nutrient delivery systems for food applications: improving bioactive dispersibility, stability, and bioavailability. *J. Food Sci.* 80: N1602–N1611.

McClements, D.J. and Rao, J. (2011). Food-grade nanoemulsions: formulation, fabrication, properties, performance, biological fate, and potential toxicity. *Crit. Rev. Food Sci. Nutr.* 51: 285–330.

Meghani, N., Dave, S., and Kumar, A. (2020). Introduction to nanofood. In: *Nano-food Engineering* (ed. U. Hebbar, S. Ranjan, N. Dasgupta and R.K. Mishra). Springer.

Meghani, N., Patel, P., Kansara, K., Ranjan, S., Dasgupta, N., Ramalingam, C., and Kumar, A. (2018). Formulation of vitamin D encapsulated cinnamon oil nanoemulsion: its potential anti-cancerous activity in human alveolar carcinoma cells. *Colloids Surf. B Biointerfaces* 166: 349–357.

Morris, V.J., Woodward, N.C., and Gunning, A.P. (2011). Atomic force microscopy as a nanoscience tool in rational food design. *J. Sci. Food Agric.* 91: 2117–2125.

Neethirajan, S. and Jayas, D.S. (2011). Nanotechnology for the food and bioprocessing industries. *Food Bioproc. Technol.* 4: 39–47.

Nickols-Richardson, S.M. (2007). Nanotechnology: implications for food and nutrition professionals. *J. Am. Diet. Assoc.* 107: 1494–1497.

Oberdörster, G., Oberdörster, E., and Oberdörster, J. (2005). Nanotoxicology: an emerging discipline evolving from studies of ultrafine particles. *Environ. Health Perspect.* 113: 823–839.

Ocampo-Salinas, I., Tellez-Medina, D., Jimenez-Martinez, C., and Davila-Ortiz, G. (2016). Application of high pressure homogenization to improve stability and decrease droplet size in emulsion-flavor systems. *Int. J. Environ. Agric. Biotech.* 1.

Ojagh, S.M., Rezaei, M., Razavi, S.H., and Hosseini, S.M.H. (2010). Effect of chitosan coatings enriched with cinnamon oil on the quality of refrigerated rainbow trout. *Food Chem.* 120: 193–198.

Patel, P., Meghani, N., Kansara, K., and Kumar, A. (2019). Nanotherapeutics for the treatment of cancer and arthritis. *Curr. Drug Metab.* 20: 430–445.

Polychniatou, V. and Tzia, C. (2014). Study of formulation and stability of co-surfactant free water-in-olive oil nano-and submicron emulsions with food grade non-ionic surfactants. *J. Am. Oil Chem. Soc.* 91: 79–88.

Porras, M., Solans, C., Gonzalez, C., and Gutierrez, J. (2008). Properties of water-in-oil (W/O) nano-emulsions prepared by a low-energy emulsification method. *Colloids Surf. A: Physicochem. Eng. Asp.* 324: 181–188.

Qian, C., Decker, E.A., Xiao, H., and McClements, D.J. (2012). Physical and chemical stability of β-carotene-enriched nanoemulsions: influence of pH, ionic strength, temperature, and emulsifier type. *Food Chem.* 132: 1221–1229.

Ranjan, S., Dasgupta, N., Chakraborty, A.R., Samuel, S.M., Ramalingam, C., Shanker, R., and Kumar, A. (2014). Nanoscience and nanotechnologies in food industries: opportunities and research trends. *J. Nanoparticle Res.* 16: 2464.

Ranjan, S., Dasgupta, N., Ramalingam, C., and Kumar, A. (2016). Nanoemulsions in food science and nutrition. In: *Nanotechnology in Nutraceuticals: Production to Consumption* (ed. S. Sen, and Y. Pathak), 135–164, CRC Press.

Roco, M.C. (2001). International strategy for nanotechnology research. *J. Nanopart. Res.* 3: 353–360.

Salem, M.A. and Ezzat, S.M. (2018a). Nanoemulsions in food industry. In: *Dispersed Food Systems* (ed. J. Milani). IntechOpen.

Salem, M.A. and Ezzat, S.M. (2018b). Nanoemulsions in food industry. In: *Some New Aspects of Colloidal Systems in Foods* (ed. J. Milani). IntechOpen.

Sánchez-González, L., Vargas, M., González-Martínez, C., Chiralt, A., and Chafer, M. (2011). Use of essential oils in bioactive edible coatings: a review. *Food Eng. Rev.* 3: 1–16.

Sarangi, M.K. and Padhi, S. (2018). Novel herbal drug delivery system: an overview. *Arch. Med. Health Sci.* 6: 171.

Sekhon, B.S. (2010). Food nanotechnology – an overview. *Nanotechnol. Sci. Appl.* 3: 1.

Shaker, D.S., Ishak, R.A., Ghoneim, A., and Elhuoni, M.A. (2019). Nanoemulsion: a review on mechanisms for the transdermal delivery of hydrophobic and hydrophilic drugs. *Sci. Pharm.* 87: 17.

Sharifzadeh Baei, M. (2013). Studying the enrichment of ice cream with alginate nanoparticles including Fe and Zn salts. *J. Nanopart.* 2013.

Sharma, C., Dhiman, R., Rokana, N., and Panwar, H. (2017). Nanotechnology: an untapped resource for food packaging. *Front. Microbiol.* 8: 1735.

Silva, H.D., Cerqueira, M.Â., and Vicente, A.A. (2012). Nanoemulsions for food applications: development and characterization. *Food Bioproc. Tech.* 5: 854–867.

Singh, T., Shukla, S., Kumar, P., Wahla, V., Bajpai, V.K., and Rather, I.A. (2017a). Application of nanotechnology in food science: perception and overview. *Front. Microbiol.* 8: 1501.

Singh, Y., Meher, J.G., Raval, K., Khan, F.A., Chaurasia, M., Jain, N.K., and Chourasia, M.K. (2017b). Nanoemulsion: concepts, development and applications in drug delivery. *J. Control. Release* 252: 28–49.

Sonneville-Aubrun, O., Simonnet, J.-T., and L'alloret, F. (2004). Nanoemulsions: a new vehicle for skincare products. *Adv. Colloid Interface Sci.* 108: 145–149.

Souza, V.G.L. and Fernando, A.L. (2016). Nanoparticles in food packaging: biodegradability and potential migration to food—A review. *Food Packag. Shelf Life* 8: 63–70.

Sugumar, S., Mukherjee, A., and Chandrasekaran, N. (2015). Nanoemulsion formation and characterization by spontaneous emulsification: investigation of its antibacterial effects on Listeria monocytogenes. *Asian J. Pharm.* 9.

Sutradhar, K.B. and Amin, M.L. (2013). Nanoemulsions: increasing possibilities in drug delivery. *Eur. J. Nanomed.* 5: 97–110.

Tadros, T., Izquierdo, P., Esquena, J., and Solans, C. (2004). Formation and stability of nano-emulsions. *Adv. Colloid Interface Sci.* 108: 303–318.

Tajkarimi, M., Ibrahim, S.A., and Cliver, D. (2010). Antimicrobial herb and spice compounds in food. *Food Control* 21: 1199–1218.

Tamine, A. and Robinson, R. (2007). *Yoghurt: Science and Technology*, 3rd rev. New York: CRC Press.

Thiruvengadam, M., Rajakumar, G., and Chung, I.-M. (2018). Nanotechnology: current uses and future applications in the food industry. *3 Biotech* 8: 74.

Uboldi, C., Giudetti, G., Broggi, F., Gilliland, D., Ponti, J., and Rossi, F. (2012). Amorphous silica nanoparticles do not induce cytotoxicity, cell transformation or genotoxicity in Balb/3T3 mouse fibroblasts. *Mutat. Res. Genet. Toxicol. Environ. Mutagen.* 745: 11–20.

Udapurkar, P., Bhusnure, O., Kamble, S., and Biyani, K. (2016). Phyto-phospholipid complex vesicles for phytoconstituents and herbal extracts: a promising drug delivery system. *Int. J. Herbal Med.* 4: 14–20.

Vidhyalakshmi, R., Bhakyaraj, R., and Subhasree, R. (2009). Encapsulation "the future of probiotics" – a review. *Adv. Biol. Res.* 3: 96–103.

Weiss, J., Takhistov, P., and McClements, D.J. (2006). Functional materials in food nanotechnology. *J. Food Sci.* 71: R107–R116.

Xu, Z., Hua, N., and Godber, J.S. (2001). Antioxidant activity of tocopherols, tocotrienols, and γ-Oryzanol components from rice bran against cholesterol oxidation accelerated by 2, 2 '-Azobis (2-methylpropionamidine) dihydrochloride. *J. Agric. Food Chem.* 49: 2077–2081.

Yin, L.-J., Chu, B.-S., Kobayashi, I., and Nakajima, M. (2009). Performance of selected emulsifiers and their combinations in the preparation of β-carotene nanodispersions. *Food Hydrocoll.* 23: 1617–1622.

Yu, H., Huang, Y., and Huang, Q. (2009). Synthesis and characterization of novel antimicrobial emulsifiers from ε-polylysine. *J. Agric. Food Chem.* 58: 1290–1295.

Yu, H., Huang, Y., and Huang, Q. (2010). Synthesis and characterization of novel antimicrobial emulsifiers from ε-polylysine. *J. Agric. Food Chem* 58: 1290–1295.

Yuan, Y., Gao, Y., Zhao, J., and Mao, L. (2008). Characterization and stability evaluation of β-carotene nanoemulsions prepared by high pressure homogenization under various emulsifying conditions. *Food Res. Int.* 41: 61–68.

10

Adverse Epigenetic Effects of Environmental Engineered Nanoparticles as Drug Carriers

*Yingxue Zhang, Eid Alshammari, Nouran Yonis, and Zhe Yang**

Department of Biochemistry, Microbiology, and Immunology, Wayne State University School of Medicine, Detroit, Michigan, USA
** Corresponding author*

Introduction

Environmental engineered nanoparticles (ENPs) are defined as solid, colloidal particles with sizes ranging from 1 to 100 nm. Owing to their unique properties, particularly their high surface-area-to-mass ratio, ENPs have been exploited in a wide range of applications in diverse areas, such as medicine, cosmetics, textiles, agriculture, and many others (Baetke et al. 2015; Elveny et al. 2021). In the medical field, ENPs have led to the development of novel devices for the early detection of malignant tumors and efficient systems for the targeted delivery of drugs, genes, or proteins (Elveny et al. 2021; Ng et al. 2018; Sun et al. 2014). In recent years, the clinical applications of ENPs as drug carriers are increasing. Compared with the conventional drug-delivery systems, the ENP-based drug-delivery systems are superior in several aspects, including enhanced solubility, bioavailability, *in vivo* stability, and target-specific delivery (Mitchell et al. 2021; Rizvi and Saleh 2018). These superiorities are largely attributed to their high surface-area-to-mass ratio and easy-to-modify surface. Because of the large surface area, ENPs can efficiently adsorb and carry a large number of small molecules and therapeutic peptides. Because of the easy-to-modify surface, ENPs can enable cell- and organelle-specific targeting through surface modification with bio-responsive targeting agents (Mitchell et al. 2021; Patra et al. 2018). The hydrophilic molecules attached to the surface of ENPs can improve ENPs' solubility and promote ENPs to cross various biological barriers, such as the blood-brain barrier, the pulmonary alveolar-capillary membrane, and the tight junctions of endothelial cells of the skin (Kohane 2007; Rizvi and Saleh 2018). The linked targeting agents can also enable site-directed delivery of ENPs and thus enhance the bioavailability of encapsulated cargos and reduce the risk of side effects. Therefore, ENPs are being embraced as ideal materials for drug-delivery applications.

Current ENP drug-delivery systems can be cataloged into three main classes based on their chemical compositions, including lipid-based ENPs, polymeric ENPs, and inorganic ENPs (Mitchell et al. 2021). Lipid-based ENPs usually have a spherical structure, consisting of at least one lipid bilayer surrounding internal aqueous compartments (Hou et al. 2021; Mitchell et al. 2021). They are the most common drug-delivery systems in FDA-approved nanomedicines, owing to many advantageous properties, such as self-assembly, biocompatibility, and the ability to carry large payloads (Hua et al. 2018; Sarfraz et al. 2018). A liposome, typically composed of phospholipids, is the most widely used lipid-based ENP, which has been used in many clinical applications due to its cell-like structure (Sercombe et al. 2015). Compared with the lipid-based ENPs, polymeric ENPs are usually larger in size, ranging from 5 to 100 nm (Rideau et al. 2018). These ENPs can be made

Impact of Engineered Nanomaterials in Genomics and Epigenomics, First Edition. Edited by Saura C. Sahu.

of various kinds of polymers, including natural or synthetic polymers. The most widely used synthetic polymers are polylactide, polycaprolactone, and polyacrylate (Liu et al. 2020). As these polymers are very different in structure and chemistry, polymeric ENPs made of these polymers can have very distinct characteristics and structures. This makes polymeric ENPs ideal for multipurpose co-delivery applications and allows for the precise control of ENP features. The easy surface modification also made polymeric ENPs well-suited for targeted delivery, which ameliorates the adverse effects of capsulated drugs (Ahmed et al. 2021). The last type of ENP drug-delivery system is inorganic ENPs, which are synthesized using inorganic materials, such as silica, gold, or iron. The unique physical, electrical, optical, and magnetic properties of these inorganic materials distinguish inorganic ENPs from the other two classes and make inorganic ENPs uniquely qualified for diagnostics and imaging therapies (Mitchell et al. 2021; Paul and Sharma 2020).

Although the ENP drug-delivery systems are advantageous over the conventional chemotherapies, the potential toxicities associated with the ENPs should not be ignored. The toxic effects of ENP exposure have been investigated in various *in vitro* and *in vivo* assays (De Jong and Borm 2008; Lewinski et al. 2008). Among the frequently observed toxicities are cytotoxicity, peroxidation of lipids, micronuclei formation, apoptosis, genotoxicity, and immunotoxicity (Lewinski et al. 2008; Musolino et al. 2022; Najahi-Missaoui et al. 2020). One possible molecular mechanism of these toxic effects is that ENPs can alter gene or protein expression, affecting many major cellular pathways that regulate cell survival and stress response. Another possible mechanism is the change in epigenetic regulation, a fundamental cell regulatory mechanism that especially acts at the interface between the genome and the environment. Epigenetics refers to the heritable changes in gene expression that do not involve alterations in the actual genetic sequence. Epigenetic regulation can occur through DNA methylation, histone modification, and small noncoding RNA (ncRNA) regulation. These epigenetic regulations are prone to environmental chemicals, such as metals, air pollutants, or organic materials (Torano et al. 2016). Disturbed epigenetic processes can cause various deleterious consequences in cellular functions, which can lead to human diseases, especially cancers. It has been shown that ENPs can induce significant epigenetic changes and can grossly affect gene or protein expression (Pogribna and Hammons 2021; Smolkova et al. 2017). For example, the application of silver NPs (AgNPs) can cause a global elevation in the DNA methylation levels in human lung cancer epithelial cells (Blanco et al. 2017), while exposure of pregnant mice to AgNPs was found to result in DNA hypomethylation at the *ZAC1* gene promoter in the placenta tissue (Zhang et al. 2015). It has also been shown that organic ENPs are more epigenetically toxic than lipid-based or polymeric ENPs, especially in formulation with heavy metals. Significant epigenetic effects, including the changes in DNA methylation, histone modification, and ncRNA regulation, have been observed when the application of metal-fused organic ENPs, such as gold NPs, iron oxide NPs, and silica NPs (Li et al. 2011; Patil et al. 2019; Seidel et al. 2017). Therefore, understanding the effects of ENPs on epigenetic regulation is essential to address the health concerns of the ENP-based drug-delivery systems. In this chapter, we summarize those ENPs commonly used in drug-delivery applications and discuss their adverse epigenetic effects, aiming to inform future research toward ameliorating their toxic effects and eradicating the bottleneck of their clinical applications.

ENP-Based Drug-Delivery Systems

Lipid-Based ENPs

Lipid-based ENPs mainly refer to liposomes and lipid nanoparticles (LNPs). As the most commonly used drug-delivery system, these ENPs are the least toxic, with several advantageous properties, such as formulation simplicity, high bioavailability, payload flexibility, and high encapsulation efficiency (Mitchell et al. 2021). Liposomes are most often made up of phospholipids and can form unilamellar and multilamellar vesicular structures (Figure 10.1). Because of the amphiphilicity of phospholipids, liposomes have strong emulsifying and wetting characteristics that give liposomes the ability to efficiently deliver a variety of physiochemically different drugs, such as hydrophilic, hydrophobic, or lipophilic drugs, and they can even entrap hydrophilic and lipophilic drugs within the same system (Allen and Cullis 2013). As phospholipids are the main components of the cell membrane, liposomes possess excellent biocompatibility and do not produce significant toxic or immunological responses when exposed to living tissues (Allen and Cullis 2013; Puri et al. 2009). However, it is because of this biocompatible property, the liposome-based drug-delivery systems suffer from rapid clearance by the reticuloendothelial system, adversely affecting their bioavailability. To prolong their circulation lifetime, liposomes are often surface modified. One of the common surface modification methods is the addition of neutral, hydrophilic polymers such as polyethylene glycol (PEG) to the outer surface (Khan et al. 2020; Sercombe et al. 2015). This strategy can effectively block opsonin binding and delay clearance, thereby increasing the circulation time of modified liposomes. The PEGylated liposomes also possess other desirable drug-delivery properties, such as improved pharmacokinetics and a higher drug-loading capacity (Khan et al. 2020). It has been shown that PEGylated liposomes, when used for the encapsulation of the hydrophilic drugs irinotecan and floxuridine, significantly improved the pharmacokinetic profiles of these drugs, leading to a marked increase in the progression-free survival of patients suffering from colorectal cancer (Batist et al. 2009).

Other molecules that have been exploited for surface modification include vitamins, carbohydrates, peptides, antibodies, and single-stranded nucleic acids (Khan et al. 2020). Liposomes

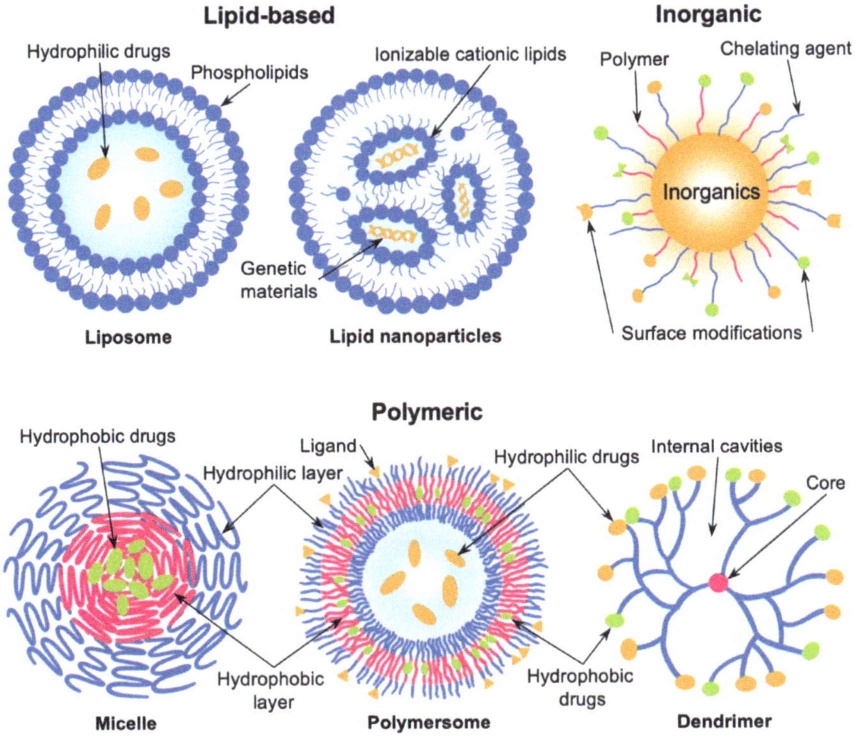

Figure 10.1 Commonly used environmental engineered nanoparticles drug-delivery systems.

modified by these molecules can exhibit greater solubility, higher stability, or specific drug targeting properties. However, the properties of liposomes can also be improved by supplementing with functional additives during their synthesis. Liposomal vesicles have a tendency to self-aggregate, which results in their sedimentation and loss of the encapsulated active molecules (Puri et al. 2009). One common solution to this problem is to increase the concentration of cholesterol in liposomes, which can strengthen the lipid membrane and lower membrane fluidity (Puri et al. 2009). Liposomes are also susceptible to oxidative damage due to the chemically unsaturated phospholipids being prone to oxidation. Adding antioxidants, such as α-tocopherol or butylated hydroxyl toluene, to liposomes can prevent free radical-induced damage to phospholipids, thus enhancing the stability of liposomes for drug delivery (Chmykh and Nadeau 2020; Khan et al. 2020).

Lipid nanoparticles (LNPs) are different from traditional liposomes in several aspects. LNPs are specialized in the delivery of nucleic acids since the major component of LNPs is ionizable cationic lipids, which can form complexes with negatively charged genetic materials (Hou et al. 2021; Leung et al. 2015) (Figure 10.1). Another difference is that LNPs form micellar structures within the particle core, which is different from liposomes that have only the outer layer of lipids, though both have spherical structures (Hou et al. 2021). LNPs have found many applications in gene therapy due to their efficacy for nucleic acid delivery, and also due to their simple synthesis, small size, and serum stability (Yonezawa et al. 2020). One of the most widely used LNP systems is Lipofectin, which is suitable for transfecting DNA, RNA, and oligonucleotides into many mammalian cells (Cheng and Lee 2016). Lipofectin consists of two different lipids: 1,2-di-O-octadecenyl-3-trimethylammonium-propane (DOTMA) and 1,2-dioleoyl-sn-glycero-3-phosphoethanolamine (DOPE). DOTMA is a quaternary ammonium lipid and has a cationic head group; DOPE is a neutral phospholipid that helps DOTMA in gene delivery by reducing the surface charge density of lipoplexes. Other cationic lipids often used for nucleic acid delivery include (2S)-2,5-bis(3-aminopropylamino)-N-[2-(dioctadecylamino)acetyl] pentanamide (DOGS; Transfectam) and DC-Cholesterol (Behr et al. 1989; Cheng and Lee 2016; Kulkarni et al. 2019). These lipids are an ideal platform for the delivery of nucleic acid therapies since they have a near-neutral charge at the physiological pH but become positively charged in acidic endosomal compartments, promoting endosomal escape for intracellular delivery (Patel et al. 2019; Vhora et al. 2019). Notably, the two recently authorized vaccines for coronavirus disease 2019 (COVID-19), mRNA-1273 and BNT162b, utilize this type of LNPs to deliver the antigen mRNA (Thi et al. 2021). Many different LNP-nucleic acid formulations have been developed and some of them are currently under clinical trials for the prevention and treatment of viral infections, cancer, or genetic diseases (Anselmo and Mitragotri 2019; Fenton et al. 2018).

Polymeric ENPs

Polymeric ENPs can be synthesized from natural or synthetic materials and have a wide variety of possible structures and characteristics. As a highly versatile drug-delivery system, polymeric ENPs deliver drugs in various ways. Drugs can be encapsulated within the ENP core, entrapped in the polymer matrix, chemically conjugated to the polymer, or bound to the ENP surface. Various payloads can be delivered by polymeric ENPs, including hydrophobic and hydrophilic compounds and cargos with different molecular weights, such as small molecules, DNA, RNA, proteins, and vaccines (Mitchell et al. 2021). The most common forms of polymeric ENPs are nanocapsules and nanospheres. They differ in their structures. Nanocapsules contain a cavity surrounded by a polymer membrane, whereas nanospheres are a solid polymer matrix system. Drugs are usually confined in the cavity of nanocapsules, whereas in nanospheres, they are dispersed in the polymer matrix. Polymeric ENPs can be synthesized in different shapes, such as micelles, polymersomes, and dendrimers (Mitchell et al. 2021). Polymeric micelles are self-assembled colloidal nanoparticles formed by amphiphilic block copolymers in an aqueous environment. They typically consist of both hydrophilic and hydrophobic units, with the hydrophobic portion forming the core of micelles while the hydrophilic portion forming the micelles' shell (Figure 10.1) (Ghezzi et al. 2021). Because of this core-shell structure and the core being hydrophobic, polymeric micelles are considered an excellent drug-delivery system for poorly water-soluble drugs, and nonpolar drugs can be solubilized within the hydrophobic core of the micelles (Ghezzi et al. 2021). In addition, polymeric micelles possess high structural stability due to the interactions between the polymeric chains in the core-forming hydrophobic blocks, which enhances the retention of encapsulated drugs and further increases drug bioavailability (Jhaveri and Torchilin 2014).

Like polymeric micelles, polymersomes are also an excellent system in drug-delivery applications, with several advantageous properties. Polymersomes are vesicular, tiny hollow spheres with a membrane made of amphiphilic block copolymers (Lee and Feijen 2012). They typically contain an aqueous solution in their core surrounded by a bilayer membrane. Such a structure allows polymersomes to simultaneously carry hydrophilic drugs inside the core and hydrophobic drugs within the bilayer membrane (Figure 10.1). The development of stimuli-responsive polymersomes has made polymersomes one of the most desirable systems in drug delivery. In stimuli-responsive polymersomes, the release of drugs can be controlled by changing the stability and permeability of the membrane (Meng et al. 2009; Rideau et al. 2018). Dependent on the types of block copolymers, the membrane can undergo those changes in response to different stimuli, such as pH, temperature, redox conditions, light, magnetic field, ionic strength, and glucose concentration (Meng et al. 2009). Other advantageous features of polymersomes are structural stability and high drug-delivery efficiency. Polymersomes are reported to have improved stability and cargo-retention efficiency compared with liposomes (Rideau et al. 2018). The higher retention efficiency is attributed to the presence of a thicker and more robust hydrophobic membrane in polymersomes, while the structural stability of their membrane reduces the leakage of incorporated compounds, further contributing to their drug-delivery efficiency (Anajafi and Mallik 2015). Similar to the surface modification that alters liposomes' properties, the properties of polymersomes can be modified and improved by copolymerization of different polymers, such as that among PEG, poly-dimethylsiloxane, or polylactide (PLA) (Discher et al. 2007). Through copolymerization with PEG, for example, PLA can be improved in hydrophilicity, degradation rate, and crystallization (Danafar et al. 2014).

Dendrimers have the most complex architectures compared with polymeric micelles and polymersomes. They are hyperbranched nanoparticles, often with a spherical three-dimensional structure and a radial symmetry (Abbasi et al. 2014). A typical dendrimer is composed of three main components: an initiating core molecule, interior repetitive branching units attached to the central core, and terminal modifiable functional groups (Figure 10.1) (Hsu et al. 2017). The terminal functional groups enable conjugation of biomolecules or contrast agents to the surface, while drugs can be loaded onto the interior repetitive branching units. Dendrimers are most commonly used for the delivery of nucleic acids and small molecules, though they have been used to carry many other types of cargoes (Chauhan 2018; Hsu et al. 2017). One of the commonly used dendrimers is poly(amidoamine) (PAMAM). PAMAM is a class of charged dendrimers, often with the primary amines, such as ethylenediamine, ammonia, and cystamine, as the initiating core, and with the repetitive tertiary amides, which are structurally very similar to polyamide fibers, as branches (Li et al. 2018). The number of available surface groups increases exponentially with the increasing diameter of dendrimers. Dendrimers are considered excellent candidates for drug delivery due to their versatile, advantageous properties, such as high drug-loading capacity, high water solubility, and biocompatibility (Chauhan 2018). Several dendrimer-based products are currently in clinical trials, where they serve as theranostic agents, transfection agents, topical gels, and contrast agents (Kannan et al. 2014; Ray et al. 2018).

The major concern for the use of polymeric ENPs in drug delivery is cytotoxicity. The high surface area of polymeric ENPs and the strong attractive interaction between the particles can cause their aggregation or agglomeration, which may block blood capillaries (Ashraf et al. 2018). Additionally, dendrimers with the positively charged surface groups can interact nonspecifically with the negatively charged cell membrane, leading to membrane destabilization, nanohole formation, cytosolic enzyme release, or cell death (Jain et al. 2010). Although the modifiable surface chemistry of polymeric ENPs is a convenient strategy to improve their pharmaceutical applications, the impurities in decoration reagents themselves can pose different risks of toxicity. For example, PEG, a commonly used decoration reagent, can reduce ENPs' protein binding in the blood and avoid their recognition by the reticular endothelial system, thus prolonging the circulation time of ENPs in the human body (D'souza and Shegokar 2016). However, the impurities in PEG, including ethylene oxide and 1,4-dioxane, are the known carcinogens and respiratory irritants (Gi et al. 2018; Stayner et al. 1993). Long-term exposure to ethylene oxide can cause serious health consequences, including damage to the nervous system (Brashear et al. 1996). To date, only a small number of polymeric nanomedicines are currently approved by the FDA, partly due to their cytotoxicity.

Inorganic ENPs

Inorganic ENPs are made of inorganic materials, which can be classified based on the properties of the chemical elements used. There are three major classes of inorganic ENPs: noble metal nanoparticles, magnetic nanoparticles, and fluorescence nanoparticles (Mitchell et al. 2021). The noble metal nanoparticles have high chemical stability and are prepared from noble metals such as gold and silver. Magnetic nanoparticles are made of Ni, Co, Fe, Fe_3O_4, or FePt, which are superparamagnetic with large magnetic moments in a magnetic field. Fluorescent nanoparticles, such as quantum dots and SiO_2, are widely used for fluorescent imaging of cells and tissues (Ghaderi et al. 2011; Vangijzegem et al. 2019; Wang et al. 2015) (Figure 10.1). Among the noble metal nanoparticles, gold NPs (AuNPs) are the most studied, due to their high stability, ease of chemical synthesis, and a high degree of surface functionalization. AuNPs have been used in various architectural forms, such as nanospheres, nanorods, nanostars, nanoshells, and nanocages (Wang et al. 2020). One interesting finding is that the gold-dendrimer nanoparticles exhibit high levels of uptake and selective targeting to certain organs, even though no specific targeting moieties are placed on their surface (Kesharwani et al. 2019). However, the surface of AuNPs can actually be easily functionalized, granting them additional properties and delivery capabilities. When coated with the human serum albumin, AuNPs can deliver drugs across the blood-brain barrier, and this ability is attributed to the low surface charge of the particles and their ability to absorb a large amount of creatine in the presence of the albumin layer (Lopez-Viota et al. 2009). Additionally, AuNPs possess photothermal properties due to the free electrons at their surface. Since the frequency of oscillation of these electrons is dependent on the size and shape of the particles, the photothermal efficiency of AuNP is size- or shape-tunable (Yang et al. 2019).

Iron oxide, Fe_3O_4 or Fe_2O_3, is the most commonly researched material for magnetic nanoparticles. Initially, iron oxide ENPs were prepared as contrast agents for magnetic resonance imaging because of their response to the magnetic fields. However, due to several other remarkable superparamagnetic properties, such as high field irreversibility, a high saturation field, extra anisotropy, or shifted loops after field cooling, iron oxide ENPs have been extended to many other applications, such as magnetic hyperthermia, thermoablation bioseparations, and biosensing (Paul and Sharma 2020). Their finite size and surface effects also made iron oxide ENPs excellent drug-delivery vehicles, and they make up the majority of FDA-approved inorganic nanomedicines (Bobo et al. 2016). It has been shown that chlorotoxin, when conjugated with iron oxide ENPs via the amine-functionalized PEG silane, exhibited substantially enhanced cellular uptake due to the tumor cell-specific binding, and the invasion inhibition rate was significantly increased compared with unconjugated chlorotoxin (Sreeram et al. 2009). Another study revealed that when combined with the superparamagnetic iron oxide nanocore, liposomes showed high biocompatibility and enhanced drug-delivery capabilities (Soenen et al. 2009).

Quantum dots (QDs), typically made of semiconducting materials, are primarily used in *in vitro* imaging applications and *in vivo* diagnostics, due to their optical and electronic properties (Wagner et al. 2019). They also show promise as drug-delivery systems and can act as nanocarriers or as fluorescent labels in a more complex architecture (Matea et al. 2017). For example, a drug-delivery system developed based on ZnO QDs has been used for intracellular controlled release of drugs (Cai et al. 2016). This system was developed to specifically deliver doxorubicin to cancer cells. To render cancer cell-specific targeting, hyaluronic acid was conjugated to ZnO QDs, which acts as a targeting ligand and can specifically bind to the overexpressed glycoprotein CD44 in cancer cells (Cai et al. 2016). One advantageous feature of this drug-delivery platform is pH-sensitivity. After uptake by cancer cells, the acidic endosomal environment can trigger the disassociation of doxorubicin from ZnO QDs, leading to a controlled release (Cai et al. 2016). QDs have also been exploited for gene delivery (Zhou et al. 2016). Carbon QDs (CQDs), when doped with S, N, and Gd, have a strong capacity to condense plasmid DNA and have a high DNA transfection efficiency (Chiu et al. 2016).

Other inorganic ENPs, such as calcium phosphate-based ENPs, have also been used successfully for gene and drug delivery. Nano-sized calcium phosphates have been studied as the drug-delivery systems in the intestinal delivery of insulin or antibiotics, and they have been shown to possess exceptional biocompatibility and bioactivity (Degli Esposti et al. 2018; Geuli et al. 2017; Uskokovic and Uskokovic 2011). Generally, inorganic ENPs are qualified for many applications, such as diagnostics, imaging, and drug delivery. Their unique magnetic, radioactive, or plasmonic properties find applications where any required properties are unattainable by organic materials. However, inorganic ENPs are limited in clinical applications due to their low solubility and toxicity concerns, especially when they are in formulation with heavy metals. Overexposure to the ENPs containing heavy metals, such as gold, may cause skin rashes, bone marrow depression, stomach and intestinal bleeding, focal or generalized continuous fine vibrating muscle movements, or yellowing of the skin (Boisselier and Astruc 2009; Kim et al. 2019; Schrand et al. 2010). Thus, further modifications are often needed for inorganic ENPs to reduce their toxicity.

Adverse Epigenetic Effects of ENPs

Overview of Epigenetic Toxicity of ENPs

Epigenetic regulation is a fundamental cellular mechanism that plays a role in many aspects of cellular functions by regulating chromatin structure and gene expression (Skvortsova et al. 2018). It has been shown that ENP exposure can induce epigenetic alterations, through which ENPs impact several normal and pathological cellular functions, including stress response, DNA damage repair, cell cycle control, oxidative stress, apoptosis, and inflammatory response (Pogribna and

Hammons 2021; Smolkova et al. 2017). The epigenetic changes caused by ENPs could have the potential to lead to serious health implications, since numerous diseases have been associated with epigenetic dysregulation, such as cancer, diabetes, or neurodegenerative disorders (Shyamasundar et al. 2015). As a result, there is increasing interest in understanding how ENPs cause epigenetic changes. Understanding their epigenetic toxicity is important for the safe application of ENPs as carriers for drug delivery.

Three main mechanisms by which ENPs may cause epigenetic toxicity are the change in global or locus-specific DNA methylation, the change in histone posttranslational modifications, and the change in noncoding RNA (ncRNA) expression (Gedda et al. 2019). It has been shown that ENPs can induce significant global or locus-specific changes in DNA methylation profiles in many types of mammalian cells, and these DNA methylation changes showed a profound impact on the remodeling of the chromatin and grossly affected the expression of the methylated genes (Sooklert et al. 2019). DNA methylation, a covalent addition of a methyl group to the 5-carbon of cytosine (5mC), is the most important epigenetic mark that usually locks genes in the silenced state (Moore et al. 2013). In the genome, DNA methylation is often found within the CpG islands at the gene promoter regions. There are three DNA methyltransferases (DNMTs) in mammalian cells, namely DNMT1, DNMT3A, and DNMT3B, and these enzymes catalyze DNA methylation using S-adenosyl methionine (SAM) as a methyl donor. Administration of ENPs has been found not only to affect the level of DNA methylation but also the expression level of the DNA methyltransferases (Gedda et al. 2019; Smolkova et al. 2017). Histone modification, another important epigenetic mark, can also be disturbed by ENPs (Pogribna and Hammons 2021). Histone proteins are subject to multiple posttranslational covalent modifications, such as phosphorylation, methylation, and acetylation (Lennartsson and Ekwall 2009). These modifications are dynamic and reversible, usually controlled by pairs of enzymes with opposing activities. Histone kinases and histone phosphatases regulate the level of histone phosphorylation. Histone methyltransferases (HMTs) and histone demethylases (HDMs) regulate the level of histone methylation. Histone acetyltransferases (HATs) and histone deacetylases (HDACs) regulate the level of histone acetylation (Lennartsson and Ekwall 2009). ENPs have been found to induce changes in all these modifications. Since these histone modifications play roles in DNA damage response, gene transcription, or mitosis, it is possible that ENPs could cause epigenetic toxicity by perturbing these critical cellular processes. Recent studies showed that ENPs can also affect the expression of ncRNAs (Gedda et al. 2019). As another important epigenetic mechanism, ncRNA regulation plays pivotal roles in many cellular processes, such as cell proliferation, differentiation, and survival (Paul et al. 2018). They are usually categorized according to their length: short ncRNAs (<200 nucleotides) and long ncRNAs (>200 nucleotides). Short ncRNAs include microRNAs (miRNAs), small interfering RNAs (siRNAs), and Piwi-interacting RNAs (piRNAs). To date, miRNAs, with a length of about 20 nucleotides, are the most studied ncRNAs. miRNAs regulate gene expression by affecting mRNA stability and by targeting mRNA for degradation, and about one-quarter of the human genes are known to be regulated by miRNAs (Kaikkonen et al. 2011; Paul et al. 2018; Portela and Esteller 2010). It has been shown that ENPs disturb the expression of the miRNAs involved in blood clotting processes, oxidative stress, and apoptosis, which has suggested possible mechanisms for ENP-induced epigenetic toxicity (Balansky et al. 2013; Brzoska et al. 2019; Gedda et al. 2019; Halappanavar et al. 2011).

The epigenetic effects of ENPs have been investigated on various cell types and tissues in both *in vitro* and *in vivo* studies (Tables 10.1–10.3). Most of these studies have focused on inorganic ENPs, especially those formulated with metals, such as gold, silver, titanium, and zinc. These metallic ENPs were found to cause changes in DNA methylation, histone modification, and ncRNA expression. Nonmetallic inorganic ENPs, including carbon nanoparticles, silicon ENPs, and graphene

Table 10.1 Epigenetic effects of common metallic environmental engineered nanoparticles.

Environmental engineered nanoparticles	Study subject	Concentration	Exposure time	Epigenetic effect	References
AuNP					
	Human normal skin fibroblasts; A375	62.5 μg/mL	24, 48 h	Global DNA hypomethylation	Patil et al. 2019
	BALB/c mice	0.25, 2.5 mg/mL	48 h	Hypermethylation at *Atm*, *Cdk*, and *Gsr* promoters; hypomethylation at *Gpx* promoter in lung tissue	Tabish et al. 2017
	MDA-MB-231; MDA-MB-468	100–500 μg/mL	24 h	Increased γ-H2AX. Negatively charged AuNPs: increased MKP-1 expression; dephosphorylation of H3S10; deacetylation of H3K9/H3K14. Positively charged AuNPs: decreased MKP-1 expression; increased acetylation of H3K9/H3K14	Surapaneni et al. 2018
	Small airway epithelial cells	1 nM	72 h	Decreased H3K27me3	Shyamasundar et al. 2015
	HepG2	12, 24 μg/mL	24, 48 h	Decreased HDAC activity; decreased HDAC1 and HDAC2 levels	Bishayee et al. 2015
	Male Wistar rats	3 mg/rat	7, 60 d	21 miRNAs dysregulated; miR-298 upregulated	Chew et al. 2012
	MRC5	1 nM	48, 72 h	Chromatin condensation; miR-155 upregulation caused *PROS1* downregulation	Ng et al. 2011
	Adult Swiss albino mice	3.3 mg/kg (gestation)	10, 12, 14, 17 d	Fetus lung: 28 miRNAs dysregulated; let-7 and miR-183 upregulation. Fetus liver: 5 miRNAs dysregulated; let-7 and miR-183 upregulation	Balansky et al. 2013
AgNP					
	HT22	1–20 μg/mL	48 h	Global DNA hypermethylation; increased levels of DNMT1, DNMT3a, and DNMT3b	Mytych et al. 2017

(Continued)

Table 10.1 (Continued)

Environmental engineered nanoparticles	Study subject	Concentration	Exposure time	Epigenetic effect	References
	BEAS-2B	1 μg/mL	6 wk	Altered DNA methylation at *LINC02200* promoter: 6 CpG sites and 5 tiling regions	Gliga et al. 2018
	A549	10–200 μg/mL	48, 72 h	Global DNA hypermethylation	Blanco et al. 2017
	Mice with ICR	1 mg/kg	6.5 d	DNA hypomethylation at *ZAC1* promoter in placenta tissue	Zhang et al. 2015
	NIH3T3	5–25 μg/mL	24 h	Altered DNA methylation	Gurunathan et al. 2018
	A549	10–200 μg/mL	48, 72 h	Global H3 deacetylation; elevation of total H3 level; increased γ-H2AX	Blanco et al. 2017
	A549; MCF7	1 μg/mL	4 h	Increased γ-H2AX	Zhao et al. 2016
	A549; MCF7; HaCaT	0.3 μg/ mL	24 h	Increased phosphorylation of H3S10 via activation of aurora kinase	Zhao et al. 2017
	A549	1 μg/mL	10 h	Increased H3S10 phosphorylation, independent of DNA damage	Zhao et al. 2019
	Mouse erythroleukemia cells	8 μg/mL	72 h	Decreased H3K4me3 and H3K79me1 on the β-globin locus	Qian et al. 2015
	Human Jurkat T cells; Jurkat clone E6-1 cells	0.2 mg/L	24 h	Altered expression of 63 miRNAs; *MT1F* and *TRIB3* gene expression negatively correlated with miR-219-5p	Eom et al. 2014
	MC3T3-E1	20 μg/ml	24 h	Altered expression of miRNAs associated with bone formation	Mahmood et al. 2011
TiO_2 ENP					
	MRC5	0.5, 4 μg/mL	24, 48 h	Global DNA hypomethylation; decrease in total DNMT activity	Patil et al. 2016
	A549	40 μg/cm^2	48, 72 h	Global DNA hypomethylation; demethylation of *LINE1* repetitive elements	Stoccoro et al. 2017

Table 10.1 (Continued)

Environmental engineered nanoparticles	Study subject	Concentration	Exposure time	Epigenetic effect	References
	SAEC	0.5, 30 μg/mL	24 h	Demethylation of *SINE B1* repetitive elements	Lu et al. 2016
	Human skin A-431 cells; HepG2; human colon CaCo-2 cells	100 μg/mL	24, 72 h	Global DNA hypomethylation; hypermethylation of *CDKN1A*, *DNAJC15*, *GADD45A*, *GDF15*, *INSIG1*, *SCARA3*, *TP53*, and *BNIP3* genes; altered expression of *DNMT1*, *DNMT3A*, *DNMT3B*, *MBD2*, and *UHRF* genes	Pogribna et al. 2020
	A549	6.25–100 μg/mL	24 h	Hypermethylation of *PARP1* promoter	Bai et al. 2015
	A549	50–1000 μg/mL	1, 24, 48 h	Increased γ-H2AX, independent of oxidative stress	Toyooka et al. 2012
	THP-1; HPMEC-ST1.6R	5–800 μg/mL	4, 24 h	Increased γ-H2AX	Hanot-Roy et al. 2016
	Human dermal fibroblasts	1–100 μg/mL	24 h	Increased γ-H2AX	Prasad et al. 2013
	Human skin fibroblasts	10–1000 μg/mL	24 h	Increased γ-H2AX	Setyawati et al. 2013
	Human dermal fibroblasts	1–100 μg/mL	24 h	Increased ATM and Chk2 phosphorylation	Prasad et al. 2013
	A549	800 μg/mL	24 h	Decreased HDAC9 expression, dependent on ROS	Jayaram and Payne 2020
	C57BL/6 female mice	$1.7 \times 10^6/cm^3$	1 h/d; 11 d	Upregulation of miR-1, miR-449a, and miR-135b in lung tissue	Halappanavar et al. 2011
	A549	20–100 μg/mL	24 h	Downregulation of miR-21 and miR-30a along with altered miR-155 expression	Alinovi et al. 2017
ZnO ENP					
	MRC5	4, 8 μg/mL	24, 48 h	Global DNA hypomethylation; decrease in total DNMT activity; reduction in expression of DNMT1 and DNMT3a	Patil et al. 2016

(Continued)

Table 10.1 (Continued)

Environmental engineered nanoparticles	Study subject	Concentration	Exposure time	Epigenetic effect	References
	HEK293	25, 50 μg/mL	48 h	Global DNA hypomethylation; increase in 5-hmC associated with increased expression of *TET1* and *TET2* genes	Choudhury et al. 2017
	Hens	10–200 mg/kg	24 wk	Increased γ-H2AX	Liu et al. 2017a
	Chicken ovarian cells	10–200 mg/kg	24 wk	Increased γ-H2AX	Liu et al. 2017a
	CHO-K1	1–100 μg/mL	4 h	Increased γ-H2AX	Toduka et al. 2012
	HaCaT	20, 50 μg/mL	24 h	Deacetylation of H4K3; increased methylation of H3K9me2; upregulation of *G9a* and *GLP* genes; downrcgulation of *HATs GCN5*, *P300*, and *CBP* genes	Gao et al. 2016
	T24	10 μg/mL	48 h	Decrease in global H3K27me3 level at *RUNX3* gene promoter	Zhang et al. 2020

A375: human skin melanoma cells; A549: human lung adenocarcinoma epithelial cells; BEAS-2B: human non-tumorigenic lung epithelial cells; CHO-K1: Chinese hamster ovary cells; HaCaT: human keratinocytes; Hela: human cervical cancer cells; HEK293: human embryonic kidney cells; HepG2: human hepatocellular carcinoma cells; HPMEC-ST1.6R: human pulmonary microvascular endothelial cells; HT22: mouse hippocampal neuronal cells; ICR: imprinting control region; MC3T3-E1: mouse osteoblastic cells; MCF7: human breast adenocarcinoma cells; MDA-MB-231 and MDA-MB-468: human triple-negative breast cancer cells; MRC5: human lung fibroblast cells; NIH3T3: mouse embryonic fibroblast cells; PC12: rat pheochromocytoma cells; ROS: reactive oxidative species; SAEC: human small airway epithelial cells; SK-Hep-1: human hepatocellular carcinoma cells; T24: human bladder cancer cells; THP-1: human leukemia monocytic cells; γ-H2AX: H2AX phosphorylation.

Table 10.2 Epigenetic effects of additional metallic ENPs.

Environmental engineered nanoparticles	Study subject	Concentration	Exposure time	Epigenetic effect
CuO ENP	SK-Hep-1	1–100 μg/mL	24 h	Increased γ-H2AX (Kung et al. 2015)
	CHO-K1	1–100 μg/mL	4 h	Increased γ-H2AX (Toduka et al. 2012)
	A549	200 μg	36 h	Decrease in total HDAC activity; reduction in the levels of HDAC1, HDAC2, HDAC3, HDAC5, HDAC9, and HDAC11 mRNAs (Kalaiarasi et al. 2018)
CuNP	Male albino Wistar rats	3.25, 6.5 mg/kg	4 wk	Global DNA hypomethylation (Ognik et al. 2019)

Table 10.2 (Continued)

Environmental engineered nanoparticles	Study subject	Concentration	Exposure time	Epigenetic effect
Fe_2O_3	NIH3T3	50–150 μg/mL	24 h	Genome-wide changes in miRNA expression profiles (Li et al. 2011)
SPION	PC12	214 μg/mL	24 h	Genome-wide changes in miRNA expression profiles (Sun et al. 2015)
	HepG2	5 μg/mL	24 h	Altered miR-34a expression (Balasubramanian et al. 2020)
IONP	Human*	0.04–0.24 mg/m^3	2 y	Peripheral blood samples: elevated 5-hmC concurrent with reduced *sTfR* expression and independent of oxidant markers (Yu et al. 2020a)
CeO_2 ENP	Human peripheral blood lymphocytes	6, 12, 18 μg/mL	3–24 h	Increased γ-H2AX (Konen-Adiguzel and Ergene 2018)
As_2O_3 ENP	HEK293; HeLa	0.2–0.8 μM	24, 48, 72 h	Decrease in global H4K16ac level; increased HDAC4 expression (Liu et al. 2015a)
CdTe-QD	MCF7	5 μg/ mL	4, 24 h	Global histone hypoacetylation; chromatin decondensation due to H3 deacetylation (Choi et al. 2008)
	NIH3T3	15–45 μg/mL	12, 24 h	Genome-wide changes in miRNA expression profiles (Li et al. 2011)

*Human: 23 people occupationally exposed to IONPs for 2 years.
A549: human lung adenocarcinoma epithelial cells; CHO-K1: Chinese hamster ovary cells; Hela: human cervical cancer cells; HEK293: human embryonic kidney cells; HepG2: human hepatocellular carcinoma cells; MCF7: human breast adenocarcinoma cells; NIH3T3: mouse embryonic fibroblast cells; PC12: rat pheochromocytoma cells; SK-Hep-1: human hepatocellular carcinoma cells.

Table 10.3 Epigenetic effects of nonmetallic environmental engineered nanoparticles.

Environmental engineered nanoparticle	Study subject	Concentration	Exposure time	Epigenetic effect
CNP	Zebrafish	10, 30 μg/mL	60 d	Global DNA hypermethylation; hypomethylation of *LEBP*, *CD248b*, and *IL11a* (Zhou et al. 2019)
GQD	Zebrafish	2, 10, 50 mg/L	7 d	Tissue-specific global DNA hypermethylation (Hu et al. 2019)
SiO_2 ENP	HaCaT	2.5–10 μg/mL	24 h	Global DNA hypomethylation; decreased protein expression of DNMT1, DNMT3a, and MBD2 (Gong et al. 2010)
	HaCaT	10, 100 μg/mL	24 h	Reduced methylation of *Alu* repetitive elements (Sooklert et al. 2019)

(Continued)

Table 10.3 (Continued)

Environmental engineered nanoparticle	Study subject	Concentration	Exposure time	Epigenetic effect
	HaCaT	10 μg/mL	24 h	Promoter hypermethylation and decreased expression of PARP1 protein (Gong et al. 2012)
	Bhas 42	15–25 $\mu g/cm^2$	48 h	~80% decrease in global DNA methylation; increased levels of DNMT3a and DNMT3b proteins (Seidel et al. 2017)
	BEAS-2B	5 μg/mL	30 passages	Global DNA hypermethylation; hypermethylation of *CREB3L1* and *BCL-2* gene promoters (Zou et al. 2016)
	Human*	N/A	2.7 y	Global DNA hypomethylation in peripheral white blood cells (Liou et al. 2017)
	Human#	20% (wt) of $2.6 \times 10^4/cm^3$	14.5, 0.5 y	Long-term exposure caused changes in DNA methylation: 341 CpG sites hypomethylated and 364 CpG sites hypermethylated. Short-term exposure did not affect DNA methylation (Rossnerova et al. 2020)
	CaCo-2	4–64 μg/mL	24 h	Increased γ-H2AX (Tarantini et al. 2015)
	Bhas 42	15–25 $\mu g/cm^2$	48 h	Increase in H3K4ac, H3K4me3, H3K9ac, and H3K27ac levels; increased HDAC2 expression; decreased HDAC1 and HDAC6 expression (Seidel et al. 2017)
	A549	50 μg/mL	3–24 h	Decreased SIRT6 expression; upregulation of FST due to suppressed deacetylation of H3K9 and H3K56 at *FST* promoter (Zhang et al. 2018)
	BALB/c mice	10–40 mg/kg	4, 8, 24, 72 h	Upregulated miR-122 and miR-192 (Nagano et al. 2013)

*Human: 31 people exposed to SiO_2 for 2.7 years (14.8 hours per week).
#Human: 20 people occupationally exposed to SiO_2 for 14.5 years or half a year.
HaCaT: human keratinocytes; Bhas 42: mouse Bhas 42 cells; BEAS-2B: human non-tumorigenic lung bronchial epithelial cells; CaCo-2: human colorectal adenocarcinoma cells; A549: human lung adenocarcinoma epithelial cells; MBD: methyl GpG binding protein 2.

QDs (GQDs), were also shown to cause epigenetic changes, though to a lesser extent. However, the epigenetic effects of lipid-based and polymeric ENPs were rarely studied. One possible reason for this lack of study is that these non-inorganic ENPs are generally considered safer, especially for liposomes, due to their cell-like structure (Scioli Montoto et al. 2020).

Epigenetic Toxicity of Metallic ENPs

Altered DNA Methylation

Metallic ENPs can induce significant changes in DNA methylation at both global and locus-specific levels. They can induce DNA hypomethylation or DNA hypermethylation, and in most cases, the hypomethylation and hypermethylation co-occur at different gene loci in the same affected genome. Hypomethylation of CpG dinucleotides usually activates gene expression, whereas DNA hypermethylation results in gene silencing (Moore et al. 2013). Thus, the functional consequences of the metallic ENP-induced changes in DNA methylation can be very complicated, involving both gene activation and gene silencing and occurring at either the global level or at various specific gene sequences. Their differential effects on DNA methylation appear to be cell-type specific, even the same type of ENPs can have the opposite effects in different cells. AuNPs were shown to induce global DNA hypomethylation in human skin fibroblasts and human skin melanoma A375 cells, whereas in human lung adenocarcinoma epithelial cells, they caused global DNA hypermethylation (Patil et al. 2019). These global changes in DNA methylation were correlated with the corresponding changes in the expression levels of DNA methyltransferases, with the DNA hypomethylation correlated with decreased expression of DNMT1, DNMT3A, and DNMT3B, while the DNA hypermethylation correlated with increased expression of these enzymes (Patil et al. 2019). AuNPs can also alter the DNA methylation level at specific gene loci. Mice intratracheally exposed to AuNPs were hypermethylated at the promoters of several genes in the lung, including the genes coding for ataxia-telangiectasia mutated (ATM), cyclin-dependent kinase, and glutathione reductase (Tabish et al. 2017). In addition to the promoter hypermethylation, AuNPs were found to induce hypomethylation at the promoter of the glutathione peroxidase gene in the same tissue (Tabish et al. 2017). All these genes are involved in oxidative stress response, suggesting a potential link between the AuNP-induced DNA methylation alteration and the stress response signaling pathways.

Other metallic ENPs have been investigated for their effects on DNA methylation, including the ENPs formulated with titanium dioxide (TiO_2), zinc oxide (ZnO), or silver (Ag). Like AuNPs, these ENPs also exhibited complex effects and can affect DNA methylation at both global and gene-specific levels. It was shown that TiO_2 ENPs can cause global DNA hypomethylation in human fetal lung fibroblast MRC5 cells, and these DNA methylation changes were concomitant with the decreased expression of DNA methyltransferases (Patil et al. 2016). It was also shown that TiO_2 ENPs can significantly increase the methylation level of the gene coding for poly(ADP-ribose) polymerase 1 (PARP1) in human lung adenocarcinoma epithelial cells (Bai et al. 2015). PARP1 is a highly conserved DNA-binding protein involved in DNA single- and double-strand break repairs (Ray Chaudhuri and Nussenzweig 2017). The increased methylation of the *PARP1* gene suggests that TiO_2 ENPs might downregulate the DNA damage response pathways, thus implying possible cytotoxicity.

ZnO ENPs were found to induce global DNA hypomethylation in human lung fibroblasts and human embryonic kidney 293 cells (Choudhury et al. 2017; Patil et al. 2016). They were also found to cause increased expression of ten-eleven translocation (*TET*) genes and an increase in DNA 5-hydroxymethylcytosine (5hmC) levels (Choudhury et al. 2017). These results suggest that the global DNA hypomethylation induced by ZnO ENPs might be due to the activation of an active DNA demethylation process since TET proteins catalyze the conversion of 5mC to 5hmC and play essential roles in active DNA demethylation (Choudhury et al. 2017). AgNPs were found to induce global DNA hypermethylation in human lung adenocarcinoma epithelial cells, and the increased DNA methylation was associated with increased expression of DNA methyltransferases (Blanco et al. 2017; Mytych et al. 2017). AgNPs can also cause locus-specific changes in DNA methylation. In the placental tissue, AgNP exposure significantly decreased DNA methylation at the *ZAC1* gene promoter and slightly increased the DNA methylation level of the *IGF2r* gene (Zhang et al. 2015). The *ZAC1* and *IGF2r* genes are two important imprinted genes involved in the control of embryonic growth. It has been shown that the change in the expression of these placental genes is associated with the accelerated fetal growth during the third trimester, due to the prematurely stopped proliferation of primordial germ cells and their early entry into the meiotic prophase (Zhang et al. 2015). Thus, AgNPs may produce developmental toxicity by altering the expression of the *ZAC1* and *IGF2r* genes and disturbing the first meiosis progression in the placental tissue.

Altered Histone Modification

Metallic ENPs have been shown to induce the changes in several histone modifications, through which they may cause epigenetic toxicity, or which act as indicators of ENP-induced cytotoxicity. One of the most frequent alterations induced by metallic ENPs is the phosphorylation of histone H2AX at serine 139 (Pogribna and Hammons 2021). This phosphorylation plays a key role in DNA double-strand break repair and is required for the assembly of DNA repair machineries at the damaged sites (Kopp et al. 2018). Several ENPs have been shown to trigger H2AX phosphorylation. AgNP exposure was found to induce H2AX phosphorylation in human lung adenocarcinoma A549 cells and human breast adenocarcinoma MCF7 cells (Zhao et al. 2016). Treatment with TiO_2 ENPs was reported to induce H2AX phosphorylation in human dermal fibroblasts and human lung adenocarcinoma A549 cells (Prasad et al. 2013; Toyooka et al. 2012). H2AX phosphorylation always follows DNA double-strand breaks and is an early step in DNA damage response pathways. In response to DNA damage, H2AX is phosphorylated by kinases such as ATM in the PI3K pathway (Davis et al. 2010). AgNPs were found to induce DNA double-strand breaks and showed synergistic toxicity with ultraviolet A exposure (Zhao et al. 2016), while TiO_2 ENPs were found to increase the phosphorylation level of ATM and its activation in response to DNA damage (Prasad et al. 2013). Incubation with TiO_2 ENPs was also found to inhibit the overall rate of DNA synthesis and the frequency of replicon initiation events (Prasad et al. 2013). These results support the idea that AgNPs and TiO_2 ENPs are genotoxic, causing DNA double-strand breaks and triggering H2AX phosphorylation.

The metallic ENP-induced changes in histone methylation or acetylation may exert epigenetic toxicity through altering chromatin condensation and transcription of genes involved in cell cycle control. In human epidermal keratinocytes HaCaT cells, ZnO ENP exposure resulted in cell cycle arrest at the G2/M checkpoint, followed by p53-Bax mitochondrial pathway-mediated apoptosis (Gao et al. 2016). In human bladder cancer T24 cells, treatment with low concentrations of ZnO resulted in cell cycle arrest at the S phase and promoted the late phase of apoptosis (Zhang et al. 2020). These cytotoxic effects were found to be associated with histone modification-induced chromatin condensation and histone modification-induced changes in gene expression. In the HaCaT cells, ZnO ENP exposure resulted in chromatin condensation, accompanied by increased H3K9 methylation and decreased H4K5 acetylation (Gao et al. 2016). Consistent with these epigenetic changes, ZnO ENP treatment also resulted in the upregulation of the H3K9 methyltransferase G9a and the downregulation of the histone acetyltransferases GCN5, P300, and CBP. In the T24 cells, the cell growth-inhibiting effects of ZnO might be attributed to increased expression of RUNX3, due to the decreased level of the histone H3K27 methyltransferase EZH2 leading to reduced H3K27 methylation at the RUNX3 promoter (Zhang et al. 2020). Changes in histone methylation also link metallic ENPs to erythroid toxicity. It was shown that AgNP treatment caused a significant reduction of the global H3 methylation level in mouse erythroleukemia cells (Qian et al. 2015). At the gene level, one significant change induced by AgNPs is reduced methylation of H3K4 and H3K79 at the β-globin locus. β-globin is a component of hemoglobin, involved in heme binding and oxygen transport. It was found that the reduced H3K4 and H3K79 methylations were associated with decreased expression of β-globin (Qian et al. 2015). The expression levels of the histone methyltransferases MLL and DOT1L, which are respectively responsible for H3K4 methylation and H3K79 methylation, were also reduced upon AgNP treatment. These results suggest that through the epigenetic downregulation of β-globin, AgNP exposure could potentially lead to erythroid toxicity and cause a deficiency in the oxygen-carrying protein hemoglobin.

The premise of epigenetic toxicity due to metallic ENP-induced histone modification depends on the specific functions of affected histone modifications and the specific gene loci affected by histone modification. The level, type, and location of histone modification together define their specific impacts on what and how genes will be influenced, which could affect different critical pathways for cell proliferation and survival. However, the histone modification changes caused by metallic ENPs have been demonstrated to be very complicated. Even the same type of ENP, but with different charges, can lead to different histone modifications. In human breast cancer MDA-MB-231 cells, the negatively charged AuNPs resulted in deacetylation of H3K9/H3K14 and dephosphorylation of H3S10, whereas treatment of the same cells with the positively charged AuNPs caused increased H3K9/H3K14 acetylation and H3S10 phosphorylation (Surapaneni et al. 2018). Another example of the charge-specific effect is cadmium telluride quantum dots (CdTe-QDs). When exposed to human leukemia monocytic THP-1 cells, the charged CdTe-QDs bound to the core histones, which changed the physiochemical properties of the histones and increased histone aggregation (Conroy et al. 2008), while the uncharged CdTe-QDs caused global H3 hypoacetylation and resulted in chromatin decondensation in human breast adenocarcinoma MCF7 cells (Choi et al. 2008). Thus, the changes in histone modification depend not only on the type of metals used to make ENPs but also on the electrostatic properties of the ENPs.

Altered miRNA Expression

Metallic ENP exposure has been shown to alter the expression of many miRNAs, which represents another major epigenetic mechanism underlying the toxicity of metallic ENPs. It was found that intravenous injection of AuNPs into the rats altered the expression of miR-155 and its putative target genes in the lung (Ng et al. 2011). One such target is tumor protein 53 inducible nuclear protein 1 (TP53INP1), which is known to induce cell growth arrest and apoptosis by modulating p53 transcriptional activity (Liu et al. 2015b). Another target gene is the protein S (*PROS1*) gene, which codes for protein S, an anticoagulant protein that regulates hemostasis and the blood clotting process (Ng et al. 2011). These findings suggest several pathological mechanisms by which AuNPs might lead to lung toxicity. Through modulating miR-155 expression, AuNPs could lead to enhanced cell apoptosis by elevating *TP53INP1* expression or could contribute to thrombosis in the pulmonary vasculature by modulating *PROS1* expression. TiO_2 ENPs were also suggested to cause lung toxicity by altering the expression of miRNAs. Mice exposed to TiO_2 ENPs by inhalation showed increased expression of 16 miRNAs (Halappanavar et al. 2011). Among the highest induced miRNAs are miR-1, miR-449a, and miR-135b, and the induction of these miRNAs was associated with the changes in the expression of genes associated with the acute phase inflammation-immune response. While the exact role of these miRNAs in the acute phase response is unknown, TiO_2 ENP inhalation might induce not only pulmonary toxicity but also systemic effects, since the acute phase response in the lung is strongly associated with systemic effects, which might lead to a dysfunction of multiple organs (Halappanavar et al. 2011). The toxicity of metallic ENPs with respect to altered miRNA expression has also been examined at the cellular level. It was shown that AgNP exposure to Jurkat T cells altered the expression of 63 miRNAs (Eom et al. 2014). Among these miRNAs, the putative target genes of miR-504, miR-33, and miR-302 were found to be involved in DNA damage and apoptosis, while hsa-miR-219-5p, whose expression was negatively correlated with the expression of metallothionein 1F (MT1F) and tribbles homolog 3 (TRIB3), was found to be associated with oxidative stress, cell cycle, and apoptosis (Eom et al. 2014). To date, all available studies about the effects of metallic ENPs on ncRNAs focused only on miRNAs, and how the expression of other ncRNAs, such as long ncRNA, siRNA, and piRNA, undergoes changes in response to metallic ENP exposure, and their associated toxicities are unknown.

Epigenetic Toxicity of Nonmetallic ENPs

Like metallic ENPs, nonmetallic ENPs can also induce DNA hypomethylation or DNA hypermethylation at both global and locus-specific levels. Their impact on the expression of genes associated with inflammation, DNA damage response, and apoptosis suggests that nonmetallic ENPs could also induce epigenetic toxicity through altering DNA methylation. It was shown that there was global DNA hypermethylation in various organs and tissues, including the heart, liver, and intestine, after zebrafish were exposed to carbon nanoparticles (CNPs) or graphene quantum dots (GQDs) (Hu et al. 2019; Zhou et al. 2019). In fish intestine, DNA methylation was found to be very sensitive to GQDs, suggesting that the intestine might be the main target for GQDs (Hu et al. 2019). However, in the heart, in addition to global DNA hypermethylation, several gene-specific loci, including *LEBP*, *CD248b*, and *IL11a*, showed hypomethylation upon CNP exposure, and the decreased promoter methylation of these genes was found to be associated with their transcriptional activation (Zhou et al. 2019). Since these genes are known to play roles in host defense by promoting the release of cytokines in response to bacterial lipopolysaccharide, the CNP-induced changes in their expression suggest that CNP exposure has triggered the innate immune response (Zhou et al. 2019). On the other hand, the mixed pattern of global DNA hypermethylation and locus-specific DNA hypomethylation is consistent with the differential expression of DNA methyltransferases, where upon CNP treatment, there was a marked reduction in the mRNA level of DNMT1, whereas the mRNA level of DNMT3a/b was upregulated. Silicon dioxide (SiO_2) ENPs also cause global and locus-specific changes in DNA methylation when exposed to cells, but they usually lead to global DNA hypomethylation (Gong et al. 2010; Seidel et al. 2017). It was shown that SiO_2 ENPs caused a dramatic decrease in global DNA methylation in human HaCaT cells and mouse Bhas 42 cells, while locus-specific hypermethylation was found at the promoter regions of several stress response-related genes, including *PARP1*, *CREB3L1*, and *BCL-2* (Chiu et al. 2021; Zou et al. 2016). PARP1 plays a key role in DNA repair in response to DNA damage, while CREB3L1 is involved in unfolded protein response and protects cells from ER stress-induced cell death. BCL-2 regulates cell death by either inhibiting or inducing apoptosis. In agreement with DNA hypermethylation, the expression of these genes was downregulated upon SiO_2 ENP treatment (Zou et al. 2016). These findings suggest the potential cytotoxic effects of SiO_2 ENPs on cellular responses to stresses, which could lead to the induction of cell-death programs. SiO_2 ENPs were also shown to cause DNA hypomethylation in the white blood cells of people working at the nanomaterial manufacturing/handling factories (Liou et al. 2017; Rossnerova et al. 2020). Long-term occupational exposure to SiO_2-containing nanomaterials was found to be associated with the substantial changes in DNA methylation in the white blood cells; however, the short-term exposure did not affect the DNA methylation patterns (Rossnerova et al. 2020).

To date, only a few studies have investigated the effects of nonmetallic ENPs on histone modification and ncRNA expression. It was shown that SiO_2 ENPs altered the global histone H4 acetylation status in mouse Bhas 42 cells and significantly increased the acetylation of H3 at the promoter of *c-Myc* (Seidel et al. 2017). The latter acetylation, including H3K4ac, H3K9ac, and H3K27ac, marked the actively transcribed chromatin and was associated with increased *c-Myc* expression. As a result of *c-Myc* overexpression, SiO_2 ENPs could potentially contribute to c-Myc-mediated cell transformation and promote cancer cell development (Seidel et al. 2017). Another study showed that SiO_2 ENP exposure to human A549 cells caused an increase in both H3K9ac and H3K56ac levels at the follistatin promoter region, and this increase in H3 acetylation was found to be associated with the reduced expression of the histone deacetylase SIRT6 (Zhang et al. 2018). Follistatin plays an important role in various processes, such as cell growth, development, and differentiation (Hemmati-Brivanlou et al. 1994). The activation of follistatin transcription during SiO_2 ENP exposure was found to play a protective role that suppresses SiO_2 ENP-induced apoptosis (Zhang et al. 2018). SiO_2 ENPs were also reported to induce H2AX phosphorylation in human intestinal CaCo-2 cells (Tarantini et al. 2015). Since the H2AX phosphorylation is an indicator of the induction of DNA damage response, this result has indicated that SiO_2 ENPs are genotoxic and could induce adverse effects on the intestinal epithelium (Tarantini et al. 2015). The extent of chromosome damage and cell death following exposure to SiO_2 ENPs was found to be dependent on the size and concentration of SiO_2 ENPs used. There is only one study that investigated ncRNA alteration upon nonmetallic ENP treatment. SiO_2 ENPs were shown to increase the serum level of miR-122 in mice and were associated with liver damage (Nagano et al. 2013). miR-122 regulates the mRNA levels of several genes, such as *CAT1*, *AldoA*, and *Hjv*, and through these regulations, miR-122 modulates fatty-acid metabolism and systemic iron homeostasis in the liver (Nagano et al. 2013). It was proposed that miR-122 could be a useful biomarker for SiO_2 ENP-induced liver damage, since its sensitivity for liver damage is similar to that of the conventional hepatic biomarkers, such as alanine aminotransferase (ALT) and aspartate aminotransferase (AST).

ENP-Induced Epigenetic Toxicity Likely Mediated by ROS

Various epigenetic toxicities caused by ENPs highlight the importance of understanding the molecular mechanisms underlying how ENPs induce epigenetic alterations, which, however, remain largely unknown. ROS could be a mechanism that mediates the effects of ENPs on epigenetic alternations, because of their link to both ENP exposure and epigenetic modification. As drug carriers, ENPs usually make the first contact with the phagocytes of the immune system. One typical biological reaction to ENP exposure is the activation of the innate immune response, which is accompanied by the induction of ROS (Fu et al. 2014; Liu et al. 2017b). ROS generation and the subsequent induction of oxidative stress have been reported as a predominant mechanism of leading to nanotoxicity, such as DNA damage, disturbed cellular signaling, cell cycle dysregulation, cancer promotion, and apoptosis (Manke et al. 2013). ROS are also known to cause epigenetic changes, either through directly modifying DNA bases or indirectly through regulating the expression and activity of enzymes responsible for the epigenetic modification, or through influencing the metabolic pathways involved in the biogenesis of the methyltransferase cofactor SAM (Cencioni et al. 2013; Kietzmann et al. 2017; Mentch et al. 2015). These pieces of evidence support the idea that the epigenetic toxicity of ENPs could be mediated by ROS and that ROS could act as intermediate effector molecules to transduce the effects of ENPs to the changes in epigenetic modification (Yu et al. 2020b).

ROS are chemically reactive molecules that play important roles in the innate immune response. During phagocytosis, ROS, such as hydrogen peroxide (H_2O_2), reactive superoxide anion radicals (O^{2-}), and hydroxyl radicals (•OH), are a major mechanism used by the phagocytes to suppress microbial growth and kill pathogens (Mittler 2017). However, excessive ROS can damage DNA, proteins, and lipids, leading to malignant cell transformation or apoptosis. Both metallic and nonmetallic ENPs are known to induce excessive ROS. Metal ions released by metallic ENPs can lead to intracellular ROS burst by participating in the redox cycling via the Fenton reaction or the Fenton-like reaction (Fu et al. 2014; Yu et al. 2020b). For instance, AgNPs were found to cause a marked increase in ROS, likely due to the Ag ions released from AgNPs, which subsequently resulted in membrane permeability disruption, DNA damage, genomic instability, and cell cycle arrest (Li et al. 2017). The level of ROS induced by metallic ENPs appeared to be associated with their size, and the larger size of selenium (Se) ENPs has been shown to induce more ROS than the smaller size of particles (Huang et al. 2019). Although nonmetallic ENPs do not have metal ions to release, their size, shape, and polarity have been found to influence their compatibility with the immune system. Studies showed that the nonmetabolic ENPs can initiate the activation of the complement system and promote ROS production and the release of inflammatory cytokines (Liu et al. 2017b; McCarthy et al. 2012; Wibroe et al. 2017). For example, the levels of ROS produced by carboxylated polystyrene ENPs were found to be dependent on the shape of particles. All shapes, including the sphere, rod, and disk, can activate the complement components C3a and C5a, but the rod- and disk-shaped particles have more pronounced complement activation than the spherical ones (Wibroe et al. 2017). C5a activation can robustly increase ROS production, which induces aberrant DNA methylation in human renal tubular epithelial cells, particularly in the gene regions involved in cell cycle control and DNA damage repair (Castellano et al. 2019). Another study showed that SiO_2 ENPs can enhance ROS generation and the release of interleukin (IL)-6 and IL-8 in human lung submucosal Calu-3 cells, leading to apoptosis and decreased cell survival (McCarthy et al. 2012).

Cellular ROS can induce epigenetic changes by affecting DNA methylation and histone modification, either directly or indirectly. The direct oxidation of 5mC by hydroxyl radicals was found to result in the demethylation of cytosine in CpG nucleotides (Branco et al. 2011; Madugundu et al. 2014). One main product of this oxidation is 5hmC, which either arises from hydroxyl radical addition at the 5,6-double bond of 5mC, or arises from hydrogen abstraction from the methyl group of 5mC (Madugundu et al. 2014). It was previously proposed that 5hmC can block the proper inheritance of the DNA methylation pattern by interfering with the action of DNMT1 and causing the demethylation of DNA (Branco et al. 2011). In agreement with this notion, ENP exposure has been shown to cause global DNA hypomethylation in many cases (Table 10.1). ROS can indirectly affect histone methylation by regulating the expression or activity of histone methyltransferases. Exposure of H_2O_2 upregulated the expression of SET and MYND domain-containing protein 1 (SMYD1) in cardiomyoblasts, an H3K4 methyltransferase (Liang et al. 2020). ROS can also regulate SMYD2 activity through oxidative stress-induced SMYD2 glutathionylation (Munkanatta Godage et al. 2018). Another mechanism by which ROS induce epigenetic changes is that ROS affect the biosynthesis of the methyltransferase cofactor SAM. Under oxidative stress, methionine, which is required for SAM biosynthesis, is mostly used for the synthesis of cysteine to produce the antioxidant glutathione. As a result of this, the level of SAM is reduced, which limits the activity of DNA or histone methyltransferases (Cyr and Domann 2011). In addition, DNA single-strand or double-strand breaks caused by ROS usually trigger rapid H2AX phosphorylation during the DNA damage response (Podhorecka et al. 2010).

ENP-induced ROS, in some cases, have been suggested to be the direct cause of epigenetic changes associated with ENP treatment. TiO_2 ENPs are known to induce ROS and DNA hypermethylation at the promoter region of *PARP1* in human lung adenocarcinoma A549 cells (Bai et al. 2015). These two events appeared to be related, as the DNMT inhibitor 5-aza-2′-deoxycytidine not only decreased *PARP1* methylation but also decreased the level of ROS induced by TiO_2 ENPs. More importantly, the ROS scavenger α-lipoic acid was shown to decrease the DNA hypermethylation of *PARP1* associated with the ENP exposure (Bai et al. 2015). These findings suggest that the TiO_2 ENP-induced ROS are likely the cause of increased *PARP1* methylation. Several other studies on histone modification provide additional evidence that the ENP-induced ROS mediate epigenetic changes. ROS induced by ZnO and CuO ENPs have been shown to be associated with H2AX phosphorylation and the activation of the DNA damage response (Toduka et al. 2012). A piece of supporting evidence for the ENP-induced ROS being responsible for this H2AX phosphorylation came from an experiment that showed the antioxidant N-acetylcysteine (NAD) partially suppressed H2AX phosphorylation (Toduka et al. 2012). Another study demonstrated that the TiO_2 ENP-induced ROS might be responsible for decreased HDAC9 expression and increased histone acetylation (Jayaram and Payne 2020). TiO_2 ENPs were shown to cause time-dependent production of intracellular superoxide, which was associated with the decreased expression of HDAC9. The causal link between these two events was demonstrated by the use of superoxide dismutase (SOD), which not only inhibited the production of intracellular superoxide but also maintained the expression level of HDAC9 (Jayaram and Payne 2020). These studies together established the link between the ENP-induced ROS and the ROS-induced epigenetic changes, with ROS acting as intermediate effector molecules.

Conclusion

Existing studies have demonstrated that ENP-based drug carriers offer many advantages, such as improved stability, favorable biodistribution profiles, slower drug-release kinetics, and cell-specific targeting. To further expand their clinical applications, it is important to thoroughly understand their toxicity profiles and how the immune response and oxidative stress induced by them contribute to their genotoxicity and epigenetic toxicity. Many epigenetic changes have been associated with ENP exposure, including DNA methylation, histone methylation, histone acetylation, histone phosphorylation, and ncRNA expression. Although the underlying mechanisms of how the ENPs induce these epigenetic alterations remain not well understood, ROS, often generated by ENP exposure, could be the major source that contributes to ENP-induced epigenetic toxicity. Excessive ROS are toxic and can cause many adverse effects and ultimately cell death. Understanding how ENPs induce ROS and how the induced ROS cause epigenetic changes could help develop novel strategies to reduce toxicity associated with ENP exposure. Recent advances have made promising progress in reducing ROS triggered by ENPs (Singh et al. 2019). There was almost no ROS-related DNA damage observed in mercaptopropionic acid-coated ENPs (Rispail et al. 2014), and ENPs coated with polyvinylpyrrolidone showed good immunological biocompatibility and fewer toxic effects (Zhi et al. 2013). In addition, platinum ENPs were shown to inhibit DNA damage by degrading ROS (Aioub et al. 2017; Panikkanvalappil et al. 2013), while cerium oxide ENPs could prevent lung cells from oxidative attack with their ROS-scavenging capacity (Rubio et al. 2016). These encouraging results warrant future research in exploiting additional new strategies to reduce oxidative stress induced by ENPs, with an ultimate goal of accelerating their clinical and commercial translation.

References

Abbasi, E., Aval, S.F., Akbarzadeh, A., Milani, M., Nasrabadi, H.T., Joo, S.W., Hanifehpour, Y., Nejati-Koshki, K., and Pashaei-Asl, R. (2014). Dendrimers: synthesis, applications, and properties. *Nanoscale Res. Lett.* 9: 247.

Ahmed, A., Sarwar, S., Hu, Y., Munir, M.U., Nisar, M.F., Ikram, F., Asif, A., Rahman, S.U., Chaudhry, A.A., and Rehman, I.U. (2021). Surface-modified polymeric nanoparticles for drug delivery to cancer cells. *Expert Opin. Drug Deliv.* 18: 1–24.

Aioub, M., Panikkanvalappil, S.R., and El-Sayed, M.A. (2017). Platinum-coated gold nanorods: efficient reactive oxygen scavengers that prevent oxidative damage toward healthy, untreated cells during plasmonic photothermal therapy. *ACS Nano.* 11: 579–586.

Alinovi, R., Goldoni, M., Pinelli, S., Ravanetti, F., Galetti, M., Pelosi, G., De Palma, G., Apostoli, P., Cacchioli, A., Mutti, A., and Mozzoni, P. (2017). Titanium dioxide aggregating nanoparticles induce autophagy and under-expression of microRNA 21 and 30a in A549 cell line: a comparative study with cobalt(II, III) oxide nanoparticles. *Toxicol. In Vitro* 42: 76–85.

Allen, T.M. and Cullis, P.R. (2013). Liposomal drug-delivery systems: from concept to clinical applications. *Adv. Drug Deliv. Rev.* 65: 36–48.

Anajafi, T. and Mallik, S. (2015). Polymersome-based drug-delivery strategies for cancer therapeutics. *Ther. Deliv.* 6: 521–534.

Anselmo, A.C. and Mitragotri, S. (2019). Nanoparticles in the clinic: an update. *Bioeng. Transl. Med.* 4: e10143.

Ashraf, M.A., Peng, W., Zare, Y., and Rhee, K.Y. (2018). Effects of size and aggregation/agglomeration of nanoparticles on the interfacial/interphase properties and tensile strength of polymer nanocomposites. *Nanoscale Res. Lett.* 13: 214.

Baetke, S.C., Lammers, T., and Kiessling, F. (2015). Applications of nanoparticles for diagnosis and therapy of cancer. *Br. J. Radiol.* 88: 20150207.

Bai, W., Chen, Y., and Gao, A. (2015). Cross talk between poly(ADP-ribose) polymerase 1 methylation and oxidative stress involved in the toxic effect of anatase titanium dioxide nanoparticles. *Int. J. Nanomed.* 10: 5561–5569.

Balansky, R., Longobardi, M., Ganchev, G., Iltcheva, M., Nedyalkov, N., Atanasov, P., Toshkova, R., De Flora, S., and Izzotti, A. (2013). Transplacental clastogenic and epigenetic effects of gold nanoparticles in mice. *Mutat. Res.* 751-752: 42–48.

Balasubramanian, S., Gunasekaran, K., Sasidharan, S., Jeyamanickavel Mathan, V., and Perumal, E. (2020). MicroRNAs and xenobiotic toxicity: an overview. *Toxicol. Rep.* 7: 583–595.

Batist, G., Gelmon, K.A., Chi, K.N., Miller, W.H., Jr, Chia, S.K., Mayer, L.D., Swenson, C.E., Janoff, A.S., and Louie, A.C. (2009). Safety, pharmacokinetics, and efficacy of CPX-1 liposome injection in patients with advanced solid tumors. *Clin. Cancer Res.* 15: 692–700.

Behr, J.P., Demeneix, B., Loeffler, J.P., and Perez-Mutul, J. (1989). Efficient gene transfer into mammalian primary endocrine cells with lipopolyamine-coated DNA. *Proc. Natl. Acad. Sci. U S A.* 86: 6982–6986.

Bishayee, K., Khuda-Bukhsh, A.R., and Huh, S.O. (2015). PLGA-loaded gold-nanoparticles precipitated with quercetin downregulate HDAC-Akt activities controlling proliferation and activate p53-ROS crosstalk to induce apoptosis in hepatocarcinoma cells. *Mol. Cells* 38: 518–527.

Blanco, J., Lafuente, D., Gomez, M., Garcia, T., Domingo, J.L., and Sanchez, D.J. (2017). Polyvinyl pyrrolidone-coated silver nanoparticles in a human lung cancer cells: time- and dose-dependent influence over p53 and caspase-3 protein expression and epigenetic effects. *Arch. Toxicol.* 91: 651–666.

Bobo, D., Robinson, K.J., Islam, J., Thurecht, K.J., and Corrie, S.R. (2016). Nanoparticle-based medicines: a review of FDA-approved materials and clinical trials to date. *Pharm. Res.* 33: 2373–2387.

Boisselier, E. and Astruc, D. (2009). Gold nanoparticles in nanomedicine: preparations, imaging, diagnostics, therapies and toxicity. *Chem. Soc. Rev.* 38: 1759–1782.

Branco, M.R., Ficz, G., and Reik, W. (2011). Uncovering the role of 5-hydroxymethylcytosine in the epigenome. *Nat. Rev. Genet.* 13: 7–13.

Brashear, A., Unverzagt, F.W., Farber, M.O., Bonnin, J.M., Garcia, J.G., and Grober, E. (1996). Ethylene oxide neurotoxicity: a cluster of 12 nurses with peripheral and central nervous system toxicity. *Neurology* 46: 992–998.

Brzoska, K., Gradzka, I., and Kruszewski, M. (2019). Silver, gold, and iron oxide nanoparticles alter miRNA expression but do not affect DNA methylation in HepG2 cells. *Materials (Basel)* 12.

Cai, X., Luo, Y., Zhang, W., Du, D., and Lin, Y. (2016). pH-sensitive ZnO quantum dots-doxorubicin nanoparticles for lung cancer targeted drug delivery. *ACS Appl. Mater. Interfaces* 8: 22442–22450.

Castellano, G., Franzin, R., Sallustio, F., Stasi, A., Banelli, B., Romani, M., De Palma, G., Lucarelli, G., Divella, C., Battaglia, M., Crovace, A., Staffieri, F., Grandaliano, G., Stallone, G., Ditonno, P., Cravedi, P., Cantaluppi, V., and Gesualdo, L. (2019). Complement component C5a induces aberrant epigenetic modifications in renal tubular epithelial cells accelerating senescence by Wnt4/betacatenin signaling after ischemia/reperfusion injury. *Aging (Albany NY)* 11: 4382–4406.

Cencioni, C., Spallotta, F., Martelli, F., Valente, S., Mai, A., Zeiher, A.M., and Gaetano, C. (2013). Oxidative stress and epigenetic regulation in ageing and age-related diseases. *Int. J. Mol. Sci.* 14: 17643–17663.

Chauhan, A.S. (2018). Dendrimers for drug delivery. *Molecules* 23.

Cheng, X. and Lee, R.J. (2016). The role of helper lipids in lipid nanoparticles (LNPs) designed for oligonucleotide delivery. *Adv. Drug. Deliv. Rev.* 99: 129–137.

Chew, W.S., Poh, K.W., Siddiqi, N.J., Alhomida, A.S., Yu, L.E., and Ong, W.Y. (2012). Short- and long-term changes in blood miRNA levels after nanogold injection in rats–potential biomarkers of nanoparticle exposure. *Biomarkers* 17: 750–757.

Chiu, L.Y., Wu, N.L., Hung, C.F., Bai, P., Dai, Y.S., and Lin, W.W. (2021). PARP-1 involves in UVB-induced inflammatory response in keratinocytes and skin injury via regulation of ROS-dependent EGFR transactivation and p38 signaling. *FASEB J.* 35: e21393.

Chiu, S.H., Gedda, G., Girma, W.M., Chen, J.K., Ling, Y.C., Ghule, A.V., Ou, K.L., and Chang, J.Y. (2016). Rapid fabrication of carbon quantum dots as multifunctional nanovehicles for dual-modal targeted imaging and chemotherapy. *Acta Biomater.* 46: 151–164.

Chmykh, Y.G. and Nadeau, J.L. (2020). Characterization of retinol stabilized in phosphatidylcholine vesicles with and without antioxidants. *ACS Omega* 5: 18367–18375.

Choi, A.O., Brown, S.E., Szyf, M., and Maysinger, D. (2008). Quantum dot-induced epigenetic and genotoxic changes in human breast cancer cells. *J. Mol. Med. (Berl)* 86: 291–302.

Choudhury, S.R., Ordaz, J., Lo, C.L., Damayanti, N.P., Zhou, F., and Irudayaraj, J. (2017). From the cover: zinc oxide nanoparticles-induced reactive oxygen species promotes multimodal cyto- and epigenetic toxicity. *Toxicol. Sci.* 156: 261–274.

Conroy, J., Byrne, S.J., Gun'ko, Y.K., Rakovich, Y.P., Donegan, J.F., Davies, A., Kelleher, D., and Volkov, Y. (2008). CdTe nanoparticles display tropism to core histones and histone-rich cell organelles. *Small* 4: 2006–2015.

Cyr, A.R. and Domann, F.E. (2011). The redox basis of epigenetic modifications: from mechanisms to functional consequences. *Antioxid. Redox Signal.* 15: 551–589.

D'souza, A.A. and Shegokar, R. (2016). Polyethylene glycol (PEG): a versatile polymer for pharmaceutical applications. *Expert Opin. Drug Deliv.* 13: 1257–1275.

Danafar, H., Rostamizadeh, K., Davaran, S., and Hamidi, M. (2014). PLA-PEG-PLA copolymer-based polymersomes as nanocarriers for delivery of hydrophilic and hydrophobic drugs: preparation and evaluation with atorvastatin and lisinopril. *Drug Dev. Ind. Pharm.* 40: 1411–1420.

Davis, A.J., So, S., and Chen, D.J. (2010). Dynamics of the PI3K-like protein kinase members ATM and DNA-PKcs at DNA double strand breaks. *Cell Cycle* 9: 2529–2536.

De Jong, W.H. and Borm, P.J. (2008). Drug delivery and nanoparticles: applications and hazards. *Int. J. Nanomed.* 3: 133–149.

Degli Esposti, L., Carella, F., Adamiano, A., Tampieri, A., and Iafisco, M. (2018). Calcium phosphate-based nanosystems for advanced targeted nanomedicine. *Drug Dev. Ind. Pharm.* 44: 1223–1238.

Discher, D.E., Ortiz, V., Srinivas, G., Klein, M.L., Kim, Y., Christian, D., Cai, S., Photos, P., and Ahmed, F. (2007). Emerging applications of polymersomes in delivery: from molecular dynamics to shrinkage of tumors. *Prog. Polym. Sci.* 32: 838–857.

Elveny, M., Khan, A., Nakhjiri, A.T., and Albadarin, A.B. (2021). A state-of-the-art review on the application of various pharmaceutical nanoparticles as a promising technology in cancer treatment. *Arab. J. Chem.* 14.

Eom, H.J., Chatterjee, N., Lee, J., and Choi, J. (2014). Integrated mRNA and micro RNA profiling reveals epigenetic mechanism of differential sensitivity of Jurkat T cells to AgNPs and Ag ions. *Toxicol. Lett.* 229: 311–318.

Fenton, O.S., Olafson, K.N., Pillai, P.S., Mitchell, M.J., and Langer, R. (2018). Advances in biomaterials for drug delivery. *Adv. Mater.* e1705328.

Fu, P.P., Xia, Q., Hwang, H.M., Ray, P.C., and Yu, H. (2014). Mechanisms of nanotoxicity: generation of reactive oxygen species. *J. Food Drug. Anal.* 22: 64–75.

Gao, F., Ma, N., Zhou, H., Wang, Q., Zhang, H., Wang, P., Hou, H., Wen, H., and Li, L. (2016). Zinc oxide nanoparticles-induced epigenetic change and G2/M arrest are associated with apoptosis in human epidermal keratinocytes. *Int. J. Nanomed.* 11: 3859–3874.

Gedda, M.R., Babele, P.K., Zahra, K., and Madhukar, P. (2019). Epigenetic aspects of engineered nanomaterials: is the collateral damage inevitable? *Front Bioeng. Biotechnol.* 7: 228.

Geuli, O., Metoki, N., Zada, T., Reches, M., Eliaz, N., and Mandler, D. (2017). Synthesis, coating, and drug-release of hydroxyapatite nanoparticles loaded with antibiotics. *J. Mater. Chem. B* 5: 7819–7830.

Ghaderi, S., Ramesh, B., and Seifalian, A.M. (2011). Fluorescence nanoparticles "quantum dots" as drug delivery system and their toxicity: a review. *J. Drug Target.* 19: 475–486.

Ghezzi, M., Pescina, S., Padula, C., Santi, P., Del Favero, E., Cantu, L., and Nicoli, S. (2021). Polymeric micelles in drug delivery: an insight of the techniques for their characterization and assessment in biorelevant conditions. *J. Control. Release* 332: 312–336.

Gi, M., Fujioka, M., Kakehashi, A., Okuno, T., Masumura, K., Nohmi, T., Matsumoto, M., Omori, M., Wanibuchi, H., and Fukushima, S. (2018). In vivo positive mutagenicity of 1,4-dioxane and quantitative analysis of its mutagenicity and carcinogenicity in rats. *Arch. Toxicol.* 92: 3207–3221.

Gliga, A.R., Di Bucchianico, S., Lindvall, J., Fadeel, B., and Karlsson, H.L. (2018). RNA-sequencing reveals long-term effects of silver nanoparticles on human lung cells. *Sci. Rep.* 8: 6668.

Gong, C., Tao, G., Yang, L., Liu, J., Liu, Q., Li, W., and Zhuang, Z. (2012). Methylation of PARP-1 promoter involved in the regulation of nano-SiO2-induced decrease of PARP-1 mRNA expression. *Toxicol. Lett.* 209: 264–269.

Gong, C., Tao, G., Yang, L., Liu, J., Liu, Q., and Zhuang, Z. (2010). SiO(2) nanoparticles induce global genomic hypomethylation in HaCaT cells. *Biochem. Biophys. Res. Commun.* 397: 397–400.

Gurunathan, S., Qasim, M., Park, C., Yoo, H., Choi, D.Y., Song, H., Park, C., Kim, J.H., and Hong, K. (2018). Cytotoxicity and transcriptomic analysis of silver nanoparticles in mouse embryonic fibroblast cells. *Int. J. Mol. Sci.* 19.

Halappanavar, S., Jackson, P., Williams, A., Jensen, K.A., Hougaard, K.S., Vogel, U., Yauk, C.L., and Wallin, H. (2011). Pulmonary response to surface-coated nanotitanium dioxide particles includes induction of acute phase response genes, inflammatory cascades, and changes in microRNAs: a toxicogenomic study. *Environ. Mol. Mutagen.* 52: 425–439.

Hanot-Roy, M., Tubeuf, E., Guilbert, A., Bado-Nilles, A., Vigneron, P., Trouiller, B., Braun, A., and Lacroix, G. (2016). Oxidative stress pathways involved in cytotoxicity and genotoxicity of titanium dioxide (TiO2) nanoparticles on cells constitutive of alveolo-capillary barrier in vitro. *Toxicol. In Vitro* 33: 125–135.

Hemmati-Brivanlou, A., Kelly, O.G., and Melton, D.A. (1994). Follistatin, an antagonist of activin, is expressed in the Spemann organizer and displays direct neuralizing activity. *Cell* 77: 283–295.

Hou, X., Zaks, T., Langer, R., and Dong, Y. (2021). Lipid nanoparticles for mRNA delivery. *Nat. Rev. Mater.* 6: 1078–1094.

Hsu, H.J., Bugno, J., Lee, S.R., and Hong, S. (2017). Dendrimer-based nanocarriers: a versatile platform for drug delivery. *Wiley Interdiscip. Rev. Nanomed. Nanobiotechnol.* 9.

Hu, J., Lin, W., Lin, B., Wu, K., Fan, H., and Yu, Y. (2019). Persistent DNA methylation changes in zebrafish following graphene quantum dots exposure in surface chemistry-dependent manner. *Ecotoxicol. Environ. Saf.* 169: 370–375.

Hua, S., De Matos, M.B.C., Metselaar, J.M., and Storm, G. (2018). Current trends and challenges in the clinical translation of nanoparticulate nanomedicines: pathways for translational development and commercialization. *Front Pharmacol.* 9: 790.

Huang, T., Holden, J.A., Heath, D.E., O'brien-simpson, N.M., and O'connor, A.J. (2019). Engineering highly effective antimicrobial selenium nanoparticles through control of particle size. *Nanoscale* 11: 14937–14951.

Jain, K., Kesharwani, P., Gupta, U., and Jain, N.K. (2010). Dendrimer toxicity: let's meet the challenge. *Int. J. Pharm.* 394: 122–142.

Jayaram, D.T. and Payne, C.K. (2020). Intracellular generation of superoxide by TiO2 nanoparticles decreases histone deacetylase 9 (HDAC9), an epigenetic modifier. *Bioconjug. Chem.* 31: 1354–1361.

Jhaveri, A.M. and Torchilin, V.P. (2014). Multifunctional polymeric micelles for delivery of drugs and siRNA. *Front Pharmacol.* 5: 77.

Kaikkonen, M.U., Lam, M.T., and Glass, C.K. (2011). Non-coding RNAs as regulators of gene expression and epigenetics. *Cardiovasc. Res.* 90: 430–440.

Kalaiarasi, A., Sankar, R., Anusha, C., Saravanan, K., Aarthy, K., Karthic, S., Mathuram, T.L., and Ravikumar, V. (2018). Copper oxide nanoparticles induce anticancer activity in A549 lung cancer cells by inhibition of histone deacetylase. *Biotechnol. Lett.* 40: 249–256.

Kannan, R.M., Nance, E., Kannan, S., and Tomalia, D.A. (2014). Emerging concepts in dendrimer-based nanomedicine: from design principles to clinical applications. *J. Intern. Med.* 276: 579–617.

Kesharwani, P., Choudhury, H., Meher, J.G., Pandey, M., and Gorain, B. (2019). Dendrimer-entrapped gold nanoparticles as promising nanocarriers for anticancer therapeutics and imaging. *Prog. Mater. Sci.* 103: 484–508.

Khan, A.A., Allemailem, K.S., Almatroodi, S.A., Almatroudi, A., and Rahmani, A.H. (2020). Recent strategies towards the surface modification of liposomes: an innovative approach for different clinical applications. *3 Biotech.* 10: 163.

Kietzmann, T., Petry, A., Shvetsova, A., Gerhold, J.M., and Gorlach, A. (2017). The epigenetic landscape related to reactive oxygen species formation in the cardiovascular system. *Br. J. Pharmacol.* 174: 1533–1554.

Kim, J.J., Kim, Y.S., and Kumar, V. (2019). Heavy metal toxicity: an update of chelating therapeutic strategies. *J. Trace Elem. Med. Biol.* 54: 226–231.

Kohane, D.S. (2007). Microparticles and nanoparticles for drug delivery. *Biotechnol. Bioeng.* 96: 203–209.

Konen-Adiguzel, S. and Ergene, S. (2018). In vitro evaluation of the genotoxicity of CeO2 nanoparticles in human peripheral blood lymphocytes using cytokinesis-block micronucleus test, comet assay, and gamma H2AX. *Toxicol. Ind. Health* 34: 293–300.

Kopp, B., Dario, M., Zalko, D., and Audebert, M. (2018). Assessment of a panel of cellular biomarkers and the kinetics of their induction in comparing genotoxic modes of action in HepG2 cells. *Environ. Mol. Mutagen.* 59: 516–528.

Kulkarni, J.A., Witzigmann, D., Leung, J., Tam, Y.Y.C., and Cullis, P.R. (2019). On the role of helper lipids in lipid nanoparticle formulations of siRNA. *Nanoscale* 11: 21733–21739.

Kung, M.L., Hsieh, S.L., Wu, C.C., Chu, T.H., Lin, Y.C., Yeh, B.W., and Hsieh, S. (2015). Enhanced reactive oxygen species overexpression by CuO nanoparticles in poorly differentiated hepatocellular carcinoma cells. *Nanoscale* 7: 1820–1829.

Lee, J.S. and Feijen, J. (2012). Polymersomes for drug delivery: design, formation and characterization. *J. Control. Release* 161: 473–483.

Lennartsson, A. and Ekwall, K. (2009). Histone modification patterns and epigenetic codes. *Biochim. Biophys. Acta* 1790: 863–868.

Leung, A.K., Tam, Y.Y., Chen, S., Hafez, I.M., and Cullis, P.R. (2015). Microfluidic mixing: a general method for encapsulating macromolecules in lipid nanoparticle systems. *J. Phys. Chem. B* 119: 8698–8706.

Lewinski, N., Colvin, V., and Drezek, R. (2008). Cytotoxicity of nanoparticles. *Small* 4: 26–49.

Li, J., Liang, H., Liu, J., and Wang, Z. (2018). Poly (amidoamine) (PAMAM) dendrimer mediated delivery of drug and pDNA/siRNA for cancer therapy. *Int. J. Pharm.* 546: 215–225.

Li, S., Wang, H., Qi, Y., Tu, J., Bai, Y., Tian, T., Huang, N., Wang, Y., Xiong, F., Lu, Z., and Xiao, Z. (2011). Assessment of nanomaterial cytotoxicity with SOLiD sequencing-based microRNA expression profiling. *Biomaterials* 32: 9021–9030.

Li, Y., Qin, T., Ingle, T., Yan, J., He, W., Yin, J.J., and Chen, T. (2017). Differential genotoxicity mechanisms of silver nanoparticles and silver ions. *Arch. Toxicol.* 91: 509–519.

Liang, Q., Cai, M., Zhang, J., Song, W., Zhu, W., Xi, L., and Tian, Z. (2020). Role of muscle-specific histone methyltransferase (Smyd1) in exercise-induced cardioprotection against pathological remodeling after myocardial infarction. *Int. J. Mol. Sci.* 21.

Liou, S.H., Wu, W.T., Liao, H.Y., Chen, C.Y., Tsai, C.Y., Jung, W.T., and Lee, H.L. (2017). Global DNA methylation and oxidative stress biomarkers in workers exposed to metal oxide nanoparticles. *J. Hazard. Mater.* 331: 329–335.

Liu, D., Wu, D., Zhao, L., Yang, Y., Ding, J., Dong, L., Hu, L., Wang, F., Zhao, X., Cai, Y., and Jin, J. (2015a). Arsenic trioxide reduces global histone H4 acetylation at lysine 16 through direct binding to histone acetyltransferase hMOF in human cells. *PLoS One* 10: e0141014.

Liu, F., Kong, X., Lv, L., and Gao, J. (2015b). MiR-155 targets TP53INP1 to regulate liver cancer stem cell acquisition and self-renewal. *FEBS Lett.* 589: 500–506.

Liu, J., Zhao, Y., Ge, W., Zhang, P., Liu, X., Zhang, W., Hao, Y., Yu, S., Li, L., Chu, M., Min, L., Zhang, H., and Shen, W. (2017a). Oocyte exposure to ZnO nanoparticles inhibits early embryonic development through the gamma-H2AX and NF-kappaB signaling pathways. *Oncotarget* 8: 42673–42692.

Liu, S., Qin, S.H., He, M., Zhou, D.F., Qin, Q.D., and Wang, H. (2020). Current applications of poly(lactic acid) composites in tissue engineering and drug delivery. *Compos. B. Eng.* 199.

Liu, Y., Hardie, J., Zhang, X., and Rotello, V.M. (2017b). Effects of engineered nanoparticles on the innate immune system. *Semin. Immunol.* 34: 25–32.

Lopez-Viota, J., Mandal, S., Delgado, A.V., Toca-Herrera, J.L., Moller, M., Zanuttin, F., Balestrino, M., and Krol, S. (2009). Electrophoretic characterization of gold nanoparticles functionalized with human serum albumin (HSA) and creatine. *J. Colloid Interface Sci.* 332: 215–223.

Lu, X., Miousse, I.R., Pirela, S.V., Melnyk, S., Koturbash, I., and Demokritou, P. (2016). Short-term exposure to engineered nanomaterials affects cellular epigenome. *Nanotoxicology* 10: 140–150.

Madugundu, G.S., Cadet, J., and Wagner, J.R. (2014). Hydroxyl-radical-induced oxidation of 5-methylcytosine in isolated and cellular DNA. *Nucleic Acids Res.* 42: 7450–7460.

Mahmood, M., Li, Z., Casciano, D., Khodakovskaya, M.V., Chen, T., Karmakar, A., Dervishi, E., Xu, Y., Mustafa, T., Watanabe, F., Fejleh, A., Whitlow, M., Al-Adami, M., Ghosh, A., and Biris, A.S. (2011). Nanostructural materials increase mineralization in bone cells and affect gene expression through miRNA regulation. *J. Cell Mol. Med.* 15: 2297–2306.

Manke, A., Wang, L., and Rojanasakul, Y. (2013). Mechanisms of nanoparticle-induced oxidative stress and toxicity. *Biomed. Res. Int.* 2013: 942916.

Matea, C.T., Mocan, T., Tabaran, F., Pop, T., Mosteanu, O., Puia, C., Iancu, C., and Mocan, L. (2017). Quantum dots in imaging, drug delivery and sensor applications. *Int. J. Nanomed.* 12: 5421–5431.

McCarthy, J., Inkielewicz-Stepniak, I., Corbalan, J.J., and Radomski, M.W. (2012). Mechanisms of toxicity of amorphous silica nanoparticles on human lung submucosal cells in vitro: protective effects of fisetin. *Chem. Res. Toxicol.* 25: 2227–2235.

Meng, F., Zhong, Z., and Feijen, J. (2009). Stimuli-responsive polymersomes for programmed drug delivery. *Biomacromolecules* 10: 197–209.

Mentch, S.J., Mehrmohamadi, M., Huang, L., Liu, X., Gupta, D., Mattocks, D., Gomez Padilla, P., Ables, G., Bamman, M.M., Thalacker-Mercer, A.E., Nichenametla, S.N., and Locasale, J.W. (2015). Histone methylation dynamics and gene regulation occur through the sensing of one-carbon metabolism. *Cell Metab.* 22: 861–873.

Mitchell, M.J., Billingsley, M.M., Haley, R.M., Wechsler, M.E., Peppas, N.A., and Langer, R. (2021). Engineering precision nanoparticles for drug delivery. *Nat. Rev. Drug Discov.* 20: 101–124.

Mittler, R. (2017). ROS are good. *Trends Plant Sci.* 22: 11–19.

Moore, L.D., Le, T., and Fan, G. (2013). DNA methylation and its basic function. *Neuropsychopharmacology* 38: 23–38.

Munkanatta Godage, D.N.P., Vanhecke, G.C., Samarasinghe, K.T.G., Feng, H.Z., Hiske, M., Holcomb, J., Yang, Z., Jin, J.P., Chung, C.S., and Ahn, Y.H. (2018). SMYD2 glutathionylation contributes to degradation of sarcomeric proteins. *Nat. Commun.* 9: 4341.

Musolino, E., Pagiatakis, C., Serio, S., Borgese, M., Gamberoni, F., Gornati, R., Bernardini, G., and Papait, R. (2022). The Yin and Yang of epigenetics in the field of nanoparticles. *Nanoscale Adv.* 4: 979–994.

Mytych, J., Zebrowski, J., Lewinska, A., and Wnuk, M. (2017). Prolonged effects of silver nanoparticles on p53/p21 pathway-mediated proliferation, DNA damage response, and methylation parameters in HT22 hippocampal neuronal cells. *Mol. Neurobiol.* 54: 1285–1300.

Nagano, T., Higashisaka, K., Kunieda, A., Iwahara, Y., Tanaka, K., Nagano, K., Abe, Y., Kamada, H., Tsunoda, S., Nabeshi, H., Yoshikawa, T., Yoshioka, Y., and Tsutsumi, Y. (2013). Liver-specific microRNAs as biomarkers of nanomaterial-induced liver damage. *Nanotechnology* 24: 405102.

Najahi-Missaoui, W., Arnold, R.D., and Cummings, B.S. (2020). Safe nanoparticles: are we there yet? *Int. J. Mol. Sci.* 22.

Ng, C.T., Baeg, G.H., Yu, L.E., Ong, C.N., and Bay, B.H. (2018). Biomedical applications of nanomaterials as therapeutics. *Curr. Med. Chem.* 25: 1409–1419.

Ng, C.T., Dheen, S.T., Yip, W.C., Ong, C.N., Bay, B.H., and Lanry Yung, L.Y. (2011). The induction of epigenetic regulation of PROS1 gene in lung fibroblasts by gold nanoparticles and implications for potential lung injury. *Biomaterials* 32: 7609–7615.

Ognik, K., Cholewinska, E., Juskiewicz, J., Zdunczyk, Z., Tutaj, K., and Szlazak, R. (2019). The effect of copper nanoparticles and copper (II) salt on redox reactions and epigenetic changes in a rat model. *J. Anim. Physiol. Anim. Nutr. (Berl).* 103: 675–686.

Panikkanvalappil, S.R., Mahmoud, M.A., Mackey, M.A., and El-Sayed, M.A. (2013). Surface-enhanced Raman spectroscopy for real-time monitoring of reactive oxygen species-induced DNA damage and its prevention by platinum nanoparticles. *ACS Nano.* 7: 7524–7533.

Patel, S., Ryals, R.C., Weller, K.K., Pennesi, M.E., and Sahay, G. (2019). Lipid nanoparticles for delivery of messenger RNA to the back of the eye. *J. Control. Release* 303: 91–100.

Patil, N.A., Gade, W.N., and Deobagkar, D.D. (2016). Epigenetic modulation upon exposure of lung fibroblasts to TiO2 and ZnO nanoparticles: alterations in DNA methylation. *Int. J. Nanomed.* 11: 4509–4519.

Patil, Y.M., Rajpathak, S.N., and Deobagkar, D.D. (2019). Characterization and DNA methylation modulatory activity of gold nanoparticles synthesized by *Pseudoalteromonas* strain. *J. Biosci.* 44.

Patra, J.K., Das, G., Fraceto, L.F., Campos, E.V.R., Rodriguez-Torres, M.D.P., Acosta-Torres, L.S., Diaz-Torres, L.A., Grillo, R., Swamy, M.K., Sharma, S., Habtemariam, S., and Shin, H.S. (2018). Nano based drug delivery systems: recent developments and future prospects. *J. Nanobiotechnol.* 16: 71.

Paul, P., Chakraborty, A., Sarkar, D., Langthasa, M., Rahman, M., Bari, M., Singha, R.S., Malakar, A.K., and Chakraborty, S. (2018). Interplay between miRNAs and human diseases. *J. Cell Physiol.* 233: 2007–2018.

Paul, W. and Sharma, C.P. (2020). Inorganic nanoparticles for targeted drug delivery. In: *Biointegration of Medical Implant Materials*, 2e, Sharma, C.P. (ed), 333–373. Elsevier.

Podhorecka, M., Skladanowski, A., and Bozko, P. (2010). H2AX phosphorylation: its role in DNA damage response and cancer therapy. *J. Nucleic Acids* 2010.

Pogribna, M. and Hammons, G. (2021). Epigenetic effects of nanomaterials and nanoparticles. *J Nanobiotechnology* 19: 2.

Pogribna, M., Koonce, N.A., Mathew, A., Word, B., Patri, A.K., Lyn-Cook, B., and Hammons, G. (2020). Effect of titanium dioxide nanoparticles on DNA methylation in multiple human cell lines. *Nanotoxicology* 14: 534–553.

Portela, A. and Esteller, M. (2010). Epigenetic modifications and human disease. *Nat. Biotechnol.* 28: 1057–1068.

Prasad, R.Y., Chastain, P.D., Nikolaishvili-Feinberg, N., Smeester, L., Kaufmann, W.K., and Fry, R.C. (2013). Titanium dioxide nanoparticles activate the ATM-Chk2 DNA damage response in human dermal fibroblasts. *Nanotoxicology* 7: 1111–1119.

Puri, A., Loomis, K., Smith, B., Lee, J.H., Yavlovich, A., Heldman, E., and Blumenthal, R. (2009). Lipid-based nanoparticles as pharmaceutical drug carriers: from concepts to clinic. *Crit. Rev. Ther. Drug Carrier Syst.* 26: 523–580.

Qian, Y., Zhang, J., Hu, Q., Xu, M., Chen, Y., Hu, G., Zhao, M., and Liu, S. (2015). Silver nanoparticle-induced hemoglobin decrease involves alteration of histone 3 methylation status. *Biomaterials* 70: 12–22.

Ray Chaudhuri, A. and Nussenzweig, A. (2017). The multifaceted roles of PARP1 in DNA repair and chromatin remodelling. *Nat. Rev. Mol. Cell Biol.* 18: 610–621.

Ray, S., Li, Z., Hsu, C.H., Hwang, L.P., Lin, Y.C., Chou, P.T., and Lin, Y.Y. (2018). Dendrimer- and copolymer-based nanoparticles for magnetic resonance cancer theranostics. *Theranostics* 8: 6322–6349.

Rideau, E., Dimova, R., Schwille, P., Wurm, F.R., and Landfester, K. (2018). Liposomes and polymersomes: a comparative review towards cell mimicking. *Chem. Soc. Rev.* 47: 8572–8610.

Rispail, N., De Matteis, L., Santos, R., Miguel, A.S., Custardoy, L., Testillano, P.S., Risueno, M.C., Perez-De-Luque, A., Maycock, C., Fevereiro, P., Oliva, A., Fernandez-Pacheco, R., Ibarra, M.R., De La Fuente, J.M., Marquina, C., Rubiales, D., and Prats, E. (2014). Quantum dot and superparamagnetic nanoparticle interaction with pathogenic fungi: internalization and toxicity profile. *ACS Appl. Mater. Interfaces* 6: 9100–9110.

Rizvi, S.A.A. and Saleh, A.M. (2018). Applications of nanoparticle systems in drug delivery technology. *Saudi Pharm. J.* 26: 64–70.

Rossnerova, A., Honkova, K., Pelclova, D., Zdimal, V., Hubacek, J.A., Chvojkova, I., Vrbova, K., Rossner, P., Jr., Topinka, J., Vlckova, S., Fenclova, Z., Lischkova, L., Klusackova, P., Schwarz, J.,

Ondracek, J., Ondrackova, L., Kostejn, M., Klema, J., and Dvorackova, S. (2020). DNA methylation profiles in a group of workers occupationally exposed to nanoparticles. *Int. J. Mol. Sci.* 21.

Rubio, L., Annangi, B., Vila, L., Hernandez, A., and Marcos, R. (2016). Antioxidant and anti-genotoxic properties of cerium oxide nanoparticles in a pulmonary-like cell system. *Arch. Toxicol.* 90: 269–278.

Sarfraz, M., Afzal, A., Yang, T., Gai, Y., Raza, S.M., Khan, M.W., Cheng, Y., Ma, X., and Xiang, G. (2018). Development of dual drug loaded nanosized liposomal formulation by a reengineered ethanolic injection method and its pre-clinical pharmacokinetic studies. *Pharmaceutics* 10.

Schrand, A.M., Rahman, M.F., Hussain, S.M., Schlager, J.J., Smith, D.A., and Syed, A.F. (2010). Metal-based nanoparticles and their toxicity assessment. *Wiley Interdiscip. Rev. Nanomed. Nanobiotechnol.* 2: 544–568.

Scioli Montoto, S., Muraca, G., and Ruiz, M.E. (2020). Solid lipid nanoparticles for drug delivery: pharmacological and biopharmaceutical aspects. *Front Mol. Biosci.* 7: 587997.

Seidel, C., Kirsch, A., Fontana, C., Visvikis, A., Remy, A., Gate, L., Darne, C., and Guichard, Y. (2017). Epigenetic changes in the early stage of silica-induced cell transformation. *Nanotoxicology* 11: 923–935.

Sercombe, L., Veerati, T., Moheimani, F., Wu, S.Y., Sood, A.K., and Hua, S. (2015). Advances and challenges of liposome assisted drug delivery. *Front Pharmacol.* 6: 286.

Setyawati, M.I., Khoo, P.K., Eng, B.H., Xiong, S., Zhao, X., Das, G.K., Tan, T.T., Loo, J.S., Leong, D.T., and Ng, K.W. (2013). Cytotoxic and genotoxic characterization of titanium dioxide, gadolinium oxide, and poly(lactic-co-glycolic acid) nanoparticles in human fibroblasts. *J. Biomed. Mater Res. A* 101: 633–640.

Shyamasundar, S., Ng, C.T., Yung, L.Y., Dheen, S.T., and Bay, B.H. (2015). Epigenetic mechanisms in nanomaterial-induced toxicity. *Epigenomics* 7: 395–411.

Singh, N., Savanur, M.A., Srivastava, S., D'silva, P., and Mugesh, G. (2019). A manganese oxide nanozyme prevents the oxidative damage of biomolecules without affecting the endogenous antioxidant system. *Nanoscale* 11: 3855–3863.

Skvortsova, K., Iovino, N., and Bogdanovic, O. (2018). Functions and mechanisms of epigenetic inheritance in animals. *Nat. Rev. Mol. Cell Biol.* 19: 774–790.

Smolkova, B., Dusinska, M., and Gabelova, A. (2017). Nanomedicine and epigenome. Possible health risks. *Food Chem. Toxicol.* 109: 780–796.

Soenen, S.J., Hodenius, M., and De Cuyper, M. (2009). Magnetoliposomes: versatile innovative nanocolloids for use in biotechnology and biomedicine. *Nanomedicine (Lond)* 4: 177–191.

Sooklert, K., Nilyai, S., Rojanathanes, R., Jindatip, D., Sae-Liang, N., Kitkumthorn, N., Mutirangura, A., and Sereemaspun, A. (2019). N-acetylcysteine reverses the decrease of DNA methylation status caused by engineered gold, silicon, and chitosan nanoparticles. *Int. J. Nanomed.* 14: 4573–4587.

Sreeram, K.J., Nidhin, M., and Nair, B.U. (2009). Synthesis of aligned hematite nanoparticles on chitosan-alginate films. *Colloids Surf. B Biointerfaces* 71: 260–267.

Stayner, L., Steenland, K., Greife, A., Hornung, R., Hayes, R.B., Nowlin, S., Morawetz, J., Ringenburg, V., Elliot, L., and Halperin, W. (1993). Exposure-response analysis of cancer mortality in a cohort of workers exposed to ethylene oxide. *Am. J. Epidemiol.* 138: 787–798.

Stoccoro, A., Di Bucchianico, S., Coppede, F., Ponti, J., Uboldi, C., Blosi, M., Delpivo, C., Ortelli, S., Costa, A.L., and Migliore, L. (2017). Multiple endpoints to evaluate pristine and remediated titanium dioxide nanoparticles genotoxicity in lung epithelial A549 cells. *Toxicol. Lett.* 276: 48–61.

Sun, B., Liu, R., Ye, N., and Xiao, Z.D. (2015). Comprehensive evaluation of microRNA expression profiling reveals the neural signaling specific cytotoxicity of superparamagnetic iron oxide nanoparticles (SPIONs) through N-methyl-D-aspartate receptor. *PLoS One* 10: e0121671.

Sun, T., Zhang, Y.S., Pang, B., Hyun, D.C., Yang, M., and Xia, Y. (2014). Engineered nanoparticles for drug delivery in cancer therapy. *Angew. Chem. Int. Ed. Engl.* 53: 12320–12364.

Surapaneni, S.K., Bashir, S., and Tikoo, K. (2018). Gold nanoparticles-induced cytotoxicity in triple negative breast cancer involves different epigenetic alterations depending upon the surface charge. *Sci. Rep.* 8: 12295.

Tabish, A.M., Poels, K., Byun, H.M., Luyts, K., Baccarelli, A.A., Martens, J., Kerkhofs, S., Seys, S., Hoet, P., and Godderis, L. (2017). Changes in DNA methylation in mouse lungs after a single intra-tracheal administration of nanomaterials. *PLoS One* 12: e0169886.

Tarantini, A., Lanceleur, R., Mourot, A., Lavault, M.T., Casterou, G., Jarry, G., Hogeveen, K., and Fessard, V. (2015). Toxicity, genotoxicity and proinflammatory effects of amorphous nanosilica in the human intestinal Caco-2 cell line. *Toxicol. In Vitro* 29: 398–407.

Thi, T.T.H., Suys, E.J.A., Lee, J.S., Nguyen, D.H., Park, K.D., and Truong, N.P. (2021). Lipid-based nanoparticles in the clinic and clinical trials: from cancer nanomedicine to COVID-19 vaccines. *Vaccines (Basel)* 9.

Toduka, Y., Toyooka, T., and Ibuki, Y. (2012). Flow cytometric evaluation of nanoparticles using side-scattered light and reactive oxygen species-mediated fluorescence-correlation with genotoxicity. *Environ. Sci. Technol.* 46: 7629–7636.

Torano, E.G., Garcia, M.G., Fernandez-Morera, J.L., Nino-Garcia, P., and Fernandez, A.F. (2016). The impact of external factors on the epigenome: in utero and over lifetime. *Biomed. Res. Int.* 2016: 2568635.

Toyooka, T., Amano, T., and Ibuki, Y. (2012). Titanium dioxide particles phosphorylate histone H2AX independent of ROS production. *Mutat. Res.* 742: 84–91.

Uskokovic, V. and Uskokovic, D.P. (2011). Nanosized hydroxyapatite and other calcium phosphates: chemistry of formation and application as drug and gene delivery agents. *J. Biomed. Mater. Res. B Appl. Biomater.* 96: 152–191.

Vangijzegem, T., Stanicki, D., and Laurent, S. (2019). Magnetic iron oxide nanoparticles for drug delivery: applications and characteristics. *Expert Opin. Drug Deliv.* 16: 69–78.

Vhora, I., Lalani, R., Bhatt, P., Patil, S., and Misra, A. (2019). Lipid-nucleic acid nanoparticles of novel ionizable lipids for systemic BMP-9 gene delivery to bone-marrow mesenchymal stem cells for osteoinduction. *Int. J. Pharm.* 563: 324–336.

Wagner, A.M., Knipe, J.M., Orive, G., and Peppas, N.A. (2019). Quantum dots in biomedical applications. *Acta Biomater.* 94: 44–63.

Wang, J., Potocny, A.M., Rosenthal, J., and Day, E.S. (2020). Gold nanoshell-linear tetrapyrrole conjugates for near infrared-activated dual photodynamic and photothermal therapies. *ACS Omega* 5: 926–940.

Wang, Y., Zhao, Q., Han, N., Bai, L., Li, J., Liu, J., Che, E., Hu, L., Zhang, Q., Jiang, T., and Wang, S. (2015). Mesoporous silica nanoparticles in drug delivery and biomedical applications. *Nanomedicine* 11: 313–327.

Wibroe, P.P., Anselmo, A.C., Nilsson, P.H., Sarode, A., Gupta, V., Urbanics, R., Szebeni, J., Hunter, A.C., Mitragotri, S., Mollnes, T.E., and Moghimi, S.M. (2017). Bypassing adverse injection reactions to nanoparticles through shape modification and attachment to erythrocytes. *Nat. Nanotechnol.* 12: 589–594.

Yang, W.J., Liang, H.Z., Ma, S.H., Wang, D., and Huang, J. (2019). Gold nanoparticle based photothermal therapy: development and application for effective cancer treatment. *Sustain. Mater. Technol.* 22.

Yonezawa, S., Koide, H., and Asai, T. (2020). Recent advances in siRNA delivery mediated by lipid-based nanoparticles. *Adv. Drug Deliv. Rev.* 154-155: 64–78.

Yu, M., Zhou, X., Ju, L., Yu, M., Gao, X., Zhang, M., and Tang, S. (2020a). Characteristics of iron status, oxidation response, and DNA methylation profile in response to occupational iron oxide nanoparticles exposure. *Toxicol. Ind. Health* 36: 170–180.

Yu, Z., Li, Q., Wang, J., Yu, Y., Wang, Y., Zhou, Q., and Li, P. (2020b). Reactive oxygen species-related nanoparticle toxicity in the biomedical field. *Nanoscale Res. Lett.* 15: 115.

Zhang, L., Han, B., Xiang, J., Liu, K., Dong, H., and Gao, X. (2018). Silica nanoparticle releases SIRT6-induced epigenetic silencing of follistatin. *Int. J. Biochem. Cell Biol.* 95: 27–34.

Zhang, T., Du, E., Liu, Y., Cheng, J., Zhang, Z., Xu, Y., Qi, S., and Chen, Y. (2020). Anticancer effects of zinc oxide nanoparticles through altering the methylation status of histone on bladder cancer cells. *Int. J. Nanomed.* 15: 1457–1468.

Zhang, X.F., Park, J.H., Choi, Y.J., Kang, M.H., Gurunathan, S., and Kim, J.H. (2015). Silver nanoparticles cause complications in pregnant mice. *Int. J. Nanomed.* 10: 7057–7071.

Zhao, X., Rao, Y., Liang, J., Lin, S., Wang, X., Li, Z., and Huang, J. (2019). Silver nanoparticle-induced phosphorylation of histone H3 at serine 10 involves MAPK pathways. *Biomolecules* 9.

Zhao, X., Takabayashi, F., and Ibuki, Y. (2016). Coexposure to silver nanoparticles and ultraviolet A synergistically enhances the phosphorylation of histone H2AX. *J. Photochem. Photobiol. B* 162: 213–222.

Zhao, X., Toyooka, T., and Ibuki, Y. (2017). Silver nanoparticle-induced phosphorylation of histone H3 at serine 10 is due to dynamic changes in actin filaments and the activation of Aurora kinases. *Toxicol. Lett.* 276: 39–47.

Zhi, X., Fang, H., Bao, C., Shen, G., Zhang, J., Wang, K., Guo, S., Wan, T., and Cui, D. (2013). The immunotoxicity of graphene oxides and the effect of PVP-coating. *Biomaterials* 34: 5254–5261.

Zhou, J., Deng, W., Wang, Y., Cao, X., Chen, J., Wang, Q., Xu, W., Du, P., Yu, Q., Chen, J., Spector, M., Yu, J., and Xu, X. (2016). Cationic carbon quantum dots derived from alginate for gene delivery: one-step synthesis and cellular uptake. *Acta Biomater.* 42: 209–219.

Zhou, W., Tian, D., He, J., Yan, X., Zhao, J., Yuan, X., and Peng, S. (2019). Prolonged exposure to carbon nanoparticles induced methylome remodeling and gene expression in zebrafish heart. *J. Appl. Toxicol.* 39: 322–332.

Zou, Y., Li, Q., Jiang, L., Guo, C., Li, Y., Yu, Y., Li, Y., Duan, J., and Sun, Z. (2016). DNA hypermethylation of CREB3L1 and Bcl-2 associated with the mitochondrial-mediated apoptosis via PI3K/Akt pathway in human BEAS-2B cells exposure to silica nanoparticles. *PLoS One* 11: e0158475.

11

Engineered Nanoparticles Adversely Impact Glucose Energy Metabolism

*Yingxue Zhang, Alexander Yang, and Zhe Yang**

Department of Biochemistry, Microbiology, and Immunology, Wayne State University School of Medicine, Detroit, Michigan, USA
** Corresponding author*

Introduction

Engineered nanoparticles are nano-sized materials that have significantly different properties from bulk materials due to the small size and the large surface area to volume ratio. Generally, nanoparticles refer to particles of 1–100 nm in size in at least one dimension. On this scale, the properties of particles are significantly different from those of bulk materials. Quantum effects can occur on this scale, which drastically impacts the optical, electrical, and magnetic properties of particles. When the size gets smaller, opaque substances can become transparent, while solids can become liquid. Additionally, the relatively large surface area of nanoparticles leads to a high fraction of molecules on the surface and makes nanoparticles more chemically reactive. Nanoparticles can be synthesized from various organic or inorganic materials and have found many applications in a wide range of sectors including medicine. One of their clinical applications is the use of magnetic nanoparticles as an anticancer strategy to destroy tumors via hyperthermia (Kobayashi et al. 2014). When exposed to high-frequency alternating magnetic fields, magnetic nanoparticles can induce localized tumor heating, destroying tumor tissues. Another important clinical application is the use of nanoparticles as drug carriers (Mitchell et al. 2021). Depending on the type of nanoparticles, a drug can be carried within the core of the nanoparticles, or it can be chemically conjugated to the surface. The surface modification of nanoparticles is also a powerful strategy in their drug delivery applications, as not only does it permit the construction of biocompatible nanoparticles, but also it can improve drug delivery efficiency, achieving specific targeting of pathogenic cells (Mitchell et al. 2021). Nanoparticles have also been exploited in fluorescence imaging and used for early-stage disease diagnosis and for monitoring real-time therapeutic effects (Sun et al. 2021). Advantages of nanoparticle-based fluorescence probes, such as photostability and strong signal strength, make them very effective for sensitive cancer imaging, revolutionizing the field of image-guided surgery.

However, the use of nanoparticles in medicine has been receiving increased attention for their potential adverse effects on human health. Recent studies have shown that exposure to nanoparticles is indeed associated with cytotoxicity, including genotoxicity that causes DNA or chromosomal damage and elevated inflammation that causes cell or tissue injury (Mohammapdour and Ghandehari 2022; Singh et al. 2009). There are also an increasing number of studies indicating that nanoparticle-induced toxicity is associated with mitochondria malfunction (Asharani et al. 2009). Nanoparticle exposure has been found to cause significant changes in multiple mitochondria-related metabolic pathways, particularly glucose energy metabolism. In humans, the most

Impact of Engineered Nanomaterials in Genomics and Epigenomics, First Edition. Edited by Saura C Sahu.

important metabolic pathways are glycolysis, tricarboxylic acid cycle (TCA), and oxidative phosphorylation. These three pathways represent a three-stage process that is central in the breakdown of glucose in the presence of oxygen, as a major source of energy for all cells, harvesting the energy for the cells to grow and divide. Nanoparticles have been found to affect all these pathways, leading to significantly altered energy metabolism and dysregulation of genes and proteins essential for these pathways (Babele et al. 2019; Carrola et al. 2016; Lee et al. 2016; Shin et al. 2019). Deficiencies in glycolysis, TCA, and oxidative phosphorylation are associated with many metabolic disorders, such as diabetes, encephalopathies, and mitochondrial complex I deficiency. Organs that have high-energy demands, such as the brain and liver, are particularly vulnerable to defective glucose metabolism. A shortage of energy in these tissues can lead to cell death, causing neurological and liver problems. Here we discuss the impact of engineered nanoparticles on key metabolic pathways and their potential implications on human health. Our aim is to highlight important safety considerations for nanotoxicity when using nanoparticles for therapeutic or diagnostic purposes.

Biological Toxicity of Engineered Nanoparticles

Engineered nanoparticles have become highly valuable materials in biomedicine given their unique properties and biocompatibility. A variety of materials, including lipids, inorganics, and polymers, have been used to produce nanoparticles. Because of their intrinsically different physiochemical properties, the resulting nanoparticles can have many different characteristics in terms of chemical, electrical, optical, and magnetic properties, thus conferring different advantages and disadvantages in medical applications, as well as with different toxicity profiles when they come in contact with biological systems (Mitchell et al. 2021). Liposomes are the most commonly used nanoparticles in biomedicine. Most often composed of phospholipids, liposomes have a spherical structure consisting of one or more concentric lipid bilayers (Garcia-Pinel et al. 2019). As an effective drug delivery system, liposomes have been successfully applied in many biomedical areas and can encapsulate both lipophilic and hydrophilic drugs. Lipid nanoparticles, another type of lipid-based nanoparticle, are made of ionizable cationic lipids, thus often used for delivering mRNA vaccines and DNA vaccines (Hou et al. 2021). In terms of physiochemical properties, inorganic nanoparticles are the most diverse. Nanoparticles prepared from noble metals, such as gold and silver, have high chemical stability. Magnetic nanoparticles, such as those formed by Fe_3O_4, are superparamagnetic with a fast response to magnetic fields. Fluorescent nanoparticles, typically made of semiconducting materials, such as quantum dots, are widely used for fluorescent imaging of cells and tissues in *in vivo* diagnostics. Another major type of nanoparticles are polymer-based nanoparticles, such as nanocapsules (Mora-Huertas et al. 2010). Nanocapsules have many medical applications; as a vesicular system, drugs and nutraceuticals are placed in the inside core cavity, which provides protection in adverse environments.

Among all known nanoparticles, lipid-based nanoparticles are the least toxic to biological systems (Mitchell et al. 2021). As often prepared from natural phospholipids, liposomes are highly biocompatible, with minimal toxicity when exposed to living tissues. For polymeric nanoparticles, their toxic effects appear to be associated with their high surface area and strong inter-particle attractive interactions. Because of these, polymeric nanoparticles tend to aggregate or agglomerate, which may cause the blockage of blood capillaries (Ashraf et al. 2018). Additionally, their nonspecific interaction with the cell membrane could lead to membrane destabilization and cytosolic enzyme release (Jain et al. 2010). Inorganic nanoparticles appear to be the most toxic among the known nanoparticles. The clinical applications of inorganic nanoparticles are currently limited, primarily due to toxicity concerns. It has been shown that exposure to nanoparticles containing heavy metals causes bone marrow depression and intestinal bleeding (Ferdous and Nemmar 2020; Schrand et al. 2010). Silver nanoparticles (AgNPs), one of the well-characterized inorganic nanoparticles, showed *in vivo* toxicity in a variety of different organs, such as the liver, skin, brain, and reproductive organs (Ferdous and Nemmar 2020). AgNPs are excellent drug delivery carriers, capable of delivering various payloads, such as small drug molecules or large biomolecules. They are also widely used in metal-enhanced fluorescence for biosensing applications and effective disease diagnosis (Lee and Jun 2019). Compared with other inorganic nanoparticles, AgNPs are advantageous as biosensors due to higher extinction coefficients, sharper extinction bands, and high field enhancements (Lee and Jun 2019). However, exposure to AgNPs has been associated with inflammatory, oxidative, genotoxic, and cytotoxic consequences, often leading to liver damage as they primarily accumulate in the liver (Park et al. 2010). The mechanism of the toxicity of AgNPs appeared to be derived from oxidative stress and inflammation that were caused by the generation of reactive oxygen species (ROS) (Asharani et al. 2009; Carlson et al. 2008). AgNPs have also been shown to impact cellular metabolism and their toxicity appears to be relevant to mitochondria dysfunction (Asharani et al. 2009). Several of the most important metabolic pathways, including glycolysis, TCA cycle, and oxidative phosphorylation (Figure 11.1), were significantly dysregulated after AgNP exposure (Babele et al. 2019; Lee et al. 2016; Shin et al. 2019), implicating their potential adverse effects on human health by impairing cellular respiration. Thus, there is increasing concern regarding the potential toxicity of AgNPs and other inorganic nanoparticles with respect to cell metabolism.

Engineered Nanoparticles Alter Glucose Metabolism

As a major source of energy for all cells, glucose metabolism is critical for cell growth and survival. Four major pathways regulate glucose metabolism, including glycolysis, gluconeogenesis, glycogenolysis, and glycogenesis. Additionally, the pentose phosphate pathway utilizes glucose to

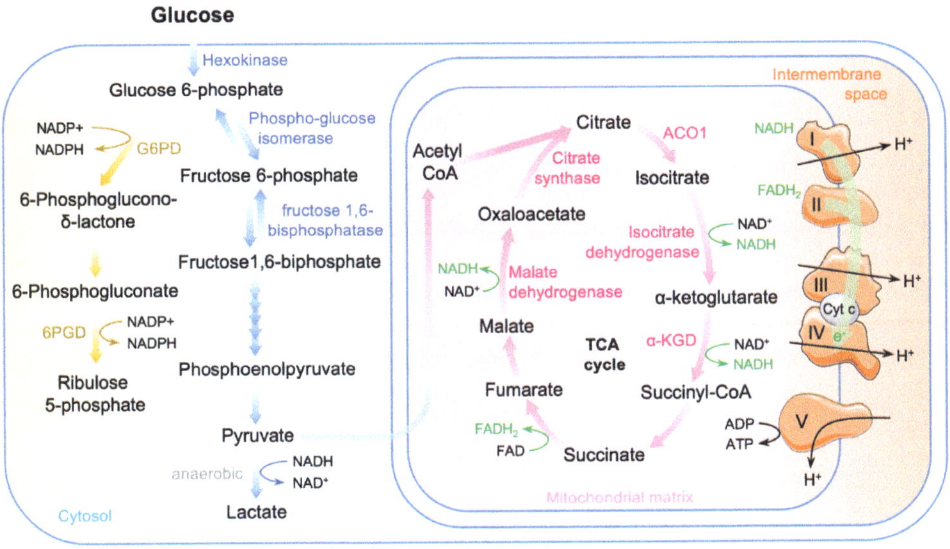

Figure 11.1 Glucose energy metabolism. Pentose phosphate pathway, glycolysis, the TCA cycle, and oxidative phosphorylation are shaded in yellow, blue, pink, and orange, respectively. G6PD: glucose 6-phosphate dehydrogenase; 6PGD: 6-phosphogluconate dehydrogenase; ACO1: aconitase 1; α-KGD: α-ketoglutarate dehydrogenase; Cyc c: cytochrome c.

generate NADPH, pentoses, and ribose 5-phosphate. Dysregulation of these pathways can disturb glucose homeostasis and lead to numerous metabolic disorders, such as diabetes, the most common metabolic disease. Nanoparticles have been shown to impact glucose metabolism. In one study, silver nanoparticles (AgNPs) were found to affect glucose consumption and lactate production, and these effects were found to be dependent on the size of nanoparticles used and the type of cells treated (Lee et al. 2016). It was shown that 5 nm AgNPs, and not 100 nm AgNPs, significantly decreased glucose consumption in HepG2 cells, but glucose consumption was not significantly affected in Huh7 cells by the size of the particles. However, lactate production was significantly reduced in both those cells after treatment with 5 nm AgNPs. The reduced lactate production suggests that the activity of certain proteins involved in regulating glycolysis, gluconeogenesis, or the pentose phosphate pathway might have been affected by AgNP treatment. Lactate production is dependent on the flux of glucose to glycolysis and the activity of gluconeogenesis. The pentose phosphate pathway also affects lactate production, since it changes the amount of glucose available to glycolysis. Thus, the reduced lactate production can be caused by the reduced activity of glycolysis, which converts glucose to lactate in anaerobic conditions, the increased activity of gluconeogenesis, which consumes lactate to generate glucose, or the increased activity of the pentose phosphate pathway, which reduces the flux of glucose to glycolysis. To determine how AgNPs affected lactate production, the expression levels of several genes involved in those pathways were examined (Lee et al. 2016). It was shown that the mRNA levels of hexokinase 2, PEPCK-C, and PEPCK-M were not significantly changed in both HepG2 and Huh7 cells treated with AgNPs, but the nuclear factor erythroid 2-related factor 2 (NRF2) mRNA levels were significantly reduced. These results indicate that AgNPs may not affect the first obligatory step of glycolysis, the conversion of glucose to glucose-6-phosphate, catalyzed by hexokinase 2. Since glucose-6-phosphate is the initial substrate of the pentose phosphate pathway, AgNPs may not affect the flux of glucose-6-phosphate to the pentose phosphate pathway either. However, the reduced NRF2 levels suggest that several steps in the pentose phosphate pathway could have been affected by AgNP treatment. NRF2 is known to regulate the pentose phosphate pathway by regulating the expression of enzymes involved in this pathway, such as G6PD and 6PGD (Figure 11.1) (Mitsuishi et al. 2012). G6PD is the rate-limiting enzyme in the pentose phosphate pathway that catalyzes the oxidation of glucose-6-phosphate, while 6PGD converts 6-phosphogluconate to ribulose 5-phosphate during the oxidative phase of this pathway. It has been shown that upon NRF2 activation, glucose is preferentially consumed through the pentose phosphate pathway (Mitsuishi et al. 2012). Thus, the reduced NRF2 levels caused by AgNP treatment are in agreement with reduced glucose consumption but do not explain the decrease in lactate production.

To further understand how AgNPs affected glucose metabolism, the kinetics of glucose consumption and lactate production were examined during AgNP treatment (Lee et al. 2016). It was shown that the decrease in glucose consumption occurred later than the decrease in lactate production (Lee et al. 2016). In other words, the reduced lactate production caused by AgNPs occurred at an earlier time before the decrease in glucose consumption could be detected. This result suggests a possibility of increased gluconeogenesis, which accelerates the depletion of lactate followed by glucose accumulation. Currently, there is no evidence to support this hypothesis. Cytoplasmic and mitochondrial PEPCK-C and PEPCK-M catalyze the conversion of oxaloacetate to phosphoenolpyruvate at the early steps of gluconeogenesis; however, no significant change in their expression levels after AgNP treatment suggests that AgNPs do not affect the concentration of these enzymes in gluconeogenesis (Lee et al. 2016). It remains unknown whether AgNPs affect the expression levels of fructose-1,6-bisphosphatase, the rate-limiting enzyme in gluconeogenesis that converts fructose-1,6-bisphosphate to fructose-6-phosphate. It also remains unknown whether

AgNPs affect the levels of glycerol and alanine, the two other known substrates of gluconeogenesis. Glycerol and alanine could affect the flux of lactate to gluconeogenesis. It has been shown that the gluconeogenic rate of a substrate when it is alone differs from its rates when it is in a mixture with other substrates (Shah and Wondisford 2020). If AgNP treatment causes changes in glycerol or alanine levels, the gluconeogenic flux rate from lactate could potentially be affected. It has also been shown that the levels of alanine were significantly disturbed after treatment with titanium dioxide (TiO_2) nanoparticles (Chen et al. 2019). Nevertheless, current evidence suggests that treatment with AgNPs affects glucose metabolism, likely through altering the expression of NRF2 and impacting the pentose phosphate pathway. An important area for future investigation is to understand whether and how gluconeogenesis is impacted by AgNP treatment.

Another study showed that silica-coated nanoparticles (SiO_2-NPs) affect glucose metabolism in HEK293 cells (Shin et al. 2019). In this study, an integrated multi-omics approach examined the metabolomic and transcriptomic effects of SiO_2-NPs. Several metabolites that are associated with glucose metabolism were found to have altered expression, and these changes were found to be associated with defective glucose uptake that affected the level of glucose in cells. Among the altered metabolites are pyruvate, 3-hydroxybutyrate, succinate, citrate, oxaloacetate, and 2-hydroxyglutarate (Shin et al. 2019). Pyruvate showed the largest alteration among all those organic acids, and its expression level was significantly increased after SiO_2-NP exposure. Pyruvate is the final product of glycolysis under aerobic conditions. The accumulation of pyruvate suggests that SiO_2-NP exposure might have induced mitochondrial defects that have impaired pyruvate transport into mitochondria. Changes in amino acid profiles can also signal homeostatic deregulation of glucose metabolism. It was found that after SiO_2-NP treatment, the levels of leucine, glutamic acid, lysine, and tyrosine were significantly increased, while the levels of glycine, serine, cysteine, and glutamine were significantly decreased (Shin et al. 2019). Cysteine and lysine have been associated with the protein expression of glucose transporters, while tyrosine has been shown to inhibit glucose transport (Gazit et al. 2003; Katsumata 2011). Thus, the altered expression of these amino acids might be associated with defective glucose uptake under the SiO_2-NP treatment condition. The ingenuity pathway analysis based on the altered metabolic profiles further suggests disturbed glucose metabolism. The metabolomics network derived from this analysis revealed a close metabolic relationship between the organic acids and amino acids with altered expression, where the suppression of glucose uptake was predicted as a result of the correlated changes in their expression. The fact that SiO_2-NP treatment caused altered glucose metabolism was further confirmed by the *in vitro* validation experiments (Shin et al. 2019). It was shown that SiO_2-NP treatment decreased glucose uptake and the amount of intracellular glucose by 40% and 30%, respectively. This result suggests that decreased glucose uptake might be responsible for the reduction in the level of intracellular glucose.

The transcriptomic analysis also indicated glucose metabolic dysfunction after treatment with SiO_2-NPs (Shin et al. 2019). The expression levels of 36 genes were changed in the SiO_2-NP-treated cells compared with untreated control cells. Among these genes, 21 genes were upregulated and 15 genes were downregulated. The highest differentially expressed genes are HNF1A and NR4A1. Both genes were shown to be associated with glucose metabolism, mainly through regulating glucose uptake. HNF1A, a transcriptional regulator of glucose transporters (Bae et al. 2010), was significantly downregulated after treatment with SiO_2-NPs. NR4A1, which regulates glucose transportation and glycolysis (Corrocher et al. 2018), showed significantly increased expression. Also, the gene expression levels of PRKCA and SLC2A2 were significantly increased after SiO_2-NP treatment. PRKCA is a kinase involved in the regulation of cell surface localization of glucose transporters, while SLC2A2 is a facilitative hexose transporter that mediates the bidirectional transport of glucose across the plasma membrane of hepatocytes (Lee et al. 2015; Mueckler and Thorens 2013). Thus, the transcriptomic network is interconnected with the metabolomic network, and the two networks have close relationships in glucose metabolic dysfunction.

In both the above studies, the production of ROS was elevated after treatment with nanoparticles (Lee et al. 2016; Shin et al. 2019). Increased ROS production represents a very common response of cells to nanoparticle treatment, likely due to the activation of the innate immune system (Liu et al. 2017). In the first study, AgNPs led to increased ROS production in both HepG2 and Huh7 cells, which was associated with altered glucose metabolism (Lee et al. 2016). Since ROS and cellular glucose metabolism have close interactions (Asharani et al. 2009), it is possible that increased ROS production was responsible for the decrease in glucose consumption and lactate production in the AgNP-treated cells. To test this hypothesis, the ROS scavenger N-acetylcysteine (NAC) was added to the cell culture before treatment with AgNPs (Lee et al. 2016). It was found that the addition of NAC restored reduced glucose consumption and lactate production caused by AgNP treatment. This result suggests that AgNPs disturb glucose consumption and lactate production likely via ROS-induced changes in glucose metabolism. One potential mechanism is that ROS stimulate cellular glucose uptake by upregulating the protein expression of glucose transporters and enhancing glucose transportation (Shin et al. 2019). In the second study, the altered metabolomic and transcriptomic profiles after SiO_2-NP treatment are highly related to ROS production, and there is a close relationship between increased ROS production and glucose metabolic dysfunction (Shin et al. 2019). It is well known that ROS and free amino acids have close interactions. ROS can directly oxidize tryptophan, tyrosine, histidine, and cysteine, while glycine protects the cell against oxidative stress by reducing the level of superoxide anions (Droge 2002; Ruiz-Ramirez et al. 2014). The oxidation of cysteine has the potential to disturb glucose metabolism since cysteine regulates glucose homeostasis through its enzymatic breakdown into hydrogen sulfide, which could be affected by ROS-induced oxidation (Droge 2002; Shin et al. 2019). Additionally, the differentially expressed NR4A1 and SLC2A2 are transcriptionally and translationally sensitive to oxidative stress (Shin et al. 2019). Thus, the combined metabolome and transcriptome network suggests the disturbance in glucose metabolism is likely due to increased ROS production in the SiO_2-NP-treatment condition.

Engineered Nanoparticles Alter TCA Cycle

The TCA cycle is the key metabolic pathway responsible for converting the chemical energy of acetyl coenzyme A (acetyl-CoA) into the reducing power of NADH. As one of the major routes of generating acetyl-CoA, the formation of acetyl-CoA from pyruvate, the final product of glycolysis, links the TCA cycle with glucose metabolism (Figure 11.1). Several studies have shown that exposure of cells to nanoparticles affects the TCA cycle. In one study, the effects of AgNPs on the metabolic pathways were examined in human dermal fibroblasts (Huang et al. 2020). By using metabolomics analysis, 22 metabolites were shown to be differentially expressed after 24 h of AgNP treatment. The key metabolic pathway affected by AgNPs was identified as the TCA cycle pathway. Two significant metabolic changes found in this pathway are the downregulation of malic acid and the upregulation of citrate (Huang et al. 2020). These two metabolic changes appear to be closely related. Malic acid is an important intermediate in the TCA cycle, and it can accelerate the TCA cycle by increasing the activity of malate dehydrogenase. However, as a downstream metabolite, the metabolism of malic acid can be affected by citrate, and increased citrate concentration can reduce the level of malic acid by inhibiting the activity of citrate synthase, the rate-limiting enzyme in the TCA cycle (Huang et al. 2020). Other studies using AgNPs also showed that exposure to nanoparticles affected the TCA cycle. AgNPs reduced the activity of the TCA cycle in human keratinocytes, while treatment with AgNPs caused the downregulation of genes involved in the TCA cycle in HepG2 cells (Bastos et al. 2016; Rank Miranda et al. 2020). Another metabolomics study found that the expression levels of five metabolites, including malic acid and citrate, were significantly altered in LO2 cells after AgNP treatment (Sun et al. 2018). Also, in an *in vivo* metabolic study, the TCA cycle was significantly downregulated in the lung of AgNP-injected mice (Jarak et al. 2017). These studies together indicate that AgNPs impact the TCA cycle by affecting the expression of its key metabolites. One potential consequence of decreased expression of malic acid is reduced ATP production, since malic acid and ATP generation are known to be closely related. To test this hypothesis, ATP content was measured after AgNP treatment (Huang et al. 2020). The ATP content in AgNP-treated cells was found to be significantly lower than that in untreated cells.

One study using an untargeted ^{1}H NMR-based metabolomics approach showed that AgNPs caused significant changes in the expression of metabolites in many metabolic pathways including the TCA cycle in yeast cells (Babele et al. 2019). Based on the chemical shift data and coupling patterns, 55 metabolites were identified to be significantly differentially expressed. These metabolites were classified by the molecular pathway analysis into amino acid metabolism, glycolysis, the TCA cycle, nucleotide metabolism, the urea cycle, and lipid metabolism. The differentially expressed metabolites linked to glycolysis and the TCA cycle include glucose, pyruvate, succinate, fumarate, citrate, and oxaloacetate, which were significantly downregulated, and lactate and acetate, which were significantly upregulated. The changes in the expression of these metabolites suggest that the cells were stressed and that there was an imbalance in carbohydrate metabolism after AgNP treatment. Since succinate, an essential intermediate of the TCA cycle, also plays an important role in ROS homeostasis (Cao et al. 2013), its downregulation could disturb the balance between ROS generation and scavenging, leading to oxidative stress. Given the upregulation of lactate and downregulation of glucose, there might be a shift in metabolism toward glycolysis to provide energy for cell survival. Also, increased acetate levels further indicate cellular stress, where it might be used as a source of energy (Babele et al. 2019; Jiang et al. 2013). Transcriptome analysis revealed that several genes involved in the TCA cycle were significantly downregulated, including ACO1 and α-KGD (Figure 11.1) (Babele et al. 2019). ACO1 is an essential enzyme in the TCA cycle and catalyzes the isomerization of citrate to isocitrate. α-KGD catalyzes the conversion of α-ketoglutarate to succinyl-CoA and is a key enzyme in determining the metabolic flux through the TCA cycle. Taken together, these results showed that AgNPs have a significant impact on the metabolic profiles of yeast cells and could cause disequilibrium in carbohydrate metabolism, leading to cytotoxicity by altering cellular transcriptome and metabolome. Consistently, another ^{1}H NMR-based metabolomic study showed that AgNPs significantly downregulated several metabolites in glycolysis and the TCA cycle, including citrate, succinate, and α-ketoglutarate, in human skin keratinocytes (Carrola et al. 2016). Not limited to AgNPs, a study showed that exposure of a buffalo rat liver cell line to SiO_2 nanoparticles caused a decrease in the expression of several enzymes in the TCA cycle (Xue et al. 2014). SiO_2 nanoparticles inhibited the expression of both isocitrate dehydrogenase and citrate synthase. This finding suggests that SiO_2 nanoparticles, like AgNPs, also have the potential to disturb the TCA cycle.

Engineered Nanoparticles Alter Oxidative Phosphorylation

Oxidative phosphorylation is the final stage in cellular respiration, which uses NADH generated in the TCA cycle to drive ATP synthesis by transferring the electrons of NADH to molecular oxygen through the mitochondrial electron transport chain (Figure 11.1). Nanoparticles have been shown to affect oxidative phosphorylation. One study used a combined transcriptomics and proteomics approach to assess the response of human monocytic cells to gold nanoparticle (AuNP) treatment (Gallud et al. 2019). Exposure to AuNPs was found to cause significant downregulation of several genes coding for subunits of the electron transport chain. The most significantly downregulated genes include cytochrome c, an essential protein in oxidative phosphorylation that functions as an electron shuttle between complexes III and IV, and NDUFAF3, which is required for the function of the electron transport chain involved in the assembly of complex I. The only upregulated gene was SOD2 (Gallud et al. 2019), a mitochondrial superoxide dismutase that provides protection against cell death by converting toxic superoxide, a byproduct of oxidative phosphorylation, to hydrogen peroxide and oxygen. Mutations in this gene have been shown to be associated with mitochondrial complex I deficiency (Gallud et al. 2019). Proteomics analysis corroborates altered oxidative phosphorylation in the AuNP-treatment condition. Pathway analysis of the significantly differentially expressed proteins showed that the dysregulated pathways were linked to mitochondrial dysfunction, oxidative phosphorylation, gluconeogenesis, and protein ubiquitination. The protein level of cytochrome c was found to be significantly downregulated, in agreement with the transcriptomic analysis. Another notable change is the downregulation of apoptosis-inducing factor (AIF). This protein is a mitochondrial oxidoreductase that is involved not only in the assembly of the electron transport chain but also in inducing caspase-independent apoptosis (Vahsen et al. 2004). Further experiments showed that those transcriptomic and proteomic changes were associated with mitochondrial dysfunction (Gallud et al. 2019). Mitochondrial functions in THP-1 cells exposed to AuNPs were assessed, including mitochondrial respiration, mitochondrial membrane potential, and mitochondrial superoxide production. AuNPs were found to significantly inhibit mitochondrial respiration based on the assessment of oxygen consumption rate. It was also found that AuNPs caused a marked loss of the mitochondrial membrane potential. Increased mitochondrial superoxide production was also observed after AuNP treatment (Gallud et al. 2019). These data together indicated that AuNPs impaired mitochondrial function, likely through altering the expression of mitochondrial proteins involved in oxidative phosphorylation that could eventually lead to mitochondria-dependent cell death.

Several other studies also showed that nanoparticles affect oxidative phosphorylation. One study investigated the effects of cadmium telluride quantum dots (CdTe-QDs) on mitochondrial function and oxidative phosphorylation in HepG2 cells (Nguyen et al. 2015). CdTe-QD treatment was found to result in the enlargement of mitochondria as examined with transmission electron microscopy and confocal microscopy. Functional analysis of mitochondria revealed that CdTe-QD treatment disrupted the mitochondrial membrane potential and decreased cellular respiration and ATP synthesis. Additionally, there was also a significant decrease in the rate of oxygen consumption. These results clearly suggest that CdTe-QDs affect the functioning of oxidative phosphorylation. In agreement with this notion, CdTe-QDs were found to affect the function of the electron transport chain complexes (Nguyen et al. 2015). The levels of complexes II, III, and IV were significantly downregulated after CdTe-QD treatment, whereas the level of complex V was significantly upregulated (Nguyen et al. 2015). It was also found that CdTe-QDs caused a significant decrease in the activity of complexes II, III, and IV but had little effect on the activity of complexes I and V. As complexes I and III are the primary sites of ROS production, the impairment of electron transfer through complexes I and III may induce ROS generation. It has been shown that CdTe-QDs can cause significantly elevated levels of ROS in AML12 cells and Huh7 cells (Katubi et al. 2019; Zhang et al. 2015). It has also been suggested that CdTe-QDs exert their cytotoxic effects by stimulating ROS generation (Katubi et al. 2019). Thus, the effects of CdTe-QDs on complex III might represent an underlying mechanism by which CdTe-QDs induce ROS production. Another study determined the potential impact of cadmium sulfide quantum dots (CdS-QDs) on yeast cells using proteomics analysis (Gallo et al. 2020). The significantly downregulated proteins after CdS-QD treatment were found to mainly function in oxidative phosphorylation, including the SDHB subunit of complex II, QCR6, QCR7, and RIP1 subunits of complex III, COX4 and COX6 subunits of complex IV, and ATP3 and ATP16 subunits of complex V. Another study focused on the potential toxic effects of SiO_2 nanoparticles on mitochondrial energy metabolism in BRL cells (Xue et al. 2014). It was found that exposure of SiO_2 nanoparticles to the cells induced cytotoxicity and mitochondrial damage, which was associated with a decrease in the activity of complexes I, III, and IV. It was also shown that SiO_2 nanoparticles activated Kupffer cells, which led to the release of TNF-α and ROS from these cells and subsequently inhibited complex I activity in BRL cells.

Conclusion

The widespread use of engineered nanoparticles poses a risk of harmful effects that may adversely affect human health. There are growing concerns about their clinical applications due to serious biosafety issues that have arisen during their direct contact with biological systems. Potential toxic effects of engineered nanoparticles have been revealed at the molecular level in multiple aspects of biology including perturbation of cellular metabolism. Providing energy necessary for cells to grow and survive, impaired cellular metabolism could result in significant health consequences, causing cell death, and eventually leading to metabolic disorders. Technological advances have allowed the examination of the adverse effects of nanoparticles at genome-wide levels using metabolomic signatures to understand nanotoxicity. Numerous changes identified in multiple metabolic pathways indicate the broad impact of nanoparticles on cellular metabolism, not only the glucose energy mechanism discussed in this chapter, but also altered metabolism of amino acids, nucleotides, and lipids, as well as the elevation of DNA damage-related metabolites. It has been shown that the nanoparticle-induced metabolic alterations are particle type- and dose-dependent, with each type of nanoparticle inducing distinct changes of certain metabolites, and with high doses inducing more metabolomic alterations. Although it has become clear about nanoparticle exposure being associated with metabolic perturbations, the link of the effects of nanoparticles on cellular metabolism with dysregulation of cellular functions remains largely unknown. Future investigation should focus on how the metabolomic alterations caused by nanoparticles can predict and reflect abnormal changes in cellular functions. With accumulating evidence of their adverse effects on human biology, the most pressing goal is to develop methods and strategies that can reduce their potential health risk to achieve maximum benefits in biomedical applications.

References

Asharani, P.V., Low Kah Mun, G., Hande, M.P., and Valiyaveettil, S. (2009). Cytotoxicity and genotoxicity of silver nanoparticles in human cells. *ACS Nano* 3: 279–290.

Ashraf, M.A., Peng, W., Zare, Y., and Rhee, K.Y. (2018). Effects of size and aggregation/agglomeration of nanoparticles on the interfacial/interphase properties and tensile strength of polymer nanocomposites. *Nanoscale Res. Lett.* 13: 214.

Babele, P.K., Singh, A.K., and Srivastava, A. (2019). Bio-inspired silver nanoparticles impose metabolic and epigenetic toxicity to *Saccharomyces cerevisiae*. *Front. Pharmacol.* 10: 1016.

Bae, J.S., Kim, T.H., Kim, M.Y., Park, J.M., and Ahn, Y.H. (2010). Transcriptional regulation of glucose sensors in pancreatic beta-cells and liver: an update. *Sensors (Basel)* 10: 5031–5053.

Bastos, V., Ferreira de Oliveira, J.M., Brown, D., Jonhston, H., Malheiro, E., Daniel-da-silva, A.L., Duarte, I.F., Santos, C., and Oliveira, H. (2016). The influence of citrate or PEG coating on silver nanoparticle toxicity to a human keratinocyte cell line. *Toxicol. Lett.* 249: 29–41.

Cao, J., Barbosa, J.M., Singh, N.K., and Locy, R.D. (2013). GABA shunt mediates thermotolerance in *Saccharomyces cerevisiae* by reducing reactive oxygen production. *Yeast* 30: 129–144.

Carlson, C., Hussain, S.M., Schrand, A.M., Braydich-Stolle, L.K., Hess, K.L., Jones, R.L., and Schlager, J.J. (2008). Unique cellular interaction of silver nanoparticles: size-dependent generation of reactive oxygen species. *J. Phys. Chem. B* 112: 13608–13619.

Carrola, J., Bastos, V., Ferreira De Oliveira, J.M., Oliveira, H., Santos, C., Gil, A.M., and Duarte, I.F. (2016). Insights into the impact of silver nanoparticles on human keratinocytes metabolism through NMR metabolomics. *Arch. Biochem. Biophys.* 589: 53–61.

Chen, Z., Zhou, D., Han, S., Zhou, S., and Jia, G. (2019). Hepatotoxicity and the role of the gut-liver axis in rats after oral administration of titanium dioxide nanoparticles. *Part Fibre Toxicol.* 16: 48.

Corrocher, F.A., Bueno De Paiva, L., Duarte, A.S.S., Ferro, K.P., Silveira, L.D.R., De Lima, T.I., Olalla Saad, S.T., and Lazarini, M. (2018). Reduced expression of NR4A1 activates glycolytic pathway in acute promyelocytic leukemia cells. *Leuk. Lymphoma* 59: 1501–1504.

Droge, W. (2002). Free radicals in the physiological control of cell function. *Physiol. Rev.* 82: 47–95.

Ferdous, Z. and Nemmar, A. (2020). Health impact of silver nanoparticles: a review of the biodistribution and toxicity following various routes of exposure. *Int. J. Mol. Sci.* 21.

Gallo, V., Srivastava, V., Bulone, V., Zappettini, A., Villani, M., Marmiroli, N., and Marmiroli, M. (2020). Proteomic analysis identifies markers of exposure to cadmium sulphide quantum dots (CdS QDs). *Nanomaterials (Basel)* 10.

Gallud, A., Kloditz, K., Ytterberg, J., Ostberg, N., Katayama, S., Skoog, T., Gogvadze, V., Chen, Y.Z., Xue, D., Moya, S., Ruiz, J., Astruc, D., Zubarev, R., Kere, J., and Fadeel, B. (2019). Cationic gold nanoparticles elicit mitochondrial dysfunction: a multi-omics study. *Sci. Rep.* 9: 4366.

Garcia-Pinel, B., Porras-Alcala, C., Ortega-Rodriguez, A., Sarabia, F., Prados, J., Melguizo, C., and Lopez-Romero, J.M. (2019). Lipid-based nanoparticles: application and recent advances in cancer treatment. *Nanomaterials (Basel)* 9.

Gazit, V., Ben-Abraham, R., Vofsi, O., and Katz, Y. (2003). L-cysteine increases glucose uptake in mouse soleus muscle and SH-SY5Y cells. *Metab. Brain Dis.* 18: 221–231.

Hou, X., Zaks, T., Langer, R., and Dong, Y. (2021). Lipid nanoparticles for mRNA delivery. *Nat. Rev. Mater.* 6: 1078–1094.

Huang, Y., Lu, X., Chen, R., and Chen, Y. (2020). Comparative study of the effects of gold and silver nanoparticles on the metabolism of human dermal fibroblasts. *Regen Biomater.* 7: 221–232.

Jain, K., Kesharwani, P., Gupta, U., and Jain, N.K. (2010). Dendrimer toxicity: let's meet the challenge. *Int. J. Pharm.* 394: 122–142.

Jarak, I., Carrola, J., Barros, A.S., Gil, A.M., Pereira, M.L., Corvo, M.L., and Duarte, I.F. (2017). From the cover: metabolism modulation in different organs by silver nanoparticles: an NMR metabolomics study of a mouse model. *Toxicol. Sci.* 159: 422–435.

Jiang, L., Gulanski, B.I., De Feyter, H.M., Weinzimer, S.A., Pittman, B., Guidone, E., Koretski, J., Harman, S., Petrakis, I.L., Krystal, J.H., and Mason, G.F. (2013). Increased brain uptake and oxidation of acetate in heavy drinkers. *J. Clin. Invest.* 123: 1605–1614.

Katsumata, M. (2011). Promotion of intramuscular fat accumulation in porcine muscle by nutritional regulation. *Anim. Sci. J.* 82: 17–25.

Katubi, K.M., Alzahrani, F.M., Ali, D., and Alarifi, S. (2019). Dose- and duration-dependent cytotoxicity and genotoxicity in human hepato carcinoma cells due to CdTe QDs exposure. *Hum. Exp. Toxicol.* 38: 914–926.

Kobayashi, T., Kakimi, K., Nakayama, E., and Jimbow, K. (2014). Antitumor immunity by magnetic nanoparticle-mediated hyperthermia. *Nanomedicine (Lond)* 9: 1715–1726.

Lee, E.E., MA, J., Sacharidou, A., Mi, W., Salato, V.K., Nguyen, N., Jiang, Y., Pascual, J.M., North, P.E., Shaul, P.W., Mettlen, M., and Wang, R.C. (2015). A protein kinase C phosphorylation motif in GLUT1 affects glucose transport and is mutated in GLUT1 deficiency syndrome. *Mol. Cell* 58: 845–853.

Lee, M.J., Lee, S.J., Yun, S.J., Jang, J.Y., Kang, H., Kim, K., Choi, I.H., and Park, S. (2016). Silver nanoparticles affect glucose metabolism in hepatoma cells through production of reactive oxygen species. *Int. J. Nanomed.* 11: 55–68.

Lee, S.H. and Jun, B.H. (2019). Silver nanoparticles: synthesis and application for nanomedicine. *Int. J. Mol. Sci.* 20.

Liu, Y., Hardie, J., Zhang, X., and Rotello, V.M. (2017). Effects of engineered nanoparticles on the innate immune system. *Semin. Immunol.* 34: 25–32.

Mitchell, M.J., Billingsley, M.M., Haley, R.M., Wechsler, M.E., Peppas, N.A., and Langer, R. (2021). Engineering precision nanoparticles for drug delivery. *Nat. Rev. Drug Discov.* 20: 101–124.

Mitsuishi, Y., Taguchi, K., Kawatani, Y., Shibata, T., Nukiwa, T., Aburatani, H., Yamamoto, M., and Motohashi, H. (2012). Nrf2 redirects glucose and glutamine into anabolic pathways in metabolic reprogramming. *Cancer Cell* 22: 66–79.

Mohammapdour, R. and Ghandehari, H. (2022). Mechanisms of immune response to inorganic nanoparticles and their degradation products. *Adv. Drug Deliv. Rev.* 180: 114022.

Mora-Huertas, C.E., Fessi, H., and Elaissari, A. (2010). Polymer-based nanocapsules for drug delivery. *Int. J. Pharm.* 385: 113–142.

Mueckler, M. and Thorens, B. (2013). The SLC2 (GLUT) family of membrane transporters. *Mol. Aspects Med.* 34: 121–138.

Nguyen, K.C., Rippstein, P., Tayabali, A.F., and Willmore, W.G. (2015). Mitochondrial toxicity of cadmium telluride quantum dot nanoparticles in mammalian hepatocytes. *Toxicol. Sci.* 146: 31–42.

Park, E.J., Bae, E., Yi, J., Kim, Y., Choi, K., Lee, S.H., Yoon, J., Lee, B.C., and Park, K. (2010). Repeated-dose toxicity and inflammatory responses in mice by oral administration of silver nanoparticles. *Environ. Toxicol. Pharmacol.* 30: 162–168.

Rank Miranda, R., Pereira Da Fonseca, M., Korzeniowska, B., Skytte, L., Lund Rasmussen, K., and Kjeldsen, F. (2020). Elucidating the cellular response of silver nanoparticles as a potential combinatorial agent for cisplatin chemotherapy. *J. Nanobiotechnol.* 18: 164.

Ruiz-Ramirez, A., Ortiz-Balderas, E., Cardozo-Saldana, G., Diaz-Diaz, E., and El-hafidi, M. (2014). Glycine restores glutathione and protects against oxidative stress in vascular tissue from sucrose-fed rats. *Clin. Sci.* 126: 19–29.

Schrand, A.M., Rahman, M.F., Hussain, S.M., Schlager, J.J., Smith, D.A., and Syed, A.F. (2010). Metal-based nanoparticles and their toxicity assessment. *Wiley Interdiscip. Rev. Nanomed. Nanobiotechnol.* 2: 544–568.

Shah, A.M. and Wondisford, F.E. (2020). Tracking the carbons supplying gluconeogenesis. *J. Biol. Chem.* 295: 14419–14429.

Shin, T.H., Seo, C., Lee, D.Y., ji, M., Manavalan, B., Basith, S., Chakkarapani, S.K., Kang, S.H., Lee, G., Paik, M.J., and Park, C.B. (2019). Silica-coated magnetic nanoparticles induce glucose metabolic dysfunction in vitro via the generation of reactive oxygen species. *Arch. Toxicol.* 93: 1201–1212.

Singh, N., Manshian, B., Jenkins, G.J., Griffiths, S.M., Williams, P.M., Maffeis, T.G., Wright, C.J., and Doak, S.H. (2009). NanoGenotoxicology: The DNA damaging potential of engineered nanomaterials. *Biomaterials* 30: 3891–3914.

Sun, A., Ban, Z., Mu, L., and HU, X. (2018). Screening small metabolites from cells as multifunctional coatings simultaneously improves nanomaterial biocompatibility and functionality. *Adv. Sci. (Weinh)* 5: 1800341.

Sun, Y., Wang, Y., Liu, Y., Weng, B., Yang, H., Xiang, Z., Ran, J., Wang, H., and Yang, C. (2021). Intelligent tumor microenvironment-activated multifunctional nanoplatform coupled with turn-on and always-on fluorescence probes for imaging-guided cancer treatment. *ACS Appl. Mater. Interfaces* 13: 53646–53658.

Vahsen, N., Cande, C., Briere, J.J., Benit, P., Joza, N., Larochette, N., Mastroberardino, P.G., Pequignot, M.O., Casares, N., Lazar, V., Feraud, O., Debili, N., Wissing, S., Engelhardt, S., Madeo, F., Piacentini, M., Penninger, J.M., Schagger, H., Rustin, P., and Kroemer, G. (2004). AIF deficiency compromises oxidative phosphorylation. *EMBO J.* 23: 4679–4689.

Xue, Y., Chen, Q., Ding, T., and Sun, J. (2014). SiO(2) nanoparticle-induced impairment of mitochondrial energy metabolism in hepatocytes directly and through a Kupffer cell-mediated pathway in vitro. *Int. J. Nanomed.* 9: 2891–2903.

Zhang, T., Hu, Y., Tang, M., Kong, L., Ying, J., Wu, T., Xue, Y., and Pu, Y. (2015). Liver toxicity of cadmium telluride quantum dots (CdTe QDs) due to oxidative stress in vitro and in vivo. *Int. J. Mol. Sci.* 16: 23279–23299.

12

Artificial Intelligence and Machine Learning of Single-Cell Transcriptomics of Engineered Nanoparticles

*Alexander Yang, Yingxue Zhang, and Zhe Yang**

Department of Biochemistry, Microbiology, and Immunology, Wayne State University School of Medicine, Detroit, Michigan, USA
** Corresponding author*

Introduction

Engineered nanoparticles have a wide range of applications in the biomedical field, with significant promising results in drug delivery and cancer imaging. Owing to their tiny size, ranging from 1 to 100 nm, nanoparticles exhibit very different physicochemical properties compared with bulk materials. The exploding growth in the use of nanoparticles has led to an increase in both intended and unintended human exposure. There are growing concerns about their biosafety, as justified by accumulating evidence that nanoparticle exposure can lead to toxic effects in many aspects of biology (Navya and Daima 2016). Nanoparticles are known to induce cytotoxicity and tissue injury, associated with the release of proinflammatory cytokines (Elsabahy and Wooley 2013; Liu et al. 2017). It has also been widely acknowledged that the excessive production of reactive oxygen species (ROS) is a common response associated with nanoparticle exposure (Fu et al. 2014). However, many puzzles exist about the mechanism by which nanoparticles induce toxicity. The pressing questions are what properties of nanoparticles make them toxic and how those properties interact with biological systems, leading to toxicity. One assumption is that different physicochemical properties of nanoparticles activate different proteins and cellular pathways that, in turn, govern their exact impact on the cell and ultimately on human health (Mitchell et al. 2021). Furthermore, despite a wide range of biological effects identified for nanoparticle exposure, the primary effects, secondary effects, or indirect effects of nanoparticles and how the complex biological responses induced by nanoparticles interact with each other remain elusive. The relationship between the biological responses and the properties of nanoparticles is one of the main issues in biomedical research of nanoparticles; however, it is far from being understood. Understanding the complex interactions between nanoparticles and the biological systems is necessary for improving their biosafety and utilization in biomedical applications.

Transcriptomic approaches are increasingly used for the assessment of the biosafety of nanoparticles and have provided global information on disturbed gene expression upon nanoparticle exposure (Horstmann et al. 2021). In response to nanoparticle exposure, numerous genes show significantly altered expression, affecting numerous critical cellular pathways, such as energy metabolism, oxidative stress, inflammation, and apoptosis (Horstmann et al. 2021; Karpeta-Kaczmarek et al. 2016; Mohammadinejad et al. 2019). The complex transcriptomic responses induced by nanoparticle exposure depict not only the activation of cell survival programs but also potential cell damage that eventually leads to cell death (Gurunathan et al. 2018). Traditional

Impact of Engineered Nanomaterials in Genomics and Epigenomics, First Edition. Edited by Saura C. Sahu.

transcriptomic studies measure gene expression in response to nanoparticle exposure in the entire cell population. Results obtained from such studies are changes in gene expression averaged over a large number of cells. Thus, the traditional transcriptomic assays are unable to describe the heterogeneity of the interactions between nanoparticles and cells. Also, it is difficult for the bulk cell-based transcriptomic analysis to describe dynamic changes in gene expression upon nanoparticle exposure and to discern the complex relationships between the induced transcriptomic responses. However, knowledge of the heterogeneous nature of the nanoparticle–cell interactions is essential for understanding the transcriptomic plasticity in the cellular response to nanoparticle exposure, thus gaining a higher resolution view of transcriptomic alterations.

Single-cell RNA sequencing (scRNA-seq), though not yet widely used in studying the biosafety of nanoparticles, has revolutionized our fundamental understanding of cellular transcriptome at the resolution of an individual cell. Cell-to-cell variability, that would otherwise be averaged in bulk experiments, can be revealed by scRNA-seq by profiling cell-specific transcriptional networks. Thus, scRNA-seq is a powerful tool to analyze cellular heterogeneity and to reveal the remodeling dynamics of gene expression networks during cell differentiation. It has been used to study intratumor heterogeneity and tumor microenvironment heterogeneity, to identify cell subsets responsible for drug resistance, and to reconstruct tumor evolution through analysis of transcriptional heterogeneity (Lee et al. 2014; Yan et al. 2022; Zhang et al. 2021). It has also been used to infer the differentiation trajectories of hematopoietic stem cells and reveal the transcriptomic signatures underlying lineage specification and commitment during hematopoiesis (Ranzoni et al. 2021). These scRNA-seq studies have provided an unprecedented understanding of the gene expression differences at the single-cell level. Only a few studies have investigated the transcriptomic effects of nanoparticles using scRNA-seq. One such study characterized the heterogenous responses of cells to different doses of quantum dots (QDs) (Mitchell et al. 2016). Because achieving uniform cellular uptake of nanoparticles is experimentally difficult, bulk RNA-seq is unable to reveal unique responses to different doses, due to signal averaging across many cells, each carrying different amounts of nanoparticles. However, this problem is not an issue for scRNA-seq since the gene expression analysis is conducted in a single cell with a defined amount of nanoparticles. It was found that the cellular responses to QDs are dose-dependent, with a higher dose causing mostly suppression of cellular processes whereas a lower dose triggering the activation of cellular defense or repair pathways (Mitchell et al. 2016). Thus, scRNA-seq technology has opened up unique opportunities to understand the transcriptomic changes at the single-cell resolution after nanoparticle exposure. In addition, capturing the transcriptome of thousands of single cells may allow us to reveal the complexity of dynamic nanoparticle–cell interactions, thereby mapping the transcriptional routes that lead to toxicity.

A vast amount of gene expression data yielded by scRNA-seq, as well as the cell-to-cell variability in gene expression due to stochastic transcriptional bursting, raise substantial challenges for the analysis of scRNA-seq data. The fact that cells are complex, dynamic, and adaptive biological systems make those challenges even more challenging. To tackle the data complexity issues in the scRNA-seq method, artificial intelligence (AI) and machine learning (ML) are increasingly used in data analysis and have been successfully applied in many areas of scRNA-seq research, from raw data analysis to inference of functional pathways and quantification of cellular differentiation states (Aevermann et al. 2021; Biancalani et al. 2021; Petegrosso et al. 2020). For example, a convolutional neural network (CNN)-based deep-learning method has been developed to mine gene–gene relationships from scRNA-seq data, which has resulted in a significant improvement in predicting targets of transcription factors and in identifying disease-related genes and inferring causality (Wang et al. 2021; Yuan and Bar-Joseph 2019) (Figure 12.1). Another study using a

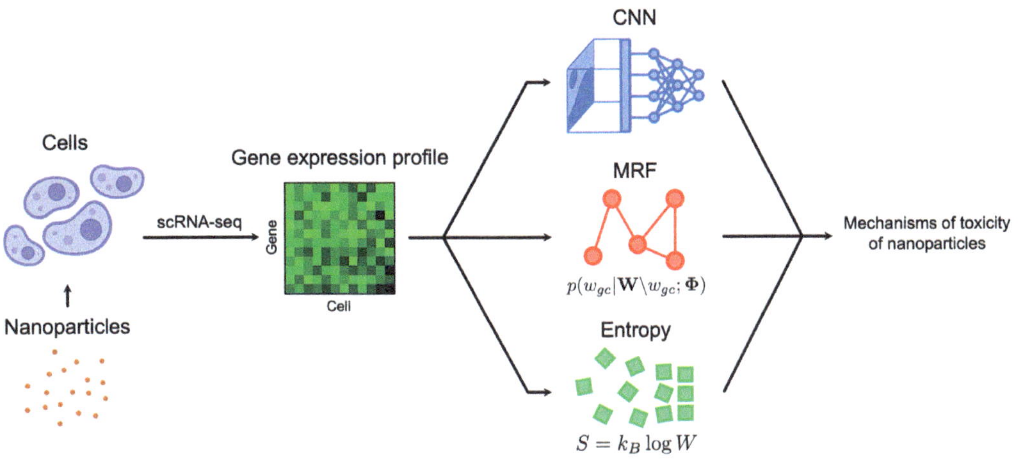

Figure 12.1 Prospective application of artificial intelligence and machine learning in determining the mechanisms of toxicity of nanoparticles from scRNA-seq data.

Markov random field (MRF) model accommodates gene network information in scRNA-seq data analysis, which has led to a significant increase in statistical power for identifying cell-specific differentially expressed genes from scRNA-seq data (Zhu et al. 2022). A different study, borrowing an idea from the information entropy theory, utilizes scRNA-seq data to compute single-cell entropy, which allows to quantitatively measure cellular differentiation states and predict cell-differentiation trajectories (Chen et al. 2021). These successful applications have showcased the power of AI and ML in scRNA-seq data analysis and their potential in uncovering the complex biological relationships from scRNA-seq data. It is conceivable that if these methods are used to study changes in tens of thousands of single-cell transcriptomes after nanoparticle exposure, the toxicity pathways that govern the toxicological activity of nanoparticles, as well as the complex interactions between these pathways, will be revealed. Here, we discuss the application of scRNA-seq in assessing the toxicological impact of nanoparticles and highlight the use of AI and ML in scRNA-seq data analysis. The purpose of these discussions is to make a proposition that the integration between scRNA-seq and those computational methods will allow an unprecedented understanding of nanoparticle-induced toxicity at a global genome-wide level.

Impact of Nanoparticles on Single-Cell Transcriptomics and Response Heterogeneity

Overview of Engineered Nanoparticles

Engineered nanoparticles, with superior physicochemical properties due to their small size, have been widely applied in various biomedical fields, such as drug delivery, tissue engineering, and cancer imaging. The wide usability of nanoparticles is also attributed to the availability of various types of nanoparticles with different surface properties, stability, and reactivity. For example, QDs have been generating a growing interest in medical diagnostics due to their high photostability, high quantum yield, and high molar extinction coefficients (Matea et al. 2017). Another example is titanium dioxide nanoparticles (TiO_2-NPs), which have gained wide attention in anticancer and antibacterial therapy owing to their chemical stability and high oxidizing abilities (Ziental et al. 2020). Silver nanoparticles (AgNPs) are another commonly used nanoparticles for biomedical applications and have been widely tested for the development of therapeutic interventions as drug carriers due to their high-density surface ligand attachment and high transmembrane delivery abilities (Xu et al. 2020). However, the extensive use of nanoparticles has raised concerns regarding their possible health hazards on humans. A growing body of evidence implicates the potential adverse effects of nanoparticles. Several toxic events have been shown to be caused by QDs, such as ROS generation, DNA damage, mitochondrial dysfunction, and cell death (Yong et al. 2013). The accumulation of TiO_2-NPs in the internal organs has been shown to be associated with intestinal tumor formation and inflammatory disorders such as inflammatory bowel disease and obesity (Grande and Tucci 2016). Extensive studies have been conducted to assess the toxicological effect of AgNPs on biological systems, which have shown that AgNPs exert cardiovascular toxicity due to their ability to enter systemic circulation (Xu et al. 2020). However, although extensive efforts have been made for understanding the toxicity of nanoparticles, the molecular mechanisms underlying their biological toxicity remain largely unknown; in particular, the toxicity mechanisms from a transcriptomic perspective. The main challenges that have hindered our understanding of transcriptomic responses associated with toxicity are the complexity of interactions between nanoparticles and the biological systems as well as the heterogeneity of cell populations due to the presence of a variety of distinct cell types or of cells in distinct states. With the recent development in scRNA-seq technology, it becomes possible for the detection of transcriptomic heterogeneity at the single-cell level. Several scRNA-seq investigations have successfully revealed the transcriptomic signatures and heterogeneous responses after nanoparticle exposure. These studies have filled the gap in our knowledge of how distinct nanoparticle properties contribute to unique cellular responses.

Dose-Dependent Heterogeneous Transcriptomic Responses to Quantum Dots

A study utilizing scRNA-seq demonstrated that cells respond to different doses of QDs through distinct transcriptomic strategies (Mitchell et al. 2016). However, before the time when this study was conducted, dose-dependent transcriptomic responses were not well characterized due to experimental difficulties in preparing cells with uniform nanoparticle loads. To overcome this challenge, this study developed a strategy that enabled the cells with defined loads of nanoparticles to be sorted out. The premise of this strategy is that if the amount of nanoparticle loads is known for individual cells, dose-dependent transcriptomic responses can be revealed by scRNA-seq analysis of these cells. Alveolar epithelial cells were chosen for the study because of their relevance to airborne nanoparticle exposure, and therefore this choice highlights the significance of this study in understanding the toxicological effect of nanoparticles on human health. Aminated and carboxylated QDs were chosen for the study because their bright fluorescence allows individual cells to be sorted out by their nanoparticle load (Mitchell et al. 2016). This choice is the foundation of the entire study, which enabled scRNA-seq analysis to reveal dose-dependent transcriptomic responses after nanoparticle exposure. It was shown that there was a wide distribution of nanoparticle loads, as much as two orders of magnitudes, justifying the necessity of using scRNA-seq, rather than bulk RNA-seq, to deduce unique transcriptomic responses to different loads of nanoparticles. Principal component analysis (PCA), based on all 9,233 genes identified in the scRNA-seq analysis, indicated that the cells exhibited distinct transcriptomic responses, which were not only dose-dependent but also depended on the type of QD (Mitchell et al. 2016). There were clearly different transcriptomic signatures between the cells with low and high loads of QDs and between the cells treated with aminated and carboxylated QDs. A subsequent classification tree-based analysis identified genes involved in stress response and cell cycle being primarily responsible for those observed transcriptomic differences.

It is not surprising for the observation that QD exposure caused the stress response of the cell, as the cellular stress response is a normal reaction that helps the cell to defend against the insult. However, of interest here is that the cell appears to respond to stress in different ways when exposed to different types and doses of QDs. To gain an in-depth understanding of mechanistic differences in the stress response pathways, this study performed gene set enrichment analysis using differentially expressed genes (DEGs) (Mitchell et al. 2016). It was found that the low level of aminated QDs caused the upregulation of many stress-related proteins and processes, including heat shock proteins, unfolded protein binding, cell cycle, apoptosis, and cell redox homeostasis. Consistent with these changes, the ubiquitin conjugation function was also uniquely upregulated by the low level of aminated QDs. The ubiquitin conjugation function plays important roles at all levels of stress response. One important function of the ubiquitination-related genes is the removal of proteins damaged by stressors to avoid accumulation as potentially harmful aggregates. Interestingly, the high level of aminated QDs caused the downregulation of most of the aforementioned stress-related processes, except for cell cycle (Mitchell et al. 2016). Considering that the high level of aminated QDs is more toxic than the low level of aminated QDs, the differences they caused in the cellular stress responses might reflect how the cell responds to different levels of toxicity. At the low level of aminated QDs, the cell appears to be capable of managing adverse effects and actively repairing damage, whereas at the high level of aminated QDs, the overly stressed cell diverts its effort from stress response to cell cycle control for survival. This notion is in line with another finding that aminated and carboxylated QDs caused similar cellular responses, including the upregulation of genes involved in cell cycle when they were at the high level. However, the low level of carboxylated QDs showed some unique responses, such as increased expression of genes in RNA binding. Together, these results support an idea that the high level of QDs, regardless of their type, leads to the convergence of the cellular response mechanisms to focus on survival, whereas the low level of QDs allows the cell to explore diverse stress response pathways for damage control and repair.

It turned out that the QD-triggered cellular responses were highly heterogeneous, where the cells responded with different strategies even to the same QD type with similar loads (Mitchell et al. 2016). Thanks to scRNA-seq that has allowed the understanding of the divergent behavior at the single-cell level. For example, one cell carrying the low level of aminated QDs showed a much greater level of stress response than other cells in the same group. In addition to this difference, several other functions were found to be upregulated in this cell, including protein translation, RNA binding, and ATP synthesis. Another cell used a different strategy, with a similar level of stress response but a marked decrease in the expression of genes involved in cell cycle, ATP binding, and translation initiation. Interestingly, the upregulation of cellular respiration was shared across all divergent cells carrying the low level of QDs (Mitchell et al. 2016). Cellular respiration is essential for cell survival because it provides the energy for the cell to perform all necessary functions. Given that the low level of QDs caused the activation of many stress response pathways, the upregulation of cellular respiration might be in line with a greater need for energy for the cell to manage adverse effects and to defend against the insult. Thus, this study provides evidence that individual cells can exhibit significantly distinct behaviors in response to nanoparticles, while demonstrating the use of scRNA-seq that has allowed the observation of response heterogeneity, which could not be achieved by bulk RNA-seq-based studies due to signal averaging.

TiO_2 Nanoparticles of Different Sizes Elicit Heterogeneous Transcriptomic Responses

By taking advantage of the sensitivity of scRNA-seq, a study demonstrated that exposure of TiO_2-NPs to human bone marrow stromal cells (BMSCs) elicited size-dependent heterogeneous transcriptomic responses in those cells (Xiang et al. 2022). This study identified diverse cellular pathways altered by TiO_2-NP exposure, including DNA repair, cell cycle, and oxidative stress, which has suggested the potential mechanisms by which TiO_2-NPs exert their toxicity on the bone marrow. The choice of BMSCs was largely because the bone marrow was considered as a major site of TiO_2-NP accumulation. Additionally, this choice represents a novel contribution to the toxicological studies of TiO_2-NPs, as most of previous toxicity studies only focused on the bone marrow-derived red and white blood cells. BMSCs are a heterogeneous, pluripotent group of cells that are indispensable for maintaining hematopoiesis through establishing the bone marrow microenvironment for hematopoietic stem cell differentiation. In this study, the interactions between TiO_2-NPs and BMSCs were investigated by scRNA-seq, which has the potential to reveal the heterogeneity and complexity of those interactions. Because size is one of the most important factors in defining nanoparticles' properties, the main aim of this study was to elucidate distinct cellular response mechanisms when BMSCs were exposed to different sizes of TiO_2-NPs. Five sizes of TiO_2-NPs were chosen in the study, ranging from 5 to 228 nm. After scRNA-seq analysis, PCA was performed with all 14,220 genes detected in the analysis, which led to the identification of two major clusters of cells: the cells exposed to 5 and 14 nm TiO_2-NPs and the cells exposed to other sizes of TiO_2-NPs (54, 135, and 228 nm). Thus, this clustering result clearly indicates that TiO_2-NPs elicited size-dependent transcriptomic responses.

All sizes of TiO_2-NPs were found to cause profound transcriptional changes in BMSCs (Xiang et al. 2022). About 15% of genes were differentially expressed in the cells exposed to the small size of TiO_2-NPs compared with about 25% of genes to the large size of TiO_2-NPs. Gene ontology (GO) pathway analysis using these DEGs revealed that the cellular responses to TiO_2-NP exposure were very complex, consisting of both size-dependent and -independent mechanisms. Several common cellular responses, including DNA repair, cell cycle, response to virus, and DNA replication, were observed in all sizes of TiO_2-NPs. This observation is consistent with the fact that about 25–35% of DEGs were shared between different sizes of TiO_2-NPs. To understand the size-dependent cellular responses, this study focused on a subset of genes (195 genes) that exhibited the largest changes in expression between the small and large sizes of TiO_2-NPs. Based on pathway analysis of these genes, it was concluded that the small size of TiO_2-NPs triggered the cellular defense response to virus and the cytokine and interferon signaling pathway, while the large size of TiO_2-NPs induced the expression of genes involved in cell proliferation and differentiation.

Like QDs discussed above, TiO_2-NPs were also found to induce cell-specific heterogeneous responses, the two together highlighting the complexity in cellular responses to the insult as well as the necessity of using scRNA-seq to resolve the diverse nature of such complexity. In this study, divergent cellular responses were identified when the levels of gene expression were compared at the single-cell level (Xiang et al. 2022). Several cells were found to have very different transcriptomic profiles compared with the rest of the cells in the same treatment group. One cell showed a significant downregulation of genes involved in the carbohydrate metabolic process and transcription regulation, whereas the metabolic process of carbohydrate was upregulated in another cell. Another cell showed extensive downregulation of cellular activities, including gene expression, the ATP catabolic process, as well as the small molecule metabolic process. In other cases, the apoptotic process was downregulated in one cell but upregulated in another, even though these two cells were exposed to the same size of TiO_2-NPs. The enrichment of the apoptosis pathway suggests that some of the cells were subjected to apoptosis after TiO_2-NP exposure. While the mechanism of heterogeneity in cellular responses to nanoparticles remains elusive, cell heterogeneity has been associated with a number of cellular factors, such as different stages of the cell cycle, circadian rhythm, and the level of ATP (Ha et al. 2020). Another possible reason in this case is the intrinsic heterogeneity of the BMSC population. BMSCs show various shapes and expressions of different cell markers, and can differentiate into various lineages, such as endothelial cells, fibroblasts, adipocytes, macrophages, osteocytes, and muscle cells (Xiang et al. 2022). While this study did not consider any potential contributions of BMSC heterogeneity to the heterogeneous cellular responses, the divergent cell behaviors might represent a complex response to TiO_2-NP perturbation, being determined by many factors that may interact with each other.

AI and ML in scRNA-Seq Data Analysis

Overview of AI and ML in Bioinformatics

The increased use of scRNA-seq in addressing the toxicity of nanoparticles has led to the appreciation of significant heterogeneity in cellular responses to nanoparticles, a multitude of cellular pathways perturbed by nanoparticle exposure, where even the same dose and size of the same type of nanoparticle can elicit opposite changes in the same cellular pathways. Thus, the use of scRNA-seq has the potential to lead to the identification of novel cellular response mechanisms that only occur in a small population of cells, allowing for the tracking of the toxicity development through the inference of transcriptional dynamics from cell to cell, which could otherwise be averaged and smoothed out in bulk RNA-seq experiments. However, the current scRNA-seq studies of nanoparticle toxicity did not fully harness the power of this single-cell based method, and even though with the large quantities of data generated by this method, their data analysis was limited to the pathway enrichment analysis that is traditionally used for bulk RNA-seq data. As a result, instead of the inference of causal relationships, complex gene–gene interactions, and dynamic transcriptional trajectories, the current analysis only yielded knowledge of statistical associations between the cellular responses elicited by nanoparticles, with limited insight into the spatiotemporal relationships between these events. To exploit the dynamics and heterogeneity of scRNA-seq data for model discovery, AI- and ML-based models are increasingly used in data analysis, with tremendous success in many biological fields, such as cell-type classification, trajectory inference, and gene-regulatory network inference.

Several notable ML-based methods that have been successfully used in scRNA-seq data analysis and other biological fields include CNN-based deep learning, probability-based MRF, and information or thermodynamic entropy that measures randomness of a system (Figure 12.1). CNN is a deep neural network originally designed for image analysis. In the field of biology, one successful application of CNN is accurate protein structure prediction; a prediction accuracy greater than 90% was achieved using the AlphaFold algorithm developed based on CNN (Jumper et al. 2021; Tunyasuvunakool et al. 2021). Deep learning is also transforming medical diagnostics and uses convolution and pooling for lesion detection and segmentation, which has allowed greater extraction of features from medical images for health-related problems (Levine et al. 2019). MRF is a powerful probability-based undirected graphical model, which can be used to infer spatial correlations among neighboring nodes and define potentials for biological interactions within a cell and across cells (Yang et al. 2022). The concept of entropy has been widely explored in many scientific disciplines, from statistical mechanics to thermodynamics and information theory. As a measure of disorder or uncertainty in a system, entropy is also a useful measure of cellular heterogeneity and has been used to define cell differentiation states based on patterns of gene expression (Teschendorff and Enver 2017). These ML-based models are increasingly used in inferring biological knowledge from scRNA-seq data, leading to an in-depth understanding of the dynamic complexity in gene interaction networks.

MRF in Differential Expression Analysis of scRNA-Seq Data

The transcriptional noise and biological variability across cells have raised analytical challenges for reliably detecting DEGs from scRNA-seq data. In the pursuit of an effective method for identifying DEGs, the MRF single-cell RNA-sequencing (MRFscRNA-seq) method was developed, which increases statistical power to identify cell-specific DEGs from scRNA-seq data (Li et al. 2021; Zhu et al. 2022). MRFscRNA-seq utilizes a two-sample t-test basis, where the test is run on each gene for each cell type between disease and control groups, and where the evidence of a gene being differentially expressed is represented by a z-score derived from the t-test statistic. Instead of simply using the z-score for inference of DEGs, this method constructs an MRF graph network model that incorporates prior knowledge of cell-type dependencies and gene network information into DEG detection. In the graph, the nodes correspond to cell-type specific genes, with their values denoting cell-type specific differential expression (DE) status, while the edges represent the relationships between the nodes, obtained from known gene–gene interactions and cell–cell dependencies. The main concept here is that the DE status of a gene is influenced by its neighbors in the network. Intuitively, if a gene is a DEG, the probability of its neighboring genes being a DEG is higher, or vice versa. In the MRF graph, these local influences are modeled by the potential of interaction between the nodes, where the sum of the potentials of all interactions defines the energy or probability of a given configuration of DE status. The primary objective of this model is to infer a configuration of DE status with the highest joint probability. For parameter estimation of the model, the expectation-maximization (EM) algorithm is used to approximate the interaction potential of the edges based on the observed evidence of z-scores. This is where the gene expression data sees the model, which makes the parameters of the model data dependent. After the model parameters are obtained from the EM algorithm, Gibbs sampling is used to sample the posterior probability of the DE status of each gene across all cell types.

In a simulation experiment comparing MRFscRNA-seq against other methods, such as the two-sample t-test and Wilcoxon rank sum test, MRFscRNA-seq was found to possess a significantly higher power than other methods, correctly identifying a much higher proportion of DEGs, while still retaining an acceptable false discovery rate. MRFscRNA-seq was also found to be robust, maintaining accuracy in highly varied populations for gene connections and DEG distributions. The effectiveness of MRFscRNA-seq was also demonstrated in studying scRNA-seq data of idiopathic pulmonary fibrosis (IPF) (Li et al. 2021). When MRFscRNA-seq was applied to the IPF study, MRFscRNA-seq was able to correctly identify numerous DEGs that other methods did not. Furthermore, ingenuity pathways analysis of DEGs showed that the method, in addition to accurately identifying pathways found by multiple other models, also revealed pathways related to IPF that other methods did not detect. These pathways include the Th2 activation pathway, which studies have shown promotes IPF progression, and the VDR/RXR activation pathway, which was found to be pathogenically related to IPF in previous studies (Li et al. 2021). The outcomes of the MRFscRNA-seq studies show promise in the value in applying ML-based methods to scRNA-seq data analysis to improve upon preexisting methods. By utilizing gene–gene and cell–cell networks, this method increases statistical power to detect DEGs from scRNA-seq data.

Deep Learning for Inferring Gene Relationships from scRNA-Seq Data

The CNN for coexpression (CNNC), a recently developed deep-learning framework, infers gene interactions from scRNA-seq data through the utilization of a CNN (Yuan and Bar-Joseph 2019). In determining gene relationships from gene expression data, while there are various successful methods, such as Pearson correlation and mutual information, those methods often suffer from significant variability, causing them to erroneously predict genes that have no relationship as being related and genes that are related as not being related (Yuan and Bar-Joseph 2019). As a result, the development of a more accurate computational method that improves upon preexisting methods has become more prudent given the advances in AI, and this prospect of potential improvement resulted in the creation of CNNC. In CNNC, gene expression data provided by scRNA-seq is transformed into a series of images, with each image representing the coexpression profile of a pair of genes across all cell samples and being depicted by a two-dimensional histogram, wherein each of the two axes represents the expression range of a gene, and each entry on the histogram represents the counts of the number of cell samples that have the respective expression levels of both genes. The model is trained using the gene expression datasets labeled with a vector that is defined by some aspect of functional interactions of genes and whose dimension is based upon the goal of the study. For example, when the model is trained to predict a causal relationship between two genes a and b, a three-dimensional vector is used, the first dimension representing a lack of an interaction, the second representing an interaction of a with b, and the third representing an interaction of b with a.

Tests determining the robustness of CNNC, including its ability to predict target genes of transcription factors, regulator–target gene pairs, and causal gene interactions, reveals that CNNC, in most cases, performs better than the preexisting methods, including count statistics, Pearson correlation, and GO, with a significantly higher median AUPRC/AUROC, two evaluation metrics that measure a method's ability to accurately find positive cases and discriminate between positive and negative cases, respectively (Yuan and Bar-Joseph 2019). In both the tests for prediction of regulator–target gene pairs and casual gene interactions, CNNC was able to attain an exceptional median AUROC value of 0.9949, nearing the perfect case of 1. Additionally, CNNC has also been successfully used to assign functions to genes with greater accuracy than other methods, such as guilt by association. Such an ability is valuable to drug design as it can reveal potential drug targets by showing which genes are critical to the pathogenesis of the disease. When tested on two lung diseases, asthma and chronic obstructive pulmonary disease, CNNC identified many genes that had undergone previous research due to their possible importance in the treatment of the disease. One of these genes identified by CNNC was the tyrosine–protein kinase Lck, which had recently been determined to be a possible drug target for the treatment of asthma. Thus, CNNC shows that through the union of scRNA-seq and deep learning, large amounts of data can be analyzed without sacrificing the ability to discern the major functions of genes and intricacies of their relationships. CNNC, whose applications excel in both depth and breadth, also demonstrates the power of scRNA-seq in fostering in-depth understanding of gene–gene interactions, highlighting the relevant mechanistic details that would otherwise be imperceptible with conventional techniques.

Determining Cell Differentiation and Lineage Based on Single-Cell Entropy

scRNA-seq offers the capability to determine cell differentiation and lineage, possibly allowing a deeper understanding of organ formation and function. However, the several existing methods that had been developed to reconstruct cell differentiation trajectories from scRNA-seq data, including Wanderlust and Monocle, require the use of external knowledge, such as gene marker or temporal information, in order to determine the directions of cell differentiation (Sagar and Grun 2020; Sun et al. 2017). By introducing the concept of entropy, a new method, named single-cell lineage inference using cell expression similarity and entropy (SLICE), was developed, which unleashes the potential of scRNA-seq in determining cell differentiation and lineage without the need for any preexisting knowledge (Guo et al. 2017). SLICE quantifies cell differentiation states by computing single-cell entropy (scEntropy), a numeric value that measures the level of uncertainty of functional diversity reflected by gene expression patterns in individual cells. Specifically, scEntropy is calculated using Boltzmann's entropy equation rewritten to be in terms of a probability distribution of genes in all functional groups, where the more skewed and concentrated the distribution, the lower the entropy. In this context, entropy describes the level of specialization and organization present in the cell. Cells with lower entropy generally are more specialized toward a few particular functions, and thus more differentiated, while those with higher entropy are less specialized and more diverse with respect to cellular functions. In the development of a cell lineage, cells tend to become more and more differentiated, becoming more specialized over time, and thus show a decreasing scEntropy. SLICE infers cell lineages by identifying cells that possess a local minimum entropy, which are considered to be in a stable state. Through clustering of the cells into distinct states, a lineage is then traced between two given stable state cells from the higher entropy cell to the lower entropy cell, using the shortest possible path through the other stable state cells between the two given cells. Finally, for every set of two neighboring stable state cells in this lineage, the transitional cells are interpolated, thus yielding a complete lineage between the two chosen cells.

SLICE was shown to correctly determine the cell lineage when applied to three scRNA-seq datasets with a known cell lineage, including the datasets collected from mouse lung alveolar type 2 cells, human skeletal muscle myoblasts, and human early embryonic cells (Guo et al. 2017). In these studies, SLICE demonstrates an approximately equivalent or greater accuracy than other methods in every category of functional inference, including the correct cell clustering, identification of DEGs, and correlation of scEntropy to cell differentiation. A direct comparison with Monocle reveals that SLICE performs similarly for relatively homogeneous populations, such as the population for human early embryonic cells, but outperforms Monocle for heterogeneous populations, shown by its more accurate cell lineage reconstruction for mouse lung alveolar type 2 cells. Furthermore, in a test using embryonic mouse lung cells, SLICE was also shown to correctly determine the lung mesenchymal cell lineage relationships that contain branched pathways, where two or more distinct cell clusters originate from a single cell cluster, while Monocle, in the absence of time information for the cells in the dataset, instead constructed a linear cell lineage for the same data. Thus, SLICE demonstrates the power of combining the entropy theory and scRNA-seq in accurately determining cellular differentiation states and reconstructing cell differentiation lineages.

Conclusion

The increased use of scRNA-seq in characterizing the toxicity of nanoparticles has provided an unprecedented understanding of the complexity and heterogeneity in the cellular responses to nanoparticles. The power of scRNA-seq lies in the fact that it allows the transcriptome of each cell to be measured individually. Unlike the bulk RNA-seq that only yields a single averaged transcriptomic profile of a population of cells, the scRNA-seq allows the analysis of tens of thousands of

cells in a single experiment, where each cell represents a unique transcriptomic response, thus allowing the revelation of the heterogeneous cellular behaviors in response to nanoparticles. Given the breadth of scRNA-seq, our understanding of the toxicity of nanoparticles has never been clearer, which has helped us develop a greater appreciation of their heterogeneity, such that some cells respond to nanoparticles by the activation of pathways that promote survival, while others, if overwhelmed by nanoparticles, elicit programmed cell death that eliminates themselves. The use of the conventional gene set enrichment analysis in scRNA-seq data analysis has allowed the identification of cellular processes and molecular functions affected by nanoparticles and provided a qualitative understanding of the response heterogeneity that can either be dependent on or independent of the dose and size of nanoparticles. However, the depth in our understanding of the toxicity of nanoparticles is limited with such an analysis as the analysis is unable to reveal any causal relationships among the affected genes and cellular pathways, and thus it is unable to fully unleash the power from a large amount of data offered by scRNA-seq. Recent studies that applied AI and ML to scRNA-seq data analysis have demonstrated a great potential for the use of these computational methods in uncovering the hidden patterns in scRNA-seq data, and it has proven to be robust in many areas of functional inference from scRNA-seq data ranging from the accurate identification of DEGs, functional annotation, disease gene prediction, to the determination of cell differentiation trajectories. This successful application not only highlights the great value in data generated by scRNA-seq, but also showcases the effectiveness of AI and ML in revealing the concealed mechanistic information from scRNA-seq data, which could not otherwise be revealed by a conventional analysis of gene expression data. Therefore, it is conceivable that recent advances in AI and its applications in general scRNA-seq data analysis, if applied to analyze the toxicity of nanoparticles, should provide a more comprehensive understanding of the toxicity profile of nanoparticles, how the complex biological responses induced by nanoparticles interact with each other, the trajectory of toxicity development, as well as the causal gene regulatory networks that relate the primary effects, secondary effects, and indirect effects of nanoparticles.

References

Aevermann, B., Zhang, Y., Novotny, M., Keshk, M., Bakken, T., Miller, J., Hodge, R., Lelieveldt, B., Lein, E., and Scheuermann, R.H. (2021). A machine learning method for the discovery of minimum marker gene combinations for cell type identification from single-cell RNA sequencing. *Genome. Res.* 31: 1767–1780.

Biancalani, T., Scalia, G., Buffoni, L., Avasthi, R., Lu, Z., Sanger, A., Tokcan, N., Vanderburg, C.R., Segerstolpe, A., Zhang, M., Avraham-Davidi, I., Vickovic, S., Nitzan, M., MA, S., Subramanian, A., Lipinski, M., Buenrostro, J., Brown, N.B., Fanelli, D., Zhuang, X., Macosko, E.Z., and Regev, A. (2021). Deep learning and alignment of spatially resolved single-cell transcriptomes with Tangram. *Nat. Methods* 18: 1352–1362.

Chen, Z., Yang, Z., Yuan, X., Zhang, X., and Hao, P. (2021). scSensitiveGeneDefine: sensitive gene detection in single-cell RNA sequencing data by Shannon entropy. *BMC Bioinformatics* 22: 211.

Elsabahy, M. and Wooley, K.L. (2013). Cytokines as biomarkers of nanoparticle immunotoxicity. *Chem. Soc. Rev.* 42: 5552–5576.

Fu, P.P., Xia, Q., Hwang, H.M., Ray, P.C., and Yu, H. (2014). Mechanisms of nanotoxicity: generation of reactive oxygen species. *J. Food Drug Anal.* 22: 64–75.

Grande, F. and Tucci, P. (2016). Titanium dioxide nanoparticles: a risk for human health? *Mini Rev. Med. Chem.* 16: 762–769.

Guo, M., Bao, E.L., Wagner, M., Whitsett, J.A., and Xu, Y. (2017). SLICE: determining cell differentiation and lineage based on single cell entropy. *Nucleic Acids Res.* 45: e54.

Gurunathan, S., Qasim, M., Park, C., Yoo, H., Choi, D.Y., Song, H., Park, C., Kim, J.H., and Hong, K. (2018). Cytotoxicity and transcriptomic analysis of silver nanoparticles in mouse embryonic fibroblast cells. *Int. J. Mol. Sci.* 19.

Ha, M.K., Kwon, S.J., Choi, J.S., Nguyen, N.T., Song, J., Lee, Y., Kim, Y.E., Shin, I., Nam, J.W., and Yoon, T.H. (2020). Mass cytometry and single-cell RNA-seq profiling of the heterogeneity in human peripheral blood mononuclear cells interacting with silver nanoparticles. *Small* 16: e1907674.

Horstmann, C., Davenport, V., Zhang, M., Peters, A., and Kim, K. (2021). Transcriptome profile alterations with carbon nanotubes, quantum dots, and silver nanoparticles: a review. *Genes (Basel)* 12.

Jumper, J., Evans, R., Pritzel, A., Green, T., Figurnov, M., Ronneberger, O., Tunyasuvunakool, K., Bates, R., Zidek, A., Potapenko, A., Bridgland, A., Meyer, C., Kohl, S.A.A., Ballard, A.J., Cowie, A., Romera-Paredes, B., Nikolov, S., Jain, R., Adler, J., Back, T., Petersen, S., Reiman, D., Clancy, E., Zielinski, M., Steinegger, M., Pacholska, M., Berghammer, T., Bodenstein, S., Silver, D., Vinyals, O., Senior, A.W., Kavukcuoglu, K., Kohli, P., and Hassabis, D. (2021). Highly accurate protein structure prediction with alphafold. *Nature* 596: 583–589.

Karpeta-Kaczmarek, J., Dziewiecka, M., Augustyniak, M., Rost-Roszkowska, M., and Pawlyta, M. (2016). Oxidative stress and genotoxic effects of diamond nanoparticles. *Environ. Res.* 148: 264–272.

Lee, M.C., Lopez-Diaz, F.J., Khan, S.Y., Tariq, M.A., Dayn, Y., Vaske, C.J., Radenbaugh, A.J., Kim, H.J., Emerson, B.M., and Pourmand, N. (2014). Single-cell analyses of transcriptional heterogeneity during drug tolerance transition in cancer cells by RNA sequencing. *Proc. Natl. Acad. Sci. U. S. A.* 111: E4726–35.

Levine, A.B., Schlosser, C., Grewal, J., Coope, R., Jones, S.J.M., and Yip, S. (2019). Rise of the machines: advances in deep learning for cancer diagnosis. *Trends Cancer* 5: 157–169.

Li, H., Zhu, B., Xu, Z., Adams, T., Kaminski, N., and Zhao, H. (2021). A Markov random field model for network-based differential expression analysis of single-cell RNA-seq data. *BMC Bioinform.* 22: 524.

Liu, Y., Hardie, J., Zhang, X., and Rotello, V.M. (2017). Effects of engineered nanoparticles on the innate immune system. *Semin. Immunol.* 34: 25–32.

Matea, C.T., Mocan, T., Tabaran, F., Pop, T., Mosteanu, O., Puia, C., Iancu, C., and Mocan, L. (2017). Quantum dots in imaging, drug delivery and sensor applications. *Int. J. Nanomed.* 12: 5421–5431.

Mitchell, H.D., Markillie, L.M., Chrisler, W.B., Gaffrey, M.J., Hu, D., Szymanski, C.J., Xie, Y., Melby, E.S., Dohnalkova, A., Taylor, R.C., Grate, E.K., Cooley, S.K., Mcdermott, J.E., Heredia-Langner, A., and Orr, G. (2016). Cells respond to distinct nanoparticle properties with multiple strategies as revealed by single-cell RNA-seq. *ACS Nano* 10: 10173–10185.

Mitchell, M.J., Billingsley, M.M., Haley, R.M., Wechsler, M.E., Peppas, N.A., and Langer, R. (2021). Engineering precision nanoparticles for drug delivery. *Nat. Rev. Drug Discov.* 20: 101–124.

Mohammadinejad, R., Moosavi, M.A., Tavakol, S., Vardar, D.O., Hosseini, A., Rahmati, M., Dini, L., Hussain, S., Mandegary, A., and Klionsky, D.J. (2019). Necrotic, apoptotic and autophagic cell fates triggered by nanoparticles. *Autophagy* 15: 4–33.

Navya, P.N. and Daima, H.K. (2016). Rational engineering of physicochemical properties of nanomaterials for biomedical applications with nanotoxicological perspectives. *Nano Converg.* 3: 1.

Petegrosso, R., Li, Z., and Kuang, R. (2020). Machine learning and statistical methods for clustering single-cell RNA-sequencing data. *Brief Bioinform.* 21: 1209–1223.

Ranzoni, A.M., Tangherloni, A., Berest, I., Riva, S.G., Myers, B., Strzelecka, P.M., Xu, J., Panada, E., Mohorianu, I., Zaugg, J.B., and Cvejic, A. (2021). Integrative single-cell RNA-seq and ATAC-seq analysis of human developmental hematopoiesis. *Cell Stem. Cell* 28 (472–487): e7.

Sagar and Grun, D. (2020). Deciphering cell fate decision by integrated single-cell sequencing analysis. *Annu. Rev. Biomed. Data Sci.* 3: 1–22.

Sun, N., Yu, X., Li, F., Liu, D., Suo, S., Chen, W., Chen, S., Song, L., Green, C.D., Mcdermott, J., Shen, Q., Jing, N., and Han, J.J. (2017). Inference of differentiation time for single cell transcriptomes using cell population reference data. *Nat. Commun.* 8: 1856.

Teschendorff, A.E. and Enver, T. (2017). Single-cell entropy for accurate estimation of differentiation potency from a cell's transcriptome. *Nat. Commun.* 8: 15599.

Tunyasuvunakool, K., Adler, J., Wu, Z., Green, T., Zielinski, M., Zidek, A., Bridgland, A., Cowie, A., Meyer, C., Laydon, A., Velankar, S., Kleywegt, G.J., Bateman, A., Evans, R., Pritzel, A., Figurnov, M., Ronneberger, O., Bates, R., Kohl, S.A.A., Potapenko, A., Ballard, A.J., Romera-Paredes, B., Nikolov, S., Jain, R., Clancy, E., Reiman, D., Petersen, S., Senior, A.W., Kavukcuoglu, K., Birney, E., Kohli, P., Jumper, J., and Hassabis, D. (2021). Highly accurate protein structure prediction for the human proteome. *Nature* 596: 590–596.

Wang, T., Bai, J., and Nabavi, S. (2021). Single-cell classification using graph convolutional networks. *BMC Bioinform.* 22: 364.

Xiang, Y., Ran, Q., Wu, C., Zhou, L.P., Zhang, W.W., Li, J.X., Xiang, L.X., Xiao, Y.N., Chen, L., Chen, Y., Chen, X.L., Stucky, A., Li, S.C., Zhong, J.F., Li, Z.J., and Cai, K.Y. (2022). Single-cell transcriptomics uncovers the impacts of titanium dioxide nanoparticles on human bone marrow stromal cells. *Chem. Eng. J.* 440.

Xu, L., Wang, Y.Y., Huang, J., Chen, C.Y., Wang, Z.X., and Xie, H. (2020). Silver nanoparticles: synthesis, medical applications and biosafety. *Theranostics* 10: 8996–9031.

Yan, H., Shi, J., Dai, Y., Li, X., Wu, Y., Zhang, J., Gu, Z., Zhang, C., and Leng, J. (2022). Technique integration of single-cell RNA sequencing with spatially resolved transcriptomics in the tumor microenvironment. *Cancer Cell Int.* 22: 155.

Yang, Y., Shi, X., Liu, W., Zhou, Q., Chan Lau, M., Chun Tatt Lim, J., Sun, L., Ng, C.C.Y., Yeong, J., and Liu, J. (2022). SC-MEB: spatial clustering with hidden Markov random field using empirical Bayes. *Brief Bioinform.* 23.

Yong, K.T., Law, W.C., Hu, R., Ye, L., Liu, L., Swihart, M.T., and Prasad, P.N. (2013). Nanotoxicity assessment of quantum dots: from cellular to primate studies. *Chem. Soc. Rev.* 42: 1236–1250.

Yuan, Y. and Bar-Joseph, Z. (2019). Deep learning for inferring gene relationships from single-cell expression data. *Proc. Natl. Acad. Sci. U. S. A.* 116: 27151–27158.

Zhang, M., Hu, S., Min, M., Ni, Y., Lu, Z., Sun, X., Wu, J., Liu, B., Ying, X., and Liu, Y. (2021). Dissecting transcriptional heterogeneity in primary gastric adenocarcinoma by single cell RNA sequencing. *Gut* 70: 464–475.

Zhu, B., Li, H., Zhang, L., Chandra, S.S., and Zhao, H. (2022). A Markov random field model-based approach for differentially expressed gene detection from single-cell RNA-seq data. *Brief Bioinform.* 23: bbac166.

Ziental, D., Czarczynska-Goslinska, B., Mlynarczyk, D.T., Glowacka-Sobotta, A., Stanisz, B., Goslinski, T., and Sobotta, L. (2020). Titanium dioxide nanoparticles: prospects and applications in medicine. *Nanomater. (Basel)* 10: 387.

13

Toxicogenomics and Toxicological Mechanisms of Engineered Nanomaterials

*Eid Alshammari, Yingxue Zhang, Alexander Yang, and Zhe Yang**

Department of Biochemistry, Microbiology, and Immunology, Wayne State University School of Medicine, Detroit, Michigan, USA
** Corresponding author*

Introduction

With recent advances in nanotechnology, engineered nanomaterials (ENMs) have emerged as a promising branch of materials with numerous advantages for different purposes, ranging from water purification to drug delivery. Since ENMs exhibit novel and excellent physicochemical properties, they have been widely used in numerous industrial, pharmaceutical, food, and clinical applications. However, the extensive production and use of ENMs have raised safety concerns regarding human health and environmental impacts. ENMs may lead to unexpected harmful consequences due to intentional and unintentional human exposure via inhalation, ingestion, intravenous injection, or skin contact (Figure 13.1). Nano-ecotoxicity could occur via different routes, such as the release of ENMs to the surrounding environments and ecosystems during the manufacturing of ENMs (Leudjo Taka et al. 2021; Seaton et al. 2010). Indeed, many investigations have shown the harmful toxicological effect of ENMs on human health (Ge et al. 2019; Pietroiusti 2012; Singh et al. 2017) and on the safety of the environment (Boyes et al. 2017; Kansara et al. 2022; Suhendra et al. 2020). These studies have revealed the complex interaction of ENMs with both biotic and abiotic environments as well as the behavior, transport, and fate of ENMs. However, the underlying mechanisms and adverse effects of ENMs' nano–bio–eco interactions are still not fully understood (He et al. 2018). Consequently, increasing efforts have been made to propose safe and sustainable utilization of ENMs and develop effective approaches for the risk assessment of ENMs to ameliorate their toxic effects (Atluri and Jensen 2017; Cockburn et al. 2012; Forest 2022; He et al. 2014; Hwang et al. 2018). One group of approaches that has been increasingly used for understanding the toxicological mechanisms of ENMs comprises toxicogenomic approaches. Unlike conventional toxicological approaches, toxicogenomic approaches have provided global information on the relationships between changes in intracellular molecular profiles and exposure to ENMs. With the identification of abnormal changes in global gene expression profiles, our knowledge of ENM-induced toxicities has been considerably expanded by toxicogenomics, with an unprecedented understanding of disturbed molecular pathways at a genome-wide level (Lan et al. 2014). Here, we discuss the application of toxicogenomics in addressing the toxicity of ENMs and the cellular and systemic effects of ENMs.

Impact of Engineered Nanomaterials in Genomics and Epigenomics, First Edition. Edited by Saura C. Sahu.

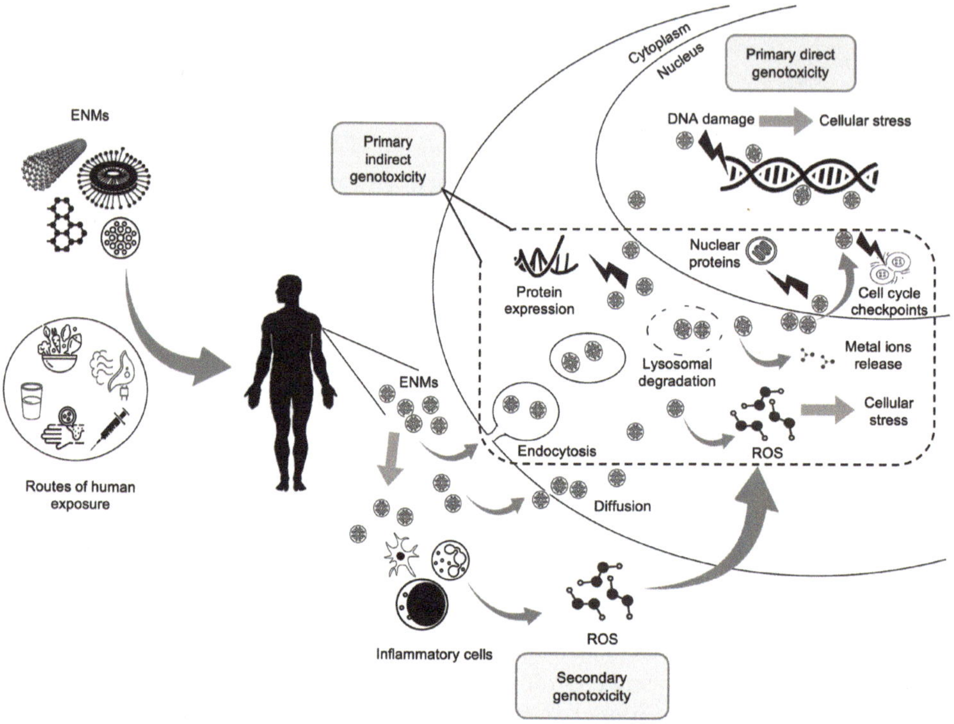

Figure 13.1 Mechanisms of toxicity of engineered nanomaterials (ENMs). A scheme of routes of human exposure to ENMs, including ingestion, inhalation, dermal contact, and injection. Cell entry mechanisms of ENMs include active diffusion and endocytosis. ENM-containing vesicles are fused to lysosomes and degraded in cytosol to release ENMs. ENMs enter the nucleus and cause primary direct genotoxicity through direct contact with DNA, causing DNA damage that may lead to cellular stress and apoptosis. ENMs can also cause primary indirect genotoxicity by interfering with the cell cycle process, affecting nuclear proteins and gene expression, and prompting the metal ion release and reactive oxygen species (ROS) generation. Inflammatory cells are recruited to the site of ENMs and induce ROS production, which leads to cellular stress, causing secondary genotoxicity.

Genomic Responses to ENMs

Over the past few decades, there has been a revolution in the field of toxicological sciences by introducing omics technologies, which has resulted in a new branch of toxicology termed "toxicogenomics." This new toxicological approach integrates toxicology with bioinformatics and functional genomics, including transcriptomics, proteomics, epigenomics, and metabolomics (Afshari et al. 2011). With these genomic analysis techniques, toxicogenomics provides a comprehensive study of the toxic effect of a substance, which represents a powerful tool to identify important cellular components and pathways associated with toxicity responses. While conventional toxicogenetics examines the expression of a single gene in response to toxic substances, toxicogenomics extends the scope for a broader effect of such substances on expression levels of tens of thousands of genes. Thus, toxicogenomics is a global assessment of the biological effect of chemical exposure and can provide valuable mechanistic insights into the mode of action of toxins (Hayes and Bradfield 2005). The application of this powerful approach in assessing the toxicity of ENMs has led to the identification of genomic and transcriptomic responses, which have provided the basis for a broader understanding of genome-wide toxicity profiles associated with the use of ENMs.

Using a quantitative toxicogenomics-based testing system with conventional phenotype-based bioassays, one study tested the genomic toxicity of seven carbon-based nanomaterials (CNMs) with different physicochemical and structural properties at sub-cytotoxic concentration levels (Jiang et al. 2021). The CNMs tested in this study include fullerenes, carbon black, carbon nanotubes, graphene, graphene oxide, nanodiamonds, and carbon-based quantum dots. Based on the toxicogenomic analysis, this study concluded that DNA damage, oxidative stress, and unfolded protein stress were the dominant mechanisms of action for all seven CNMs, with the severity of their toxicity associated with their structural properties. Several other studies also demonstrated that CNM-induced genotoxicity was associated with DNA damage. In a study, mammalian cells were found to be more susceptible to genotoxicity by three types of CNMs, namely carbon black, single-walled carbon nanotube, and fullerene, that induced severe DNA double-strand breaks (Lan et al. 2014). Another study found the genotoxicity and carcinogenic potential of several CNMs, indicated by DNA-strand break, p53 downregulation, and acquisition of cancer stem cell-like properties in human small airway epithelial cells (Kiratipaiboon et al. 2019). In addition, exposure to single-walled carbon nanotubes (SWCNTs) has been shown to induce oxidative stress and DNA damage based on the assessment of formamidopyrimidine DNA glycosylase sensitive sites in human hepatocytes (Vesterdal et al. 2014). Moreover, SWCNTs have been found to cause higher cytotoxicity and genotoxicity than multi-walled carbon nanotubes (MWCNTs), as well as to promote hypomethylation of genes involved in several signaling pathways (Oner et al. 2017). Some CNMs' properties, such as length, functional groups, and electronic structures, can greatly affect their behaviors and thereby their potential toxicological effect on humans. A recent study found that for SWCNT-induced DNA damage and oxidative stress, the short SWCNTs exhibited higher toxicity levels than the long SWCNTs (Jiang et al. 2020). In terms of surface functional groups, carboxylated and hydroxylated SWCNT functionalization resulted in higher levels of DNA damage compared with unmodified SWCNTs. Therefore, the current evidence indicates that the genotoxic effects of SWCNTs occur through the induction of DNA damage or the disruption of the DNA repair machinery.

Graphene-based nanomaterials, another major family of CNMs, have also received scrutiny regarding their potential genomic toxicity, as they are being widely used in biomedical applications as drug carriers and in improving the therapeutic effect of anticancer agents (Cui et al. 2021; de Melo-Diogo et al. 2019). One study found that graphene quantum dots caused DNA damage, demonstrated by the release of reactive oxygen species (ROS) and increased expression of DNA damage-related proteins, such as p53, RAD51, and OGG1 in fibroblast cell lines (Wang et al. 2015). Other graphene-based nanomaterials, such as pristine, carboxylated, and aminated graphene as well as single- and few-layer graphene oxide, have been shown to cause single-strand DNA damage, possibly by impacting the nucleotide excision repair and non-homologous end-joining repair mechanisms (Chatterjee et al. 2016). Moreover, single- and few-layer graphene oxide exposure led to an increase in global DNA methylation, while exposure to pristine, carboxylated (COOH), or aminated (NH_2) graphene nanoplatelets caused hypomethylation by suppressing the induction of DNA methyltransferase 3B (DNMT3B) (Chatterjee et al. 2016). These studies indicated that those graphene-based nanomaterials could result in differential gene expression by altering DNA methylation patterns. Another study revealed a destructive effect of graphene oxide nanosheets on DNA stability, likely due to their ability to promote DNA melting and interfere with DNA synthesis and chromosome segregation (Ivask et al. 2015). It was also shown that graphene oxide could interfere with DNA replication and alter the mRNA expression of over 100 genes involved in DNA damage repair, apoptosis, cell cycle, and metabolism. In addition, it has been demonstrated both *in vitro* and *in vivo* that graphene oxide can induce DNA mutagenesis (Liu et al. 2013). A recent study further confirmed the harmful impact of graphene oxide nanoparticles at the molecular level. Graphene oxide treatment at different doses significantly promoted DNA breakages in a dose-dependent manner and led to genomic instability, induction of oxidative stress, mutations in the p53 and the catalytic component of γ-secretase complex presenilin genes, and elevated levels of p53 protein in liver and brain tissues of a mouse model (Mohamed et al. 2020). Therefore, the toxicity of graphene-based nanomaterials depends on many factors, including dose, lateral size, and surface charge, and their physicochemical properties, such as shape, particle size, layer count, and surface functionalization, which together govern their behaviors on cells and affect different cellular responses (Xiaoli et al. 2020).

Metal-based nanomaterials are another important nanotechnology sector commonly used in various biomedical fields. Silver, copper, gold, titanium, zinc, and iron are among many metal nanoparticles (MeNPs) that have found their use in medical applications, such as drug delivery, antimicrobial agents, and radiosensitizers (Sanchez-Lopez et al. 2020; Zhang et al. 2019). However, there is accumulating evidence that MeNPs are involved in genotoxicity and carcinogenesis through stimulation of ROS generation (Mortezaee et al. 2019). One potential mechanism by which MeNPs contribute to ROS production is their ability to cause a reduction of water molecules in cells. Some MeNPs can release metal ions that interact with water molecules, thereby stimulating the generation of free radicals (Mortezaee et al. 2019). The increased levels of free radicals due to exposure to MeNPs have been shown to increase the intracellular calcium levels and lead to oxidative stress and upregulation of apoptotic pathways (Huang et al. 2010). Moreover, the increased production of free radicals and apoptosis after exposure to MeNPs can trigger reduction/oxidation reactions, which can result in continuous endogenous ROS generation in a positive feedback loop (Mortezaee et al. 2019). Several reports have pointed out that MeNPs, such as silver (Ag) and copper oxide (CuO), are involved in DNA damage through the induction of endogenous production of nitric oxide (NO), possibly by affecting the activity of endothelial nitric oxide synthase (eNOS) and inducible nitric oxide synthase (iNOS) (Anavi and Tirosh 2020; Arancibia et al. 2016; Zhen et al. 2008; Zielinska et al. 2016). NO, a known immune response mediator, has been shown to interfere with the cellular repair process, leading to oxidative DNA damage and tumorigenesis (Cheng et al. 2014). Another investigation found that gold nanoparticles (AuNPs) increased ROS production and programmed cell death and decreased the mitochondrial membrane potential in human kidney cells (Enea et al. 2020). One study that aimed to investigate the effect of AuNPs on lung cancer cell lines revealed that the cellular exposure to AuNPs decreased the intracellular glutathione levels and resulted in DNA damage in the form of alkali-labile lesions (May et al. 2018). However, this AuNP-induced DNA damage was transient and was repaired by highly controlled repair mechanisms. Using an air–liquid interface cell culture system to assess the toxicity of copper- and titanium-based nanoparticles, a study found that titanium dioxide nanoparticles (TiO_2-NPs) did not induce any cytotoxicity or genomic instability while CuO-NP altered the expression patterns of metal homeostasis-associated genes, such as *MT1X* and *MT2A*. CuO-NP also induced cellular stress, ROS generation, and DNA damage, as indicated by the oxidative stress markers *HMOX1* and *HSPA1A* and the DNA damage response-associated genes *DDIT3* and *GADD45A* (Hufnagel et al. 2020). Long-term and repetitive exposure of human endothelial cells to zinc oxide NPs has been shown to trigger DNA damage and impact cell proliferation and angiogenesis (Poier et al. 2020). Therefore, these data indicate the health consequences of metal-based nanomaterial exposure and the need for a thorough nanotoxicology assessment to improve their safe use in biomedical applications.

Transcriptomic Responses to ENMs

While cytological and histological evaluation is an essential initial step to evaluate the ENM-induced cellular and immune responses, evaluating the global transcriptome provides more detailed information about specific molecular responses and how different biological pathways respond to ENM exposure. This information can be used for the characterization of gene expression signatures associated with unique ENM properties and for comprehensive mapping of any molecular events consequently causing ENM toxicity (Alexander-Dann et al. 2018). A recent study has reported that exposure of animal models to 16 ENMs led to the transcriptomic alteration of long noncoding RNAs (lncRNAs) (Ndika et al. 2021). When mice were exposed to CuO, MWCNTs, spherical TiO_2, or rod-like TiO_2 particles, each with four different surface functionalizations, via oropharyngeal aspiration, the type-, shape-, and surface chemistry-dependent toxicological effect was observed on lncRNA expression. It was shown that the altered lncRNAs are implicated in the gene regulation of critical cellular processes, like cell division and DNA damage response, based on mRNA expression analysis. This study also found that NH_2 surface-containing ENMs were more toxic than ENMs with pristine, COOH, or polyethylene glycol (PEG) surface functional groups. In addition to altered lncRNA expression, the same study identified a significant amount of differentially expressed genes (DEGs) by comparing the gene expression profiles from the lung tissue samples of ENM-exposed mice and corresponding controls (Ndika et al. 2021). It was found that ENMs with CuO and MWCNT cores caused the greatest transcriptome changes, with 3,599 and 1,544 DEGs, respectively. Other ENMs, including silver, spherical TiO_2, and rod-like TiO_2, had an intermediate effect on the gene expression profiles. In terms of the toxicological impact of surface functionalizations, the NH_2-group caused the most changes in the number of DEGs, followed by the COOH and PEG groups. Another study showed that NH_2 and COOH surface-functionalized ENMs shared more than 50% of DEGs, or 618 DEGs, which led to the conclusion that surface functionalization alone cannot induce strong transcriptomic alteration if the chemistry of the ENM's core is inactive or nontoxic (Kinaret et al. 2021).

A study investigated the transcriptomic changes after pulmonary exposure to two graphene-based nanomaterials (Poulsen et al. 2021). Graphene oxide (GO) and reduced graphene oxide (rGO) share similar physical properties but have different surface chemistry. Both GO and rGO were shown to induce transcriptomic responses in the lung and liver after their exposure to mice. They also induced ROS production and genotoxicity and impacted the canonical pathways in lung and liver tissues. However, GO exposure was more potent in inducing the transcriptomic changes associated with pulmonary inflammation, innate immune response, and fibrosis, and solely activated the acute phase response in the lung. In addition, GO exposure also increased the risk of atherosclerosis by affecting hepatic lipid homeostasis. It was hypothesized that the robust transcriptomic responses after GO exposure could be attributed to a larger number of hydroxyl groups on the surface of GO (Poulsen et al. 2021). Although GO nanomaterials show no acute cytotoxicity and are relatively biocompatible with human umbilical vein endothelial cells (HUVECs), omics techniques and bioinformatics analysis revealed that GO can alter several functional pathways related to the immune systems and the response to viruses (Luo et al. 2021). Moreover, GO exposure can significantly reduce the levels of Toll-like receptor 3 (TLR3) and interleukin 8 proteins, which might indicate an immune-suppressive effect of GO through the inhibition of TLR3 (Luo et al. 2021). This also suggests that even without acute cytotoxicity, some nanomaterials may induce other adverse direct or indirect toxicological effects.

Carbon nanotubes (CNTs) have also been shown to cause significant transcriptomic changes. Previous research showed that due to their small size, CNTs have a greater capability of interacting with cells and thereby smaller CNTs have a greater toxicological effect on cellular transcriptome (Zhao et al. 2019). When HUVECs were treated with two modified MWCNTs, endoplasmic reticulum (ER) stress was induced, as indicated by the mRNA level of *DDIT3* and *XBP1*, the gene biomarkers for ER stress (Chang et al. 2018; Zhao et al. 2019). It was shown that shorter MWCNTs decreased the expression of *DDIT3* while treatment of longer MWCNTs significantly increased *DDIT3* levels. As DDIT3 regulates inflammatory cytokines, including IL-6, dysregulation of this gene would lead to augmented ER stress and inflammatory response (Chang et al. 2018). This study has also found a dose-dependent effect of pristine and carboxylated MWCNTs on the induction of ER stress. These two MWCNTs induced cytotoxicity and alteration in the expression of genes associated with ER stress, such as *HSPA5*, *DDIT3*, *XBP1*, *BECN1*, and *ATG12*, in HUVECs and A549/THP-1 macrophages (Chang et al. 2018). Another study showed that unmodified and COOH-functionalized MWCNTs affected the innate immune system and triggered the inflammatory response through the activation of the NLRP3 inflammasome in mice lungs (Sager et al. 2014). Transcriptomics studies that assessed the effect of CNTs in bacteria found that even at a low concentration (10 μg/ml) of SWCNTs, *Escherichia coli* showed altered gene expression while high concentration (100 μg/ml) of SWCNTs induced the expression of the phage shock protein (Psp) operon and led to more severe cellular perturbation and significant gene downregulation (Anh Le et al. 2019).

Quantum dots (QDs) are another type of ENMs and have been widely utilized in molecular imaging and as biosensors and drug carriers (Hardman 2006). However, due to their diverse physicochemical properties, such as size, charge, composition, concentration, and outer coating bioactivity, there is a high chance that some combinations of these properties may bring about toxic effects on human health (Manshian et al. 2017). The toxic effect of glutathione (GSH)-capped cadmium telluride (CdTe-GSH) QDs on *E. coli* has been demonstrated to be size-dependent, with red QDs (5 nm) found to be more toxic than green QDs (3 nm). The *E. coli*'s global transcriptomic response to these QDs revealed a change in expression of a total of 95 genes in response to red QD treatment compared with only 42 genes when green QDs were used. Some of the altered genes of both QD types were found to be involved in transport pathways, biosynthesis, and metabolism (Monras et al. 2014). Another study showed that the transcriptomic response of *Pseudomonas stutzeri* to cadmium selenide/cadmium zinc sulfide (CdSe/CdZnS) QDs altered seven genes involved in the denitrification and upregulation of superoxide dismutase in response to QD-induced ROS generation (Yang et al. 2011). Exposure of CdSe QDs to *Pseudomonas aeruginosa* PAO1 has been found to impact the expression of genes implicated in heavy metal and oxidative stress responses (Yang et al. 2012). Using whole-genome microarray, a study found that treatment with red, blue, and green QDs affected the expression of many genes in human embryonic kidney fibroblast cells (HEK293) (Jeong et al. 2010). Green and blue QDs induced more upregulated genes while red QDs showed more downregulated genes. Gene ontology analysis of these DEGs showed responses to wounding, cell stress, apoptosis, and defense, which were enriched in all three QDs.

Conclusion

Toxicological studies in the ENM field are evolving and adapting new methodologies such as toxicogenomics approaches. Unlike conventional toxicity methods that deal with extensive observations of phenotypic endpoints, toxicogenomics integrates conventional toxicology with omics assays for better characterization of the mechanism of action of ENMs. ENMs have been shown to influence several genomic events like induction of DNA damage through direct interaction with the DNA or through the adverse effect of ROS production (Figure 13.1). The former is considered a primary direct genotoxicity effect, while the latter is an indirect or secondary genotoxicity effect, and these effects may consequently lead to serious health consequences. ENMs also impact gene expression homeostasis, which results in the dysregulation of crucial pathways and cellular processes. The use of toxicogenomics including genomics and transcriptomics has helped understand the toxicity pathways that underlie the action of ENMs as well as cell susceptibility to distinct ENMs. Therefore, toxicogenomics-based assays have proved to be a promising approach for a comprehensive assessment of the adverse effects of ENMs on human health.

References

Afshari, C.A., Hamadeh, H.K., and Bushel, P.R. (2011). The evolution of bioinformatics in toxicology: advancing toxicogenomics. *Toxicol. Sci.* 120 (Suppl 1): S225–37.

Alexander-Dann, B., Pruteanu, L.L., Oerton, E., Sharma, N., Berindan-Neagoe, I., Modos, D., and Bender, A. (2018). Developments in toxicogenomics: understanding and predicting compound-induced toxicity from gene expression data. *Mol. Omics* 14: 218–236.

Anavi, S. and Tirosh, O. (2020). iNOS as a metabolic enzyme under stress conditions. *Free Radic. Biol. Med.* 146: 16–35.

Anh Le, T.T., Thuptimdang, P., McEvoy, J., and Khan, E. (2019). Phage shock protein and gene responses of *Escherichia coli* exposed to carbon nanotubes. *Chemosphere* 224: 461–469.

Arancibia, S., Barrientos, A., Torrejon, J., Escobar, A., and Beltran, C.J. (2016). Copper oxide nanoparticles recruit macrophages and modulate nitric oxide, proinflammatory cytokines and PGE2 production through arginase activation. *Nanomed. (Lond)* 11: 1237–1251.

Atluri, R. and Jensen, K.A. (2017). Engineered nanomaterials: their physicochemical characteristics and how to measure them. *Adv. Exp. Med. Biol.* 947: 3–23.

Boyes, W.K., Thornton, B.L.M., Al-Abed, S.R., Andersen, C.P., Bouchard, D.C., Burgess, R.M., Hubal, E.A.C., Ho, K.T., Hughes, M.F., Kitchin, K., Reichman, J.R., Rogers, K.R., Ross, J.A., Rygiewicz, P.T., Scheckel, K.G., Thai, S.F., Zepp, R.G., and Zucker, R.M. (2017). A comprehensive framework for evaluating the environmental health and safety implications of engineered nanomaterials. *Crit. Rev. Toxicol.* 47: 767–810.

Chang, S., Zhao, X., Li, S., Liao, T., Long, J., Yu, Z., and Cao, Y. (2018). Cytotoxicity, cytokine release and ER stress-autophagy gene expression in endothelial cells and alveolar-endothelial co-culture exposed to pristine and carboxylated multi-walled carbon nanotubes. *Ecotoxicol. Environ. Saf.* 161: 569–577.

Chatterjee, N., Yang, J., and Choi, J. (2016). Differential genotoxic and epigenotoxic effects of graphene family nanomaterials in human bronchial epithelial cells. *Mutat. Res. Genet. Toxicol. Environ. Mutagen* 798-799: 1–10.

Cheng, H., Wang, L., Mollica, M., Re, A.T., Wu, S., and Zuo, L. (2014). Nitric oxide in cancer metastasis. *Cancer Lett.* 353: 1–7.

Cockburn, A., Bradford, R., Buck, N., Constable, A., Edwards, G., Haber, B., Hepburn, P., Howlett, J., Kampers, F., Klein, C., Radomski, M., Stamm, H., Wijnhoven, S., and Wildemann, T. (2012). Approaches to the safety assessment of engineered nanomaterials (ENM) in food. *Food Chem. Toxicol.* 50: 2224–2242.

Cui, G., Wu, J., Lin, J., Liu, W., Chen, P., Yu, M., Zhou, D., and Yao, G. (2021). Graphene-based nanomaterials for breast cancer treatment: promising therapeutic strategies. *J. Nanobiotechnology* 19: 211.

de Melo-Diogo, D., Lima-Sousa, R., Alves, C.G., and Correia, I.J. (2019). Graphene family nanomaterials for application in cancer combination photothermal therapy. *Biomater. Sci.* 7: 3534–3551.

Enea, M., Pereira, E., de Almeida, M.P., Araujo, A.M., de Lourdes Bastos, M., and Carmo, H. (2020). Gold nanoparticles induce oxidative stress and apoptosis in human kidney cells. *Nanomater. (Basel)* 10.

Forest, V. (2022). Experimental and computational nanotoxicology-complementary approaches for nanomaterial hazard assessment. *Nanomater. (Basel)* 12.

Ge, D., Du, Q., Ran, B., Liu, X., Wang, X., Ma, X., Cheng, F., and Sun, B. (2019). The neurotoxicity induced by engineered nanomaterials. *Int. J. Nanomed.* 14: 4167–4186.

Hardman, R. (2006). A toxicologic review of quantum dots: toxicity depends on physicochemical and environmental factors. *Environ. Health Perspect* 114: 165–172.

Hayes, K.R. and Bradfield, C.A. (2005). Advances in toxicogenomics. *Chem. Res. Toxicol.* 18: 403–414.

He, X., Aker, W.G., Leszczynski, J., and Hwang, H.M. (2014). Using a holistic approach to assess the impact of engineered nanomaterials inducing toxicity in aquatic systems. *J. Food Drug Anal.* 22: 128–146.

He, X., Fu, P., Aker, W.G., and Hwang, H.M. (2018). Toxicity of engineered nanomaterials mediated by nano–bio–eco interactions. *J. Environ. Sci. Health C Environ. Carcinog. Ecotoxicol. Rev.* 36: 21–42.

Huang, C.C., Aronstam, R.S., Chen, D.R., and Huang, Y.W. (2010). Oxidative stress, calcium homeostasis, and altered gene expression in human lung epithelial cells exposed to ZnO nanoparticles. *Toxicol. In Vitro* 24: 45–55.

Hufnagel, M., Schoch, S., Wall, J., Strauch, B.M., and Hartwig, A. (2020). Toxicity and gene expression profiling of copper- and titanium-based nanoparticles using air–liquid interface exposure. *Chem. Res. Toxicol.* 33: 1237–1249.

Hwang, R., Mirshafiee, V., Zhu, Y., and Xia, T. (2018). Current approaches for safer design of engineered nanomaterials. *Ecotoxicol. Environ. Saf.* 166: 294–300.

Ivask, A., Voelcker, N.H., Seabrook, S.A., Hor, M., Kirby, J.K., Fenech, M., Davis, T.P., and Ke, P.C. (2015). DNA melting and genotoxicity induced by silver nanoparticles and graphene. *Chem. Res. Toxicol.* 28: 1023–1035.

Jeong, S.-H., Lee, S.-E., Hong, S.-W., Kim, S., and Lee, D.-K. (2010). Evaluation of toxicity and gene expression changes triggered by quantum dots. *Bull. Korean Chem. Soc.* 31: 1555–1560.

Jiang, T., Amadei, C.A., Gou, N., Lin, Y., Lan, J., Vecitis, C.D., and Gu, A.Z. (2020). Toxicity of single-walled carbon nanotubes (SWCNTs): effect of lengths, functional groups and electronic structures revealed by a quantitative toxicogenomics assay. *Environ. Sci. Nano* 7: 1348–1364.

Jiang, T., Lin, Y., Amadei, C.A., Gou, N., Rahman, S.M., Lan, J., Vecitis, C.D., and Gu, A.Z. (2021). Comparative and mechanistic toxicity assessment of structure-dependent toxicity of carbon-based nanomaterials. *J. Hazard Mater.* 418: 126282.

Kansara, K., Bolan, S., Radhakrishnan, D., Palanisami, T., Al-Muhtaseb, A.H., Bolan, N., Vinu, A., Kumar, A., and Karakoti, A. (2022). A critical review on the role of abiotic factors on the transformation, environmental identity and toxicity of engineered nanomaterials in aquatic environment. *Environ. Pollut.* 296: 118726.

Kinaret, P.A.S., Ndika, J., Ilves, M., Wolff, H., Vales, G., Norppa, H., Savolainen, K., Skoog, T., Kere, J., Moya, S., Handy, R.D., Karisola, P., Fadeel, B., Greco, D., and Alenius, H. (2021). Toxicogenomic profiling of 28 nanomaterials in mouse airways. *Adv. Sci. (Weinh)* 8: 2004588.

Kiratipaiboon, C., Stueckle, T.A., Ghosh, R., Rojanasakul, L.W., Chen, Y.C., Dinu, C.Z., and Rojanasakul, Y. (2019). Acquisition of cancer stem cell-like properties in human small airway epithelial cells after a long-term exposure to carbon nanomaterials. *Environ. Sci. Nano* 6: 2152–2170.

Lan, J., Gou, N., Gao, C., He, M., and Gu, A.Z. (2014). Comparative and mechanistic genotoxicity assessment of nanomaterials via a quantitative toxicogenomics approach across multiple species. *Environ. Sci. Technol.* 48: 12937–12945.

Leudjo Taka, A., Tata, C.M., Klink, M.J., Mbianda, X.Y., Mtunzi, F.M., and Naidoo, E.B. (2021). A review on conventional and advanced methods for nanotoxicology evaluation of engineered nanomaterials. *Molecules* 26.

Liu, Y., Luo, Y., Wu, J., Wang, Y., Yang, X., Yang, R., Wang, B., Yang, J., and Zhang, N. (2013). Graphene oxide can induce in vitro and in vivo mutagenesis. *Sci. Rep.* 3: 3469.

Luo, Y., Wang, X., and Cao, Y. (2021). Transcriptomic-based toxicological investigations of graphene oxide with modest cytotoxicity to human umbilical vein endothelial cells: changes of Toll-like receptor signaling pathways. *Toxicol. Res. (Camb)* 10: 1104–1115.

Manshian, B.B., Martens, T.F., Kantner, K., Braeckmans, K., De Smedt, S.C., Demeester, J., Jenkins, G.J.S., Parak, W.J., Pelaz, B., Doak, S.H., Himmelreich, U., and Soenen, S.J. (2017). The role of intracellular trafficking of CdSe/ZnS QDs on their consequent toxicity profile. *J. Nanobiotechnology* 15: 45.

May, S., Hirsch, C., Rippl, A., Bohmer, N., Kaiser, J.P., Diener, L., Wichser, A., Burkle, A., and Wick, P. (2018). Transient DNA damage following exposure to gold nanoparticles. *Nanoscale* 10: 15723–15735.

Mohamed, H.R.H., Welson, M., Yaseen, A.E., and El-Ghor, A.A. (2020). Estimation of genomic instability and mutation induction by graphene oxide nanoparticles in mice liver and brain tissues. *Environ. Sci. Pollut. Res. Int.* 27: 264–278.

Monras, J.P., Collao, B., Molina-Quiroz, R.C., Pradenas, G.A., Saona, L.A., Duran-Toro, V., Ordenes-Aenishanslins, N., Venegas, F.A., Loyola, D.E., Bravo, D., Calderon, P.F., Calderon, I.L., Vasquez, C.C., Chasteen, T.G., Lopez, D.A., and Perez-Donoso, J.M. (2014). Microarray analysis of the *Escherichia coli* response to CdTe-GSH Quantum Dots: understanding the bacterial toxicity of semiconductor nanoparticles. *BMC Genom.* 15: 1099.

Mortezaee, K., Najafi, M., Samadian, H., Barabadi, H., Azarnezhad, A., and Ahmadi, A. (2019). Redox interactions and genotoxicity of metal-based nanoparticles: a comprehensive review. *Chem. Biol. Interact* 312: 108814.

Ndika, J., Karisola, P., Kinaret, P., Ilves, M., and Alenius, H. (2021). Profiling non-coding RNA changes associated with 16 different engineered nanomaterials in a mouse airway exposure model. *Cells* 10.

Oner, D., Moisse, M., Ghosh, M., Duca, R.C., Poels, K., Luyts, K., Putzeys, E., Cokic, S.M., Van Landuyt, K., Vanoirbeek, J., Lambrechts, D., Godderis, L., and Hoet, P.H. (2017). Epigenetic effects of carbon nanotubes in human monocytic cells. *Mutagenesis* 32: 181–191.

Pietroiusti, A. (2012). Health implications of engineered nanomaterials. *Nanoscale* 4: 1231–1247.

Poier, N., Hochstoger, J., Hackenberg, S., Scherzad, A., Bregenzer, M., Schopper, D., and Kleinsasser, N. (2020). Effects of zinc oxide nanoparticles in HUVEC: cyto- and genotoxicity and functional impairment after long-term and repetitive exposure in vitro. *Int. J. Nanomed.* 15: 4441–4452.

Poulsen, S.S., Bengtson, S., Williams, A., Jacobsen, N.R., Troelsen, J.T., Halappanavar, S., and Vogel, U. (2021). A transcriptomic overview of lung and liver changes one day after pulmonary exposure to graphene and graphene oxide. *Toxicol. Appl. Pharmacol.* 410: 115343.

Sager, T.M., Wolfarth, M.W., Andrew, M., Hubbs, A., Friend, S., Chen, T.H., Porter, D.W., Wu, N., Yang, F., Hamilton, R.F., and Holian, A. (2014). Effect of multi-walled carbon nanotube surface modification on bioactivity in the C57BL/6 mouse model. *Nanotoxicology* 8: 317–327.

Sanchez-Lopez, E., Gomes, D., Esteruelas, G., Bonilla, L., Lopez-Machado, A.L., Galindo, R., Cano, A., Espina, M., Ettcheto, M., Camins, A., Silva, A.M., Durazzo, A., Santini, A., Garcia, M.L., and Souto, E.B. (2020). Metal-based nanoparticles as antimicrobial agents: an overview. *Nanomater. (Basel)* 10.

Seaton, A., Tran, L., Aitken, R., and Donaldson, K. (2010). Nanoparticles, human health hazard and regulation. *J. R. Soc. Interface* 7 (Suppl 1): S119–S129.

Singh, N., Nelson, B.C., Scanlan, L.D., Coskun, E., Jaruga, P., and Doak, S.H. (2017). Exposure to engineered nanomaterials: impact on DNA repair pathways. *Int J Mol Sci.* 18.

Suhendra, E., Chang, C.H., Hou, W.C., and Hsieh, Y.C. (2020). A review on the environmental fate models for predicting the distribution of engineered nanomaterials in surface waters. *Int. J. Mol. Sci.* 21.

Vesterdal, L.K., Danielsen, P.H., Folkmann, J.K., Jespersen, L.F., Aguilar-Pelaez, K., Roursgaard, M., Loft, S., and Moller, P. (2014). Accumulation of lipids and oxidatively damaged DNA in hepatocytes exposed to particles. *Toxicol. Appl. Pharmacol.* 274: 350–360.

Wang, D., Zhu, L., Chen, J.F., and Dai, L. (2015). Can graphene quantum dots cause DNA damage in cells? *Nanoscale* 7: 9894–9901.

Xiaoli, F., Qiyue, C., Weihong, G., Yaqing, Z., Chen, H., Junrong, W., and Longquan, S. (2020). Toxicology data of graphene-family nanomaterials: an update. *Arch. Toxicol.* 94: 1915–1939.

Yang, Y., Mathieu, J.M., Chattopadhyay, S., Miller, J.T., Wu, T., Shibata, T., Guo, W., and Alvarez, P.J. (2012). Defense mechanisms of *Pseudomonas aeruginosa* PAO1 against quantum dots and their released heavy metals. *ACS Nano* 6: 6091–6098.

Yang, Y., Zhu, H., Colvin, V.L., and Alvarez, P.J. (2011). Cellular and transcriptional response of *Pseudomonas stutzeri* to quantum dots under aerobic and denitrifying conditions. *Environ. Sci. Technol.* 45: 4988–4994.

Zhang, C., Yan, L., Gu, Z., and Zhao, Y. (2019). Strategies based on metal-based nanoparticles for hypoxic-tumor radiotherapy. *Chem. Sci.* 10: 6932–6943.

Zhao, X., Chang, S., Long, J., Li, J., LI, X., and Cao, Y. (2019). The toxicity of multi-walled carbon nanotubes (MWCNTs) to human endothelial cells: the influence of diameters of MWCNTs. *Food Chem. Toxicol.* 126: 169–177.

Zhen, J., Lu, H., Wang, X.Q., Vaziri, N.D., and Zhou, X.J. (2008). Upregulation of endothelial and inducible nitric oxide synthase expression by reactive oxygen species. *Am. J. Hypertens.* 21: 28–34.

Zielinska, E., Tukaj, C., Radomski, M.W., and Inkielewicz-Stepniak, I. (2016). Molecular mechanism of silver nanoparticles–induced human osteoblast cell death: protective effect of inducible nitric oxide synthase inhibitor. *PLoS One* 11: e0164137.

14

Carbon Nanotubes Alter Metabolomics Pathways Leading to Broad Ecological Toxicity

*Nouran Yonis, Eid Alshammari, and Zhe Yang**

Department of Biochemistry, Microbiology, and Immunology, Wayne State University School of Medicine, Detroit, Michigan, USA
** Corresponding author*

Introduction

Carbon nanotubes (CNTs) are hollow, cylindrical tubes made of carbon atoms with diameters typically measured in nanometers. In their natural state, CNTs have a plain surface due to the carbon atoms in CNTs bonding to each other via the sp^2-hybridized orbitals, forming connected hexagonal planes (Laurila et al. 2017). Because of their unique physical, chemical, and electrical properties, CNTs have attracted a lot of attention in biomedical applications, including tissue engineering, drug delivery, biosensors, and gene therapy. With a large surface area, CNTs exhibit high drug-loading capability, regardless of whether the drug is adsorbed inside the tube or bound to the outer surface (Zare et al. 2021). In line with this ability, CNTs show the potential to improve drug encapsulation efficiency and stability while lowering drug-release rates and cytotoxicity. CNTs also have strong tissue permeability and can pass through a variety of biological barriers (Chen et al. 2019). However, the biomedical applications of CNTs remain challenging due to the fact that pristine CNTs are very hydrophobic, and if directly administered into the human body, CNTs could cause a variety of unwanted and severe toxicities, such as cell-cycle arrest, apoptosis, and necrosis. It has been shown that agglomeration, size, contaminants, and length are all contributing factors to CNT toxicity (Thakur et al. 2022). To improve the biocompatibility of CNTs, an effective means is functionalizing their surfaces (Raissi and Mollania 2014; Zhao and Liu 2012). Surface functionalization of CNTs also improves water solubility, dispersion, and effectiveness of insoluble medicines, as well as reducing nonspecific interactions with cell membranes and their associated toxicities. Depending on the application, functionalization is accomplished via covalent or noncovalent bonding of various functional groups on the surface or at the open ends of CNTs (Mody et al. 2014). Covalent functionalization of CNTs involves several chemical modifications, such as oxidation, hydrogenation, amidation, halogenation, thiolation, and the addition of carbenes or nitrene. These modifications produce a shift in carbon atom hybridization from sp^2 to sp^3, while also preventing the electron transition, which is beneficial for pharmaceutical drug delivery since it provides a gradual release of medicinal substances. As a result, covalent functionalized CNTs are seen to be the most promising for diagnostics and medicine administration (Hassani et al. 2022; Kumar et al. 2017). Types of noncovalent functionalization include supramolecular adhesion of polymers, biomolecules, or amphiphilic surfactants on the CNT surface through van der Waals, π–π stacking, or hydrophobic interactions. Noncovalent functionalization can improve CNT solubility and dispersibility, but unlike covalent functionalization, it has limited drug transport capabilities due to the

Impact of Engineered Nanomaterials in Genomics and Epigenomics, First Edition. Edited by Saura C. Sahu.

weak intermolecular forces. Proteins, nucleic acids, and small medicinal molecules have been successfully delivered to cell-specific targets using functionalized CNTs (Al-Qattan et al. 2018).

CNTs are divided into two types: single-walled carbon nanotubes (SWCNTs) and multi-walled carbon nanotubes (MWCNTs) (Figure 14.1). SWCNTs have a single cylinder, while MWCNTs have many cylinders that are nested with coinciding centers. SWCNTs have diameters ranging from 0.4 to 3.0 nm and lengths ranging from 20 to 1,000 nm, whereas MWCNTs have a central cylindrical tube diameter of 1–3 nm and an exterior cylindrical tube diameter of 2–100 nm, with lengths ranging from one to several micrometers (Al-Qattan et al. 2018). Both SWCNTs and MWCNTs have been extensively investigated for drug delivery. SWCNTs are capable of delivering antibacterial agents and cytotoxic drug molecules, while MWCNTs have been shown to be a suitable carrier for poorly soluble drugs (Assali et al. 2021). However, the wide use of these CNTs in biomedical applications raises increasing safety concerns associated with their potential toxicological effects and risks to human health. It has been shown that DNA damage and oxidative stress are the major mechanisms for SWCNTs to exert their toxicological effects, with the level of toxicity associated with their length, functionalization, and electronic structure. For MWCNTS, their toxicological effects have recently been assessed in animals including zebrafish (*Danio rerio*), copepods (*Tigriopus japonicus*), frogs (*Xenopus laevis*), rats, and mice, where the liver, spleen, and kidney were studied after MWCNT exposure, which has provided a thorough understanding of MWCNT-induced toxicity (Ganesan et al. 2021). MWCNTs influenced apoptosis, DNA damage, oxidative

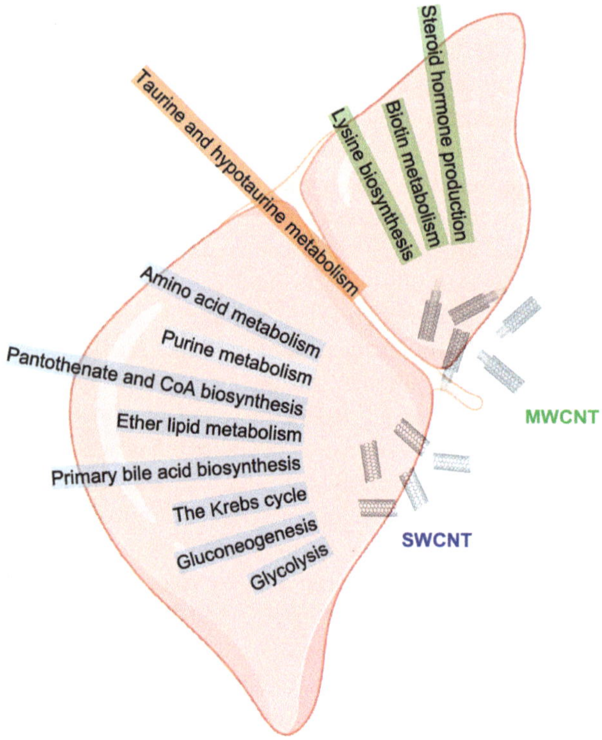

Figure 14.1 Carbon nanotubes impact cellular metabolism. Metabolic pathways uniquely affected by single-walled carbon nanotubes and multi-walled carbon nanotubes are colored in blue and green respectively, and the shared one is colored in orange. The background is a diagram of the liver since the liver is the primary site for regulating metabolic homeostasis of the body.

stress, and inflammation, as well as protein and gene expression. Additionally, both SWCNTs and MWCNTs have been shown to influence metabolic profiles at the proteome level and have effects on different metabolic pathways (Haniu et al. 2010; Phuyal et al. 2018). Some of these metabolic pathways are related to energy metabolism such as glycolysis and the Krebs cycle. As cellular metabolism is fundamentally important for cell survival and growth, understanding the metabolomic effects of CNTs is essential to address the health concerns of their metabolomic toxicity. Here, we summarize CNTs commonly used in biomedical applications and discuss their metabolomic toxicity, aiming for promoting the future safe use of CNTs in their clinical applications.

Biomedical Application and Toxicity of Carbon Nanotubes

Single-Walled Carbon Nanotubes

Because of their superior mechanical, electrical, thermal, and spectroscopic capabilities, SWCNTs have sparked a lot of interest in biomedical applications. SWCNTs have a large carrying capacity, allowing several copies of different biological ligands to be loaded at the same time. They are capable of delivering genes, peptides, oligonucleotides, antibacterial agents, and cytotoxic drug molecules (Beg et al. 2018). Because of their tiny size with a hollow tube structure, SWCNTs can effectively transport macromolecules into cells by serving as a nano-needle that passes through the cell membrane without causing any membrane damage. Additionally, SWCNTs are uniquely suited to transport antigens across biological membranes, allowing for the effective use of antigens that had previously failed to elicit adequate or appropriate responses, as well as significant means of enhancing and modulating immune responses (Kafa et al. 2015; C. Zhang et al. 2020). By conjugating bioactive compounds or ligands to the surface of SWCNTs, SWCNTs increase solubility, bioavailability, residence duration, and therapeutic efficacy of drugs at specific target locations (Kafa et al. 2015). According to studies, SWCNTs may form chemical bonds with drugs or proteins and transport them into cells while being processed by the kidney and eliminated through the urine (Murugesan et al. 2006). It was also shown that SWCNTs have cavities that can hold medications and protect them from destruction during transit (Kam and Dai 2005). Thus, SWCNT is an excellent carrier with many benefits.

Sequence-specific interactions between SWCNTs and DNA, although the molecular mechanisms of these interactions are still not being fully understood, have enabled a wide range of biomedical applications, especially as a carrier of DNA vaccines. The structure and characteristics of SWCNT–DNA hybrids depend on DNA sequence. If SWCNT is assembled with a DNA containing its specific binding sequence, the formed SWCNT–DNA hybrids have greater dispersion quality and a higher concentration of monodispersed particles (Zhao et al. 2020). The stronger the binding between SWCNT and the DNA, the tighter conformation of the DNA on the SWCNT surface and the greater zeta-potential of the SWCNT–DNA hybrids. With the stronger interaction, the dispersions of the specific SWCNT–DNA hybrids are more stable against salt and have a higher fluorescence emission intensity, thus contributing to the high-quality dispersion of the particles. Because of this interesting feature, SWCNTs have been exploited as a potential DNA vaccine carrier. A study showed that SWCNTs carrying varying concentrations of the *MCP* gene as DNA vaccine, when they were inoculated in young mandarin fish, elicited much stronger immune responses in muscle, spleen, and kidney tissues (Golubewa et al. 2021). In comparison with fish vaccinated with *MCP* alone, the immune response, including immune-related gene expression, serum antibody production, and C3 content, was considerably boosted in fish vaccinated with SWCNT–*MCP*. After the challenge, the relative percentage survival of the SWCNT–*MCP* group was higher than the naked MCP group.

In addition to the use as a carrier of DNA vaccines, SWCNTs have also been shown to have an excellent ability to deliver protein subunit vaccines into fish, which have significantly improved bath vaccination efficiency (Yan et al. 2020; Zhu et al. 2014). A study showed that SWCNTs carrying a subunit vaccine against grass carp reovirus offered 90% protection (Sun et al. 2012). A subunit vaccine against spring viremia of carp virus carried by SWCNTs showed similar efficacy. A different study, built on a recombinant MCP subunit covalently connected to SWCNTs, showed that the MCP subunit carried by SWCNTs resulted in a more significant increase in the immune response compared with the naked MCP subunit, where numerous immunological biomarkers showed increased activities and expression levels, such as increased enzyme activities of acid phosphatase, superoxide dismutase, and alkaline phosphatase in serum, as well as increased mRNA expression of many immune genes, including *IgM*, *Mx1*, *IL1-β*, *TNFα*, *MHCI*, and *MHCII* (Ke et al. 2020). These studies suggest that SWCNTs are a good carrier of subunit vaccines against fish viral diseases.

Multi-Walled Carbon Nanotubes

MWCNTs are easier to synthesize than SWCNTs, which has sped up their commercial manufacturing and utilization. However, the unavoidable release of MWCNTs during manufacture, use, or disposal has raised growing concerns regarding their ecological safety (Zhao et al. 2021). MWCNTs have been shown to cause inflammation, oxidative stress, lung fibrosis, mesothelioma-like effects, and cardiovascular consequences *in vitro* and *in vivo* (Kuempel et al. 2017). A study has investigated the impact of long-term chronic exposure of human lung cells to MWCNTs on cellular proteins and lipids (Phuyal et al. 2018). A normal bronchial epithelial cell line model was chosen since the lung is the primary target organ of MWCNT exposure. In the study, cells were continually exposed to two nontoxic and low dosages of MWCNT to imitate exposure to produced nanomaterials in occupational settings. It was shown that MWCNT exposure elevated reactive oxygen species (ROS) levels in cells without increasing oxidative DNA damage, resulting in differential expression of many anti- and pro-apoptotic proteins. According to the proteomic analysis, more than 200 proteins were differentially expressed in the MWCNT-exposed cells. Additionally, the lipidomic analysis demonstrated increased levels of various classes of lipids as well. These results indicate that long-term MWCNT exposure of the human normal lung cells at occupationally relevant low doses alters both the proteome and lipidome profiles of the target epithelial cells in the lung (Phuyal et al. 2018).

Several epidemiological investigations have provided data on biological changes in occupationally exposed people, involving the lungs and immune system, as a result of MWCNT exposure. A study found that occupational exposure to MWCNTs induced oxidative stress in exhaled breath condensate, though there were no observed impacts on lung function or hematological parameters (Hamilton et al. 2018). Another research investigated 11 workers who had been exposed to MWCNT for more than a year and 14 nonexposed controls found significant correlations between MWCNT exposure and sputum levels of interleukin (IL)-1, IL-4, IL-5, IL-6, IL-8, tumor necrosis factor (TNF), and Krebs von den Lungen-6 (KL-6), as well as significant correlations between MWCNT exposure and serum levels of IL-1, IL-4, IL-10, and TNF (Vlaanderen et al. 2017). The same study also found differences in mRNA and noncoding RNA expression across the genome between people exposed to MWCNTs and nonexposed controls. Different research assessed potential MWCNT toxicity was conducted between people occupationally exposed to MWCNTs at a MWCNT manufacturing plant and nonexposed controls (Khaliullin et al. 2020). This study examined the relationship between MWCNT exposure and a number of indicators of lung health and early perturbations of the immune system, including complete blood cell counts (CBC), 51 circulating inflammatory markers, 3 pneumoproteins, lung function, including forced expiratory volume and forced vital capacity, and fractional exhaled nitric oxide (FENO). It was found that after exposure to MWCNT, there was a significant increase in the levels of several immune markers, such as C-C motif ligand 20, basic fibroblast growth factor, and soluble IL-1 receptor II (Dong 2020). It was also found that exposed and nonexposed people showed differences in FENO and several CBC parameters, but their lung function or levels of pneumoproteins are similar.

Metabolomics Toxicity of Carbon Nanotubes

A Brief of Metabolomic Techniques Used for CNT Toxicity Profiling

Metabolomics is a high-throughput global metabolite analysis platform that has been increasingly used for nanotoxicity analysis by monitoring the metabolic alterations of numerous metabolites upon nanoparticle exposure. As the levels of metabolites are strictly regulated in metabolomic networks, metabolomics investigation can yield significant information on the toxicological effects of nanoparticles by assessing any abnormal metabolic changes. The currently available techniques to assess metabolomics pathways include ^{1}H nuclear magnetic resonance (^{1}H-NMR) spectroscopy and liquid chromatography-mass spectrometry (LC-MS) (Emwas 2015; Emwas et al. 2013). ^{1}H-NMR spectroscopy is a robust technique for metabolomics research because of its lack of need for derivatization and ability to measure and identify both known and undiscovered chemicals. The chemical input, biosynthesis intermediates, and end products of biological activity can all be described using ^{1}H-NMR spectroscopy in metabolic profiling (Baghdadchi et al. 2018). LC-MS is another widely used metabolomics approach due to its high sensitivity, efficient detection, and amazing universality. Because of its use for nontargeted or targeted metabolomics research, LC-MS allows hundreds of metabolites to be detected in complicated samples (Zhang et al. 2020).

NMR-Based Metabolomic Profiling

A study using ^{1}H-NMR spectroscopy investigated the metabolic effects of SWCNTs in zebrafish (Ganesan et al. 2021). SWCNT exposure had a substantial influence on metabolic profiles of zebrafish and induced considerable phenotypic alterations in numerous metabolites. Among the top affected metabolites are malate, fumarate, oxalacetate, phenylalanine, taurine, sn-glycero-3-phosphate, glycine, lactate, ATP, AMP, valine, pyruvate, ADP, serine, and niacinamide. SWCNT exposure significantly increased the levels of asparagine, glycine, histidine, malate, and oxalacetate, whereas there was a considerable reduction in the levels of dimethylamine, fumarate, and isoleucine. Metabolic pathway enrichment analysis of the affected metabolites showed that SWCNT exposure disturbed a wide range of metabolic pathways, including glycolysis, gluconeogenesis, the Krebs cycle, primary bile acid biosynthesis, taurine and hypotaurine metabolism, ether lipid metabolism, pantothenate and CoA biosynthesis, purine metabolism, as well as amino acid metabolism, including phenylalanine, tyrosine, cysteine, methionine, glycine, serine, threonine, and tryptophan biosynthesis. It is worth noting that SWCNT exposure also resulted in a considerable reduction in the NADPH level. NADPH is the reducing power required for the synthesis of fatty acids, sterols, nucleotides, and nonessential amino acids. Reduced NADPH levels are associated with a significant increase in oxidative stress due to the production of ROS (Yuan et al. 2012). Previous studies showed that the uptake of SWCNTs induced oxidative stress and produced both mitochondrial and cytoplasmic stress (Hilton et al. 2015; Zhu et al. 2018). Therefore, these metabolic changes caused by SWCNT exposure support the notion that SWCNTs exert a broad impact on cellular metabolism and significant disruption of the metabolic networks of zebrafish, which may be associated with ROS induction and oxidative stress.

Another study using ^{1}H-NMR in conjunction with biochemistry and histopathology assessed the metabolomic effects of MWCNTs in mice (Baghdadchi et al. 2018). Steroid hormone production, biotin (vitamin B7) metabolism, lysine biosynthesis, and taurine and hypotaurine metabolism were shown to be the most altered pathways after MWCNT treatment. The steroid hormones play a wide range of physiological roles, such as regulating metabolism, inflammation, immunological function, salt and water balance, sexual characteristics development, and stress tolerance (West et al. 2001). This study examined 70 steroid hormone biosynthesis-related metabolites, among which 23 showed changes in levels following exposure to MWCNTs. When compared with the control groups, cholesterol, a precursor to all steroid hormones, showed higher levels in the medium-dose and high-dose groups. The precursors of testosterone and estrogens, including dehydroepiandrosterone (DHEA), androstenedione, and androstanedione, were also altered in their levels. The high-dose group had higher levels of androstenedione, whereas the medium-dose and low-dose groups had lower levels of this hormone. All dose groups had higher levels of cholesterol, DHEA, androstanedione, testosterone, and etiocholanolone. Additionally, the levels of the primary type of estrogenic steroids estradiol, estrone, and estriol were increased in the medium-dose and high-dose groups, while all dose groups showed elevated levels of their metabolites, including 2-methoxyestradiol, 2-hydroxyestradiol, 16a-hydroxyestrone, 2-methoxyestrone, and 2-hydroxyestrone. The fact that androgens are precursors of estrogens suggests that the change in the levels of estrogenic metabolites is a result of the change in androgenic metabolites. Additionally, the progesterone levels were increased following MWCNT exposure (Almeida et al. 2017). In steroid hormone biosynthesis, progesterone is a key precursor for the production of cortisol, estrogen, testosterone, and DHEA. Because the mice in this study were all male, it is assumed that the progesterone levels are likely to have risen as a result of the higher amounts of the aforementioned hormones (Baghdadchi et al. 2018). Other changes include elevated levels of aldosterone and its precursor deoxycorticosterone in all dose groups and elevated levels of intermediates of the corticosterone pathway, such as 17-hydroxypregnenolone, 17-hydroxyprogestrone, and 17, 20-dihydroxypregn-4en-3on as well as corticosterone and cortisone in the high-dose group.

In addition to steroid hormone production, MWCNT exposure also impacted the lysine biosynthesis process and biotin metabolism (Baghdadchi et al. 2018). Three of four metabolites in the lysine biosynthesis process showed changes in levels following MWCNT treatment. L-lysine levels were higher in all dose groups, while saccharopine levels were higher in the low-dose group and lower in the medium-dose and high-dose groups. The medium-dose group showed a decrease in aminoadipate levels, whereas the low-dose and high-dose groups saw an increase in this metabolite. For biotin metabolism, the levels of biotin became higher in the low-dose group, but lower in the medium-dose and high-dose groups. Biotin has a variety of roles in metabolic pathways, including fatty acid, glucose, and amino acid metabolism. This vitamin is required for food to be converted into energy (Zempleni et al. 2008). As a result, any change in biotin concentration might affect energy metabolism. Biotin is also a significant cofactor for carboxylase enzymes in metabolic activities including fatty acid production and gluconeogenesis, such as acetyl-CoA carboxylase and pyruvate carboxylase. As a result, decreasing biotin levels in the medium-dose and high-dose groups may cause enzymatic abnormalities in those metabolic pathways (Baghdadchi et al. 2018).

^{1}H-NMR spectroscopy-based metabolomics has shown that both SWCNTs and MWCNTs induced significant changes in cellular metabolomic profiles (Figure 14.1). SWCNTs primarily altered the levels of metabolites in energy metabolism, including processes such as glycolysis and the Krebs cycle, while MWCNTs altered the levels of numerous steroid hormones and their related metabolites, such as androgens, estrogens, corticosterone, and aldosterone. One metabolic process affected by both these CNTs is taurine and hypotaurine metabolism. Hypotaurine is the key intermediate for taurine synthesis, which might play a cytoprotective role and act as an antioxidant to scavenge reactive hydroxyl radicals and hypochlorous acids (Surai et al. 2021). Thus, the change in this metabolism following CNT treatment might represent a common stress response mechanism in response to CNT-induced oxidative stress. In conclusion, these studies led to new insights into metabolic disturbance caused by CNTs and the interactions between CNTs and metabolic pathways at a global metabolomic level.

LC-MS-Based Metabolomic Profiling

A study using liquid chromatography-mass spectrometry (LC-MS) examined the metabolomic effect of SWCNTs on human lung A549 cells (Zhang et al. 2020). There were considerable metabolomic changes caused by SWCNTs, mainly on four metabolic pathways, including alanine, aspartate, and glutamate metabolism; pyrimidine metabolism; glycine, serine, and threonine metabolism; and cysteine and methionine metabolism. The changes in amino acid metabolism could potentially affect cellular energy metabolism, especially the Krebs cycle. The downregulation of alanine following SWCNT treatment could disturb the glucose–alanine cycle and biosynthesis of acetyl-CoA since alanine can be oxidized to generate acetyl-CoA to enter the Krebs cycle. Dysregulation of glutamate and aspartate could also contribute to the deregulation of the Krebs cycle. Glutamate can be converted to 2-oxoglutarate by glutamate dehydrogenase, while aspartate can be deaminated to oxaloacetate either in the cytoplasm or in the mitochondria. Both 2-oxoglutarate and oxaloacetate are the key intermediates in the Krebs cycle. In addition to cellular energy metabolism, the dysregulation of glutamate and aspartate by SWCNTs could also affect the DNA or RNA metabolic process, since these amino acids are known to be involved in the generation of purine and pyrimidine bases, nucleosides, and nucleotides, which serve as the primary constituent of DNA and RNA. Additionally, SWCNT could cause dysregulation of cellular redox status by altering methionine levels. Methionine can be converted to S-adenosylmethionine and cysteine, where cysteine is required for the *de novo* synthesis of glutathione, an important antioxidant that regulates cellular redox state (Eriksson et al. 2015). Thus, these results revealed the primary effects of SWCNTs on cellular metabolism, including amino acids metabolism, energy metabolism, and nucleotides biosynthesis.

Another study using LC-MS found that absorption of MWCNTs by tomato fruits significantly affected total fruit metabolome (McGehee et al. 2017). Numerous metabolites were dysregulated in MWCNT-exposed fruits, including 93 upregulated and 70 downregulated metabolites. Kyoto Encyclopedia of Genes and Genomes pathway enrichment analysis led to the discovery of 71 unique pathways affected by dysregulated metabolites in MWCNT-exposed tomato fruits. Interestingly, many dysregulated metabolites are secondary metabolites. Secondary metabolites are compounds produced by plants that are not essential for plant growth and reproduction but can confer adaptative advantages by protecting plants from pathogens. Several secondary metabolites derived from the flavone and flavonol biosynthetic pathway, including luteolin 7-O-glucuronide and quercetin 3-O-glucoside, were upregulated. Luteolin has diverse biological activities, including antioxidant and anti-inflammatory properties (Bhaskarachary and Joshi 2018), while quercetin exhibits high antioxidant and phytotoxic activity and may act as an allelopathic agent to regulate plant growth and survival (Zaïri et al. 2020). Other secondary metabolites that were upregulated following MWCNT treatment, include gossypol and senecionine. Gossypol, produced by the terpenoid pathway, acts as a natural defensive agent against predators, provoking infertility in insects (Gadelha et al. 2014). Senecionine, a product of the alkaloid pathway, exhibits high phytotoxic activity (Loo 2009). In addition to secondary metabolites, numerous primary metabolites that are involved in pyrimidine metabolism, purine metabolism, and amino sugar and nucleotide sugar metabolism were affected by MWCNT exposure. These results indicate a significant impact of MWCNTs on the tomato metabolome.

Conclusion

Carbon nanotubes have been extensively explored as novel drug and gene delivery carriers due to their unique cylindrical structure and superior mechanical strength. The ease of functionalization with various chemical modifications further expands the range of their biomedical applications with improved pharmacological properties and biocompatibility. However, there are growing concerns regarding a multitude of toxicological effects caused by carbon nanotubes, including oxidative stress, inflammatory responses, DNA damage, malignant transformation, and apoptosis, which has significantly dampened enthusiasm for their biomedical applications. Additionally, carbon nanotubes are potentially toxic to cellular metabolism, with numerous metabolic pathways found to be impacted by carbon nanotubes, including glycolysis and the Krebs cycle, the most central metabolic pathways in energy metabolism required for cell growth and survival. Not limited to humans, carbon nanotubes have a wide range of ecological effects, causing significant dysregulation of metabolic pathways in fish and plants. Therefore, the future use of carbon nanotubes is largely dependent on safety issues. To make carbon nanotubes more biocompatible, less cytotoxic, and ideally biodegradable, it is important for future research to aim at developing novel strategies for safety-enhancing functionalization, with a thorough characterization of their potential toxicological consequences.

References

Almeida, M. et al. (2017). Estrogens and androgens in skeletal physiology and pathophysiology. *Physiol. Rev.* American Physiological Society, 97 (1): 135. doi: 10.1152/PHYSREV.00033.2015.

Al-Qattan, M.N., Deb, P.K., and Tekade, R.K. (2018). Molecular dynamics simulation strategies for designing carbon-nanotube-based targeted drug delivery. *Drug Discov. Today*. Elsevier Current Trends, 23 (2): 235–250. doi: 10.1016/J.DRUDIS.2017.10.002.

Assali, M. et al. (2021). Dual covalent functionalization of single-walled carbon nanotubes for effective targeted cancer therapy. *Nanotechnology*. IOP Publishing, 32 (20): 205101. doi: 10.1088/1361-6528/ABE48C.

Baghdadchi, Y. et al. (2018). The assessment of metabolite alteration induced by –OH functionalized multi-walled carbon nanotubes in mice using NMR-based metabonomics. *BioImpacts: BI*. Tabriz University of Medical Sciences, 8 (2): 107. doi: 10.15171/BI.2018.13.

Beg, S. et al.(2018). Emergence in the functionalized carbon nanotubes as smart nanocarriers for drug delivery applications. In: *Fullerens, Graphenes and Nanotubes*. EditorGrumezescu, A.M. Publisher: Elsevier, 105. doi: 10.1016/B978-0-12-813691-1.00004-X.

Bhaskarachary, K. and Joshi, A.K.R. (2018). Natural bioactive molecules with antidiabetic attributes: insights into structure–activity relationships. *Stud. Nat. Prod. Chem.* Elsevier B.V., 57: 353–388. doi: 10.1016/B978-0-444-64057-4.00011-9.

Chen, G. et al. (2019). Isoniazid-loaded chitosan/carbon nanotubes microspheres promote secondary wound healing of bone tuberculosis. *J. Biomater. Appl.* SAGE Publications Ltd, 33 (7): 989–996. doi: 10.1177/0885328218814988.

Dong, J. (2020). Signaling pathways implicated in carbon nanotube-induced lung inflammation. *Front. Immunol.* Frontiers Media SA, 11. doi: 10.3389/FIMMU.2020.552613.

Emwas, A.H.M. et al. (2013). NMR-based metabolomics in human disease diagnosis: applications, limitations, and recommendations. *Metabolomics*. Springer, 9 (5): 1048–1072. doi: 10.1007/S11306-013-0524-Y.

Emwas, A.H.M. (2015). The strengths and weaknesses of NMR spectroscopy and mass spectrometry with particular focus on metabolomics research. *Methods Mol. Biol.* Humana Press Inc., 1277: 161–193. doi: 10.1007/978-1-4939-2377-9_13/COVER.

Eriksson, S. et al. (2015). Dietary methionine can sustain cytosolic redox homeostasis in the mouse liver. *Nat. Commun.* Nature Publishing Group, 6 (1): 1–9. doi: 10.1038/ncomms7479.

Gadelha, I.C.N. et al. (2014). Gossypol toxicity from cottonseed products. *Sci. World J.* The Scientific World Journal. doi: 10.1155/2014/231635.

Ganesan, R. et al. (2021). Carbon nanotubes induce metabolomic profile disturbances in zebrafish: NMR-based metabolomics platform. *Front. Mol. Biosci.* Frontiers Media S.A., 8. doi: 10.3389/FMOLB.2021.688827/FULL.

Golubewa, L. et al. (2021). Rapid and delayed effects of single-walled carbon nanotubes in glioma cells. *Nanotechnology*. IOP Publishing, 32 (50): 505103. doi: 10.1088/1361-6528/AC28DA.

Hamilton, R.F. et al. (2018). Length, but not reactive edges, of cup-stack MWCNT is responsible for toxicity and acute lung inflammation. *Toxicol. Pathol.* NIH Public Access, 46 (1): 62. doi: 10.1177/0192623317732303.

Haniu, H. et al. (2010). Proteomics-based safety evaluation of multi-walled carbon nanotubes. *Toxicol. Appl. Pharmacol.* Academic Press, 242 (3): 256–262. doi: 10.1016/J.TAAP.2009.10.015.

Hassani, M. et al. (2022). Robust antibacterial activity of functionalized carbon nanotube-levofloxacine conjugate based on in vitro and in vivo studies. *Sci. Rep.* Nature Publishing Group, 12 (1): 1–19. doi: 10.1038/s41598-022-14206-w.

Hilton, G.M. et al. (2015). Toxicoproteomic analysis of pulmonary carbon nanotube exposure using LC-MS/MS. *Toxicology*. Toxicology, 329: 80–87. doi: 10.1016/J.TOX.2015.01.011.

Kafa, H. et al. (2015). The interaction of carbon nanotubes with an in vitro blood–brain barrier model and mouse brain in vivo. *Biomaterials*. Biomaterials, 53: 437–452. doi: 10.1016/J.BIOMATERIALS.2015.02.083.

Kam, N.W.S. and Dai, H. (2005). Carbon nanotubes as intracellular protein transporters: generality and biological functionality. *J. Am. Chem. Soc.* American Chemical Society, 127 (16): 6021–6026. doi: 10.1021/JA050062V/ASSET/IMAGES/MEDIUM/JA050062VN00001.GIF.

Ke, F. et al. (2020). Dispersion quality of single-walled carbon nanotubes reveals the recognition sequence of DNA. *Nanotechnology*. Nanotechnology, 31 (25). doi: 10.1088/1361-6528/AB7DE3.

Khaliullin, T.O. et al. (2020). Comparative analysis of lung and blood transcriptomes in mice exposed to multi-walled carbon nanotubes. *Toxicol. Appl. Pharmacol.* Academic Press Inc., 390: 114898–114898. doi: 10.1016/J.TAAP.2020.114898.

Kuempel, E.D. et al. (2017). Evaluating the mechanistic evidence and key data gaps in assessing the potential carcinogenicity of carbon nanotubes and nanofibers in humans. *Crit. Rev. Toxicol.* 47 (1): 1–58. doi: 10.1080/10408444.2016.1206061.

Kumar, S. et al. (2017). Carbon nanotubes: a novel material for multifaceted applications in human healthcare. *Chem. Soc. Rev.* Royal Society of Chemistry, 46 (1): 158–196. doi: 10.1039/C6CS00517A.

Laurila, T., Sainio, S., and Caro, M. (2017). Hybrid carbon based nanomaterials for electrochemical detection of biomolecules. *Prog. Mater. Sci.* Pergamon, 88: 499–594. doi: 10.1016/J.PMATSCI.2017.04.012.

Loo, M. (2009). Herbs and biological agents. *Integr. Med. Children*. Elsevier, 106–130. doi: 10.1016/B978-141602299-2.10007-6.

McGehee, D.L. et al. (2017). Multiwalled carbon nanotubes dramatically affect the fruit metabolome of exposed tomato plants. *ACS Appl. Mater. Interfaces*. American Chemical Society, 9 (38): 32430–32435. doi: 10.1021/ACSAMI.7B10511/SUPPL_FILE/AM7B10511_SI_001.PDF.

Mody, N. et al. (2014). Dendrimer, liposomes, carbon nanotubes and PLGA nanoparticles: one platform assessment of drug delivery potential. *AAPS Pharm. Sci. Tech.* Springer New York LLC, 15 (2): 388–399. doi: 10.1208/S12249-014-0073-3/TABLES/4.

Murugesan, S. et al. (2006). Blood compatible carbon nanotubes–nano-based neoproteoglycans. *Langmuir*, 22 (8): 3461–3463. doi: 10.1021/LA0534468.

Phuyal, S., Kasem, M., Knittelfelder, O., Sharma, A., Fonseca, D. de M. et al. (2018). Characterization of the proteome and lipidome profiles of human lung cells after low dose and chronic exposure to multiwalled carbon nanotubes. *Nanotoxicology*. 12 (2): 138–152. doi: 10.1080/17435390.2018.1425500.

Raissi, H. and Mollania, F. (2014). Immunosuppressive agent leflunomide: a SWNTs-immobilized dihydroortate dehydrogenase inhibitory effect and computational study of its adsorption properties on zigzag single walled (6, 0) carbon and boron nitride nanotubes as controlled drug delivery devices. *Eur. J. Pharm. Sci.* Elsevier, 56 (1): 37–54. doi: 10.1016/J.EJPS.2014.02.006.

Sun, Y. et al. (2012). Construction and comparative study of monovalent and multivalent DNA vaccines against *Streptococcus iniae*. *Fish Shellfish Immunol.* Academic Press, 33 (6): 1303–1310. doi: 10.1016/J.FSI.2012.10.004.

Surai, P.F., Earle-payne, K., and Kidd, M.T. (2021). Taurine as a natural antioxidant: from direct antioxidant effects to protective action in various toxicological models. *Antioxidants*. Multidisciplinary Digital Publishing Institute (MDPI), 10 (12). doi: 10.3390/ANTIOX10121876.

Thakur, A., Bharti, R., and Sharma, R. (2022). Carbon nanotubes: types, synthesis, cytotoxicity and applications in biomedical. *Mater. Today: Proc.* Elsevier, 50: 2256–2268. doi: 10.1016/J.MATPR.2021.10.002.

Vlaanderen, J. et al. (2017). A cross-sectional study of changes in markers of immunological effects and lung health due to exposure to multi-walled carbon nanotubes. *Nanotoxicology*. Taylor and Francis Ltd, 11 (3): 395–404. doi: 10.1080/17435390.2017.1308031/SUPPL_FILE/INAN_A_1308031_SM0032.DOCX.

West, L.A. et al. (2001). Steroidogenic acute regulatory protein and peripheral-type benzodiazepine receptor associate at the mitochondrial membrane. *Endocrinology*. 142 (1): 502–505. doi: 10.1210/ENDO.142.1.8052.

Yan, Y. *et al.* (2020). β-glucan and anisodamine can enhance the immersion immune efficacy of inactivated cyprinid herpesvirus 2 vaccine in *Carassius auratus gibelio*. *Fish Shellfish Immunol.* Fish Shellfish Immunol, 98: 285–295. doi: 10.1016/J.FSI.2020.01.025.

Yuan, C.Y., Lee, Y.J., and Hsu, G.S.W. (2012). Aluminum overload increases oxidative stress in four functional brain areas of neonatal rats. *J. Biomed. Sci.* 19: 1. doi: 10.1186/1423-0127-19-51.

Zaïri, A. et al. (2020). Phytochemical profile, cytotoxic, antioxidant, and allelopathic potentials of aqueous leaf extracts of *Olea europaea*. *Food Sci. Nutr.* Wiley-Blackwell, 8 (9): 4805. doi: 10.1002/FSN3.1755.

Zare, H. et al. (2021). Carbon nanotubes: smart drug/gene delivery carriers. *Int. J. Nanomed.* Dove Press, 16: 1681. doi: 10.2147/IJN.S299448.

Zempleni, J., Hassan, Y.I., and Wijeratne, S.S.K. (2008). Biotin and biotinidase deficiency. *Expert Rev. Endocrinol. Metab.* NIH Public Access, 3 (6): 715. doi: 10.1586/17446651.3.6.715.

Zhang, C. et al. (2020). Application of antigen presenting cell-targeted nanovaccine delivery system in rhabdovirus disease prophylactics using fish as a model organism. *J. Nanobiotechnol.* BioMed Central, 18 (1): 1–13. doi: 10.1186/S12951-020-0584-X/FIGURES/7.

Zhang, D. et al. (2020). Investigating biological effects of multidimensional carboxylated carbon-based nanomaterials on human lung A549 cells revealed via non-targeted metabolomics approach. *Nanotechnology* IOP Publishing, 32 (1): 015704. doi: 10.1088/1361-6528/ABB55B.

Zhao, D. et al. (2021). Multi-walled carbon nanotubes prevent high temperature–induced damage by activating the ascorbate-glutathione cycle in *Paeonia ostii* T. Hong et J. X. Zhang. *Ecotoxicol. Environ. Saf.* Academic Press, 227: 112948. doi: 10.1016/J.ECOENV.2021.112948.

Zhao, X. and Liu, R. (2012). Recent progress and perspectives on the toxicity of carbon nanotubes at organism, organ, cell, and biomacromolecule levels. *Environ. Int.* Pergamon, 40 (1): 244–255. doi: 10.1016/J.ENVINT.2011.12.003.

Zhao, Z. et al. (2020). Single-walled carbon nanotubes as delivery vehicles enhance the immunoprotective effect of an immersion DNA vaccine against infectious spleen and kidney necrosis virus in mandarin fish. *Fish Shellfish Immunol.* Academic Press, 97: 432–439. doi: 10.1016/J.FSI.2019.12.072.

Zhu, B. et al. (2014). Single-walled carbon nanotubes as candidate recombinant subunit vaccine carrier for immunization of grass carp against grass carp reovirus. *Fish Shellfish Immunol.* Academic Press, 41 (2): 279–293. doi: 10.1016/J.FSI.2014.09.014.

Zhu, S. et al. (2018). Biocompatibility assessment of single-walled carbon nanotubes using *Saccharomyces cerevisiae* as a model organism. *J. Nanobiotechnol.* BioMed Central, 16 (1): 44. doi: 10.1186/S12951-018-0370-1.

15

Assessment of the Biological Impact of Engineered Nanomaterials Using Mass Spectrometry-Based MultiOmics Approaches

*Nicholas Day, Tong Zhang, Matthew J. Gaffrey, Brian D. Thrall, and Wei-Jun Qian**

Biological Sciences Division, Pacific Northwest National Laboratory, Richland, WA
** Corresponding author*

Introduction

Engineered nanomaterials (ENMs) have found extensive applications from food packaging [1, 2] to drug delivery [3–5] due to their favorable physiochemical properties. They also pose a potential risk to human health, making characterization of the cellular responses to ENMs a prerequisite in evaluating their biological impact. While conventional assays assess cellular phenotypes, such as cell proliferation or the activity of several selected pathways, they are unable to reveal an overall picture of ENM–cell interactions [6]. Nonetheless, it is clear that ENMs can interact with cells at multiple levels by dysregulating cell signaling, gene expression, protein turnover, and metabolism, calling for alternative methods that can provide a comprehensive and detailed view of cellular responses to ENMs [7–9]. In addition, the outcomes of ENM–cell interactions depend on multiple ENM parameters, such as the chemical composition, size, charge, and shape, indicating different types of cellular machinery are involved to cope with a diverse range of ENMs. As a result, it is challenging to predict the biological impact and cellular mechanisms governing ENM–cellular interactions. Thus, comprehensive profiling of the biological responses to ENMs is important for both safety assessment and mechanistic understanding of ENM–cell interactions.

The development of advanced higher information content technologies such as high-throughput sequencing has made valuable contributions toward system-wide characterization of cellular responses to ENM exposure. As genomic and epigenomic studies resolve details about changes in gene sequences or expression, application of such technology alone is insufficient to capture and deconstruct the complexity of a biological system at the protein network and metabolism levels [10]. Thus, additional studies are still needed to understand how ENM exposure disrupts biological processes beyond the genetic level. Analysis of other biomolecules like proteins and their protein posttranslational modifications (PTMs), as well as many types of lipids and metabolites enables a broader examination of the physiological status of the cell. Proteins are essential to facilitate the many processes of a cell, including energy production, metabolism, replication, transport, and structural support. Lipids are an integral part of membrane architecture in cells, serve as a form of energy storage, and have roles in signaling pathways. Metabolites are small molecules that are created as intermediate or end products from metabolic pathways and can serve as energy sources, stimulants, inhibitors, or signaling molecules.

Because the composition and concentration of proteins, metabolites, and lipids reflect the activity of various cellular signaling pathways and biochemical pathways, corresponding omics

Impact of Engineered Nanomaterials in Genomics and Epigenomics, First Edition. Edited by Saura C. Sahu.

measurements provide a direct readout of cellular activities. While proteomics can reveal changes in protein abundances, measurements of different metabolite or lipid species can provide clues on the status of various metabolic processes and functions. Since the synthesis and catabolism of metabolites and lipids results from the activity of specific proteins (enzymes) [11–13], the interdependence of these other biomolecules makes their studies complementary to each other when studying a perturbation to a biological system.

In this chapter, we will cover studies that demonstrate the application of proteomics, metabolomics, or lipidomics for evaluating the biological impact of ENM exposure in different biological systems. The goal of this review is to present readers the capacity of omics tools in the ENM field and highlight biological insights gained from such studies. As most current studies use primarily one omics technique, we will also discuss the integration of these other omics approaches, as it provides an opportunity to draw mechanistic links between ENM exposure and the reprograming of the genome/proteome/metabolome, as well as phenotypical changes.

Applications of MS for the Measurements of Proteins, PTMs, Lipids, and Metabolites

MS is an analytical tool where the mass-to-charge ratio (m/z) of analytes are measured to determine their molecular weight. Prior to analysis on a mass spectrometer, separations of biomolecules are needed to reduce sample complexity (Figure 15.1a) to enhance ionization efficiency and overall sensitivity during an analytical run, which typically involves chromatography with a liquid (LC) or gas (GC) mobile phase. The stationary phases of GC columns use high boiling polymers for a coating, while LC columns are usually packed with silica particles coated with different chemical groups to promote separation of specific molecules (4 carbon alkyl chain for proteins or 18 carbon alkyl chain for peptides). Molecules are carried by a gas- or liquid-based mobile phase and are separated by their varying affinities to the stationary phase. While both LC and GC accomplish separation of molecules, GC is best suited for smaller molecules like metabolites and lipids, but LC has greater versatility to separate multiple classes of biomolecules, including proteins (or their corresponding peptides), lipids, and metabolites. Mass spectrometers typically comprise three major components: (i) an ion source that ionizes molecules into gas-phase ions; (ii) a mass analyzer that can sort analytes by their m/z ratios via an electric or magnetic field; and (iii) a detector that records the relative abundances of ions that leave the mass analyzer (Figure 15.1a). Derivations of these major components have expanded the capability of mass spectrometers. For example, ion sources are available in different formats, such as electrospray ionization or matrix-assisted laser desorption/ionization, each having their own advantages for ionizing certain types of molecules [14]. Equipping a mass spectrometer with two or more mass analyzers is another highly desirable configuration, as it enables both isolation and fragmentation of ions. Analysis of the fragmented ion spectra allows for the highly accurate identification of molecules present in a sample (Figure 15.1a). There are derivations of mass analyzers beyond the typical quadrupole, such as the time of flight or orbitrap mass analyzers (Figure 15.1a), which make mass spectrometers more versatile and specialized [15]. For additional detail on the different configurations of mass spectrometers, readers are directed to a previous review [16].

Biomolecules such as proteins and peptides, lipids, and metabolites in complex samples can be analyzed using MS-based omics approaches for both identification and quantification purposes. For example, MS-based proteomics can identify and quantify thousands of proteins as well as PTMs in a biological system. Currently, most applications in the ENM field use a bottom-up approach that digests proteins into peptides for proteomic analysis. Major steps include: (i) lysis of cell culture or tissue to extract proteins; (ii) reduction and alkylation of cysteine thiol groups; (iii) digestion of proteins into peptides by a sequence-specific protease, such as trypsin; and (iv) clean-up of peptides to remove salts and detergents (Figure 15.1b). For investigation of PTMs, such as

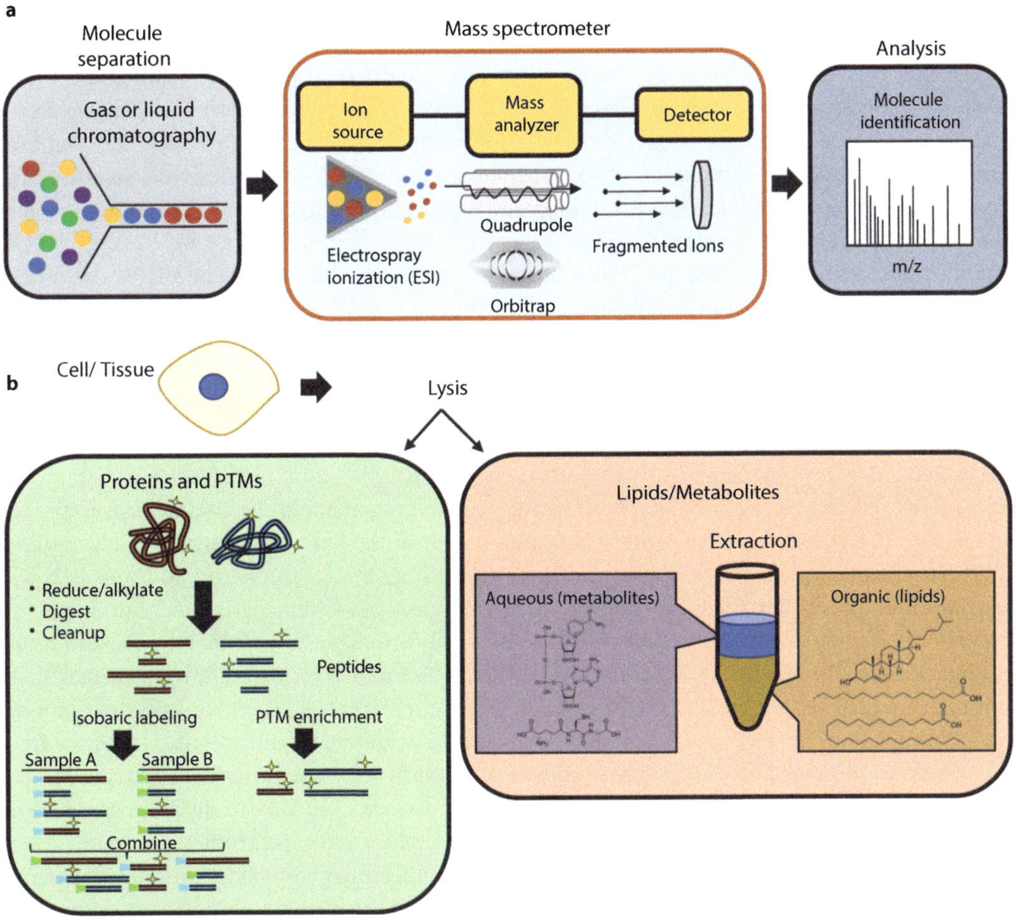

Figure 15.1 A simplified view of mass spectrometry and sample preparation of proteins, lipids, and metabolites. a) Schematic illustrating the basic principles of mass spectrometry-based analysis of molecules. Molecule separation is achieved using gas or liquid chromatography (illustrated). The mass spectrometer is made up of 3 major components. 1) An ion source that vaporizes and ionizes molecules (electrospray ionization creating electrically charged particles is illustrated). 2) A mass analyzer that measures the mass to charge ratio of ions via an electric or magnetic field (commonly used quadrupole and orbitrap mass analyzers are illustrated). Adding more mass analyzers enables fragmentation of ions (illustrated) which better resolves molecule identity. 3) A detector that measures the relative abundance of ions as they exit the mass analyzer(s). Analysis of the spectra from mass spectrometry data are used to determine what molecules were present in a sample. b) After lysis of a cell culture or a tissue, lipids and metabolites can be separated by a biphasic extraction procedure. Proteins and their PTMs (yellow star) are processed using a standard workflow, where the cysteine thiol groups of proteins are reduced and alkylated, the proteins are digested with a sequence specific protease to create peptides, such as trypsin, and cleaned up to remove any residual salts and detergents. The peptides can be further processed by enrichment strategies that capture specific PTMs (illustrated) or labeled with isobaric tags for combining multiple samples into a single mass spectrometry run (illustrated).

phosphorylation, acetylation, glycosylation, or cysteine oxidation, specialized workflows have been developed for their selective enrichment (Figure 15.1b) [17–20]. Quantification of relative abundances of peptides can be performed by either labeling with isobaric compounds such as tandem mass tags or label-free approaches. One major advantage of isobaric labeling is the ability to combine up to 18 samples into a single analytical run (Figure 15.1b), while label-free analysis must run one sample at a time.

While proteomics follows the axis of gene expression and may cross-validate genomics data, lipidomics and metabolomics offer valuable insight into the physiological status of the cell. Biomolecules like lipids, amino acids, nucleic acids, carbohydrates, and organic acids, are critical to cellular processes and their abundances are subject to change following a perturbation. Metabolites are broadly categorized as low molecular weight biomolecules (50–1500 Da) that are soluble in polar solvents, allowing them to be separated from other molecules with solubility in nonpolar solvents (i.e., lipids). As a result, biphasic separation (Figure 15.1b) of these two major classes of biomolecules is possible through different methods [21], including the well-known Folch method, which simply uses methanol, chloroform, and water to extract these biomolecules. With diverse chemical structures, the metabolomes of biological systems are complex and analytically challenging to characterize. Two main current analytical methods are nuclear magnetic resonance (NMR) and MS coupled with GC or LC with their own strengths [22, 23]. NMR-based studies are informative, but LC/GC-MS is more sensitive and provides greater throughput for metabolomics [24, 25], making it an ideal platform for toxicological studies.

Once considered a single field under the broad scope of metabolomics, lipidomics has emerged as its own distinct field and platform. The diversity and heterogeneity of the chemical structure of lipids has led to a classification of eight groups that comprise the lipidome: fatty acyls, glycerolipids, glycerophospholipids, sterol lipids, prenol lipids, sphingolipids, saccharolipids, and polyketides [26]. Like other biomolecules, the lipidome is dynamic, as changes in the abundance or composition of lipids can result from a countless array of factors, such as toxicant exposure [27], diet [28], or age [29]. Further complexity arises from differential expression of lipid species that can be dependent on tissue type [30, 31]. In addition, lipids form PTMs on proteins (N-myristoylation, S-/N-palmitoylation, farnesylation), but are also subject to oxidative modifications on their hydrocarbon chains or head groups [32]. These lipid-specific modifications can alter their function, leading to the induction of alternate biological responses or disease [33]. Due to the high number of building blocks and possible combinations used to make lipids, mass spectrometers can be tuned to accommodate the high diversity of the lipidome. One such strategy involves switching between positive and negative ionization of molecules, which expands the number of lipid species that are detectable [34]. For additional detail of the specific analytical approaches utilized in lipidomics, readers are directed to reviews found elsewhere [35–37].

MS-based multiomics studies have been increasingly used to study the biological impact of ENMs primarily because it can provide a comprehensive view of the interaction between ENMs and a biological system (Figure 15.2). Quantitative proteome-level comparison enables the identification of proteins and biological processes and/or pathways that are responding to ENM exposure [38–40]. Since many diseases or challenges to steady-state conditions are accompanied by changes in metabolite composition, metabolomics is useful for breaking down the complex responses in a biological system. Despite the complexity that is nearly comparable to that of proteins, our understanding of the lipidome is inadequate. Interest in quantifying and measuring the lipid composition of a sample is gaining traction, including in the field of nanomaterials, as viewing changes in the lipidome offers a different perspective on the toxic effects of these materials. Collectively, these high-sensitivity and large-scale MS-based studies provide additional details

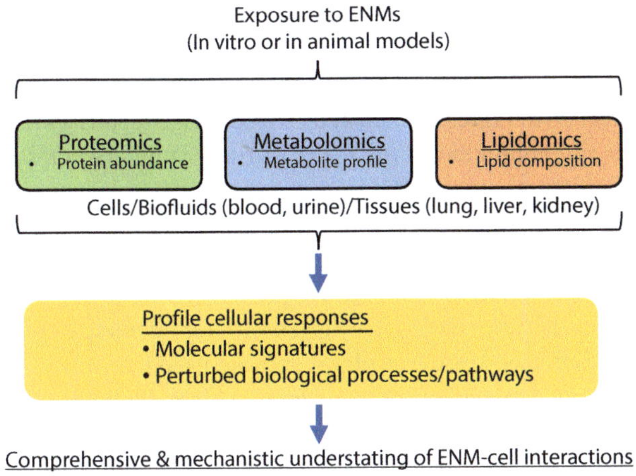

Figure 15.2 Conceptual workflow of mass spectrometry based "omics" methods to profile proteins, metabolites, and lipids.

about cellular perturbations that may not be detectable with other commonly used toxicological approaches (e.g., cell viability and proliferation) at lower and more realistic doses of materials.

It should be noted that multiomics studies also have been used to understand the composition of the biomolecules that form coronas around nanoparticles, as it plays a critical role in ENM–cell interactions. Interested readers are redirected to other excellent resources for corona profiling [41–49]. This review serves as an extension of the application of proteomics in this field since our previous reviews [40, 50]; therefore, we will cover relevant proteomics literature primarily within the past three years. Besides proteomics, we will also showcase the capacity for MS-based metabolomics and lipidomics to profile the biological responses to ENMs in an array of biological systems. Finally, we make the case for the need for integrative multiomics studies in future investigations of ENM toxicity.

Multiomics Investigation of ENM Exposure to Microorganisms

With the small size and high surface to volume ratio of ENMs, microscopic organisms found ubiquitously in the environment, such as bacteria and fungi, are at risk of exposure to nanoparticles [51]. Indeed, ENMs have advantageous uses in curbing bacterial infections in medical applications, but the presence of these materials in wastewater discharged into aquatic environments poses a threat to resident microorganisms. MS-based omics studies have been used to answer fundamental questions about ENM interactions with microorganisms. For instance, to understand the molecular targets of silver nanoparticles in inhibiting *Staphylococcus aureus*, a common cause of hospital-acquired infections, Wang et al. used a metalloproteomics approach to identify 38 silver (Ag^{+})-binding proteins [52]. They further validated the proteomics findings by showing that Ag binds to the catalytic site and thus inhibits the activity of 6-phosphogluconate dehydrogenase. Thus, a discovery-proteomics approach, coupled with orthogonal studies, helped to confirm the cellular targets of silver nanoparticles in combating *S. aureus*. The antimicrobial effect of nanoparticles can be dampened if bacteria become resistant. Salas-Orozco et al. showed that *Enterococcus faecalis*, a facultative Gram-positive bacterium with high plasticity, acquired resistance under sublethal concentrations of silver nanoparticles [53]. The authors took a proteomics approach to identify proteins and pathways that are activated in the resistant strain, where they found that biofilm-related proteins were upregulated to increase survivability. Other significantly altered processes include efflux of cations, reactive oxygen species (ROS) regulation, and DNA damage repair, which were likely the mechanisms by which *E. faecalis* counteracted the toxicity of silver nanoparticles.

Saccharomyces cerevisiae, a fungus also known as single-celled yeast, is another model system that has recently been used to study ENM toxicity. Lipidomics was recently used to better understand the effects of carboxymethyl cellulose copper nanoparticle damage to the surface of yeast cells revealed by scanning electron microscopy. Lipids such as phosphatidylcholine and phosphatidylethanolamine showed increased abundances, while triacylglycerols were depleted [54]. This indicated upregulation of the Kennedy pathway, since triacylglycerols are used to synthesize phosphatidyl – choline or – ethanolamine lipids, which may suggest altered membrane lipid composition. Upregulation of ROS generation following ENM exposure was also detected, which likely contributed to the high level of lipid peroxidation, a phospholipid modification that compromises membrane integrity. Cellulose nanostructures have been used to stabilize metallic nanoparticles as a strategy to expand their use and effectiveness. A previous study used proteomics to study how cellulosic cupric nanoparticles affected the growth of various strains of *S. cerevisiae* to determine the potential of the material as an antifungal agent. Increased expression of zinc transporter proteins was identified as the cause for copper tolerance in the resistant strains compared with copper-sensitive strains [55]. Interestingly, a combination of the hybrid nanomaterial with a low dose of zinc inhibits the growth of copper-resistant strains of yeast, highlighting interdependence of metals in creating antimicrobial or antifungal agents.

The toxicity of ENM exposure for microorganisms in aquatic ecosystems and subsequent ecological effects remains unexplored or is poorly understood. Graphene oxide nanosheets (1–5 μM) and quantum dots (20–50 nm) were previously found to impact the metabolome of the single-cell microalga *Chlorella vulgaris*, which is a major contributor to the energy and biomass of the aquatic environment. Despite the identical composition of the material, the quantum dots appeared to exert a greater effect on abundances of metabolites, such as linolenic acid, and amino acids, such as isoleucine and glutamine, suggesting perturbed amino acid metabolism [56]. This study shows how other properties, such as the shape of ENMs, can determine their biological impact, which can be resolved by a comparative analysis of MS-based omics data. Another study utilized metabolomics to resolve the effects of silver nanoparticles internalized by *Poterioochromonas malhamensis*, a different species of alga. A time course study of metabolite composition revealed changes in amino acid metabolism and the TCA cycle, which supported physiological assays that found elevated ROS production, increased lipid peroxidation, and altered photosynthetic activity [57]. Notably, this study showed how metabolomics offers an early and sensitive assessment of ENM toxicity and stress before physiological changes are detected.

As the number of MS-based omics datasets continues to increase, synthesizing data from published studies may identify what common and unique cellular pathways/mechanisms are utilized by microorganisms in response to a perturbation by ENMs. For example, Mortimer et al. found that membrane transport, signal transduction, and energy metabolism are among the most commonly deregulated pathways in bacteria interacting with different ENMs in more than a dozen proteomics studies, despite the great variation in ENM type, size, and concentration, as well as bacterial strains used [58]. By categorizing ENMs into inhibitory, subinhibitory, and stimulatory groups depending on the effect on bacterial growth and survival, the authors found pathways that are modulated in the subinhibitory compared with inhibitory ENM concentrations. This study exemplifies the power of systematic analysis in integrating all available studies for a "big data" view. The benefit is clear as most studies only test a limited number of samples covering specific ENM types and biological systems.

Multiomics Investigation of ENM Exposure Using In Vitro Cell Culture Models

Besides microorganisms, mammalian cell culture based biological systems are a more conventional platform for studying material responses relevant to human health. Tissue culture offers an isolated system to investigate the effects of ENM exposure before scaling up to more complex models (i.e. in vivo mouse or rat). Furthermore, utilization of cell lines derived from specific organs and organ systems can provide insight into how ENMs might disrupt those types of tissues. For example, culturing cells derived from lung tissue can model how the pulmonary system responds to ENMs. As the respiratory tract is a route of nanoparticle exposure, a number of studies have focused on the metabolic changes in the lung. The metabolic profiles of human lung bronchial epithelial cells exposed to silica nanoparticles was recently reported, which bolstered accompanying phenotypic data showing mitochondrial dysfunction. Metabolites with significantly changing abundances, such as downregulation of glutathione and pantothenic acid as well as upregulation of 8-hydroxyguanosine, suggested oxidative stress and disrupted energy metabolism, respectively [59]. Altered mitochondrial structure along with dysregulated ROS or ATP production, respiration, and proton leak supported the metabolite data, leading to a better understanding of how silica nanoparticles can disrupt multiple pathways in lung tissue.

Following oral ingestion, ENMs have the potential to reach the digestive system, making research of these tissues another area of interest. Soria et al. performed an in-depth toxicological study of copper oxide nanomaterials, which revealed dose-dependent effects on lipid composition [27]. Lipidomics analysis of human colon carcinoma HCT-116 cells exposed to copper oxide revealed the most dramatic differences in membrane lipids, such as those belonging to the sphingolipids or glycerophospholipids classes. These significant changes in lipid composition are thought to be the result of membrane remodeling, which would promote the uptake of these ENMs.

The renal system is another area at risk of toxic ENM exposure and damage, as the organs and tissues are involved in the removal of toxins from the body. In a recent study, silica nanoparticles of different sizes (16, 29, and 51 nm) were used to treat normal rat kidney cells, which are a commonly used epithelial culture system to study the function and responses of the renal system. Of the 3 particle sizes, 60 μg/mL of 16 nm silica nanoparticles led to the activation of autophagy without compromised cell viability, where integrated transcriptomic, proteomic, and phosphoproteomic analyses provided more detail on the response to the ENM [60]. Across the three omics tools used, the CDK7-CDK4 signaling axis was identified as the major autophagy-activating kinase cascade used to mitigate toxicological responses to the material, which was validated by specific inhibition of these kinases in an orthogonal assay.

Analysis of ENM Toxicity in Liver-Based Cell Models

As numerous studies have shown the liver as a target organ where high levels of nanoparticles accumulate, liver-based tissue culture systems are ideal for identifying mechanisms of hepatotoxicity. Proteomic works by several groups have demonstrated substantial, early-stage proteomic changes with sub-toxic concentrations of silica nanoparticles in HepG2 cells, such as activation of the acute phase response and inflammation pathways, as well as disrupted lipid metabolism [61].

In a different study, label-free proteomic analysis of HepG2 cells treated with silica helped to define the detailed mechanism of toxicity over time, where early responses involved mRNA splicing and cell-cycle arrest, while later stages showed increases in ROS and immune response [62]. These studies highlight the idea that while the level of exposure is not cytotoxic and cell morphology may appear normal, omics tools can reveal the sensitivity of a cell type toward any ENM and identify early response markers. On a different note, silver nanoparticle exposure in different types of hepatocytes revealed the protein targets of silver ions (Ag^+) released from silver nanoparticles [63]. The authors used a metallomic and proteomic approach to identify six proteins including glutathione S-transferase (GST), whose activity was inhibited by Ag^+. Activity depletion of GST due to silver binding led to oxidative stress in the cell and was considered a main driver of toxicity.

Besides proteomic changes, HepG2 cells offer another vantage point for exploring lipidome changes in response to a material. Kitchin and colleagues found that cerium oxide induced significant changes in lipid concentrations, particularly in long chain fatty acids and monoacylglycerols [64]. A follow-up study by the same group confirmed these findings for cerium dioxide, but also revealed that silicon dioxide and copper oxide elevated the concentration of similar classes of lipids [65]. Concurrent metabolomics analysis revealed an increase in glycerol levels but a decline in glycerol 3-phosphate, which is necessary for the synthesis of free fatty acids into triacylglycerides. This suggested that impaired glycerol conversion into glycerol 3-phosphate dampens triacylglyceride synthesis, causing free fatty acids to accumulate and be stored. This suggests that free fatty acids are a potential biomarker for hepatocyte health, since their composition fluctuates and they are involved in various liver diseases. HepG2 cells were also recently used to evaluate the hepatotoxicity of iron oxide and silver nanoparticles, as they are speculated to alter the composition of phospholipids in cell membranes. Following exposure to iron oxide and silver nanoparticles, lipidomics analysis revealed the upregulation of phosphatidylinositol levels that was concurrent with a reduction in phosphatidylacid levels [66]. Significant differences in the levels of phosphatidylcholine species supported the necrotic phenotype of the cells and suggested that inflammation is one of the major mechanisms that makes these materials toxic to the cells.

Macrophage-Based Studies of ENM Toxicity

Macrophages, which purposefully ingest cellular debris, particles, and other macromolecules to promote inflammatory and/or immune responses, are activated by ENM exposure. As such, there are a variety of studies that have evaluated the toxicity of nanomaterials after their engulfment by phagocytosis. Titanium dioxide nanoparticles are speculated to disrupt the phospholipids of cellular membranes since they have an affinity for phosphopeptides [67]. A previous study found that a 24-hour exposure to titanium dioxide ENMs caused significant reductions in multiple cardiolipin species in macrophage cell lines [68]. Since cardiolipin is a phospholipid found primarily in the mitochondrial membrane and facilitates electron transport chain function, researchers concluded that the ENM is a mitotoxicant and causes mitochondrial dysfunction and elevated oxidative stress. Another study used lipidomics to uncover the specific mechanism that triggered cell death in RAW264.7 macrophages exposed to graphene quantum dots. About 140 lipids showed greater than a two-fold change in abundance following exposure, with a number of these being phospholipids that are involved in ferroptosis [69]. This coincided with reduced expression of lipid repair enzyme glutathione peroxidase 4 and increased lipid peroxidation, effects that contribute to ferroptosis, which was suggested to be the cause of cell death following graphene quantum dot exposure.

From a proteomics perspective, our group previously used a two-tiered strategy to assess both targeted biological endpoints and untargeted proteome profiles of human THP-1-derived macrophages exposed to 11 metal/metal-oxide ENMs. In the first tier, the MTT assay showed that many ENMs, including ZnS, CdS, Au, SiO_2, Fe_2O_3, CeO_2, MgO, TiO_2-P25, and 1% Ag-SiO_2, did not result in a significant decrease in cell viability even at a high concentration (50 μg/mL). In the second tier, proteomics data revealed these non-cytotoxic ENMs, even at a low concentration (12.5 μg/mL), can induce significant perturbations in biological processes, such as cell–cell adhesion, translation, and mRNA catabolic process [70] (Figure 15.3). In our study, biological processes responsive to non-cytotoxic ENMs were also altered by cytotoxic ENMs, but at a higher amplitude, indicating that they are adaptive cellular processes that may respond to ENMs in a "stress dosage"-dependent manner. In addition, we observed biological processes uniquely altered by cytotoxic ENMs, including the TCA cycle, response to unfolded protein, and NIK/NF-κB signaling. The data from this study also revealed protein signatures related to ENM-induced oxidative stress. In agreement with this observation, our lab also previously reported sites of protein S-glutathionylation, a cysteine thiol redox-based PTM, in response to ENM exposure and accompanying oxidative stress. Redox proteomics analysis of RAW 264.7 macrophage cells treated with cobalt oxide, silicon oxide, or iron oxide highlighted protein translation and stability and macrophage innate immune function as the most perturbed pathways [71]. These studies showcase the complexity and multifactorial nature of ENM–cell interactions that require unbiased and in-depth analysis of the proteome to understand the molecular pathways in response to ENMs.

Neuronal Cell Models Reveal Potential Mechanisms of ENM-Induced Neurotoxicity

Besides the tissues that can encounter the external environment, deep tissues are capable of being disturbed by ENM exposure. ROS-dependent molecular pathway changes upon ENM exposure are ubiquitous to any cell type, including neuronal cells. Bend3 cells were treated with silver nanoparticles in the presence or absence of the antioxidant N-acetyl L-cysteine (NAC). Using the NAC group as a control, the authors were able to identify pathways that are dependent on ENM-induced ROS through proteomic analysis, including upregulation of the NRF2 antioxidant pathway [72]. In another study, proteomics was used to determine the effect of zinc oxide (ZnO) nanoparticles on

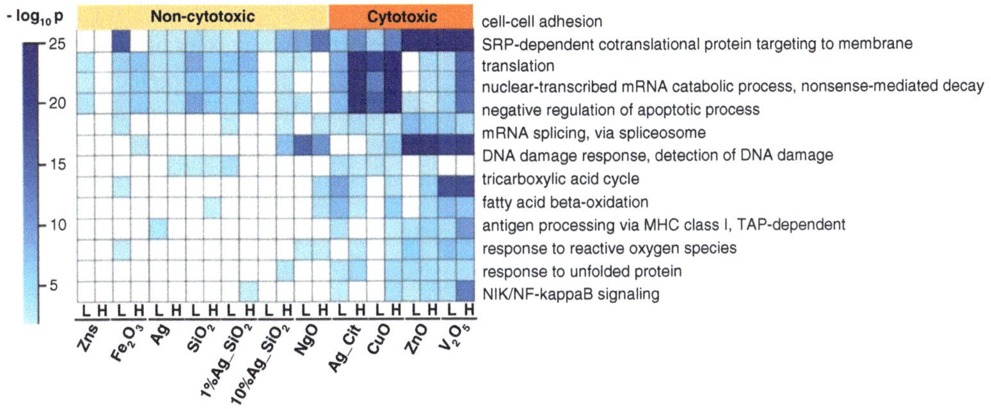

Figure 15.3 Proteomics reveals common and unique biological processes altered in response to a panel of 11 ENMs. For each ENM, both low (12.5 μg/mL, L) and high (25 μg/mL, H) doses were used to treat human THP1-derived macrophage cells. Proteins showing a significant abundance in each condition were enriched by gene ontology, and the significance was represented by the negative logarithm of the p value (0.01 as the cutoff). If a biological process was not enriched in a given condition, it is denoted as a white block. Source: Zhang et al. (2020) / With permission of Elsevier.

neuronal cell differentiation. The proteomics data revealed that ZnO was more toxic to differentiated cells compared with undifferentiated cells, which was also seen at the microRNA level [73]. The deregulated proteins pointed to pathways associated with neurodegeneration, while physiological studies revealed mitochondrial dysfunction, highlighting ZnO as a neurotoxic ENM.

Lipidomics analysis of primary cortical rat neurons exposed to graphene oxide monolayer sheets revealed significant differences in lipid composition, compared with the vehicle control [74]. Membrane phospholipids showed the most dramatic increases in abundances, suggesting that key lipid species are sensitive to ENM exposure, even when compared with other forms of structural lipids like cholesterol, which did not significantly change. These changes in membrane composition were speculated to contribute to attenuated network function and reduced excitatory synaptic contacts and transmissions, which depend on proper membrane architecture to propagate a signal. A follow-up proteomic and lipidomics study by the same group found that graphene oxide and few-layer graphene flakes differentially disrupt rat astrocyte function. Graphene oxide is more disruptive toward membrane cholesterol levels and calcium binding proteins, which are critical for calcium-based signaling of astrocytes [75]. Three-dimensional human brain organoids have also been used to model the neurotoxicity of graphene oxide sheets. Lipidomics analysis of brain organoids revealed upregulation of the ceramide class of lipids, which is known to be cytotoxic as it impairs the functions of organelles and correlated with transcriptomics analysis [76]. Few studies have used 3D organoids to study ENM toxicity; however, this demonstrates its utility as a valuable tool in identifying responses that may better resemble exposure in humans.

Multiomics Studies Reveal Organ-Specific Toxicity at the Organismal Level

While studies using tissue culture offer valuable insight into ENM toxicity at the cellular level, more complex biological responses can be learned from experiments at the organismal level. Investigation into the toxicological effects of ENM exposure are valuable when applied to model systems, such as the mouse or rat, which more accurately represent the complex modes of exposure encountered by humans. Compared with histological studies, MS-based omics approaches offer more content and detail regarding organ targeting and toxicity. As the roles of the liver and kidneys are to metabolize and excrete toxicants, it is not surprising that they are targeted by ENMs, but other organs are also susceptible to ENM-induced perturbations. In the following section, we will discuss how MS-based omics has improved our understanding of biological responses toward ENMs in the lung, gut, and embryo of different whole animal systems.

Mechanisms of ENM-Induced Toxicity in the Lung

Using lung tissue from a mouse model, intratracheal instillations of polyvinylpyrrolidone-coated silver nanoparticles (5 versus 50 nm) yielded a number of perturbations. Metabolites related to the TCA cycle (succinate and malonate) and amino acid metabolism (glycine, glutamine, and lysine) exhibited dynamic changes in abundances as a function of dosage frequency or particle size (to a lesser extent), while creatine exhibited abundance changes exclusive to the 50 nm particle size [77]. From a metabolic perspective, these findings suggest that energy and amino acid metabolism processes are the most perturbed in the lung following exposure to silver nanoparticles. By contrast, ENM exposure may yield no significant differences in metabolite profiles, as was reported for both human lung bronchial epithelial cells and rat lung tissue after exposure to iron oxide particles at levels comparable with the environment [78]. This suggested that material dosage at sub- or low- cytotoxic levels does not induce significant metabolic changes or strong pulmonary toxicity.

An early application of lipidomics in the ENM field was reported by Tyurina et al. [79], which investigated the lipid profile of mouse lung tissue following inhalation of single-wall carbon nanotubes. While the composition of pulmonary phospholipids in the lung did not significantly differ, specific classes were subject to peroxidation, such as cardiolipin, which is central to mitochondrial function and induces apoptosis. The study also demonstrated that ENM exposure can induce oxidation of specific classes of lipids, rather than stochastic nonspecific oxidation. In contrast, cerium oxide nanoparticles were found to upregulate expression of glycerophospholipids in the rat lung, specifically glycerophosphocholines and glycerophosphoethanolamines [80]. These types of lipids typically exist in low abundance in the lung, but the marked changes observed after cerium oxide exposure suggest that these lipids are biologically relevant for specific cells or pathways that respond to the material.

In a recent study, integrated MS-based lipidomic and proteomic analysis of bronchoalveolar lavage fluid (BALF) from healthy and metabolic syndrome (MetS) mice was used to profile protein and lipid interactions with iron oxide nanoparticles. Formation of biomolecule coronas derived from MetS mice showed different protein and lipid compositions compared with healthy mice. Enrichment of apolipoprotein c-III and glycolysis enzyme alpha-enolase proteins were unique to the MetS mice, while lipidomics revealed elevated phosphatidylcholine and sphingolipid species in MetS mice compared with the healthy control [81]. An assessment of the toxicity of these iron oxide particles with their MetS – or healthy control – BALF biocoronas on alveolar macrophages revealed that MetS-based biocorona induced inflammation and upregulation of the Jak/Stat pathway. This study demonstrated that subpopulations with underlying diseases such as MetS may induce more acute responses to ENMs, due to the differential enrichment of proteins and lipids compared with that of healthy lungs. Surprisingly, another recent study revealed that nanoparticulate inhalation can impact liver health in the absence of a protein that is critical for fat metabolism. Proteomics analysis of apolipoprotein E knockout mice exposed to silica nanoparticles revealed downregulation of protein synthesis machinery and mitochondrial proteins, compared with the water inhalation control [82]. Decreased expression of these proteins are thought to be part of the molecular mechanism that causes liver steatosis and nonalcoholic fatty liver disease, drawing a link between particulate matter exposure and potential damage in susceptible populations.

Elucidation of Response Pathways Following Ingestion of ENMs

Beyond inhalation, oral ingestion of ENMs is another complex mode of exposure that has been investigated in murine models. Metabolomics, proteomics, and transcriptomics have been used to profile the level of hepatotoxicity induced by ENMs delivered to mice orally (200 mg/kg/day) or intravenously (20 mg/kg/day). Mesoporous silica nanoparticles delivered to mice revealed that the different forms of material administration led to the induction of dissimilar response pathways. Histopathology of liver tissue revealed intravenous injection as a more damaging mode of exposure, which was supported by more robust changes in the metabolite, protein, and transcript profiles [83]. Intravenous exposure stimulated glycolysis and glutathione synthesis, while downregulating TCA and oxidative phosphorylation pathways, suggesting oxidative stress. In contrast, oral uptake yielded less dramatic toxicological effects even with a higher dosage of material, with pathways similar to the intravenous exposure only being mildly perturbed.

Assessment of changes in lipid metabolism for rats after oral ingestion of titanium dioxide nanoparticles was previously reported. Following a dose of 0, 2, 10, or 50 mg of material per kilogram of rat body weight for 90 days, the lipid composition in the sera of the rats were evaluated, revealing stark differences in the abundance of different classes of lipids [84]. Most notably, blood lipid analysis revealed a significant reduction in the level of triglycerides treated with 10 or 50 mg of titanium dioxide, which coincided with weight loss. A more in-depth analysis indicated that other lipid species and their derivatives, such as phosphatidylcholine and phosphatidylethanolamine, show reduced abundances. Pathway analysis reflected this, by emphasizing glycerophospholipid metabolism as significantly disrupted. Since the liver is a major site of the biogenesis of glycerophospholipids, such as phosphatidylcholine, this supported titanium dioxide as a hepatotoxic ENM. A follow-up study by the same group used the same material and experimental conditions to probe the lipid composition of the gut–liver axis after treatment. Lipidomic analysis of rat blood, liver tissue, and feces yielded glycerophospholipid metabolism and phosphatidylcholine lipid abundances as the most perturbed features found common to all sample types [85]. Additionally, upregulated lipid peroxidation was detected in these samples, which was considered an indicator of oxidative stress and an instigator of disordered lipid metabolism, leading ultimately to hepatoxicity.

ENM-Induced Metabolic Changes in the Gut: Involvement of Multiple Biological Systems

Among the range of organs where ENMs are toxic, the gut and the accompanying microbiome have gained interest in recent years. As nanomaterials possess antimicrobial properties, they have the potential to negatively impact the commensal microorganisms that reside within the gastrointestinal tract. Metabolites generated by bacteria as end products of their own metabolism are beneficial to the host through regulation of immune system, mucosal wall integrity, and energy metabolism. Likewise, metabolites created by the host can be important for the microbiome, thereby supporting a symbiotic relationship. Given the importance of microorganisms within the gastrointestinal tract, profiling the metabolic changes in response to ENMs can reveal how they influence gut-associated metabolism.

Metabolite analysis of rat feces exposed to 29 ± 9 nm titanium dioxide nanoparticles for 30 days revealed 25 metabolites with significant changes in abundances, many of which were organic acids or organoheterocyclic compounds [86]. Interestingly, *N*-acetylhistamine, an intermediate of histamine metabolism, showed the most significant increase, while its precursor, L-histidine showed a significant decrease, suggesting both disrupted histamine and histidine metabolism. A supplementary study using similar conditions and the same material at 20 and 250 nm further expanded the range of toxicity associated with the specific ENM. The study uncovered both dose- and particle size-dependent differences in the metabolomes, where glutathione and propionic acid metabolism pathways were the most perturbed pathways for both particle sizes, while biosynthesis of unsaturated fatty acids and D-glutamate or glutamine metabolism were uniquely perturbed in the 250 and 20 nm groups, respectively [87]. Interestingly, correlation analysis between the metabolite data and the gut microbiota revealed pyroglutamic acid and L-glutamic acid as key metabolites impacting the abundance of different genera of bacteria, such as *Anaerotruncus*, *Lactobacillus*, or *Prevotella*.

Broad profiling of plasma metabolites can also yield relevant information about the gut, as was shown recently in a study that examined the effects of silver or silicon dioxide nanoparticle exposure. A notable observation was a reduction in indole-3-acetic acid abundance after ENM exposure, an anti-inflammatory metabolite created by *Lactobacillus*-mediated metabolism of tryptophan, which coincided with a reduction in *Lactobacillus* bacteria in the microbiota data [88]. In support of the prior study mentioned above, this study identified an increase in the abundance of *Prevotella* following exposure to either nanoparticle, which can trigger gut inflammation. Taken together, these studies provide a molecular explanation of how the symbiotic relationship between the gut microbiome and host can be disrupted by ENM exposure.

ENM-Induced Metabolic Changes During Embryo Development

With many biological changes that take place during pregnancy and early development, these stages of life are more susceptible to toxicants and often at high risk for abnormalities. Consequently, ENM exposure poses a risk to these sensitive phases of life, making it an important area of focus for toxicological studies. To evaluate the risk of maternal–fetal transfer, polyvinylpyrrolidone-coated silver nanoparticles (20 and 110 nm) were delivered orally or intravenously in pregnant rats. The study revealed that these materials accumulate throughout the maternal tissue as well as placental and fetal tissues, while the metabolome data of maternal urine indicated that carbohydrate and amino acid metabolism pathways are the most perturbed [89]. Citrate, allantoin, and glutamine were among the key metabolites that showed most distinguished changes in abundance. From a different perspective of the maternal to offspring axis, the effects of ENM exposure on lactation and breastfeeding has also been recently explored. A combined MS-based metabolomics and redox proteomics study was used to determine the effects of titanium dioxide and zinc oxide via inhalation on lactating mice, which reduced milk production and destroyed mammary gland structure. Metabolomics revealed increases in glycolysis and fatty acid biosynthesis, which were speculated to be part of the compensatory response toward the increased apoptosis [90]. Altered glutathione metabolism was another observation, which suggested increased oxidative stress and prompted the authors to investigate glutathionylation as a redox-based PTM in the proteome. Redox proteomics revealed induced glutathionylation of proteins involved in an array of processes such as neutrophil degranulation, cell respiration, redox homeostasis, vesicle or fatty acid transport, and carbohydrate biosynthesis. The authors also noted glutathionylation sites that were specific to a particular ENM, suggesting material-dependent induction of PTMs.

Another report revealed metabolic changes in zebrafish embryos treated with 20 nm polystyrene nanoparticles that caused morphological defects. Among 28 metabolites showing significant changes, 3-methoxytyramine and inosine monophosphate are involved in dopamine synthesis and energy metabolism, respectively, and are important for the proper development of the embryo [91]. A related study using the same particles of a larger size (500 nm) found metabolic effects during the embryonic and larval stages as well as dysbiosis of gut microbiota and defective reproduction in the adult stage. A number of significantly changing metabolites in this study also pointed toward perturbed energy metabolism, especially for fatty acids [92]. Similarly, the impact of copper oxide nanoparticle exposure on sea urchin embryos has been evaluated by metabolomics. Significant dose-dependent changes in abundance of choline, *N*-acetyl serotonin, and glycine were revealed, suggesting neurotoxic effects as well as alterations in skeleton formation [93]. As a whole, these studies provide a small molecule perspective of what metabolic processes are most susceptible to the toxic effects of an ENM during developmental stages.

Probing the Relationship Between Particle Size and Toxicity in Whole Animal Systems

Beyond identification of physiological and organ-targeted effects of ENM exposure, metabolomics offers an alternate perspective for comparative studies of responses to ENMs beyond typical analysis of gene products, histology, or clinical chemistry parameters. Some of the studies described above have alluded to the potential of MS-based omics to explore the relationship between size and the biological responses for the same type of material. As the size of a nanomaterial can play a role in its biocompatibility, it is important to study this relationship using high-throughput omics approaches for a global view of all biological changes, rather than a panel of specific parameters.

Silicon dioxide nanoparticles (30, 70, or 300 nm) were previously investigated using metabolomics to determine whether the biological response is size-dependent. Analysis of the metabolite composition of liver and blood serum derived from mice treated with materials of different sizes revealed profiles distinct from the control but similar to each other, where surface area as a function of dosage had the greatest influence on toxicity [94]. Interestingly, a follow-up study by the same group reported that amino or carboxyl surface modifications had a higher impact on the metabolite profile and biological response than particle size [95]. On the contrary, a transcriptomics study using plain surface silicon dioxide nanoparticles by the same group revealed particle size-dependent gene expression profile differences, which converged to similar cellular response pathways [96]. This suggests that multiple omics studies are ideal to understand the complex biological responses after ENM exposure and help decipher whether a material is toxic at the gene, gene product, or post-gene product level in a system.

In contrast to silicon dioxide, a comprehensive metabolomics study of amino-functionalized, silica-coated iron (II,III) oxide nanoparticles revealed size-dependent particle toxicity. Exposure of different sizes of this ENM (10, 20, or 40 nm) to mice revealed metabolite abundance changes in kidney, liver, lung, or spleen tissue that were specific to or dependent on particle size [97]. Some notable metabolites with altered abundances as a function of particle size were formate and glutathione, suggesting perturbed nucleotide and energy metabolism as well as oxidative stress. Additionally, the blood serum biochemistry data revealed that lactate dehydrogenase activity was reduced as particle size increased, which was supported by reduced nicotinamide adenine dinucleotide cofactor levels in the metabolomics data.

Conclusions and Perspectives

Proteomic, metabolomic, or lipidomic approaches are powerful tools in revealing the molecular details by which cells respond to ENMs. As these studies demonstrate the complexity of ENM–cell interactions, further in-depth research is expected to reveal novel cellular response pathways. Through examination of significant changes in the abundances of metabolites and lipids, disruptions to a biological system are revealed that would otherwise remain unresolved at the genomic, epigenomic, or even proteomic level. The complementation of metabolomics and/or lipidomics data with other omics, biochemical, biological, phenotypic, or traditional clinical chemistry data is another valuable aspect. This can bolster inferences made by other forms of omics or phenotypic data, leading to development of more robust explanatory models of toxicity. By identifying specific classes of lipids and their corresponding organelles, pathways, or signaling functions, details about the lipid composition of a sample can further inform scientists on how ENM exposure disrupts cellular processes. Additionally, lipidomics data can also relate back to the physicochemical properties of ENMs by revealing what types of lipids are most disrupted by a specific ENM. This can also guide the molecular dynamics field, which focuses on computational modeling to visualize lipid–ENM interactions [98] or even show how protein function can be modulated in the presence of an ENM [99]. Given the large number of ENMs that may pose adverse effects on human health, application of omics tools in a high-throughput and systematic manner will also be expected. Below, we discuss a few key points that should be addressed in future MS-based omics studies of ENMs.

Many studies include only a small set of samples and thus cannot reveal the heterogeneity among samples. This is especially relevant to the clinical applications of nanoparticles, as a large degree of variation among individuals may exist in responding to nanoparticles for antimicrobial or antifungal purposes. For instance, the corona composition between nanoparticles and blood can be largely dependent on the donor with the specific enrichment of immunoglobulin or complement proteins [100]. Such variations in the biomolecule composition of the nanoparticle corona can further lead to differences in the interactions between nanoparticles and blood immune cells. Similarly, variations in the cellular responses to nanoparticles among individuals should be expected. The extent of variation should be assessed in further studies by including a reasonably larger number of samples. Second, the response to long-term nanoparticle exposure remains largely unknown as most studies focused on acute responses in cell or animal models. Third, a comprehensive understanding of the ENM's physicochemical properties and how they alter the proteome, metabolome, or lipidome is still lacking. Collection of sufficient data is needed to model the cellular outcomes and regulatory pathways as a function of the physiochemical properties of ENMs. Such information can be used to guide the synthesis of more effective and safer ENMs in the future. Lastly, the PTM layer of the proteome in response to ENM challenges remains largely unexplored. Several studies have already shown that reversible protein oxidation is altered upon exposure to ENMs [101, 102], supporting the concept that ROS can mediate ENM effects by changing the redox proteome. More work is needed to extend PTM-level proteome analysis to gauge the contribution of ROS in ENM–cell interactions. Coupled with analysis of global protein abundance, it is possible to pinpoint regulatory proteins that respond to ENMs by changes in protein abundance or redox status.

While we have reviewed MS-based omics studies in an array of model systems, it should be noted that these systems offer specific advantages. Indeed, whole animal systems are costly compared with cell culture systems, but more accurately reflect the complex physiological mechanisms in humans. Furthermore, whole animal systems like rats and mice offer multiple points of interest for comprehensive toxicological studies via biofluids (blood, urine, spinal fluid) and tissues (muscle, lung, gut, brain, liver, kidney, etc.). In contrast, cell culture systems are isolated, meaning that researchers know the cell type identity of their model system, rather than a complex milieu of cell types in a specific organ or the animal system as a whole. This is advantageous for profiling cell type-specific responses, but ultimately more complex model systems (in vivo animal models) are needed to better understand biological responses and complement findings. Perhaps the use of 3D cell culture may help to bridge the gap between cell culture and whole animal systems, as organoid systems can coculture multiple cell types and hold promise for toxicological studies [103]. The challenge remains for studies such as those that explore gut–microbiome relationships, where an organ system coexists with a complex population of microorganisms, which may be difficult to recapitulate in vitro. ENMs are known to impact human gut microbiomes [104, 105], thus there is a need to characterize the mechanisms of toxicity. A number of studies have used simplified human microbiome model systems to investigate the adverse effects of ENM exposure; however, there are inconsistencies in experimental design and modes of analysis [106]. Development of more accurate model systems will be necessary, concomitantly with sensitive multiomic methods that can evaluate the health of the human microbiome from minimally invasive sources (blood, urine, fecal matter) or tissue biopsies.

Although we review studies that use individual omics tools in the ENM field, there is a trend of integrating multiomics data in a single study. Such multiomics approaches have been used in the assessment of cigarette aerosols in mouse lung [39, 107], peroxisome disorders [108], and neuroinflammation [109]. In the ENM field, this integration would enable a systems toxicology approach for evaluating and understanding mechanisms of ENM toxicity (Figure 15.4). Currently, a limited number of ENM studies have incorporated multiomics methods, such as an investigation of the toxicity of ingested carbon dots [110]. However, most may only use one or two omics methods, leading to a potentially incomplete understanding of the responses in a biological system.

To facilitate the integration of multiple omics methods, extraction of proteins, metabolites, and lipids from the same sample is ideal. This involves use of solvents that target their physicochemical properties to strategically separate and capture them from a heterogeneous mixture of all types of biomolecules [111]. The development of the metabolite, protein, and lipid extraction (MPLEx) protocol [112] or related "3-in-1" method [113] have paved the way for multiomics measurements of all biomolecule types from the same sample to reduce variability and improve throughput. Given the importance and volume of information from these multiple sample types, their incorporation into a combined analytical workflow will likely be a baseline for toxicological studies in the future.

Methods for analysis of integrated omics data are continuously developing, with the goal of leading to a more comprehensive understanding of how a biological system responds to a perturbation. Identifying trends between genomics and proteomics data is a fairly straightforward process, as these omics data follow the gene and gene product axis. However, the challenge arises when integrating metabolomics and lipidomics data, which require sufficient knowledge of corresponding pathways to enable crosstalk between the different data types. The development of these tools remains complex and not user-friendly to individuals with little to no bioinformatics experience. As the datasets become larger and more complex, integration strategies have been developed with

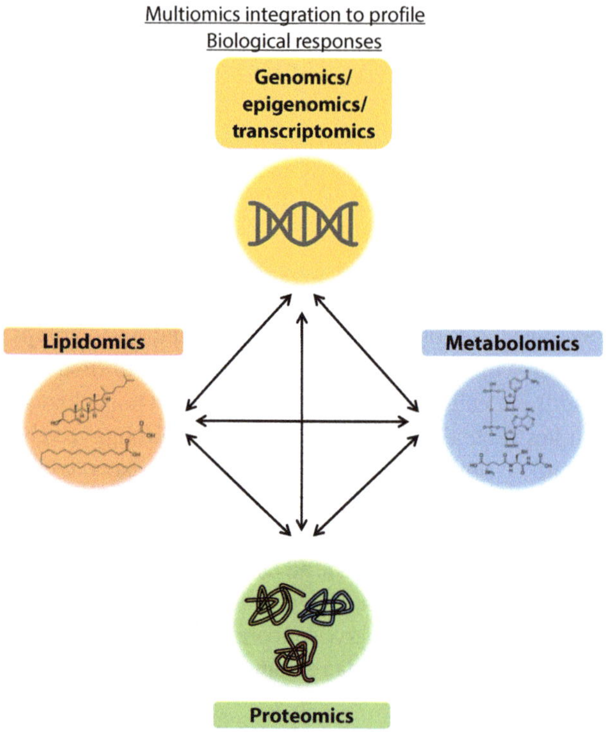

Figure 15.4 Schematic of multiple omics approaches used in understanding the biological impacts of ENMs. Arrows illustrate how each omic approach can support the findings of another omic method, where no single method has centrality or greater importance over the others. Structural formulas shown are NAD+ and glutathione (Metabolomics); cholesterol, palmitic acid, and cerotic acid (Lipidomics).

the help of machine learning [114]. A myriad of other bioinformatic tools developed for handling multiomics data are reported, including Causal Oriented Search of Multiomics Space [115] or STATegra [116]. With increased demand for more comprehensive multiomics experiments, the development of more standardized integrative analysis procedures will help shape the direction of the field.

Acknowledgments

Portions of the work were supported by NIH grants U01ES027292 as part of the Nanotechnology Health Implications Research (NHIR) Consortium, and R01HL139335.

Compliance with Ethical Standards

The authors declare no conflict of interest.

References

Hajipour, M.J., Fromm, K.M., Ashkarran, A.A., de Aberasturi, D.J., de Larramendi, I.R., Rojo, T., Serpooshan, V., Parak, W.J., and Mahmoudi, M. (2013). Antibacterial properties of nanoparticles (vol 30, pg 499, 2012). *Trends Biotechnol.* 31 (1): 61–62.

Sharma, C., Dhiman, R., Rokana, N., and Panwar, H. (2017). Nanotechnology: an untapped resource for food packaging. *Front Microbiol.* 8: 22.

Parveen, S., Misra, R., and Sahoo, S.K. (2012). Nanoparticles: a boon to drug delivery, therapeutics, diagnostics and imaging. *Nanomed. Nanotechnol Biol. Med.* 8 (2): 147–166.

Wong, I.Y., Bhatia, S.N., and Toner, M. (2013). Nanotechnology: emerging tools for biology and medicine. *Genes Dev.* 27 (22): 2397–2408.

Carabineiro, S.A.C. (2017). Applications of gold nanoparticles in nanomedicine: recent advances in vaccines. *Molecules* 22 (5): 16.

Nel, A., Xia, T., Meng, H., Wang, X., Lin, S., Ji, Z., and Zhang, H. (2013). Nanomaterial toxicity testing in the 21st century: use of a predictive toxicological approach and high-throughput screening. *Acc. Chem. Res.* 46 (3): 607–621.

Shang, L., Nienhaus, K., and Nienhaus, G.U. (2014). Engineered nanoparticles interacting with cells: size matters. *J. Nanobiotechnol.* 12 (1): 5.

Villanueva-Flores, F., Castro-Lugo, A., Ramírez, O.T., and Palomares, L.A. (2020). Understanding cellular interactions with nanomaterials: towards a rational design of medical nanodevices. *Nanotechnology* 31 (13): 132002.

Runa, S., Hussey, M., and Payne, C.K. (2018). Nanoparticle-cell interactions: relevance for public health. *J. Phys. Chem. B* 122 (3): 1009–1016.

Karczewski, K.J. and Snyder, M.P. (2018). Integrative omics for health and disease. *Nat. Rev. Genet.* 19 (5): 299–310.

Noda-Garcia, L., Liebermeister, W., and Tawfik, D.S. (2018). Metabolite-enzyme coevolution: from single enzymes to metabolic pathways and networks. *Annu. Rev. Biochem.* 87 (1): 187–216.

Vance, J.E. (2015). Phospholipid synthesis and transport in mammalian cells. *Traffic* 16 (1): 1–18.

Jacquemyn, J., Cascalho, A., and Goodchild, R.E. (2017). The ins and outs of endoplasmic reticulum-controlled lipid biosynthesis. *EMBO Rep.* 18 (11): 1905–1921.

Nadler, W.M., Waidelich, D., Kerner, A., Hanke, S., Berg, R., Trumpp, A., and Rösli, C. (2017). MALDI versus ESI: the impact of the ion source on peptide identification. *J. Proteome Res.* 16 (3): 1207–1215.

Haag, A.M. (2016). Mass analyzers and mass spectrometers. In: *Modern Proteomics – Sample Preparation, Analysis and Practical Applications*, edn. (ed. H. Mirzaei and M. Carrasco), 157–169. Cham: Springer International Publishing.

Di Girolamo, F., Lante, I., Muraca, M., and Putignani, L. (2013). The role of mass spectrometry in the "omics" era. *Curr. Org. Chem.* 17 (23): 2891–2905.

Ino, Y., Kinoshita, E., Kinoshita-Kikuta, E., Akiyama, T., Nakai, Y., Nishino, K., Osada, M., Ryo, A., Hirano, H., Koike, T. et al. (2022). Evaluation of four phosphopeptide enrichment strategies for mass spectrometry-based proteomic analysis. *Proteomics* 22 (7): 2100216.

Schilling, B., Jg, M., Wei, L., Ott, M., and Verdin, E. (2019). High-resolution mass spectrometry to identify and quantify acetylation protein targets. *Methods Mol. Biol.* 1983: :3–16.

Morgenstern, D., Wolf-Levy, H., Tickotsky-Moskovitz, N., Cooper, I., Buchman, A.S., Bennett, D.A., Beeri, M.S., and Levin, Y. (2022). Optimized glycopeptide enrichment method – it is all about the sauce. *Anal. Chem.* 94 (29): 10308–10313.

Guo, J., Gaffrey, M.J., Su, D., Liu, T., Camp, D.G., 2nd, Smith, R.D., and Qian, W.J. (2014). Resin-assisted enrichment of thiols as a general strategy for proteomic profiling of cysteine-based reversible modifications. *Nat. Protoc.* 9 (1): 64–75.

Ulmer, C.Z., Jones, C.M., Yost, R.A., Garrett, T.J., and Bowden, J.A. (2018). Optimization of folch, bligh-dyer, and matyash sample-to-extraction solvent ratios for human plasma-based lipidomics studies. *Anal. Chim. Acta* 1037: :351–357.

Tyagi, R., Kumar, P., and Sharma, U. (2021). Chapter 1 – metabolomics techniques: a brief update. In: *Epigenetics and Metabolomics*, 28e (ed. P.K. Agrawala and P. Rana), 1–29. Academic Press.

Gallart-Ayala, H., Teav, T., and Ivanisevic, J. (2020). Metabolomics meets lipidomics: assessing the small molecule component of metabolism. *BioEssays* 42 (12): 2000052.

Ren, J.-L., Zhang, A.-H., Kong, L., and Wang, X.-J. (2018). Advances in mass spectrometry–based metabolomics for investigation of metabolites. *RSC Adv.* 8 (40): 22335–22350.

Liu, X., Zhou, L., Shi, X., and Xu, G. (2019). New advances in analytical methods for mass spectrometry–based large-scale metabolomics study. *TrAC Trends in Analy. Chem.* 121: 115665.

Fahy, E., Subramaniam, S., Brown, H.A., Glass, C.K., Merrill, A.H., Murphy, R.C., Raetz, C.R.H., Russell, D.W., Seyama, Y., Shaw, W. et al. (2005). A comprehensive classification system for lipids. *J. Lipid Res.* 46 (5): 839–861.

Chavez Soria, N.G., Aga, D.S., and Atilla-Gokcumen, G.E. (2019). Lipidomics reveals insights on the biological effects of copper oxide nanoparticles in a human colon carcinoma cell line. *Molecular Omics* 15 (1): 30–38.

Levental, K.R., Malmberg, E., Symons, J.L., Fan, -Y.-Y., Chapkin, R.S., Ernst, R., and Levental, I. (2020). Lipidomic and biophysical homeostasis of mammalian membranes counteracts dietary lipid perturbations to maintain cellular fitness. *Nat. Commun.* 11 (1): 1339.

Eum, J.Y., Lee, J.C., Yi, S.S., Kim, I.Y., Seong, J.K., and Moon, M.H. (2020). Aging-related lipidomic changes in mouse serum, kidney, and heart by nanoflow ultrahigh-performance liquid chromatography-tandem mass spectrometry. *J. Chromatogr. A* 1618: 460849.

Bhaduri, A., Neumann, E.K., Kriegstein, A.R., and Sweedler, J.V. (2021). Identification of lipid heterogeneity and diversity in the developing human brain. *JACS Au.* 1 (12): 2261–2270.

Al-Sari, N., Suvitaival, T., Mattila, I., Ali, A., Ahonen, L., Trost, K., Henriksen, T.F., Pociot, F., Dragsted, L.O., and Legido-Quigley, C. (2020). Lipidomics of human adipose tissue reveals diversity between body areas. *Plos One* 15 (6): e0228521.

Spickett, C.M. (2020). Formation of oxidatively modified lipids as the basis for a cellular epilipidome. *Front Endocrinol (Lausanne)* 11.

Agmon, E., Solon, J., Bassereau, P., and Stockwell, B.R. (2018). Modeling the effects of lipid peroxidation during ferroptosis on membrane properties. *Sci. Rep.* 8 (1): 5155.

Breitkopf, S.B., Ricoult, S.J.H., Yuan, M., Xu, Y., Peake, D.A., Manning, B.D., and Asara, J.M. (2017). A relative quantitative positive/negative ion switching method for untargeted lipidomics via high resolution LC-MS/MS from any biological source. *Metabolomics* 13 (3).

Züllig, T., Trötzmüller, M., and Köfeler, H.C. (2020). Lipidomics from sample preparation to data analysis: a primer. *Anal. Bioanal. Chem.* 412 (10): 2191–2209.

Wu, Z., Bagarolo, G.I., Thoröe-Boveleth, S., and Jankowski, J. (2020). "Lipidomics": mass spectrometric and chemometric analyses of lipids. *Adv. Drug Deliv. Rev.* 159: 294–307.

Züllig, T. and Köfeler, H.C. (2021). High resolution mass spectrometry in lipidomics. *Mass Spectrom. Rev.* 40 (3): 162–176.

Nath Roy, D., Goswami, R., and Pal, A. (2017). Nanomaterial and toxicity: what can proteomics tell us about the nanotoxicology? *Xenobiotica* 47 (7): 632–643.

Frohlich, E. (2017). Role of omics techniques in the toxicity testing of nanoparticles. *J. Nanobiotechnol.* 15 (1): 84.

Zhang, T., Gaffrey, M.J., Thrall, B.D., and Qian, W.J. (2018). Mass spectrometry–based proteomics for system-level characterization of biological responses to engineered nanomaterials. *Anal. Bioanal. Chem.* 410 (24): 6067–6077.

Wang, C., Chen, B., He, M., and Hu, B. (2021). Composition of intracellular protein corona around nanoparticles during internalization. *ACS Nano* 15 (2): 3108–3122.

Zhou, M., Tang, M., Li, S., Peng, L., Huang, H., Fang, Q., Liu, Z., Xie, P., Li, G., and Zhou, J. (2018). Effective lock-in strategy for proteomic analysis of corona complexes bound to amino-free ligands of gold nanoparticles. *Nanoscale* 10 (26): 12413–12423.

Pinals, R.L., Yang, D., Rosenberg, D.J., Chaudhary, T., Crothers, A.R., Iavarone, A.T., Hammel, M., and Landry, M.P. (2020). Quantitative protein corona composition and dynamics on carbon nanotubes in biological environments. *Angew. Chem. Int. Ed. Engl.* 59 (52): 23668–23677.

Chetwynd, A.J. and Lynch, I. (2020). The rise of the nanomaterial metabolite corona, and emergence of the complete corona. *Environ. Sci. Nano.* 7 (4): 1041–1060.

Chetwynd, A.J., Zhang, W., Thorn, J.A., Lynch, I., and Ramautar, R. (2020). The nanomaterial metabolite corona determined using a quantitative metabolomics approach: a pilot study. *Small* 16 (21): 2000295.

Fleury, J.-B., Werner, M., Guével, X.L., and Baulin, V.A. (2021). Protein corona modulates interaction of spiky nanoparticles with lipid bilayers. *J. Colloid. Interface Sci.* 603: 550–558.

Francia, V., Yang, K., Deville, S., Reker-Smit, C., Nelissen, I., and Salvati, A. (2019). Corona composition can affect the mechanisms cells use to internalize nanoparticles. *ACS Nano* 13 (10): 11107–11121.

Lee, H. (2021). Effect of protein corona on nanoparticle–lipid membrane binding: the binding strength and dynamics. *Langmuir* 37 (12): 3751–3760.

Lima, T., Bernfur, K., Vilanova, M., and Cedervall, T. (2020). Understanding the lipid and protein corona formation on different sized polymeric nanoparticles. *Sci. Rep.* 10 (1): 1129.

Zhang, T., Gaffrey, M.J., Qian, W.-J., and Thrall, B.D. (2020). Oxidative stress and redox modifications in nanomaterial–cellular interactions. In: *Interaction of Nanomaterials with the Immune System*, edn (ed. J.C. Bonner and J.M. Brown), 127–148. Cham: Springer International Publishing.

Niazi, J.H. and Gu, M.B. (2009). Toxicity of metallic nanoparticles in microorganisms – a review. In: *Atmospheric and Biological Environmental Monitoring*, edn. (ed. Y.J. Kim, U. Platt, M.B. Gu, and H. Iwahashi), 193–206. Dordrecht: Springer Netherlands.

Wang, H., Wang, M., Xu, X., Gao, P., Xu, Z., Zhang, Q., Li, H., Yan, A., Kao, R.Y., and Sun, H. (2021). Multi-target mode of action of silver against *Staphylococcus aureus* endows it with capability to combat antibiotic resistance. *Nat. Commun.* 12 (1): 3331.

Salas-Orozco, M.F., Niño-Martínez, N., Martínez-Castañón, G.A., Méndez, F.T., Morán, G.M.M., Bendaña-Piñeiro, A.E., Ruiz, F., and Bach, H. (2022). Proteomic analysis of an *Enterococcus faecalis* mutant generated against the exposure to silver nanoparticles. *J. Appl. Microbiol.* 132 (1): 244–255.

Winans, M.J. and Gallagher, J.E.G. (2020). Metallomic and lipidomic analysis of *S. cerevisiae* response to cellulosic copper nanoparticles uncovers drivers of toxicity. *Metallomics* 12 (5): 799–812.

Rong-Mullins, X., Winans, M.J., Lee, J.B., Lonergan, Z.R., Pilolli, V.A., Weatherly, L.M., Carmenzind, T.W., Jiang, L., Cumming, J.R., Oporto, G.S. et al. (2017). Proteomic and genetic analysis of the response of *S. cerevisiae* to soluble copper leads to improvement of the antimicrobial function of cellulosic copper nanoparticles. *Metallomics* 9 (9): 1304–1315.

Ouyang, S., Hu, X., and Zhou, Q. (2015). Envelopment–internalization synergistic effects and metabolic mechanisms of graphene oxide on single-cell *Chlorella vulgaris* are dependent on the nanomaterial particle size. *ACS Appl. Mater. Interfaces* 7 (32): 18104–18112.

Liu, W., Majumdar, S., Li, W., Keller, A.A., and Slaveykova, V.I. (2020). Metabolomics for early detection of stress in freshwater alga *Poterioochromonas malhamensis* exposed to silver nanoparticles. *Sci. Rep.* 10 (1): 20563.

Mortimer, M., Wang, Y., and Holden, P.A. (2021). Molecular mechanisms of nanomaterial–bacterial interactions revealed by omics – the role of nanomaterial effect level. *Front. Bioeng. Biotechnol.* 9: 683520.

Zhao, X., Abulikemu, A., Lv, S., Qi, Y., Duan, J., Zhang, J., Chen, R., Guo, C., Li, Y., and Sun, Z. (2021). Oxidative stress- and mitochondrial dysfunction-mediated cytotoxicity by silica nanoparticle in lung epithelial cells from metabolomic perspective. *Chemosphere* 275: 129969.

Ruan, C., Wang, C., Gong, X., Zhang, Y., Deng, W., Zhou, J., Huang, D., Wang, Z., Zhang, Q., Guo, A. et al. (2021). An integrative multi-omics approach uncovers the regulatory role of CDK7 and CDK4 in autophagy activation induced by silica nanoparticles. *Autophagy* 17 (6): 1426–1447.

Lorscheidt, S., Shetab Boushehri, M.A., Klaschik, S., and Lamprecht, A. (2019). Sub-cytotoxic doses of pharmaceutical silica nanoparticles show significant impact on the proteome of HepG2 cells. *J. Control. Release* 306: 1–14.

Lee, S.Y., Kim, I.Y., Heo, M.B., Moon, J.H., Son, J.G., and Lee, T.G. (2021). Global proteomics to study silica nanoparticle-induced cytotoxicity and its mechanisms in HepG2 cells. *Biomolecules* 11 (3).

Xu, M., Yang, Q., Xu, L., Rao, Z., Cao, D., Gao, M., and Liu, S. (2019). Protein target identification and toxicological mechanism investigation of silver nanoparticles-induced hepatotoxicity by integrating proteomic and metallomic strategies. *Part Fibre Toxicol.* 16 (1): 46.

Kitchin, K.T., Grulke, E., Robinette, B.L., and Castellon, B.T. (2014). Metabolomic effects in HepG2 cells exposed to four TiO_2 and two CeO_2 nanomaterials. *Environ. Sci. Nano* 1 (5): 466–477.

Kitchin, K.T., Stirdivant, S., Robinette, B.L., Castellon, B.T., and Liang, X. (2017). Metabolomic effects of CeO(2), SiO(2) and CuO metal oxide nanomaterials on HepG2 cells. *Part Fibre Toxicol.* 14 (1): 50.

Adeyemi, J.A., Sorgi, C.A., Machado, A.R.T., Ogunjimi, A.T., Gardinassi, L.G.A., Nardini, V., Faccioli, L.H., Antunes, L.M.G., and Barbosa, F. (2020). Phospholipids modifications in human hepatoma cell lines (HepG2) exposed to silver and iron oxide nanoparticles. *Arch. Toxicol.* 94 (8): 2625–2636.

Thingholm, T.E., Tjd, J., Jensen, O.N., and Larsen, M.R. (2006). Highly selective enrichment of phosphorylated peptides using titanium dioxide. *Nat. Protoc.* 1 (4): 1929–1935.

Chen, Q., Wang, N., Zhu, M., Lu, J., Zhong, H., Xue, X., Guo, S., Li, M., Wei, X., Tao, Y. et al. (2018). TiO(2) nanoparticles cause mitochondrial dysfunction, activate inflammatory responses, and attenuate phagocytosis in macrophages: a proteomic and metabolomic insight. *Redox Biol.* 15: :266–276.

Shao, Y., Wang, L., Chen, J., Hunag, Y., Huang, Y., Wang, X., Zhou, D., Zhang, J., Wu, W., Zhang, Q. et al. (2021). Increased lipid peroxidation by graphene quantum dots induces ferroptosis in macrophages. *NanoImpact* 23: 100334.

Zhang, T., Gaffrey, M.J., Thomas, D.G., Weber, T.J., Hess, B.M., Weitz, K.K., Piehowski, P.D., Petyuk, V.A., Moore, R.J., W-j, Q. et al. (2020). A proteome-wide assessment of the oxidative stress paradigm for metal and metal-oxide nanomaterials in human macrophages. *NanoImpact* 17: 100194.

Duan, J., Zhang, T., Gaffrey, M.J., Weitz, K.K., Moore, R.J., Li, X., Xian, M., Thrall, B.D., and Qian, W.J. (2020). Stochiometric quantification of the thiol redox proteome of macrophages reveals subcellular compartmentalization and susceptibility to oxidative perturbations. *Redox Biol.* 36: 101649.

Miranda, R.R., Oliveira, A.C.S., Skytte, L., Rasmussen, K.L., and Kjeldsen, F.: Proteome-wide analysis reveals molecular pathways affected by AgNP in a ROS-dependent manner. (1743-5404 (Electronic)).

Srivastava, A.K., Yadav, S.S., Mishra, S., Yadav, S.K., Parmar, D., and Yadav, S. (2020). A combined microRNA and proteome profiling to investigate the effect of ZnO nanoparticles on neuronal cells. *Nanotoxicology* 14 (6): 757–773.

Bramini, M., Sacchetti, S., Armirotti, A., Rocchi, A., Vázquez, E., León Castellanos, V., Bandiera, T., Cesca, F., and Benfenati, F. (2016). Graphene oxide nanosheets disrupt lipid composition, Ca2+ homeostasis, and synaptic transmission in primary cortical neurons. *ACS Nano* 10 (7): 7154–7171.

Bramini, M., Chiacchiaretta, M., Armirotti, A., Rocchi, A., Kale, D.D., Martin, C., Vázquez, E., Bandiera, T., Ferroni, S., Cesca, F. et al. (2019). An increase in membrane cholesterol by graphene oxide disrupts calcium homeostasis in primary astrocytes. *Small* 15 (15): 1900147.

Liu, X., Yang, C., Chen, P., Zhang, L., and Cao, Y. (2022). The uses of transcriptomics and lipidomics indicated that direct contact with graphene oxide altered lipid homeostasis through ER stress in 3D human brain organoids. *Sci. Total Environ.* 849: 157815.

Rosário, F., Duarte, I.F., Pinto, R.J.B., Santos, C., Hoet, P.H.M., and Oliveira, H. (2021). Biodistribution and pulmonary metabolic effects of silver nanoparticles in mice following acute intratracheal instillations. *Environ. Sci. Pollut. Res.* 28 (2): 2301–2314.

Guo, C., Weber, R.J.M., Buckley, A., Mazzolini, J., Robertson, S., Delgado-Saborit, J.M., Rappoport, J.Z., Warren, J., Hodgson, A., Sanderson, P. et al. (2021). Environmentally relevant iron oxide nanoparticles produce limited acute pulmonary effects in rats at realistic exposure levels. *Int. J. Mol. Sci.* 22 (2).

Tyurina, Y.Y., Kisin, E.R., Murray, A., Tyurin, V.A., Kapralova, V.I., Sparvero, L.J., Amoscato, A.A., Samhan-Arias, A.K., Swedin, L., Lahesmaa, R. et al. (2011). Global phospholipidomics analysis reveals selective pulmonary peroxidation profiles upon inhalation of single-walled carbon nanotubes. *ACS Nano* 5 (9): 7342–7353.

Guo, C., Robertson, S., Weber, R.J.M., Buckley, A., Warren, J., Hodgson, A., Rappoport, J.Z., Ignatyev, K., Meldrum, K., Römer, I. et al. (2019). Pulmonary toxicity of inhaled nano-sized cerium oxide aerosols in Sprague-Dawley rats. *Nanotoxicology* 13 (6): 733–750.

Xia, L., Alqahtani, S., Ferreira, C.R., Aryal, U.K., Biggs, K., and Shannahan, J.H. (2022). Modulation of pulmonary toxicity in metabolic syndrome due to variations in iron oxide nanoparticle-biocorona composition. *Nanomater. (Basel)* 12 (12).

Stachyra, K., Kiepura, A., Suski, M., Ulatowska-Białas, M., Kuś, K., Wiśniewska, A., Czepiel, K., Majka, G., and Olszanecki, R. (2022). Changes in the liver proteome in apoE knockout mice exposed to inhalation of silica nanoparticles indicate mitochondrial damage and impairment of ER stress responses associated with microvesicular steatosis. *Environ. Sci. Pollut. Res. Epub ahead of print.*

Li, J., Sun, R., Xu, H., and Wang, G. (2022). Integrative metabolomics, proteomics and transcriptomics analysis reveals liver toxicity of mesoporous silica nanoparticles. *Front Pharmacol.* 13: 835359.

Chen, Z., Han, S., Zheng, P., Zhou, D., Zhou, S., and Jia, G. (2020). Effect of oral exposure to titanium dioxide nanoparticles on lipid metabolism in Sprague-Dawley rats. *Nanoscale* 12 (10): 5973–5986.

Chen, Z., Han, S., Zheng, P., Zhang, J., Zhou, S., and Jia, G. (2022). Landscape of lipidomic metabolites in gut-liver axis of Sprague-Dawley rats after oral exposure to titanium dioxide nanoparticles. *Part Fibre Toxicol.* 19 (1): 53.

Chen, Z., Han, S., Zhou, D., Zhou, S., and Jia, G. (2019). Effects of oral exposure to titanium dioxide nanoparticles on gut microbiota and gut-associated metabolism in vivo. *Nanoscale* 11 (46): 22398–22412.

Yan, J., Chen, Q., Tian, L., Li, K., Lai, W., Bian, L., Han, J., Jia, R., Liu, X., and Xi, Z. (2022). Intestinal toxicity of micro- and nano-particles of foodborne titanium dioxide in juvenile mice: disorders of gut microbiota–host co-metabolites and intestinal barrier damage. *Sci. Total Environ.* 821: 153279.

Landsiedel, R., Hahn, D., Ossig, R., Ritz, S., Sauer, L., Buesen, R., Rehm, S., Wohlleben, W., Groeters, S., Strauss, V. et al. (2022). Gut microbiome and plasma metabolome changes in rats after oral gavage of nanoparticles: sensitive indicators of possible adverse health effects. *Part Fibre Toxicol.* 19 (1): 21.

Fennell, T.R., Mortensen, N.P., Black, S.R., Snyder, R.W., Levine, K.E., Poitras, E., Harrington, J.M., Wingard, C.J., Holland, N.A., Pathmasiri, W. et al, (2017). Disposition of intravenously or orally administered silver nanoparticles in pregnant rats and the effect on the biochemical profile in urine. *J. Appl. Toxicol.* 37 (5): 530–544.

Cai, J., Zang, X., Wu, Z., Liu, J., and Wang, D. (2021). Altered protein S-glutathionylation depicts redox imbalance triggered by transition metal oxide nanoparticles in a breastfeeding system. *NanoImpact* 22: 100305.

Sulukan, E., Şenol, O., Baran, A., Kankaynar, M., Yıldırım, S., Kızıltan, T., Bolat, İ., and Ceyhun, S.B. (2022). Nano-sized polystyrene plastic particles affect many cancer-related biological processes even in the next generations; zebrafish modeling. *Sci. Total Environ.* 838 (Pt 3): 156391.

Marana, M.H., Poulsen, R., Thormar, E.A., Clausen, C.G., Thit, A., Mathiessen, H., Jaafar, R., Korbut, R., Hansen, A.M.B., Hansen, M. et al. (2022). Plastic nanoparticles cause mild inflammation, disrupt metabolic pathways, change the gut microbiota and affect reproduction in zebrafish: a full generation multi-omics study. *J. Hazard. Mater.* 424: 127705.

Cappello, T., Vitale, V., Oliva, S., Villari, V., Mauceri, A., Fasulo, S., and Maisano, M. (2017). Alteration of neurotransmission and skeletogenesis in sea urchin *Arbacia lixula* embryos exposed to copper oxide nanoparticles. *Comp. Biochem. Physiol. C Toxicol. Pharmacol.* 199: 20–27.

Lu, X., Tian, Y., Zhao, Q., Jin, T., Xiao, S., and Fan, X. (2010). Integrated metabonomics analysis of the size–response relationship of silica nanoparticles-induced toxicity in mice. *Nanotechnology* 22 (5): 055101.

Lu, X., Ji, C., Jin, T., and Fan, X. (2015). The effects of size and surface modification of amorphous silica particles on biodistribution and liver metabolism in mice. *Nanotechnology* 26 (17): 175101.

Lu, X., Jin, T., Jin, Y., Wu, L., Hu, B., Tian, Y., and Fan, X. (2013). Toxicogenomic analysis of the particle dose– and size–response relationship of silica particles–induced toxicity in mice. *Nanotechnology* 24 (1): 015106.

Li, J., Yuan, Z., Liu, H., Feng, J., and Chen, Z. (2019). Size-dependent tissue-specific biological effects of core-shell structured Fe(3)O(4)@SiO(2)-NH(2) nanoparticles. *J. Nanobiotechnol.* 17 (1): 124.

Tsukanov, A.A. and Vasiljeva, O. (2021). Nanomaterials interaction with cell membranes: computer simulation studies. In: *Multiscale Biomechanics and Tribology of Inorganic and Organic Systems: In Memory of Professor Sergey Psakhie*, edn. (ed. G.-P. Ostermeyer, V.L. Popov, E.V. Shilko, and O.S. Vasiljeva), 189–210. Cham: Springer International Publishing.

Foreman-Ortiz, I.U., Liang, D., Laudadio, E.D., Calderin, J.D., Wu, M., Keshri, P., Zhang, X., Schwartz, M.P., Hamers, R.J., Rotello, V.M. et al. (2020). Anionic nanoparticle-induced perturbation to phospholipid membranes affects ion channel function. *Proc. Natl. Acad. Sci.* 117 (45): 27854–27861.

Ju, Y., Kelly, H.G., Dagley, L.F., Reynaldi, A., Schlub, T.E., Spall, S.K., Bell, C.A., Cui, J., Mitchell, A.J., Lin, Z. et al. (2020). Person-specific biomolecular coronas modulate nanoparticle interactions with immune cells in human blood. *ACS Nano* 14 (11): 15723–15737.

Duan, J., Kodali, V.K., Gaffrey, M.J., Guo, J., Chu, R.K., Camp, D.G., Smith, R.D., Thrall, B.D., and Qian, W.J. (2016). Quantitative profiling of protein S-glutathionylation reveals redox-dependent regulation of macrophage function during nanoparticle-induced oxidative stress. *ACS Nano* 10 (1): 524–538.

Holmila, R., Wu, H., Lee, J., Tsang, A.W., Singh, R., and Furdui, C.M. (2021). Integrated redox proteomic analysis highlights new mechanisms of sensitivity to silver nanoparticles. *Mol. Cell Proteomics* 20: 100073.

Nabi, S.U., Ali, S.I., Rather, M.A., Sheikh, W.M., Altaf, M., Singh, H., Mumtaz, P.T., Mishra, N.C., Nazir, S.U., and Bashir, S.M. (2022). Organoids: a new approach in toxicity testing of nanotherapeutics. *J. Appl. Toxicol.* 42 (1): 52–72.

Ghebretatios, M., Schaly, S., and Prakash, S. (2021). Nanoparticles in the food industry and their impact on human gut microbiome and diseases. *Int. J. Mol. Sci.* 22 (4).

Yu, R., Ahmed, T., Jiang, H., Zhou, G., Zhang, M., Lv, L., and Li, B. (2021). Impact of zinc oxide nanoparticles on the composition of gut microbiota in healthy and autism spectrum disorder children. *Materials (Basel)* 14 (19).

Utembe, W., Tlotleng, N., and Kamng'ona, A.W. (2022). A systematic review on the effects of nanomaterials on gut microbiota. *Curr. Res. Microbial. Sci.* 3: 100118.

Matysiak, M., Kapka-Skrzypczak, L., Brzoska, K., Gutleb, A.C., and Kruszewski, M. (2016). Proteomic approach to nanotoxicity. *J. Proteomics* 137: 35–44.

Garikapati, V., Colasante, C., Baumgart-Vogt, E., and Spengler, B. (2022). Sequential lipidomic, metabolomic, and proteomic analyses of serum, liver, and heart tissue specimens from peroxisomal biogenesis factor 11α knockout mice. *Anal. Bioanal. Chem.* 414 (6): 2235–2250.

Puris, E., Kouřil, Š., Najdekr, L., Auriola, S., Loppi, S., Korhonen, P., Gómez-Budia, M., Fricker, G., Kanninen, K.M., Malm, T. et al. (2022). Metabolomic, lipidomic and proteomic characterisation of lipopolysaccharide-induced inflammation mouse model. *Neuroscience* 496: 165–178.

Cao, X., Pan, X., Couvillion, S.P., Zhang, T., Tamez, C., Bramer, L.M., White, J.C., Qian, W.-J., Thrall, B.D., Ng, K.W. et al. (2021). Fate, cytotoxicity and cellular metabolomic impact of ingested nanoscale carbon dots using simulated digestion and a triculture small intestinal epithelial model. *NanoImpact* 23: 100349.

Wang, R., Li, B., Lam, S.M., and Shui, G. (2020). Integration of lipidomics and metabolomics for in-depth understanding of cellular mechanism and disease progression. *J. Genet. Genomics.* 47 (2): 69–83.

Nakayasu, E.S., Nicora, C.D., Sims, A.C., Burnum-Johnson, K.E., Kim, Y.-M., Kyle, J.E., Matzke, M.M., Shukla, A.K., Chu, R.K., Schepmoes, A.A. et al. (2016). MPLEx: a robust and universal protocol for single-sample integrative proteomic, metabolomic, and lipidomic analyses. *mSystems* 1 (3): e00043–00016.

Kang, J., David, L., Li, Y., Cang, J., and Chen, S. (2021). Three-in-one simultaneous extraction of proteins, metabolites and lipids for multi-omics. *Front Genet.* 12: 635971.

Picard, M., Scott-Boyer, M.-P., Bodein, A., Périn, O., and Droit, A. (2021). Integration strategies of multi-omics data for machine learning analysis. *Comput. Struct. Biotechnol. J.* 19: :3735–3746.

Dugourd, A., Kuppe, C., Sciacovelli, M., Gjerga, E., Gabor, A., Emdal, K.B., Vieira, V., Bekker-Jensen, D.B., Kranz, J., Bindels, E.M.J. et al. (2021). Causal integration of multi-omics data with prior knowledge to generate mechanistic hypotheses. *Mol. Syst. Biol.* 17 (1): e9730.

Planell, N., Lagani, V., Sebastian-Leon, P., van der Kloet, F., Ewing, E., Karathanasis, N., Urdangarin, A., Arozarena, I., Jagodic, M., Tsamardinos, I. et al. (2021). STATegra: multi-omics data integration – a conceptual scheme with a bioinformatics pipeline. *Front Genet.* 12.

16

Current Scenario and Future Trends of Plant Nano-Interaction to Mitigate Abiotic Stresses: A Review

Farhat Yasmeen[1], *Ghazala Mustafa*[2], *Hafiz Muhammad Jhanzab*[3], *and Setsuko Komatsu*[4]

[1] *Department of Biosciences, University of Wah, Wah Cantt, 47040, Pakistan*
[2] *Department of Plant Sciences, Quaid-i-Azam University, Islamabad 45320, Pakistan*
[3] *Department of Agronomy, The University of Agriculture Dera Ismail Khan, Khyber Pakhtunkhwa, 29111, Pakistan*
[4] *Faculty of Environmental and Information Sciences, Fukui University of Technology, Fukui 910–8505, Japan*

Abbreviations

Aluminum oxide Al_2O_3
Carbon C
Catalase CAT
Cadmium Cd
Copper oxide CuO
Gold Au
Iron Fe
Nanoparticles NPs
Peroxidase POD
Reactive oxygen species ROS
Silver Ag
Superoxide dismutase SOD
Silicon Si
Titanium oxide TiO_2
Zinc oxide ZnO

Introduction

Engineered nanoparticles (NPs) are believed to have potential applications as insecticides, growth enhancers, fertilizers, and pesticides in agriculture (Landa 2021). Nanotechnology triggered the demand for meticulous research on ultrafine nanomaterials, which improve the productivity and quality of agricultural goods (Shang et al. 2019). The quantity of fabricated nanomaterials that eventually penetrate living systems has increased significantly due to nanotechnology's rapid growth and widespread use (Aslani et al. 2014). Nanotechnology emerges as an enthusiastically evolving field, with numerous applications in materials, energy, manufacturing, and medical diagnosis (Neto 2014; Safari and Zarnegar 2014). There are about 800 different types of nanomaterial goods on the market, which are the by-products of nanotechnology generated with NPs (Dutschk et al. 2014: Zhou et al. 2014). Nanotechnology holds good promises for solving the issues related to abiotic stresses ensuring

Impact of Engineered Nanomaterials in Genomics and Epigenomics, First Edition. Edited by Saura C. Sahu.

sustainable agriculture. Therefore, innovative and resource conservation technologies are necessary to fulfill the future demands of increasing population.

NPs have considerable impacts on the synthesis of proteins, carbohydrates, fats, and vitamins. They raise the chlorophyll content in leaves, improving the process of photosynthesis, which indicates that metal NPs can neutralize various toxic substances in the soil (El-Baz et al. 2015; Jiang et al. 2015). Plants, on exposure to NPs, become more resistant to adverse conditions like flooding, drought, disease, and pests (Mustafa and Komatsu 2016b). Studies revealed that NPs significantly stimulate the germination and growth of plant at early stages of ontogenesis (Barabanov et al. 2018; Krishnaraj et al. 2012). Booming agricultural production through the application of novel NPs for sustainability, depicts potential output by promotion of plant growth and development through nano-fertilizers, nano-herbicides, and nano-pesticides as well as inducing stress resilience (Manjunatha et al. 2019; Saxena et al. 2016). Efficient application of nutrients can only be achieved through nano-encapsulated fertilizer to reach sustainable production and significant reduction in soil toxicity and leaching losses (Li et al. 2016a). However, NPs with a positive influence on the biological system should be explored to boost plant growth.

Technological advancement in the agriculture sector is a prerequisite for improving crop production. Biotic and abiotic stresses include pathogen, virus, fungi, salinity, drought, flooding, water logging, and metal toxicity/inadequacies, which limit crop growth and productivity (Tripathi et al. 2012, 2015). Foliar application of zinc oxide (ZnO) NPs promoted photosynthetic function along with activity of enzymes to mitigate the negative effects of stress. Plant growth increased as the efficiency of the photosystem and stress tolerance increased (Alabdallah and Alzahrani 2020). Various antioxidant enzymes like peroxidase (POD), catalase (CAT), and superoxide dismutase (SOD) enhanced their activity on exposure to NPs (Laware and Raskar 2014). Therefore, NPs are proven to boost plant vigor and development, leading to enhancement in yield ability and stress tolerance. Hence, more investigation is needed to explore the strategies to induce stress tolerance in plants through the development of new nano-based materials.

Synthesis of Nanoparticles

In nanotechnology, NPs synthesis is promising, which supplies many solutions to the agriculture sector (Bahrulolum et al. 2021). During the process of synthesis, mechanical, electrical, and physiochemical properties are controlled to produce finer particles (Khan et al. 2020). Bulk material and NPs are differentiated based on shape and size that are controlled through pH, temperature, and concentration (Maynard et al. 2006; Wiesner et al. 2006). NPs could be synthesized through physical, chemical, and biological methods (Figure 16.1). NPs are synthesized using two methods: the top-down and bottom-up approaches. In the bottom-up approach, size reduction was conducted through physical and chemical methods (Mourdikoudis et al. 2018) while the bottom-up approach involves the reduction/oxidation process (Gawande et al. 2015). Moreover, plant extracts are also used for NPs synthesis due to the presence of reducing agents, such as terpenoids, phenols, carbohydrates, and alkaloids that stabilize NPs (Ovais et al. 2018). Efficient method for NPs synthesis should be followed to obtain finer particles for inducing growth improvement and tolerance in plants (Table 16.1).

Silver Nanoparticles

Silver (Ag) NPs have diverse applications in several industries, like catalysis, electronics, optics, environment, and biotechnology (Abbasi et al. 2016). One of the most favored strategies for NPs synthesis is reduction through chemicals with the help of organic and inorganic reducing agents

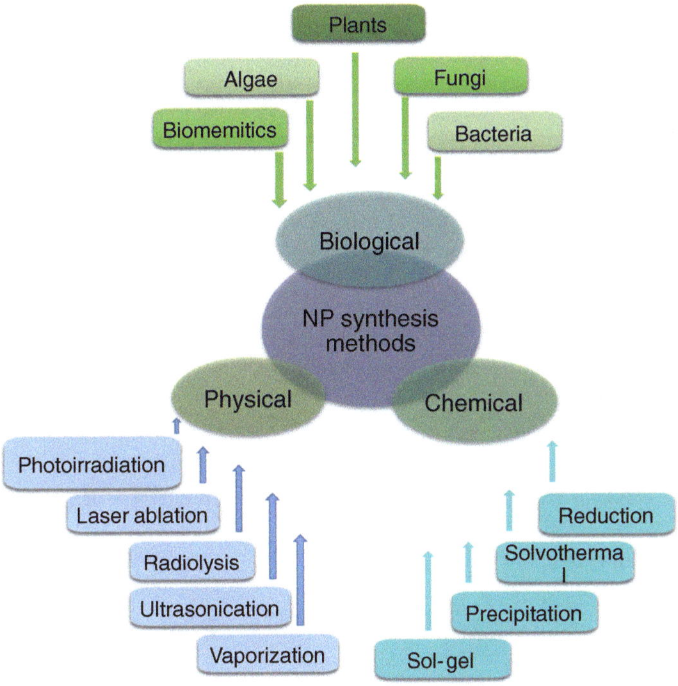

Figure 16.1 Nanoparticles (NPs) synthesis approaches: the figure summarizes various methods that can be used for NPs synthesis. These methods are based on either of the three main approaches i.e., biological, chemical, or physical.

Table 16.1 List of nanoparticles synthesized through biological and physicochemical processes.

Nanoparticle	Phase of synthesis	Method of synthesis	Source	Size (nm)	References
Ag	Solid	Biological	Fungi, *Fusarium oxysporum*	5–15	Ahmad et al. (2003)
	Solid	Biological	Fungi, *Aspergillus fumigatus*	5–25	Bhainsa and D'Souza (2006)
	Solid	Biological	Algae, *Portieria hornemannii*	25–30	Fatima et al. (2021)
	Solid	Biological	Algae, *Gracilaria birdiae*	27	Passos de Aragão et al. (2019)
	Solid	Biological	Plant, *Parkia speciosa* leaves	31–35	Ravichandran et al. (2019)
	Solid	Biological	Plant, *Eclipta alba* leaves	30–35	Premasudha et al. (2015)
	Solid	Biological	Plant, *Artocarpus heterophyllus* fruit latex	15	Sundararajan et al. (2021)
	Solid	Biological	Plant *Protium serratum* leaves	75	Mohanta et al. (2016)
	Solid	Biological	Plant, *Passiflora caerulea* leaves	16	Santhoshkumar et al. (2019)
	Solid	Biological	Plant, *Semecarpus anacardium* leaves	63	Kishore et al. (2020)
	Solid	Biological	Plant, *Camellia sinensis* leaves	35	Goswami et al. (2015)

(Continued)

Table 16.1 (Continued)

Nanoparticle	Phase of synthesis	Method of synthesis	Source	Size (nm)	References
	Solid	Biological	Plant, *Acalypha wilkesiana* leaves	20	Ahmed et al. (2021)
	Solid	Biological	Plant, *Atalantia monophylla* leaves	35	Mahadevan et al. (2017)
	Solid	Biological	Plant, *Lippia citriodora* leaves	24	Elemike et al. (2017)
	Solid	Biological	Plant, *Psidium guajava* leaves	25	Wang et al. (2018)
	Solid	Biological	Plant, *Abelmoschus esculentus* leaves	9	Elmusa et al. (2021)
	Solid	Biological	Plant, *Rosa santana* petals	15	Jahan et al. (2021)
	Solid	Biological	Plant, *Parthenium hysterophorus* leaves	25	Ahsan et al. (2020)
	Solid	Biological	Plant, *Origanum majorana* aerial parts	72	Erenler and Dag (2022)
	Solid	Biological	Plant, *Passiflora edulis f. flavicarpa* leaves	35	Thomas et al. (2019)
	Solid	Biological	Plant, *Abelmoschus esculentus* pulp	21	Mollick et al. (2019)
	Solid	Biological	Plant, *Lytraria acaulis* leaves	100	Aswini et al. (2021)
	Solid	Biological	Plant, *Cannabis sativa* stem	30	Singh et al. (2018)
	Solid	Biological	Plant, *Madhuca longifolia* flower	40	Patil et al. (2018)
	Solid	Biological	Plant, *Momordica charantia* peel	25	Amruta et al. (2019)
	Solid	Biological	Plant, *Camellia sinensis* stem waste	20	Cheng et al. (2020)
	Solid	Biological	Plant, *Ginkgo Biloba* leaves	13	Ren et al. (2016)
	Solid	Biological	Plant, *Quercus acerifolia* fruit hull	19	Sorbiun et al. (2018)
	Solid	Biological	Plant, *Asparagus officinalis* root	16	Tripathi et al. (2019)
	Solid	Biological	Plant, *Camellia sinensis* tea extract	25	Rolim et al. (2019)
Al_2O_3	Solid	Biological	Plant, *Cymbopogon citratus leaves*	35	Ansari et al. (2015)
	Solid	Physical	Hydrothermal treatment	20	Zhang et al. (2018)
CuO	Solid	Biological	Bacterial strains	30	John et al. (2021)
	Solid	Biological	Fungi, *Aspergillus niger* strain STA9	500	Noor et al. (2020)
	Solid	Biological	Plant, *Camilla japonica* leaves	25	Rajivgandhi et al. (2019)
	Solid	Biological	Plant, *Caesalpinia bonducella* seed	20	Sukumar et al. (2020)
	Solid	Biological	Plant, *Vaccinium myrtillus* waste	25	Benassai et al. (2021)
	Solid	Biological	Plant, *Moringa oleifera* leaves	130	Bai et al. (2022)
	Solid	Biological	Plant, *Punica granatum* fruit rind	55	Padma et al. (2018)
	Solid	Biological	Plant, *Aloe Barbadensis* leaves	6	Sharma et al. (2021)

Table 16.1 (Continued)

Nanoparticle	Phase of synthesis	Method of synthesis	Source	Size (nm)	References
	Solid	Biological	Plant, *Millettia pinnata* flower	23	Thiruvengadam et al. (2019)
	Solid	Biological	Plant, *Tinospora cardifolia* leaves	63	Sharma et al. (2019)
	Solid	Physical	Microwave-assisted approach	200	Długosz et al. (2021)
	Solid	Physical	Microwave-assisted approach	90	Henam et al. (2019)
	Liquid	Chemical	Reduction	45	Kruk et al. (2015)
FeO	Solid	Biological	Bacteria, D-4 strain of *Microbacterium marinilacus*	*25*	Mukherjee (2017)
	Solid	Biological	Bacteria, Strain LS4, genus *Desulfovibrio*	*18*	Das et al. (2018)
	Solid	Biological	Fungi, *Pleurotus spp.*	*200*	Mazumdar and Haloi (2011b)
	Solid	Biological	Fungi, *Aspergillus niger*	50	Abdeen et al. (2016)
	Solid	Biological	Fungi, *Alternaria alternata*	*75*	Sarkar et al. (2017)
	Solid	Biological	Fungi, *Saccharomyces cerevisiae*	10	Mehrotra et al. (2017)
	Solid	Biological	Plant, *Camellia sinensis* tea extract	45	Lin et al. (2020)
	Solid	Biological	Plant, *Moringa oleifera* leaves	35	Silveira et al. (2017)
	Solid	Biological	Plant, *Parthenium hysteropherous* leaves	45	Rawat et al. (2021)
	Solid	Biological	Plant, *Camellia sinensis* waste	99	Gautam et al. (2018)
	Solid	Biological	Plant, *Piper nigrum* seed	25	Khaghani et al. (2017)
	Solid	Biological	Plant, *Rosa indica* flower	25	Afsheen et al. (2018)
	Solid	Biological	Plant, *Lantana camara* leaves	60	Nithya et al. (2021)
	Solid	Biological	Plant, *Punica granatum* seed	55	Bibi et al. (2019)
	Solid	Biological	Plant, *Magnifera indica* peel	35	Desalegn et al. (2019)
	Solid	Biological	Plant, *Cynara cardunculus* leaves	13	Ruíz-Baltazar et al. (2019)
	Solid	Biological	Plant, *Tamarix aphylla* stem	25	Ahmad et al. (2020a)
	Solid	Biological	Plant, *Moringa oleifera* leaves	20	Aisida et al. (2021)
	Solid	Biological	Plant, *Rhamnella gilgitica* leaves	21	Iqbal et al. (2020)
	Solid	Biological	Plant, *Terminalia bellirica* fruit	21	Jagadeesan et al. (2019)
	Solid	Biological	Plant, *Stevia rebaudiana* leaves	20	Khatami et al. (2019)
	Solid	Biological	Plant, *Centella asiatica* whole plant	30	Poka et al. (2019)
	Solid	Biological	Plant, *Cocos nucifera* husk	35	Sebastian et al. (2018)
	Solid	Biological	Plant, *Avecinnia marina* flower	45	Karpagavinayagam and Vedhi (2019)
	Solid	Biological	Plant, *Camellia sinensis* leaves	5	Liu et al. (2019)
	Solid	Biological	Plant, *Calliandra haematocephala* leaves	85	Sirdeshpande et al. (2018)

(Continued)

Table 16.1 (Continued)

Nanoparticle	Phase of synthesis	Method of synthesis	Source	Size (nm)	References
	Solid	Biological	Plant, *Trigonella foenum-graecum* seed	14	Radini et al. (2018)
	Solid	Biological	Plant, *Rheum emodi* root	30	Sharma et al. (2020)
	Solid	Biological	Plant, *Citrus maxima* peel	100	Wei et al. (2017)
	Solid	Biological	Plant, *Ruellia tuberosa* leaves	80	Vasantharaj et al. (2019)
	Solid	Biological	Plant, *Lawsonia inermis* whole plant	150	Chauhan and Upadhyay (2019)
	Solid	Biological	Plant, *Moringa oleifera* leaves	50	Katata-Seru et al. (2018)
	Solid	Biological	Plant, *Upressus sempervirens* leaves	19	Ebrahiminezhad et al. (2017)
	Solid	Chemical	Microwave assisted	45	Williams et al. (2016)
	Solid	Chemical	Microwave assisted	9	Lastovina et al. (2017)
	Solid	Chemical	Microwave assisted	25	Gonzalez-Moragas et al., (2015)
	Solid	Chemical	Microwave assisted	35	Mahmoud et al. (2016)
	Solid	Chemical	Ball milling	10	Pithawala and Deevi (2004)
	Liquid	Chemical	Microemulsion	30	Huber (2005)
	Liquid	Chemical	Hydrothermal	20	Edward et al. (2011)
	Vapour	Chemical	Vapor condensation	35	Tavakoli et al. (2007)
C	Solid	Biological	Animal, bovine serum albumin	3.8	Pichardo-Molina et al. (2020)
	Solid	Chemical	Vapor deposition	80	Hlekelele et al. (2016)
	Solid	Physical	Sol–gel	15	Tripathi et al. (2016)
	Solid	Chemical	Microwave assisted	5	Zhou et al. (2007)
	Solid	Chemical	Microwave assisted	20	Tripathi et al. (2017)
	Solid	Chemical	Microwave assisted	35	Hou et al. (2015)
	Solid	Chemical	Microwave assisted	51	Liu et al. (2016b)
	Solid	Chemical	Oxidation	85	Ray et al. (2009)
	Solid	Chemical	Oxidation	70	Peng and Sejdic (2009)
	Solid	Physical	Laser ablation	65	Yu et al. (2015)
	Solid	Physical	Laser ablation	55	Nguyen et al. (2015)
	Solid	Physical	Ultrasonic treatment	45	Wang et al. (2015)
	Solid	Physical	Ultrasonic treatment	25	Dang et al. (2016)
Au	Solid	Biological	Bacteria, *Bacillus subtilis*	20	He et al. (2006)
	Solid	Biological	Bacteria, *Pseudoalteromonas* Sp. Bac178	25	Patil et al. (2019)
	Solid	Biological	Fungi, *Cladosporium cladosporioides*	60	Joshi et al. (2017)

Table 16.1 (Continued)

Nanoparticle	Phase of synthesis	Method of synthesis	Source	Size (nm)	References
	Solid	Biological	Fungi, *Cladosporium oxysporum*	75	Bhargava et al. (2016)
	Solid	Biological	Fungi, *Fusarium oxysporum*	55	Naimi-Shamel et al. (2019)
	Solid	Biological	Plant, *Scutellaria barbata* whole plant	30	Wang et al. (2010)
	Solid	Biological	Plant, *Lonicera japonica* flower extract	40	Patil et al. (2019)
	Solid	Biological	Plant, Bo*mbyx mori* coccons	35	Ranjana et al. (2020)
	Solid	Biological	Plant, *Artemisia vulgaris* leaves	25	Sundararajan and Kumari (2017)
	Solid	Biological	Plant, *Bauhinia purpurea* leaves	45	Vijayan et al. (2019)
	Solid	Biological	Plant, *Sansevieria roxburghiana* leaves	65	Kumar et al. (2019)
Pd	Solid	Biological	Plant, *Aspalathus linearis* whole plant	22	Ismail et al. (2017)
	Solid	Biological	Plant, *Citrus sinensis* peel	4	Karim et al. (2019)
ZnO	Solid	Biological	Plant, *Melia azedarach* leaves	35	Dhandapani et al. (2020)
	Solid	Biological	Plant, *Syzygium cumini* leaves	25	Arumugam et al. (2021)
NiO	Solid	Biological	Plant, *Zea mays* Silk	20	Nwanya et al. (2020)
MgO	Solid	Chemical	Sol–gel	14	Fardood et al. (2018)
ZnO	Solid	Biological	Plant, *Quercus acerifolia* fruit hull	34	Sorbiun et al. (2018)
	Solid	Chemical	Sol–gel	10	Fardood et al. (2017)
SiO	Solid	Biological	Plant, *Oryza sativa* husk	85	Zarei et al. (2021)

(Iravani et al. 2014). Ag NPs could be synthesized chemically by reducing Ag nitrate with trisodium citrate dihydrate (Kulkarni 2007). Different reducing agents, such as ascorbate, sodium citrate, sodium borohydride, dimethylformamide, and elemental hydrogen are applied for the reduction of Ag ions in an aqueous environment to produce metallic Ag (Merga et al. 2007). Microemulsion technique is also used for the synthesis of a uniform and controllable size of Ag NPs (Krutyakov et al. 2008). However, the physicochemical methods of manufacturing Ag NPs results in the production of various detrimental organic solvents as by-products (Zhang et al. 2016). Ag NPs synthesized via the green approach using algae, fungi, actinomycetes, bacteria, and plants extend numerous benefits compared with physical and chemical synthesis approaches (Oves et al. 2018, 2019; Qayyum et al. 2017). Biosynthesis of Ag NPs is considered an eco-friendly, and low-cost process (Wypij et al. 2021). Ag NPs having a size range of 25 nm were used to induce green synthesis through the leaf extract of *Artemisia vulgaris* without adding external reducing/capping agent (Rasheed et al. 2017). Ag NPs were synthesized through the bio-reduction of Ag nitrate using leaf extracts of *Parkia speciosa* (Ravichandran et al. 2019) and *Origanum majorana* (Erenler and Dag 2022). It is suggested that green synthesis and organic reduction are cheaper methods for obtaining quality NPs.

Aluminum Oxide Nanoparticles

Aluminum oxide (Al_2O_3) NPs have gained greater importance due to their extensive applications in biomedical and antimicrobial fields (Galfetti et al. 2006). Al NPs can be synthesized through chemical and physical processes and the green method (Guo et al. 2004). Al NPs were synthesized using Al nitrate and lemon extract as they are eco-friendly and nontoxic. The salt solution and lemon extract were radiated using a microwave. Sodium citrate was used as a stabilizing agent that is environmentally friendly (Narayanan and Rakesh 2018). In the solution method, a combination of Al nitrate and citric acid was heated at high temperatures to form an amorphous gel that was then dried (Mohammed and Safwat. 2020). The Al_2O_3 NPs were prepared using leaf extracts of lemongrass and its antibacterial activities were also evaluated (Ansari et al. 2015). Al NPs are synthesized by adopting different methods; however, more eco-friendly strategies are preferred compared with the physical and chemical approaches.

Copper Nanoparticles

Among the metallic nanomaterials, copper (Cu) NPs are used in electronics, wind/hydraulic filtration, ceramics, wood preservatives, bioactive coverings, textile, makeups, and lubricants (Yang et al. 2009). Cu NPs are synthesized by the reduction of Cu sulfate pentahydrate with onion extract. The translucent yellowish green color indicated the conversion of Cu ions to Cu NPs (Yasmeen et al. 2018). A simplistic reduction approach in which sodium borohydride is used as the reducing agent and starch as the stabilizer leads to monodisperse Cu NPs in aqueous medium at ambient conditions (Suramwar et al. 2016). Cu NPs are synthesized by reducing an aqueous solution of Cu (II)-nitrilotriacetic acid complex with hydrazine at room temperature under air (Kamikoriyama et al. 2019). Cu NPs have gained greater importance due to their growth-promoting effects; however, new reducing materials should be explored for finer NPs synthesis.

Synthesis of metal NPs applying microorganisms is a clean, eco-friendly, and viable strategy that is complementary to physical and chemical methods. Bacterial strains isolated from *Antarctic consortium* was utilized for Cu NPs synthesis that resulted in spherical, monodispersed, and well-separated NPs with an average size of 30 nm (Noor et al. 2020). CuO NPs are biologically manufactured using leaf extracts of *Camilla japonica* with 15–25 nm size (Rajivgandhi et al. 2019) and *Annona muricata* with 30–40 nm size (Sukumar et al. 2020). Hierarchical CuO microspheres are prepared with leaf extract of *Moringa oleifera* using the green synthesis method (Kalaiyan et al. 2021). Cu NPs are biologically synthesized with the aqueous extract of the flower *Millettia pinnata* that resulted in extremely durable and spherical particles with an average size of 23 nm (Thiruvengadam et al. 2019). Microwave-aided approach is used for obtaining Cu (II) oxide and iron (Fe) (III) oxide NPs using leaf extract of *Euphorbia helioscopia*. NPs of Fe and Cu oxide are obtained in a spherical shape with sizes of 7–10 nm diameter (Henam et al. 2019). NPs with characteristic sizes can be prepared using different plant extracts that could be exploited for their possible applications in crop plants.

Iron Nanoparticles

Nowadays, a number of methods are used for the preparation of Fe nanomaterials, such as borohydride reduction (Liu et al. 2015), solution technique (Xu et al. 2007), electrochemical deposition (Ali et al. 2016), and ultrasonic treatment (Zhang et al. 2007). The particle sizes and morphologies of the nanomaterials synthesized by different methods may vary depending on the synthesis conditions (Wang et al. 2010). Fe NPs was synthesized by the chemical reduction of ferric chloride hexahydrate using onion extract through stepwise reduction, leading to the synthesis of reddish-brown solution (Jhanzab et al. 2022; Yasmeen et al. (2017a). Fe NPs are prepared through variable synthesis strategies using its different salts.

Synthesis of Fe NPs from anaerobic bacteria grown in anaerobic fluidized bed reactors showed amorphous distribution of NPs on the bioparticles surface (Romero-Cedillo et al. 2020). Magnetic Fe oxide NPs with 32–48 nm size were synthesized with the help of *Microbacterium marinilacus* strains (Mukherjee 2017). Moreover, magnetic and magnetite Fe NPs were synthesized using *Aspergillus niger* through fungi-based intracellular method (Abdeen et al. 2016). Different plant parts and leaf extracts, such as *Azardirachta indica* (Bolade et al. 2018; Devatha et al. 2018), *Camellia sinensis* (Zhu et al. 2018), *Syzygium aromaticum* (Afsheen et al. 2018), *Citrus maxima* (Li et al. 2018b), *Lantana camera* (Rajiv et al. 2017), *Eucalyptus globulus* (Weng et al. 2017), *Cymbopogon citratus* (Bolade et al. 2018), *Rosa damascene*, *Thymus vulgaris*, and *Urtica dioica* (Fazlzadeh et al. 2017), and *Canna indica* (Bolade et al. 2019) were employed as bio-reducing agents for NPs synthesis. Fe NPs are synthesized using *Citrus maxima* peels in aqueous solution through the reduction of Fe chloride (Wei et al. 2016). Fe NPs are also synthesized through fresh plant extracts of *Thymus vulgaris*, *Rosa damascene*, and *Urtica dioica*, and dried green extracts (Borja et al. 2015; Fazlzadeh et al. 2017). Plant extracts as reducing agents have the potential to yield high-quality particles that could be used for improving plant growth through stress tolerance.

Carbon Nanoparticles

Carbon (C) NPs are widely used in bioimaging (Tao et al. 2012), chemical sensing (Sharma et al. 2020), biomedical applications (Chen et al. 2017), and diagnostics (Liu et al. 2016a). Therefore, various physical, chemical, and biological techniques are used for the synthesis of CNPs, such as electrochemical oxidation (Zhou et al. 2014), ultrasonic treatment (Wang et al. 2015), hydrothermal synthesis (Yang et al. 2011), microwave-aided synthesis (Zhai et al. 2012), and thermal decomposition (Wang et al. 2016a). Fluorescent CNPs were synthesized through C soot oxidation of nitric acid (Ray et al. 2009). Fluorescent CNPs were synthesized through microwave irradiation of phosphoric acid with sucrose (Chandra et al. 2011). For the first time, CNPs are also obtained from commercial honey, resulting in surface-coated particles with size 7 nm (Wu et al. 2013). Fluorescent CNPs were synthesized through the hydrothermal process utilizing cheap wastes of pomelo peel as C source (Lu et al. 2012). C NPs having fluorescent properties are prepared through different strategies that might have potential applications in crop improvements.

CNPs are also prepared through the hydrothermal treatment of grape juice, devoid of salts, acids, bases, and organic solvents (Huang et al. 2014). Similarly, novel C quantum dots are synthesized from caffeic acid through the rapid hydrothermal method (Zhang et al. 2021). CNPs altered maize growth, enhancing, and reducing growth at low and at high concentrations, respectively. CNPs enhanced the plant physiological parameters leading to elevated plant growth (Zhao et al. 2021). Biological synthesis of NPs is a dire need for eco-friendly growth-promoting chemicals with no adverse effects.

Synthesis of Other Metal Nanoparticles

Due to their unique physical properties and technological applications, metal NPs are synthesized with different size, shape, chemical composition, and controlled monodispersity (Aromal and Philip 2012). The TiO_2 NPs of 40–60 nm are bio-synthesized through *Bacillus mycoides* (Órdenes-Aenishanslins et al. 2019), and *Bacillus subtilis* with a size range of 10–30 nm (Dhandapani et al. 2012). Spherical Au NPs are synthesized using the aqueous extract of *Macrotyloma uniflorum* with a size range of 14–17 nm (Aromal and Philip 2012). Cerium oxide NPs of 23 nm size were prepared using honey as a cheap material (Darroudi et al. 2014) and using *Gloriosa superba* (Arumugam et al. 2015). Size-controlled Cd sulfide NPs were successfully synthesized by immobilized *Rhodobacter sphaeroides* (Bai et al. 2009). Leaf extracts of *Hibiscus rosa-sinensis* was used in the synthesis of ZnO NPs (Devi and Gayathri 2014), and *Aloe vera* (Sangeetha et al. 2011) as well as through reproducible bacteria, *Aeromonas hydrophila* (Jayaseelan et al. 2012). Ruthenium NPs synthesis was done by applying an aqueous solution of *Gloriosa superba* with a size range of 25–90 nm (Gopinath et al. 2014). Further, research is needed for the exploration of plants with high reduction potential for the synthesis and production of NPs.

Morphophysiological Effects of Nanoparticles on Plant

Nanotechnology has great capability in enhancing plant growth because of very unique properties of NPs. The small size of NPs facilitates penetration and absorption of nonmaterial through plasma membrane and cell wall (Valletta et al. 2014). Large surface-to-volume ratio enables nonmaterial to act as a carrier and vector for nutrients and agrochemical absorption (Liu et al. 2016b). Seed priming of tomato with graphene augmented the vitamin C, chlorophyll, β-carotene, and hydrogen peroxide content (López-Vargas et al. 2020). Nano-blended formulations increase the efficient extent of adhesion and deposition of leaf (Yu et al. 2017). Once NPs enter plant cells, they cross different cellular barriers like plasmodesmata and vascular bundles. Inside cells, NPs raise biomass measure, photosynthetic rate, and antioxidants. NPs regulated the genes related to biotic and abiotic stress tolerance (Singh et al. 2021). NPs are magic bullets, which increased metabolic activities of plants; however, a lot of research work is in process to explore and utilize full capacity of NPs in the improvement of plant health. However, a more comprehensive analysis is needed for the correct mode of action using NPs (Table 16.2).

Table 16.2 Morphophysiological response of plants on nanoparticles treatment.

Crops	Nanoparticles	Morphophysiological response	References
Arabidopsis	Ag	Enhanced root growth and activated gene expression involved in cellular events	Syu et al. (2014)
		Upregulated the genes responsible for the activities of antioxidant enzymes	Kohan-Baghkheirati and Geisler-Lee (2015)
		Activated gene expression in response to metals and oxidative stress and downregulated the genes involved in response to pathogens and hormonal signals	Kaveh et al. (2013)
		Induced the secondary metabolites and accumulation of phytoalexins	Kruszka et al. (2020)
		Activated gene expression involved in cellular events, including cell proliferation, metabolism, and hormone signaling pathways	Syu et al. (2014)
	CuO	Increase in POD and SOD	Costa and Sharma (2016)
		Upregulated the gene expression at higher concentrations related to oxidative stress responses, sulfur assimilation, glutathione, and proline biosynthesis	Nair and Chung (2014)

Table 16.2 (Continued)

Crops	Nanoparticles	Morphophysiological response	References
		Deferentially expressed genes involved in root growth and ROS regulation	Wang et al. (2016b)
		Increase in root length and fresh weight	Shaw and Hossain (2013)
	FeO	Elongation of root by cell wall loosening	Kim et al. (2014)
	SiO	Reduced growth and development of seedlings and caused chlorosis in leaves	Slomberg and Schoenfisch (2012)
Rice	Ag	Increased root length and changed the endoplasmic reticulum and vacuole proteins	Vannini et al. (2013)
		Significant increase in length of roots	Mirzajani et al. (2013)
	CNP	Suspended flowering resulting in declined productivity	Lin et al. (2009)
	CeO	Exponential rate of growth	Divya and Jisha (2018)
	CuO	Decrease in the germination rate and plant weight	Shaw and Hossain (2013)
	$Ca(PO_4)$ NPs	Increased the growth rate and affect the physiology of the plants	Upadhyaya et al. (2017)
	FeO	Improved plant length and biomass with improved level of proline, glutathione, and phyto-chelatins	Bidi et al. (2021)
		Increased biomass with improved photosynthetic potential and nutrient uptake	Ahmed et al. (2021)
	Nano phosphorus	Improved photosynthetic activity and nutrient use efficiency, ultimately enhancing the quality and yield	Miranda-Villagomez et al. (2019)
	Nano nitrogen	Increased plant height, number of tillers per plant, and plant dry weight	Rathnayaka et al. (2018)
	SiO	Increased seed germination, speed of germination, root length, and shoot length	Patil et al. (2018)
		Enhanced the root length, root volume, and dry matter weight of shoot and root	Adhikari et al. (2013)
	TiO_2	Increased 51% total biomass and decreased 38% Cd in shoot	Rizwan et al. (2019)
	ZnO	Significant increase in yield, dry matter accumulation, quality, and grain Zn contents	Zhang et al. (2021)
		Effective resistance to arsenic toxicity by increasing germination, and biomass	Wu et al. (2020)
Wheat	Ag	Increased leaf area, shoot fresh and dry weight, root biomass, and chlorophyll content	Razzaq et al. (2015)
		Enhanced germination rate and germination percentage, increased antioxidant enzyme activities, and mitigated the levels of oxidative stress	Wahid et al. (2020)

(Continued)

Table 16.2 (Continued)

Crops	Nanoparticles	Morphophysiological response	References
		Increased proline and carbohydrates content	Mohamed et al. (2017)
		Increased growth, protein content, and antioxidant enzyme activities	Jhanzab et al. (2022)
		Decline in seedling weight	Singh et al. (2015)
		Elevated antioxidative enzymes activities	Mohamed et al. (2017)
		Increase in shoot and root fresh weight, chlorophyll a, b, and carotenoid contents	Abou-Zeid, and Ismai (2018)
	Al_2O_3	Significant improvement in length of root	Riahi-Madvar et al. (2012)
	CeO	Reduction in plant weight	Du et al. (2015)
		Risen the growth, shoot weight, and productivity	Rico et al. (2013)
	C	Elevated length of root	Wang et al. (2012)
	CuO	Declined growth of seedling	Hawthorne et al. (2012)
		Decrease in growth of root and seedling	Lee et al. (2008)
		Increase in plant growth and weight with improved nutrient uptake	Noman et al. (2020a)
		Inhibited growth and structure of root	Tang et al. (2016)
		Declined growth of root and seedling	Lee et al. (2008)
		Increased plant growth, chlorophyll contents, and yield	Hafeez et al. (2015)
		Increased growth and biomass of plants, Fe uptake, and antioxidant defense system	Noman et al. (2020b)
	FeO	Enhanced rate of germination and development of plant	Feizi et al. (2013)
		Increased spike weight, 1000 grain weight, grain yield, and protein content	Bakhtiari et al. (2015)
		Increase in plant height with spike length and improved the biomass and productivity	Rizwan et al. (2019)
		Increase in SOD and POD content	Konate et al. (2017)
		Enhanced growth parameters, photosynthetic activities, and uptake of Fe	Adrees et al. (2020)
		Increase in plant biomass with enhanced antioxidants and pigment content	Manzoor et al. (2021)
		Increase in plant growth with improved biomass and leaf surface index	Moghadam et al. (2012)
		Increase in lengths of seedling, weight of shoot, root and proteins related to photosynthesis, and protein metabolism	Yasmeen et al. (2016b)
	Si	Improvement in chlorophyll and reduced the malondialdehyde content	Khan et al. (2020)
		Improved seedling height and chlorophyll	Jiang et al. (2021)

Table 16.2 (Continued)

Crops	Nanoparticles	Morphophysiological response	References
	TiO_2	Elevated root length with no effect on germination and plant weight	Larue et al. (2012)
		Enhanced length of shoot	Rafique et al. (2014)
		Declined plant overall weight	Jacob et al. (2013)
	ZnO	Declined plant weight	Du et al. (2015)
		Declined plant weight	Jacob et al. (2013)
		Improved growth and productivity with elevated chlorophyll	Adrees et al. (2021)
		Induced plant growth and biomass	Rizwan et al. (2019)
		Delay in the panicle initiation time with rise in grain yield	Dimkpa et al. (2020)
Soybean	Al_2O_3	Increased hypocotyl growth and level of protein in membrane and glycolysis process	Mustafa et al. (2015b)
		Increased growth of plants, protein, and enzyme activities	Mustafa and Komatsu (2016a)
		Elevated survival percentage and growth of root	Yasmeen et al. (2016a)
	CNP	Increased rate of germination and growth	Lahiani et al. (2013)
	FeO	Enhanced biomass of leaf, pod weight, and yield	Sheykhbaglou et al. (2010)
		Increased leaf phosphorus content, translocation of Fe to leaf tissues, and uptake of Fe	Burke et al. (2015)
		Increased chlorophyll content and improved plant growth, evidenced by a greater root biomass system and increased shoot biomass	Groppa et al. (2022)
	ZnO	Increased seed germination and plant growth	Sedghi et al. (2013)
		Increased germination ratio, seedlings root, and shoot development	Montanha et al. (2020)
		Significant increase in enzymes involved in the ascorbate–glutathione cycle including SOD, CAT, ascorbate peroxidase, and glutathione reductase	Ahmad et al. (2020b)
Common beans and maize	Ag	Increase in shoot length, root length, leaf area, chlorophyll contents, carbohydrates, and protein contents	Salama (2012)
Sorghum	FeO	Increase in seed germination, growth of plants, total chlorophyll content, relative water contents and photosynthetic efficiency	Maswada et al. (2018)
Peppermint	FeO	Increased leaf biomass, uptake of macro and micro nutrients	Askary et al. (2016)
Maize	CuO	Increase in biomass, chlorophyll and Fe contents, reduced oxidative stress through antioxidant defense system	Van-Nguyen et al. (2022)

(Continued)

Table 16.2 (Continued)

Crops	Nanoparticles	Morphophysiological response	References
Maize	ZnO	Rise in levels of superoxide and SOD activity	Wang et al. (2016a)
Egg plant	ZnO	Increased growth, photosynthetic efficiency, and fruit yield	Semida et al. (2021)
Cotton	ZnO	Increased SOD and POX activities with decrease in lipid peroxidation content	Venkatachalam et al. (2017)
Brassica	Ag	Enhanced root growth	Sharma et al. (2012)
Saffron	Ag	Enhanced root growth by blocking ethylene signaling	Rezvani et al. (2012)
Maize	TiO_2	Reduced the size of cell-wall pores of roots and inhibited hydraulic conductivity and transpiration	Asli and Neumann (2009)
Cotton	ZnO	Increased growth, biochemical parameters, and biomass	Priyanka and Venkatachalam (2016)
Fenugreek	Ag	Increased germination, root length, and root biomass	Hojjat (2015)
Fenugreek	Ag	Increase in growth, photosynthetic pigments, and yield	Sadak (2019)
Asparagus	Ag	Increase in plant growth and biochemical attributes, such as protein and carbohydrates formation, antioxidant enzyme activities, and chlorophyll contents	Sharma et al. (2012)
Bacopa	Ag	Increase in seed germination, growth, protein contents, carbohydrates synthesis, and SOD and CAT activities	Krishnaraj et al. (2012)

Arabidopsis

Arabidopsis is an effective system to study stress interactions due to its short life cycle (Milewska-Hende et al. 2019). *Arabidopsis* seedling and root length were reduced on exposure to 25 mg/L FeNPs (Bombin et al. 2015). However, the growth of root tip cells was reduced due to the accumulation of Cu contents on exposure to CuO NPs (Jia et al. 2020; Yuan et al. 2016). ZnO NPs accumulated in the root tip, restricting their translocation toward the shoot (Nair and Chung 2017). Most of the applied Ag was stored in roots when evaluated using an inductively coupled plasma-atomic emission spectrometer (Xu et al. 2020). Ag NPs treatment significantly inhibited plant growth and chlorophyll synthesis (Li et al. 2018a). The shape of the NPs also determines their effect as the spherical Ag NPs reduced root growth promotion and anthocyanin was accumulated in *Arabidopsis* seedlings (Syu et al. 2014). Cd NPs exposure caused oxidative stress by damaging the respiration and reduction of chlorophyll and carotenoids (Marmiroli et al. 2020). Gold (Au) NPs adhered to the surface of *Arabidopsis* roots (Milewska-Hendel et al. 2019). For these reasons, novel nano-safety approaches to assess the environmental impact of engineered nanomaterials on plants are required.

Rice

Rice is a major crop of economic importance due to its extended consumption worldwide (Kovach et al. 2007). However, various environmental stresses influence the growth and yield of rice (Cramer et al. 2011). High concentration of CuO NPs and bulk-CuO treatment diminished grain production and reduced yield (Deng et al. 2022). Under CuO NPs exposure, Cu uptake was increased in rice seedlings, which enhanced seedling growth (Liu et al. 2018). In the hydroponic system, long-term exposure of CuO NPs quelled rice seedlings growth (Yang et al. 2020). Exposure to lower concentrations of CuO NPs enhanced rice growth, while higher concentrations significantly decreased growth (Tiwari et al. 2016). Higher concentration of CuO NPs reduced germination and seedling growth, which might be linked to the accumulation of Cu facilitating the process of lignification (Wang et al. 2020). Phytostimulatory and inhibitory effects of nanoparticles depend upon size and concentration; however, a lower concentration with a finer particle size may stimulate plant growth and development.

Soybean

Soybean is a significant crop due to its high protein contents. Because of the predominant nutritional quantities, soybean is deemed as a health-benefit food (Wang and Komatsu 2017). In soybean, NPs-mediated responses at the morphological level were comprehensively reviewed (Mustafa and Komatsu 2016a). Al_2O_3 NPs did not change plant morphology, while Al_2O_3 NPs altered the appearance of the root surface, imparting crack formation and damaging the root cap (Almeida et al. 2021). Fe_2O_3 NPs increased chlorophyll levels in soybean without causing any prominent toxicity (Ghafariyan et al. 2013). Fe_3O_4 NPs are readily absorbed without translocation within the plant and increased root weight (Iannone et al. 2021). Chromium oxide NPs' suspensions reduced soybean growth by decreasing root/shoot biomass (Li et al. 2018a). The growth of the root was increased, whereas the growth of the shoot was decreased (Li et al. 2020). Soybean development was dented with ZnO NPs (Yoon et al. 2014). Various NPs interact differentially with plant species; however, information regarding size, concentration, and extent of NPs to increase soybean growth is limited.

Wheat

Wheat is the major staple food due to its adaptability to a broad range of climatic conditions (Shewry 2009). Wheat seeds primed with 10 mg/L ZnO NPs indicated a significant positive influence on germination and vigor index as compared with unprimed and hydroprimed seeds (Rai-Kalal and Jajoo 2021). The silicon (Si) NPs enhanced plant growth and photosynthesis, while lowered the Cd concentrations without or with drought stress (Khan et al. 2020). Ag NPs exposure to wheat increased germination rate, enzymes, and lowered oxidative stress (Wahid et al. 2020). ZnO NPs treatment boosted grain yield and delayed panicle initiation (Dimkpa et al. 2020). Growth, protein contents, and antioxidant enzyme activities were increased under Ag NPs (Jhanzab et al. 2022). Fe NPs increased the yield attributes in wheat (Yasmeen et al. 2016b). Growth and productivity increased with increased level of chlorophyll contents by ZnO NPs treatment (Adrees et al. 2020). Therefore, application of NPs promoted the growth of wheat; however, detailed metabolomic analysis is critical to comprehend the fate of nonmaterials.

Other Plants

High concentration of bulk CuO and CuO NPs improved Fe in seeds; however, very low concentration of CuO NPs does not influence nutritional quality of green peas (Ochoa et al. 2018). CuO NPs raised Cu uptake in rice seedlings and improved growth (Liu et al. 2018). CuO NPs enhanced nutrient levels, suggesting their possible use as nanofertilizer for onion production (Wang et al. 2020). At greater concentrations, CuO NPs inhibited germination and radicle growth in lettuce (Pelegrino et al. 2020). Fe_2O_3 NPs at low concentration promoted root elongation by increasing germination and vigor indexes. However, high concentrations of Fe_2O_3 NPs remarkably decreased root length in corn (Li et al. 2016b). Plants are the basic unit of food chain; however, mechanisms of plant nano-interaction is needed to find the useful impact of NPs on important crop species.

Molecular Mechanism Altered by Nanoparticles

The application of NPs upregulated genes that encode for proteins is crucial for maintaining reactive oxygen species (ROS) balance (Ruotolo et al. 2018). After treatment with CuO NPs suspension, rice roots were seriously damaged, while malondialdehyde, SOD, POD, and hydrogen peroxide contents increased (Wang et al. 2019). During flooding stress, Al_2O_3 NPs exposure of soybean improved the recovery response through the regulation of different proteins (Yasmeen et al. 2016a). NPs help plants reduce the level of stress through increasing the antioxidant enzyme activities; however, the time of application of NPs varied for different plant species (Table 16.3) (Figure 16.2).

Table 16.3 Alteration in molecular mechanism of plants on nanoparticles application.

Molecular mechanism	Nanoparticles	Crops	Response	References
Oxidative stresses	Ag	*Brassica*	Increased levels of antioxidant enzymes, guaiacol peroxide, CAT, and ascorbate peroxidase, which resulted in decreased levels of ROS	Sharma et al. (2012)
		Rice	Altered transcription and protein degradation, oxidative stress response pathway, and calcium signaling pathway	Mirzajani et al. (2014)
		Saffron	Blocking of ethylene signaling, promotion of root growth	Rezvani et al. (2012)
		Salad	Changes in the proteins related to redox regulation, disrupting cellular homeostasis	Vannini et al. (2013)
		Soybean	Changed proteins related to stress, signaling, and cell metabolism	Mustafa et al. (2015a)
		Rice	Increased oxidative stress response pathway, Ca^{2+} regulation signaling, transcription, protein degradation, cell-wall synthesis, cell division, and apoptosis	Mirzajani et al. (2014)
	Al_2O_3	Soybean	Change in proteins related to protein synthesis/degradation, glycolysis, and lipid metabolism	Mustafa et al. (2015b)

Table 16.3 (Continued)

Molecular mechanism	Nanoparticles	Crops	Response	References
	C	Tomato	Influenced seed germination and seedling growth by upregulating stress-related gene expression	Khodakovskaya et al. (2009)
		Tomato	Upregulation in the stress and water channel-related genes	Khodakovskaya et al. (2011)
	CeO	Cotton	Alleviated ROS accumulation	An et al. (2020)
		Cotton	Help to maintain Na +/K + ratio	Liu et al. (2021)
	CuO	Tomato	Increased content of glutathione	Perez-Labrada et al. (2019)
		Tomato	Increased content of phenols, β-carotene, ascorbic acid, and lycopene	Hernandez-Hernández et al. (2018)
		Radish, ryegrass	Mediated DNA damage and plant growth restriction	Atha et al. (2012)
	Fe_3O_4	Tomato	Improved Fe content	Shankramma et al. (2016)
		Moringa	Reduction in Na^+ and Cl^- contents, increasing N, P, K^+, Fe, Zn, total chlorophyll, carotenoids, proline, carbohydrates, crude protein, and enzymatic/nonenzymatic antioxidants	Soliman et al. (2015)
	Au	*Brassica*	Increased activities of enzymes, such as SOD, CAT guaiacol peroxidase, ascorbate peroxidase, and glutathione reductase	Gunjan and Zaidi (2014)
	Mn_3O_4	Cucumber	Scavenged overaccumulated ROS to improve salt tolerance	Lu et al. (2020)
	SiO_2	Strawberry	Maintain chlorophyll content and leaf relative water content/protection	Avestan et al. (2019)
		Cucumber	Increased K^+ content of seedlings by maintaining K^+/Na^+ ratio	Alsaeedi et al. (2019)
		Rice	Regulated jasmonic acid signal	Abdel-Haliem et al. (2017)
	ZnO	Tomato	Modulated the activities of antioxidant enzymes	Faizan et al. (2021)
		Moringa	Reduction in Na^+ and Cl^- contents, increasing N, P, K +, Fe, Zn, total chlorophyll, carotenoids, proline, carbohydrates, crude protein, and enzymatic/nonenzymatic antioxidants	Soliman et al. (2015)
		Sunflower	Increasing growth, net CO_2 assimilation rate, sub-stomatal CO_2 content, chlorophyll content, Fv/Fm and Zn content, and decreasing Na^+ content	Torabian et al. (2018)
Energy regulation	Cu–Zn	Wheat	Increased activity of antioxidative enzymes, reduced the level of accumulation of thiobarbituric acid reactive substances and stabilized content of photosynthetic pigments	Taran et al. (2017)

(Continued)

Table 16.3 (Continued)

Molecular mechanism	Nanoparticles	Crops	Response	References
	SiO_2	Hawthorn	Improvement of photosynthesis rate and stomatal conductance	Ashkavand et al. (2015)
		Sorghum	Improved root growth and maintenance of photosynthetic rate	Hattori et al. (2005)

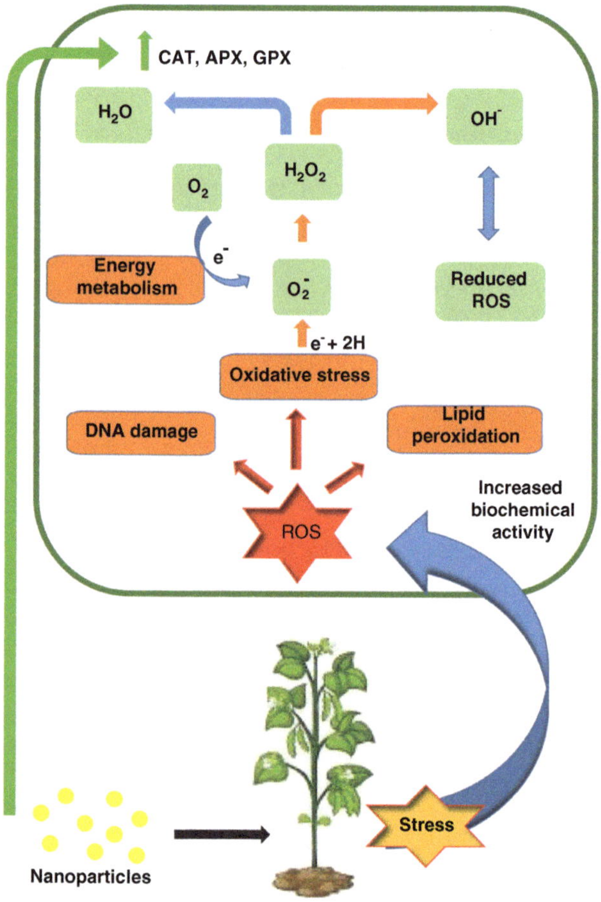

Figure 16.2 Role of nanoparticles (NPs) during stress. Stressful conditions cause an excessive reactive oxygen species (ROS) production that results in oxidative stress, DNA damage, and lipid peroxidation. NPs are responsible for the increase in antioxidant enzymatic activity that is involved in the ROS scavenging, thus protecting plants from the damaging effects of the ROS.

Oxidative Stresses

Exposure of NPs altered the ROS levels inside cells by regulating the levels of nicotinamide adenine dinucleotide phosphate oxidase, SOD, glutathione *S*-transferases, and POD (Ruotolo et al. 2018). Entry of NPs into the cell can disrupt the electron transport chain of chloroplast and mitochondria causing oxidative rupture due to increased ROS concentration (Hossain et al.

2015). ROS includes hydroxyl radical, superoxide free radical, nonradical singlet oxygen, and hydrogen peroxidase (Dimkpa et al. 2013). Excess level of ROS inside cells caused adversative impacts known as oxidative stress (Faisal et al. 2013; Gill and Tuteja 2010; Jiang et al. 2014; Pakrashi et al. 2014). Oxidative stress induced destruction of DNA, oxidation of protein, lipid peroxidation, and membrane damage that led to cell death (Van Breusegem and Dat 2006). Exposure to NPs increased ROS production through enhanced antioxidant activity to withstand adverse environments.

ROS triggered oxidative destruction of chloroplasts and mitochondria, causing dysfunction and complete inactivation of certain proteins (Atha et al. 2012). Under Ag NPs treatment, ROS production was increased, which regulated the antioxidant defense mechanism in *Ricinus communis* seed (Belava et al. 2017). Ag NPs treatment increased antioxidant enzymes, like CAT and ascorbate peroxidase in *Brassica juncea*, leading to decreased ROS activity (Sharma et al. 2012). Proteins related to different metabolic functions were altered under Ag NPs (Mirzajani et al. 2014). NPs perform a significant part in regulating ROS production and controlling the redox homeostasis of cell under stress conditions.

CAT and SOD activities decreased under CuO NPs, while SOD activity increased in rice leaves (Yang et al. 2020). It was found that hydrogen peroxide and proline contents increased under AuNPs application. The activity of glutathione reductase and ascorbate peroxidase is stimulated in the presence of Au NPs (Mazumdar and Ahmed 2011a). Therefore, NPs can increase plant growth by regulating the redox homeostasis and metabolic activity under adverse environmental conditions.

Energy Regulation

The goal of any scientific development is to increase the well-being of human health. Green nanotechnology uses the concepts of green chemistry (Verma et al. 2019). NPs possess very unique properties and therefore have the ability to impose variable impacts on biochemistry of plant metabolism (Adekanmi et al. 2022). Chemo-blended Ag NPs increased the proteins related to photosynthesis and protein synthesis, while recox and mitochondrial electron transport chain related proteins were decreased (Jhanzab et al. 2019). Antioxidant enzyme activities help regulate energy metabolism through the regulation of glycolysis-associated proteins in response to Ag NPs.

Transportation of Fe through leaves, stems, and roots via Fe NPs application changed the antioxidant enzymatic activities with ultimate chlorophyll degradation, leading to reduced photosynthesis and biomass (Lu et al. 2020). Fe_3O_4 NPs reduced seed germination and root elongation in *Cucumis sativus* (Mushtaq 2011). Ag NPs significantly promoted photosynthesis by increasing chlorophyll contents in cucumber (Zhang et al. 2021). The largest quantum yield of photosystem-II as well as the photochemical quenching decreased, showing that the photosynthetic system was damaged on treatment with chromium oxide NPs (Li et al. 2018a). Under cerium oxide NPs, chlorophyll synthesis was inhibited, while the maximum quantum yield and photochemical extinction of the photosystem-II complex decreased in soybean (Li et al. 2020). Cerium oxide NPs hindered C assimilation, leading to reduced RuBisCo (Cao et al. 2018). Under Al_2O_3 NPs, energy metabolism decreased compared with flooding stress in soybean (Mustafa et al. 2015b). Si NPs heightened the gaseous leaf exchange attributes and chlorophyll concentrations, while decreased oxidative stress through the enhancement of SOD and POD activities (Ali et al. 2019). NPs help plants regulate plant growth and development through energy regulation and increased level of antioxidant enzyme activities.

Nanoparticles Interaction with Plants

Ag NPs' interaction with crops relies upon plant species, size, and concentration. Application of Ag NPs enhanced root growth in *Brassica juncea* (Sharma et al. 2012), *Crocus sativus* (Rezvani et al. 2012), and *Arabidopsis* (Syu et al. 2014). Exposure of Ag NPs increased the germination of *Boswellia ovaliofoliolata* (Savithramma et al. 2012), while reduced the plant biomass, chlorophyll contents, and tissue nitrogen contents of *Sprodela polyrhiza* (Jiang et al. 2014). In wheat, Ag NPs increased plant growth (Razzaq et al. 2016), while *Phaselous radiatus*' growth was not affected (Lee et al. 2012). Earlier studies suggest that engineered NPs can encourage plant growth; however, the pathways and mode of translocation of NPs to food chain need more exploration (Figure 16.3).

Nanoparticles Interaction with Soybean

In soybean, SiO_2 and TiO_2 NPs increased germination and growth by promoting water and fertilizer-use efficiency, while Fe_2O_3 NPs increased grain dry weight and yield (Lu et al. 2001; Sheykhbaglou et al. 2010). Soybean treated with Ag NPs experienced less oxygen scarcity stress that led to improved growth

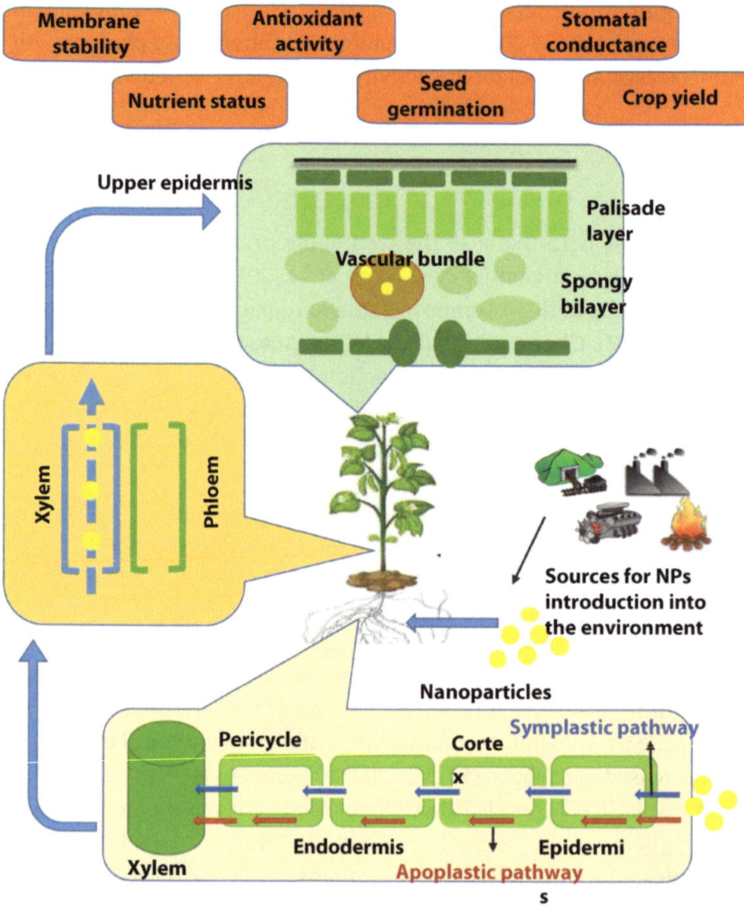

Figure 16.3 Interactions of nanoparticles (NPs) with plants. NPs first come in contact with roots. NPs can enter the plant root through symplastic or apoplastic pathways. These are then transported through xylem to the aerial parts of the plant. NPs alter several metabolic processes such as seed germination, stomatal conductance, nutrient uptake, and membrane stability that affect plant growth and crop yield.

performance under flooding (Mustafa et al. 2015a). Different sizes of Ag NPs improved soybean growth through the regulation of the level of proteins related to amino acid synthesis and wax formation (Mustafa et al. 2016). Al_2O_3 NPs increased hypocotyl growth and regulated the level of proteins related to cell death energy and metabolism in soybean (Mustafa and Komatsu 2016c; Mustafa et al. 2015b). Al_2O_3 NPs helped soybean recover from flooding stress through the enhancement of root length and weight (Yasmeen et al. 2016b). ZnO NPs promoted the germination and health of soybean (Sedghi et al. 2013). Fe_3O_4 NPs enhanced chlorophyll and improved plant growth, indicated through greater root/shoot biomass system of soybean (Groppa et al. 2022). Multi-walled C nanotubes significantly increased the growth of soybean (Lahiani et al. 2013). ZnO NPs boosted the germination ratio and seedlings development, depicting a workable substitute to Zn supply (Montanha et al. 2020). Fe_3O_4 NPs raised leaf phosphorus content, movement of Fe to leaf tissues, and uptake of Fe by soybean plants (Burke et al. 2015). Growth of soybean increased significantly in response to nanoparticles; however, limited information about the interaction of other NPs is available.

Nanoparticles Interaction with Wheat

Ag NPs interaction with plants has been reported to cause both positive and negative impacts. Among different NPs, Ag NPs are widely used as antibacterial, antifungal, and coating material and for water purification (Duran et al. 2007). Cu NPs enriched the yield of wheat through a number of grains (Hafeez et al. 2015; Yasmeen et al. 2017a; Yasmeen et al. 2018). Growth and ascorbic contents of wheat remained unaffected under Ag NPs (Liu et al. 2017). Ag NPs enhanced germination rate, increased antioxidant enzyme activities, and mitigated the levels of oxidative stress in wheat (Wahid et al. 2020), while proline and carbohydrate contents also increased in wheat (Mohamed et al. 2017). Wheat growth, chlorophyll contents, protein contents, and antioxidant enzyme activities increased with chemo-blended nanoparticles (Jhanzab et al. 2022). Fe NPs exposure in wheat enhanced growth parameters, photosynthetic activities, and uptake of Fe (Adrees et al. 2020; Yasmeen et al. 2017b). Growth and development of wheat plants is improved; however, more elaborative research is vital to improve the quality of food through the application of NPs.

Nanoparticles Interaction with Other Plants

CuO NPs interacted with actin protein and decreased the growth rate of actin filaments that disturbed the subcellular localization and auxin distribution in root tips in *Arabidopsis* (Jia et al. 2020). Single-walled C NPs increased photosystem-II activity, C metabolism, and encouraged glycolysis and tricarboxylic acid cycle to spawn energy in *Sophora alopecuroides* (Wan et al. 2020). Chitosan NPs reduced hydrogen peroxide and malondialdehyde contents but increased chlorophyll contents in *Vigna radiata* (Sen et al. 2020). Si NPs improved photosynthesis and chlorophyll content in *Musa acuminate* (Mahmoud et al. 2020). NPs caused variable growth impacts on plant metabolism; therefore, NPs mediated stress response mechanisms, which needs to be explored in detail to exploit the beneficial side of nanomaterials.

Conclusion and Future Prospects

Nanotechnology has accomplished a tremendous position in the modern era owing to its exclusive functions in different areas of the agricultural sector. However, the unintentional delivery of artificially synthesized nanomaterials into the environment has evoked worldwide concern. Concern about the potential release of NPs into the environment with ultimate influence on living organisms has increased. Considerable focus has been on the schemes of NPs synthesis, contact, and possible destiny in the environment. NPs are manufactured via physical and chemical methods; however, synthesis through biological methods is environmentally friendly and less toxic. Ag, Al, Cu, and Fe NPs are synthesized through different strategies and their interaction with crop plants imparts changes in the morphological and biochemical levels. NP treatments on plants mainly impact physiological, metabolic, and biochemical processes, which inhibit plant productivity in positive and negative ways. To understand the influence of NPs on plants, it is essential to be aware that certain mechanisms are adopted by plants, which enhance their tolerance. These NPs regulate the oxidative stresses and energy regulations under abiotic stresses. Moreover, these NPs interacted with soybean and altered the energy metabolism and helped the plant to combat stress conditions. In wheat, these NPs improved yield by regulating antioxidant systems and photosynthetic activities. Considering all these studies, more thorough studies incorporating various omics, genomics, proteomics, transcriptomics, and metabolomics, might be favorable to a comprehensive uncovering of plant–NPs interaction mechanisms.

References

Abbasi, E., Milani, M., Fekri Aval, S., Kouhi, M., Akbarzadeh, A., Tayefi Nasrabadi, H., Nikasa, P., Joo, S.W., Hanifehpour, Y., Nejati-Koshki, K., and Samiei, M. (2016). Silver nanoparticles: synthesis methods, bio-applications and properties. *Crit. Rev. Microbiol.* 42 (2): 173–180.

Abdeen, M., Sabry, S., Ghozlan, H., El-Gendy, A.A., and Carpenter, E.E. (2016). Microbial-physical synthesis of Fe and Fe_3O_4 magnetic nanoparticles using *Aspergillus niger* YESM1 and supercritical condition of ethanol. *J. Nanomater.* 2016 1–7.

Abdel-Haliem, M.E.F., Hegazy, H.S., Hassan, N.S., and Naguib, D.M. (2017). Effect of silica ions and nano silica on rice plants under salinity stress. *Ecol. Eng.* 99: 282–289.

Abou-Zeid, H. and Ismail, G. (2018). The role of priming with biosynthesized silver nanoparticles in the response of *Triticum aestivum* L to salt stress. *Egypt. J. Bot.* 58: 73–85.

Adekanmi, A.A., Akinkunmi, O.O., Ajewole, I.D., Kehinde, L.I., Micheal, O.G., and Young, A.R. (2022). Characterization and antimicrobial property of nickel nanoparticle synthesized using leaves extract of *Launaea taraxacifolia* (African lettuce). *Int. J. Research Sci. Engineering.* 2(2): 56–69.

Adhikari, T., Kundu, S., and Rao, A.S. (2013). Impact of SiO_2 and Mo nanoparticles on seed germination of rice (*Oryza sativa* L.). *Int. J. Agric. Food Sci. Technol.* 4 (8): 809–816.

Adrees, M., Khan, Z.S., Ali, S., Hafeez, M., Khalid, S., Rehman, M.Z.U., Hussain, A., Hussain, K., Chatha, S.A.S., and Rizwan, M. (2020). Simultaneous mitigation of cadmium and drought stress in wheat by soil application of iron nanoparticles. *Chemosphere* 238: 124681.

Adrees, M., Khan, Z.S., Hafeez, M., Rizwan, M., Hussain, K., Asrar, M., Alyemeni, M.N., Wijaya, L., and Ali, S. (2021). Foliar exposure of zinc oxide nanoparticles improved the growth of wheat (*Triticum aestivum* L.) and decreased cadmium concentration in grains under simultaneous Cd and water deficient stress. *Ecotoxicol. Environ. Safe.* 208: 111627.

Afsheen, S., Bilal, M., Iqbal, T., Liaqat, A., and Abrar, M. (2018). Green synthesis and characterization of novel iron particles by using different extracts. *J. Alloys Compd* 732: 935–944.

Ahmad, A., Senapati, S.S., Khan, M.I., Kumar, R., Ramani, R., Srinivas, V., and Sastry, M. (2003). Intracellular synthesis of gold nanoparticles by a novel alkalotolerant actinomycete, *Rhodococcus* species. *Nanotechnology* 14: 824.

Ahmad, P., Alyemeni, M.N., Al-Huqail, A.A., Alqahtani, M.A., Wijaya, L., Ashraf, M., and Bajguz, A. (2020b). Zinc oxide nanoparticles application alleviates arsenic (As) toxicity in soybean plants by restricting the uptake of as and modulating key biochemical attributes, antioxidant enzymes, ascorbate-glutathione cycle and glyoxalase system. *Plants* 9: 825.

Ahmad, W., Khan, A.U., Shams, S., Qin, L., Yuan, Q., Ahmad, A., Wei, Y., Khan, Z.U.H., Ullah, S., and Rahman, A.U. (2020a). Eco-benign approach to synthesize spherical iron oxide nanoparticles: a new insight in photocatalytic and biomedical applications. *J. Photochem. Photobiol. B: Biol.* 205: 111821.

Ahmed, A.H., Hanfy, A.E., Ansari, M.J., Dessoky, E.S., Attia, A.O., Elshaer, M.M., Gaber, M.K., Kordy, A.S., Doma, A., Abdelkhalek, A., and Behiry, S.I. (2021). Green biosynthesized silver nanoparticles using *Acalypha wilkesiana* extract control root-knot nematode. *J. King Saud Univ.-Sci.* 33: 101516.

Ahsan, A., Farooq, M.A., Ahsan Bajwa, A., and Parveen, A. (2020). Green synthesis of silver nanoparticles using *Parthenium hysterophorus*: optimization, characterization and in vitro therapeutic evaluation. *Molecules* 25 (15): 3324.

Aisida, S.O., Ugwu, K., Akpa, P., Nwanya, A.C., Nwankwo, U., Bashir, A.K.H., Madiba, I., Ahmed, I., and Ezema, F. (2021). Synthesis and characterization of iron oxide nanoparticles capped with *Moringa oleifera*: the mechanisms of formation effects on the optical, structural, magnetic and morphological properties. *Mater. Today: Proc.* 36: 214–218.

Alabdallah, N.M. and Alzahrani, H.S. (2020). The potential mitigation effect of ZnO nanoparticles on (*Abelmoschus esculentus* L. Moench) metabolism under salt stress conditions. *Saudi J. Biol. Sci.* 27: 3132–3137.

Ali, M.A., Al-Mokaram, A.A., Yahya, R., Abdi, M.M., and Ekramul Mahmud, H.N.M. (2016). One-step electrochemical deposition of Polypyrrole-Chitosan-Iron oxide nanocomposite films for non-enzymatic glucose biosensor. *Mater. Lett.* 183: 90–93.

Ali, S., Rizwan, M., Hussain, A., Zia ur Rehman, M., Ali, B., Yousaf, B., Wijaya, L., Alyemeni, M.N., and Ahmad, P. (2019). Silicon nanoparticles enhanced the growth and reduced the cadmium accumulation in grains of wheat (*Triticum aestivum* L.). *Plant Physiol. Biochem.* 140: 1–8.

Almeida, G.H.G., Siqueira-Soares, R.C., Mota, T.R., Oliveira, D.M., Abrahão, J., Foletto-Felipe, M.P., Dos Santos, W.D., Ferrarese-Filho, O., and Marchiosi, R. (2021). Aluminum oxide nanoparticles affect the cell-wall structure and lignin composition slightly altering the soybean growth. *Plant Physiol. Biochem.* 159: 335–346.

Alsaeedi, A., El-Ramady, H., Alshaal, T., El-Garawany, M., Elhawat, N., and Al-Otaibi, A. (2019). Silica nanoparticles boost growth and productivity of cucumber under water deficit and salinity stresses by balancing nutrients uptake. *Plant Physiol. Biochem.* 139: 1–10.

Amruta, S., Jaiprakash, S., Shampa, C., Ajay, S., Patil, V., and Suresh, G. (2019). Helminthicidal and larvicidal potentials of biogenic silver nanoparticles synthesized from medicinal plant *Momordica charantia*. *Med. Chem.* 15: 781–789.

An, J., Hu, P., Li, F., Wu, H., Shen, Y., and White, J.C. (2020). Emerging investigator series: molecular mechanisms of plant salinity stress tolerance improvement by seed priming with cerium oxide nanoparticles. *Environ. Sci. Nano.* 7: 2214–2228.

Ansari, M.A., Khan, H.M., Alzohairy, M.A., Jalal, M., Ali, S.G., Pal, R., and Musarrat, J. (2015). Green synthesis of Al_2O_3 nanoparticles and their bactericidal potential against clinical isolates of multi-drug resistant *Pseudomonas aeruginosa*. *World J. Microbiol. Biotechnol.* 31: 153–164.

Aromal, S.A. and Philip, D. (2012). Green synthesis of gold nanoparticles using *Trigonella foenum-graecum* and its size-dependent catalytic activity. *Spectrochim. Acta Part A: Mol. Biomol. Spectrosc.* 971–975.

Arumugam, A., Karthikeyan, C., Hameed, A.S.H., Gopinath, K., Gowri, S., and Karthika, V. (2015). Synthesis of cerium oxide nanoparticles using *Gloriosa superba* L. leaf extract and their structural, optical and antibacterial properties. *Mater. Sci. Eng.: C* 49: 408–415.

Arumugam, M., Natarajan, T.S., Saelee, T., Praserthdam, S., Ashokkumar, M., and Praserthdam, P. (2021). Recent developments on bismuth oxyhalides (BiOX; X = Cl, Br, I) based ternary nanocomposite photocatalysts for environmental applications. *Chemosphere* 282: 131054.

Ashkavand, P., Tabari, M., Zarafshar, M., Tomaskova, I., and Struve, D. (2015). Effect of SiO_2 nanoparticles on drought resistance in hawthorn seedlings. *For. Res. Pap.* 76: 350–359.

Askary, M., Talebi, S.M., Amini, F., and DoustiBaloutBangan, A. (2016). Effect of NaCl and Iron oxide nanoparticles on *Mentha piperita* essential oil composition. *Environ. Exp. Biol.* 14: 27–32.

Aslani, F., Bagheri, S., Muhd Julkaplimailto, N., Juraimi, A.S., Hashemi, F.S.G., and Baghdadi, A. (2014). Effects of engineered nanomaterials on plants growth: an overview. *Sci. World J.* 2014: 641759.

Asli, S. and Neumann, P.M. (2009). Colloidal suspensions of clay or titanium dioxide nanoparticles can inhibit leaf growth and transpiration via physical effects on root water transport. *Plant, Cell Environ.* 32: 577–584.

Aswini, R., Murugesan, S., and Kannan, K. (2021). Bio-engineered TiO2 nanoparticles using *Ledebouria revoluta* extract: larvicidal, histopathological, antibacterial and anticancer activity. *Int. J. Environ. Anal. Chem.* 101: 2926–2936.

Atha, D.H., Wang, H., Petersen, E.J., Cleveland, D., Holbrook, R.D., Jaruga, P., and Nelson, B.C. (2012). Copper oxide nanoparticle mediated DNA damage in terrestrial plant models. *Environ. Sci. Technol.* 46: 1819–1827.

Avestan, S., Ghasemnezhad, M., Esfahani, M., and Byrt, C.S. (2019). Application of nano-silicon dioxide improves salt stress tolerance in strawberry plants. *Agronomy* 9: 246.

Bahrulolum, H., Nooraei, S., Javanshir, N., Tarrahimofrad, H., Mirbagheri, V.S., Easton, A.J., and Ahmadian, G. (2021). Green synthesis of metal nanoparticles using microorganisms and their application in the agrifood sector. *J. Nanobiotechnol.* 19: 1–26.

Bai, B., Saranya, S., Dheepaasri, V., Muniasamy, S., Alharbi, N.S., Selvaraj, B., Undal, V.S., and Gnanamangai, B.M. (2022). Biosynthesized copper oxide nanoparticles (CuO NPs) enhances the anti-biofilm efficacy against *K. pneumoniae* and *S. aureus*. *J. King Saud Univ-Sci.* 34: 102120.

Bai, H.J., Zhang, Z.M., Guo, Y., and Yang, G.E. (2009). Biosynthesis of cadmium sulfide nanoparticles by photosynthetic bacteria *Rhodopseudomonas palustris*. *Colloids Surf. B: Biointerfaces* 70: 142–146.

Bakhtiari, M., Moaveni, P., and Sani, B. (2015). The effect of iron nanoparticles spraying time and concentration on wheat. *Biol. Forum* 7: 679–683.

Barabanov, P.V., Gerasimov, A.V., Blinov, A.V., Kravtsov, A.A., and Kravtsov, V.A. (2018). Influence of nanosilver on the efficiency of *Pisum sativum* crops germination. *Ecotoxicol. Environ. Saf.* 147: 715–719.

Belava, V.N., Panyuta, O.O., Yakovleva, G.M., Pysmenna, Y.M., and Volkogon, M.V. (2017). The effect of silver and copper nanoparticles on the wheat–*Pseudocercosporellah erpotrichoides* pathosystem. *Nanoscale Res. Lett.* 12: 1–10.

Benassai, E., Bubba, M.D., Ancillotti, C., Colzi, I., Gonnelli, C., Calisi, N., Salvatici, M.C., Casalone, E., and Ristori, S. (2021). Green and cost-effective synthesis of copper nanoparticles by extracts of non-edible and waste plant materials from Vaccinium species: characterization and antimicrobial activity. *Mater. Sci. Eng.: C* 119: 111453.

Bhainsa, K.C. and D'Souza, S.F. (2006). Extracellular biosynthesis of silver nanoparticles using the fungus *Aspergillus fumigatus*. *Colloids Surf. B: Biointerfaces* 47: 160–164.

Bhargava, A., Jain, N., Khan, M.A., Pareek, V., Dilip, R.V., and Panwar, J. (2016). Utilizing metal tolerance potential of soil fungus for efficient synthesis of gold nanoparticles with superior catalytic activity for degradation of rhodamine B. *J. Environ. Manage.* 183: 22–32.

Bibi, I., Nazar, N., Ata, S., Sultan, M., Ali, A., Abbas, A., Jilani, K., Kamal, S., Sarim, F.M., Khan, M.I., Jalal, F., and Iqbal, M. (2019). Green synthesis of iron oxide nanoparticles using pomegranate seeds extract and photocatalytic activity evaluation for the degradation of textile dye. *J. Mater. Res. Technol.* 8: 6115–6124.

Bidi, H., Fallah, H., Niknejad, Y., and Tari, D.B. (2021). Iron oxide nanoparticles alleviate arsenic phytotoxicity in rice by improving iron uptake, oxidative stress tolerance and diminishing arsenic accumulation. *Plant Physiol. Biochem.* 163: 348–357.

Bolade, O.P., Akinsiku, A.A., Adeyemi, A.O., Jolayemi, G.E., Williams, A.B., and Benson, N.U. (2019). Qualitative analysis, total phenolic content, FT-IR and GC-MS characterization of *Canna indica*: bioreducing agent for nanoparticles synthesis. *J. Phys. Conf. Ser.* 129: 012135.

Bolade, O.P., Akinsiku, A.A., Adeyemi, A.O., Williams, A.B., and Benson, N.U. (2018). Dataset on phytochemical screening, FTIR and GC-MS characterisation of *Azadirachta indica* and *Cymbopogon citratus* as reducing and stabilising agents for nanoparticles synthesis. *Data Brief* 38: 107407.

Bombin, S., LeFebvre, M., Sherwood, J., Xu, Y., Bao, Y., and Ramonell, K.M. (2015). Developmental and reproductive effects of iron oxide nanoparticles in *Arabidopsis thaliana*. *Int. J. Mol. Sci.* 16: 24174–24193.

Borja, J.Q., Ngo, M.A.S., Saranglao, C.C., Tiongco, R.P.M., Roque, E.C., and Dugos, N.P. (2015). Synthesis of green zero-valent iron using polyphenols from dried green tea extract. *J. Eng. Sci. Technol.* 10: 22–31.

Burke, D.J., Pietrasiak, N., Situ, S.F., Abenojar, E.C., Porche, M., Kraj, P., Lakliang, Y., and Samia, A.C.S. (2015). Iron oxide and titanium dioxide nanoparticle effects on plant performance and root associated microbes. *Int. J. Mol. Sci.* 16: 23630–23650.

Cao, Z., Rossi, L., Stowers, C., Zhang, W., Lombardini, L., and Ma, X. (2018). The impact of cerium oxide nanoparticles on the physiology of soybean (*Glycine max* Merr.) under different soil moisture conditions. *Environ. Sci. Pollut. Res. Int.* 25: 930–939.

Chandra, S., Das, P., Bag, S., Laha, D., and Pramanik, P. (2011). Synthesis, functionalization and bioimaging applications of highly fluorescent carbon nanoparticles. *Nanoscale* 3: 1533–1540.

Chauhan, S. and Upadhyay, L.S.B. (2019). Biosynthesis of iron oxide nanoparticles using plant derivatives of *Lawsonia inermis* (Henna) and its surface modification for biomedical application. *Nanotechnol. Environ. Eng.* 4: 8.

Chen, F., Gao, W., Qiu, X., Zhang, H., Liu, L., Liao, P., Fu, W., and Luo, Y. (2017). Graphene quantum dots in biomedical applications: recent advances and future challenges. *Front. Lab. Med.* 1: 192–199.

Cheng, T., Yang, Z., Tang, R.C., and Zhai, A.D. (2020). Functionalization of silk by silver nanoparticles synthesized using the aqueous extract from tea stem waste. *J. Mater. Res. Technol.* 9: 4538–4549.

Costa, M.V.J. and Sharma, P.K. (2016). Effect of copper oxide nanoparticles on growth, morphology, photosynthesis, and antioxidant response in *Oryza sativa*. *Photosynthetica* 54: 110–119.

Cramer, G.R., Urano, K., Delrot, S., Pezzotti, M., and Shinozaki, K. (2011). Effects of abiotic stress on plants: a systems biology perspective. *BMC Plant Biol.* 11: 163.

Dang, H., Huang, L.K., Zhang, Y., Wang, C.F., and Chen, S. (2016). Large-scale ultrasonic fabrication of white fluorescent carbon dots. *Ind. Eng. Chem. Res.* 55: 5335–5341.

Darroudi, M., Hoseini, S.J., Oskuee, R.K., Hosseini, H.A., Gholami, L., and Gerayli, S. (2014). Food-directed synthesis of cerium oxide nanoparticles and their neurotoxicity effects. *Ceramics Int.* 40: 7425–7430.

Das, K.R., Kowshik, M., Praveen Kumar, M.K., Kerkar, S., Shyama, S.K., and Mishra, S. (2018). Native hypersaline sulphate reducing bacteria contributes to iron nanoparticle formation in saltpan sediment: a concern for aquaculture. *J. Environ. Manage.* 206: 556–564.

Deng, C., Wang, Y., Navarro, G., Sun, Y., Cota-Ruiz, K., Hernandez-Viezcas, J.A., Niu, G., Li, C., White, J.C., and Gardea-Torresdey, J. (2022). Copper oxide (CuO) nanoparticles affect yield, nutritional quality, and auxin associated gene expression in weedy and cultivated rice (*Oryza sativa* L.) grains. *Sci. Total Environ.* 810: 152260.

Desalegn, B., Megharaj, M., Chen, Z., and Naidu, R. (2019). Green synthesis of zero valent iron nanoparticle using mango peel extract and surface characterization using XPS and GC-MS. *Heliyon* 5: e01750.

Devatha, C.P., Jagadeesh, K., and Patil, M. (2018). Effect of green synthesized iron nanoparticles by *Azardirachta indica* in different proportions on antibacterial activity. *Environm. Nanotechnol. Monit. Manag.* 9: 85–94.

Devi, R.S. and Gayathri, R. (2014). Green synthesis of zinc oxide nanoparticles by using *Hibiscus rosa-sinensis*. *Int. J. Curr. Eng. Technol.* 4: 2444–2446.

Dhandapani, P., Maruthamuthu, S., and Rajagopal, G. (2012). Bio-mediated synthesis of TiO_2 nanoparticles and its photocatalytic effect on aquatic biofilm. *J. Photochem. Photobiol. B* 110: 43–49.

Dhandapani, P., Prakash, A.A., AlSalhi, M.S., Maruthamuthu, S., Devanesan, S., and Rajasekar, A. (2020). Ureolytic bacteria mediated synthesis of hairy ZnO nanostructure as photocatalyst for decolorization of dyes. *Mater. Chem. Phy.* 122619.

Dimkpa, C.O., Andrews, J., Sanabria, J., Bindraban, P.S., Singh, U., Elmer, W.H., Gardea-Torresdey, J.L., and White, J.C. (2020). Interactive effects of drought, organic fertilizer, and zinc oxide nanoscale and bulk particles on wheat performance and grain nutrient accumulation. *Sci. Total Environ.* 722: 137808.

Dimkpa, C.O., McLean, J.E., Martineau, N., Britt, D.W., Haverkamp, R., and Anderson, A.J. (2013). Silver nanoparticles disrupt wheat (*Triticum aestivum* L.) growth in a sand matrix. *Environ. Sci. Technol.* 47: 1082–1090.

Divya, K. and Jisha, M.S. (2018). Chitosan nanoparticles preparation and applications. *Environ. Chem. Lett.* 16: 101–112.

Długosz, O., Sochocka, M., Ochnik, M., and Banach, M. (2021). Metal and bimetallic nanoparticles: flow synthesis, bioactivity and toxicity. *J. Coll. Interf. Sci.* 586: 807–818.

Du, W., Gardea-Torresdey, J.L., Ji, R., Yin, Y., Zhu, Z., Peralta-Videa, J.R., and Guo, H. (2015). Physiological and biochemical changes imposed by CeO_2 nanoparticles on wheat: a life cycle field study. *Environ. Sci. Technol.* 49: 11884–11893.

Duran, N., Marcato, P.D., De Souza, G.I.H., Alves, O.L., and Esposito, E. (2007). Antibacterial effect of silver nanoparticles produced by fungal process on textile fabrics and their effluent treatment. *J. Biomed. Nanotechnol.* 3: 203–208.

Dutschk, V., Karapantsios, T., Liggieri, L., McMillan, N., Miller, R., and Starov, V.M. (2014). Smart and green interfaces: from single bubbles/drops to industrial environmental and biomedical applications. *Adv. Coll. Interf. Sci.* 209: 109–126.

Ebrahiminezhad, A., Taghizadeh, S., Ghasemi, Y., and Berenjian, A. (2017). Green synthesized nanoclusters of ultra-small zero valent iron nanoparticles as a novel dye removing material. *Sci. Total Environ.* 621: 1527–1532.

Edwards, E.R., Antunes, E.F., Botelho, E.C., Baldan, M.R., and Corat, E.J. (2011). Evaluation of residual iron in carbon nanotubes purified by acid treatments. *Appl. Surf. Sci.* 258: 641–648.

El-Baz, F.K., Mahmoud, K., El-Senousy, W.M., Darwesh, O.M., and ElGohary, A.E. (2015). Antiviral–antimicrobial and schistosomicidal activities of *Eucalyptus camaldulensis* essential oils. *Int. J. Pharm. Sci. Rev. Res.* 31: 262–268.

Elemike, E.E., Onwudiwe, D.C., Ekennia, A.C., Ehiri, R.C., and Nnaji, N.J. (2017). Phytosynthesis of silver nanoparticles using aqueous leaf extracts of *Lippia citriodora*: antimicrobial, larvicidal and photocatalytic evaluations. *Mater. Sci. Eng.: C* 75: 980–989.

Elmusa, F., Aygun, A., and Gulbagca, F. (2021). Investigation of the antibacterial properties of silver nanoparticles synthesized using *Abelmoschus esculentus* extract and their ceramic applications. *Int. J. Environ. Sci. Technol.* 18: 849–860.

Erenler, E. and Dag, B. (2022). Biosynthesis of silver nanoparticles using *Origanum majorana* L. and evaluation of their antioxidant activity. *Inorg. Nano-Metal Chem.* 52: 485–492.

Faisal, M., Saquib, Q., Alatar, A.A., Al-Khedhairy, A.A., Hegazy, A.K., and Musarrat, J. (2013). Phytotoxic hazards of NiO-nanoparticles in tomato: a study on mechanism of cell death. *J. Hazard. Mater.* 250: 318–332.

Faizan, M., Bhat, J.A., Chen, C., Alyemeni, M.N., and Yu, F. (2021). Zinc oxide nanoparticles (ZnO-NPs) induce salt tolerance by improving the antioxidant system and photosynthetic machinery in tomato. *Plant Physiol. Biochem.* 161: 122–130.

Fardood, S.T., Ramazani, A., and Joo, S.W. (2017). Green synthesis of zinc oxide nanoparticles using arabic gum and photocatalytic degradation of direct blue 129 dye under visible light. *J. Mater. Sci.: Mater. Electron* 28: 13596–13601.

Fardood, S.T., Ramazani, A., and Joo, S.W. (2018). Eco-friendly synthesis of magnesium oxide nanoparticles using arabic gum. *J. Appl. Chem. Res.* 12: 8–15.

Fatima, F., Hashim, A., and Anees, S. (2021). Efficacy of nanoparticles as nanofertilizer production: a review. *Environ. Sci. Poll. Res.* 28: 1292–1303.

Fazlzadeh, M., Rahmani, K., Zarei, A., Abdoallahzadeh, H., Nasiri, F., and Khosravi, R. (2017). A novel green synthesis of zero valent iron nanoparticles (NZVI) using three plant extracts and their efficient application for removal of Cr (VI) from aqueous solutions. *Adv. Powder Technol.* 28: 122–130.

Feizi, H., Kamali, M., Jafari, L., and Moghaddam, P.R. (2013). Phytotoxicity and stimulatory impacts of nanosized and bulk titanium dioxide on fennel (*Foeniculum vulgare* Mill). *Chemosphere* 91: 506–511.

Galfetti, L., De Luca, L.T., Severini, F., Meda, L., Marra, G., Marchetti, M., and Bellucci, S. (2006). Nanoparticles for solid rocket propulsion. *J. Phy. Cond. Matter* 18: S1991.

Gautam, A., Rawat, S., Verma, L., Singh, J., Sikarwar, S., Yadav, B.C., and Kalamdhad, A.S. (2018). Green synthesis of iron nanoparticle from extract of waste tea: an application for phenol red removal from aqueous solution. *Environ. Nanotechnol. Monit. Manag.* 10: 377–387.

Gawande, M.B., Goswami, A., Asefa, T., Guo, H., Biradar, A.V., Peng, D.L., and Varma, R.S. (2015). Core-shell NPs: synthesis and applications in catalysis and electrocatalysis. *Chem. Soc. Rev.* 44: 7540–7590.

Ghafariyan, M.H., Malakouti, M.J., Dadpour, M.R., Stroeve, P., and Mahmoudi, M. (2013). Effects of magnetite nanoparticles on soybean chlorophyll. *Environ. Sci. Technol.* 47: 10645–10652.

Gill, S.S. and Tuteja, N. (2010). Reactive oxygen species and antioxidant machinery in abiotic stress tolerance in crop plants. *Plant Physiol. Biochem.* 48: 909–930.

Gonzalez-Moragas, L., Yu, S., Carenza, E., Laromaine, A., and Roig, A. (2015). Protective effects of bovine serum albumin on superparamagnetic iron oxide nanoparticles evaluated in the nematode *Caenorhabditis elegans*. *ACS Biomater. Sci. Eng.* 11: 1129–1138.

Gopinath, K., Karthika, V., Gowri, S., Senthilkumar, V., Kumaresan, S., and Arumugam, A. (2014). Antibacterial activity of ruthenium nanoparticles synthesized using *Gloriosa superba* L. leaf extract. *J. Nanostruct. Chem.* 4: 83.

Goswami, S.R., Sahareen, T., Singh, M., and Kumar, S. (2015). Role of biogenic silver nanoparticles in disruption of cell–cell adhesion in *Staphylococcus aureus* and *Escherichia coli* biofilm. *J. Ind. Eng. Chem.* 26: 73–80.

Groppa, M.D., Zawoznik, M.S., Benavides, M.P., and annone, M.F. (2022). Beneficial effects of magnetite nanoparticles on soybean-*Bradyrhizobium japonicum* and alfalfa-*Sinorhizobium meliloti* associations. *Plant Physiol. Biochem.* 180: 42–49.

Gunjan, B. and Zaidi, M. (2014). Impact of gold nanoparticles on physiological and biochemical characteristics of *Brassica juncea*. *J. Plant Biochem. Physiol.* 2: 67–73.

Guo, Z., Kumar, C.S., Henry, L.L., Doomes, E.E., Hormes, J., and Podlaha, E.J. (2004). Displacement synthesis of Cu shells surrounding Co NPs. *J. Electrochem. Soc.* 152: 1.

Hafeez, A., Razzaq, A., Mahmood, T., and Jhanzab, H.M. (2015). Potential of copper NPs to increase growth and yield of wheat. *J. Nanosci. Adv. Technol.* 1: 6–11.

Hattori, T., Inanaga, S., Araki, H., An, P., Morita, S., Luxova, M., and Lux, A. (2005). Application of silicon enhanced drought tolerance in Sorghum bicolor. *Physiol. Plantarum* 123: 459–466.

Hawthorne, J., Musante, C., Sinha, S.K., and White, J.C. (2012). Accumulation and phytotoxicity of engineered nanoparticles to *Cucurbita pepo*. *Int. J. Phytoremediation* 14: 429–442.

He, Y., Yuan, J., Su, F., Xing, X., and Shi, G. (2006). *Bacillus subtilis* assisted assembly of gold nanoparticles into long conductive nodous ribbons. *J. Phys. Chem. B.* 110 (36): 17813–17818.

Henam, S.D., Ahmad, F., Shah, M.A., Parveen, S., and Wani, A.H. (2019). Microwave synthesis of nanoparticles and their antifungal activities. *Spectrochim. Acta A: Mol. Biomol. Spectrosc.* 2213: 337–341.

Hernandez-Hernandez, H., González-Morales, S., Benavides-Mendoza, A., Ortega-Ortiz, H., Cadenas-Pliego, G., and Juarez-Maldonado, A. (2018). Effects of chitosan–PVA and cu nanoparticles on the growth and antioxidant capacity of tomato under saline stress. *Molecules* 23: 178.

Hlekelele, L., Frankly, P.J., Tripathi, P.K., and Durbach, S.H. (2016). Morphological and crystallinity differences in nitrogen-doped carbon nanotubes grown by chemical vapour deposition decomposition of melamine over coal fly ash. *RSC Adv.* 6: 76773–76779.

Hojjat, S.S. (2015). Effect of nano silver on seed germination and seedling growth in fenugreek seed. *Int. J. Food Eng.* 1: 106–110.

Hossain, Z., Mustafa, G., and Komatsu, S. (2015). Plant responses to nanoparticle stress. *Int. J. Mol. Sci.* 16: 26644–26653.

Hou, Y., Lu, Q., Deng, J., Li, H., and Zhang, Y. (2015). One-pot electrochemical synthesis of functionalized fluorescent carbon dots and their selective sensing for mercury ion. *Anal. Chem. Acta* 866: 69–74.

Huang, H., Xu, Y., Tang, C., Chen, J., Wang, A., and Feng, J. (2014). Facile and green synthesis of photoluminescent carbon nanoparticles for cellular imaging. *N. J. Chem.* 38: 784–789.

Huber, D.L. (2005). Synthesis, properties, and applications of iron nanoparticle. *Nano, Micro, Small* 1: 482–501.

Iannone, M.F., Groppa, M.D., Zawoznik, M.S., Coral, D.F., Fernandez, van Raap, M.B., and Benavides, M.P. (2021). Magnetite nanoparticles coated with citric acid are not phytotoxic and stimulate soybean and alfalfa growth. *Ecotoxicol. Environ. Saf.* 211 111942.

Iqbal, J., Abbassi, B.A., Ahmad, R., Shahbaz, A., Zahra, S.A., Kanwal, S., Rabbani, A., and Mahmood, T. (2020). Biogenic synthesis of green and cost-effective iron nanoparticles and evaluation of their potential biomedical properties. *J. Mol. Struct.* 1199: 126979.

Iravani, S., Korbekandi, H., Mirmohammadi, S.V., and Zolfaghari, B. (2014). Synthesis of silver nanoparticles: chemical, physical and biological methods. *Res. Pharm. Sci.* 9: 385–406.

Ismail, E., Khenfouch, M.M., Dhlamini, M., Dube, S., and Maaza, M. (2017). Green palladium and palladium oxide nanoparticles synthesized via *Aspalathus linearis* natural extract. *J. Alloys Compds.* 695: 3632–3638.

Jacob, D.L., Borchardt, J.D., Navaratnam, L., Otte, M.L., and Bezbaruah, A.N. (2013). Uptake and translocation of Ti from nanoparticles in crops and wetland plants. *Int. J. Phytoremediation* 15: 142–153.

Jagadeesan, G., Srimathi, K., Srinivas, N.S., Manishkanna, S., and Vignesh, D. (2019). Green synthesis of iron oxide nanoparticles using *Terminalia bellirica* and *Moringa oleifera* fruit and leaf extracts: antioxidant, antibacterial and thermoacoustic properties. *Biocatal. Agric. Biotechnol.* 21: 101354.

Jahan, S., Karim, M.E., and Chowdhury, E.H. (2021). Nanoparticles targeting receptors on breast cancer for efficient delivery of chemotherapeutics. *Biomedicines* 9: 114.

Jayaseelan, C., Abdul Rahuman, A., Kirthi, A.V., Marimuthu, S., Santhoshkumar, T., Bagavan, A., Gaurav, K., Karthik, L., and Rao, K.V.B. (2012). Novel microbial route to synthesize ZnO nanoparticles using *Aeromonas hydrophila* and their activity against pathogenic bacteria and fungi. *Spectrochim. Acta Part A: Mol. Biomol. Spectrosc.* 90: 78–84.

Jhanzab, H.M., Qayyum, A., Bibi, Y., Sher, A., Hayat, M.T., Iqbal, J., and Askary, A.E. (2022). Chemo-blended Ag and Fe nanoparticles effect on growth, physiochemical and yield traits of wheat (*Triticum aestivum*). *Agronomy* 12: 757.

Jhanzab, H.M., Razzaq, A., Bibi, Y., Yasmeen, F., Yamaguchi, H., Hitachi, K., and Komatsu, S. (2019). Proteomic analysis of the effect of inorganic and organic chemicals on silver nanoparticles in wheat. *Int. J. Mol. Sci.* 20: 825.

Jia, H., Chen, S., Wang, X., Shi, C., Liu, K., Zhang, S., and Li, J. (2020). Copper oxide nanoparticles alter cellular morphology via disturbing the actin cytoskeleton dynamics in *Arabidopsis* roots. *Nanotoxicol.* 14: 127–144.

Jiang, C., Zu, C., Shen, J., Shao, F., and Li, T. (2015). Effects of selenium on the growth and photosynthetic characteristics of flue-cured tobacco (*Nicotiana tabacum* L.). *Acta Soc. Bot. Pol.* 84 (1).

Jiang, H.S., Qiu, X.N., Li, G.B., Li, W., and Yin, L.Y. (2014). Silver nanoparticles induced accumulation of reactive oxygen species and alteration of antioxidant systems in the aquatic plant *Spirodela polyrhiza*. *Environ. Toxicol. Chem.* 33 (6): 1398–40540.

Jiang, L., Jia, F., Han, Y., Meng, X., Xiao, Y., and Bai, S. (2021). Development and characterization of zein edible films incorporated with catechin/β-cyclodextrin inclusion complex nanoparticles. *Carbohydr. Polym.* 261: 117877.

John, M.S., Nagoth, J.A., Zannotti, M., Giovannetti, R., Mancini, A., Ramasamy, K.P., Miceli, C., and Pucciarelli, S. (2021). Biogenic synthesis of copper nanoparticles using bacterial strains isolated from an Antarctic consortium associated to a psychrophilic marine ciliate: characterization and potential application as antimicrobial agents. *Mar. Drugs* 19 (5): 263.

Joshi, C.G., Danagoudar, A., Poyya, J., Kudva, A.K., and Dhananjaya, B.L. (2017). Biogenic synthesis of gold nanoparticles by marine endophytic fungus–*Cladosporium cladosporioides* isolated from seaweed and evaluation of their antioxidant and antimicrobial properties. *Process Biochem.* 63: 137–140.

Kalaiyan, G., Suresh, S., Prabu, K.M., Thambidurai, S., Kandasamy, M., Pugazhenthiran, N., Kumar, S.K., and Muneeswaran, T. (2021). Bactericidal activity of *Moringa oleifera* leaf extract assisted green synthesis of hierarchical copper oxide microspheres against pathogenic bacterial strains. *J. Environm. Chem. Eng.* 9: 104847.

Kamikoriyama, Y., Imamura, H., and Muramatsu, A. (2019). Ambient aqueous-phase synthesis of copper nanoparticles and nanopastes with low-temperature sintering and ultra-high bonding abilities. *Sci. Rep.* 9: 899.

Karim, N.A., Rubinsin, N.J., Burukan, M.A.A., and Kamarudin, S.K. (2019). Sustainable route of synthesis platinum nanoparticles using orange peel extract. *Int. J. Green Energy* 16: 1518–1526.

Karpagavinayagam, P. and Vedhi, C. (2019). Green synthesis of iron oxide nanoparticles using *Avicennia marina* flower extract. *Vacuum* 160: 286–292.

Katata-Seru, L., Moremedi, T., Aremu, O.S., and Bahadur, I. (2018). Green synthesis of iron nanoparticles using *Moringa oleifera* extracts and their applications: removal of nitrate from water and antibacterial activity against *Escherichia coli*. *J. Mol. Liq.* 256: 296–304.

Kaveh, R., Li, Y., Ranjbar, S., Tehrani, R., Brueck, C.L., and Aken, B.V. (2013). Changes in *Arabidopsis thaliana* gene expression in response to silver nanoparticles and silver ions. *Environ. Sci. Technol.* 47 (18): 10637–10644.

Khaghani, S., Ghanbari, D., and Khaghani, S. (2017). Green synthesis of iron oxide–palladium nanocomposites by pepper extract and its application in removing of colored pollutants from water. *Arak Univ. Technol.* 175–182.

Khan, Z.S., Rizwan, M., Hafeez, M., Ali, S., Adrees, M., Qayyum, M.F., Khalid, S., Ur Rehman, M.Z., and Sarwar, M.A. (2020). Effects of silicon nanoparticles on growth and physiology of wheat in cadmium contaminated soil under different soil moisture levels. *Environ. Sci. Poll. Res. Int.* 27 (5): 4958–4968.

Khatami, M., Alijani, H.Q., Fakheri, B., Mobasseri, M., Heydarpour, M., Farahani, Z.K., and Khan, A.U. (2019). Superparamagnetic iron oxide nanoparticles (SPIONs): greener synthesis using *Stevia* plant and evaluation of its antioxidant properties. *J. Clean Prod.* 208: 1171–1177.

Khodakovskaya, M., Dervishi, E., Mahmood, M., Yang, X., Li, Z., and Fumiya, W. (2009). Carbon nanotubes are able to penetrate plant seed coat and dramatically affect seed germination and plant growth. *Am. Chem. Soc. Nano* 3: 3221–3227.

Khodakovskaya, M.V., de Silva, K., Nedosekin, D.A., Dervishi, E., Biris, A.S., Shashkov, E.V., Galanzha, E.I., and Zharov, V.P. (2011). Complex genetic, photothermal, and photoacoustic analysis of nanoparticle–plant interactions. *Proceedings of the National Academy of Sciences*, 108, 1028–1033.

Kim, J.H., Lee, Y., Kim, E.J., Gu, S., Sohn, E.J., Seo, Y.S., and Chang, Y.S. (2014). Exposure of iron nanoparticles to *Arabidopsis thaliana* enhances root elongation by triggering cell wall loosening. *Environ. Sci. Technol.* 48 (6): 3477–3485.

Kishore, M.Y., Kunal, B., Kumar, J.S., Abeer, H., A Fathi, A.E., and Kumar, M.T. (2020). Anti-biofilm and antibacterial activities of silver nanoparticles synthesized by the reducing activity of phytoconstituents present in the Indian medicinal plants. *Front. Microbiol.* 11: 01143.

Kohan-Baghkheirati, E. and Geisler-Lee, J. (2015). Gene expression, protein function and pathways of *Arabidopsis thaliana* responding to silver nanoparticles in comparison to silver ions, cold, salt, drought, and heat. *Nanomater.* 5: 436–467.

Konate, A., He, X., Zhang, Z., Ma, Y., Zhang, P., Alugongo, G.M., and Rui, Y. (2017). Magnetic (Fe_3O_4) nanoparticles reduce heavy metals uptake and mitigate their toxicity in wheat seedling. *Sustainability* 9 (5): 790.

Kovach, M.J., Sweeney, M.T., and McCouch, S.R. (2007). New insights into the history of rice domestication. *Trends Genet.* 23: 578–587.

Krishnaraj, C., Jagan, E.G., Ramachandran, R., Abirami, S.M., Mohan, N., and Kalaichelvan, P.T. (2012). Effect of biologically synthesized silver NPs on *Bacopa monnieri* (Linn.) Wettst. plant growth metabolism. *Process Biochem.* 47 (4): 651–658.

Kruk, T., Szczepanowicz, K., Stefańska, J., Socha, R.P., and Warszyński, P. (2015). Synthesis and antimicrobial activity of monodisperse copper nanoparticles. *Colloids Surf. B Biointerf.* 128: 17–22.

Kruszka, D., Sawikowska, A., Selvakesavan, R.K., Krajewski, P., Kachlicki, P., and Franklin, G. (2020). Silver nanoparticles affect phenolic and phytoalexin composition of *Arabidopsis thaliana*. *Sci. Total Environ.* 716: 135–361.

Krutyakov, Y.A., Kudrinskiy, A.A., Olenin, A.Y., and Lisichkin, G.V. (2008). Synthesis and properties of silver nanoparticles: advances and prospects. *Russ. Chem. Rev.* 77: 233.

Kulkarni, S.K. (2007). *Nanotechnology: Principles and Practices*. New Delhi, India: Capital Pub. Co.

Kumar, P.V., Kala, S.M.J., and Prakash, K. (2019). Green synthesis of gold nanoparticles using *Croton caudatus* Geisel leaf extract and their biological studies. *Mater. Lett.* 236: 19–22.

Lahiani, M.H., Dervishi, E., Chen, J., Nima, Z., Gaume, A., Biris, A.S., and Khodakovskaya, M.V. (2013). Impact of carbon nanotube exposure to seeds of valuable crops. *ACS Appl. Mater. Interf.* 5: 7965–7973.

Landa, P. (2021). Positive effects of metallic nanoparticles on plants: overview of involved mechanisms. *Plant Physiol. Biochem.* 161: 12–24.

Larue, C., Veronesi, G., Flank, A., Surble, S., Herlin-Boime, N., and Carrière, M. (2012). Comparative uptake and impact of TiO_2 nanoparticles in wheat and rapeseed. *J. Toxicol. Environ. Health, Part A* 75: 13–15.

Lastovina, T.A., Budnyk, A.P., Soldatov, M.A., Rusalev, Y.V., Guda, A.A., Bogdan, A.S., and Soldatov, A.V. (2017). Microwave-assisted synthesis of magnetic iron oxide nanoparticles in oleylamine–oleic acid solutions. *Mendeleev Commun.* 27: 487–489.

Laware, S.L. and Raskar, S. (2014). Effect of titanium dioxide NPs on hydrolytic and antioxidant enzymes during seed germination in onion. *Int. J. Curr. Microbiol. Appl. Sci.* 3: 749–760.

Lee, H., Lytton-Jean, A., Chen, Y., Love, K.T., Park, A.I., Karagiannis, E.D., and Sehgal, A. (2012). Molecularly self-assembled nucleic acid nanoparticles for targeted *in vivo* siRNA delivery. *Nat. Nanotechnol.* 7: 389–393.

Lee, Y., Choi, J., Lee, K.J., Stott, N.E., and Kim, D. (2008). Large-scale synthesis of copper nanoparticles by chemically controlled reduction for applications of inkjet-printed electronics. *Nanotechnol.* 19: 415604.

Li, H., Shan, C., Zhang, Y., Cai, J., Zhang, W., and Pan, B. (2016a). Arsenate adsorption by hydrous ferric oxide nanoparticles embedded in cross-linked anion exchanger: effect of the host pore structure. *ACS Appl. Mat. Interf.* 8: 3012–3020.

Li, J., Hu, J., Ma, C., Wang, Y., Wu, C., Huang, J., and Xing, B. (2016b). Uptake, translocation and physiological effects of magnetic iron oxide (γ-Fe_2O_3) nanoparticles in corn (*Zea mays* L.). *Chemosphere* 159: 326–334.

Li, J., Mu, Q., Du, Y., Luo, J., Liu, Y., and Li, T. (2020). Growth and photosynthetic inhibition of cerium oxide nanoparticles on soybean (*Glycine max*). *Bull. Environ. Contam. Toxicol.* 105 (1): 119–126.

Li, J., Song, Y., Wu, K., Tao, Q., Liang, Y., and Li, T. (2018a). Effects of Cr_2O_3 nanoparticles on the chlorophyll fluorescence and chloroplast ultrastructure of soybean (*Glycine max*). *Environ. Sci. Poll. Res. Int.* 25 (20): 19446–19457.

Li, X., Ke, M., Zhang, M., Peijnenburg, W.J.G.M., Fan, X., Xu, J., Zhang, Z., Lu, T., Fu, Z., and Qian, H. (2018b). The interactive effects of diclofop-methyl and silver nanoparticles on *Arabidopsis thaliana*: growth, photosynthesis and antioxidant system. *Environ. Poll.* 232: 212–219.

Lin, J., Chen, J., Huang, S., Ko, J., Wang, Y., Chen, T., and Wang, L. (2009). Folic acid–Pluronic F127 magnetic nanoparticle clusters for combined targeting, diagnosis, and therapy applications. *Biomater.* 30: 5114–5124.

Lin, Z., Weng, X., Owens, G., and Chen, Z. (2020). Simultaneous removal of Pb(II) and rifampicin from wastewater by iron nanoparticles synthesized by a tea extract. *J. Clean. Prod.* 242: 118476.

Liu, A., Liu, J., and Zhang, W. (2015). Transformation and composition evolution of nanoscale zero valent iron (nZVI) synthesized by borohydride reduction in static water. *Chemosphere* 119: 1068–1074.

Liu, D., Yang, F., Xiong, F., and Gu, N. (2016a). The smart drug delivery system and its clinical potential. *Theranostics* 6: 1306–1323.

Liu, G., Zhang, M., Jin, Y. Fan, X., Xu, J., Zhu, Y., Fu, Z., Pan, X., and Qain, H. (2017). The effects of low concentrations of silver nanoparticles on wheat growth, seed quality, and soil microbial communities. *Water Air Soil Pollut.* 228: 348.

Liu, H., Sun, Y., Yu, T., Zhang, J., Zhang, X., Zhang, H., Zhao, K., and Wei, J. (2019). Plant-mediated biosynthesis of iron nanoparticles–calcium alginate hydrogel membrane and its eminent performance in removal of Cr(VI). *Chem. Eng. J.* 378: 122120.

Liu, J., Fu, C., Li, G., Khan, M.N., and Wu, H. (2021). ROS homeostasis and plant salt tolerance: plant nanobiotechnology updates. *Sustainability* 13: 3552.

Liu, J., Simms, M., Song, S., King, R.S., and Cobb, G.P. (2018). Physiological effects of copper oxide nanoparticles and arsenic on the growth and life cycle of rice (*Oryza sativa* japonica 'Koshihikari'). *Environ. Sci. Technol.* 52 (23): 13728–13737.

Liu, M., Xu, Y., Niu, F., Gooding, J.J., and Liu, J. (2016b). Carbon quantum dots directly generated from electrochemical oxidation of graphite electrodes in alkaline alcohols and the applications for specific ferric ion detection and cell imaging. *Analyst* 141 (9): 2657–2664.

López-Vargas, E.R., González-García, Y., Pérez-Álvarez, M., Cadenas-Pliego, G., González-Morales, S., Benavides-Mendoza, A., Cabrera, R.I., and Juárez-Maldonado, A. (2020). Seed priming with carbon nanomaterials to modify the germination, growth, and antioxidant status of tomato seedlings. *Agronomy* 10: 639.

Lu, C., Zhang, C., Wen, J., Wu, G., and Tao, M. (2001). Research of the effect of nanometer materials on germination and growth enhancement of *Glycine max* and its mechanism. *Soybean Sci.* 21: 168–171.

Lu, G., Li, S., Guo, Z., Farah, O.K., Hauser, B.G., Qi, X., Wang, Y., Wang, X., and Han, S. (2012). Imparting functionality to a metal–organic framework material by controlled nanoparticle encapsulation. *Nat. Chem.* 4: 310–316.

Lu, L., Huang, M., Huang, Y., Corvini, P.F.X., Ji, R., and Zhao, L. (2020). Mn_3O_4 nanozymes boost endogenous antioxidant metabolites in cucumber (*Cucumis sativus*) plant and enhance resistance to salinity stress. *Environ. Sci. Nano* 7: 1692–1703.

Mahadevan, S., Vijayakumar, S., and Arulmozhi, P. (2017). Green synthesis of silver nano particles from *Atalantia monophylla* (L) Correa leaf extract, their antimicrobial activity and sensing capability of H_2O_2. *Microb. Pathol.* 113: 445–450.

Mahmoud, A.W.M., Abdeldaym, E.A., Abdelaziz, S.M., El-Sawy, M.B., and Mottaleb, S.A. (2020). Synergetic effects of zinc, boron, silicon, and zeolite nanoparticles on confer tolerance in potato plants subjected to salinity. *Agronomy* 10: 19.

Mahmoud, O., Din, Vryzas, N., Zisis, H.A., and Kelessidis, V.Z. (2016). Nanoparticle-Based drilling fluids for minimizing formation damage in HP/HT applications. In: *Paper presented at the SPE International Conference and Exhibition on Formation Damage Control*. Lafayette, Louisiana, USA, February 2016. doi: https://doi.org/10.2118/178949-MS

Manjunatha, R.L., Naik, D., and Usharani, K.V. (2019). Nanotechnology application in agriculture: a review. *J. Pharmacol. Phytochem.* 8: 1073–1083.

Manzoor, N., Ahmed, T., Noman, M., Shahid, M., Nazir, M., Ali, L., Alnusaire, T.S., Li, B., Schulin, R., and Wang, G. (2021). Iron oxide nanoparticles ameliorated the cadmium and salinity stresses in wheat plants, facilitating photosynthetic pigments and restricting cadmium uptake. *Sci. Total Environ.* 769: 145221.

Marmiroli, M., Mussi, F., Pagano, L., Imperiale, D., Lencioni, G., Villani, M., Zappettini, A., White, J.C., and Marmiroli, N. (2020). Cadmium sulfide quantum dots impact *Arabidopsis thaliana* physiology and morphology. *Chemosphere* 124856.

Maswada, H.F., Djanaguiraman, M., and Prasad, P.V.V. (2018). Seed treatment with nano-iron (III) oxide enhances germination, seeding growth and salinity tolerance of sorghum. *J. Agron. Crop Sci.* 204: 577–587.

Maynard, A.D., Aitken, R.J., Butz, T., Colvin, V., Donaldson, K., Oberdorster, G., and Warheit, D.B. (2006). Safe handling of nanotechnology. *Nature* 444: 267–269.

Mazumdar, H. and Ahmed, G.U. (2011a). Phytotoxicity effect of silver nanoparticles on *Oryza sativa*. *Int. J. Chem. Technol. Res.* 3: 1494–1500.

Mazumdar, H. and Haloi, N. (2011b). A study on biosynthesis of iron nanoparticles by *Pleurotus* sp. *J. Microbiol. Biotechnol. Res.* 1: 39–49.

Mehrotra, N., Tripathi, R.M., Zafar, F., and Singh, M.P. (2017). Catalytic degradation of dichlorvos using biosynthesized zero valent iron nanoparticles. *IEEE Trans. Nano Biosci.* 16: 280–286.

Merga, G., Wilson, R., Lynn, G., Milosavljevic, B., and Meisel, D. (2007). Redox catalysis on "naked" silver NPs. *J. Phys. Chem.* 111: 12220–12206.

Milewska-Hendel, A., Zubko, M., Stroz, D., and Kurczynska, E.U. (2019). Effect of nanoparticles surface charge on the *Arabidopsis thaliana* (L.) roots development and their movement into the root cells and protoplasts. *Int. J. Mol. Sci.* 20: 1650.

Miranda-Villagomez, E., Trejo-Tellez, L.I., Gomez-Merino, F.C., Sandoval-Villa, M., Sanchez-Garcia, P., and Aguilar-Mendez, M.A. (2019). Nanophosphorus fertilizer stimulates growth and photosynthetic activity and improves P status in rice. *J. Nanomater.* 2019: 5368027.

Mirzajani, F., Askari, H., Hamzelou, S., Farzaneh, M., and Ghassempour, A. (2013). Effect of silver nanoparticles on *Oryza sativa* L. and its rhizosphere bacteria. *Ecotoxicol. Environm. Saf.* 88: 48–54.

Mirzajani, F., Askari, H., Hamzelou, S., Schober, Y., Rompp, A., Ghassempour, A., and Spengler, B. (2014). Proteomics study of silver nanoparticles toxicity on *Bacillus thuringiensis*. *Ecotoxicol. Environm. Saf.* 100: 122–130.

Moghadam, B.Y., Hou, W.C., Corredor, C., Westerhoff, P., and Posner, J.D. (2012). Role of nanoparticle surface functionality in the disruption of model cell membranes. *Langmuir* 28: 16318–16326.

Mohamed, A.K.S., Qayyum, M.F., Abdel-Hadi, A.M., Rehman, R.A., Ali, S., and Rizwan, M. (2017). Interactive effect of salinity and silver nanoparticles on photosynthetic and biochemical parameters of wheat. *Arch. Agron. Soil Sci.* 63: 1736–1747.

Mohammed, E.T. and Safwat, G.M. (2020). Grape seed proanthocyanidin extract mitigates titanium dioxide nanoparticle (TiO2-NPs)–induced hepatotoxicity through TLR-4/NF-κB signaling pathway. *Biol. Trace Elem. Res.* 196: 579–589.

Mohanta, Y., Panda, S.K., Biswas, K., Tamang, A., De Bandyopadhyay, J.D., Mohanta, D., and Bastia, A.K. (2016). Biogenic synthesis of silver nanoparticles from *Cassia fistula* (Linn.): *in vitro* assessment of their antioxidant, antimicrobial and cytotoxic activities. *IET Nanobiotechnol.* 10: 438–444.

Mollick, M.M.R., Rana, D., Dash, S.K., Chattopadhyay, S., Bhowmick, B., Maity, D., Mondal, D., Pattanayak, S., Roy, S., Chakraborty, M., and Chattopadhyay, D. (2019). Studies on green synthesized silver nanoparticles using *Abelmoschus esculentus* (L.) pulp extract having anticancer (in vitro) and antimicrobial applications. *Arab. J. Chem.* 12: 2572–2584.

Montanha, G.S., Rodrigues, E.S., Marques, J.P.R., de Almeida, E., Colzato, M., and Pereira de Carvalho, H.W. (2020). Zinc nanocoated seeds: an alternative to boost soybean seed germination and seedling development. *SN Appl. Sci.* 2 (5): 1–11.

Mourdikoudis, S., Pallares, R.M., and Thanh, N.T. (2018). Characterization techniques for NPs: comparison and complementarity upon studying nanoparticle properties. *Nanoscale* 10: 12871–12934.

Mukherjee, P. (2017). *Stenotrophomonas* and *Microbacterium*: mediated biogenesis of copper, silver and iron nanoparticles-proteomic insights and antibacterial properties versus biofilm formation. *J. Clust. Sci.* 28: 331–358.

Mushtaq, Y.K. (2011). Effect of nanoscale Fe_3O_4, TiO_2 and carbon particles on cucumber seed germination. *J. Environ. Sci.* 46: 1732–1735.

Mustafa, G. and Komatsu, S. (2016a). Nanoparticles mediated soybean response mechanism at morphological, physiological, and proteomic level. *Curr. Proteom.* 13: 1–10.

Mustafa, G. and Komatsu, S. (2016b). Toxicity of heavy metals and metal-containing nanoparticles on plants. *Biochim. et Biophys. Acta (BBA)-Proteins Proteom* 1864: 932–944.

Mustafa, G. and Komatsu, S. (2016c). Insights into the response of soybean mitochondrial proteins to various sizes of aluminum oxide nanoparticles under flooding stress. *J. Proteome Res.* 15: 4464–4475.

Mustafa, G., Sakata, K., Hossain, Z., and Komatsu, S. (2015a). Proteomic study on the effects of silver nanoparticles on soybean under flooding stress. *J. Proteomics* 122: 100–118.

Mustafa, G., Sakata, K., and Komatsu, S. (2015b). Proteomic analysis of flooded soybean root exposed to aluminum oxide nanoparticles. *J. Proteomics* 128: 280–297.

Mustafa, G., Sakata, K., and Komatsu, S. (2016). Proteomic analysis of soybean root exposed to varying sizes of silver nanoparticles under flooding stress. *J. Proteomics* 148: 113–125.

Naimi-Shamel, N., Pourali, P., and Dolatabadi, S. (2019). Green synthesis of gold nanoparticles using *Fusarium oxysporum* and antibacterial activity of its tetracycline conjugant. *J. Mycol. Med.* 29: 7–13.

Nair, P.M. and Chung, I.M. (2014). Impact of copper oxide nanoparticles exposure on *Arabidopsis thaliana* growth, root system development, root lignification, and molecular level changes. *Environ. Sci. Poll. Res. Int.* 21: 12709–12022.

Nair, P.M.G. and Chung, I.M. (2017). Regulation of morphological, molecular and nutrient status in *Arabidopsis thaliana* seedlings in response to ZnO nanoparticles and Zn ion exposure. *Sci. Total Environ.* 575: 187–198.

Narayanan, M.V. and Rakesh, S.G. (2018). Synthesis of colloidal alumina nanoparticles using green method. Conference Series Material Science Engineering, 402, 012150.

Neto, O.P.V. (2014). Intelligent computational nanotechnology: the role of computational intelligence in the development of nanoscience and nanotechnology. *J. Comput. Theor. Nanosci.* 11 (4): 928–944.

Nguyen, V., Yan, L., Si, J., and Hou, X. (2015). Femtosecond laser-induced size reduction of carbon nanodots in solution: effect of laser fluence, spot size, and irradiation time. *J. Appl. Phy.* 117 (8): 084304.

Nithya, K., Sathish, A., and Sivamani, S. (2021). In situ synthesis of mesostructured iron oxide nanoparticles embedded in *L. camara*: adsorption insights and modeling studies. *Biomass Convers. Biorefin.* doi: https://doi.org/10.1007/s13399-021-01764-8.

Noman, M., Ahmed, T., Hussain, S., Niazi, M.B.K., Shahid, M., and Song, F. (2020a). Biogenic copper nanoparticles synthesized by using a copper-resistant strain *Shigella flexneri* SNT22 reduced the translocation of cadmium from soil to wheat plants. *J. Hazard. Mater.* 398: 123175.

Noman, M., Shahid, M., Ahmed, T., Tahir, M., Naqqash, T., Muhammad, S., and Aslam, Z. (2020b). Green copper nanoparticles from a native *Klebsiella pneumoniae* strain alleviated oxidative stress impairment of wheat plants by reducing the chromium bioavailability and increasing the growth. *Ecotoxicol. Environ. Saf.* 192: 110303.

Noor, S., Shah, Z., Javed, A., Ali, A., Hussain, S.B., Zafar, S., Ali, H., and Muhammad, S.A. (2020). A fungal based synthesis method for copper nanoparticles with the determination of anticancer, antidiabetic and antibacterial activities. *J. Microbiol. Methods* 174: 105966.

Nwanya, A.C., Ndipingwi, M.M., Ikpo, C.O., Obodo, R.M., Nwanya, S.C., Botha, S., Ezema, F.I., Iwuoha, E.I., and Maaza, M. (2020). *Zea mays* lea silk extract mediated synthesis of nickel oxide nanoparticles as positive electrode material for asymmetric supercabattery. *J. Alloys Compds.* 822: 153581.

Ochoa, L., Zuverza-Mena, N., Medina-Velo, I.A., Flores-Margez, J.P., Peralta-Videa, J.R., and Gardea-Torresdey, J.L. (2018). Copper oxide nanoparticles and bulk copper oxide, combined with indole-3-acetic acid, alter aluminum, boron, and iron in *Pisum sativum* seeds. *Sci. Total Environ.* 634: 1238–1245.

Órdenes-Aenishanslins, N.A., Saona, L.A., and Durán-Toro, V.M. (2019). Use of titanium dioxide nanoparticles biosynthesized by *Bacillus mycoides* in quantum dot sensitized solar cells. *Microb. Cell Fact.* 13: 90.

Ovais, M., Khalil, A.T., Islam, N.U., Ahmad, I., Ayaz, M., Saravanan, M., and Mukherjee, S. (2018). Role of plant phytochemicals and microbial enzymes in biosynthesis of metallic NPs. *Appl. Microbiol. Biotechnol.* 102: 6799–6814.

Oves, M., Aslam, M., Rauf, M.A., Qayyum, S., Qari, H.A., and Khan, M.S. (2018). Antimicrobial and anticancer activities of silver nanoparticles synthesized from the root hair extract of *Phoenix dactylifera*. *Mater. Sci. Eng. C Mater. Biol. Appl.* 89: 429–443.

Oves, M., Rauf, M.A., Hussain, A., Qari, H.A., Khan, A.A.P., Muhammad, P., Rehman, M.T., Alajmi, M.F., and Ismail, I.I.M. (2019). Antibacterial silver nanomaterial synthesis from *Mesoflavibacter zeaxanthinifaciens* and targeting biofilm formation. *Front. Pharmacol.* 10: 801.

Padma, P.N., Banu, S., and Kumari, S.C. (2018). Studies on green synthesis of copper nanoparticles using *Punica granatum*. *Ann. Res. Rev. Biol.* 23 (1): 1–10.

Pakrashi, S., Jain, N., Dalai, S., Jayakumar, J., Chandrasekaran, P.T., Raichur, A.M., and Mukherjee, A. (2014). *In vivo* genotoxicity assessment of titanium dioxide nanoparticles by *Allium cepa* root tip assay at high exposure concentrations. *PloS One* 9 (2): e87789.

Passos de Aragão, A., Maria de Oliveira, T., Quelemes, P.V., Perfeito, M.L.G., Araújo, M.C., Santiago, J.A.S., Cardoso, V.S., Quaresma, P., Souza de Almeida Leite, J.R., and Alves da Silva, D. (2019). Green synthesis of silver nanoparticles using the seaweed *Gracilaria birdiae* and their antibacterial activity. *Arab. J. Chem.* 12: 4182–4188.

Patil, D.S., Shetti, N.P., Nayak, D.S., and Revankar, R.S. (2019). Fabrication of multi-walled carbon nanotubes and ZnO nanoparticles composite electrode as a sensor for paracetamol. *Mater. Today Proc.* 18: 1124–1131.

Patil, M.P., Singh, R.D., Koli, P.B., Patil, K.T., Jagdale, B.S., Tipare, A.R., and Kim, G.D. (2018). Antibacterial potential of silver nanoparticles synthesized using *Madhuca longifolia* flower extract as a green resource. *Microb. Pathog.* 121: 184–189.

Pelegrino, M.T., Kohatsu, M.Y., Seabra, A.B., Monteiro, L.R., Gomes, D.G., Oliveira, H.C., Rolim, W.R., de Jesus, T.A., Batista, B.L., and Lange, C.N. (2020). Effects of copper oxide nanoparticles on growth of lettuce (*Lactuca sativa* L.) seedlings and possible implications of nitric oxide in their antioxidative defense. *Environ. Monit. Assess.* 192 (4): 232.

Peng, H. and Sejdic, J.T. (2009). Simple aqueous solution route to luminescent carbogenic dots from carbohydrates. *Chem. Mater.* 21 (23): 5563–5565.

Perez-Labrada, F., López-Vargas, E.R., Ortega-Ortiz, H., Cadenas-Pliego, G., Benavides-Mendoza, A., and Juarez-Maldonado, A. (2019). Responses of tomato plants under saline stress to foliar application of copper nanoparticles. *Plants* 8 (6): 151.

Pichardo-Molina, J.L., Cardoso-Avila, P.E., Flores-Villavicencio, L.L., Gomez-Ortiz, N.M., and Rodriguez-Rivera, M.A. (2020). Fluorescent carbon nanoparticles synthesized from bovine serum albumin nanoparticles. *Int. J. Biol. Macromol.* 142: 724–731.

Pithawalla, Y.B. and Deevi, S. (2004). Chemical synthesis of iron aluminide [FeAl] and iron aluminum carbide [Fe3AlC0. 5] nanopowders. *Mater. Res. Bull.* 39 (14–15): 2303–2316.

Poka, L.P., Krishna, M.G., Venkateswara, K.R., and Shanker, K. (2019). Biosynthesis, characterization and acute oral toxicity studies of synthesized iron oxide nanoparticles using ethanolic extract of *Centella asiatica* plant. *Mater. Lett.* 236: 256–259.

Premasudha, P., Venkataramana, M., Abirami, M., Vanathi, P., Krishna, K., and Rajendran, R. (2015). Biological synthesis and characterization of silver nanoparticles using *Eclipta alba* leaf extract and evaluation of its cytotoxic and antimicrobial potential. *Bull. Mater. Sci.* 38 (4): 965–973.

Priyanka, N. and Venkatachalam, P. (2016). Biofabricated zinc oxide NPs coated with phycomolecules as novel micronutrient catalysts for stimulating plant growth of cotton. *Adv. Nat. Sci.: Nanoscience Nanotechnol.* 7 (4): 045018.

Qayyum, S., Oves, M., and Khan, A.U. (2017). Obliteration of bacterial growth and biofilm through ROS generation by facilely synthesized green silver nanoparticles. *PLoS One* 12: 0181363.

Radini, I.A., Hansan, N., Malik, M.A., and Khan, Z. (2018). Biosynthesis of iron nanoparticles using *Trigonella foenum-graecum* seed extract for photocatalytic methyl orange dye degradation and antibacterial applications. *J. Photochem. Photobiol. B: Biol.* 183: 154–163.

Rafique, R., Arshad, M., Khokhar, M.F., Qazi, I.A., Hamza, A., and Virk, N. (2014). Growth response of wheat to titania nanoparticles application. *NUST J. Eng. Sci.* 7 (1): 42–46.

Rai-Kalal, P. and Jajoo, A. (2021). Priming with zinc oxide nanoparticles improve germination and photosynthetic performance in wheat. *Plant Physiol. Biochem.* 160: 341–351.

Rajiv, P., Bavadharani, B., Kumar, M.N., and Vanathi, P. (2017). Synthesis and characterization of biogenic iron oxide nanoparticles using green chemistry approach and evaluating their biological activities, *Biocatalysis Agri. Biotechnol.* 12: 45–49.

Rajivgandhi, G., Maruthupandy, M., Muneeswaran, T., Ramachandran, G., Manoharan, N., Quero, F., Anand, M., and Song, J. (2019). Biologically synthesized copper oxide nanoparticles enhanced intracellular damage in ciprofloxacin resistant ESBL producing bacteria. *Microb. Pathog.* 127: 267–276.

Ranjana, R., Parushuram, N., Harisha, K.S., Asha, S., and Sangappa, Y. (2020). Silk fibroin a bio-template for synthesis of different shaped gold nanoparticles: characterization and ammonia detection application. *Mater. Today: Proc.* 27: 434–439.

Rasheed, T., Bilal, M., Iqbal, H.M.N., and Li, C. (2017). Green biosynthesis of silver nanoparticles using leaves extract of *Artemisia vulgaris* and their potential biomedical applications. *Colloids Surf. B Biointerfaces* 158: 408–415.

Rathnayaka, R.M.N.N., Iqbal, Y.B., and Rifnas, L.M. (2018). Influence of urea and nano-nitrogen fertilizers on the growth and yield of rice (*Oryza sativa* L.) cultivar. *Int. J. Res. Publications* 5 (2): 7–7.

Ravichandran, V., Vasanthi, C., Shalini, S., Shah, S., Tripathy, M., and Paliwal, N. (2019). Green synthesis, characterization, antibacterial, antioxidant and photocatalytic activity of *Parkia speciosa* leaves extract mediated silver nanoparticles. *Results Phys.* 15: 102565.

Rawat, S., Samreen, K., Nayak, A.K., Singh, J., and Koduru, J.R. (2021). Fabrication of iron nanoparticles using Parthenium: a combinatorial eco-innovative approach to eradicate crystal violet dye and phosphate from the aqueous environment. *Environ. Nanotechnol. Monit. Manag.* 15: 100426.

Ray, S.C., Saha, A., and Jana, N.R. (2009). Fluorescent carbon nanoparticles: synthesis, characterization, and bioimaging application. *J. Phys. Chem. C* 113 (43): 18546–18551.

Razzaq, A., Ammara, R., Jhanzab, H.M., Mahmood, T., Hafeez, A., and Hussain, S. (2016). A novel nanomaterial to enhance growth and yield of wheat. *J. Nanosci. Technol.* 55–58.

Ren, Y.Y., Yang, H., Wang, T., and Wang, C. (2016). Green synthesis and antimicrobial activity of monodisperse silver nanoparticles synthesized using *Ginkgo biloba* leaf extract. *Phys. Lett. A* 380 (45): 3773–3777.

Rezvani, N., Sorooshzadeh, A., and Farhadi, N. (2012). Effect of nano-silver on growth of saffron in flooding stress. *World Acad. Sci. Eng. Technol.* 1: 517–522.

Riahi-Madvar, A., Rezaee, F., and Alili, V. (2012). Effects of alumina nanoparticles on morphological properties and antioxidant system of *Triticum aestivum*. *Iran. J. Plant Pathol.* 3: 595–603.

Rico, C.M., Morales, M., Barrios, A.C., McCreary, R., Hong, J., Lee, W.Y., Nunez, J., Peralta-Videa, J.R., and Gardea-Torresdey, J.L. (2013). Effect of cerium oxide nanoparticles on the quality of rice (*Oryza sativa* L.) grains. *J. Agri. Food Chem.* 61: 11278–11285.

Rizwan, M., Ali, S., Malik, S., Adrees, M., Qayyum, M.F., Alamri, S.A., and Ahmad, P. (2019). Effect of foliar applications of silicon and titanium dioxide nanoparticles on growth, oxidative stress, and cadmium accumulation by rice (*Oryza sativa*). *Acta Physiol. Plantarum* 41 (3): 1–12.

Rolim, W.R., Pelegrino, M.T., Lima, B.A., Ferraz, L.S., Costa, F.N., Bernardes, J.S., Rodigues, T., Brocchi, M., and Seabra, A.B. (2019). Green tea extract mediated biogenic synthesis of silver nanoparticles: characterization, cytotoxicity evaluation and antibacterial activity. *Appl. Surf. Sci.* 463: 66–74.

Romero-Cedillo, L., Poggi-Varaldo, H.M., Santoyo, J., Escamilla-Alvarado, C., Matsumoto-Kuwabara, Y., Ponce-Noyola, M.T., Bretón-Deval, L., and García-Rocha, M. (2020). Biological synthesis of iron nanoparticles using hydrolysates from a waste-based biorefinery. *Environ. Sci. Poll. Res. Int.* 23: 28649–28669.

Ruíz-Baltazar, Á.J., Reyes-López, S.Y., MondragónSánchez, M.L., Robles-Cortés, A.I., and Pérez, R. (2019). Eco-friendly synthesis of Fe3O4 nanoparticles: evaluation of their catalytic activity in methylene blue degradation by kinetic adsorption models. *Results Phys.* 7: 989–995.

Ruotolo, R., Maestri, E., Pagano, L., Marmiroli, M., White, J.C., and Marmiroli, N. (2018). Plant response to metal-containing engineered nanomaterials: an omics-based perspective. *Environ. Sci. Technol.* 52 (5): 2451–2467.

Sadak, M.S. (2019). Impact of silver nanoparticles on plant growth, some biochemical aspects, and yield of fenugreek plant (*Trigonella foenum-graecum*). *Bull. Nat. Res. Cent.* 43 (1): 1–6.

Safari, J. and Zarnegar, Z. (2014). Advanced drug delivery systems: nanotechnology of health design A review. *J. Saudi Chem. Soc.* 18 (2): 85–99.

Salama, H.M. (2012). Effects of silver nanoparticles in some crop plants, common bean (*Phaseolus vulgaris* L.) and corn (*Zea mays* L.). *Int. Res. J. Biotechnol.* 3 (10): 190–197.

Sangeetha, G., Rajeshwari, S., and Venckatesh, R. (2011). Green synthesis of zinc oxide nanoparticles by *Aloe barbadensis* miller leaf extract: structure and optical properties. *Mat. Res. Bull.* 46 (12): 2560–2566.

Santhoshkumar, J., Sowmya, B., Venkat Kumar, S., and Rajeshkumar, S. (2019). Toxicology evaluation and antidermatophytic activity of silver nanoparticles synthesized using leaf extract of *Passiflora caerulea*. *S. Afr. J. Chem. Eng.* 29: 17–23.

Sarkar, J., Mollick, M.M.R., Chattopadhyay, D., and Acharya, K. (2017). An eco-friendly route of γ-Fe2O3 nanoparticles formation and investigation of the mechanical properties of the HPMC-γ-Fe2O3 nanocomposites. *Bioprocess Biosyst. Eng.* 40 (3): 351–359.

Savithramma, N., Ankanna, S., and Bhumi, G. (2012). Effect of nanoparticles on seed germination and seedling growth of *Boswellia ovalifoliolata* an endemic and endangered medicinal tree taxon. *Nano Vision* 2 (1): 2.

Saxena, R., Tomar, R.S., and Kumar, M. (2016). Exploring nanobiotechnology to mitigate abiotic stress in crop plants. *J. Pharm. Sci. Res.* 8 (9): 974.

Sebastian, A., Nangia, A., and Prasad, M.N.V. (2018). A green synthetic route to phenolics fabricated magnetite nanoparticles from coconut husk extract: implications to treat metal contaminated water and heavy metal stress in *Oryza sativa* L. *J. Clean Prod.* 174: 355–366.

Sedghi, M., Hadi, M., and Toluie, S.G. (2013). Effect of nano zinc oxide on the germination parameters of soybean seeds under drought stress. Annales of West University of Timisoara. *Series of Biol.* 16 (2): 73.

Semida, W.M., Abdelkhalik, A., Mohamed, G., El-Mageed, A., Taia, A., El-Mageed, A., and Ali, E.F. (2021). Foliar application of zinc oxide nanoparticles promotes drought stress tolerance in eggplant (*Solanum melongena* L.). *Plants* 10 (2): 421.

Sen, S.K., Chouhan, D., Das, D., Ghosh, R., and Mandal, P. (2020). Improvisation of salinity stress response in mung bean through solid matrix priming with normal and nanosized chitosan. *Int. J. Biol. Macromol.* 145: 108–123.

Shang, Y., Hasan, M.K., Ahammed, G.J., Li, M., Yin, H., and Zhou, J. (2019). Applications of nanotechnology in plant growth and crop protection: a review. *Molecules* 24: 2558.

Shankramma, K., Yallappa, S., Shivanna, M.B., and Manjanna, J. (2016). Fe_2O_3 magnetic nanoparticles to enhance *S. lycopersicum* (tomato) plant growth and their biomineralization. *Appl. Nanosci.* 6 (7): 983–990.

Sharma, D., Ledwani, L., Mehrotra, T., Kumar, N., Pervaiz, N., and Kumar, R. (2020). Biosynthesis of hematite nanoparticles using *Rheum emodi* and their antimicrobial and anticancerous effects *in vitro*. *J. Photochem. Photobiol. B: Biol.* 2020 206.

Sharma, P., Bhatt, D., Zaidi, M., Saradhi, P.P., Khanna, P., and Arora, S. (2012). Silver nanoparticle-mediated enhancement in growth and antioxidant status of *Brassica juncea*. *Appl. Biochem. Biotechnol.* 167: 2225–2233.

Sharma, P., Pant, S., Dave, V., Tak, K., Sadhu, V., and Reddy, K.R. (2019). Green synthesis and characterization of copper nanoparticles by *Tinospora cardifolia* to produce nature-friendly copper nano-coated fabric and their antimicrobial evaluation. *J. Microbiol. Methods* 60: 107–116.

Sharma, S., Yadav, D.K., Chawla, K., Lal, N., and Lal, C. (2021). Synthesis and characterization of CuO nanoparticles by *Aloe barbadensis* leaves. *Quantum J. Eng. Sci. Technol.* 2 (5): 1–9.

Shaw, A.K. and Hossain, Z. (2013). Impact of nano-CuO stress on rice (*Oryza sativa* L.) seedlings. *Chemosphere* 93 (6): 906–915.

Shewry, P.R. (2009). Wheat. *J. Exp. Bot.* 60 (6): 1537–1553.

Sheykhbaglou, R., Sedghi, M., Shishevan, M.T., and Sharifi, R.S. (2010). Effects of nano-iron oxide particles on agronomic traits of soybean. *Not. Sci. Biol.* 2 (2): 112–113.

Silveira, C., Shimabuku, Q.L., Fernandes Silva, M., and Bergamasco, R. (2017). Iron-oxide nanoparticles by green synthesis method using *Moringa oleifera* leaf extract for fluoride removal. *Environ. Technol.* 3330: 1–40.

Singh, A., Tiwari, S., Pandey, J., Lata, C., and Singh, I.K. (2021). Role of nanoparticles in crop improvement and abiotic stress management. *J. Biotechnol.* 337: 57–70.

Singh, J., Mehta, A., Rawat, M., and Basu, S. (2018). Green synthesis of silver nanoparticles using sun dried tulsi leaves and its catalytic application for 4-Nitrophenol reduction. *J. Environ. Chem. Eng.* 6: 1468–1474.

Singh, R., Shedbalkar, U.U., Wadhwani, S.A., and Chopade, B.A. (2015). Bacteriagenic silver nanoparticles: synthesis, mechanism, and applications. *Appl. Microbiol. Biotechnol.* 99 (11): 4579–4593.

Sirdeshpande, K.D., Sridhar, A., Cholkar, K.M., and Selvaraj, R. (2018). Structural characterization of mesoporous magnetite nanoparticles synthesized using the leaf extract of *Calliandra haematocephala* and their photocatalytic degradation of malachite green dye. *Appl. Nanosci.* 8: 675–683.

Slomberg, D.L. and Schoenfisch, M.H. (2012). Silica nanoparticle phytotoxicity to *Arabidopsis thaliana*. *Environ. Sci. Technol.* 46 (18): 10247–10254.

Soliman, M.G., Pelaz, B., Parak, W.J., and Del Pino, P. (2015). Phase transfer and polymer coating methods toward improving the stability of metallic nanoparticles for biological applications. *Chem. Mater.* 27 (3): 990–997.

Sorbiun, M., Mehr, S.E., Ramazani, A., and Fardood, S.T. (2018). Biosynthesis of Ag, ZnO and bimetallic Ag/ZnO alloy nanoparticles by aqueous extract of oak fruit hull (Jaft) and investigation of photocatalytic activity of ZnO and bimetallic Ag/ZnO for degradation of basic violet 3 dye. *J. Mater. Sci.: Mater. Electron.* 29: 2806–2814.

Sukumar, K., Sengottaiyan, A., Selvankumar, T., Senthilkumar, B., Sudhakar, C., and Selvam, K. (2020). Eco-friendly cost-effective approach for synthesis of copper oxide nanoparticles for enhanced photocatalytic performance. *Optik* 202: 163507.

Sundararajan, B. and Kumari, B.R. (2017). Novel synthesis of gold nanoparticles using *Artemisia vulgaris* L. leaf extract and their efficacy of larvicidal activity against dengue fever vector *Aedes aegypti* L. *J. Trace Elem. Med. Biol.* 43: 187–196.

Suramwar, N.V., Thakare, S.R., and Khaty, N.T. (2016). One pot synthesis of copper nanoparticles at room temperature and its catalytic activity. *Arabian J. Chem.* 9(2): S1807–S1812.

Sundararajan, B., Mahendran, G., Thamaraiselvi, R., and Kumari, B.D.R. (2021). Biological activities of synthesized silver nanoparticles from *Cardiospermum halicacabum* L. *Bull. Mater. Sci.* 39: 423–431.

Syu, Y.Y., Hung, J.H., Chen, J.C., and Chuang, H.W. (2014). Impacts of size and shape of silver nanoparticles on *Arabidopsis* plant growth and gene expression. *Plant Physiol. Biochem.* 83: 57–64.

Tang, H., Wei, J., Liu, F., Qiao, B., Pan, X., Li, L., and Zhang, T. (2016). Strong metal–support interactions between gold nanoparticles and nonoxides. *J. Am. Chem. Soc.* 138 (1): 56–59.

Tao, H., Yang, K., Ma, Z., Wan, J., Zhang, Y., Kang, Z., and Liu, Z. (2012). In vivo NIR fluorescence imaging, biodistribution, and toxicology of photoluminescent carbon dots produced from carbon nanotubes and graphite. *Small* 8 (2): 281–290.

Taran, N., Storozhenko, V., Svietlova, N., Batsmanova, L., Shvartau, V., and Kovalenko, M. (2017). Effect of zinc and copper nanoparticles on drought resistance of wheat seedlings. *Nanoscale Res. Lett.* 12: 60.

Tavakoli, A., Sohrabi, M., and Kargari, A. (2007). A review of methods for synthesis of nanostructured metals with emphasis on iron compounds. *Chem. Pap.* 61 (3): 151–170.

Thiruvengadam, M., Chung, I.M., Gomathi, T., Ansari, M.A., Gopiesh, K.V., Babu, V., and Rajakumar, G. (2019). Synthesis, characterization and pharmacological potential of green synthesized copper nanoparticles. *Bioprocess Biosyst. Eng.* 42: 1769–1777.

Thomas, B., Vithiya, S.M., Prasad, T.A.A., Mohamed, S.B., Magdalane, C.M., Kaviyarasu, K., and Maaza, M. (2019). Antioxidant and photocatalytic activity of aqueous leaf extract mediated green synthesis of silver nanoparticles using *Passiflora edulis f. flavicarpa*. *J. Nanosci. Nanotechnol.* 19 (5): 2640–2648.

Tiwari, M., Krishnamurthy, S., Shukla, D., Kiiskila, J., Jain, A., Datta, R., and Sahi, S.V. (2016). Comparative transcriptome and proteome analysis to reveal the biosynthesis of gold NPs in *Arabidopsis*. *Sci. Rep.* 6: 21733.

Torabian, S., Zahedi, M., and Khoshgoftarmanesh, A. (2018). Effect of foliar spray of zinc oxide on some antioxidant enzymes activity of sunflower under salt stress. *J. Agri. Sci. Technol.* 18: 1013–1025.

Tripathi, D., Modi, A., Narayan, G., and Rai, S.P. (2019). Green and cost effective synthesis of silver nanoparticles from endangered medicinal plant *Withania coagulans* and their potential biomedical properties. *Mater. Sci. Eng.: C* 100: 152–164.

Tripathi, D.K., Singh, S., Singh, S., Mishra, S., Chauhan, D.K., and Dubey, N.K. (2015). Micronutrients and their diverse role in agricultural crops: advances and future prospective. *Acta Physiol. Plantarum* 37 (7): 1–14.

Tripathi, D.K., Singh, S., Singh, S., Pandey, R., Singh, V.P., Sharma, N.C., and Chauhan, D.K. (2017a). An overview on manufactured nanoparticles in plants: uptake, translocation, accumulation and phytotoxicity. *Plant Physiol. Biochem.* 110: 2–12.

Tripathi, D.K., Singh, V.P., Kumar, D., and Chauhan, D.K. (2012). Impact of exogenous silicon addition on chromium uptake, growth, mineral elements, oxidative stress, antioxidant capacity, and leaf and root structures in rice seedlings exposed to hexavalent chromium. *Acta Physiol. Plantarum* 34 (1): 279–289.

Tripathi, P.K., Gan, L., and Liu, M. (2016). Synthesis of large pore carbon nanoparticles for removal of malachite green. *J. Nanosci. Nanotechnol.* 16 (1): 892–897.

Upadhyaya, H., Begum, L., Dey, B., Nath, P.K., and Panda, S.K. (2017). Impact of calcium phosphate nanoparticles on rice plant. *J. Plant Sci. Phytopathol.* 1: 1–10.

Valletta, A., Chronopoulou, L., Palocci, C., Baldan, B., Donati, L., and Pasqua, G. (2014). Poly (lactic-co-glycolic) acid nanoparticles uptake by *Vitis vinifera* and grapevine-pathogenic fungi. *J. Nanoparticle Res.* 16: 2744.

Van Breusegem, F. and Dat, J.F. (2006). Reactive oxygen species in plant cell death. *Plant Physiol.* 141: 384–390.

Vannini, C., Domingo, G., Onelli, E., Prinsi, B., Marsoni, M., Espen, L., and Bracale, M. (2013). Morphological and proteomic responses of *Eruca sativa* exposed to silver nanoparticles or silver nitrate. *PloS One* 8 (7): 68752.

Vasantharaj, S., Sathiyavimal, S., Saravanan, M., Senthilkumar, P., Gnanasekaran, K., Shanmugavel, M., and Pugazhendhi, A. (2019). Synthesis of ecofriendly copper oxide nanoparticles for fabrication over textile fabrics: characterization of antibacterial activity and dye degradation potential. *J. Photochem. Photobiol. B: Biol.* 191: 143–149.

Verma, A., Gautam, S.P., Bansal, K.P., Prabhakar, N., and Rosenholm, J.M. (2019). Green nanotechnology: advancement in phytoformulation research. *Medicines* 6 (1): 10.3390/ medicines6010039

Venkatachalam, P., Priyanka, N., Manikandan, K., Ganeshbabu, I., Indiraarulselvi, P., Geetha, N., and Sahi, S.V. (2017). Enhanced plant growth promoting role of phycomolecules coated zinc oxide nanoparticles with P supplementation in cotton (*Gossypium hirsutum* L.). *Plant Physiol. Biochem.* 110: 118–127.

Vijayan, R., Joseph, S., and Mathew, B. (2019). Anticancer, antimicrobial, antioxidant, and catalytic activities of green synthesized silver and gold nanoparticles using *Bauhinia purpurea* leaf extract. *Bioprocess Biosyst. Eng.* 42: 305–319.

Wahid, I., Kumari, S., Ahmad, R., Hussain, S.J., Alamri, S., Siddiqui, M.H., and Khan, M.I.R. (2020). Silver nanoparticle regulates salt tolerance in wheat through changes in ABA concentration, ion homeostasis, and defense systems. *Biomolecules* 10: 1506.

Wan, J., Wang, R., Bai, H., Wang, Y., and Xu, J. (2020). Comparative physiological and metabolomics analysis reveals that single-walled carbon nanohorns and ZnO nanoparticles affect salt tolerance in *Sophora alopecuroides*. *Environ. Sci.: Nano* 7: 2968–2981.

Wang, F., Banerjee, D., Liu, Y., Chen, X., and Liu, X. (2010). Upconversion nanoparticles in biological labeling, imaging, and therapy. *Analyst* 135 (8): 1839–1854.

Wang, F., Liu, X., Shi, Z., Tong, R., Adams, C.A., and Shi, X. (2016a). Arbuscular mycorrhizae alleviate negative effects of zinc oxide nanoparticle and zinc accumulation in maize plants – a soil microcosm experiment. *Chemosphere* 147: 88–97.

Wang, H., Lee, H.W., Deng, Y., Lu, Z., Hsu, P.C., Liu, Y., and Cui, Y. (2015). Bifunctional non-noble metal oxide nanoparticle electrocatalysts through lithium-induced conversion for overall water splitting. *Nature Commun.* 6 (1): 1–8.

Wang, L., Wu, Y., Xie, J., Wu, S., and Wu, Z. (2018). Characterization, antioxidant and antimicrobial activities of green synthesized silver nanoparticles from *Psidium guajava* L. leaf aqueous extracts. *Mater. Sci. Eng.: C* 86: 1–8.

Wang, R., Liu, H., Wang, X., Li. X., Gu, X., and Zheng, Z. (2020). Plasmon-enhanced furfural hydrogenation catalyzed by stable carbon-coated copper nanoparticles driven from metal–organic frameworks. *Catal. Sci. Technol.* 10: 6483–6494.

Wang, X., Han, H., Liu, X., Gu, X., Chen, K., and Lu, D. (2012). Multi-walled carbon nanotubes can enhance root elongation of wheat (*Triticum aestivum*) plants. *J. Nanoparticle Res.* 14: 841.

Wang, X. and Komatsu, S. (2017). Improvement of soybean products through the response mechanism analysis using proteomic technique. *Adv. Food Nutr. Res.* 82: 117–148.

Wang, X., Sun, W., and Ma, X. (2019). Differential impacts of copper oxide nanoparticles and copper (II) ions on the uptake and accumulation of arsenic in rice (Oryza sativa). *Environ. Pollution*, 252: 967–973.

Wang, Z., Xu, L., Zhao, J., Wang, X., White, J.C., and Xing, B. (2016b). CuO nanoparticle interaction with *Arabidopsis thaliana*: toxicity, parent-progeny transfer, and gene expression. *Environ. Sci. Technol.* 50 (11): 6008–6016.

Wei, Y., Fang, Z., Zheng, L., and Pokeung, E. (2017). Biosynthesized iron nanoparticles in aqueous extracts of *Eichhornia crassipes* and its mechanism in the hexavalent chromium removal. *Appl. Surf. Sci.* 399: 322–329.

Wei, Y., Fang, Z., Zheng, L., Tan, L., and Tsang, E.P. (2016). Green synthesis of Fe nanoparticles using *Citrus maxima* peels aqueous extracts. *Mat. Lett.* 185: 384–386.

Weng, X., Guo, M., Luo, F., and Chen, Z. (2017). One-step green synthesis of bimetallic Fe/Ni nanoparticles by eucalyptus leaf extract: biomolecules identification, characterization and catalytic activity. *Chem. Eng. J.* 308: 904–911.

Wiesner, M.R., Lowry, G.V., Alvarez, P., Dionysiou, D., and Biswas, P. (2006). Assessing the risks of manufactured nanomaterials. *Environ. Sci. Technol.* 40: 4336–4345.

Williams, M.J., Sánchez, E., Aluri, E.R., Douglas, F.J., MacLaren, D.A., and Collins, O.M. (2016). Microwave-assisted synthesis of highly crystalline, multifunctional iron oxide nanocomposites for imaging applications. *RSC Adv.* 6 (87): 83520–83528.

Wu, F., Fang, Q., Yan, S., Pan, L., Tang, X., and Ye, W. (2020). Effects of zinc oxide nanoparticles on arsenic stress in rice (*Oryza sativa* L.): germination, early growth, and arsenic uptake. *Environ. Sci. Poll. Res.* 27 (21): 26974–26981.

Wu, L., Cai, X., and Nelson, K. (2013). A green synthesis of carbon nanoparticles from honey and their use in real-time photoacoustic imaging. *Nano Res.* 6: 312–325.

Wypij, M., Jędrzejewski, T., Trzcińska-Wencel, J., Ostrowski, M., Rai, M., and Golinska, P. (2021). Green synthesized silver nanoparticles: antibacterial and anticancer activities, biocompatibility, and analyses of surface-attached proteins. *Front. Microbiol.* 12: 888.

Xu, H., Yu, T., Fu, Y., Dang, Z., Wang, L., Xie, S., and Ren, Q. (2020). Biosynthesis of Ag nanoparticles and two-dimensional element distribution in *Arabidopsis*. *IET Nanobiotechnol.* 14 (4): 325–330.

Xu, X.Y., McGrath, S.P., and Zhao, F.J. (2007). Rapid reduction of arsenate in the medium mediated by plant roots. *New Phytol.* 176 (3): 590–599.

Yang, A.H., Moore, S.D., Schmidt, B.S., Klug, M., Lipson, M., and Erickson, D. (2009). Optical manipulation of nanoparticles and biomolecules in sub-wavelength slot waveguides. *Nature* 457 (7225): 71–75.

Yang, Z., Xiao, Y., Jiao, T., Zhang, Y., Chen, J., and Gao, Y. (2020). Effects of copper oxide nanoparticles on the growth of rice (*Oryza sativa* L.) seedlings and the relevant physiological responses. *Int. J. Environ. Res. Public Health* 17 (4): 1260.

Yang, Z.C., Wang, M., Yong, A.M., Wong, S.Y., Zhang, X.H., Tan, H., Chang, A.Y., Li, X., and Wang, J. (2011). Intrinsically fluorescent carbon dots with tunable emission derived from hydrothermal treatment of glucose in the presence of monopotassium phosphate. *Chem. Commun.* 47 (42): 11615–11617.

Yasmeen, F., Raja, N.I., Ilyas, N., and Komatsu, S. (2018). Quantitative proteomic analysis of shoot in stress tolerant wheat varieties on copper nanoparticle exposure. *Plant Mol. Biol. Rep.* 36 (2): 326–340.

Yasmeen, F., Raja, N.I., Mustafa, G., Sakata, K., and Komatsu, S. (2016a). Quantitative proteomic analysis of post-flooding recovery in soybean root exposed to aluminum oxide nanoparticles. *J. Proteom.* 143: 136–150.

Yasmeen, F., Raja, N.I., Razzaq, A., and Komatsu, S. (2016b). Gel-free/label-free proteomic analysis of wheat shoot in stress tolerant varieties under iron nanoparticles exposure. *Biochim. et Biophys. Acta-Proteins and Proteom.* 1864 (11): 1586–1598.

Yasmeen, F., Raja, N.I., Razzaq, A., and Komatsu, S. (2017a). Proteomic and physiological analyses of wheat seeds exposed to copper and iron nanoparticles. *Biochim. Biophys. Acta (BBA): Proteins and Proteomics*, 1865 (1): 28–42.

Yasmeen, F., Raja, N.I., Razzaq, A., and Komatsu, S. (2017b). Gel-free/label-free proteomic analysis of wheat shoot in stress tolerant varieties under iron nanoparticles exposure. *Biochim. Biophys. Acta*, 1864 (2016): 1586–1598.

Yoon, S.J., Kwak, J.I., Lee, W.M., Holden, P.A., and An, Y.J. (2014). Zinc oxide nanoparticles delay soybean development: a standard soil microcosm study. *Ecotoxicol. Environ. Saf.* 100: 131–137.

Yu, H., Li, X., Zeng, X., and Lu, Y. (2015). Preparation of carbon dots by non-focusing pulsed laser irradiation in toluene. *Chem. Commun.* 52 (4): 819–822.

Yu, M., Yao, J., Liang, J., Zeng, Z., Cui, B., Zhao, X., and Cui, H. (2017). Development of functionalized abamectin poly (lactic acid) nanoparticles with regulatable adhesion to enhance foliar retention. *RSC Adv.* 7: 11271–11280.

Yuan, J., He, A., Huang, S., Hua, J., and Sheng, G.D. (2016). Internalization and phytotoxic effects of CuO nanoparticles in *Arabidopsis thaliana* as revealed by fatty acid profiles. *Environ. Sci. Technol.* 50: 10437–10447.

Zarei, M., Karimi, E., Oskoueian, E., Es-Haghi, A., and Yazdi, M.E.T. (2021). Comparative study on the biological effects of sodium citrate-based and apigenin-based synthesized silver nanoparticles. *Nutr. Cancer* 73 (8): 1511–1519.

Zhai, X., Zhang, P., Liu, C., Bai, T., Li, W., Da, L., and Liu, W. (2012). Highly luminescent carbon nanodots by microwave-assisted pyrolysis. *Chem. Commun.* 48 (64): 7955–7957.

Zhang, H., Wang, R., Chen, Z., Cui, P., Lu, H., Yang, Y., and Zhang, H. (2021). The effect of zinc oxide nanoparticles for enhancing rice (*Oryza sativa* L.) yield and quality. *Agriculture* 11: 1247.

Zhang, Q., Dehaini, D., Zhang, Y., Zhou, J., Chen, X., Zhang, L., and Zhang, L. (2018). Neutrophil membrane-coated nanoparticles inhibit synovial inflammation and alleviate joint damage in inflammatory arthritis. *Nat. Nanotechnol.* 13 (12): 1182–1190.

Zhang, W., Qiao, X., and Chen, J. (2007). Synthesis of nanosilver colloidal particles in water/oil microemulsion. *Colloids Surf. A: Physicochem. Eng. Asp.* 299: 22–28.

Zhang, X.F., Liu, Z.G., Shen, W., and Gurunathan, S. (2016). Silver nanoparticles: synthesis, characterization, properties, applications, and therapeutic approaches. *Int. J. Mol. Sci.* 17 (9): 1534.

Zhu, S., Huang, X., Wang, D., Wang, L., and Ma, F. (2018). Enhanced hexavalent chromium removal performance and stabilization by magnetic iron nanoparticles assisted biochar in aqueous solution: mechanisms and application potential. *Chemosphere*, 207: 50–59.

Zhao, F., Xin, X., Cao, Y., Su, D., Ji, P., Zhu, Z., and He, Z. (2021). Use of carbon nanoparticles to improve soil fertility, crop growth and nutrient uptake by corn (*Zea mays* L.). *Nanomaterials* 11 (10): 2717.

Zhou, J., Booker, C., Li, R., Zhou, X., Sun, X., and Ding, Z. (2007). An electrochemical avenue to blue luminescent nanocrystals from multi-walled carbon nanotubes (MWCNTs). *J. Chem. Soc.* 129 (4): 744–745.

Zhou, X., Torabi, M., Lu, J., Shen, R., and Zhang, K. (2014). Nanostructured energetic composites: synthesis, ignition/combustion modeling, and applications. *ACS Appl. Mater. Interf.* 6: 3058–3074.

17

Latest Insights on Genomic and Epigenomic Mechanisms of Nanotoxicity

Vratko Himič[1,2], Nikolaos Syrmos[3], Gianfranco K.I. Ligarotti[4], and Mario Ganau[1,]*

[1] *Nuffield Department of Clinical Neurosciences, University of Oxford, UK*
[2] *University of Oxford Medical School, Oxford, UK*
[3] *School of Medicine, Aristotle University of Thessaloniki, Greece*
[4] *Institute of AeroSpace Medicine, Milan*
[*] *Corresponding author*

Introduction

While there is a large body of evidence concerning the cytotoxicity of engineered nanomaterials (ENMs), relatively little is known about their adverse effects at the genomic and epigenomic level. The substantial increase in the presence of ENMs in all spheres of life science, from everyday exposure to cutting-edge drug delivery and therapeutic design, also brings with it a responsibility to evaluate their effects through the lens of genomics and epigenomics [1–4]. While the study of genetics concerns individual genes to high degrees of detail, genomics is concerned with the expression patterns and interactions of all genes that make up a person's genome. Similarly, epigenomics is the study of the various proteins and compounds that influence the genome through signaling and regulation.

These questions are important in order to contextualize the place of nanomedicines in today's world, from the laboratories to the hospital environment where diagnostics tools and surgery in the realm of nanometers are becoming a reality [2, 3, 5, 6]. We will see that while significant unknowns remain, novel work has uncovered, and in some cases quantified, the genotoxic effects of ENMs to cells and tissues.

In this chapter, we will look at the impact that ENMs have on genes and their expression, and how genetic and epigenetic mediators could underscore several of their nanotoxic side effects. After discussing the mechanisms of direct and indirect genotoxicity, we will review the contribution of ENMs to DNA damage as well as their effects on epigenetic pathways. We will close with a comparison of how the different properties that convey ENM genotoxicity (like size and charge) can eventually be exploited to suit modern therapeutic needs.

Mechanisms of Genotoxicity

In the universe of genomics, one of the leading research trends aims at exploring the mechanisms of damage to the DNA. Epigenomics studies the changes that occur through the superimposition of chemical and enzyme-mediated changes that act on DNA and RNA to up- or downregulate gene expression. There are several mechanisms through which this genotoxicity can be induced: namely, we defined direct and indirect genotoxicity, the former occurring where ENMs directly bind the DNA, the latter affecting gene expression via a secondary cascade of events (see Table 17.1). In fact, when

Impact of Engineered Nanomaterials in Genomics and Epigenomics, First Edition. Edited by Saura C. Sahu.

Table 17.1 Summary of mechanisms of genotoxicity.

Mechanism	Principles
Direct genotoxicity	Access to DNA • during mitosis stage, when nuclear membrane has not enveloped genetic material • pass through nuclear pores directly if small enough • damage nuclear membrane to gain access Binding to DNA • direct binding • electrostatic forces
Indirect genotoxicity	
Oxidative stress	Generation of reactive oxygen species through mitochondrial dysfunction that damage DNA or target epigenetic processes
Inflammatory changes	Cytokine cocktail release form macrophages that engulf and process the ENM May have broad-ranging and longer-term side effects like tumorigenesis
DNA processing and repair	Targeting DNA replication, transcription, and the proteins involved in the DNA damage response, turning temporary genotoxicity into permanent genomic changes
Hypoxia	Biomarkers of hypoxia, including hypoxia-inducible factor seem to be upregulated; however, evidence is not clear
Epigenetic changes	DNA methylation • affecting the extent to which genes are transcribed, based on the balance of addition or removal of a methyl group to nucleotides Histone modifications • like DNA methylation, ENMs can affect the extent to which histone proteins that compact DNA into nucleosomes are acetylated • acetylated state relaxes chromatin and increases transcription, deacetylation preventing transcription and access of DNA repair proteins Noncoding RNAs • vast and complex set of regulatory RNAs that affect the expression of other protein-coding genes, of each other, and on intracellular signaling pathways

ENMs enter nuclei, they can immediately exert many effects on DNA; on the other hand, oxidative stress and a pro-inflammatory state can lead to genomic and epigenomic changes via indirect damage that may occur if there is a perturbation in the DNA repair pathways normally employed by cells to continuously repair and reverse any unwanted lesions [7].

Direct Genotoxicity

An initial evidence of direct genotoxic lesions to DNA was provided when scientists started investigating the role of silver nanoparticles (AgNPs) in DNA damage through cell-cycle arrest [8, 9]; since then, several other ways through which ENMs enact their damaging effects on DNA have been identified (see Figure 17.1). More specifically, ENMs can do so by damaging the DNA during the mitosis stage, when the nuclear membrane has not yet enveloped the genetic material (obviously this can happen only if they are small enough and can be internalized through the nuclear pores); or alternatively, they can interact with the DNA having previously caused any sort of damage the nuclear membrane (either intentionally or by accident) to gain access to the nucleus [7, 10–12]. There is also

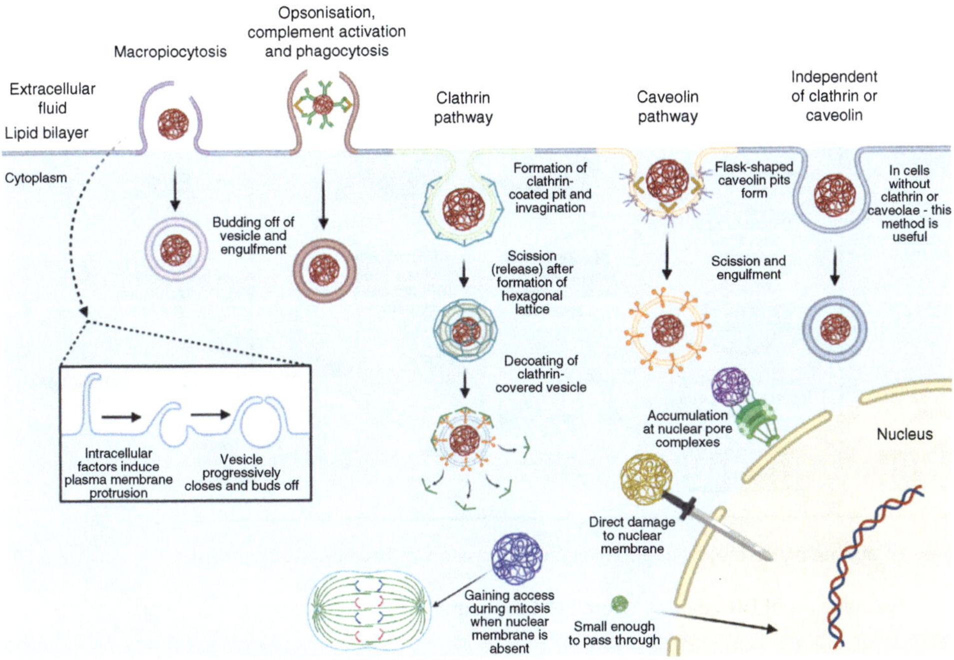

Figure 17.1 Cellular and nuclear uptake of ENMs.

a growing evidence base regarding the mechanisms that the ENMs adopt to bind the DNA and exert their genotoxic effects. Some work has suggested a direct binding process with DNA [13, 14], others, via electrostatic forces [15, 16]. Therefore, both the passage of ENMs and their interactions with genetic material are potential mechanisms of genotoxic effect.

Indirect Genotoxicity

On the other hand, ENMs can cause indirect genotoxicity through any of the following: oxidative stress, inflammation, DNA processing and repair, hypoxia, and epigenetic changes [7]. While it may be tempting to characterize these separately, it is clear that these pathways are interlinked (see Figure 17.2); for example, inflammation can directly induce cytotoxicity and apoptosis via oxidative stress [17], but it can also activate various genetic and epigenetic pathways that may be independent or may well influence and reinforce each other. Those aspects will be briefly summarized below:

i) **Oxidative stress.** Several ENMs can induce the generation of reactive oxygen species (ROS), which, as a downstream effect, damage DNA or affect epigenetic processes. ROS can be produced through mitochondrial dysfunction, which was firstly demonstrated through a very low-dose exposure of AgNPs in rats [18]. This oxidative stress can have effects that are directly genotoxic or induce epigenetic changes without direct genotoxicity. Several lines of evidence suggest that ENMs cause mitochondrial dysfunction and induce the production of ROS from aberrant functioning of the electron transport chain of the mitochondria (potentially by decreasing the mitochondrial membrane potential as shown in ferric oxide and zinc oxide NPs) [19–21].

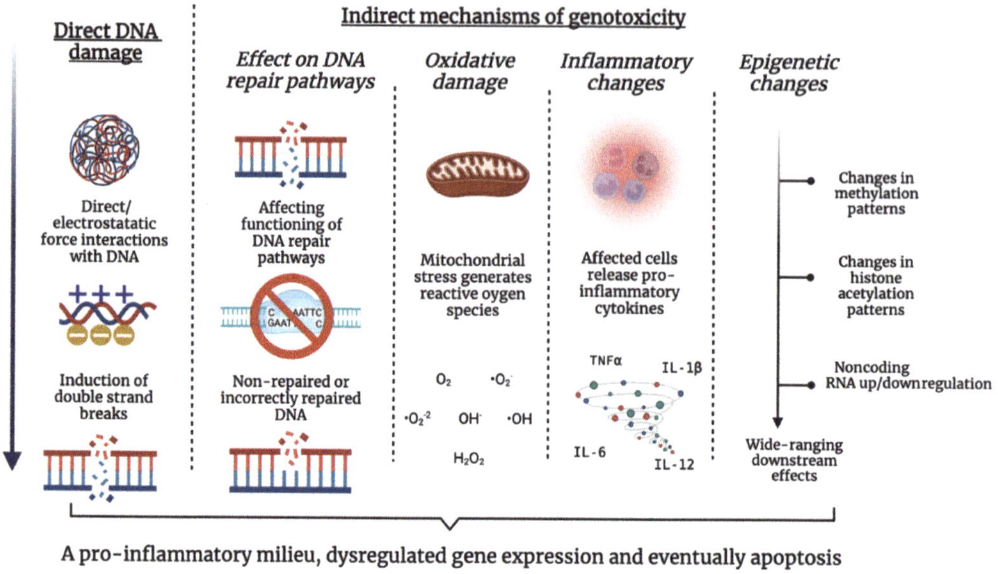

Figure 17.2 Mechanisms of DNA damage from ENM exposure.

ii) **Inflammatory changes.** ENMs can also convey their cytotoxic effects via the release of a cytokine cocktail. These inflammatory mediators are released by the macrophages that process the ENMs, particularly those in the liver and spleen. This pro-inflammatory state can also lead to secondary effects, such as increasing the likelihood of tumor formation.

iii) **DNA processing and repair.** ENMs can also exert their genotoxic effects on gene expression by targeting the different components of DNA processing, such as DNA replication, transcription and, crucially, the processes that are used in DNA repair (also called DNA damage response [DDR]). DDR is central to combating and reversing some of the direct forms of genotoxicity, and DDR genes are important in preventing permanent gene changes that can lead to the development of dysregulated growth and, eventually, cancerous lesions [22].

iv) **Hypoxia.** Acting through other pathways, ENMs can induce the activation of certain factors and signaling markers (hypoxia-inducible factor, HIF) [9], although this effect has rarely been documented and is not clearly understood [7].

v) **Epigenetic changes.** Lastly, as well as influencing the underlying genome, ENMs are also being shown to have an impact on epigenetic processes, albeit the latter represents a developing and relatively novel research interest. Several epigenetic modifications have been demonstrated, including DNA methylation, histone modifications, and the alteration of the levels of noncoding RNAs. These modifications will be discussed in detail later in this chapter.

Genomic Consequences of ENM Exposure

Over the years, there has been growing concern about the effects of exposure to ENMs on our underlying genetic information, namely DNA, RNA, and exosomes, in any industry where nanomaterials have been introduced and are currently employed. The seminal finding that cobalt nanoparticle exposure can cause DNA damage in the human lungs through oxidative stress has catalyzed an interest in characterizing the different ways (see Table 17.2) through which each subclass of ENMs can contribute to DNA damage [23]. The consequences of the interactions between ENMs and DNA can result in a broad spectrum of lesions in the DNA, such as strand breaks (be they single, double, within, or between strands), the formation of modified bases, and rearrangements of the genetic material. Ultimately, depending on the severity and scope of the alterations caused by ENM, this can result in aberrant gene expression (with the production of mutant protein structures), cell death, or even cancerous changes within cells [24, 25].

Direct DNA Damage

Direct DNA genotoxicity occurs after exposure to several different ENMs that manage to gain access to the genetic material. And while delivering NPs to the nucleus of cells may be advantageous and even a desired mechanism of delivering cancer therapies [26, 27], the unintended consequences of DNA exposure to ENMs could be deleterious to healthy cells. For example, carbon black, a type of carbon nanoparticle (CNP), induced double strand breaks (DSBs) by electrostatically binding to DNA in mouse livers after only 24 hours of co-culture [13]. A similar effect was

Table 17.2 Genomic effects of various engineered nanomaterials.

Mechanism	Engineered nanomaterials implicated
Direct DNA damage	• Carbon black NP-induced double strand breaks (DSBs) in mouse hepatocyte model after 24 h [13] • Gold nanoparticles induction of DSBs is exacerbated when coated with polyvinylpyrrolidone [14] • Graphene quantum dots can cause early apoptosis and induce cross-linking of DNA [30]
Oxidative damage	• Silicon dioxide NPs cause DSBs in lung cells [31] • Citrated silver nanoparticles induce oxidative stress and DNA damage in a dose- and time-dependent manner [32] • Graphene oxide (GO) NPs induce mutational changes to DNA *in vitro* and in mice [34] • GO quantum dot induction of autophagy is dependent on reactive oxygen species levels [36] • Zinc oxide nanoparticles induce oxidative stress and DNA damage in a gastrointestinal helminth parasite, showing potential use as an anti-parasitic treatment [39] • Gold nanoparticles-induced DNA damage via oxidative stress; however, this seemed to be transient in nature and majority of the damage was repaired [40] • Hydroxyapatite-loaded cadmium NPs induce DNA damage in zebrafish liver [41]
Inflammatory changes	• ENMs emitted from printers induce pro-inflammatory cytokine release in a variety of human cell lines [42] • Nickel oxide NPs induce inflammation via the MAPK cascade, affecting cell cycle and DNA repair pathways [43] • Multiwalled carbon nanotubes induce IL-1β and TNF-α production in murine lung cells [44]
Impact on DNA repair pathways	• Difficult to elucidate causal relationships as DNA repair pathways can be the target of ENMs but also a biomarker of underlying ENM-induced DNA damage [25] • The base-excision repair mechanisms are upregulated by some and downregulated by other ENMs

seen after 96 hours of exposure of fish to gold nanoparticles (AuNPs). Interestingly, the induction of these DNA strand breaks seemed to be more severe when AuNPs are coated with polyvinylpyrrolidone (rather than citrate) in a dose-dependent manner – suggesting that specific coatings can affect the genotoxic properties of ENMs [14]. The graphene family of nanomaterials (GFNs) has relatively less evidence for genotoxicity relative to more traditional ENMs, like metals and metal oxides [28]. Some GFNs can directly pierce the cell membrane [29] and others can enter the nucleus through other means [30]. Graphene quantum dots are an example of such activity: alongside causing early apoptosis, they accumulate in the nucleus (perhaps via specific nuclear pore complexes) and induced cleavage and cross-linking of DNA [30]. This mechanism does not occur in isolation, as ROS and caspase upregulation also occur concomitantly; demonstrating that the splitting genotoxicity mechanisms into either direct or indirect may not help distinguish or define the effect of ENMs on an accurate or clinically relevant level. While several mechanisms and consequences have been reported in the literature for indirect DNA damage mechanisms, the evidence for direct interactions between DNA and ENMs is less clear and further exploration is definitely required to reach a quorate conclusion [7].

Oxidative Damage

Moving forward, oxidative stress, and the generation of ROS have been shown to be key contributors to indirect genotoxicity. Oxidative damage has been strongly suggested as the mechanisms of DNA damage after short-term inhalation of silicon dioxide NPs (SiO_2-NPs) caused DNA strand breaks in the lungs of rat models [31]. Similarly, exposing rabbit liver tissue to a single injection of citrated AgNP causes a dose-dependent and time-dependent change in oxidative stress and subsequent DNA damage [32]. Similarly, graphene oxide (GO) NPs have recently been shown to also induce oxidative damage [33]. It was initially demonstrated that GO has the ability to affect DNA replication and to induce DNA mutagenesis *in vitro* and *in vivo* [34]. From work in mice, we now know that these effects are time- and dose-dependent [35]. Furthermore, GO quantum dots can induce autophagy (a key step in stem cell differentiation and in normal cell recycling and turnover) in a way that was dependent on ROS levels [36]. While further evidence has shown that varying doses of GO cause genomic instability and mutations via what are probably oxidative stress mechanisms, this line of evidence is not entirely clear as a few studies did not report any genotoxicity despite doses reaching 200 μg/mL [37, 38]. Certainly the role of oxidative stress in DNA damage seems to be very clear, as demonstrated in AuNPs, ZnO-NPs, and hydroxyapatite-loaded cadmium NPs [39–41].

Inflammatory Changes

The effects of ENMs on inflammatory pathways are complex to elucidate for several reasons. Firstly, inflammation has an essential role in the normal functioning of healthy cells and the organs they are part of, and secondly, inflammation is interlinked with many other processes, including oxidative stress and a wide variety of diseases, including cancerous changes in tissue. However, specific inflammatory factors can also contribute to DNA damage in the context of genomic changes. For example, ENMs that are emitted from printers have been shown to induce pro-inflammatory cytokine release alongside ROS generation in macrophages, lymphoblasts, and airway epithelial cells [42]. Similarly, nickel oxide nanoparticles (NiO-NPs) can induce a pro-inflammatory state with cytokine release via the MAPK cascade, along with cell-cycle dysfunction and DDR suggesting underlying DNA damage [43]. CNPs, such as multiwalled carbon nanotubes

(MWCNTs), also contribute to a pro-inflammatory state through IL-1β and TNF-α production [44]. For all of the above, we can certainly affirm that inflammation is the most investigated consequence of ENM exposure.

Impact on DNA Repair Pathways

As mentioned above, ENMs can also have an impact on DNA repair processes. These pathways are complex and varied, but, ultimately, their role is to inactivate, prevent, or even reverse damage done to DNA. In this way, DNA repair pathways can both be the target of ENMs, or they can be used by diagnostic assays measuring the underlying DNA damage process that is being caused by ENMs. The various methods of DNA repair include base-excision repair, DNA damage signaling, mismatch repair, nonhomologous end joining repair among many others, excellently summarized elsewhere [25]. It is, at this stage, difficult to evaluate what the effect of ENMs is on DNA repair pathways as these pathways work together and are cross-regulated across several shared enzymes and other gene products. For example, while some ENMs upregulate the apurinic/apyrimidinic endonuclease pathway in the base-excision repair mechanism, some downregulate it [25].

A Primer on Epigenetic Processes

Before we discuss the epigenetic mechanisms of ENM nanotoxicity, it is important to be aware of the three key processes through which ENMs can exert their effects – DNA methylation, histone modifications, and noncoding RNAs. This section serves as a propaedeutic to the epigenomic consequences of ENM exposure further in the chapter.

DNA Methylation

One of the epigenetic processes that ENMs affect are those of DNA methylation, a key cellular self-regulating mechanism that regulates to which extent certain genes are transcribed. Methylation is carried out by DNA methyltransferases (DNMT) that covalently add a methyl (CH_3) group to nucleotides (for example, converting a cytosine [C] to a 5-methylcytosine [5mC]). Hypermethylation of cytosine–guanine dinucleotide islands on promoter regions suppresses gene expression while hypomethylation stimulates it. Although there are other forms of methylation, this is the most well understood [45]. This methylation can occur *de novo* or in a maintenance state to suppress the expression of specific genes. Concomitantly, there is a process of demethylation that is carried out by DNA demethylases; however, this process mostly occurs during early embryogenesis and gametogenesis, after which a global DNA methylation pattern is established (although the levels of methylation thereafter vary between different tissues). Relatively little is known about DNA demethylation, but oxidation or deamination of 5mC has been shown to play a significant role [46]. It is important to note, however, that DNA methylation of the gene *body*, as opposed to the CpG islands, promotes gene expression [47].

Histone Modifications

Another form of epigenetic regulation affected by ENMs is through histone modifications. Histones are proteins that are used in the efficient and compact packaging of DNA. They form into nucleosomes that are linked by an intermediary linker DNA sequence of a few 10s of bases and a histone

H1 protein. This structure is together termed chromatin, and the extent to which the chromatin is compacted or more "loose" affects the expression of genes within that region [48]. Much like DNA methylation, modifications can be made to the histones themselves via enzymatic or nonenzymatic pathways. Notably, histones also have their own methylation state regulation that affects how compacted chromatin will be depending on which sites and residues are affected. Similarly, histone acetylation is a process that involves the addition and removal of an acetyl group to lysine by either histone acetyltransferases or histone deacetylases, respectively [49]. Acetylation relaxes chromatin and enables access by transcription proteins and enzymes, while a deacetylated histone residue favors a tighter chromatin structure, preventing transcription and also DNA repair proteins from accessing that region of DNA.

ncRNAs

Lastly, NPs can affect the expression of noncoding (nc)RNAs. These ncRNAs play a role in regulating the expression of other protein-coding genes, as well as influencing intracellular signaling and developmental processes. ncRNAs can be shorter (~20–30nt long), such as micro (mi)RNAs and small interfering (si)RNAs, or they can take the form of long noncoding (lnc)RNAs that are longer (100s or 1000s of nucleotides long) [50]. To add further complexity, these ncRNAs regulate and, in some instances, are regulated by the aforementioned methylation and histone modification processes in a complex feedback loop; for this, studying these pathways in isolation to see the impact of NPs is not a trivial task [47, 51, 52].

Epigenomic Consequences of ENM Exposure

In recent years, it became clear that ENMs have a vast (perhaps previously underappreciated) impact on epigenomics, and their effects can be seen at various levels (see Table 17.3). First, we will focus on *in vitro* and *in vivo* studies, where most of the evidence base lies. The current evidence base suggests that the effect of ENMs on the epigenome is not just carried out via direct interactions with the proteins involved in methylation, histone modification, and ncRNA regulation. The most common observation of EMNs is the induction of the aforementioned oxidative stress and inflammation; these two pathways can themselves lead to DNA hypomethylation or hypermethylation, respectively [53, 54].

Apoptosis

First, EMNs can have deleterious effects on cells via their effects on autophagy, apoptosis, and even DNA repair pathways, and epigenetic processes seem to be the common denominator targeted in these processes. For example, in mouse macrophages, the miRNA miR-350 was highly expressed after exposure to titanium oxide (TiO_2-NPs); this miRNA increased apoptosis of cells via negative regulation of the *PIK3R3* gene [55]. Similarly, SiO_2-NPs increased the apoptotic rate of human bronchial epithelial cells through DNA hypermethylation of *CREB3L1* and *Bcl-2* via mitochondrially mediated pathways, even at a low dose, although over a long period [56]. ZnO-NPs have also induced apoptotic pathways through the arrest of the cell cycle at the G2/M stage with increased expression of pro-apoptotic genes [57]. Furthermore, AgNPs can also contribute to epigenetically regulated changes in the normal cell proliferation pathways; low-dose exposure and the subsequent removal of AgNPs led to DNA methylation pattern changes (specifically in DNMT levels) and cell-cycle disruption via p53/p21 mechanisms [58]. These effects can be even seen 96 hours after the

Table 17.3 Epigenomic effects of various engineered nanomaterials.

Mechanism	Engineered nanomaterials implicated
Apoptosis	• Titanium dioxide NPs-induced apoptosis through miR-350 negative regulation of *PIK3R3* [55] • Silicon dioxide NPs increased the apoptotic rate of bronchial epithelial cells through *CREB3L1* and *Bcl-2* hypermethylation [56] • Zinc oxide NPs increase pro-apoptotic gene expression through cell-cycle arrest at G2/M [57] • Silver NPs disrupt the cell cycle through p53/p21 mechanisms and via DNA methylation pattern changes that last for days after removal of the ENM [58] • Silver NPs contribute to cellular death through *Bcl-2* repression via ODRUL (noncoding RNA) action [59]

Inflammation and oxidative stress	• Via oxidative stress pathways, titanium dioxide NPs alter histone acetylation patterns and induce lesions in the central nervous system after inhalation [60–62]
	• Multiwalled carbon nanotubes induce inflammation in the lung in a hypomethylated DNA state [64]
	• Carbon nanotubes induce VEGF and platelet activation pathways and polarize macrophages to the M2 subgroup [65, 66]
	• Carbon dots (CDs) act on immune response pathways via miRNA pattern changes [68]
	• In zebrafish hearts, CDs cause a dose-dependent increase in inflammatory cytokines and DNA methylation changes [69]
	• In mouse lung, single-walled carbon nanotubes hypermethylate and repress the *Atm* promoter [70]
	• Silicon dioxide NPs upregulate miRNAs that target *PDC4*, *LIN28B*, and *CTGF* (regulators of the MAPK pathway) in rat lung [71]
	• Silicon dioxide NPs induce MAPK through upregulating *P21*, *P19*, *CFOS*, *CJUN*, and *KSR1* while downregulating *BRAF*, *MAPK11*, and *NFATC4* in human hepatoma cells (Huh7) [73, 74]
	• In humans:
	  ○ silicon dioxide NPs lead to global demethylation of DNA in white blood cells alongside increased oxidative DNA damage [94]
	  ○ carbon nanotubes induce changes in promoters of genes that code for methylation and acetylation proteins [95]

Epigenomic changes and cancer	• Over six weeks in human lung cells, silver NPs led to differential expression of 2,000 genes with evidence of epithelial–mesenchymal transition taking place [76] • Aluminum NPs in the spleen contributed to the development of metastatic disease alongside NLRP3 inflammasome activation [77]
Development and genomic imprinting	• The effects of titanium dioxide NP inhalation on epigenomic changes are much greater in young mice compared with adults [78] • Maternal inhalation of titanium dioxide NP induces epigenetic changes in offspring cardiac tissue [79] • Exposure of silver NPs to pregnant mice leads to *Zac 1* methylation pattern changes in the developing embryo [80] • Injection of silver NPs into zebrafish embryos impairs cardiovascular development and stunts growth [81] • Gold NPs act across the placenta; miRNAs involved in proliferation and apoptosis were affected in fetal lung and liver [82] • Transgenerational effects have been seen after silicon dioxide treatment of male mice through genomic imprinting in Dlk1/Dio3 domains, as well as after silver NP exposure in *C. elegans*, the effect of which could not be rescued at the F6 offspring stage [87, 89]

removal of AgNPs, suggesting these changes are prolonged and not transitory and restricted to the treatment period. AgNPs have also induced pro-apoptotic changes through the expression of ODRUL (a lncRNA) that ultimately repressed Bcl-2 and hence increased cellular death [59]. Taking the examples of AgNPs here, we can see that there are several different target methylation patterns that ultimately can give rise to a pro-apoptotic state, even through the action of the same ENM.

Taken together, this evidence, if reproduced in more robust *in vivo* models, could be quite concerning as the pathways that are in question are at the very core of gene expression regulation and cell-cycle regulation; affecting these could even underscore future cancerous changes.

Inflammation and Oxidative Stress

Epigenetic changes do not have to occur through direct interaction with ENMs but can manifest as a consequence of the effects of ROS and a pro-inflammatory environment. For example, even non-cytotoxic levels of TiO_2-NPs in the lungs caused alterations in acetylation patterns, which has been shown to be driven by the generation of superoxide [60, 61]. Several pathways have been suggested to contribute to increased oxidative stress. Lesions of the central nervous system, particularly in the hippocampus, were reported in response to inhalation of TiO_2-NP in mice, with increased glutamic and nitric oxide release, protein oxidation, and increased catalase activity being demonstrated [62, 63].

In terms of inflammation, CNPs feature prominently, and this is quite topical due to the major concerns of environmental pollution with carbon emissions. MWCNT exposure corresponds to lung inflammation while simultaneously inducing an environment of DNA hypomethylation [64]. CNPs could also be contributing to inflammatory and immunological changes. *In vitro*, both single-walled carbon nanotubes (SWCNTs) and MWCNTs, respectively induce hypomethylation in over 1,000 different genes after just 24 hours of exposure [65]. The pathways involved (vascular endothelial growth factor [VEGF] and platelet activation pathways) are heavily involved in polarizing macrophages toward the M2 subgroup. Interestingly, in further studies, this group reported that while MWCNTs induced hypomethylation of promoter regions, SWCNTs hypomethylated specific CpG islands [66, 67]. Furthermore, pathways involved in the immune response were disrupted when carbon dots (CDs; a novel <10 nm form of nanomaterial) mediated by differential expression patterns of miRNA, although, DNA methylation was not affected [68]. This effect was the same, regardless of the surface charge on the CD. *In vivo* work showed a dose-dependent expression of inflammatory cytokines in the heart of zebrafish, alongside DNA methylation changes [69]. Similarly, the *Atm* promoter, a gene that usually has protective properties in an inflammatory milieu, was hypermethylated in response to SWCNTs in the murine lung [47, 70].

Another example of toxicity caused by a response to oxidative stress is provided by silicon oxide. The inflammatory effect of SiO_2-NPs has been shown to be underscored by epigenomic changes, upregulating several miRNAs in lung tissue [71]. The target genes of these miRNAs (including *PDC4*, *LIN28B*, and *CTGF*) are involved in regulating inflammatory signaling pathways, such as the central MAPK pathway. A growing evidence base suggests that while inflammation and genotoxicity were initially thought to be separate from epigenomics, these mechanisms can now be linked across domains [72–74].

Epigenomic Changes and Cancer

While ENMs have induced changes that are pro-apoptotic, exposure to them has also been implicated in cancerous changes in cells. The epithelial–mesenchymal transition (EMT) is a hallmark feature of tumor cells attempting to invade the surrounding tissue and metastasize to distant sites

[75]. An important study evaluated the long-term effects of AgNPs over six weeks (this is particularly useful, as most studies evaluate effects over days) [76]. In a human lung cell line, almost 2,000 genes were differentially expressed after exposure and clear markers of EMT were detected, irrespective of size (10 or 75 nm). This study shed light on the potentially damaging, and so far, poorly understood, long-term epigenetic changes following chronic NP exposure. Indeed, the cancerous changes do not act in isolation, but have been correlated with an increased pro-inflammatory state; aluminum-based NPs in the spleen lead to activation of the NLRP3 inflammasome and increase the secretion of the transforming growth factor (TGF)-β, subsequently causing an increase in EMT and the development of metastatic disease [77].

Development and Genomic Imprinting

One particularly concerning data in the context of nanotoxicity of ENMs relate to their effects on younger animals and in certain developmental phases. This is relevant due to the importance of methylation and demethylation balance in embryogenesis; in fact, events occurring during embryogenesis can act as key regulators in influencing the global methylation state that continues and affects cellular function into adulthood. For example, TiO_2-NPs inhalation causes differential epigenetic changes in younger mice compared with adult mice; a decrease in global methylation was found only in the young group, particularly in the TNF-α and Thy-1 promoters, and young mice had almost 200 more genes upregulated than adult mice. The genes whose expression was altered play a role in cancer pathways [78].

Even longer-term effects have been documented: more specifically, effects at the offspring development stage. These findings raise concerns over the access of ENMs across the placenta as well as the transgenerational risks of genomic imprinting. As cardiovascular defects have been reported after exposure to ENMs, investigations into whether they can have effects on progeny have been carried out. Maternal inhalation of TiO_2-NPs aerosols showed significant epigenetic changes in the hearts of offspring [79]. More immediate effects, including the effects on embryonic development, have been reported in animal models. When mice were exposed to AgNPs, their embryos had significantly altered differential methylation patterns of *Zac1*, a gene relevant in controlling the progression of embryonic growth [80]. Similarly, when zebrafish embryos were directly injected with AgNPs, they had impaired cardiovascular development as well as shorter bodies, shown to be mediated by downregulations of Ca^{2+} signaling through the methylation of specific DNA loci relevant to myofibrillogenesis [81]. Transplacental effects have also been reported with AuNPs; certain miRNAs involved in cell proliferation and apoptosis (let-7a and miR-183) were highly altered in the fetal lung and liver [82]. Epigenetic changes in the expression of the let-7a family have also been implicated after carcinogenic exposure [83]. Interestingly, this effect was only seen with 100 nm (but not 40 nm) AuNPs, suggesting that this transplacental effect could be size-dependent, although perhaps quite paradoxically, as one may expect that smaller particles would cross the placenta with greater ease. While the authors did conclude that both sizes of NPs made their way into cells with electron microscopy, the effects of size on the biodistribution and toxicity of AuNPs has had varying results [84–86].

Engineered nanomaterials could also contribute to alterations in the imprinting of genetic expression patterns over generations. SiO_2-NP treatment of male mice had alterations in testis tissue associated with methylation pattern changes in the imprinted Dlk1/Dio3 domains (regions that are used in establishing the methylation status before birth and are maintained during future spermatogonia and spermatocytes) [87, 88]. These potentially multigenerational effects have also been described after AgNP exposure. In a *C. elegans* model, AgNPs led to increased histone methylation at the H3K4me2 marker [89]. Interestingly, the first three generations were exposed to the treatment, and when rescue was carried out at F6, it was unsuccessful. This evidence, albeit in a relatively rudimentary organism, suggests that unexposed offspring, many generations after exposure, could have epigenetic biomarkers of the exposure of their ancestors to ENMs.

Importance of Duration and Dose of Exposure

Three aspects play a role in nanotoxicity: duration, dose, and continuity of exposure. The **duration** of exposure to NPs seems to affect the degree of the severity of their effects; however, data are somehow conflicting. We know, for instance, that short-term exposure does not affect methylation in the same way as longer-term exposure does [90, 91]; what is more difficult to figure out is how to define short exposure (few hours or few minutes?). For ZnO-NPs, we have evidence of an increased generation of ROS at 6 hours of treatment, and DNA damage at 24 hours, suggesting that this is not a binary, but rather a cumulative and gradual process [57].

The **dose** is also relevant. Higher concentrations of AgNP induce more DNA methylation and histone deacetylation across the board [92]. Furthermore, the effects of SiO_2-NPs on global methylation levels have been shown to be dose-dependent [93], but even in this case, it is more difficult to establish how dose and duration interact and potentiate each other.

Interestingly, some of the epigenetic effects vary based on the **continuity** of exposure, as well as the duration and dose. When the effect of AgNP exposure on tumor suppressor genes and apoptosis pathways are evaluated in a cell line, the methylation and expression of certain genes vary based on different exposure patterns. Daily exposure at a low dose or single exposure at a high dose decreased p53, p21, and caspase-3 protein expression. Conversely, discontinuous low dose or a low, single-dose exposure led to increases in the levels of expression in those genes [92].

Evidence in Humans

The effects of exposure to ENMs have been most extensively studied with *in vitro* and *in vivo* models. There is relatively less evidence for their effects in humans [53]. Of note, exposure to SiO_2-NPs led to global DNA demethylation in white blood cells and a concomitant increase in a marker of oxidative DNA damage (8-hydroxy-2′-deoxyguanosine [8-OHdG]) in those same cells [94]. CNT, on the other hand, did not induce changes in DNA methylation levels globally but rather identified individual changes in promoters of genes coding for proteins involved in methylation and acetylation [95]. Additionally, the finding that changes in DNA methylation patterns are induced by long-term (rather than short-term) exposure have also been confirmed in human studies [96].

Is There a Need for Epigenetic Testing of ENMs?

From the available evidence, it is crystal clear how important it is to assess the toxicological profile of ENMs that are trying to make it to the market. What is more difficult to ascertain is what exactly the testing is for: for now, most of the evidence points to up- or downregulation of several epigenetic pathways to some extent. However, because of the complex interplay between these pathways, and the fact that much of the evidence measures global as opposed to locus-specific changes, the true effect is not clear. For example, AgNPs, AuNPs, and INPs were shown to affect miRNA expression but not DNA methylation in a cell line, each to a different extent [97]. However, other studies we have discussed do show changes to DNA methylation profiles from these nanomaterials. It may be that the effects of these ENMs are highly dependent on the cell type studied and the formulation used (alongside the other properties discussed previously). Furthermore, most of the evidence has been obtained during studies lasting hours, days, and weeks. Truly long-term effects at the macro scale will require more robust evidence from human studies.

Importance of Properties of ENMs

The specific properties of ENMs manifest with a range of different effects on the genome and epigenome. ENMs have their properties defined in several ways, be it through chemical composition, physical properties (size and shape), structural or colloidal properties, their hydrophobicity and their functional groups, and charge [98]. These properties are not just of academic interest; they have consequences with respect to their effects on tissues and are used in conjunction with nanotoxicological analysis to investigate the safety of ENMs. Superimposed on top of the properties of the ENM in isolation is the specific cell type or cell-cycle phase during which the ENM enters, as well as being affected by the experimental parameters. Crucially, ENMs are typically coated with proteins to form a protein corona in the extracellular fluid, which then determines the interaction of the ENM with the cell membrane and determines whether the transport into the cell is active or passive [99]; the mechanism of ENM endocytosis then, is also of relevance (be it clathrin-/caveolin-mediated, independent of the two, phagocytosis or micropinocytosis [100, 101]). Here, we first briefly cover the key relevant properties of ENMs and then discuss how taking control of, and selecting for, these parameters can be exploited for improved delivery or therapeutic effects in the context of affecting genomic and epigenomic processes.

The chemical composition of ENMs is a key influence on their nanotoxic effect and their compatibility with human tissues. Most noble metals, alongside ENMs based on biopolymers and metal oxides are relatively biocompatible and have wider applications relative to those based on heavy toxic metals like lead or cadmium [102–105]. Size, shape, and size distribution of ENMs is also relevant, as it dictates how they will traverse tissues along their journey from mucosa or skin through to the extracellular matrix and through the cellular (and possibly nuclear) membrane [106]. In the same way, ENMs can be filtered during entry into cells by their chemical and functional group properties, the same can occur with size. Smaller ENMs, such as SWCNTs, that are only 30 mm in size, enter cells much more readily than larger variants [107]. Smaller NPs are also generally more cytotoxic as they are more readily endocytosed [106]. However, it is not as simple as "the smaller the size the easier the uptake," and we can see this in experimental data; from a selection of 14 nm, 50 nm, and 74 nm AuNPs, the 50 nm ones had the easiest entry into mammalian cells [108]. Modifications to certain functional groups can also induce greater genotoxic effects; when graphene quantum dots were aminated, they induced higher cleavage and cross-linking of DNA in alveolar macrophages [30]. With regard to the surface charge, neutral ENMs have a lower affinity for cellular entry due to the increased membrane distortion that cationic ENMs achieve in the cell membrane [98, 109]. In terms of their interactions in solution, ENMs that have a more hydrophilic surface stay in circulation longer as they can evade phagocytic engulfment, possibly mediated by the fact that the polarization of macrophages has been shown to be affected by this property [110, 111].

Sometimes, these properties convey advantageous benefits, if adjusted appropriately to fit the desired therapeutic purpose (see Table 17.4). For example, as titanium is used in medical implants, such as in neurosurgical procedures to replace parts of the cranial vault, the ability of those implants to undergo osseointegration is key. As such, titanium surfaces can be modified with nanostructures to enhance osseointegration: this modification downregulated miR-23a, which in turn enhanced the osteogenic differentiation of bone marrow-derived mesenchymal stem cells [112]. miR-23a was shown to be under repression from p53, and the nanotube-modified titanium released this repression and led to the targeting of *CXCL12* via miR-23a, activating the osteogenic differentiation pathways. This is a pertinent example of how specific surface modifications, such as the sandblasted, large-grit, acid-etched method can be used to convey ENM properties in a beneficial way in well-defined clinical scenarios. In this way, ENMs have found traction with several conditions affecting the brain, as nanodrugs for gliomas, nanofabricated contrast agents, and as part of the diagnostic framework [113–118]. The charge and hydrophobic properties of ENMs are also being exploited to generate mesoporous silica nanoparticles; those that are cationic and lipophilic have shown close interactions with the membrane of mitochondria and have enabled the targeted delivery of doxorubicin to mitochondria [119]. Similarly, smaller AuNPs (1 nm) can pass the nuclear membrane and directly interact with DNA and, while this may seem like an adverse property, given the direct genotoxicity mechanisms discussed earlier, this could pave the way for ENMs being inert nanomaterials for the attachment and stable delivery of various nucleic acids to the nucleus [120, 121]. Indeed, this type of nuclear targeting has already been used to induce DNA damage and cause apoptosis in cancer cells [26]. Modifying various functional groups on the surface of ENMs can also be utilized clinically; when AuNPs are conjugated with a cationic polyelectrolyte, they confer protection to DNA against nuclease degradation in the context of nonviral gene delivery [122, 123]. In terms of hydrophobicity of ENMs, this property is key in the drug delivery context when the drug in question is insoluble in water [124]. In the context of genomics, when

Table 17.4 Influence of ENM properties on genomic and epigenomic challenges and potential future scenarios.

Property	Challenge	Use
Chemical composition	The initial/intrinsic genotoxicity of ENMs is governed by their composition Heavy toxic metals are more cytotoxic than noble metals like gold or platinum [102]	Gold, silver, and platinum nanoparticles are being trialed extensively as delivery vehicles from nucleic acids to proteins [103–105]
Size and shape	Smaller ENMs are (generally) more readily endocytosed [106] Smaller ENMs can avoid macrophage detection and enter the nucleus [26]	Small 1 nm gold NPs can enter the nucleus and be used as an inert surface on which to deliver nucleic acids in a stable way [120, 121]
Charge	Cationic ENMs have a higher affinity for cellular entry relative to neutral ENMs Charge and the extent of hydrophobicity are closely linked [109]	Mesoporous silica nanoparticles are cationic, and they interact closely with the mitochondrial membrane enabling targeted doxorubicin delivery [119]
Hydrophobicity/ hydrophilicity	ENMs that are hydrophilic remain in the circulation longer as they can evade phagocytosis [110] Hydrophilicity affects macrophage polarization [111]	Modifying glycol chitosan to be more hydrophobic enables self-assembly and packaging of hydrophobic DNA into a nanocarrier with improved transfection [125]
Functional groups	Amination of graphene dots induces higher rates of cleaves and cross-linking in the DNA of exposed alveolar macrophages [30]	Conjugating gold NPs to a cationic polyelectrolyte like polydiallyldimethyl-ammonium chloride confers protection from nuclease degradation of packaged DNA [122]

glycol chitosan was modified to be hydrophobic, it readily self-assembled into a hydrophobic DNA nanocarrier with improved transfection efficiencies [125].

It is important to appreciate then that the properties of ENMs play a pivotal role in their genomic and epigenomic effects, as well as their potential uses in the clinical setting; finding a sweet spot between usability and optimally reduced nanotoxicity could be used to define what the "ideal" ENM traits are for a given real-world use.

Future Perspectives

It is becoming clear that it will be difficult to study these individual processes in isolation if each ENM will execute its genotoxic side effects through a specific pathway that involves genomic or epigenomic processes. However, through the mounting evidence provided in this chapter, it is possible to slowly start picking out pathways and mechanisms, as we have tried to do here. From an experimental perspective, in many cases, it may be helpful to simply deduce which pathways ENMs are *not* using to narrow down the true culprits. By way of example, the genotoxic and carcinogenic side effects of nickel NPs have recently been shown to be mediated via the HIF-1α/miR-210/Rad52 pathway; here, a single type of ENM triggers inflammatory, epigenomic, and DNA repair pathways [126]. Further, more studies where several steps of inhibitors and assays are used to clarify which processes are involved will provide new insights into genotoxicity, potentially demystifying the real consequences of ENMs on genes and their products [126]. Particularly, the transplacental and transgenerational effects will need a careful examination in longer observational studies. While most of this work has so far been carried out in cell lines and in some early preclinical animal models, more evidence from human studies is needed to understand whether these changes can be replicated, and if they have any significant effect.

References

Ganau, M., Prisco, L., and Ganau, L. (2015). Ethics and nanoethics. In: *Commercializing Nanomedicine* (ed. L.Escoffier, M.Ganau, J.Wong). Jenny Stanford Publishing.

Ganau, M., Foroni, R.I., and Ambu, R. (2015). Surgery in the realm of nanometers. In: *Commercializing Nanomedicine* (ed. L.Escoffier, M.Ganau, J.Wong). Jenny Stanford Publishing.

Ganau, M., Bosco, A., Parisse, P., and Casalis, L. (2015). Diagnostic challenges of nanomedicine. In: *Commercializing Nanomedicine* (ed. L.Escoffier, M.Ganau, J.Wong). Jenny Stanford Publishing.

Ganau, M., Paris, M., Nicassio, N., and Ligarotti, G.K. (2015). Nanomedicine: the dawn of a new era. In: *Commercializing Nanomedicine* (ed. L.Escoffier, M.Ganau, J.Wong). Jenny Stanford Publishing.

Ganau, M., Prisco, L., Syrmos, N. and Ganau, L. (2016). Principles of nanoethics: theoretical models and clinical practice. In: *Handbook of Clinical Nanomedicine* (ed. R. Bawa, G.F. Audette, and B. Reese). Jenny Stanford Publishing.

Ganau, M., Foroni, R.I., Soddu, A. and Ambu, R. (2016). Biomedical engineering and nanoneurosurgery: from the laboratory to the operating room. In: *Handbook of Clinical Nanomedicine* (ed. R. Bawa, G.F. Audette, and I. Rubinstein). Jenny Stanford Publishing.

Huang, R., Zhou, Y., Hu, S., and Zhou, P.-K. (2019). Targeting and non-targeting effects of nanomaterials on DNA: challenges and perspectives. *Rev. Environ. Sci. Biotechnol.* 18: 617–634.

Huk, A. et al. (2015). Impact of nanosilver on various DNA lesions and HPRT gene mutations – effects of charge and surface coating. *Part. Fibre Toxicol.* 12: 25.

Brzóska, K., Grądzka, I., and Kruszewski, M. (2018). Impact of silver, gold, and iron oxide nanoparticles on cellular response to tumor necrosis factor. *Toxicol. Appl. Pharmacol.* 356: 140–150.

Roxbury, D., Jagota, A., and Mittal, J. (2011). Sequence-specific self-stitching motif of short single-stranded DNA on a single-walled carbon nanotube. *J. Am. Chem. Soc.* 133: 13545–13550.

Samanta, A. and Medintz, I.L. (2016). Nanoparticles and DNA – a powerful and growing functional combination in bionanotechnology. *Nanoscale* 8: 9037–9095.

Thongkumkoon, P. et al. (2014). Direct nanomaterial-DNA contact effects on DNA and mutation induction. *Toxicol. Lett.* 226: 90–97.

Zhang, R., Zhang, X., Jia, C., Pan, J., and Liu, R. (2019). Carbon black induced DNA damage and conformational changes to mouse hepatocytes and DNA molecule: a combined study using comet assay and multi-spectra methods. *Ecotoxicol. Environ. Saf.* 170: 732–738.

Barreto, A. et al. (2019). Genotoxicity of gold nanoparticles in the gilthead seabream (*Sparus aurata*) after single exposure and combined with the pharmaceutical gemfibrozil. *Chemosphere* 220: 11–19.

Vellampatti, S., Chandrasekaran, G., Mitta, S.B., Lakshmanan, V.-K., and Park, S.H. (2018). Metallo-curcumin-conjugated DNA complexes induces preferential prostate cancer cells cytotoxicity and pause growth of bacterial cells. *Sci. Rep.* 8: 1–11.

Nguyen, K.T., Shukla, K.P., Moctezuma, M., and Tang, L. (2007). Cellular and molecular responses of smooth muscle cells to surface nanotopography. *J. Nanosci. Nanotechnol.* 7: 2823–2832.

Wallach, D., Kang, T.-B., and Kovalenko, A. (2014). Concepts of tissue injury and cell death in inflammation: a historical perspective. *Nat. Rev. Immunol.* 14: 51–59.

Teodoro, J.S. et al. (2016). Low-dose, subchronic exposure to silver nanoparticles causes mitochondrial alterations in Sprague-Dawley rats. *Nanomed.* 11: 1359–1375.

Atha, D.H. et al. (2017). Quantifying engineered nanomaterial toxicity: comparison of common cytotoxicity and gene expression measurements. *J. Nanobiotechnol.* 15: 79.

Saquib, Q. et al. (2016). Genotoxicity of ferric oxide nanoparticles in *Raphanus sativus*: deciphering the role of signaling factors, oxidative stress and cell death. *J. Environ. Sci.* 47: 49–62.

Roy, K., Kanwar, R.K., and Kanwar, J.R. (2015). LNA aptamer based multi-modal, Fe3O4-saturated lactoferrin (Fe3O4-bLf) nanocarriers for triple positive (EpCAM, CD133, CD44) colon tumor targeting and NIR, MRI and CT imaging. *Biomaterials* 71: 84–99.

Broustas, C.G. and Lieberman, H.B. (2014). DNA damage response genes and the development of cancer metastasis. *Radiat. Res.* 181: 111–130.

Wan, R. et al. (2012). DNA damage caused by metal nanoparticles: involvement of oxidative stress and activation of ATM. *Chem. Res. Toxicol.* 25: 1402–1411.

Dogan-Topal, B., Uslu, B., and Ozkan, S.A. (2018). Chapter 14 – Detection of DNA damage induced by nanomaterials. In: *Nanoscale Fabrication, Optimization, Scale-Up and Biological Aspects of Pharmaceutical Nanotechnology* (ed. A.M. Grumezescu), 547–577. William Andrew Publishing. doi: 10.1016/B978-0-12-813629-4.00014-0.

Singh, N. et al. (2017). Exposure to engineered nanomaterials: impact on DNA repair pathways. *Int. J. Mol. Sci.* 18: 1515.

Kang, B., Mackey, M.A., and El-Sayed, M.A. (2010). Nuclear targeting of gold nanoparticles in cancer cells induces DNA damage, causing cytokinesis arrest and apoptosis. *J. Am. Chem. Soc.* 132: 1517–1519.

Qi, L. et al. (2021). Biocompatible nucleus-targeted graphene quantum dots for selective killing of cancer cells via DNA damage. *Commun. Biol.* 4: 1–12.

Wu, K., Zhou, Q., and Ouyang, S. (2021). Direct and indirect genotoxicity of graphene family nanomaterials on DNA—a review. *Nanomaterials* 11: 2889.

Lammel, T., Boisseaux, P., Fernández-Cruz, M.-L., and Navas, J.M. (2013). Internalization and cytotoxicity of graphene oxide and carboxyl graphene nanoplatelets in the human hepatocellular carcinoma cell line Hep G2. *Part. Fibre Toxicol.* 10: 27.

Xu, L. et al. (2018). Graphene quantum dots in alveolar macrophage: uptake-exocytosis, accumulation in nuclei, nuclear responses and DNA cleavage. *Part. Fibre Toxicol.* 15: 45.

Brandão, F. et al. (2021). Genotoxicity and gene expression in the rat lung tissue following instillation and inhalation of different variants of amorphous silica nanomaterials (aSiO2 NM). *Nanomater. Basel Switz.* 11: 1502.

Kim, Y.J. et al. (2019). Assessment of in vivo genotoxicity of citrated-coated silver nanoparticles via transcriptomic analysis of rabbit liver tissue. *Int. J. Nanomed.* 14: 393–405.

Rhazouani, A. et al. (2021). Synthesis and toxicity of graphene oxide nanoparticles: a literature review of in vitro and in vivo studies. *BioMed Res. Int.* 2021: e5518999.

Liu, Y. et al. (2013). Graphene oxide can induce in vitro and in vivo mutagenesis. *Sci. Rep.* 3: 3469.

El-Yamany, N.A. et al. (2017). Graphene oxide nanosheets induced genotoxicity and pulmonary injury in mice. *Exp. Toxicol. Pathol.* 69: 383–392.

Li, X. et al. (2020). Graphene oxide quantum dots-induced mineralization via the reactive oxygen species-dependent autophagy pathway in dental pulp stem cells. *J. Biomed. Nanotechnol.* 16: 965–974.

Bengtson, S. et al. (2016). No cytotoxicity or genotoxicity of graphene and graphene oxide in murine lung epithelial FE1 cells in vitro. *Environ. Mol. Mutagen.* 57: 469–482.

Mohamed, H.R.H., Welson, M., Yaseen, A.E., and EL-Ghor, A.A. (2020). Estimation of genomic instability and mutation induction by graphene oxide nanoparticles in mice liver and brain tissues. *Environ. Sci. Pollut. Res.* 27: 264–278.

Esmaeilnejad, B., Samiei, A., Mirzaei, Y., and Farhang-Pajuh, F. (2018). Assessment of oxidative/nitrosative stress biomarkers and DNA damage in *Haemonchus contortus*, following exposure to zinc oxide nanoparticles. *Acta Parasitol.* 63: 563–571.

May, S. et al. (2018). Transient DNA damage following exposure to gold nanoparticles. *Nanoscale* 10: 15723–15735.

Gao, M., Yang, Y., Lv, M., Song, W., and Song, Z. (2018). Oxidative stress and DNA damage in zebrafish liver due to hydroxyapatite nanoparticles-loaded cadmium. *Chemosphere* 202: 498–505.

Pirela, S.V. et al. (2016). Effects of laser printer–emitted engineered nanoparticles on cytotoxicity, chemokine expression, reactive oxygen species, DNA methylation, and DNA damage: a comprehensive in vitro analysis in human small airway epithelial cells, macrophages, and lymphoblasts. *Environ. Health Perspect.* 124: 210–219.

Capasso, L., Camatini, M., and Gualtieri, M. (2014). Nickel oxide nanoparticles induce inflammation and genotoxic effect in lung epithelial cells. *Toxicol. Lett.* 226: 28–34.

Fukai, E., Sato, H., Watanabe, M., Nakae, D., and Totsuka, Y. (2018). Establishment of an in vivo simulating co-culture assay platform for genotoxicity of multi-walled carbon nanotubes. *Cancer Sci.* 109: 1024–1031.

Gujar, H., Weisenberger, D.J., and Liang, G. (2019). The roles of human DNA methyltransferases and their isoforms in shaping the epigenome. *Genes* 10: 172.

Bochtler, M., Kolano, A., and Xu, G.-L. (2017). DNA demethylation pathways: additional players and regulators. *BioEssays News Rev. Mol. Cell. Dev. Biol.* 39: 1–13.

Moreira, L., Costa, C., Pires, J., Teixeira, J.P., and Fraga, S. (2021). How can exposure to engineered nanomaterials influence our epigenetic code? A review of the mechanisms and molecular targets. *Mutat. Res. Mutat. Res.* 788: 108385.

Bannister, A.J. and Kouzarides, T. (2011). Regulation of chromatin by histone modifications. *Cell Res.* 21: 381–395.

Gräff, J. and Tsai, L.-H. (2013). Histone acetylation: molecular mnemonics on the chromatin. *Nat. Rev. Neurosci.* 14: 97–111.

Wei, J.-W., Huang, K., Yang, C., and Kang, C.-S. (2017). Non-coding RNAs as regulators in epigenetics (Review). *Oncol. Rep.* 37: 3–9.

Yao, Q., Chen, Y., and Zhou, X. (2019). The roles of microRNAs in epigenetic regulation. *Curr. Opin. Chem. Biol.* 51: 11–17.

Yang, X. et al. (2020). Epigenetic modulations of noncoding RNA: a novel dimension of cancer biology. *Mol. Cancer* 19: 64.

Pogribna, M. and Hammons, G. (2021). Epigenetic effects of nanomaterials and nanoparticles. *J. Nanobiotechnol.* 19: 2.

Valinluck, V. et al. (2004). Oxidative damage to methyl-CpG sequences inhibits the binding of the methyl-CpG binding domain (MBD) of methyl-CpG binding protein 2 (MeCP2). *Nucleic Acids Res.* 32: 4100–4108.

Sui, J. et al. (2018). Molecular mechanism for miR-350 in regulating of titanium dioxide nanoparticles in macrophage RAW264.7 cells. *Chem. Biol. Interact.* 280: 77–85.

Zou, Y. et al. (2016). DNA hypermethylation of CREB3L1 and Bcl-2 associated with the mitochondrial-mediated apoptosis via PI3K/Akt pathway in human BEAS-2B cells exposure to silica nanoparticles. *PLoS ONE* 11: e0158475.

Gao, F. et al. (2016). Zinc oxide nanoparticles-induced epigenetic change and G2/M arrest are associated with apoptosis in human epidermal keratinocytes. *Int. J. Nanomed.* 11: 3859–3874.

Mytych, J., Zebrowski, J., Lewinska, A., and Wnuk, M. (2017). Prolonged effects of silver nanoparticles on p53/p21 pathway-mediated proliferation, DNA damage response, and methylation parameters in HT22 hippocampal neuronal cells. *Mol. Neurobiol.* 54: 1285–1300.

Gao, M. et al. (2017). Nrf-2-driven long noncoding RNA ODRUL contributes to modulating silver nanoparticle-induced effects on erythroid cells. *Biomaterials* 130: 14–27.

Jayaram, D.T. et al. (2019). TiO_2 nanoparticles generate superoxide and alter gene expression in human lung cells. *RSC Adv.* 9: 25039–25047.

Jayaram, D.T. and Payne, C.K. (2020). Intracellular generation of superoxide by TiO2 nanoparticles decreases histone deacetylase 9 (HDAC9), an epigenetic modifier. *Bioconjug. Chem.* 31: 1354–1361.

Wang, J. et al. (2008). Potential neurological lesion after nasal instillation of TiO(2) nanoparticles in the anatase and rutile crystal phases. *Toxicol. Lett.* 183: 72–80.

Rashid, M.M., Forte Tavčer, P., and Tomšič, B. (2021). Influence of titanium dioxide nanoparticles on human health and the environment. *Nanomaterials* 11: 2354.

Brown, T.A. et al. (2016). Alterations in DNA methylation corresponding with lung inflammation and as a biomarker for disease development after MWCNT exposure. *Nanotoxicology* 10: 453–461.

Öner, D. et al. (2017). Epigenetic effects of carbon nanotubes in human monocytic cells. *Mutagenesis* 32: 181–191.

Öner, D. et al. (2018). Differences in MWCNT- and SWCNT-induced DNA methylation alterations in association with the nuclear deposition. *Part. Fibre Toxicol.* 15: 11.

Öner, D. et al. (2020). Induction and recovery of CpG site specific methylation changes in human bronchial cells after long-term exposure to carbon nanotubes and asbestos. *Environ. Int.* 137: 105530.

Sima, M. et al. (2020). The differential effect of carbon dots on gene expression and DNA methylation of human embryonic lung fibroblasts as a function of surface charge and dose. *Int. J. Mol. Sci.* 21: 4763.

Zhou, W. et al. (2019). Prolonged exposure to carbon nanoparticles induced methylome remodeling and gene expression in zebrafish heart. *J. Appl. Toxicol.* 39: 322–332.

Tabish, A.M. et al. (2017). Changes in DNA methylation in mouse lungs after a single intra-tracheal administration of nanomaterials. *PLOS ONE* 12: e0169886.

Yang, H. et al. (2017). Altered microRNA expression profiles in lung damage induced by nanosized SiO2. *Bioengineered* 8: 45–54.

Lee, K.-I. et al. (2020). Ultrafine silicon dioxide nanoparticles cause lung epithelial cells apoptosis via oxidative stress-activated PI3K/Akt-mediated mitochondria- and endoplasmic reticulum stress-dependent signaling pathways. *Sci. Rep.* 10: 9928.

Christen, V. and Fent, K. (2016). Silica nanoparticles induce endoplasmic reticulum stress response and activate mitogen activated kinase (MAPK) signalling. *Toxicol. Rep.* 3: 832–840.

Almansour, M., Alarifi, S., and Jarrar, B. (2018). In vivo investigation on the chronic hepatotoxicity induced by intraperitoneal administration of 10-nm silicon dioxide nanoparticles. *Int. J. Nanomed.* 13: 2685–2696.

Kalluri, R. and Weinberg, R.A. (2009). The basics of epithelial-mesenchymal transition. *J. Clin. Invest.* 119: 1420–1428.

Gliga, A.R., Di Bucchianico, S., Lindvall, J., Fadeel, B., and Karlsson, H.L. (2018). RNA-sequencing reveals long-term effects of silver nanoparticles on human lung cells. *Sci. Rep.* 8: 6668.

Manshian, B.B. et al. (2018). Nanoparticle-induced inflammation can increase tumor malignancy. *Acta Biomater.* 68: 99–112.

Ma, Y. et al. (2019). Different effects of titanium dioxide nanoparticles instillation in young and adult mice on DNA methylation related with lung inflammation and fibrosis. *Ecotoxicol. Environ. Saf.* 176: 1–10.

Stapleton, P.A. et al. (2018). Maternal engineered nanomaterial inhalation during gestation alters the fetal transcriptome. *Part. Fibre Toxicol.* 15: 3.

Zhang, X.-F. et al. (2015). Silver nanoparticles cause complications in pregnant mice. *Int. J. Nanomed.* 10: 7057–7071.

Xu, Q.-H. et al. (2018). Silver nanoparticles impair zebrafish skeletal and cardiac myofibrillogenesis and sarcomere formation. *Aquat. Toxicol.* 200: 102–113.

Balansky, R. et al. (2013). Transplacental clastogenic and epigenetic effects of gold nanoparticles in mice. *Mutat. Res. Mol. Mech. Mutagen.* 751–752: 42–48.

Boyerinas, B., Park, S.-M., Hau, A., Murmann, A.E., and Peter, M.E. (2010). The role of let-7 in cell differentiation and cancer. *Endocr. Relat. Cancer* 17: F19–F36.

Sonavane, G., Tomoda, K., and Makino, K. (2008). Biodistribution of colloidal gold nanoparticles after intravenous administration: effect of particle size. *Colloids Surf. B Biointerfaces* 66: 274–280.

Lasagna-Reeves, C. et al. (2010). Bioaccumulation and toxicity of gold nanoparticles after repeated administration in mice. *Biochem. Biophys. Res. Commun.* 393: 649–655.

Zhang, X.-D. et al. (2011). Size-dependent in vivo toxicity of PEG-coated gold nanoparticles. *Int. J. Nanomed.* 6: 2071–2081.

Yuan, B., Zhang, H., Wang, X., Pan, Y., and Jiang, J. (2018). Effect of nano-SiO2 on expression and aberrant methylation of imprinted genes in lung and testis. *Nanoscale Res. Lett.* 13: 266.

Zeng, T.-B. et al. (2014). DNA methylation dynamics of a maternally methylated DMR in the mouse Dlk1–Dio3 domain. *FEBS Lett.* 588: 4665–4671.

Wamucho, A., Heffley, A., and Tsyusko, O.V. (2020). Epigenetic effects induced by silver nanoparticles in *Caenorhabditis elegans* after multigenerational exposure. *Sci. Total Environ.* 725: 138523.

Stoccoro, A. et al. (2017). Multiple endpoints to evaluate pristine and remediated titanium dioxide nanoparticles genotoxicity in lung epithelial A549 cells. *Toxicol. Lett.* 276: 48–61.

Pogribna, M. et al. (2020). Effect of titanium dioxide nanoparticles on DNA methylation in multiple human cell lines. *Nanotoxicology* 14: 534–553.

Blanco, J. et al. (2017). Polyvinyl pyrrolidone-coated silver nanoparticles in a human lung cancer cells: time- and dose-dependent influence over p53 and caspase-3 protein expression and epigenetic effects. *Arch. Toxicol.* 91: 651–666.

Gong, C. et al. (2010). SiO2 nanoparticles induce global genomic hypomethylation in HaCaT cells. *Biochem. Biophys. Res. Commun.* 397: 397–400.

Pilger, A. and Rüdiger, H.W. (2006). 8-Hydroxy-2′-deoxyguanosine as a marker of oxidative DNA damage related to occupational and environmental exposures. *Int. Arch. Occup. Environ. Health* 80: 1–15.

Ghosh, M. et al. (2017). Changes in DNA methylation induced by multi-walled carbon nanotube exposure in the workplace. *Nanotoxicology* 11: 1195–1210.

Rossnerova, A. et al. (2020). DNA methylation profiles in a group of workers occupationally exposed to nanoparticles. *Int. J. Mol. Sci.* 21: 2420.

Brzóska, K., Grądzka, I., and Kruszewski, M. (2019). Silver, gold, and iron oxide nanoparticles alter miRNA expression but do not affect DNA methylation in HepG2 cells. *Materials* 12: 1038.

Augustine, R. et al. (2020). Cellular uptake and retention of nanoparticles: insights on particle properties and interaction with cellular components. *Mater. Today Commun.* 25: 101692.

Lynch, I., Salvati, A., and Dawson, K.A. (2009). What does the cell see? *Nat. Nanotechnol.* 4: 546–547.

Gustafson, H.H., Holt-Casper, D., Grainger, D.W., and Ghandehari, H. (2015). Nanoparticle uptake: the phagocyte problem. *Nano Today* 10: 487–510.

Behzadi, S. et al. (2017). Cellular uptake of nanoparticles: journey inside the cell. *Chem. Soc. Rev.* 46: 4218–4244.

Zhou, H.S., Honma, I., Komiyama, H., and Haus, J.W. (1993). Coated semiconductor nanoparticles; the cadmium sulfide/lead sulfide system's synthesis and properties. *J. Phys. Chem.* 97: 895–901.

Yamada, M., Foote, M., and Prow, T.W. (2015). Therapeutic gold, silver, and platinum nanoparticles. *WIREs Nanomed. Nanobiotechnol.* 7: 428–445.

Sharifi, M. et al. (2020). Plasmonic and chiroplasmonic nanobiosensors based on gold nanoparticles. *Talanta* 212: 120782.

Augustine, R. and Hasan, A. (2020). Emerging applications of biocompatible phytosynthesized metal/metal oxide nanoparticles in healthcare. *J. Drug Deliv. Sci. Technol.* 56: 101516.

Shang, L., Nienhaus, K., and Nienhaus, G.U. (2014). Engineered nanoparticles interacting with cells: size matters. *J. Nanobiotechnol.* 12: 5.

Donkor, D.A. and Tang, X.S. (2014). Tube length and cell type-dependent cellular responses to ultra-short single-walled carbon nanotube. *Biomaterials* 35: 3121–3131.

Chithrani, B.D., Ghazani, A.A., and Chan, W.C.W. (2006). Determining the size and shape dependence of gold nanoparticle uptake into mammalian cells. *Nano Lett.* 6: 662–668.

Verma, A. and Stellacci, F. (2010). Effect of surface properties on nanoparticle–cell interactions. *Small* 6: 12–21.

Otsuka, H., Nagasaki, Y., and Kataoka, K. (2012). PEGylated nanoparticles for biological and pharmaceutical applications. *Adv. Drug Deliv. Rev.* 64: 246–255.

Hotchkiss, K.M. et al. (2016). Titanium surface characteristics, including topography and wettability, alter macrophage activation. *Acta Biomater.* 31: 425–434.

Zhuang, X.-M., Zhou, B., and Yuan, K.-F. (2019). Role of p53 mediated miR-23a/CXCL12 pathway in osteogenic differentiation of bone mesenchymal stem cells on nanostructured titanium surfaces. *Biomed. Pharmacother.* 112: 108649.

Ganau, M. (2014). Tackling gliomas with nanoformulated antineoplastic drugs: suitability of hyaluronic acid nanoparticles. *Clin. Transl. Oncol. Off. Publ. Fed. Span. Oncol. Soc. Natl. Cancer Inst. Mex.* 16: 220–223.

Ganau, M. et al. (2017). Enhancing contrast agents and radiotracers performance through hyaluronic acid-coating in neuroradiology and nuclear medicine. *Hell. J. Nucl. Med.* 20: 166–168.

Ganau, L., Prisco, L., Ligarotti, G.K.I., Ambu, R., and Ganau, M. (2018). Understanding the pathological basis of neurological diseases through diagnostic platforms based on innovations in biomedical engineering: new concepts and theranostics perspectives. *Med. Basel Switz.* 5: E22.

Ganau, M. et al. (2015). A DNA-based nano-immunoassay for the label-free detection of glial fibrillary acidic protein in multicell lysates. *Nanomed. Nanotechnol. Biol. Med.* 11: 293–300.

Ganau, M. et al. (2018). How nanotechnology and biomedical engineering are supporting the identification of predictive biomarkers in neuro-oncology. *Med. Basel Switz.* 5: E23.

Ganau, M. et al. (2018). Current and future applications of biomedical engineering for proteomic profiling: predictive biomarkers in neuro-traumatology. *Med. Basel Switz.* 5: E19.

Qu, Q., Ma, X., and Zhao, Y. (2015). Targeted delivery of doxorubicin to mitochondria using mesoporous silica nanoparticle nanocarriers. *Nanoscale* 7: 16677–16686.

Sandström, P., Boncheva, M., and Åkerman, B. (2003). Nonspecific and thiol-specific binding of DNA to gold nanoparticles. *Langmuir* 19: 7537–7543.

DeLong, R.K. et al. (2010). Functionalized gold nanoparticles for the binding, stabilization, and delivery of therapeutic DNA, RNA, and other biological macromolecules. *Nanotechnol. Sci. Appl.* 3: 53–63.

Panicker, S. et al. (2019). Gold-Aryl nanoparticles coated with polyelectrolytes for adsorption and protection of DNA against nuclease degradation. *Appl. Organomet. Chem.* 33: e4803.

Xia, T., Kovochich, M., Liong, M., Zink, J.I., and Nel, A.E. (2008). Cationic polystyrene nanosphere toxicity depends on cell-specific endocytic and mitochondrial injury pathways. *ACS Nano.* 2: 85–96.

Sosnik, A. and Augustine, R. (2016). Challenges in oral drug delivery of antiretrovirals and the innovative strategies to overcome them. *Adv. Drug Deliv. Rev.* 103: 105–120.

Sang Yoo, H., Eun Lee, J., Chung, H., Chan Kwon, I., and Young Jeong, S. (2005). Self-assembled nanoparticles containing hydrophobically modified glycol chitosan for gene delivery. *J. Control. Release* 103: 235–243.

Mo, Y. et al. (2021). Nickel nanoparticle-induced cell transformation: involvement of DNA damage and DNA repair defect through HIF-1α/miR-210/Rad52 pathway. *J. Nanobiotechnol.* 19: 370.

Index

Note: Page numbers followed by *f* refer to figures and *t* refer to tables.

Impact of Engineered Nanomaterials in Genomics and Epigenomics, First Edition. Edited by Saura C. Sahu.

f

g

o

Printed in the USA/Agawam, MA
September 16, 2026

910661.190